Formulas/Equations

Distance Formula

If $P_1 = (x_1, y_1)$ and $P_2 = (x_2, y_2)$, the distance from P_1 to P_2 is
$$d(P_1, P_2) = \sqrt{(x_2 - x_1)^2 + (y_2 - y_1)^2}$$

Standard Equation of a Circle

The standard equation of a circle of radius r with center at (h, k) is
$$(x - h)^2 + (y - k)^2 = r^2$$

Slope Formula

The slope m of the line containing the points $P_1 = (x_1, y_1)$ and $P_2 = (x_2, y_2)$ is

$$m = \frac{y_2 - y_1}{x_2 - x_1} \qquad \text{if } x_1 \neq x_2$$

$$m \text{ is undefined} \qquad \text{if } x_1 = x_2$$

Point–Slope Equation of a Line

The equation of a line with slope m containing the point (x_1, y_1) is
$$y - y_1 = m(x - x_1)$$

Slope–Intercept Equation of a Line

The equation of a line with slope m and y-intercept b is
$$y = mx + b$$

Quadratic Formula

The solutions of the equation $ax^2 + bx + c = 0$, $a \neq 0$, are

$$x = \frac{-b \pm \sqrt{b^2 - 4ac}}{2a}$$

If $b^2 - 4ac > 0$, there are two real unequal solutions.
If $b^2 - 4ac = 0$, there is a repeated real solution.
If $b^2 - 4ac < 0$, there are two complex solutions that are not real.

Geometry Formulas

Circle

r = Radius, A = Area, C = Circumference
$$A = \pi r^2 \qquad C = 2\pi r$$

Triangle

b = Base, h = Altitude (Height), A = Area
$$A = \tfrac{1}{2} bh$$

Rectangle

l = Length, w = Width, A = Area, P = Perimeter
$$A = lw \qquad P = 2l + 2w$$

Rectangular Box

l = Length, w = Width, h = Height, V = Volume
$$V = lwh$$

Sphere

r = Radius, V = Volume, S = Surface area
$$V = \tfrac{4}{3}\pi r^3 \qquad S = 4\pi r^2$$

Trigonometry

Fifth Edition

MICHAEL SULLIVAN
Chicago State University

Prentice Hall, Upper Saddle River, NJ 07458

Library of Congress Cataloging-in-Publication Data

Sullivan, Michael, 1942–
    Trigonometry / Michael Sullivan.—5th ed.
      p. cm.
    Includes index.
    ISBN 0-13-095393-8
    1. Trigonometry. I. Title.
    Qa531.S85 1999
    516—dc21   98-37230
                                       CIP

Executive Editor: *Sally Yagan*
Marketing Manager: *Patrice Lumumba Jones*
Supplements Editor/Editorial Assistant: *Sara Beth Newell*
Editor-in-Chief: *Jerome Grant*
Editorial Director: *Tim Bozik*
Associate Editor-in-Chief, Development: *Carol Trueheart*
Production Editor: *Robert C. Walters*
Senior Managing Editor: *Linda Mihatov Behrens*
Executive Managing Editor: *Kathleen Schiaparelli*
Assistant Vice President of Production and Manufacturing: *David W. Riccardi*
Manufacturing Buyer: *Alan Fischer*
Manufacturing Manager: *Trudy Pisciotti*
Art Director: *Maureen Eide*
Associate Creative Director: *Amy Rosen*
Director of Creative Services: *Paula Maylahn*
Assistant to Art Director: *John Christiana*
Art Manager: *Gus Vibal*
Interior Designer: *Elm Street Publishing*
Cover Designer: *Maureen Eide*
Cover Photo: Rafters, by *Michael K. Daly/The Stock Market*
Photo Researcher: *Beth Boyd*
Photo Research Administrator: *Melinda Reo*
Art Studio: *Academy Artworks*

Printed in the United States of America

10 9 8 7 6

ISBN 0-13-095393-8

PRENTICE-HALL INTERNATIONAL (UK) LIMITED, LONDON
PRENTICE-HALL OF AUSTRALIA PTY. LIMITED, SYDNEY
PRENTICE-HALL CANADA INC. TORONTO
PRENTICE-HALL HISPANOAMERICANA, S.A., MEXICO
PRENTICE-HALL OF INDIA PRIVATE LIMITED, NEW DELHI
PRENTICE-HALL OF JAPAN, INC., TOKYO
SIMON & SCHUSTER ASIA PTE, LTD., SINGAPORE
EDITORA PRENTICE-HALL DO BRASIL, LTDA., RIO DE JANEIRO

*In memory of Joe and Rita
and my sister Maryrose*

CONTENTS

CHAPTER 1   FUNCTIONS AND THEIR GRAPHS   1

CHAPTER 2   TRIGONOMETRIC FUNCTIONS   101

CHAPTER 3   ANALYTIC TRIGONOMETRY   189

APPENDIX A: ALGEBRA TOPICS   521

APPENDIX B: GRAPHING UTILITIES   549

As a professor at an urban public university for over 30 years, I am aware of the varied needs of trigonometry students ranging from those who have little mathematical background and/or have been away from mathematics for a while to those who have a strong mathematics education and are extremely motivated. For some of your students, this may be their last course in mathematics, while others may decide to further their mathematical education. I have written this text for both groups. As the author of precalculus, engineering calculus, finite math and business calculus texts, and as a teacher, I understand what students must know if they are to be focused and successful in upper level mathematics courses. However, as a father of four, I also understand the realities of college life. I have taken great pains to insure that the text contains solid, student-friendly examples and problems, as well as a clear writing style.

In the Fifth Edition

The Fifth Edition builds upon a solid foundation by integrating new features and techniques that further enhance student interest and involvement. The elements of previous editions that have proved successful remain, while many changes, some obvious, others subtle, have been made. A huge benefit of authoring a successful series is the broad-based feedback upon which improvements and additions are ultimately based. Virtually every change to this edition is the result of thoughtful comments and suggestions made by colleagues and students who have used previous editions. I am sincerely grateful for this feedback and have tried to make changes that improve the flow and usability of the text. For example, some topics have been moved to better reflect the way teachers approach the course. In other places problems have been added where more practice was needed. The testing package has been significantly upgraded with the addition of Testpro 3. Also, the Sullivan web site, www.prenhall.com/sullivan, represents a truly useful integration of learning technology.

Changes to the Fifth Edition

- Each chapter now begins with an Internet Excursion: a real-world problem that can be analyzed utilizing the material found in the chapter and information found on the Web.
- Many new exercises, problems, and examples that utilize real data in table form have been added.
- The new four-color design allows key concepts and formulas to be presented prominently for easier recognition and review.

Specific Organizational Changes

Chapter 2: The graphs of the six trigonometric functions now appear in this chapter, followed by sinusoidal graphs and applications of sinusoidal graphs. The former section on Applications appears later in the new Chapter 4.

Chapter 3: The material of this chapter has been distributed elsewhere in the new edition to improve flow and teachability.

Chapter 4: Now Chapter 3, the section on Inverse Trigonometric Functions is presented just before Trigonometric Equations.

Chapter 5: Now Chapter 4, the chapter begins with applications involving Right Triangles and closes with applications involving Simple Harmonic Motion and Damped Motion.

The material on Polar Coordinates and Polar Equations now appears in the new Chapter 5. Also in this chapter are Complex Numbers and De Moivre's Theorem (formerly in Chapter 7). Additionally, this new Chapter 5 contains independent sections on Vectors and the Dot Product, formerly in Chapter 7. New to this edition is the inclusion of a section on Vectors in Space.

This reorganization should provide instructors with greater flexibility in selecting topics to cover. These changes are predicated on the belief that it is better to place material as close as possible to where it is actually needed than it is to cover it, set it aside, and then expect the student to remember it when it is needed.

Specific Content Changes

Chapter 1: Scatter diagrams are used in Section 1.2 to further motivate the idea of rectangular coordinates. In Section 1.4 more emphasis is placed on the average rate of change of a function.

Chapter 2: The new section on sinusoidal graphs now contains current applications to sinusoidal curve fitting.

Chapter 4: A new section on Simple Harmonic Motion now includes Damped Motion.

Chapter 7: Section 7.4, Properties of Logarithms, contains a brief discussion of data that fit exponential or logarithmic functions. Section 7.7, Growth and Decay, now contains an example of Logistic Growth.

Appendix A: New, Section A.2, Linear Curve Fitting, explores the idea of linearly related data and fitting lines to scatter diagrams. This section is optional and does not require graphing calculators.

Appendix B: This optional appendix on graphing utilities has been up-dated to include some of the more powerful features of graphing technology.

As a result of these changes this edition will be an improved teaching device for professors and a better learning tool for students.

Acknowledgments

Textbooks are written by authors, but evolve from an idea into final form through the efforts of many people. Special thanks to Don Dellen, who first suggested this book and the other books in this series. Don's extensive contributions to publishing and mathematics are well known; we all miss him dearly.

There are many people I would like to thank for their input, encouragement, patience, and support. They have my deepest thanks and appreciation. I apologize for any omissions . . .

James Africh, *College of DuPage*
Steve Agronsky, *Cal Poly State University*
Dave Anderson, *South Suburban College*
Joby Milo Anthony, *University of Central Florida*
James E. Arnold, *University of Wisconsin-Milwaukee*
Agnes Azzolino, *Middlesex County College*
Wilson P. Banks, *Illinois State University*
Dale R. Bedgood, *East Texas State University*

Beth Beno, *South Suburban College*

Carolyn Bernath, *Tallahassee Community College*

William H. Beyer, *University of Akron*

Richelle Blair, *Lakeland Community College*

Trudy Bratten, *Grossmont College*

William J. Cable, *University of Wisconsin-Stevens Point*

Lois Calamia, *Brookdale Community College*

Roger Carlsen, *Moraine Valley Community College*

John Collado, *South Suburban College*

Denise Corbett, *East Carolina University*

Theodore C. Coskey, *South Seattle Community College*

John Davenport, *East Texas State University*

Duane E. Deal, *Ball State University*

Vivian Dennis, *Eastfield College*

Guesna Dohrman, *Tallahassee Community College*

Karen R. Dougan, *University of Florida*

Louise Dyson, *Clark College*

Paul D. East, *Lexington Community College*

Don Edmondson, *University of Texas-Austin*

Christopher Ennis, *University of Minnesota*

Ralph Esparza, Jr., *Richland College*

Garret J. Etgen, *University of Houston*

W. A. Ferguson, *University of Illinois-Urbana/Champaign*

Iris B. Fetta, *Clemson University*

Mason Flake, *student at Edison Community College*

Merle Friel, *Humboldt State University*

Richard A. Fritz, *Moraine Valley Community College*

Carolyn Funk, *South Suburban College*

Dewey Furness, *Ricke College*

Dawit Getachew, *Chicago State University*

Wayne Gibson, *Rancho Santiago College*

Sudhir Kumar Goel, *Valdosta State University*

Joan Goliday, *Sante Fe Community College*

Frederic Gooding, *Goucher College*

Ken Gurganus, *University of North Carolina*

James E. Hall, *University of Wisconsin-Madison*

Judy Hall, *West Virginia University*

Edward R. Hancock, *DeVry Institute of Technology*

Julia Hassett, *DeVry Institute-Dupage*

Brother Herron, *Brother Rice High School*

Kim Hughes, *California State College-San Bernardino*

Ron Jamison, *Brigham Young University*

Richard A. Jensen, *Manatee Community College*

Sandra G. Johnson, *St. Cloud State University*

Moana H. Karsteter, *Tallahassee Community College*

Arthur Kaufman, *College of Staten Island*

Thomas Kearns, *North Kentucky University*

Teddy Koukounas, *SUNY at Old Westbury*

Keith Kuchar, *Manatee Community College*

Tor Kwembe, *Chicago State University*

Linda J. Kyle, *Tarrant Country Jr. College*

H. E. Lacey, *Texas A & M University*

Christopher Lattin, *Oakton Community College*

Adele LeGere, *Oakton Community College*

Stanley Lukawecki, *Clemson University*

Janice C. Lyon, *Tallahassee Community College*

Virginia McCarthy, *Iowa State University*

James McCollow, *DeVry Institute of Technology*

Laurence Maher, *North Texas State University*

Jay A. Malmstrom, *Oklahoma City Community College*

James Maxwell, *Oklahoma State University-Stillwater*

Carolyn Meitler, *Concordia University*

Eldon Miller, *University of Mississippi*

James Miller, *West Virginia University*

Michael Miller, *Iowa State University*

Kathleen Miranda, *SUNY at Old Westbury*

A. Muhundan, *Manatee Community College*

Jane Murphy, *Middlesex Community College*

Bill Naegele, *South Suburban College*

James Nymann, *University of Texas-El Paso*

Sharon O'Donnell, *Chicago State University*

Seth F. Oppenheimer, *Mississippi State University*

E. James Peake, *Iowa State University*

Thomas Radin, *San Joaquin Delta College*

Ken A. Rager, *Metropolitan State College*

Elsi Reinhardt, *Truckee Meadows Community College*

Jane Ringwald, *Iowa State University*

Stephen Rodi, *Austin Community College*

Howard L. Rolf, *Baylor University*

Edward Rozema, *University of Tennessee at Chattanooga*

Dennis C. Runde, *Manatee Community College*

John Sanders, *Chicago State University*

Susan Sandmeyer, *Jamestown Community College*

A. K. Shamma, *University of West Florida*

Martin Sherry, *Lower Columbia College*

Anita Sikes, *Delgado Community College*

Timothy Sipka, *Alma College*

Lori Smellegar, *Manatee Community College*

John Spellman, *Southwest Texas State University*

Becky Stamper, *Western Kentucky University*

Judy Staver, *Florida Community College-South*

Neil Stephens, *Hinsdale South High School*

Diane Tesar, *South Suburban College*

Tommy Thompson, *Brookhaven College*

Richard J. Tondra, *Iowa State University*

Marvel Townsend, *University of Florida*

Jim Trudnowski, *Carroll College*

Richard G. Vinson, *University of South Alabama*
Mary Voxman, *University of Idaho*
Darlene Whitkenack, *Northern Illinois University*

Christine Wilson, *West Virginia University*
Carlton Woods, *Auburn University*
George Zazi, *Chicago State University*

Recognition and thanks are due particularly to the following individuals for their valuable assistance in the preparation of this edition: Jerome Grant for his support and commitment; Sally Simpson, for her genuine interest and insightful direction; Bob Walters for his organizational skill as production supervisor; Patrice Lumumba Jones for his innovative marketing efforts; Tony Palermino for his specific editorial comments; the entire Prentice Hall sales staff for their confidence; and to Katy Murphy and Michael Sullivan III for checking the answers to all the exercises.

Michael Sullivan

As you begin your study of Trigonometry, you may feel overwhelmed by the number of theorems, definitions, procedures, and equations that confront you. You may even wonder whether or not you can learn all of this material in the time allotted. These concerns are normal. Keep in mind that the concepts of Trigonometry are all around us as we go through our daily routines. Many of the concepts you will learn to express mathematically, you already know intuitively. Many of you will be going on to take advanced math courses; for some this may be your final math course. Either way, this text was written with you in mind. I have taught Trigonometry courses for over thirty years. I am also the father of four college students who called home from time to time frustrated and with questions. I know what you're going through. So I have written a text that doesn't overwhelm, or unnecessarily complicate the concepts of Trigonometry, but at the same time gives you the skills and practice you need to be successful.

This text is designed to help you, the student, master the terminology and basic concepts of Trigonometry. These aims have helped to shape every aspect of the book. Many learning aids are built into the format of the text to make your study of the material easier and more rewarding. This book is meant to be a "machine for learning," one that can help you focus your efforts and get the most from the time and energy you invest.

Please do not hesitate to contact me, through Prentice Hall, with any suggestions or comments that would improve this text.

Best Wishes!

Michael Sullivan

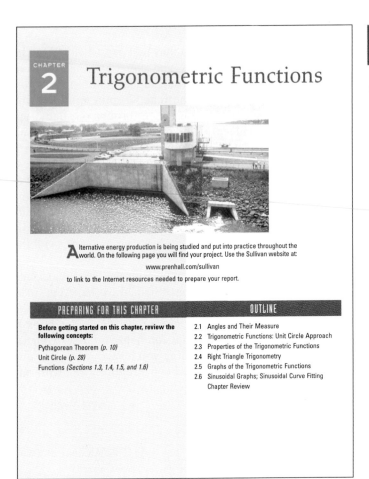

CHAPTER 2

Trigonometric Functions

Alternative energy production is being studied and put into practice throughout the world. On the following page you will find your project. Use the Sullivan website at:

www.prenhall.com/sullivan

to link to the Internet resources needed to prepare your report.

PREPARING FOR THIS CHAPTER	OUTLINE
Before getting started on this chapter, review the following concepts:	2.1 Angles and Their Measure
Pythagorean Theorem *(p. 10)*	2.2 Trigonometric Functions: Unit Circle Approach
Unit Circle *(p. 28)*	2.3 Properties of the Trigonometric Functions
Functions *(Sections 1.3, 1.4, 1.5, and 1.6)*	2.4 Right Triangle Trigonometry
	2.5 Graphs of the Trigonometric Functions
	2.6 Sinusoidal Graphs; Sinusoidal Curve Fitting
	Chapter Review

PAGE 101

CHAPTER OPENERS

Each chapter begins with Preparing for this Chapter, which serves as a "just-in-time" review, while Chapter Outlines provide students with a basic road map of the chapter.

Students will already have an understanding of which concepts are most important before beginning the chapter.

INTERNET PROJECTS

Exploration in each chapter starts right from the beginning with optional Internet Projects inviting students to use the Sullivan Web site to gather information and real data to solve mathematical problems.

Students can see what a useful and accessible tool the Internet is. Likewise, they will also be able to see algebra as a valuable tool for understanding the world outside the classroom.

Internet Projects also help foster active learning and participation in lecture.

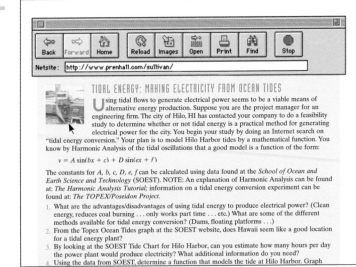

Netsite: http://www.prenhall.com/sullivan/

TIDAL ENERGY: MAKING ELECTRICITY FROM OCEAN TIDES

Using tidal flows to generate electrical power seems to be a viable means of alternative energy production. Suppose you are the project manager for an engineering firm. The city of Hilo, HI has contacted your company to do a feasibility study to determine whether or not tidal energy is a practical method for generating electrical power for the city. You begin your study by doing an Internet search on "tidal energy conversion." Your plan is to model Hilo Harbor tides by a mathematical function. You know by Harmonic Analysis of the tidal oscillations that a good model is a function of the form:

$$v = A \sin(bx + c) + D \sin(ex + f)$$

The constants for A, b, c, D, e, f can be calculated using data found at the *School of Ocean and Earth Science and Technology* (SOEST). NOTE: An explanation of Harmonic Analysis can be found at: *The Harmonic Analysis Tutorial*; information on a tidal energy conversion experiment can be found at: *The TOPEX/Poseidon Project.*

1. What are the advantages/disadvantages of using tidal energy to produce electrical power? (Clean energy, reduces coal burning . . . only works part time . . . etc.) What are some of the different methods available for tidal energy conversion? (Dams, floating platforms . . .)
2. From the Topex Ocean Tides graph at the SOEST website, does Hawaii seem like a good location for a tidal energy plant?
3. By looking at the SOEST Tide Chart for Hilo Harbor, can you estimate how many hours per day the power plant would produce electricity? What additional information do you need?
4. Using the data from SOEST, determine a function that models the tide at Hilo Harbor. Graph

PAGE 102

2.2 | TRIGONOMETRIC FUNCTIONS: UNIT CIRCLE APPROACH

1 Find the Exact Value of the Trigonometric Functions Using a Point on the Unit Circle
2 Find the Exact Value of the Trigonometric Functions of Quadrantal Angles
3 Find the Exact Value of the Trigonometric Functions of 45° Angles
4 Find the Exact Value of the Trigonometric Functions of 60° Angles
5 Find the Exact Value of the Trigonometric Functions of 30° Angles
6 Use a Calculator to Approximate the Value of the Trigonometric Functions

We are now ready to introduce trigonometric functions. As we said earlier, the approach taken uses the unit circle.

The Unit Circle

Recall that the **unit circle** is a circle whose radius is 1 and whose center is at the origin of a rectangular coordinate system. Because the radius r of the unit circle is 1, we see from the formula $s = r\theta$ that on the unit circle a central angle of θ radians subtends an arc whose length s is

$$s = \theta$$

PAGE 114

SECTION OBJECTIVES

Each section begins with a bulleted synopsis of the topics it contains.

Students have a clear idea of which concepts they will cover, enhancing their comfort level as new topics are introduced.

E X A M P L E 10

Finding the Sine Function of Best Fit

Use a graphing utility to find the sine function of best fit for the data in Table 11. Graph this function with the scatter diagram of the data.

Solution

Enter the data from Table 11 and execute the SINe REGression program. The result is shown in Figure 90.

FIGURE 90

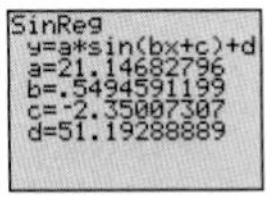

```
SinReg
y=a*sin(bx+c)+d
a=21.14682796
b=.5494591199
c=-2.35007307
d=51.19288889
```

The output that the utility provides shows us the equation

$$y = a \sin (bx + c) + d.$$

The sinusoidal function of best fit is

$$y = 21.15 \sin(0.55x - 2.35) + 51.19$$

where x represents the month and y represents the average temperature. Figure 91 shows the graph of the sinusoidal function of best fit on the scatter diagram.

FIGURE 91

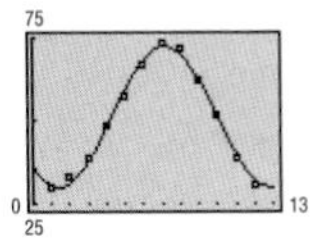

Now work Problem 79.

Since the number of hours of sunlight in a day cycles annually, the number of hours of sunlight in a day for a given location can be modeled by a sinusoidal function.

The longest day of the year (in terms of hours of sunlight) occurs on the day of the summer solstice. The summer solstice is the time when the sun is farthest north (for locations in the northern hemisphere). In 1997, the summer solstice occurs on June 21 (the 172nd day of the year) at 8:21 AM

PAGE 174

OPTIONAL GRAPHING UTILITIES AND TECHNIQUES

Graphing utilities are optional in this text and their use is clearly identified by the presence of a graphing utility icon.

At appropriate places, graphing utilities are used for Exploration, Seeing the Concept, and Checks, often accompanied by TI-83 screen shots.

Optional graphing utility exercises and examples also appear.

Graphing utilities allow students to explore algebraic concepts visually, promoting more thorough understanding of key objectives.

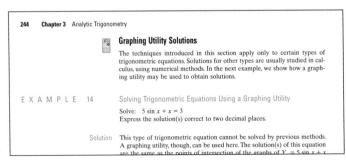

244   **Chapter 3**   Analytic Trigonometry

Graphing Utility Solutions

The techniques introduced in this section apply only to certain types of trigonometric equations. Solutions for other types are usually studied in calculus, using numerical methods. In the next example, we show how a graphing utility may be used to obtain solutions.

E X A M P L E 14

Solving Trigonometric Equations Using a Graphing Utility

Solve: $5 \sin x + x = 3$
Express the solution(s) correct to two decimal places.

Solution

This type of trigonometric equation cannot be solved by previous methods. A graphing utility, though, can be used here. The solution(s) of this equation are the same as the points of intersection of the graphs of $Y = 5 \sin x + x$

PAGE 244

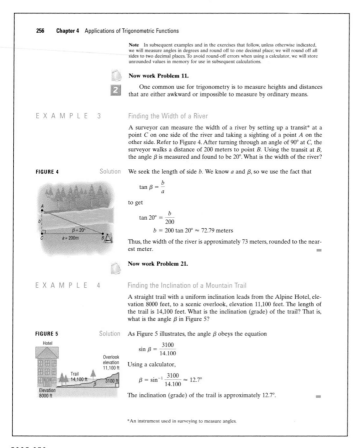

256   Chapter 4   Applications of Trigonometric Functions

Note In subsequent examples and in the exercises that follow, unless otherwise indicated, we will measure angles in degrees and round off to one decimal place; we will round off all sides to two decimal places. To avoid round-off errors when using a calculator, we will store unrounded values in memory for use in subsequent calculations.

Now work Problem 11.

One common use for trigonometry is to measure heights and distances that are either awkward or impossible to measure by ordinary means.

EXAMPLE 3   Finding the Width of a River

A surveyor can measure the width of a river by setting up a transit* at a point C on one side of the river and taking a sighting of a point A on the other side. Refer to Figure 4. After turning through an angle of 90° at C, the surveyor walks a distance of 200 meters to point B. Using the transit at B, the angle β is measured and found to be 20°. What is the width of the river?

FIGURE 4   Solution   We seek the length of side b. We know a and β, so we use the fact that

$$\tan \beta = \frac{b}{a}$$

to get

$$\tan 20° = \frac{b}{200}$$

$$b = 200 \tan 20° \approx 72.79 \text{ meters}$$

Thus, the width of the river is approximately 73 meters, rounded to the nearest meter. ∎

Now work Problem 21.

EXAMPLE 4   Finding the Inclination of a Mountain Trail

A straight trail with a uniform inclination leads from the Alpine Hotel, elevation 8000 feet, to a scenic overlook, elevation 11,100 feet. The length of the trail is 14,100 feet. What is the inclination (grade) of the trail? That is, what is the angle β in Figure 5?

FIGURE 5   Solution   As Figure 5 illustrates, the angle β obeys the equation

$$\sin \beta = \frac{3100}{14.100}$$

Using a calculator,

$$\beta = \sin^{-1} \frac{3100}{14.100} \approx 12.7°$$

The inclination (grade) of the trail is approximately 12.7°. ∎

*An instrument used in surveying to measure angles.

PAGE 256

Blue "Now Work" icons appear at appropriate places within each section, sending the student to work specific exercises as topics are introduced.

Students are able to test their understanding as they study, rather than waiting until the end of the section and trying to determine where they may have missed a concept.

A hidden benefit of the "Now Work" problems is that they enable students to ask more specific questions in class, because they know exactly in which areas they need help.

Sullivan's accessible writing style is apparent throughout, often utilizing different explanations of the same concept.

An author who writes clearly makes potentially difficult concepts intuitive. Class time is infinitely more productive when students have the support of a well-written text.

Sometimes it is helpful to think of a function f as a machine that receives as input a number from the domain, manipulates it, and outputs the value. See Figure 39.

FIGURE 39   Input x

Output
$y = f(x)$

The restrictions on this input/output machine are

1. It only accepts numbers from the domain of the function.
2. For each input, there is exactly one output (which may be repeated for different inputs).

For a function $y = f(x)$, the variable x is called the **independent variable,** because it can be assigned any of the permissible numbers from the domain. The variable y is called the **dependent variable,** because its value depends on x.

PAGE 38

MISSION POSSIBLE

Locating Lost Treasure

While scuba diving off Wreck Hill in Bermuda, a group of five entrepreneurs discovered a treasure map in a small watertight cask on a pirate schooner that had sunk in 1747. The map directed them to an area of Bermuda now known as The Flatts, but when they got there, they realized that the most important landmark on the map was gone. They called in the Mission Possible team to help them to re-create the map. They promised you 25% of whatever treasure was found.

The directions on the map read as follows:

1. From the tallest palm tree, sight the highest hill. Drop your eyes vertically until you sight the base of the hill.
2. Turn 40° clockwise from that line and walk 70 paces to the big red rock.
3. From the red rock, walk 50 paces back to the sight line between the palm tree and the hill. Dig there.

The five entrepreneurs said that they believed that they had found the red rock and the highest hill in the vicinity, but the "tallest palm tree" had long since fallen and disintegrated. It had occurred to them that the treasure must be located on a circle with radius 50 "paces" centered around the red rock, but they had decided against digging a trench 942 feet in circumference, especially since they had no assurance that the treasure was still there. (They had decided that a "pace" must be about a yard.)

1. Determine a plan to locate the position of the lost palm tree, and write out an explanation of y cedure for the entrepreneurs.
2. Unfortunately, it turns out that the entrepreneurs had more in common with the eighteenth-ce rates than you had bargained for. Once you told them the location of the lost palm tree, they all to the red rock, saying that they could take it from there. From the location of the palm tree sighted 40° counterclockwise from the rock to the hill and then ran about 50 yards to the circ they had traced about the rock and began to dig frantically. Nothing. After about an hour, the off shouting back at you, "25% of nothing is nothing!"

   Fortunately, the entrepreneurs had left the shovels. After you managed to untie yourselves, yo to the correct location and found the treasure. Where was it? How far from the palm tree? Ex
3. People who scuba dive for sunken treasure have certain legal obligations. What are they? Sh share the treasure with a lawyer, just to make sure that you get to keep the rest?

Section 4.4 The Area of a Triangle 289

PAGE 289

In every chapter, there is a full-page collaborative project, a *Mission Possible*. These multi-tasked projects, written by Hester Lewellen, one of the co-authors of the University of Chicago High School Mathematics Project, will help your students learn through and with each other.

The collaborative format has been shown to increase student motivation and interest.

79. **Monthly Temperature** The following data represent the average monthly temperatures for Juneau, Alaska.

Month, x	Average Monthly Temperature, °F
January, 1	24.2
February, 2	28.4
March, 3	32.7
April, 4	39.7
May, 5	47.0
June, 6	53.0
July, 7	56.0
August, 8	55.0
September, 9	49.4
October, 10	42.2
November, 11	32.0
December, 12	27.1

Source: U.S. National Oceanic and Atmospheric Administration.

Month, x	Average Monthly Temperature, °F
January, 1	34.6
February, 2	37.5
March, 3	47.2
April, 4	56.5
May, 5	66.4
June, 6	75.6
July, 7	80.0
August, 8	78.5
September, 9	71.3
October, 10	59.7
November, 11	49.8
December, 12	39.4

Source: U.S. National Oceanic and Atmospheric Administration.

(a) Draw a scatter diagram of the data for one period.
(b) Find a sinusoidal function of the form $y = A \sin(\omega x - \phi) + B$ that fits the data.
(c) Draw the sinusoidal function found in (b) on the scatter diagram.
(d) Use a graphing utility to find the sinusoidal function of best fit.
(e) Draw the sinusoidal function of best fit on the scatter diagram.

PAGE 180

SOURCED DATA

Sullivan provides students with applications that employ sourced real-world data.

Students find it difficult to get into hypothetical "widget" problems. Real data allows students to see the relevancy of algebra outside the classroom.

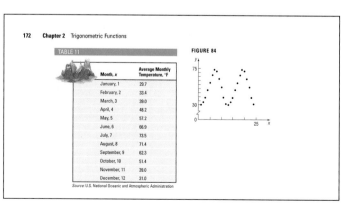

172 Chapter 2 Trigonometric Functions

TABLE 11

Month, x	Average Monthly Temperature, °F
January, 1	29.7
February, 2	33.4
March, 3	39.0
April, 4	48.2
May, 5	57.2
June, 6	66.9
July, 7	73.5
August, 8	71.4
September, 9	62.3
October, 10	51.4
November, 11	39.0
December, 12	31.0

Source: U.S. National Oceanic and Atmospheric Administration

FIGURE 84

PAGE 172

OVERVIEW

FIGURE 36

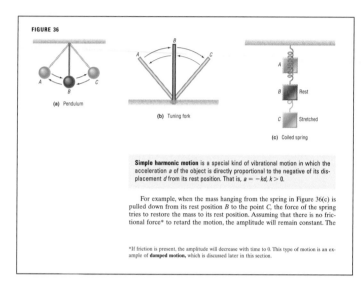

(a) Pendulum

(b) Tuning fork

(c) Coiled spring

A
B Rest
C Stretched

Simple harmonic motion is a special kind of vibrational motion in which the acceleration a of the object is directly proportional to the negative of its displacement d from its rest position. That is, $a = -kd$, $k > 0$.

For example, when the mass hanging from the spring in Figure 36(c) is pulled down from its rest position B to the point C, the force of the spring tries to restore the mass to its rest position. Assuming that there is no frictional force* to retard the motion, the amplitude will remain constant. The

*If friction is present, the amplitude will decrease with time to 0. This type of motion is an example of **damped motion**, which is discussed later in this section.

PAGE 290

MODELING

Students are asked to create and manipulate mathematical models.

The construction and interpretation of mathematical models is essential to better understanding of not only trigonometry, but virtually any mathematics course that your students are likely to take during their college careers.

END-OF-SECTION EXERCISES

Sullivan's exercises are unparalleled in terms of thorough coverage and accuracy.

Well-executed problem sets better prepare students for exams.

Each end-of-section exercise set begins with visual and concept-based problems, starting students out with the basics of the section.

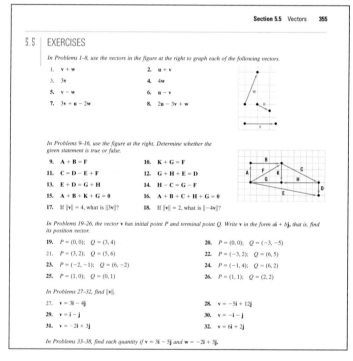

Section 5.5 Vectors 355

5.5 EXERCISES

In Problems 1–8, use the vectors in the figure at the right to graph each of the following vectors.

1. $\mathbf{v} + \mathbf{w}$
2. $\mathbf{u} + \mathbf{v}$
3. $3\mathbf{v}$
4. $4\mathbf{w}$
5. $\mathbf{v} - \mathbf{w}$
6. $\mathbf{u} - \mathbf{v}$
7. $3\mathbf{v} + \mathbf{u} - 2\mathbf{w}$
8. $2\mathbf{u} - 3\mathbf{v} + \mathbf{w}$

In Problems 9–16, use the figure at the right. Determine whether the given statement is true or false.

9. $\mathbf{A} + \mathbf{B} = \mathbf{F}$
10. $\mathbf{K} + \mathbf{G} = \mathbf{F}$
11. $\mathbf{C} = \mathbf{D} - \mathbf{E} + \mathbf{F}$
12. $\mathbf{G} + \mathbf{H} + \mathbf{E} = \mathbf{D}$
13. $\mathbf{E} + \mathbf{D} = \mathbf{G} + \mathbf{H}$
14. $\mathbf{H} - \mathbf{C} = \mathbf{G} - \mathbf{F}$
15. $\mathbf{A} + \mathbf{B} + \mathbf{K} + \mathbf{G} = 0$
16. $\mathbf{A} + \mathbf{B} + \mathbf{C} + \mathbf{H} + \mathbf{G} = 0$
17. If $\|\mathbf{v}\| = 4$, what is $\|3\mathbf{v}\|$?
18. If $\|\mathbf{v}\| = 2$, what is $\|-4\mathbf{v}\|$?

In Problems 19–26, the vector $\mathbf{v}$ has initial point P and terminal point Q. Write $\mathbf{v}$ in the form $a\mathbf{i} + b\mathbf{j}$, that is, find its position vector.

19. $P = (0,0)$; $Q = (3,4)$
20. $P = (0,0)$; $Q = (-3,-5)$
21. $P = (3,2)$; $Q = (5,6)$
22. $P = (-3,2)$; $Q = (6,5)$
23. $P = (-2,-1)$; $Q = (6,-2)$
24. $P = (-1,4)$; $Q = (6,2)$
25. $P = (1,0)$; $Q = (0,1)$
26. $P = (1,1)$; $Q = (2,2)$

In Problems 27–32, find $\|\mathbf{v}\|$.

27. $\mathbf{v} = 3\mathbf{i} - 4\mathbf{j}$
28. $\mathbf{v} = -5\mathbf{i} + 12\mathbf{j}$
29. $\mathbf{v} = \mathbf{i} - \mathbf{j}$
30. $\mathbf{v} = -\mathbf{i} - \mathbf{j}$
31. $\mathbf{v} = -2\mathbf{i} + 3\mathbf{j}$
32. $\mathbf{v} = 6\mathbf{i} + 2\mathbf{j}$

In Problems 33–38, find each quantity if $\mathbf{v} = 3\mathbf{i} - 5\mathbf{j}$ and $\mathbf{w} = -2\mathbf{i} + 3\mathbf{j}$.

PAGE 355

FIGURE 51
Constant Function

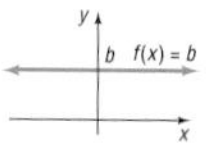

Constant Function

$$f(x) = b \quad b \text{ is a real number}$$

See Figure 51.
A **constant function** is a special linear function ($m = 0$). Its domain is the set of all real numbers; its range is the set consisting of a single number b. Its graph is a horizontal line whose y-intercept is b. The constant function is an even function whose graph is constant over its domain.

FIGURE 52
Identity Function

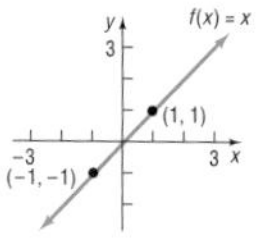

Identity Function

$$f(x) = x$$

See Figure 52.
The **identity function** is also a special linear function. Its domain and its range are the set of all real numbers. Its graph is a line whose slope is $m = 1$ and whose y-intercept is 0. The line consists of all points for which the x-coordinate equals the y-coordinate. The identity function is an odd function that is increasing over its domain. Note that the graph bisects quadrants I and III.

FIGURE 53
Square Function

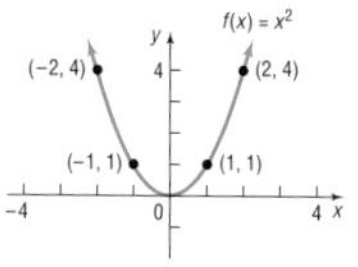

Square Function

$$f(x) = x^2$$

See Figure 53.
The domain of the **square function** f is the set of all real numbers; its range is the set of nonnegative real numbers. The graph of this function is a parabola, whose intercept is at (0, 0). The square function is an even function that is decreasing on the interval $(-\infty, 0)$ and increasing on the interval $(0, \infty)$.

FIGURE 54
Cube Function

Cube Function

$$f(x) = x^3$$

See Figure 54.
The domain and range of the **cube function** are the set of a... bers. The intercept of the graph is at (0, 0). The cube function i... increasing on the interval $(-\infty, \infty)$.

KEY CONCEPTS

Key concepts and formulas are clearly identified throughout the text.

Students are able to review more efficiently when formulas and concepts essential to the chapter material are readily identifiable.

VISUAL EXERCISES

Often students can perform mathematical operations without achieving a conceptual understanding. This text features visual exercises that encourage students to consider mathematics intuitively.

In Problems 49–62, find the domain of each function.

49. $f(x) = 3x + 4$

50. $f(x) = 5x^2 + 2$

51. $f(x) = \dfrac{x}{x^2 + 1}$

52. $f(x) = \dfrac{x^2}{x^2 + 1}$

53. $g(x) = \dfrac{x}{x^2 - 1}$

54. $h(x) = \dfrac{x}{x - 1}$

55. $F(x) = \dfrac{x - 2}{x^3 + x}$

56. $G(x) = \dfrac{x + 4}{x^3 - 4x}$

57. $h(x) = \sqrt{3x - 12}$

58. $G(x) = \sqrt{1 - x}$

59. $f(x) = \dfrac{4}{\sqrt{x - 9}}$

60. $f(x) = \dfrac{x}{\sqrt{x - 4}}$

61. $p(x) = \sqrt{\dfrac{x - 2}{x - 1}}$

62. $a(x) = \sqrt{x^2 - x - 2}$

63. Match each of the following functions with the graphs that best describe the situation.
(a) The cost of building a house as a function of its square footage.
(b) The height of an egg dropped from a 300-foot building as a function of time.
(c) The height of a human as a function of time.
(d) The demand for Big Macs as a function of price.
(e) The height of a child on a swing as a function of time.

(I)   (II)   (III)   (IV)   (V)

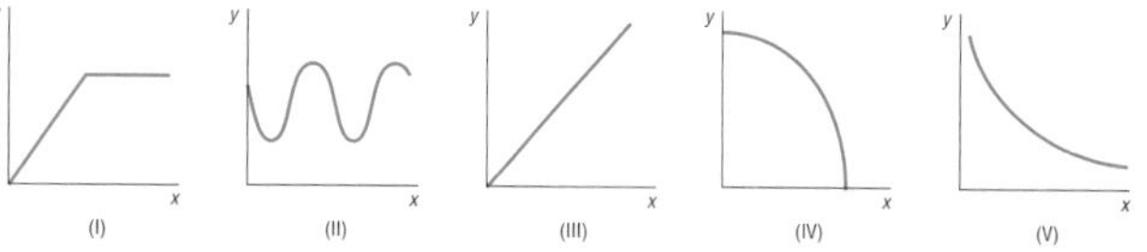

64. Consider the following scenario: Barbara decides to take a walk. She leaves home, walks 2 blocks in 5 minutes at a constant speed, and realizes that she forgot to lock the door. So Barbara runs home in 1 minute. While at her doorstep, it takes her 1 minute to find her keys and lock the door. Barbara walks 5 blocks in 15 minutes and then decides to jog home. It takes her 7 minutes to get home. Draw a graph of Barbara's distance from home (in blocks) as a function of time.

65. If $f(x) = 2x^3 + Ax^2 + 4x - 5$ and $f(2) = 5$, what is the value of A?

66. If $f(x) = 3x^2 - Bx + 4$ and $f(-1) = 12$, what is the value of B?

67. If $f(x) = (3x + 8)/(2x - A)$ and $f(0) = 2$, what is the value of A?

68. If $f(x) = (2x - B)/(3x + 4)$ and $f(2) = \frac{1}{2}$, what is the value of B?

69. If $f(x) = (2x - A)/(x - 3)$ and $f(4) = 0$, what is the value of A? Where is f not defined?

70. If $f(x) = (x - B)/(x - A)$, $f(2) = 0$, and $f(1)$ is undefined, what are the values of A and B?

71. **Demand for Jeans**   The marketing manager at Levi–Strauss wishes to find a function that relates the demand D of men's jeans and p, the price of the jeans. The following data were obtained based on a price history of the jeans.

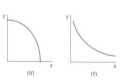

Price ($/Pair), p	Demand (Pairs of Jeans Sold per Day), D
20	60
22	57
23	56
23	53
27	52
29	49
30	44

(a) Does the relation defined by the set of ordered pairs (p, D) represent a function?
(b) Draw a scatter diagram of the data.
(c) Select two points from the data and find an equation of the line containing the points. (The answer in the back of the book uses the first and last data points.)
(d) Interpret the slope.
(e) Express the relationship found in (c) using function notation.
(f) What is the domain of the function?
(g) How many jeans will be demanded if the price is $28 a pair?
(h)  Using a graphing utility, find the line of best fit relating price and quantity demanded.

CHAPTER REVIEW

THINGS TO KNOW

Definitions

Angle in standard position	Vertex is at the origin; initial side is along the positive x-axis
Degree (1°)	$1° = \frac{1}{360}$ revolution
Radian (1)	The measure of a central angle whose rays subtend an arc whose length is the radius of the circle
Unit circle	Center at origin; radius = 1 Equation: $x^2 + y^2 = 1$
Trigonometric functions	$P = (a, b)$ is the point on the unit circle corresponding to $\theta = t$ radians:

$$\sin t = \sin \theta = b \qquad \cos t = \cos \theta = a$$

$$\tan t = \tan \theta = \frac{b}{a}, \quad a \neq 0 \qquad \cot t = \cot \theta = \frac{a}{b}, \quad b \neq 0$$

$$\csc t = \csc \theta = \frac{1}{b}, \quad b \neq 0 \qquad \sec t = \sec \theta = \frac{1}{a}, \quad a \neq 0$$

Periodic function	$f(\theta + p) = f(\theta)$, for all θ, $p > 0$, where the smallest such p is the fundamental period
Acute angle	An angle θ whose measure is $0° < \theta < 90°$ (or $0 < \theta < \pi/2$)
Complementary angles	Two acute angles whose sum is 90° ($\pi/2$)

HOW TO

Convert an angle from radian measure to degree measure

Convert an angle from degree measure to radian measure

Solve applied problems involving circular motion

Find the value of each of the remaining trigonometric functions if the value of one function and the quadrant of the angle are given

Use the theorem on cofunctions of complementary angles

Use reference angles to find the value of a trigonometric function

Use a calculator to find the value of a trigonometric function

Graph the trigonometric functions, including transformations

Find the period and amplitude of a sinusoidal function and use them to graph the function

Find a function whose sinusoidal graph is given

Find a sinusoidal function from data

FILL-IN-THE-BLANK ITEMS

1. Two rays drawn with a common vertex form a(n) _____. One of the rays is called the _____ _____; the other is called the _____ _____.

2. In the formula $s = r\theta$ for measuring the length s of arc along a circle of radius r, the angle θ must be measured in _____.

3. 180 degrees = _____ radians.

4. Two acute angles whose sum is a right angle are called _____.

5. The sine and _____ functions are cofunctions.

6. An angle is in _____ _____ if its vertex is at the origin and its initial side coincides with the positive x-axis.

REVIEW EXERCISES

Blue problem numbers indicate the author's suggestions for use in a Practice Test.

In Problems 1–4, solve each triangle.

1.

2.

3.

4.

In Problems 5–24, find the remaining angle(s) and side(s) of each triangle, if it (they) exists. If no triangle exists, say "No triangle."

5. $\alpha = 50°$, $\beta = 30°$, $a = 1$

6. $\alpha = 10°$, $\gamma = 40°$, $c = 2$

7. $\alpha = 100°$, $a = 5$, $c = 2$

8. $a = 2$, $c = 5$, $\alpha = 60°$

9. $a = 3$, $c = 1$, $\gamma = 110°$

10. $a = 3$, $c = 1$, $\gamma = 20°$

11. $a = 3$, $c = 1$, $\beta = 100°$

12. $a = 3$, $b = 5$, $\beta = 80°$

13. $a = 2$, $b = 3$, $c = 1$

14. $a = 10$, $b = 7$, $c = 8$

15. $a = 1$, $b = 3$, $\gamma = 40°$

16. $a = 4$, $b = 1$, $\gamma = 100°$

17. $a = 5$, $b = 3$, $\alpha = 80°$

18. $a = 2$, $b = 3$, $\alpha = 20°$

19. $a = 1$, $b = \frac{1}{2}$, $c = \frac{4}{3}$

The Chapter Review checks students' understanding of the chapter material in several ways. "Things to Know" gives a general overview of review topics. The "How To" section gives the student a concept-by-concept listing of the operations they are expected to perform. The "Review Exercises" then serve as a chance to practice the concepts presented within the chapter. Contained here are author suggestions for a practice test.

Students master and feel confident in knowing the chapter concepts before moving on to tackle more difficult mathematics. The Chapter Review section is also another excellent review for students when preparing for an exam or final.

Supplements

STUDENT SUPPLEMENTS

- **Student Solutions Manual**
  Worked solutions to all odd-numbered exercises from the text, including chapter review problems
  Student Solutions Manual for *Trigonometry*
  ISBN: 0-13-096122-1
- **Lecture Videos**
  The instructional tapes, in a lecture format, feature worked-out examples and exercises from each section of the text.
  ISBN: 0-13-096121-3
- *New York Times* **Themes of the Times**
  A free newspaper from Prentice Hall and *The New York Times*. Interesting and current articles on mathematics which invite discussion and writing about mathematics.
  ISBN: 0-13-973520-8
- **Mathematics on the Internet**
  Free guide providing a brief history of the Internet, discussing the use of the World Wide Web, and describing how to find your way within the Internet and how to find others on it.
  ISBN: 0-13-083998-1

INSTRUCTOR SUPPLEMENTS

- **Instructor's Solutions Manual**
  Contains complete step-by-step worked-out solutions to all even-numbered exercises in the textbook.
  Instructor's Solutions Manual for *Trigonometry*
  ISBN: 0-13-961186-X
- **Test Item File**
  Hard copy of the algorithmic computerized testing materials.
  Test Item File for *Trigonometry*
  ISBN: 0-13-020341-6
- **TestPro**
  - Algorithmically generates multiple forms of different but equivalent tests by chapter, section, or objective.
  - Includes multiple-choice and free-response options.
  - Features editing and formatting capabilities.

  For Windows and Macintosh.

  ISBN: IBM TestPro 0-13-096123-X
  ISBN: Macintosh TestPro 0-13-961194-0

http://www.prenhall.com/sulli-

THE SULLIVAN WEB SITE

For students to succeed in mathematics, they need practice in problem solving and the opportunity to explore. Prentice Hall's Companion Website for Sullivan provides both—and much more:

- Each text-specific site features easy-to-use study and exploration aids.
- Organized by chapter
- **Destinations.** Links to interesting and relevant mathematics sites.
- **Communication Center.** Links to online chat board and messaging functions.
- **Problem Solving Center.** Offers both Practice Exercises and Quizzes in multiple choice forms. Hints and page references are included. Quizzes and exercises offer students immediate feedback and the opportunity to send completed quizzes to their instructors.
- **Chapter Projects.** Linked to the Internet Projects at the beginning of each chapter.

Check out the site today!

Chapter 1	Sir Isaac Newton	Erich Lessing/Art Resource, NY
Chapter 2	Annapolis Tidal Generating Station	Nova Scotia Power
Chapter 3	A Prism	George Dodson/Uniphoto
Chapter 4	Polyhedra	Prof. George W. Hart
Chapter 5	Golden Gate Bridge	Brian Parker/Tom Stack & Associates
Chapter 6	Spirograph (drawing)	Kathy Enfalice
Chapter 7	Honeybee	SuperStock, Inc.

Functions and Their Graphs

On the page following each chapter opener is an Internet Excursion. These Excursions will place you in intriguing situations where information from various Internet resources is needed to answer thought-provoking questions. All the links and even more information on these Excursions can be found on the Sullivan website. The URL address is:

http.//www.prenhall.com/sullivan

Once at the website, find the appropriate text and explore the student's pages of the site. You will find an interactive version of the Excursion, as well as many other useful materials.

Turn the page and learn all about Isaac Newton and his mathematical discoveries. . . .

OUTLINE

THE CATALOG OF EQUATIONS

Trinity College, 1667. You have just begun your college work and one of your teachers is a strange fellow named *Isaac Newton*. His ideas are difficult to understand and very radical. Yet, when you listen to him speak, his passion makes a profound impression on you. His curious experiments and clever devices attract you to him.

After class one day, he asks you to help him with his Catalog of Equations. He quickly explains about the coordinate geometry of Descartes and how algebra and geometry are now joined together. He assigns to you the equation

$$(-2x + x^2 + y^2)^2 = x^2 + y^2$$

He gives you a manuscript for *The Witch* and asks you to do a similar work for your assigned curve. As you walk away, you feel a great sense of honor in being selected to do research for Isaac Newton.

1. Have you ever thought about such a Catalog of Equations? Take a quick tour of the famous curves at the *MacTutor Website*. Notice the Cartesian formulas (*x* and *y*) for the curves. How would you organize a catalog for the classification of algebraic curves?

2. Today a student does not hear much about a Catalog of Equations. Why do you suppose that is?

3. From looking at your equation, can you use any of the symmetry principles to get some idea of its graph? Can you graph this equation on your calculator? Are you able to graph it by making a chart of *x* and *y* values?

4. Visit the *Visual Dictionary of Plane Curves Website* and read about catacaustic curves. Can you determine the catacaustic of your curve with a light source on the edge? How could you graph it?

5. Tie together your explorations into a "manuscript."

Perhaps the most central idea in mathematics is the notion of a *function*. This important chapter deals with what a function is, how to determine characteristics of functions, how to graph functions, and how to find the inverse of certain functions.

The word *function* apparently was introduced by René Descartes in 1637. For him, a function simply meant any positive integral power of a variable *x*. Gottfried Wilhelm von Leibniz (1646–1716), who always emphasized the geometric side of mathematics, used the word *function* to denote any quantity associated with a curve, such as the coordinates of a point on the curve. Leonhard Euler (1707–1783) employed the word to mean any equation or formula involving variables and constants. His idea of a function is similar to the one most often used today in courses that precede calculus. Later, the use of functions in investigating heat flow equations led to a very broad definition, due to Lejeune Dirichlet (1805–1859), which describes a function as a rule or correspondence between two sets. It is his definition that we use here.

1.1 TOPICS FROM ALGEBRA AND GEOMETRY

1 Interval Notation
2 Absolute Value
3 Exponents
4 Pythagorean Theorem

Sets

When we want to treat a collection of similar but distinct objects as a whole, we use the idea of a **set.** For example, the set of *digits* consists of the collection of numbers 0, 1, 2, 3, 4, 5, 6, 7, 8, and 9. If we use the symbol D to denote the set of digits, then we can write

$$D = \{0, 1, 2, 3, 4, 5, 6, 7, 8, 9\}$$

In this notation, the braces { } are used to enclose the objects, or **elements,** in the set. This method of denoting a set is called the **roster method.** A second way to denote a set is to use **set-builder notation,** where the set D of digits is written as

$$D = \{ \quad x \quad | \quad x \text{ is a digit} \}$$

Read as "D is the set of all x such that x is a digit."

E X A M P L E   1

Using Set-builder Notation and the Roster Method

(a) $E = \{x | x \text{ is an even digit}\} = \{0, 2, 4, 6, 8\}$
(b) $O = \{x | x \text{ is an odd digit}\} = \{1, 3, 5, 7, 9\}$

∎

In listing the elements of a set, we do not list an element more than once because the elements of a set are distinct. Also, the order in which the elements are listed is not relevant. Thus, for example, {2, 3} and {3, 2} both represent the same set.

If every element of a set A is also an element of a set B, then we say that A is a **subset** of B. If two sets A and B have the same elements, then we say that A **equals** B. For example, {1, 2, 3} is a subset of {1, 2, 3, 4, 5}; and {1, 2, 3} equals {2, 3, 1}.

Real Numbers

Real numbers are represented by symbols such as

$$25, \quad 0, \quad -3, \quad \tfrac{1}{2}, \quad -\tfrac{5}{4}, \quad 0.125, \quad \sqrt{2}, \quad \pi, \quad \sqrt[3]{-2}, \quad 0.666\ldots$$

The set of **counting numbers,** or **natural numbers,** is the set $\{1, 2, 3, 4, \ldots\}$. (The three dots, called an **ellipsis,** indicate that the pattern continues indefinitely.) The set of **integers** is the set $\{\ldots, -3, -2, -1, 0, 1, 2, 3, \ldots\}$. A **rational number** is a number that can be expressed as a *quotient a/b* of two integers, where the integer b cannot be 0. Examples of rational numbers are $\tfrac{3}{4}, \tfrac{5}{2}, \tfrac{0}{4}$, and $-\tfrac{2}{3}$. Since $a/1 = a$ for any integer a, every integer is also a rational number. Real numbers that are not rational are called **irrational.** Examples of irrational numbers are $\sqrt{2}$ and π (the Greek letter pi), which equals the constant ratio of the circumference to the diameter of a circle. See Figure 1.

Real numbers can be represented as **decimals.** Rational real numbers have decimal representations that either **terminate** or are nonterminating with **repeating** blocks of digits. For example, $\tfrac{3}{4} = 0.75$, which terminates; and

FIGURE 1

$$\pi = \frac{C}{d}$$

$\frac{2}{3} = 0.666\ldots$, in which the digit 6 repeats indefinitely. Irrational real numbers have decimal representations that neither repeat nor terminate. For example, $\sqrt{2} = 1.414213\ldots$ and $\pi = 3.14159\ldots$. In practice, irrational numbers are generally represented by approximations. We use the symbol $\approx$ (read as "approximately equal to") to write $\sqrt{2} \approx 1.4142$ and $\pi \approx 3.1416$.

Often, letters are used to represent numbers. If the letter used is to represent *any* number from a given set of numbers, it is referred to as a **variable.** A **constant** is either a fixed number, such as 5, $\sqrt{2}$, and so on, or a letter that represents a fixed (possibly unspecified) number. In general, we will follow the practice of using letters near the beginning of the alphabet, such as *a*, *b*, and *c*, for constants and using those near the end, such as *x*, *y*, and *z*, for variables.

In working with expressions or formulas involving variables, the variables may only be allowed to take on values from a certain set of numbers, called the **domain of the variable.** For example, in the expression $1/x$, the variable *x* cannot take on the value 0, since division by 0 is not allowed.

It can be shown that there is a one-to-one correspondence between real numbers and points on a line. That is, every real number corresponds to a point on the line and, conversely, each point on the line has a unique real number associated with it. We establish this correspondence of real numbers with points on a line in the following manner.

We start with a line that is, for convenience, drawn horizontally. Pick a point on the line and label it *O*, for **origin.** Then pick another point some fixed distance to the right of *O* and label it *U*, for **unit.** The fixed distance, which may be 1 inch, 1 centimeter, 1 light-year, or any unit distance, determines the **scale.** We associate the real number 0 with the origin *O* and the number 1 with the point *U*. Refer to Figure 2. The point to the right of *U* that is twice as far from *O* as *U* is associated with the number 2. The point to the right of *U* that is three times as far from *O* as *U* is associated with the number 3. The point midway between *O* and *U* is assigned the number 0.5, or $\frac{1}{2}$. Corresponding points to the left of the origin *O* are assigned the numbers $-\frac{1}{2}$, -1, -2, -3, and so on. The real number *x* associated with a point *P* is called the **coordinate** of *P*, and the line whose points have been assigned coordinates is called the **real number line.** Notice in Figure 2 that we placed an arrowhead on the right end of the line to indicate the direction in which the assigned numbers increase. Figure 2 also shows the points associated with the irrational numbers $\sqrt{2}$ and π.

The real number line divides the real numbers into three classes: the **negative real numbers** are the coordinates of points to the left of the origin *O*; the real number **zero** is the coordinate of the origin *O*; the **positive real numbers** are the coordinates of points to the right of the origin *O*.

Let *a* and *b* be two real numbers. If the difference $a - b$ is positive, then we say that *a* is **greater than** *b* and write $a > b$. Alternatively, if $a - b$ is positive, we can also say that *b* is **less than** *a* and write $b < a$. Thus, $a > b$ and $b < a$ are equivalent statements.

On the real number line, if $a > b$, the point with coordinate *a* is to the right of the point with coordinate *b*. For example, $0 > -1$, $\pi > 3$, and $\sqrt{2} < 2$. Furthermore,

$$a > 0 \quad \text{is equivalent to } a \text{ is positive}$$
$$a < 0 \quad \text{is equivalent to } a \text{ is negative}$$

FIGURE 2
Real number line

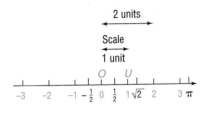

If the difference $a - b$ of two real numbers is positive or 0, that is, if $a > b$ or $a = b$, then we say that a is **greater than or equal to** b and write $a \geq b$. Alternatively, if $a \geq b$, we can also say that b is **less than or equal to** a and write $b \leq a$.

Statements of the form $a < b$ or $b > a$ are called **strict inequalities;** statements of the form $a \leq b$ or $b \geq a$ are called **nonstrict inequalities.** The symbols $>, <, \geq$, and $\leq$ are called **inequality signs.**

If x is a real number and $x \geq 0$, then x is either positive or 0. As a result, we describe the inequality $x \geq 0$ by saying that x is nonnegative.

Inequalities are useful in representing certain subsets of real numbers. In so doing, though, other variations of the inequality notation may be used.

E X A M P L E   2

FIGURE 3
$x > 4$

FIGURE 4
$x > 4$ and $x \leq 6$

Graphing Inequalities

(a) In the inequality $x > 4$, x is any number greater than 4. In Figure 3, we use a left parenthesis to indicate that the number 4 is not part of the graph.

(b) In the inequality $4 < x \leq 6$, x is any number between 4 and 6, including 6 but excluding 4. In Figure 4, we use a right bracket to indicate that 6 is part of the graph.

■

   Now work Problem 15 (in the exercise set at the end of this section).

1   Let a and b represent two real numbers with $a < b$: A **closed interval,** denoted by **[a, b],** consists of all real numbers x for which $a \leq x \leq b$. An **open interval,** denoted by **(a, b),** consists of all real numbers x for which $a < x < b$. The **half-open** or **half-closed intervals** are **(a, b],** consisting of all real numbers x for which $a < x \leq b$, and **[a, b),** consisting of all real numbers x for which $a \leq x < b$. In each of these definitions, a is called the **left endpoint** and b the **right endpoint** of the interval. Figure 5 illustrates each type of interval.

FIGURE 5

(a) Closed interval $[a,b];\ a \leq x \leq b$

(b) Open interval $(a,b);\ a < x < b$

(c) Half-open (half-closed) intervals $[a,b);\ a \leq x < b$ $(a,b];\ a < x \leq b$

The symbol ∞ (read as "infinity") is not a real number, but a notational device used to indicate unboundedness in the positive direction. The symbol $-\infty$ (read as "minus infinity") also is not a real number, but a notational device used to indicate unboundedness in the negative direction. Using the symbols ∞ and $-\infty$, we can define five other kinds of intervals:

$[a, \infty)$ consists of all real numbers x for which $x \geq a$ ($a \leq x < \infty$)
(a, ∞) consists of all real numbers x for which $x > a$ ($a < x < \infty$)
$(-\infty, a]$ consists of all real numbers x for which $x \leq a$ ($-\infty < x \leq a$)
$(-\infty, a)$ consists of all real numbers x for which $x < a$ ($-\infty < x < a$)
$(-\infty, \infty)$ consists of all real numbers x ($-\infty < x < \infty$)

Figure 6 illustrates these types of intervals.

FIGURE 6

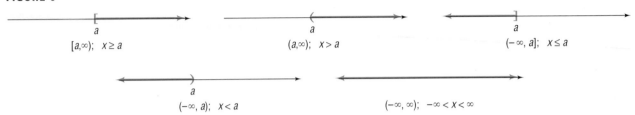

Note that ∞ and $-\infty$ are never included as endpoints, since they are not real numbers.

Now work Problems 21 and 25.

FIGURE 7

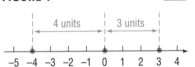

The *absolute value* of a number a is the distance from the point whose coordinate is a to the origin. For example, the point whose coordinate is -4 is 4 units from the origin. The point whose coordinate is 3 is 3 units from the origin. See Figure 7. Thus, the absolute value of -4 is 4, and the absolute value of 3 is 3.

A more formal definition of absolute value is given next.

> The **absolute value** of a real number a, denoted by the symbol $|a|$, is defined by the rules
>
> $$|a| = a \quad \text{if } a \geq 0 \quad \text{and} \quad |a| = -a \quad \text{if } a < 0$$

For example, since $-4 < 0$, then the second rule must be used to get $|-4| = -(-4) = 4$.

E X A M P L E   3   Computing Absolute Value

(a) $|8| = 8$ (b) $|0| = 0$ (c) $|-15| = 15$ ∎

Now work Problem 29.

Look again at Figure 7. The distance from the point whose coordinate is -4 to the point whose coordinate is 3 is 7 units. This distance is the difference $3 - (-4)$, obtained by subtracting the smaller coordinate from the larger. However, since $|3 - (-4)| = |7| = 7$ and $|-4 - 3| = |-7| = 7$, we can use the absolute value to calculate the distance between two points without being concerned about which coordinate is smaller.

If P and Q are two points on a real number line with coordinates a and b, respectively, the **distance between P and Q**, denoted by $d(P, Q)$, is

$$d(P, Q) = |b - a|$$

Since $|b - a| = |a - b|$, it follows that $d(P, Q) = d(Q, P)$.

E X A M P L E   4   Finding Distance on a Number Line

Let P, Q, and R be points on the real number line with coordinates $-5, 7,$ and -3, respectively. Find the distance:

(a) Between P and Q (b) Between Q and R

Solution   (a) $d(P, Q) = |7 - (-5)| = |12| = 12$ (See Figure 8.)
(b) $d(Q, R) = |-3 - 7| = |-10| = 10$

FIGURE 8

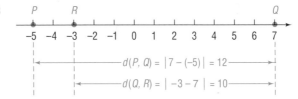

Exponents

3 Integer exponents provide a shorthand device for representing repeated multiplications of a real number.

If a is a real number and n is a positive integer, then the symbol a^n represents the product of n factors of a. That is,

$$a^n = \underbrace{a \cdot a \cdot \ldots \cdot a}_{n \text{ factors}}$$

where it is understood that $a^1 = a$. Thus, $a^2 = a \cdot a$, $a^3 = a \cdot a \cdot a$, and so on. In the expression a^n, a is called the **base** and n is called the **exponent,** or **power.** We read a^n as "a raised to the power n" or as "a to the nth power." We usually read a^2 as "a squared" and a^3 as "a cubed."

Care must be taken when parentheses are used in conjunction with exponents. For example, $-2^4 = -(2 \cdot 2 \cdot 2 \cdot 2) = -16$, whereas $(-2)^4 = (-2) \cdot (-2) \cdot (-2) \cdot (-2) = 16$. Notice the difference: The exponent applies only to the number or parenthetical expression immediately preceding it.

If $a \neq 0$, we define

$$a^0 = 1 \quad \text{if } a \neq 0$$

If $a \neq 0$ and if n is a positive integer, then we define

$$a^{-n} = \frac{1}{a^n} \qquad \text{if } a \neq 0$$

With these definitions, the symbol a^n is defined for any integer n.

The following properties, called the **laws of exponents,** can be proved using the preceding definitions. In the list, a and b are real numbers, and m and n are integers.

Laws of Exponents

$$a^m a^n = a^{m+n} \qquad (a^m)^n = a^{mn} \qquad (ab)^n = a^n b^n$$

$$\frac{a^m}{a^n} = a^{m-n} = \frac{1}{a^{n-m}}, \qquad \text{if } a \neq 0 \qquad \left(\frac{a}{b}\right)^n = \frac{a^n}{b^n}, \qquad \text{if } b \neq 0$$

E X A M P L E   5

Using the Laws of Exponents

Write each expression so that all exponents are positive.

(a) $\dfrac{x^5 y^{-2}}{x^3 y}, \quad x \neq 0, y \neq 0$ (b) $\dfrac{xy}{x^{-1} - y^{-1}}, \quad x \neq 0, y \neq 0$

Solution   (a) $\dfrac{x^5 y^{-2}}{x^3 y} = \dfrac{x^5}{x^3} \cdot \dfrac{y^{-2}}{y} = x^{5-3} \cdot y^{-2-1} = x^2 y^{-3} = x^2 \cdot \dfrac{1}{y^3} = \dfrac{x^2}{y^3}$

(b) $\dfrac{xy}{x^{-1} - y^{-1}} = \dfrac{xy}{\dfrac{1}{x} - \dfrac{1}{y}} = \dfrac{xy}{\dfrac{y - x}{xy}} = \dfrac{(xy)(xy)}{y - x} = \dfrac{x^2 y^2}{y - x}$

Now work Problem 57.

Let $n \geq 2$ be an integer.

The **principal nth root of a number a,** symbolized by $\sqrt[n]{a}$, where $n \geq 2$ is an integer, is defined as follows:

$$\sqrt[n]{a} = b \quad \text{means} \quad a = b^n \qquad \begin{array}{l} \text{where } a \geq 0 \text{ and } b \geq 0 \text{ if } n \geq 2 \text{ is even} \\ \text{and } a, b \text{ are any real numbers if } n \geq 3 \\ \text{is odd} \end{array}$$

Notice that if a is negative and n is even then $\sqrt[n]{a}$ is not defined. When it is defined, the principal nth root of a number is unique.

The symbol $\sqrt[n]{a}$ for the principal nth root of a is sometimes called a **radical**; the integer n is called the **index,** and a is called the **radicand.** If the index of a radical is 2, we call $\sqrt[2]{a}$ the **square root** of a and omit the index 2 by simply writing $\sqrt{a}$. If the index is 3, we call $\sqrt[3]{a}$ the **cube root** of a.

E X A M P L E 6

Simplifying Principal nth Roots

(a) $\sqrt[3]{8} = \sqrt[3]{2^3} = 2$ (b) $\sqrt[3]{-64} = \sqrt[3]{(-4)^3} = -4$

(c) $\sqrt[4]{\frac{1}{16}} = \sqrt[4]{(\frac{1}{2})^4} = \frac{1}{2}$ (d) $\sqrt[6]{(-2)^6} = |-2| = 2$ ∎

Here (a), (b), and (c) are examples of **perfect roots.** Thus, 8 and -64 are perfect cube roots, since $8 = 2^3$ and $-64 = (-4)^3$; $\frac{1}{2}$ is a perfect 4th root of $\frac{1}{16}$, since $\frac{1}{16} = (\frac{1}{2})^4$.

Notice the need for the absolute value in Example 6(d). If n is even, then a^n is positive whether $a > 0$ or $a < 0$. But if n is even, the principal nth root must be nonnegative. Hence, the reason for using the absolute value—it gives a nonnegative result.

In general, if $n \geq 2$ is a positive integer and a is a real number,

$$\sqrt[n]{a^n} = a \qquad \text{if } n \geq 3 \text{ is odd} \tag{1a}$$

$$\sqrt[n]{a^n} = |a| \qquad \text{if } n \geq 2 \text{ is even} \tag{1b}$$

E X A M P L E 7

Simplifying Radicals

(a) $\sqrt{8} = \sqrt{4 \cdot 2} = \sqrt{4} \cdot \sqrt{2} = 2\sqrt{2}$

(b) $\sqrt[3]{-16} = \sqrt[3]{-8 \cdot 2} = \sqrt[3]{-8} \cdot \sqrt[3]{2} = -2\sqrt[3]{2}$

(c) $\sqrt{x^2} = |x|$ ∎

Now work Problem 49.

Radicals are used to define **rational exponents.** If a is a real number and $n \geq 2$ is an integer, then

$$a^{1/n} = \sqrt[n]{a}$$

provided $\sqrt[n]{a}$ exists.

If a is a real number and m and n are integers containing no common factors with $n \geq 2$, then

$$a^{m/n} = \sqrt[n]{a^m} = (\sqrt[n]{a})^m \qquad (2)$$

provided $\sqrt[n]{a}$ exists.

In simplifying $a^{m/n}$, either $\sqrt[n]{a^m}$ or $(\sqrt[n]{a})^m$ may be used. Generally, taking the root first, as in $(\sqrt[n]{a})^m$, is preferred.

E X A M P L E   8

Using Equation (2)

(a) $8^{2/3} = (\sqrt[3]{8})^2 = 2^2 = 4$ (b) $16^{3/2} = (\sqrt{16})^3 = 4^3 = 64$

(c) $(-8x^5)^{1/3} = \sqrt[3]{-8x^3 \cdot x^2} = \sqrt[3]{(-2x)^3 \cdot x^2} = \sqrt[3]{(-2x)^3}\,\sqrt[3]{x^2}$

$$= -2x\sqrt[3]{x^2}$$

∎

   Now work Problem 47.

Pythagorean Theorem

FIGURE 9

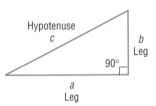

4 The *Pythagorean Theorem* is a statement about *right triangles*. A **right triangle** is one that contains a **right angle,** that is, an angle of 90°. The side of the triangle opposite the 90° angle is called the **hypotenuse;** the remaining two sides are called **legs.** In Figure 9 we have used c to represent the length of the hypotenuse and a and b to represent the lengths of the legs. Notice the use of the symbol ⌐ to show the 90° angle. We now state the Pythagorean Theorem.

> **Pythagorean Theorem**
>
> In a right triangle, the square of the length of the hypotenuse is equal to the sum of the squares of the lengths of the legs. That is, in the right triangle shown in Figure 9,
>
> $$c^2 = a^2 + b^2 \qquad (3)$$

∎

E X A M P L E   9

Finding the Hypotenuse of a Right Triangle

In a right triangle, one leg is of length 4 and the other is of length 3. What is the length of the hypotenuse?

Solution   Since the triangle is a right triangle, we use the Pythagorean Theorem with $a = 4$ and $b = 3$ to find the length c of the hypotenuse. Thus, from equation (3), we have

$$c^2 = a^2 + b^2$$
$$c^2 = 4^2 + 3^2 = 16 + 9 = 25$$
$$c = 5$$

∎

   Now work Problem 69.

The converse of the Pythagorean Theorem is also true.

Converse of the Pythagorean Theorem

In a triangle, if the square of the length of one side equals the sum of the squares of the lengths of the other two sides, then the triangle is a right triangle. The 90° angle is opposite the longest side.

E X A M P L E   10   Verifying That a Triangle Is a Right Triangle

Show that a triangle whose sides are of lengths 5, 12, and 13 is a right triangle. Identify the hypotenuse.

Solution   We square the lengths of the sides:

FIGURE 10

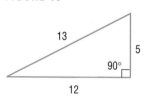

$$5^2 = 25, \qquad 12^2 = 144, \qquad 13^2 = 169$$

Notice that the sum of the first two squares (25 and 144) equals the third square (169). Hence, the triangle is a right triangle. The longest side, 13, is the hypotenuse. See Figure 10.

Now work Problem 79.

Geometry Formulas

Certain formulas from geometry are useful in solving algebra problems. We list some of these formulas next.

For a rectangle of length l and width w,

$$\text{Area} = lw \qquad \text{Perimeter} = 2l + 2w$$

For a triangle with base b and altitude h,

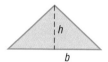

$$\text{Area} = \frac{1}{2}bh$$

For a circle of radius r (diameter $d = 2r$),

$$\text{Area} = \pi r^2 \qquad \text{Circumference} = 2\pi r = \pi d$$

For a rectangular box of length *l*, width *w*, and height *h*,

$$\text{Volume} = lwh$$

Calculators

Calculators are finite machines. As a result, they are incapable of displaying decimals that contain a large number of digits. For example, some calculators are capable of displaying only eight digits. When a number requires more than eight digits, the calculator either truncates or rounds. To see how your calculator handles decimals, divide 2 by 3. How many digits do you see? Is the last digit a 6 or a 7? If it is a 6, your calculator truncates; if it is a 7, your calculator rounds.

There are different kinds of calculators. An **arithmetic** calculator can only add, subtract, multiply, and divide numbers; therefore, this type is not adequate for this course. **Scientific** calculators have all the capabilities of arithmetic calculators and also contain **function keys** labeled ln, log, sin, cos, tan, x^y, inv, and so on. As you proceed through this text, you will discover how to use many of the function keys. **Graphing** calculators have all the capabilities of scientific calculators and contain a screen on which graphs can be displayed.

For those who have access to a graphing calculator, we have included comments, examples, and exercises marked with a 📟, indicating that a graphing calculator is required. We have also included an appendix that explains some of the capabilities of a graphing calculator. The 📟 comments, examples, and exercises may be omitted without loss of continuity, if so desired.

Arithmetic operations on your calculator are generally shown as keys labeled as follows:

$\boxed{+}$ Addition $\boxed{-}$ Subtraction $\boxed{\times}$ Multiplication $\boxed{\div}$ Division

Your calculator also has the following key:

$\boxed{\text{AC}}$ or $\boxed{\text{C}}$ To clear the memory and the display window of previous entries.

The next example illustrates a use of the keys for parentheses,

$\boxed{(}$ and $\boxed{)}$

EXAMPLE 11   Using a Calculator

Evaluate: $\dfrac{22 + 8}{8 + 2}$

Solution   Remember, we treat this expression as if parentheses enclose the numerator and the denominator.

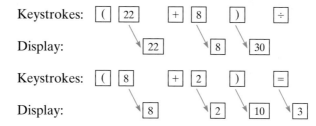

Be careful! If you work the problem in Example 11 without using parentheses, you obtain $22 + 8/8 + 2 = 22 + 1 + 2 = 25$.

   Now work Problem 93.

EXERCISES

In Problems 1–10, replace the question mark by $<$, $>$, or $=$, whichever is correct.

1. $\frac{1}{2}$? 0   **2.** 5 ? 6   **3.** -1 ? -2 **4.** -3 ? $-\frac{5}{2}$ **5.** π ? 3.14

6. $\sqrt{2}$? 1.41   **7.** $\frac{1}{2}$? 0.5   **8.** $\frac{1}{3}$? 0.33   **9.** $\frac{2}{3}$? 0.67   **10.** $\frac{1}{4}$? 0.25

11. On the real number line, label the points with coordinates $0, 1, -1, \frac{5}{2}, -2.5, \frac{3}{4}$, and 0.25.

12. Repeat Problem 11 for the coordinates $0, -2, 2, -1.5, \frac{3}{2}, \frac{1}{3}$, and $\frac{2}{3}$.

In Problems 13–20, write each statement as an inequality.

13. x is positive

14. z is negative

15. x is less than 2

16. y is greater than -5

17. x is less than or equal to 1

18. x is greater than or equal to 2

19. x is less than 5 and x is greater than 2

20. y is less than or equal to 2 and y is greater than 0

In Problems 21–24, write each inequality using interval notation, and illustrate each inequality using the real number line.

21. $0 \le x \le 4$ **22.** $-1 < x < 5$ **23.** $4 \le x < 6$ **24.** $-2 < x \le 0$

In Problems 25–28, write each interval as an inequality involving x, and illustrate each inequality using the real number line.

25. $[2, 5]$ **26.** $(1, 2)$ **27.** $[4, \infty)$ **28.** $(-\infty, 2]$

In Problems 29–32, find the value of each expression if $x = 2$ and $y = -3$.

29. $|x + y|$ **30.** $|x - y|$ **31.** $|x| + |y|$ **32.** $|x| - |y|$

In Problems 33–52, simplify each expression.

33. 3^0 **34.** 3^2 **35.** 4^{-2} **36.** $(-3)^2$

37. $\left(\frac{2}{3}\right)^2$ **38.** $\left(\frac{-4}{5}\right)^3$ **39.** $3^{-6} \cdot 3^4$ **40.** $4^{-2} \cdot 4^3$

41. $\left(\frac{2}{3}\right)^{-2}$ **42.** $\left(\frac{3}{2}\right)^{-3}$ **43.** $\dfrac{2^3 \cdot 3^2}{2 \cdot 3^{-2}}$ **44.** $\dfrac{3^{-2} \cdot 5^3}{3 \cdot 5}$

45. $9^{3/2}$ **46.** $16^{3/4}$ **47.** $(-8)^{4/3}$ **48.** $(-27)^{2/3}$

49. $\sqrt{32}$ **50.** $\sqrt[3]{24}$ **51.** $\sqrt[3]{-\frac{8}{27}}$ **52.** $\sqrt{\frac{4}{9}}$

In Problems 53–68, simplify each expression so that all exponents are positive. Whenever an exponent is negative or 0, we assume that the base does not equal 0.

53. $x^0 y^2$

54. $x^{-1} y$

55. $x^{-2} y$

56. $x^4 y^0$

57. $\dfrac{x^{-2} y^3}{xy^4}$

58. $\dfrac{x^{-2} y}{xy^2}$

59. $\left(\dfrac{4x}{5y}\right)^{-2}$

60. $(xy)^{-2}$

61. $\dfrac{x^{-1} y^{-2} z}{x^2 y z^3}$

62. $\dfrac{3x^{-2} y z^2}{x^4 y^{-3} z}$

63. $\dfrac{(-2)^3 x^4 (yz)^2}{3^2 xy^3 z^4}$

64. $\dfrac{4x^{-2}(yz)^{-1}}{(-5)^2 x^4 y^2 z^{-2}}$

65. $\dfrac{x^{-2}}{\dfrac{1}{x^2} + \dfrac{1}{y^2}}$

66. $\dfrac{x^{-1} + y^{-1}}{x^{-1} - y^{-1}}$

67. $\left(\dfrac{3x^{-1}}{4y^{-1}}\right)^{-2}$

68. $\left(\dfrac{5x^{-2}}{6y^{-2}}\right)^{-3}$

In Problems 69–78, a and b are the lengths of the legs of a right triangle and c is the length of the hypotenuse. Find the missing length.

69. $a = 5, b = 12, c = ?$

70. $a = 6, b = 8, c = ?$

71. $a = 10, b = 24, c = ?$

72. $a = 4, b = 3, c = ?$

73. $a = 7, b = 24, c = ?$

74. $a = 14, b = 48, c = ?$

75. $a = 3, c = 5, b = ?$

76. $b = 6, c = 10, a = ?$

77. $b = 7, c = 25, a = ?$

78. $a = 10, c = 13, b = ?$

In Problems 79–84, the lengths of the sides of a triangle are given. Determine which are right triangles. For those that are right triangles, identify the hypotenuse.

79. $3, 4, 5$

80. $6, 8, 10$

81. $4, 5, 6$

82. $2, 2, 3$

83. $7, 24, 25$

84. $10, 24, 26$

In Problems 85–96, use a calculator to approximate each expression. Round your answer to two decimal places.

85. $(8.51)^2$

86. $(9.62)^2$

87. $4.1 + (3.2)(8.3)$

88. $(8.1)(4.2) + 6.1$

89. $(8.6)^2 + (6.1)^2$

90. $(3.1)^2 + (9.6)^2$

91. $8.6 + \dfrac{10.2}{4.2}$

92. $9.1 - \dfrac{8.2}{10.2}$

93. $\dfrac{2.3 - 9.25}{8.91 + 5.4}$

94. $\dfrac{4.73 - 2.4}{81 + 6.39}$

95. $\dfrac{\pi + 8}{10.2 + 8.6}$

96. $\dfrac{21.3 - \pi}{6.1 + 8.8}$

97. Geometry   Find the diagonal of a rectangle whose length is 8 inches and whose width is 5 inches.

98. Geometry   Find the length of a rectangle of width 3 inches if its diagonal is 20 inches long.

99. Finding the Length of a Guy Wire   A radio transmission tower is 100 feet high. How long does a guy wire need to be if it is to connect a point halfway up the tower to a point 30 feet from the base?

100. Answer Problem 99 if the guy wire is attached to the top of the tower.

101. How Far Can You See?   The tallest inhabited building in the world is the Sears Tower in Chicago.* If the observation tower is 1450 feet above ground level, use the figure to determine how far a person standing in the observation tower can see (with the aid of a telescope). Use 3960 miles for the radius of Earth.

[**Note:** 1 mile = 5280 feet]

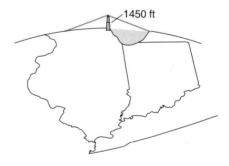

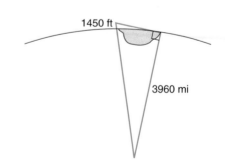

*Source: Council on Tall Buildings and Urban Habitat (1997): Sears Tower No. 1 for tallest roof (1450 ft) and tallest occupied floor (1431 ft).

In Problems 102–104, use the fact that the radius of Earth is 3960 miles.

102. How Far Can You See? The conning tower of the USS *Silversides,* a World War II submarine now permanently stationed in Muskegon, Michigan, is approximately 20 feet above sea level. How far can one see from the conning tower?

103. How Far Can You See? A person who is 6 feet tall is standing on the beach in Fort Lauderdale, Florida and looks out into the Atlantic Ocean. Suddenly, a ship appears on the horizon. How far is the ship from shore?

104. How Far Can You See? The deck of a destroyer is 100 feet above sea level. How far can a person see from the deck? How far can a person see from the bridge, which is 150 feet above sea level?

105. If $a \le b$ and $c > 0$, show that $ac \le bc$.
[**Hint:** Since $a < b$, it follows that $a - b \le 0$. Now multiply each side by c.]

106. If $a \le b$ and $c < 0$, show that $ac \ge bc$.

107. If $a < b$, show that $a < (a + b)/2 < b$. The number $(a + b)/2$ is called the **arithmetic mean** of a and b.

108. Refer to Problem 107. Show that the arithmetic mean of a and b is equidistant from a and b.

 109. Are there any real numbers that are both rational and irrational? Are there any real numbers that are neither? Explain your reasoning.

110. Explain why the sum of a rational number and an irrational number must be irrational.

111. What rational number does the repeating decimal 0.9999 . . . equal?

112. Is there a positive real number "closest" to 0?

113. I'm thinking of a number! It lies between 1 and 10; its square is rational and lies between 1 and 10. The number is larger than π. Correct to two decimal places, name the number. Now think of your own number, describe it, and challenge a fellow student to name it.

114. Write a brief paragraph that illustrates the similarities between "less than" ($<$) and "less than or equal" ($\le$).

115. The Gibb's Hill Lighthouse, Southampton, Bermuda, in operation since 1846, stands 117 feet high on a hill 245 feet high, so its beam of light is 362 feet above sea level. A brochure states that the light itself can be seen on the horizon about 26 miles distant. Verify the correctness of this information. The brochure further states that ships 40 miles away can see the light and planes flying at 10,000 feet can see it 120 miles away. Verify the accuracy of these statements. What assumption did the brochure make about the height of the ship?

120 miles

40 miles

116. You have 1000 feet of flexible pool siding and wish to construct a swimming pool. Experiment with rectangular-shaped pools with perimeters of 1000 feet. How do their areas vary? What is the shape of the rectangle with the largest area? Now compute the area enclosed by a circular pool with a perimeter (circumference) of 1000 feet. What would be your choice of shape for the pool? If rectangular, what is your preference for dimensions? Justify your choice. If your only consideration is to have a pool that encloses the most area, what shape should you use?

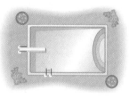

1.2 | RECTANGULAR COORDINATES; GRAPHS; CIRCLES

We locate a point on the real number line by assigning it a single real number, called the *coordinate of the point.* For work in a two-dimensional plane, we locate points by using two numbers.

We begin with two real number lines located in the same plane: one horizontal and the other vertical. We call the horizontal line the **x-axis,** the vertical line the **y-axis,** and the point of intersection the **origin O.** We assign coordinates to every point on these number lines as shown in Figure 11, using a convenient scale. In mathematics, we usually use the same scale on each axis; in applications, a different scale is often used on each axis.

FIGURE 11

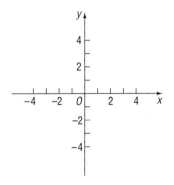

The origin O has a value of 0 on both the x-axis and the y-axis. We follow the usual convention that points on the x-axis to the right of O are associated with positive real numbers, and those to the left of O are associated with negative real numbers. Those on the y-axis above O are associated with positive real numbers, and those below O are associated with negative real numbers. In Figure 11, the x-axis and y-axis are labeled as x and y, respectively, and we have used an arrow at the end of each axis to denote the positive direction.

The coordinate system described here is called a **rectangular,** or **Cartesian* coordinate system.** The plane formed by the x-axis and y-axis is sometimes called the **xy-plane,** and the x-axis and y-axis are referred to as the **coordinate axes.**

Any point P in the xy-plane can then be located by using an **ordered pair** (x, y) of real numbers. Let x denote the signed distance of P from the y-axis (*signed* in the sense that, if P is to the right of the y-axis, then $x > 0$, and if P is to the left of the y-axis, then $x < 0$); and let y denote the signed distance of P from the x-axis. The ordered pair (x, y), also called the **coordinates** of P, then gives us enough information to locate the point P in the plane.

FIGURE 12

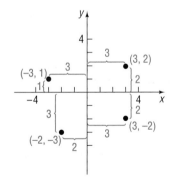

For example, to locate the point whose coordinates are $(-3, 1)$, go 3 units along the x-axis to the left of O and then go straight up 1 unit. We **plot** this point by placing a dot at this location. See Figure 12, in which the points with coordinates $(-3, 1)$, $(-2, -3)$, $(3, -2)$, and $(3, 2)$ are plotted.

The origin has coordinates $(0, 0)$. Any point on the x-axis has coordinates of the form $(x, 0)$, and any point on the y-axis has coordinates of the form $(0, y)$.

If (x, y) are the coordinates of a point P, then x is called the **x-coordinate,** or **abscissa,** of P, and y is the **y-coordinate,** or **ordinate,** of P. We identify the

---

*Named after René Descartes (1596–1650), a French mathematician, philosopher, and theologian.

FIGURE 13

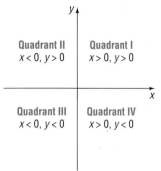

point P by its coordinates (x, y) by writing $P = (x, y)$. Usually, we will simply say "the point (x, y)" rather than "the point whose coordinates are (x, y)."

The coordinate axes divide the xy-plane into four sections, called **quadrants,** as shown in Figure 13. In quadrant I, both the x-coordinate and the y-coordinate of all points are positive; in quadrant II, x is negative and y is positive; in quadrant III, both x and y are negative; and in quadrant IV, x is positive and y is negative. Points on the coordinate axes belong to no quadrant.

Now work Problem 1.

Comment   On a graphing calculator, you can set the scale on each axis. Once this has been done, you obtain the **viewing rectangle.** See Figure 14 for a typical viewing rectangle. You should now read Section B.1, The Viewing Rectangle, in Appendix B.   ▬

FIGURE 14

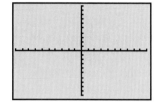

If the same units of measurement, such as inches, centimeters, and so on, are used for both the x-axis and the y-axis, then all distances in the xy-plane can be measured using this unit of measurement. The **distance formula** provides a straightforward method for computing the distance between two points.

Theorem   Distance Formula

The distance between two points $P_1 = (x_1, y_1)$ and $P_2 = (x_2, y_2)$, denoted by $d(P_1, P_2)$, is

$$d(P_1, P_2) = \sqrt{(x_2 - x_1)^2 + (y_2 - y_1)^2} \qquad (1)$$

▬

E X A M P L E   1

Finding the Distance between Two Points

Find the distance d between the points $(-4, 5)$ and $(3, 2)$.

Solution

Using the distance formula (1), the solution is obtained as follows:

$$d = \sqrt{[3 - (-4)]^2 + (2 - 5)^2} = \sqrt{7^2 + (-3)^2}$$
$$= \sqrt{49 + 9} = \sqrt{58} \approx 7.62$$

▬

Proof of the Distance Formula   Let (x_1, y_1) denote the coordinates of point P_1, and let (x_2, y_2) denote the coordinates of point P_2. Assume that the line joining P_1 and P_2 is neither horizontal nor vertical. Refer to Figure 15(a). The coordinates of P_3 are (x_2, y_1). The horizontal distance from P_1 to P_3 is the absolute value of the difference of the x-coordinates, or $|x_2 - x_1|$. The vertical distance from P_3 to P_2 is the absolute value of the difference of the y-coordinates, or $|y_2 - y_1|$. See Figure 15(b). The distance $d(P_1, P_2)$ that we seek is the length of the hypotenuse of the right triangle, so, by the Pythagorean Theorem, it follows that

$$[d(P_1, P_2)]^2 = |x_2 - x_1|^2 + |y_2 - y_1|^2$$
$$= (x_2 - x_1)^2 + (y_2 - y_1)^2$$
$$d(P_1, P_2) = \sqrt{(x_2 - x_1)^2 + (y_2 - y_1)^2}$$

FIGURE 15

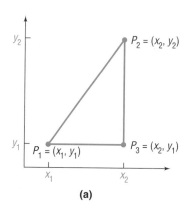

(a)

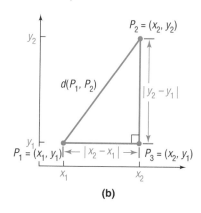

(b)

Now, if the line joining P_1 and P_2 is horizontal, then the y-coordinate of P_1 equals the y-coordinate of P_2; that is, $y_1 = y_2$. Refer to Figure 16(a). In this case, the distance formula (1) still works, because, for $y_1 = y_2$, it reduces to

$$d(P_1, P_2) = \sqrt{(x_2 - x_1)^2 + 0^2} = \sqrt{(x_2 - x_1)^2} = |x_2 - x_1|$$

A similar argument holds if the line joining P_1 and P_2 is vertical. See Figure 16(b). Thus, the distance formula is valid in all cases.

FIGURE 16

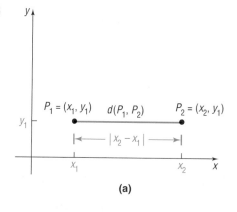

(a)

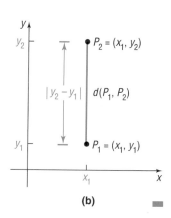

(b)

Now work Problem 25.

The distance between two points $P_1 = (x_1, y_1)$ and $P_2 = (x_2, y_2)$ is never a negative number. Furthermore, the distance between two points is 0 only when the points are identical, that is, when $x_1 = x_2$ and $y_1 = y_2$. Also, because $(x_2 - x_1)^2 = (x_1 - x_2)^2$ and $(y_2 - y_1)^2 = (y_1 - y_2)^2$, it makes no difference whether the distance is computed from P_1 to P_2 or from P_2 to P_1; that is, $d(P_1, P_2) = d(P_2, P_1)$.

The introduction to this chapter mentioned that rectangular coordinates enable us to translate geometry problems into algebra problems, and vice versa. The next example shows how algebra (the distance formula) can be used to solve geometry problems.

E X A M P L E   2   Using Algebra to Solve Geometry Problems

Consider the three points $A = (-2, 1)$, $B = (2, 3)$, and $C = (3, 1)$.

(a) Plot each point and form the triangle ABC.
(b) Find the length of each side of the triangle.
(c) Verify that the triangle is a right triangle.
(d) Find the area of the triangle.

Solution

(a) Points A, B, C and triangle ABC are plotted in Figure 17.

FIGURE 17

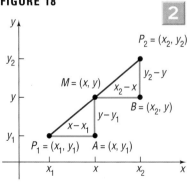

(b) $d(A, B) = \sqrt{[2 - (-2)]^2 + (3 - 1)^2} = \sqrt{16 + 4} = \sqrt{20} = 2\sqrt{5}$
$d(B, C) = \sqrt{(3 - 2)^2 + (1 - 3)^2} = \sqrt{1 + 4} = \sqrt{5}$
$d(A, C) = \sqrt{[3 - (-2)]^2 + (1 - 1)^2} = \sqrt{25 + 0} = 5$

(c) To show that the triangle is a right triangle, we need to show that the sum of the squares of the lengths of two of the sides equals the square of the length of the third side. (Why is this sufficient?) Looking at Figure 17, it seems reasonable to conjecture that the right angle is at vertex B. Thus, we shall check to see whether

$$[d(A, B)]^2 + [d(B, C)]^2 = [d(A, C)]^2$$

We find that

$$[d(A, B)]^2 + [d(B, C)]^2 = (2\sqrt{5})^2 + (\sqrt{5})^2$$
$$= 20 + 5 = 25 = [d(A, C)]^2$$

so it follows from the Converse of the Pythagorean Theorem that triangle ABC is a right triangle.

(d) Because the right angle is at B, the sides AB and BC form the base and altitude of the triangle. Its area is therefore

$$\text{Area} = \frac{1}{2}(\text{base})(\text{altitude}) = \frac{1}{2}(2\sqrt{5})(\sqrt{5}) = 5 \text{ square units}$$

Now work Problem 19.

FIGURE 18

We now derive a formula for the coordinates of the **midpoint of a line segment.** Let $P_1 = (x_1, y_1)$ and $P_2 = (x_2, y_2)$ be the end points of a line segment, and let $M = (x, y)$ be the point on the line segment that is the same distance from P_1 as it is from P_2. See Figure 18. The triangles P_1AM and MBP_2 are congruent.* [Do you see why? Angle AP_1M = angle BMP_2.† Angle P_1MA = angle

*The following statement is a postulate from geometry. Two triangles are congruent if their sides are the same length (SSS), or if two sides and the included angle are the same (SAS), or if two angles and the included side are the same (ASA).

†Another postulate from geometry states that the transversal $\overline{P_1P_2}$ forms equal corresponding angles with the parallel lines $\overline{P_1A}$ and $\overline{MB}$.

MP_2B, and $d(P_1, M) = d(M, P_2)$ are given. Thus, we have angle–side–angle.] Hence, corresponding sides are equal in length. That is,

$$x - x_1 = x_2 - x \quad \text{and} \quad y - y_1 = y_2 - y$$
$$2x = x_1 + x_2 \qquad\qquad 2y = y_1 + y_2$$
$$x = \frac{x_1 + x_2}{2} \qquad\qquad y = \frac{y_1 + y_2}{2}$$

Theorem Midpoint Formula

The midpoint (x, y) of the line segment from $P_1 = (x_1, y_1)$ to $P_2 = (x_2, y_2)$ is

$$(x, y) = \left(\frac{x_1 + x_2}{2}, \frac{y_1 + y_2}{2} \right) \qquad (2)$$

Thus, to find the midpoint of a line segment, we average the x-coordinates and the y-coordinates of the end points.

E X A M P L E   3   Finding the Midpoint of a Line Segment

Find the midpoint of a line segment from $P_1 = (-5, 3)$ to $P_2 = (3, 1)$. Plot the points P_1 and P_2 and their midpoint. Check your answer.

Solution   We apply the midpoint formula (2) using $x_1 = -5, x_2 = 3, y_1 = 3,$ and $y_2 = 1$. Then the coordinates (x, y) of the midpoint M are

$$x = \frac{x_1 + x_2}{2} = \frac{-5 + 3}{2} = -1 \quad \text{and} \quad y = \frac{y_1 + y_2}{2} = \frac{3 + 1}{2} = 2$$

FIGURE 19

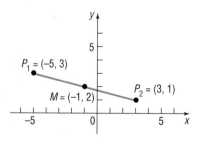

That is, $M = (-1, 2)$. See Figure 19.

Check: Because M is the midpoint, we check the answer by verifying that $d(P_1, M) = d(M, P_2)$.

$$d(P_1, M) = \sqrt{[-1 - (-5)]^2 + (2 - 3)^2} = \sqrt{16 + 1} = \sqrt{17}$$
$$d(M, P_2) = \sqrt{[3 - (-1)]^2 + (1 - 2)^2} = \sqrt{16 + 1} = \sqrt{17}$$

Now work Problem 9.

3   A **relation** is a correspondence between two variables, say x and y. When relations are written as ordered pairs (x, y), we say that x is related to y. Often, we are interested in specifying the type of relation (such as an equation) that might exist between two variables. The first step in finding this relation is to plot the ordered pairs in a Cartesian coordinate system. The resulting graph is called a **scatter diagram.**

E X A M P L E   4   Drawing a Scatter Diagram

Table 1 shows the price per share of Disney stock at the end of each month during 1997.

(a) Draw a scatter diagram of the data in Table 1.

 (b) Use a graphing utility to draw the scatter diagram.

(c) Describe what happens to the price of Disney stock from January to December 1997.

TABLE 1

Date	Closing Price	Date	Closing Price
1/31/97	$72\frac{7}{8}$	7/31/97	$80\frac{3}{4}$
2/28/97	$74\frac{1}{4}$	8/31/97	$76\frac{3}{4}$
3/31/97	$72\frac{7}{8}$	9/30/97	$80\frac{5}{8}$
4/30/97	$81\frac{3}{4}$	10/31/97	$82\frac{3}{8}$
5/31/97	$81\frac{7}{8}$	11/30/97	95
6/30/97	$80\frac{1}{4}$	12/31/97	99

Courtesy of A.G. Edwards & Sons, Inc.

Solution

(a) To draw a scatter diagram, we plot ordered pairs of the data in Table 1 using time as the x-coordinate and the price of the stock as y-coordinate. See Figure 20. Notice that the points in a scatter diagram are not connected.

(b) Figure 21 shows the scatter diagram of the data using a TI-83 graphing calculator. Notice that we use 1 for Jan., 2 for Feb., and so on.

FIGURE 20

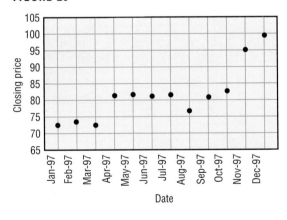

FIGURE 21

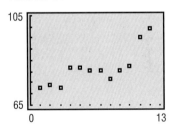

(c) From the scatter diagram, we conclude that the price of Disney stock trended upward during 1997.

Graphs

The **graph of an equation** in two variables x and y consists of the set of points in the xy-plane whose coordinates (x, y) satisfy the equation.

Let's look at some examples.

E X A M P L E   5   Graphing an Equation

Graph the equation: $y = 2x + 5$

FIGURE 22
$y = 2x + 5$

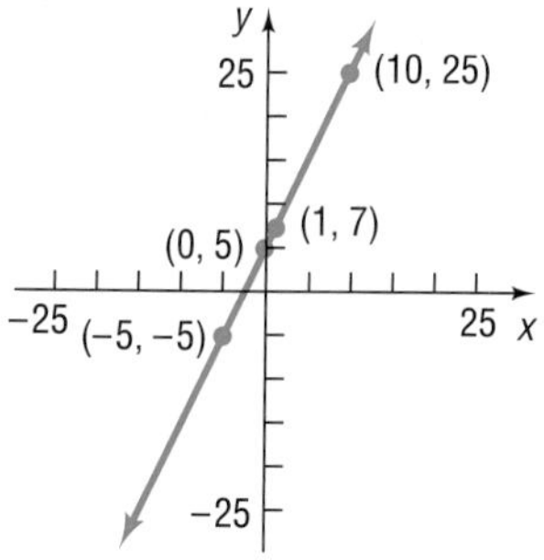

Solution   We want to find all points (x, y) that satisfy the equation. To locate some of these points (and thus get an idea of the pattern of the graph), we assign some numbers to x and find corresponding values for y.

If	Then	Point on Graph
$x = 0$	$y = 2(0) + 5 = 5$	$(0, 5)$
$x = 1$	$y = 2(1) + 5 = 7$	$(1, 7)$
$x = -5$	$y = 2(-5) + 5 = -5$	$(-5, -5)$
$x = 10$	$y = 2(10) + 5 = 25$	$(10, 25)$

By plotting these points and then connecting them, we obtain the graph of the equation (a line*), as shown in Figure 22.   ■

E X A M P L E   6   Graphing an Equation

Graph the equation: $y = x^2$

Solution   Table 2 provides several points on the graph. In Figure 23 we plot these points and connect them with a smooth curve to obtain the graph (a *parabola*).

TABLE 2		
x	**y = x²**	**(x, y)**
-4	16	$(-4, 16)$
-3	9	$(-3, 9)$
-2	4	$(-2, 4)$
-1	1	$(-1, 1)$
0	0	$(0, 0)$
1	1	$(1, 1)$
2	4	$(2, 4)$
3	9	$(3, 9)$
4	16	$(4, 16)$

FIGURE 23
$y = x^2$

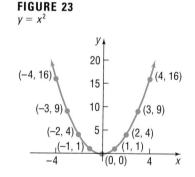

The graphs of the equations shown in Figures 22 and 23 do not show all points. For example, in Figure 22, the point $(20, 45)$ is a part of the graph of $y = 2x + 5$, but it is not shown. Since the graph of $y = 2x + 5$ could be extended out as far as we please, we use arrows to indicate that the pattern shown continues. Thus, it is important when illustrating a graph to present enough of the graph so that any viewer of the illustration will "see" the rest of it as an obvious continuation of what is actually there. This is referred to as a **complete graph.**

So, one way to obtain a complete graph of an equation is to plot a sufficient number of points on the graph until a pattern becomes evident. Then these points are connected with a smooth curve following the suggested pattern. But how many points are sufficient? Sometimes knowledge about the

*Section A.1 in Appendix A provides a review of lines.

equation tells us. For example, if an equation is of the form $y = mx + b$, then its graph is a line. In this case, two points would suffice to obtain the graph.

One purpose of this book is to investigate the properties of equations in order to decide whether a graph is complete. In this section, we shall graph equations by plotting a sufficient number of points on the graph until a pattern becomes evident; then we connect these points with a smooth curve following the suggested pattern. Shortly, we shall investigate various techniques that will enable us to graph an equation without plotting so many points.

FIGURE 24

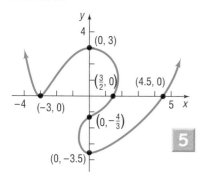

Comment: Another way to obtain the graph of an equation is to use a graphing utility. Read Section B.2, Using a Graphing Utility to Graph Equations, in Appendix B. ▬

Two techniques that reduce the number of points required to graph an equation involve finding *intercepts* and checking for *symmetry*.

The points, if any, at which a graph crosses or touches the coordinate axes are called the **intercepts.** The x-coordinate of a point at which the graph crosses or touches the x-axis is an **x-intercept,** and the y-coordinate of a point at which the graph crosses or touches the y-axis is a **y-intercept.** For example, the graph in Figure 24 has three x-intercepts, -3, $\frac{3}{2}$, and 4.5, and three y-intercepts, -3.5, $-\frac{4}{3}$, and 3.

Procedure for Finding Intercepts

> 1. To find the x-intercept(s), if any, of the graph of an equation, let $y = 0$ in the equation and solve for x.
> 2. To find the y-intercept(s), if any, of the graph of an equation, let $x = 0$ in the equation and solve for y.

E X A M P L E 7

Finding Intercepts from an Equation

Find the x-intercept(s) and the y-intercept(s) of the graph of $y = x^2 - 4$.

Solution

To find the x-intercepts, we let $y = 0$ and obtain the equation

$$x^2 - 4 = 0$$

The equation has two solutions, -2 and 2. Thus, the x-intercepts are -2 and 2.
    To find the y-intercept(s), we let $x = 0$ and obtain the equation $y = -4$. The only y-intercept is -4. ▬

Comment For many equations, finding intercepts may not be so easy. In such cases, a graphing utility can be used. Read Section B.3, Using a Graphing Utility to Locate Intercepts, in Appendix B to find out how a graphing utility locates intercepts. ▬

Now work Problem 35(a).

Another useful tool for graphing equations by hand involves *symmetry*, particularly symmetry with respect to the x-axis, the y-axis, and the origin.

A graph is said to be **symmetric with respect to the x-axis** if, for every point (x, y) on the graph, the point $(x, -y)$ is also on the graph.

A graph is said to be **symmetric with respect to the y-axis** if, for every point (x, y) on the graph, the point $(-x, y)$ is also on the graph.

A graph is said to be **symmetric with respect to the origin** if, for every point (x, y) on the graph, the point $(-x, -y)$ is also on the graph.

Figure 25 illustrates the definition. Notice that, when a graph is symmetric with respect to the x-axis, the part of the graph above the x-axis is a reflection of the part below it, and vice versa. When a graph is symmetric with respect to the y-axis, the part of the graph to the right of the y-axis is a reflection of the part to the left of it, and vice versa. Notice that symmetry with respect to the origin may be viewed in two ways:

1. As a reflection about the y-axis, followed by a reflection about the x-axis.
2. As a projection along a line through the origin so that the distances from the origin are equal.

FIGURE 25

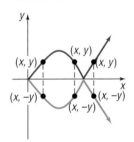

Symmetric with respect to
the x-axis

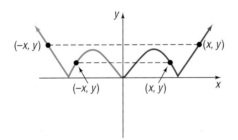

Symmetric with respect to the y-axis

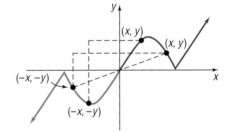

Symmetric with respect to
the origin

 Now work Problem 35(b).

When the graph of an equation is symmetric, the number of points that you need to plot in order to see the pattern is reduced. For example, if the graph of an equation is symmetric with respect to the y-axis, then, once points to the right of the y-axis are plotted, an equal number of points on the graph can be obtained by reflecting them about the y-axis. Thus, before we graph an equation, we first want to determine whether it has any symmetry. The following tests are used for this purpose.

Tests for Symmetry

> To test the graph of an equation for symmetry with respect to the:
>
> **x-axis**   Replace y by $-y$ in the equation. If an equivalent equation results, the graph of the equation is symmetric with respect to the x-axis.
>
> **y-axis**   Replace x by $-x$ in the equation. If an equivalent equation results, the graph of the equation is symmetric with respect to the y-axis.
>
> **Origin**   Replace x by $-x$ and y by $-y$ in the equation. If an equivalent equation results, the graph of the equation is symmetric with respect to the origin.

E X A M P L E   8

Graphing an Equation by Finding Intercepts and Checking for Symmetry

Graph the equation $y = x^3$. Find any intercepts and check for symmetry first.

Solution   First, we seek the intercepts. When $x = 0$, then $y = 0$; and when $y = 0$, then $x = 0$. Thus, the origin $(0, 0)$ is the only intercept. Now we test for symmetry.

x-Axis: Replace y by $-y$. Since the result, $-y = x^3$, is not equivalent to $y = x^3$, the graph is not symmetric with respect to the x-axis.

y-Axis: Replace x by $-x$. Since the result, $y = (-x)^3 = -x^3$, is not equivalent to $y = x^3$, the graph is not symmetric with respect to the y-axis.

Origin: Replace x by $-x$ and y by $-y$. Since the result, $-y = -x^3$, is equivalent to $y = x^3$, the graph is symmetric with respect to the origin.

To graph $y = x^3$, we use the equation to obtain several points on the graph. Because of the symmetry, we only need to locate points on the graph for which $x \geq 0$. See Table 3. Figure 26 shows the graph.

TABLE 3		
x	**$y = x^3$**	**(x, y)**
0	0	(0, 0)
1	1	(1, 1)
2	8	(2, 8)
3	27	(3, 27)

FIGURE 26
$y = x^3$

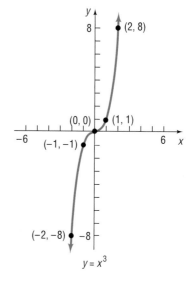

Now work **Problem 77.**

E X A M P L E   9

FIGURE 27
$x = y^2$

Graphing an Equation

Graph the equation $x = y^2$. Find any intercepts and check for symmetry first.

Solution   The lone intercept is $(0, 0)$. The graph is symmetric with respect to the x-axis. (Do you see why? Replace y by $-y$.) Figure 27 shows the graph.

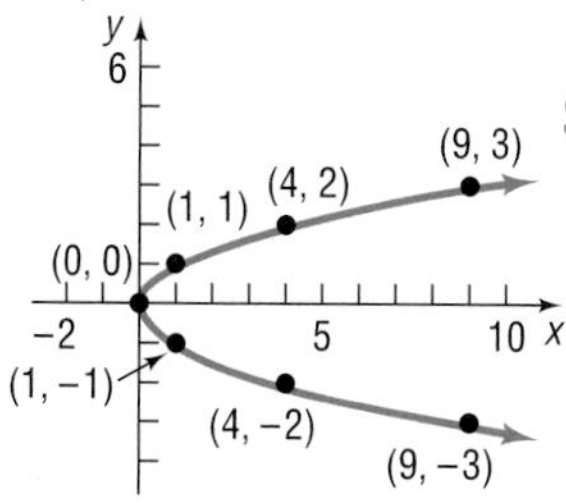

If we restrict y so that $y \geq 0$, the equation $x = y^2$, $y \geq 0$, may be written equivalently as $y = \sqrt{x}$. The portion of the graph of $x = y^2$ in quadrant I is therefore the graph of $y = \sqrt{x}$. See Figure 28.

Comment: To see the graph of the equation $x = y^2$ on a graphing calculator, you will need to graph two equations $Y_1 = \sqrt{x}$ and $Y_2 = -\sqrt{x}$. We discuss why in Section 1.3. See Figure 29.

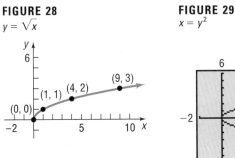

FIGURE 28
$y = \sqrt{x}$

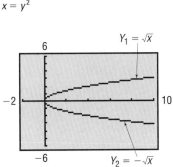

FIGURE 29
$x = y^2$

E X A M P L E   10   Graphing the Equation $y = 1/x$

Graph the equation: $y = \dfrac{1}{x}$

Find any intercepts and check for symmetry first.

Solution   We check for intercepts first. If we let $x = 0$, we obtain a 0 denominator, which is not allowed. Hence, there is no y-intercept. If we let $y = 0$, we get the equation $1/x = 0$, which has no solution. Hence, there is no x-intercept. Thus, the graph of $y = 1/x$ does not cross or touch the coordinate axes.
Next we check for symmetry.

x-Axis: Replacing y by $-y$ yields $-y = 1/x$, which is not equivalent to $y = 1/x$.

y-Axis: Replacing x by $-x$ yields $y = -1/x$, which is not equivalent to $y = 1/x$.

Origin: Replacing x by $-x$ and y by $-y$ yields $-y = -1/x$, which is equivalent to $y = 1/x$.

The graph is symmetric with respect to the origin.
Finally, we set up Table 4, listing several points on the graph. Because of the symmetry with respect to the origin, we use only positive values of x. From Table 4, we infer that if x is a large and positive number then $y = 1/x$ is a positive number close to 0. We also infer that if x is a positive number close to 0 then $y = 1/x$ is a large and positive number. Armed with this information, we can graph the equation. Figure 30 illustrates some of these points

and the graph of $y = 1/x$. Observe how the absence of intercepts and the existence of symmetry with respect to the origin were utilized.

TABLE 4

x	$y = 1/x$	(x, y)
$\frac{1}{10}$	10	$(\frac{1}{10}, 10)$
$\frac{1}{3}$	3	$(\frac{1}{3}, 3)$
$\frac{1}{2}$	2	$(\frac{1}{2}, 2)$
1	1	$(1, 1)$
2	$\frac{1}{2}$	$(2, \frac{1}{2})$
3	$\frac{1}{3}$	$(3, \frac{1}{3})$
10	$\frac{1}{10}$	$(10, \frac{1}{10})$

FIGURE 30

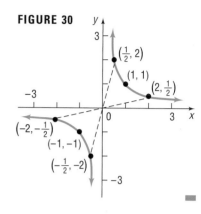

FIGURE 31

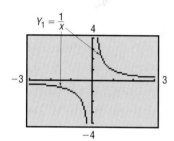

Comment Figure 31 shows the graph of $y = 1/x$ using a graphing utility. We infer from the graph that there are no intercepts. We may also infer from the graph that there is possible symmetry with respect to the origin. The TRACE feature provides further evidence of this symmetry. As the graph is TRACEd, the points $(-x, -y)$ and (x, y) are obtained. For example, the pair of points $(-0.95238, -1.05)$ and $(0.95238, 1.05)$ both lie on the graph.

Circles

7

One advantage of a coordinate system is that it enables us to translate a geometric statement into an algebraic statement, and vice versa. Consider, for example, the following geometric statement that defines a circle.

> A **circle** is a set of points in the xy-plane that are a fixed distance r from a fixed point (h, k). The fixed distance r is called the **radius,** and the fixed point (h, k) is called the **center** of the circle.

FIGURE 32

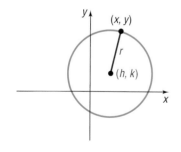

Figure 32 shows the graph of a circle. Is there an equation having this graph? If so, what is the equation? To find the equation, we let (x, y) represent the coordinates of any point on a circle with radius r and center (h, k). Then the distance between the points (x, y) and (h, k) must always equal r. That is, by the distance formula,

$$\sqrt{(x - h)^2 + (y - k)^2} = r$$

or, equivalently,

$$(x - h)^2 + (y - k)^2 = r^2$$

The **standard form of an equation of a circle** with radius r and center (h, k) is

$$(x - h)^2 + (y - k)^2 = r^2 \qquad (3)$$

The standard form of an equation of a circle with radius r and center at the origin $(0, 0)$ is

$$x^2 + y^2 = r^2$$

FIGURE 33
Unit circle $x^2 + y^2 = 1$

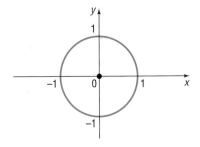

If the radius $r = 1$, the circle whose center is at the origin is called the **unit circle** and has the equation

$$x^2 + y^2 = 1$$

See Figure 33.

EXAMPLE 11   Writing the Standard Form of the Equation of a Circle

Write the standard form of the equation of the circle with radius 3 and center $(1, -2)$.

Solution   Using the form of equation (3) and substituting the values $r = 3, h = 1$, and $k = -2$, we have

$$(x - h)^2 + (y - k)^2 = r^2 \quad h = 1, k = -2, r = 3$$
$$(x - 1)^2 + (y + 2)^2 = 9$$

∎

Now work Problem 45.

 Conversely, by reversing the steps, we conclude that the graph of any equation of the form of equation (3) is that of a circle with radius r and center (h, k).

EXAMPLE 12   Graphing a Circle

Graph the equation: $(x + 3)^2 + (y - 2)^2 = 16$

Solution   By comparing the given equation to the standard form of the equation of a circle, equation (3), we conclude that the graph of the given equation is a circle. Moreover, the comparison yields information about the circle.

FIGURE 34

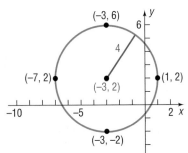

$$(x + 3)^2 + (y - 2)^2 = 16$$
$$(x - (-3))^2 + (y - 2)^2 = 4^2$$
$$\underset{\uparrow}{(x - h)^2} + \underset{\uparrow}{(y - k)^2} = \underset{\uparrow}{r^2}$$

We see that $h = -3, k = 2$, and $r = 4$. Hence, the circle has center $(-3, 2)$ and a radius of 4 units. To graph this circle, we first plot the center $(-3, 2)$. Since the radius is 4, we can locate four points on the circle by going out 4 units to the left, to the right, up, and down from the center. These four points can then be used as guides to obtain the graph. See Figure 34.

∎

Now work Problem 61.

  Comment: Now read Section B.4, Square Screens, in Appendix B. ▬

E X A M P L E   13

Using a Graphing Utility to Graph a Circle

Graph the equation: $x^2 + y^2 = 4$

FIGURE 35

Solution

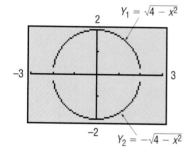

This is the equation of a circle with center at the origin and radius 2. In order to graph this equation, we must first solve for y.*

$$x^2 + y^2 = 4$$
$$y^2 = 4 - x^2$$
$$y = \pm \sqrt{4 - x^2}$$

There are two equations to graph: first, we graph $Y_1 = \sqrt{4 - x^2}$ and then $Y_2 = -\sqrt{4 - x^2}$ on the same square screen. (Your circle will appear oval if you do not use a square screen.) See Figure 35. ▬

If we eliminate the parentheses from the standard form of the equation of the circle obtained in Example 11, we get

$$(x - 1)^2 + (y + 2)^2 = 9$$
$$x^2 - 2x + 1 + y^2 + 4y + 4 = 9$$

which we find, upon simplifying, is equivalent to

$$x^2 + y^2 - 2x + 4y - 4 = 0$$

By completing the squares on both the x- and y-terms,† it can be shown that any equation of the form

$$x^2 + y^2 + ax + by + c = 0$$

has a graph that is a circle, or a point, or has no graph at all. For example, the graph of the equation $x^2 + y^2 = 0$ is the single point $(0, 0)$. The equation $x^2 + y^2 + 5 = 0$, or $x^2 + y^2 = -5$, has no graph, because sums of squares of real numbers are never negative. When its graph is a circle, the equation

$$x^2 + y^2 + ax + by + c = 0$$

is referred to as the **general form of the equation of a circle.**

---

*Some graphing utilities (e.g., TI-82, TI-83, TI-85, and TI-86) have a CIRCLE function that allows the user to enter only the coordinates of the center of the circle and its radius to graph the circle.
†The process of completing the square is reviewed in Appendix A, Section A.3.

9   The next example shows how to transform an equation in the general form to an equivalent equation in standard form. As we said earlier, the idea is to use the method of completing the square on both the x- and y-terms.

E X A M P L E   14   Graphing a Circle Whose Equation Is in General Form

Graph the equation: $x^2 + y^2 + 4x - 6y + 12 = 0$

Solution   We rearrange the equation as follows:

$$(x^2 + 4x) + (y^2 - 6y) = -12$$

Next, we complete the square of each expression in parentheses. Remember that any number added on the left must also be added on the right.

$$(x^2 + 4x + 4) + (y^2 - 6y + 9) = -12 + 4 + 9$$
$$(x + 2)^2 + (y - 3)^2 = 1$$

We recognize this equation as the standard form of the equation of a circle with radius 1 and center $(-2, 3)$. See Figure 36.

FIGURE 36
$x^2 + y^2 + 4x - 6y + 12 = 0$

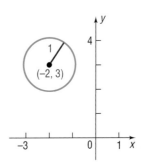

Now work Problem 63.

1.2 │ EXERCISES

In Problems 1 and 2, plot each point in the xy-plane. Tell in which quadrant or on what coordinate axis each point lies.

1. (a) $A = (-3, 2)$ (b) $B = (6, 0)$ (c) $C = (-2, -2)$
    (d) $D = (6, 5)$ (e) $E = (0, -3)$ (f) $F = (6, -3)$

2. (a) $A = (1, 4)$ (b) $B = (-3, -4)$ (c) $C = (-3, 4)$
    (d) $D = (4, 1)$ (e) $E = (0, 1)$ (f) $F = (-3, 0)$

3. Plot the points $(2, 0), (2, -3), (2, 4), (2, 1)$, and $(2, -1)$. Describe the set of all points of the form $(2, y)$, where y is a real number.

4. Plot the points $(0, 3), (1, 3), (-2, 3), (5, 3)$, and $(-4, 3)$. Describe the set of all points of the form $(x, 3)$, where x is a real number.

In Problems 5–18, find the distance $d(P_1, P_2)$ between the points P_1 and P_2. Also find the midpoint of the line segment joining P_1 and P_2.

5. 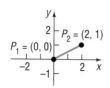 **6.** 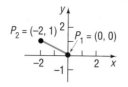 **7.** 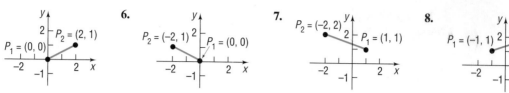 **8.**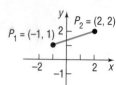

9. $P_1 = (3, -4);\quad P_2 = (3, 1)$

10. $P_1 = (-1, 0);\quad P_2 = (2, 1)$

11. $P_1 = (-3, 2);\quad P_2 = (6, 0)$

12. $P_1 = (2, -3);\quad P_2 = (4, 2)$

13. $P_1 = (4, -3);\quad P_2 = (6, 1)$

14. $P_1 = (-4, -3);\quad P_2 = (2, 2)$

15. $P_1 = (-0.2, 0.3);\quad P_2 = (2.3, 1.1)$

16. $P_1 = (1.2, 2.3);\quad P_2 = (-0.3, 1.1)$

17. $P_1 = (a, b);\quad P_2 = (0, 0)$

18. $P_1 = (a, a);\quad P_2 = (0, 0)$

In Problems 19–22, plot each point and form the triangle ABC. Verify that the triangle is a right triangle. Find its area.

19. $A = (-2, 5);\quad B = (1, 3);\quad C = (-1, 0)$

20. $A = (-2, 5);\quad B = (12, 3);\quad C = (10, -11)$

21. $A = (-5, 3);\quad B = (6, 0);\quad C = (5, 5)$

22. $A = (-6, 3);\quad B = (3, -5);\quad C = (-1, 5)$

23. Find all points having an x-coordinate of 2 whose distance from the point $(-2, -1)$ is 5.

24. Find all points having a y-coordinate of -3 whose distance from the point $(1, 2)$ is 13.

25. Find all points on the x-axis that are 5 units from the point $(2, -3)$.

26. Find all points on the y-axis that are 5 units from the point $(-4, 4)$.

In Problems 27–34, plot each point. Then plot the point that is symmetric to it with respect to:

(a) The x-axis   *(b) The y-axis*   *(c) The origin*

27. $(3, 4)$ **28.** $(5, 3)$ **29.** $(-2, 1)$ **30.** $(4, -2)$

31. $(1, 1)$ **32.** $(-1, -1)$ **33.** $(-3, -4)$ **34.** $(4, 0)$

In Problems 35–44, the graph of an equation is given.

(a) List the intercepts of the graph.

(b) Based on the graph, tell whether the graph is symmetric with respect to the x-axis, y-axis, and/or origin.

35. 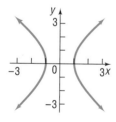 **36.** 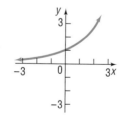 **37.**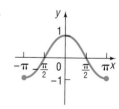

38. 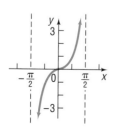 **39.** 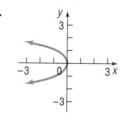 **40.**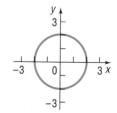

41.

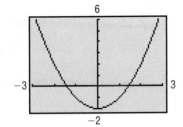

42.

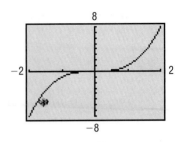

43.

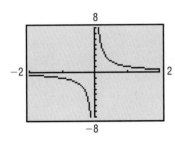

44.

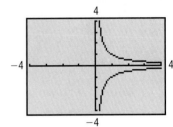

In Problems 45–50, write the standard form of the equation and the general form of the equation of each circle of radius r and center (h, k).

45. $r = 2$; $(h, k) = (0, 2)$ **46.** $r = 3$; $(h, k) = (1, 0)$ **47.** $r = 5$; $(h, k) = (4, -3)$

48. $r = 4$; $(h, k) = (2, -3)$ **49.** $r = 2$; $(h, k) = (0, 0)$ **50.** $r = 3$; $(h, k) = (0, 0)$

In Problems 51–54, find the center and radius of each circle. Write the standard form of the equation.

51. 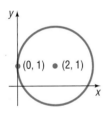   **52.** 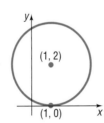   **53.** 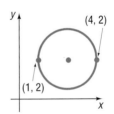   **54.**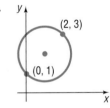

In Problems 55–58, match each graph with the correct equation.

(a) $(x - 3)^2 + (y + 3)^2 = 9$ (c) $(x + 1)^2 + (y - 2)^2 = 4$

(b) $(x - 1)^2 + (y + 2)^2 = 4$ (d) $(x + 3)^2 + (y - 3)^2 = 9$

55.

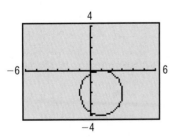

56.

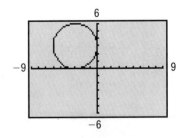

57.

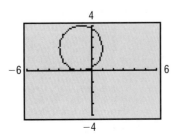

58.

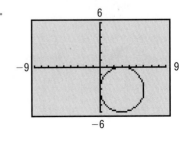

In Problems 59–68, find the center (h, k) and radius r of each circle. Graph the circle.

59. $x^2 + y^2 = 4$

60. $x^2 + (y - 1)^2 = 1$

61. $(x - 3)^2 + y^2 = 4$

62. $(x + 1)^2 + (y - 1)^2 = 2$

63. $x^2 + y^2 + 4x - 4y - 1 = 0$

64. $x^2 + y^2 - 6x + 2y + 9 = 0$

65. $x^2 + y^2 - x + 2y + 1 = 0$

66. $x^2 + y^2 + x + y - \frac{1}{2} = 0$

67. $2x^2 + 2y^2 - 12x + 8y - 24 = 0$

68. $2x^2 + 2y^2 + 8x + 7 = 0$

In Problems 69–72, use the accompanying graph.

69. Draw the graph to make it symmetric with respect to the x-axis.

70. Draw the graph to make it symmetric with respect to the y-axis.

71. Draw the graph to make it symmetric with respect to the origin.

72. Draw the graph to make it symmetric with respect to the x-axis, y-axis, and origin.

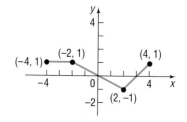

In Problems 73–86, list the intercepts and test for symmetry.

73. $x^2 = y$

74. $y^2 = x$

75. $y = 3x$

76. $y = -5x$

77. $x^2 + y - 9 = 0$

78. $y^2 - x - 4 = 0$

79. $4x^2 + 9y^2 = 36$

80. $x^2 + 4y^2 = 4$

81. $y = x^3 - 27$

82. $y = x^4 - 1$

83. $y = x^2 - 3x - 4$

84. $y = x^2 + 4$

85. $y = \dfrac{x}{x^2 + 9}$

86. $y = \dfrac{x^2 - 4}{x}$

87. Baseball A major league baseball "diamond" is actually a square, 90 feet on a side (see the figure). Overlay a rectangular coordinate system on a major league baseball diamond so that the origin is at home plate, the positive x-axis lies in the direction from home plate to first base, and the positive y-axis lies in the direction from home plate to third base.

(a) What are the coordinates of first base, second base, and third base? Use feet as the unit of measurement.

(b) If the right fielder is located at (310, 15), how far is it from there to second base?

(c) If the center fielder is located at (300, 300), how far is it from there to third base?

88. Little League Baseball The layout of a Little League playing field is a square, 60 feet on a side.* Overlay a rectangular coordinate system on a Little League baseball diamond so that the origin is at home plate, the positive x-axis lies in the direction from home plate to first base, and the positive y-axis lies in the direction from home plate to third base.

(a) What are the coordinates of first base, second base, and third base? Use feet as the unit of measurement.

(b) If the right fielder is located at (180, 20), how far is it from there to second base?

(c) If the center fielder is located at (220, 220), how far is it from there to third base?

89. An Olds Aurora and a Mack truck leave an intersection at the same time. The Aurora heads east at an average speed of 40 miles per hour, while the truck heads south at an average speed of 30 miles per hour. Find an expression for their distance apart d (in miles) at the end of t hours.

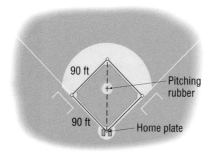

90 ft — Pitching rubber

90 ft — Home plate

*Source: Little League Baseball, Official Regulations and Playing Rules, 1998.

90. A hot-air balloon, headed due east at an average speed of 15 miles per hour and at a constant altitude of 100 feet, passes over an intersection (see the figure). Find an expression for its distance d (measured in feet) from the intersection t seconds later.

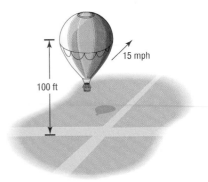

15 mph

100 ft

91. Prices of Motorola Stock   The data shown represent the closing price of Motorola, Inc., stock at the end of each month in 1997.
(a) Draw a scatter diagram.
(b) Using a graphing utility, draw a scatter diagram.
(c) How would you describe the trend in these data over time?

Month	Closing Price
January, 1997	$68\frac{1}{4}$
February, 1997	56
March, 1997	$60\frac{1}{2}$
April, 1997	$57\frac{1}{8}$
May, 1997	$66\frac{3}{8}$
June, 1997	$76\frac{1}{8}$
July, 1997	$80\frac{3}{8}$
August, 1997	$73\frac{3}{8}$
September, 1997	$71\frac{7}{8}$
October, 1997	62
November, 1997	$62\frac{7}{8}$
December, 1997	57

Courtesy of A. G. Edwards & Sons, Inc.

92. Price of Microsoft Stock   The data shown represent the closing price of Microsoft Corporation stock at the end of each month in 1997.
(a) Draw a scatter diagram.
(b) Using a graphing utility, draw a scatter diagram.
(c) How would you describe the trend in these data over time?

Month	Closing Price
January, 1997	102
February, 1997	$97\frac{1}{2}$
March, 1997	$91\frac{1}{2}$
April, 1997	$121\frac{1}{2}$
May, 1997	124
June, 1997	$128\frac{3}{8}$
July, 1997	$141\frac{3}{8}$
August, 1997	$132\frac{1}{8}$
September, 1997	$132\frac{1}{4}$
October, 1997	138
November, 1997	$141\frac{1}{2}$
December, 1997	$129\frac{3}{4}$

Courtesy of A. G. Edwards & Sons, Inc.

In Problem 93, you may use a graphing calculator, but it is not required.

93. (a) Graph $y = \sqrt{x^2}$, $y = x$, $y = |x|$, and $y = (\sqrt{x})^2$, noting which graphs are the same.
(b) Explain why the graphs of $y = \sqrt{x^2}$ and $y = |x|$ are the same.
(c) Explain why the graphs of $y = x$ and $y = (\sqrt{x})^2$ are not the same.
(d) Explain why the graphs of $y = \sqrt{x^2}$ and $y = x$ are not the same.

94. Make up an equation with the intercepts $(2, 0)$, $(4, 0)$, and $(0, 1)$. Compare your equation with a friend's equation. Comment on any similarities.

95. An equation is being tested for symmetry with respect to the x-axis, the y-axis, and the origin. Explain why, if two of these symmetries are present, the remaining one must also be present.

96. Draw a graph that contains the points $(-2, -1)$, $(0, 1)$, $(1, 3)$, and $(3, 5)$. Compare your graph with those of other students. Are most of the graphs almost straight lines? How many are "curved"? Discuss the various ways that these points might be connected.

1.3 | FUNCTIONS

1 Determine Whether a Relation Represents a Function
2 Find the Value of a Function
3 Find the Domain of a Function
4 Identify the Graph of a Function
5 Obtain Information from or about the Graph of a Function

In Section 1.2, we said that a **relation** is a correspondence between two variables, say x and y. When relations are written as ordered pairs (x, y), we say that x is related to y. Often, we are interested in specifying the type of relation (such as an equation) that might exist between the two variables.

For example, the relation between the revenue R resulting from the sale of x items selling for \$10 each may be expressed by the equation $R = 10x$. If we know how many items have been sold, then we can calculate the revenue by using the equation $R = 10x$. This equation is an example of a *function*.

As another example, suppose that an icicle falls off a building from a height of 64 feet above the ground. According to a law of physics, the distance s (in feet) of the icicle from the ground after t seconds is given (approximately) by the formula $s = 64 - 16t^2$. When $t = 0$ seconds, the icicle is $s = 64$ feet above the ground. After 1 second, the icicle is $s = 64 - 16(1)^2 = 48$ feet above the ground. After 2 seconds, the icicle strikes the ground. The formula $s = 64 - 16t^2$ provides a way of finding the distance s when the time t $(0 \leq t \leq 2)$ is prescribed. There is a correspondence between each time t in the interval $0 \leq t \leq 2$ and the distance s. We say that the distance s is a *function* of the time t because

1. There is a correspondence between the set of times and the set of distances.
2. There is exactly one distance s obtained for a prescribed time t in the interval $0 \leq t \leq 2$.

Let's now look at the definition of a function.

Definition of Function

Let X and Y be two nonempty sets of real numbers.* A **function** from X into Y is a relation that associates with each element of X a unique element of Y.

The set X is called the **domain** of the function. For each element x in X, the corresponding element y in Y is called the **value** of the function at x, or the **image** of x. The set of all images of the elements of the domain is called the **range** of the function. See Figure 37.

Since there may be some elements in Y that are not the image of some x in X, it follows that the range of a function may be a subset of Y, as shown in Figure 37.

FIGURE 37

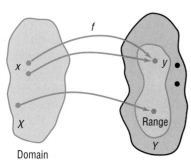

Domain

*The two sets X and Y can also be sets of complex numbers (discussed in Section 3.6), and then we have defined a complex function. In the broad definition (due to Lejeune Dirichlet), X and Y can be any two sets.

Warning   Do not confuse the two meanings given for the word range. When used in connection with a function, it means the set of all images of the elements of the domain of the function. When the word RANGE is used in connection with a graphing utility, it means the settings used for the viewing rectangle, that is, it means the WINDOW. ▬

1 Not all relations between two sets are functions.

E X A M P L E   1

Determining Whether a Relation Represents a Function

Determine whether the following relations represent functions.

(a) For this relation, the domain represents the employees of Sara's Pre-Owned Car Mart and the range represents their base salary.

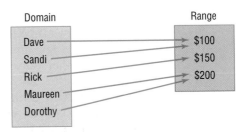

(b) For this relation, the domain represents the employees of Sara's Pre-Owned Car Mart and the range represents their phone number(s).

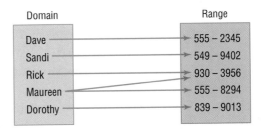

Solution   (a) The relation is a function because each element in the domain corresponds to a unique element in the range. Notice that more than one element in the domain can correspond to the same element in the range.

(b) The relation is not a function because each element in the domain does not correspond to a unique element in the range. Maureen has two telephone numbers; therefore, if Maureen is chosen from the domain, a unique telephone number cannot be assigned to her. ▬

We may think of a function as a set of ordered pairs (x, y) in which no two distinct pairs have the same first element. The set of all first elements x is the domain of the function, and the set of all second elements y is its range. Thus, there is associated with each element x in the domain a unique element y in the range.

E X A M P L E   2   Determining Whether a Relation Represents a Function

Determine whether each relation represents a function.

(a) $\{(1, 4), (2, 5), (3, 6), (4, 7)\}$
(b) $\{(1, 4), (2, 4), (3, 5), (6, 10)\}$
(c) $\{(-3, 9), (-2, 4), (0, 0), (1, 1), (-3, 8)\}$

Solution   (a) This relation is a function because there are no distinct ordered pairs with the same first element.
(b) This relation is a function because there are no distinct ordered pairs with the same first element.
(c) This relation is not a function because there is a first element, -3, that corresponds to two different second elements, 9 and 8.   ■

In Example 2(b), notice that 1 and 2 in the domain each have the same image in the range. This does not violate the definition of a function; two different first elements can have the same second element. A violation of the definition occurs when two ordered pairs have the same first element and different second elements, as in Example 2(c).

The relation referred to in the definition of a function is most often given as an equation in two variables, usually denoted x and y.

E X A M P L E   3   Example of a Function

Consider the function defined by the equation

$$y = 2x - 5 \qquad 1 \leq x \leq 6$$

The domain $1 \leq x \leq 6$ specifies that the number x is restricted to the real numbers from 1 to 6, inclusive. The equation $y = 2x - 5$ specifies that the number x is to be multiplied by 2 and then 5 is to be subtracted from the result to get y. For example, if $x = \frac{3}{2}$, then $y = 2 \cdot \frac{3}{2} - 5 = -2$.   ■

 Now work Problems 1 and 5.

Function Notation

Functions are often denoted by letters such as f, F, g, G and so on. If f is a function, then for each number x in its domain the corresponding image in the range is designated by the symbol $f(x)$, read as "f of x" or as "f at x." We refer to $f(x)$ as the **value of f at the number x.** Thus, $f(x)$ is the number that results when x is given and the function f is applied; $f(x)$ does *not* mean "f times x." For example, the function given in Example 3 may be written as $y = f(x) = 2x - 5, 1 \leq x \leq 6$. Then $f(\frac{3}{2}) = -2$.

Figure 38 illustrates some other functions. Note that in every function illustrated, for each x in the domain, there is one value in the range.

FIGURE 38

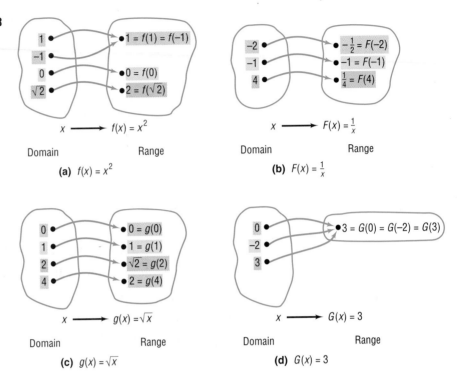

(a) $f(x) = x^2$ (b) $F(x) = \frac{1}{x}$

(c) $g(x) = \sqrt{x}$ (d) $G(x) = 3$

Sometimes it is helpful to think of a function f as a machine that receives as input a number from the domain, manipulates it, and outputs the value. See Figure 39.

FIGURE 39

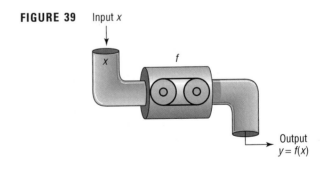

The restrictions on this input/output machine are

1. It only accepts numbers from the domain of the function.
2. For each input, there is exactly one output (which may be repeated for different inputs).

For a function $y = f(x)$, the variable x is called the **independent variable,** because it can be assigned any of the permissible numbers from the domain. The variable y is called the **dependent variable,** because its value depends on x.

Any symbol can be used to represent the independent and dependent variables. For example, if f is the *cube function,* then f can be defined by $f(x) = x^3$ or $f(t) = t^3$ or $f(z) = z^3$. All three functions are the same: Each tells us to cube the independent variable. In practice, the symbols used for the independent and dependent variables are based on common usage.

The variable x is also called the **argument** of the function. Thinking of the independent variable as an argument sometimes can make it easier to find the value of a function. For example, if f is the function defined by $f(x) = x^3$, then f tells us to cube the argument. Thus, $f(2)$ means to cube 2, $f(a)$ means to cube the number a, and $f(x + h)$ means to cube the quantity $x + h$.

E X A M P L E 4

Finding Values of a Function

For the function G defined by $G(x) = 2x^2 - 3x$, evaluate:

(a) $G(3)$ (b) $G(x) + G(3)$ (c) $G(-x)$

(d) $-G(x)$ (e) $G(x + 3)$

Solution   (a) We substitute 3 for x in the equation for G to get

$$G(3) = 2(3)^2 - 3(3) = 18 - 9 = 9$$

(b) $G(x) + G(3) = (2x^2 - 3x) + (9) = 2x^2 - 3x + 9$

(c) We substitute $-x$ for x in the equation for G:

$$G(-x) = 2(-x)^2 - 3(-x) = 2x^2 + 3x$$

(d) $-G(x) = -(2x^2 - 3x) = -2x^2 + 3x$

(e) $G(x + 3) = 2(x + 3)^2 - 3(x + 3)$ Notice the use of parentheses here.

$$= 2(x^2 + 6x + 9) - 3x - 9$$
$$= 2x^2 + 12x + 18 - 3x - 9$$
$$= 2x^2 + 9x + 9$$

Notice in this example that $G(x + 3) \neq G(x) + G(3)$ and $G(-x) \neq -G(x)$.

Now work Problem 13.

Most calculators have special keys that enable you to find the value of certain commonly used functions. For example, you should be able to find the square function, $f(x) = x^2$; the square root function, $f(x) = \sqrt{x}$; the reciprocal function, $f(x) = 1/x = x^{-1}$; and many others that will be discussed later in this book (such as $\ln x$, $\log x$, and so on). Verify the results of Example 5 on your calculator.

E X A M P L E 5

Finding Values of a Function on a Calculator

(a) $f(x) = x^2$; $f(1.234) = 1.522756$

(b) $F(x) = 1/x$; $F(1.234) = 0.8103727715$

(c) $g(x) = \sqrt{x}$; $g(1.234) = 1.110855526$

In general, when a function f is defined by an equation in x and y, we say that the function f is given **implicitly.** If it is possible to solve the equation for y in terms of x, then we write $y = f(x)$ and say that the function is given **explicitly.** In fact, we usually write "the function $y = f(x)$" when we mean "the function f defined by the equation $y = f(x)$." Although this usage is not entirely correct, it is rather common and should not cause any confusion. For example,

Implicit Form	**Explicit Form**
$3x + y = 5$	$y = f(x) = -3x + 5$
$x^2 - y = 6$	$y = f(x) = x^2 - 6$
$xy = 4$	$y = f(x) = 4/x$

Not all equations in x and y define a function $y = f(x)$. If an equation is solved for y and two or more values of y can be obtained for a given x, then the equation does not define a function. For example, consider the equation $x^2 + y^2 = 1$, which defines a circle. If we solve for y, we obtain $y = \pm\sqrt{1 - x^2}$, so two values of y result for numbers x between -1 and 1. Thus, $x^2 + y^2 = 1$ does not define a function.

 Comment The explicit form of a function is the form required by a graphing calculator. Now do you see why it is necessary to graph a circle in two "pieces"? ▬

We list below a summary of some important facts to remember about a function f.

SUMMARY OF IMPORTANT FACTS ABOUT FUNCTIONS

1. To each x in the domain of f, there is one and only one image $f(x)$ in the range.
2. f is the symbol we use to denote the function. It is symbolic of the domain and the equation that we use to get from an x in the domain to $f(x)$ in the range.
3. If $y = f(x)$, then x is called the independent variable or argument of f and y is called the dependent variable or the value of f at x.

Domain of a Function

 Often, the domain of a function f is not specified; instead, only the equation defining the function is given. In such cases, we agree that the domain of f is the largest set of real numbers for which the value $f(x)$ is a real number. Thus, the domain of f is the same as the domain of the variable x in the expression $f(x)$.

E X A M P L E   6   Finding the Domain of a Function

Find the domain of each of the following functions:

(a) $f(x) = x^2 + 5x$ (b) $g(x) = \dfrac{3x}{x^2 - 4}$ (c) $h(x) = \sqrt{4 - 3x}$

Solution   (a) The function tells us to square a number and then add five times the number. Since these operations can be performed on any real number, we conclude that the domain of f is all real numbers.

(b) The function g tells us to divide $3x$ by $x^2 - 4$. Since division by 0 is not allowed, the denominator $x^2 - 4$ can never be 0. Thus, x can never equal 2 or -2. The domain of the function g is $\{x \mid x \neq -2, x \neq 2\}$.

(c) The function h tells us to take the square root of $4 - 3x$. But only non-negative numbers have real square roots. Hence, we require that

$$4 - 3x \geq 0$$
$$-3x \geq -4$$
$$x \leq \tfrac{4}{3}$$

The domain of h is $\{x \mid x \leq \tfrac{4}{3}\}$ or the interval $(-\infty, \tfrac{4}{3}]$. ▬

Now work Problem 53.

If x is in the domain of a function f, we shall say that **f is defined at x,** or **$f(x)$ exists.** If x is not in the domain of f, we say that **f is not defined at x,** or **$f(x)$ does not exist.** For example, if $f(x) = x/(x^2 - 1)$, then $f(0)$ exists, but $f(1)$ and $f(-1)$ do not exist. (Do you see why?)

We have not said much about finding the range of a function. The reason is that when a function is defined by an equation it is often difficult to find the range. Therefore, we shall usually be content to find just the domain of a function when only the rule for the function is given. We shall express the domain of a function using inequalities, interval notation, set notation, or words, whichever is most convenient.

The Graph of a Function

In applications, a graph often demonstrates more clearly the relationship between two variables than, say, an equation or table would. For example, Table 5 shows the price per share of Disney stock at the end of each month during 1997.

If we plot the data in Table 5, using the date as the x-coordinate and the price as the y-coordinate, and then connect the points, we obtain the graph in Figure 40.

TABLE 5			
Date	**Closing Price**	**Date**	**Closing Price**
1/31/97	$72\tfrac{7}{8}$	7/31/97	$80\tfrac{3}{4}$
2/28/97	$74\tfrac{1}{4}$	8/31/97	$76\tfrac{3}{4}$
3/31/97	$72\tfrac{7}{8}$	9/30/97	$80\tfrac{5}{8}$
4/30/97	$81\tfrac{3}{4}$	10/31/97	$82\tfrac{3}{8}$
5/31/97	$81\tfrac{7}{8}$	11/30/97	95
6/30/97	$80\tfrac{1}{4}$	12/31/97	99

Courtesy of A.G. Edwards & Sons, Inc.

FIGURE 40
Monthly closing prices of Disney stock in 1997

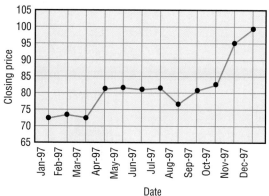

We can see from the graph that the price of the stock was fairly constant during May, June, and July and was rising during November and December. The graph also shows that the lowest price during this period occurred at the end of January and March, while the highest occurred at the end of December. Equations and tables, on the other hand, usually require some calculations and interpretation before this kind of information can be "seen."

Look again at Figure 40. The graph shows that for each time on the horizontal axis there is only one price on the vertical axis. Thus, the graph represents a function, although the exact rule for getting from time to price is not given.

When a function is defined by an equation in x and y, the **graph** of the function is the graph of the equation, that is, the set of points (x, y) in the xy-plane that satisfies the equation.

 Comment When we select a viewing rectangle to graph a function, the values of Xmin, Xmax give the domain that we wish to view, while Ymin, Ymax give the range that we wish to view. These settings usually do not represent the actual domain and range of the function. ▬

 Not every collection of points in the xy-plane represents the graph of a function. Remember, for a function, each number x in the domain has one and only one image y. Thus, the graph of a function cannot contain two points with the same x-coordinate and different y-coordinates. Therefore, the graph of a function must satisfy the following **vertical-line test.**

> **Theorem** Vertical-line Test
>
> A set of points in the xy-plane is the graph of a function if and only if every vertical line intersects the graph in at most one point.
>
> ▬

It follows that, if any vertical line intersects a graph at more than one point, the graph is not the graph of a function.

EXAMPLE 7

Identifying the Graph of a Function

Which of the graphs in Figure 41 are graphs of functions?

FIGURE 41

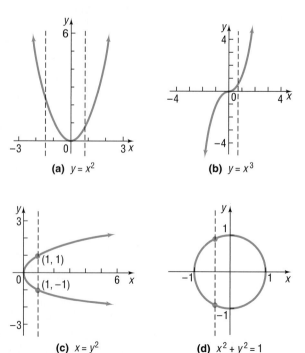

(a) $y = x^2$

(b) $y = x^3$

(c) $x = y^2$

(d) $x^2 + y^2 = 1$

Solution   The graphs in Figures 41(a) and 41(b) are graphs of functions, because a vertical line intersects each graph in at most one point. The graphs in Figures 41(c) and 41(d) are not graphs of functions, because some vertical line intersects each graph in more than one point.   ■

Now work Problem 41.

5   If (x, y) is a point on the graph of a function f, then y is the value of f at x, that is, $y = f(x)$. The next example illustrates how to obtain information about a function if its graph is given.

E X A M P L E   8   Obtaining Information from the Graph of a Function

FIGURE 42

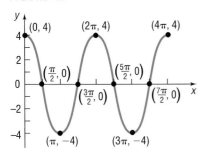

Let f be the function whose graph is given in Figure 42. The graph of f represents the distance that the bob of a pendulum is from its "at rest" position. Negative values of y mean that the pendulum is to the left of the "at rest" position and positive values of y mean that the pendulum is to the right of the "at rest" position.

(a) What is the value of the function when $x = 0$, $x = 3\pi/2$, and $x = 3\pi$?
(b) What is the domain of f?
(c) What is the range of f?
(d) List the intercepts. (Recall that these are the points, if any, where the graph crosses or touches the coordinate axes.)

Solution   (a) Since $(0, 4)$ is on the graph of f, the y-coordinate 4 is the value of f at the x-coordinate 0; that is, $f(0) = 4$. In a similar way, we find that when $x = 3\pi/2$ then $y = 0$; or $f(3\pi/2) = 0$. When $x = 3\pi$, then $y = -4$; or $f(3\pi) = -4$.
(b) To determine the domain of f, we notice that the points on the graph of f have x-coordinates between 0 and 4π, inclusive; and, for each number x between 0 and 4π, there is a point $(x, f(x))$ on the graph. Thus, the domain of f is $\{x|0 \leq x \leq 4\pi\}$ or the interval $[0, 4\pi]$.
(c) The points on the graph all have y-coordinates between -4 and 4, inclusive; and, for each such number y, there is at least one number x in the domain. Hence, the range of f is $\{y|-4 \leq y \leq 4\}$ or the interval $[-4, 4]$.
(d) The intercepts are $(0, 4)$, $(\pi/2, 0)$, $(3\pi/2, 0)$, $(5\pi/2, 0)$, and $(7\pi/2, 0)$.   ■

When the graph of a function is given, its domain may be viewed as the shadow created by the graph on the x-axis by vertical beams of light. Its range can be viewed as the shadow created by the graph on the y-axis by horizontal beams of light. Try this technique with the graph given in Figure 42.

Now work Problems 37 and 39.

E X A M P L E   9   Obtaining Information about the Graph of a Function

Consider the function: $f(x) = \dfrac{x}{x + 2}$

(a) Is the point $(1, 1/2)$ on the graph of f?
(b) If $x = -1$, what is $f(x)$? What point is on the graph of f?
(c) If $f(x) = 2$, what is x? What point is on the graph of f?

Solution (a) When $x = 1$, then $f(x) = f(1) = 1/(1 + 2) = 1/3$. The point $(1, 1/2)$ is therefore not on the graph of f.

(b) If $x = -1$, then $f(x) = f(-1) = -1/(-1 + 2) = -1$, so the point $(-1, -1)$ is on the graph of f.

(c) If $f(x) = 2$, then

$$\frac{x}{x + 2} = 2$$
$$x = 2(x + 2)$$
$$x = 2x + 4$$
$$x = -4$$

The point $(-4, 2)$ is on the graph of f. ▬

Now work Problem 33.

Applications

When we use functions in applications, the domain may be restricted by physical or geometric considerations. For example, the domain of the function f defined by $f(x) = x^2$ is the set of all real numbers. However, if f is used to obtain the area of a square when the length x of a side is known, then we must restrict the domain of f to the positive real numbers, since the length of a side can never be 0 or negative.

E X A M P L E   10   Area of a Circle

Express the area of a circle as a function of its radius.

Solution We know that the formula for the area A of a circle of radius r is $A = \pi r^2$. If we use r to represent the independent variable and A to represent the dependent variable, the function expressing this relationship is

$$A(r) = \pi r^2$$

In this setting, the domain is $\{r | r > 0\}$. (Do you see why?) ▬

E X A M P L E   11   Construction Cost

Sally, a builder of homes, is interested in finding a function that relates the cost C of building a house and x, the number of square feet of the house, so that she can easily provide estimates to clients for various sized homes. The data she obtained from constructing homes last year are listed in Table 6.

TABLE 6		
	Square Feet, x	Cost, C
	1500	165,000
	1600	176,300
	1700	183,000
	1700	189,500
	1830	201,700
	1970	217,000
	2050	220,000
	2100	237,400

(a) Does the relation defined by the set of ordered pairs (x, C) represent a function?

(b) Draw a scatter diagram of the data.

(c) Select two points from the data and find an equation of the line containing the points.

(d) Interpret the slope.

(e) Express the relationship found in (c) using function notation.

(f) What is the domain of the function?

(g) Use the function to find the cost to build a 2000 square foot house.

 (h) Using a graphing utility, find the line of best fit* relating square feet and cost. Interpret the slope.

(i) Give some reasons to explain why the cost of building a 1700 square foot house might cost $183,000 in one case and $189,500 in another.

Solution   (a) No, because the first element 1700 has, corresponding to it, two second elements, 183,000 and 189,500.

(b) See Figure 43.

FIGURE 43

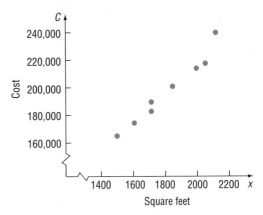

(c) We will choose the points (1500, 165,000) and (2050, 220,000). (You should select your own two points and complete the solution.) The slope of the line containing these points is

$$m = \frac{220,000 - 165,000}{2050 - 1500} = \frac{55,000}{550} = 100$$

The equation of the line containing these points is

$$y - 165,000 = 100(x - 1500)$$
$$y = 100x + 15,000$$

(d) The cost per square foot to build the house is $100. For each additional square foot of house, the cost increases by $100.

(e) Using function notation, $C(x) = 100x + 15,000$.

(f) The domain is $\{x | x > 0\}$ since a house cannot have 0 or negative square feet.

(g) The cost to build a 2000 square foot house is

$$C(2000) = 100(2000) + 15,000 = \$215,000$$

*Lines of best fit are discussed in Appendix A, Section A.2. This material is optional.

(h) The line of best fit is $y = 112.09x - 3734.02$. The cost per square foot to build a house is approximately $112.09. Thus, for each additional square foot of house being built, the cost increases by $112.09.

(i) One explanation is that the fixtures used in one house are more expensive than those used in the other house.   ▬

Observe in the solution to Example 11 that we used the symbol C in two ways: it is used to name the function, and it is used to symbolize the dependent variable. This double use is common in applications and should not cause any difficulty.

Refer back to Example 11. Notice that the data in Table 6 does not represent a function since the same first element is paired with two different second elements ((1700, 183,000), (1700, 189,500)). However, when we find a line that fits these data, we find a function that relates the data. Thus, **curve fitting** is a process whereby we find a functional relationship between two or more variables even though the data, themselves, may not represent a function.

Now work Problem 71.

Examples 10 and 11 demonstrate that some functions are determined from data, while others are based on geometric, physical, or other relations.

SUMMARY

We list here some of the important vocabulary introduced in this section, with a brief description of each term.

Function	A relation between two sets of real numbers so that each number x in the first set, the domain, has corresponding to it exactly one number y in the second set.
	A set of ordered pairs (x, y) or $(x, f(x))$ in which no two distinct pairs have the same first element.
	The range is the set of y values of the function for the x values in the domain.
	A function f may be defined implicitly by an equation involving x and y or explicitly by writing $y = f(x)$.
Unspecified domain	If a function f is defined by an equation and no domain is specified, then the domain will be taken to be the largest set of real numbers for which the equation defines a real number.
Function notation	$y = f(x)$
	f is a symbol for the function.
	x is the independent variable.
	y is the dependent variable.
	$f(x)$ is the value of the function at x, or the image of x.
Graph of a function	The collection of points (x, y) that satisfies the equation $y = f(x)$.
	A collection of points is the graph of a function provided every vertical line intersects the graph in at most one point (vertical-line test).

1.3 | EXERCISES

In Problems 1–12, determine whether each relation represents a function.

1.

2.
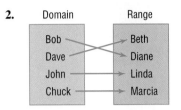

3.
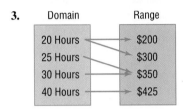

4.
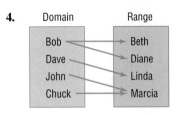

5. $\{(2, 6), (-3, 6), (4, 9), (1, 10)\}$

6. $\{(-2, 5), (-1, 3), (3, 7), (4, 12)\}$

7. $\{(1, 3), (2, 3), (3, 3), (4, 3)\}$

8. $\{(0, -2), (1, 3), (2, 3), (3, 7)\}$

9. $\{(-2, 4), (-2, 6), (0, 3), (3, 7)\}$

10. $\{(-4, 4), (-3, 3), (-2, 2), (-1, 1), (-4, 0)\}$

11. $\{(-2, 4), (-1, 1), (0, 0), (1, 1)\}$

12. $\{(-2, 16), (-1, 4), (0, 3), (1, 4)\}$

In Problems 13–20, find the following values for each function:

(a) $f(0)$;   (b) $f(1)$;   (c) $f(-1)$;   (d) $f(-x)$;   (e) $-f(x)$;   (f) $f(x + 1)$

13. $f(x) = -3x^2 + 2x - 4$

14. $f(x) = 2x^2 + x - 1$

15. $f(x) = \dfrac{x}{x^2 + 1}$

16. $f(x) = \dfrac{x^2 - 1}{x + 4}$

17. $f(x) = |x| + 4$

18. $f(x) = \sqrt{x^2 + x}$

19. $f(x) = \dfrac{2x + 1}{3x - 5}$

20. $f(x) = 1 - \dfrac{1}{(x + 2)^2}$

In Problems 21–32, use the graph of the function f given in the figure.

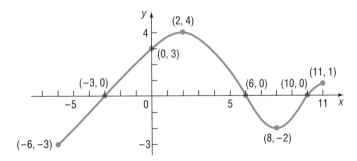

21. Find $f(0)$ and $f(-6)$.

22. Find $f(6)$ and $f(11)$.

23. Is $f(2)$ positive or negative?

24. Is $f(8)$ positive or negative?

25. For what numbers x is $f(x) = 0$?

26. For what numbers x is $f(x) > 0$?

27. What is the domain of f?

28. What is the range of f?

29. What are the x-intercepts?

30. What are the y-intercepts?

31. How often does the line $y = \frac{1}{2}$ intersect the graph?

32. How often does the line $y = 3$ intersect the graph?

In Problems 33–36, answer the questions about the given function.

33. $f(x) = \dfrac{x + 2}{x - 6}$

    (a) Is the point $(3, 14)$ on the graph of f?
    (b) If $x = 4$, what is $f(x)$? What point is on the graph of f?
    (c) If $f(x) = 2$, what is x? What point is on the graph of f?
    (d) What is the domain of f?

34. $f(x) = \dfrac{x^2 + 2}{x + 4}$

    (a) Is the point $(1, \frac{3}{5})$ on the graph of f?
    (b) If $x = 0$, what is $f(x)$? What point is on the graph of f?
    (c) If $f(x) = \frac{1}{2}$, what is x? What point is on the graph of f?
    (d) What is the domain of f?

35. $f(x) = \dfrac{2x^2}{x^4 + 1}$

    (a) Is the point $(-1, 1)$ on the graph of f?
    (b) If $x = 2$, what is $f(x)$? What point is on the graph of f?
    (c) If $f(x) = 1$, what is x? What point is on the graph of f?
    (d) What is the domain of f?

36. $f(x) = \dfrac{2x}{x - 2}$

    (a) Is the point $(\frac{1}{2}, -\frac{2}{3})$ on the graph of f?
    (b) If $x = 4$, what is $f(x)$? What point is on the graph of f?
    (c) If $f(x) = 1$, what is x? What point is on the graph of f?
    (d) What is the domain of f?

In Problems 37–48, determine whether the graph is that of a function by using the vertical-line test. If it is, use the graph to find:

(a) Its domain and range

(b) The intercepts, if any

(c) Any symmetry with respect to the x-axis, y-axis, or origin

37.
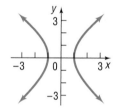

38.
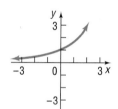

39.
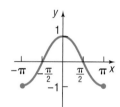

40.
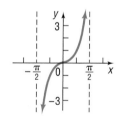

41.
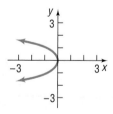

42.
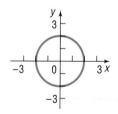

43.
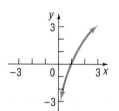

44.
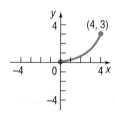

45.
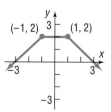

46.
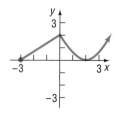

47.
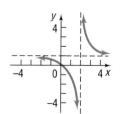

48.
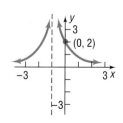

In Problems 49–62, find the domain of each function.

49. $f(x) = 3x + 4$

50. $f(x) = 5x^2 + 2$

51. $f(x) = \dfrac{x}{x^2 + 1}$

52. $f(x) = \dfrac{x^2}{x^2 + 1}$

53. $g(x) = \dfrac{x}{x^2 - 1}$

54. $h(x) = \dfrac{x}{x - 1}$

55. $F(x) = \dfrac{x - 2}{x^3 + x}$

56. $G(x) = \dfrac{x + 4}{x^3 - 4x}$

57. $h(x) = \sqrt{3x - 12}$

58. $G(x) = \sqrt{1 - x}$

59. $f(x) = \dfrac{4}{\sqrt{x - 9}}$

60. $f(x) = \dfrac{x}{\sqrt{x - 4}}$

61. $p(x) = \sqrt{\dfrac{x - 2}{x - 1}}$

62. $q(x) = \sqrt{x^2 - x - 2}$

63. Match each of the following functions with the graphs that best describe the situation.
   (a) The cost of building a house as a function of its square footage.
   (b) The height of an egg dropped from a 300-foot building as a function of time.
   (c) The height of a human as a function of time.
   (d) The demand for Big Macs as a function of price.
   (e) The height of a child on a swing as a function of time.

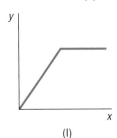

(I)

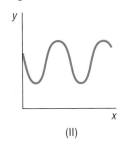

(II)

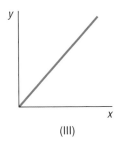

(III)

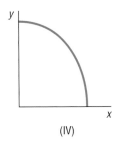

(IV)

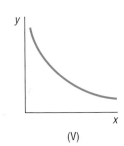

(V)

64. Consider the following scenario: Barbara decides to take a walk. She leaves home, walks 2 blocks in 5 minutes at a constant speed, and realizes that she forgot to lock the door. So Barbara runs home in 1 minute. While at her doorstep, it takes her 1 minute to find her keys and lock the door. Barbara walks 5 blocks in 15 minutes and then decides to jog home. It takes her 7 minutes to get home. Draw a graph of Barbara's distance from home (in blocks) as a function of time.

65. If $f(x) = 2x^3 + Ax^2 + 4x - 5$ and $f(2) = 5$, what is the value of A?

66. If $f(x) = 3x^2 - Bx + 4$ and $f(-1) = 12$, what is the value of B?

67. If $f(x) = (3x + 8)/(2x - A)$ and $f(0) = 2$, what is the value of A?

68. If $f(x) = (2x - B)/(3x + 4)$ and $f(2) = \frac{1}{2}$, what is the value of B?

69. If $f(x) = (2x - A)/(x - 3)$ and $f(4) = 0$, what is the value of A? Where is f not defined?

70. If $f(x) = (x - B)/(x - A)$, $f(2) = 0$, and $f(1)$ is undefined, what are the values of A and B?

71. **Demand for Jeans**   The marketing manager at Levi–Strauss wishes to find a function that relates the demand D of men's jeans and p, the price of the jeans. The following data were obtained based on a price history of the jeans.

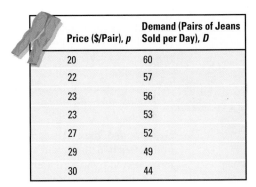

Price ($/Pair), p	Demand (Pairs of Jeans Sold per Day), D
20	60
22	57
23	56
23	53
27	52
29	49
30	44

(a) Does the relation defined by the set of ordered pairs (p, D) represent a function?
(b) Draw a scatter diagram of the data.
(c) Select two points from the data and find an equation of the line containing the points. (The answer in the back of the book uses the first and last data points.)
(d) Interpret the slope.
(e) Express the relationship found in (c) using function notation.
(f) What is the domain of the function?
(g) How many jeans will be demanded if the price is $28 a pair?
 (h) Using a graphing utility, find the line of best fit relating price and quantity demanded.

72. **Advertising and Sales Revenue** A marketing firm wishes to find a function that relates the sales S of a product and A, the amount spent on advertising the product. The data are obtained from past experience. Advertising and sales are measured in thousands of dollars.

Advertising Expenditures, A	Sales, S
20	335
22	339
22.5	338
24	343
24	341
27	350
28.3	351

(a) Does the relation defined by the set of ordered pairs (A, S) represent a function?
(b) Draw a scatter diagram of the data.
(c) Select two points from the data and find an equation of the line containing the points.
(d) Interpret the slope.
(e) Express the relationship found in (c) using function notation.
(f) What is the domain of the function?
(g) Predict sales if advertising expenditures are $25,000.
(h) Using a graphing utility, find the line of best fit relating advertising expenditures and sales.

73. **Distance and Time** Using the following data, find a function that relates the distance, s, in miles, driven by a Ford Taurus and t, the time the Taurus has been driven.

Time (Hours), t	Distance (Miles), s
0	0
1	30
2	55
3	83
4	100
5	150
6	210
7	260
8	300

(a) Does the relation defined by the set of ordered pairs (t, s) represent a function?
(b) Draw a scatter diagram of the data.
(c) Select two points from the data and find an equation of the line containing the points. (The answer in the back of the book uses the first and last data points.)
(d) Interpret the slope.
(e) Express the relationship found in (c) using function notation.
(f) What is the domain of the function?
(g) Predict the distance the car is driven after 11 hours.
(h) Using a graphing utility, find the line of best fit relating time and distance.

74. **High School versus College GPA** An administrator at Southern Illinois University wants to find a function that relates a student's college grade point average G to the high school grade point average, x. She randomly selects 8 students and obtains the following data:

High School GPA, x	College GPA, G
2.73	2.43
2.92	2.97
3.45	3.63
3.78	3.81
2.56	2.83
2.98	2.81
3.67	3.45
3.10	2.93

(a) Does the relation defined by the set of ordered pairs (x, G) represent a function?
(b) Draw a scatter diagram of the data.
(c) Select two points from the data and find an equation of the line containing the points.
(d) Interpret the slope.
(e) Express the relationship found in (c) using function notation.
(f) What is the domain of the function?
(g) Predict a student's college GPA if her high school GPA is 3.23.
(h) Using a graphing utility, find the line of best fit relating high school GPA and college GPA.

75. Geometry Express the area A of a rectangle as a function of the length x if the length is twice the width of the rectangle.

76. Geometry Express the area A of an isosceles right triangle as a function of the length x of one of the two equal sides.

77. Express the gross salary G of a person who earns $10 per hour as a function of the number x of hours worked.

78. Tiffany, a commissioned salesperson, earns $100 base pay plus $10 per item sold. Express her gross salary G as a function of the number x of items sold.

79. Page Design A page with dimensions of $8\frac{1}{2}$ inches by 11 inches has a border of uniform width x surrounding the printed matter of the page, as shown in the figure.

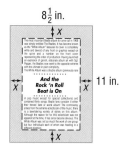

(a) Write a formula for the area A of the printed part of the page as a function of the width x of the border.
(b) Give the domain and range of A.
(c) Find the area of the printed page for borders of widths 1 inch, 1.2 inches, and 1.5 inches.
 (d) Graph the function $A = A(x)$.
(e) Use TRACE to determine what margin should be used to obtain an area of 70 square inches and of 50 square inches.

80. Cost of Trans-Atlantic Travel A Boeing 747 crosses the Atlantic Ocean (3000 miles) with an airspeed of 500 miles per hour. The cost C (in dollars) per passenger is given by

$$C(x) = 100 + \frac{x}{10} + \frac{36{,}000}{x}$$

where x is the ground speed (airspeed $\pm$ wind).
(a) What is the cost per passenger for quiescent (no wind) conditions?
(b) What is the cost per passenger with a head wind of 50 miles per hour?
(c) What is the cost per passenger with a tail wind of 100 miles per hour?
(d) What is the cost per passenger with a head wind of 100 miles per hour?
(e) Graph the function $C = C(x)$.
(f) As x varies from 400 to 600 miles per hour, how does the cost vary?

81. Period of a Pendulum The period T (in seconds) of a simple pendulum is a function of its length l (in feet) defined by the equation

$$T(l) = 2\pi\sqrt{\frac{l}{g}}$$

where $g \approx 32.2$ feet per second is the acceleration due to gravity.
(a) Use a graphing utility to graph the function $T = T(l)$.
(b) Use the TRACE function to see how the period T varies as l changes from 1 to 10.
(c) What length should be used if a period of 10 seconds is required?

82. Effect of Elevation on Weight If an object weighs m pounds at sea level, then its weight W (in pounds) at a height of h miles above sea level is given approximately by

$$W(h) = m\left(\frac{4000}{4000 + h}\right)^2$$

(a) If Amy weighs 120 pounds at sea level, how much will she weigh on Pike's Peak, which is 14,110 feet above sea level?
(b) Use a graphing utility to graph the function $W = W(h)$. Use $m = 120$ pounds.
(c) Use the TRACE function to see how weight W varies as h changes from 0 to 5 miles.
(d) At what height will Amy weigh 121 pounds?
(e) Does your answer to part (d) seem reasonable?

In Problems 83–90, tell whether the set of ordered pairs (x, y) defined by each equation is a function.

83. $y = x^2 + 2x$

84. $y = x^3 - 3x$

85. $y = \dfrac{2}{x}$

86. $y = \dfrac{3}{x} - 3$

87. $y^2 = 1 - x^2$

88. $y = \pm\sqrt{1 - 2x}$

89. $x^2 + y = 1$

90. $x + 2y^2 = 1$

91. Some functions f have the property that $f(a + b) = f(a) + f(b)$ for all real numbers a and b. Which of the following functions have this property?
(a) $h(x) = 2x$ (b) $g(x) = x^2$
(c) $F(x) = 5x - 2$ (d) $G(x) = 1/x$

92. Draw the graph of a function whose domain is $\{x | -3 \leq x \leq 8, \ x \neq 5\}$ and whose range is $\{y | -1 \leq y \leq 2, y \neq 0\}$. What point(s) in the rectangle $-3 \leq x \leq 8$, $-1 \leq y \leq 2$ cannot be on the graph? Compare your graph with those of other students. What differences do you see?

93. Are the functions $f(x) = x - 1$ and $g(x) = (x^2 - 1)/(x + 1)$ the same? Explain.

94. Describe how you would proceed to find the domain and range of a function if you were given its graph. How would your strategy change if, instead, you were given the equation defining the function?

95. How many x-intercepts can the graph of a function have? How many y-intercepts can it have?

96. Is a graph that consists of a single point the graph of a function? Can you write the equation of such a function?

97. Is there a function whose graph is symmetric with respect to the x-axis?

98. Investigate when, historically, the use of function notation $y = f(x)$ first appeared.

1.4 MORE ABOUT FUNCTIONS

1 Find the Average Rate of Change of a Function
2 Determine Where a Function Is Increasing and Decreasing
3 Determine Even or Odd Functions from a Graph
4 Identify Even or Odd Functions from the Equation
5 Graph Certain Important Functions
6 Graph Piecewise-defined Functions

Average Rate of Change

 The slope of a line can be interpreted as the average rate of change. Often, we are interested in the rate at which functions change. To find the average rate of change of a function between any two points on its graph, we calculate the slope of the line containing the two points.

EXAMPLE 1   Finding the Average Rate of Change of a Function

Suppose that you drop a ball from a cliff 1000 feet high. You measure the distance s that the ball has fallen after time t using a motion detector and obtain the data in Table 7.

(a) Draw a scatter diagram of the data, treating time as the independent variable.
(b) Draw a line from the point $(0, 0)$ to $(2, 64)$.
(c) Find the average rate of change of the ball between 0 and 2 seconds, that is, find the slope of the line in (b).
(d) Interpret the average rate of change found in (c).
(e) Draw a line from the point $(5, 400)$ to $(7, 784)$.

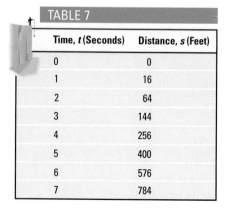

TABLE 7

Time, t (Seconds)	Distance, s (Feet)
0	0
1	16
2	64
3	144
4	256
5	400
6	576
7	784

(f) Find the average rate of change of the ball between 5 and 7 seconds.

(g) Interpret the average rate of change found in (f).

(h) What is happening to the average rate of change as time passes?

Solution   (a) We plot the ordered pairs $(0, 0)$, $(1, 16)$, and so on, using rectangular coordinates. See Figure 44.

(b) Draw the line from $(0, 0)$ to $(2, 64)$. See Figure 45.

(c) The average rate of change is found by computing the slope of the line passing through the points $(0, 0)$ and $(2, 64)$.

$$\text{Average Rate of Change} = \frac{\Delta s}{\Delta t} = \frac{64 - 0}{2 - 0} = \frac{64}{2} = 32 \text{ ft/sec}$$

(d) Since the change in distance divided by the change in time represents a speed, we would interpret the slope of the line as an average speed. The average speed of the ball between 0 and 2 seconds is 32 feet per second.

(e) Draw the line from $(5, 400)$ to $(7, 784)$. See Figure 46.

FIGURE 44

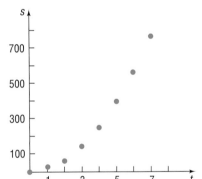

FIGURE 45

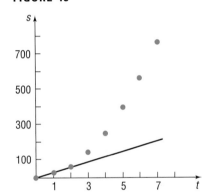

FIGURE 46

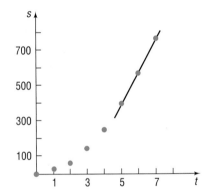

(f) The average rate of change of the ball between 5 and 7 seconds is

$$\text{Average Rate of Change} = \frac{\Delta s}{\Delta t} = \frac{784 - 400}{7 - 5} = \frac{384}{2} = 192 \text{ ft/sec}$$

(g) The average speed of the ball between 5 and 7 seconds is 192 feet per second.

(h) The average speed of the ball is increasing as time passes since the ball is accelerating due to the effect of gravity.   ▬

If we know the function f that relates the time t to the distance s that the ball has fallen, $s = f(t)$, then the average speed of the ball between 0 and 2 seconds is

$$\frac{f(2) - f(0)}{2 - 0} = \frac{64 - 0}{2} = 32 \text{ ft/sec}$$

The average speed of the ball between 5 and 7 seconds is

$$\frac{f(7) - f(5)}{7 - 5} = \frac{784 - 400}{2} = 192 \text{ ft/sec}$$

Expressions like these occur frequently in calculus.

> If c is in the domain of a function $y = f(x)$, the **average rate of change of f** between c and x is defined as
>
> $$\text{Average rate of change} = \frac{\Delta y}{\Delta x} = \frac{f(x) - f(c)}{x - c} \qquad x \neq c \qquad (1)$$
>
> This expression is also called the **difference quotient** of f at c.

The average rate of change of a function has an important geometric interpretation. Look at the graph of $y = f(x)$ in Figure 47. We have labeled two points on the graph: $(c, f(c))$ and $(x, f(x))$. The slope of the line containing these two points is

$$\frac{f(x) - f(c)}{x - c}$$

This line is called a **secant line.**

FIGURE 47

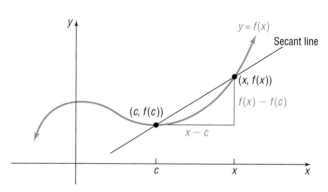

> **Theorem** **Slope of a Secant Line**
>
> The average rate of change of a function equals the slope of a secant line containing two points on its graph.

E X A M P L E   2

Finding the Average Rate of Change

The function $s(t) = -4.9t^2 + 100t + 5$ gives the height s (in meters) of a bullet fired straight up as a function of time t (in seconds).

(a) Find the average rate of change of the height of the bullet between 2 and t seconds.

(b) Using the result found in (a), find the average rate of change of the height of the bullet between 2 and 3 seconds.

Solution

(a) From expression (1), the average rate of change of the height of the bullet between 2 and t seconds is

$$\frac{\Delta s}{\Delta t} = \frac{s(t) - s(2)}{t - 2} \qquad t \neq 2$$

We begin by finding $s(2)$:

$$s(2) = -4.9(2)^2 + 100(2) + 5 = -19.6 + 200 + 5 = 185.4$$

Then the average rate of change of s between 2 and t seconds is

$$\frac{s(t) - s(2)}{t - 2} = \frac{-4.9t^2 + 100t + 5 - 185.4}{t - 2} = \frac{-4.9t^2 + 100t - 180.4}{t - 2}$$

$$= \frac{(-4.9t + 90.2)(t - 2)}{t - 2} = -4.9t + 90.2$$

(b) The average rate of change between 2 and 3 seconds is found by letting $t = 3$ in the expression found in (a). Thus, the average rate of change of the bullet between 2 and 3 seconds is $-4.9(3) + 90.2 = 75.5$ meters per second. ▄

Now work Problem 29.

Increasing and Decreasing Functions

[2] Consider the graph given in Figure 48. If you look from left to right along the graph of the function, you will notice that parts of the graph are rising, parts are falling, and parts are horizontal. In such cases, the function is described as *increasing*, *decreasing*, and *constant*, respectively.

FIGURE 48

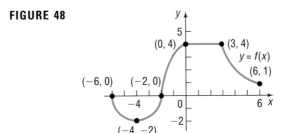

E X A M P L E   3   Determining Where a Function Is Increasing, Decreasing, or Constant

Where is the function in Figure 48 increasing? Where is it decreasing? Where is it constant?

Solution   To answer the question of where a function is increasing, where it is decreasing, and where it is constant, we use inequalities involving the independent variable x or we use open intervals* of x-coordinates. The graph in Figure 48 is rising (increasing) from the point $(-4, -2)$ to the point $(0, 4)$, so we conclude that it is increasing on the open interval $(-4, 0)$ (or for $-4 < x < 0$). The graph is falling (decreasing) from the point $(-6, 0)$ to the point $(-4, -2)$ and from the point $(3, 4)$ to the point $(6, 1)$. We conclude that the graph is decreasing on the open intervals $(-6, -4)$ and $(3, 6)$ (or for $-6 < x < -4$ and $3 < x < 6$). The graph is constant on the open interval $(0, 3)$ (or for $0 < x < 3$). ▄

More precise definitions follow.

---

*The open interval (a, b) consists of all real numbers x for which $a < x < b$. Refer to Section 1.1, if necessary.

A function f is **increasing** on an open interval I if, for any choice of x_1 and x_2 in I, with $x_1 < x_2$, we have $f(x_1) < f(x_2)$.

A function f is **decreasing** on an open interval I if, for any choice of x_1 and x_2 in I, with $x_1 < x_2$, we have $f(x_1) > f(x_2)$.

A function f is **constant** on an open interval I if, for all choices of x in I, the values $f(x)$ are equal.

Thus, the graph of an increasing function goes up from left to right, the graph of a decreasing function goes down from left to right, and the graph of a constant function remains at a fixed height. Figure 49 illustrates the definitions.

FIGURE 49

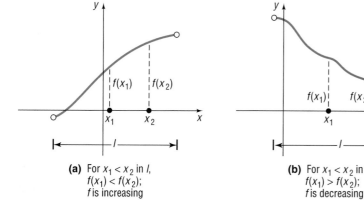

(a) For $x_1 < x_2$ in I,
$f(x_1) < f(x_2)$;
f is increasing

(b) For $x_1 < x_2$ in I,
$f(x_1) > f(x_2)$;
f is decreasing

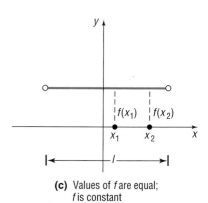

(c) Values of f are equal;
f is constant

Even and Odd Functions

A function f is **even** if for every number x in its domain the number $-x$ is also in the domain and

$$f(-x) = f(x)$$

In other words, if $y = f(x)$, then f is even if and only if, whenever the point (x, y) is on the graph of f, the point $(-x, y)$ is also on the graph.

A function *f* is **odd** if for every number *x* in its domain the number −*x* is also in the domain and

$$f(-x) = -f(x)$$

In other words, if $y = f(x)$, then *f* is odd if and only if, whenever the point (x, y) is on the graph of *f*, the point $(-x, -y)$ is also on the graph.

Refer to Section 1.2, where the tests for symmetry are listed. The following results are then evident.

Theorem

A function is even if and only if its graph is symmetric with respect to the *y*-axis. A function is odd if and only if its graph is symmetric with respect to the origin.

E X A M P L E   4

Determining Even and Odd Functions from the Graph

3

Determine whether each graph in Figure 50 is the graph of an even function, an odd function, or a function that is neither even nor odd.

FIGURE 50

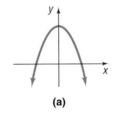
(a)

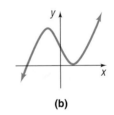
(b)

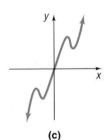
(c)

Solution

The graph in Figure 50(a) is that of an even function, because the graph is symmetric with respect to the *y*-axis. The function whose graph is given in Figure 50(b) is neither even nor odd, because the graph is neither symmetric with respect to the *y*-axis nor symmetric with respect to the origin. The function whose graph is given in Figure 50(c) is odd, because its graph is symmetric with respect to the origin.

Now work Problem 9.

4

In the next example, we show how to verify algebraically whether a function is even, odd, or neither.

E X A M P L E   5

Identifying Even and Odd Functions Algebraically

Determine whether each of the following functions is even, odd, or neither. Then determine whether the graph is symmetric with respect to the *y*-axis or with respect to the origin.

(a) $f(x) = x^2 - 5$ (b) $g(x) = x^3 - 1$
(c) $h(x) = 5x^3 - x$ (d) $F(x) = |x|$

Solution   (a) We replace x by $-x$ in $f(x) = x^2 - 5$. Then

$$f(-x) = (-x)^2 - 5 = x^2 - 5 = f(x)$$

Since $f(-x) = f(x)$, we conclude that f is an even function, and the graph is symmetric with respect to the y-axis.

(b) We replace x by $-x$. Then

$$g(-x) = (-x)^3 - 1 = -x^3 - 1$$

Since $g(-x) \neq g(x)$ and $g(-x) \neq -g(x)$, we conclude that g is neither even nor odd. The graph is not symmetric with respect to the y-axis nor with respect to the origin.

(c) We replace x by $-x$ in $h(x) = 5x^3 - x$. Then

$$h(-x) = 5(-x)^3 - (-x) = -5x^3 + x = -(5x^3 - x) = -h(x)$$

Since $h(-x) = -h(x)$, h is an odd function, and the graph of h is symmetric with respect to the origin.

(d) We replace x by $-x$ in $F(x) = |x|$. Then

$$F(-x) = |-x| = |x| = F(x)$$

Since $F(-x) = F(x)$, F is an even function, and the graph of F is symmetric with respect to the y-axis.

 Now work Problem 39.

Library of Functions

We now give names to some of the functions that we have encountered. In going through this list, pay special attention to the characteristics of each function, particularly to the shape of each graph. Knowing these graphs will lay the foundation for later graphing techniques.

Linear Functions

$$f(x) = mx + b \qquad m \text{ and } b \text{ are real numbers.}$$

The domain of a **linear function** f consists of all real numbers. The graph of this function is a nonvertical line with slope m and y-intercept b. A linear function is increasing if $m > 0$, decreasing if $m < 0$, and constant if $m = 0$.

FIGURE 51
Constant Function

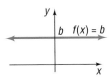

Constant Function

$$f(x) = b \qquad b \text{ is a real number}$$

See Figure 51.

A **constant function** is a special linear function ($m = 0$). Its domain is the set of all real numbers; its range is the set consisting of a single number b. Its graph is a horizontal line whose y-intercept is b. The constant function is an even function whose graph is constant over its domain.

FIGURE 52
Identity Function

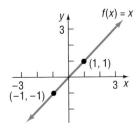

Identity Function

$$f(x) = x$$

See Figure 52.

The **identity function** is also a special linear function. Its domain and its range are the set of all real numbers. Its graph is a line whose slope is $m = 1$ and whose y-intercept is 0. The line consists of all points for which the x-coordinate equals the y-coordinate. The identity function is an odd function that is increasing over its domain. Note that the graph bisects quadrants I and III.

FIGURE 53
Square Function

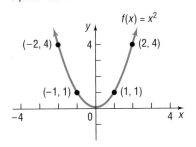

Square Function

$$f(x) = x^2$$

See Figure 53.

The domain of the **square function** f is the set of all real numbers; its range is the set of nonnegative real numbers. The graph of this function is a parabola, whose intercept is at $(0, 0)$. The square function is an even function that is decreasing on the interval $(-\infty, 0)$ and increasing on the interval $(0, \infty)$.

FIGURE 54
Cube Function

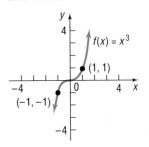

Cube Function

$$f(x) = x^3$$

See Figure 54.

The domain and range of the **cube function** are the set of all real numbers. The intercept of the graph is at $(0, 0)$. The cube function is odd and is increasing on the interval $(-\infty, \infty)$.

FIGURE 55
Square Root Function

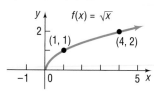

Square Root Function

$$f(x) = \sqrt{x}$$

See Figure 55.

The domain and range of the **square root function** are the set of non-negative real numbers. The intercept of the graph is at $(0, 0)$. The square root function is neither even nor odd and is increasing on the interval $(0, \infty)$.

FIGURE 56
Reciprocal Function

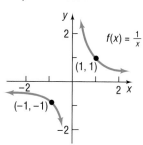

Reciprocal Function

$$f(x) = \frac{1}{x}$$

Refer to Example 10, p. 26, for a discussion of the equation $y = 1/x$. See Figure 56.

The domain and range of the **reciprocal function** are the set of all nonzero real numbers. The graph has no intercepts. The reciprocal function is decreasing on the intervals $(-\infty, 0)$ and $(0, \infty)$ and is an odd function.

FIGURE 57
Absolute Value Function

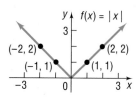

Absolute Value Function

$$f(x) = |x|$$

See Figure 57.

The domain of the **absolute value function** is the set of all real numbers; its range is the set of nonnegative real numbers. The intercept of the graph is at $(0, 0)$. If $x \geq 0$, then $f(x) = x$ and the graph of f is part of the line $y = x$; if $x < 0$, then $f(x) = -x$ and the graph of f is part of the line $y = -x$. The absolute value function is an even function; it is decreasing on the interval $(-\infty, 0)$ and increasing on the interval $(0, \infty)$.

 Check: Graph $y = |x|$ on a square screen and compare what you see with Figure 57. Note that some graphing calculators use the symbol abs (x) for absolute value. If your utility has no built-in absolute value function, you can still graph $y = |x|$ by using the fact that $|x| = \sqrt{x^2}$. ▬

The notation int(x) stands for the largest integer less than or equal to x. For example,

$$\text{int}(1) = 1 \qquad \text{int}(2.5) = 2 \qquad \text{int}(\tfrac{1}{2}) = 0 \qquad \text{int}(\tfrac{-3}{4}) = -1 \qquad \text{int}(\pi) = 3$$

This type of correspondence occurs frequently enough in mathematics that we give it a name.

Greatest Integer Function

> $f(x) = \text{int}(x) = $ Greatest integer less than or equal to x

We obtain the graph of $f(x) = \text{int}(x)$ by plotting several points. See Table 8. For values of x, $-1 \leq x < 0$, the value of $f(x) = \text{int}(x)$ is -1; for values of x, $0 \leq x < 1$, the value of f is 0. See Figure 58 for the graph.

TABLE 8

x	$y = f(x) =$ int(x)	(x, y)
-1	-1	$(-1, -1)$
$-\frac{1}{2}$	-1	$(-\frac{1}{2}, -1)$
$-\frac{1}{4}$	-1	$(-\frac{1}{4}, -1)$
0	0	$(0, 0)$
$\frac{1}{4}$	0	$(\frac{1}{4}, 0)$
$\frac{1}{2}$	0	$(\frac{1}{2}, 0)$
$\frac{3}{4}$	0	$(\frac{3}{4}, 0)$

FIGURE 58
Greatest Integer Function

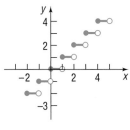

The domain of the **greatest integer function** is the set of all real numbers; its range is the set of integers. The y-intercept of the graph is at 0. The x-intercepts lie in the interval $[0, 1)$. The greatest integer function is neither even nor odd. It is constant on every interval of the form $[k, k + 1)$, for k an integer. In Figure 58, we use a solid dot to indicate, for example, that at $x = 1$ the value of f is $f(1) = 1$; we use an open circle to illustrate that the function does not assume the value of 0 at $x = 1$.

From the graph of the greatest integer function, we can see why it is also called a **step function**. At $x = 0, x = \pm 1, x = \pm 2$, and so on, this function exhibits what is called a *discontinuity;* that is, at integer values, the graph suddenly "steps" from one value to another without taking on any of the intermediate values. For example, to the immediate left of $x = 3$, the y-coordinates are 2, and to the immediate right of $x = 3$, the y-coordinates are 3.

 Comment   When graphing a function, you can choose either the **connected mode,** in which points plotted on the screen are connected, making the graph appear without any breaks, or the **dot mode,** in which only the points plotted appear. When graphing the greatest integer function with a graphing utility, it is necessary to be in the **dot mode.** This is to prevent the utility from "connecting the dots" when $f(x)$ changes from one integer value to the next. See Figure 59.

FIGURE 59
$f(x) = \text{int}(x)$

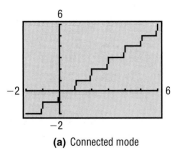

(a) Connected mode

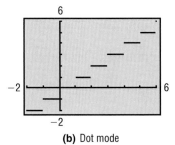

(b) Dot mode

The functions that we have discussed so far are basic. Whenever you encounter one of them, you should see a mental picture of its graph. For example, if you encounter the function $f(x) = x^2$, you should see in your mind's eye a picture like Figure 53.

Now work Problems 1–8.

Piecewise-defined Functions

Sometimes, a function is defined differently on different parts of its domain. For example, the absolute value function $f(x) = |x|$ is actually defined by two equations: $f(x) = x$ if $x \geq 0$ and $f(x) = -x$ if $x < 0$. For convenience, we generally combine these equations into one expression as

$$f(x) = |x| = \begin{cases} x & \text{if } x \geq 0 \\ -x & \text{if } x < 0 \end{cases}$$

When functions are defined by more than one equation, they are called **piecewise-defined** functions.

Let's look at another example of a piecewise-defined function.

E X A M P L E   6

Analyzing a Piecewise-defined Function

The function f is defined as:

$$f(x) = \begin{cases} -x + 1 & \text{if } -1 \leq x < 1 \\ 2 & \text{if } x = 1 \\ x^2 & \text{if } x > 1 \end{cases}$$

(a) Find $f(0)$, $f(1)$, and $f(2)$.   (b) Determine the domain of f.
(c) Graph f.   (d) Use the graph to find the range of f.

Solution   (a) To find $f(0)$, we observe that when $x = 0$ the equation for f is given by $f(x) = -x + 1$. So we have

$$f(0) = -0 + 1 = 1$$

When $x = 1$, the equation for f is $f(x) = 2$. Thus,

$$f(1) = 2$$

When $x = 2$, the equation for f is $f(x) = x^2$. So

$$f(2) = 2^2 = 4$$

FIGURE 60

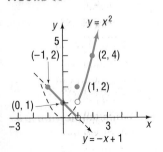

(b) To find the domain of f, we look at its definition. We conclude that the domain of f is $\{x \mid x \geq -1\}$, or $[-1, \infty)$.
(c) To graph f, we graph "each piece." Thus, we first graph the line $y = -x + 1$ and keep only the part for which $-1 \leq x < 1$. Then we plot the point $(1, 2)$ because, when $x = 1$, $f(x) = 2$. Finally, we graph the parabola $y = x^2$ and keep only the part for which $x > 1$. See Figure 60.
(d) From the graph, we conclude that the range of f is $\{y \mid y > 0\}$, or $(0, \infty)$.

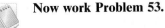

Now work Problem 53.

EXAMPLE   7

Cost of Electricity

In the winter, Commonwealth Edison Company supplies electricity to residences for a monthly customer charge of $8.91 plus 10.494¢ per kilowatt-hour (kWhr) for the first 400 kWhr supplied in the month and 7.91¢ per kWhr for all usage over 400 kWhr in the month.*

(a) What is the charge for using 300 kWhr in a month?
(b) What is the charge for using 700 kWhr in a month?
(c) If C is the monthly charge for x kWhr, express C as a function of x.

Solution

(a) For 300 kWhr, the charge is $8.91 plus 10.494¢ = $0.10494 per kWhr. Thus,

$$\text{Charge} = \$8.91 + \$0.10494(300) = \$40.39$$

(b) For 700 kWhr, the charge is $8.91 plus 10.494¢ for the first 400 kWhr plus 7.91¢ for the 300 kWhr in excess of 400. Thus,

$$\text{Charge} = \$8.91 + \$0.10494(400) + \$0.0791(300) = \$74.62$$

(c) If $0 \leq x \leq 400$, the monthly charge C (in dollars) can be found by multiplying x times $0.10494 and adding the monthly customer charge of $8.91. Thus, if $0 \leq x \leq 400$, then $C(x) = 0.10494x + 8.91$. For $x > 400$, the charge is $0.10494(400) + 8.91 + 0.0791(x - 400)$, since $x - 400$ equals the usage in excess of 400 kWhr, which costs $0.0791 per kWhr. Thus, if $x > 400$, then

$$\begin{aligned}
C(x) &= 0.10494(400) + 8.91 + 0.0791(x - 400) \\
&= 50.89 + 0.0791(x - 400) \\
&= 0.0791x + 19.25
\end{aligned}$$

The rule for computing C follows two equations:

$$C(x) = \begin{cases} 0.10494x + 8.91 & \text{if } 0 \leq x \leq 400 \\ 0.0791x + 19.25 & \text{if } x > 400 \end{cases}$$

See Figure 61 for the graph.

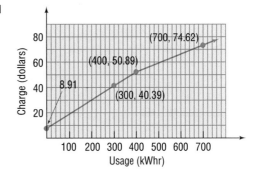

FIGURE 61

Source: Commonwealth Edison Co., Chicago, Illinois, 1998.

1.4 | EXERCISES

In Problems 1–8, match each graph to the function listed whose graph most resembles the one given.

A. *Constant function*   B. *Linear function*

C. *Square function*   D. *Cube function*

E. *Square root function*   F. *Reciprocal function*

G. *Absolute value function*   H. *Greatest integer function*

1.

2.

3.

4.

5.

6.

7.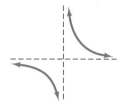

8.

In Problems 9–22, the graph of a function is given. Use the graph to find

(a) Its domain and range

(b) The intervals on which it is increasing, decreasing, or constant

(c) Whether it is even, odd, or neither

(d) The intercepts, if any

9.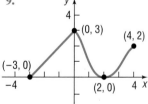

10.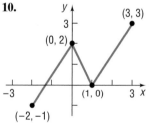

11.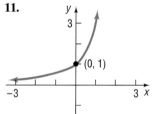

12.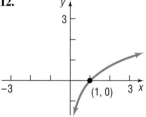

13.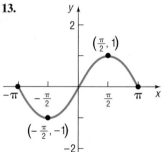

14.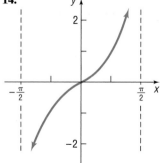

15.

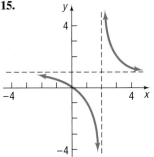

16.

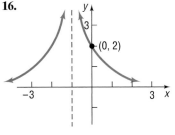

17.

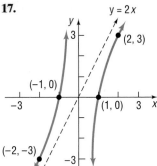

18.

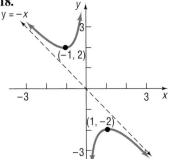

19.

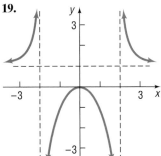

20.

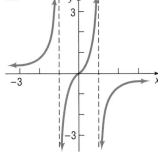

📷 *In Problems 21–22, assume that the entire graph is shown.*

21.

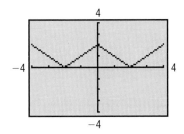

22.

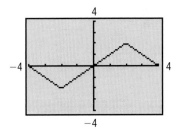

23. If

$$f(x) = \begin{cases} x^2 & \text{if } x < 0 \\ 2 & \text{if } x = 0 \\ 2x + 1 & \text{if } x > 0 \end{cases}$$

find: (a) $f(-2)$ (b) $f(0)$ (c) $f(2)$

24. If

$$f(x) = \begin{cases} x^3 & \text{if } x < 0 \\ 3x + 2 & \text{if } x \geq 0 \end{cases}$$

find: (a) $f(-1)$ (b) $f(0)$ (c) $f(1)$

25. If $f(x) = \text{int}(2x)$, find: (a) $f(1.2)$ (b) $f(1.6)$ (c) $f(-1.8)$

26. If $f(x) = \text{int}(x/2)$, find: (a) $f(1.2)$ (b) $f(1.6)$ (c) $f(-1.8)$

In Problems 27–38, find the average rate of change of f between 1 and x,

$$\frac{f(x) - f(1)}{x - 1} \qquad x \neq 1$$

for each function f. Be sure to simplify.

27. $f(x) = 3x$

28. $f(x) = -2x$

29. $f(x) = 1 - 3x$

30. $f(x) = x^2 + 1$

31. $f(x) = 3x^2 - 2x$

32. $f(x) = 4x - 2x^2$

33. $f(x) = x^3 - x$

34. $f(x) = x^3 + x$

35. $f(x) = \dfrac{2}{x + 1}$

36. $f(x) = \dfrac{1}{x^2}$

37. $f(x) = \sqrt{x}$

38. $f(x) = \sqrt{x + 3}$

In Problems 39–50, determine (algebraically) whether each function is even, odd, or neither.

39. $f(x) = 4x^3$

40. $f(x) = 2x^4 - x^2$

41. $g(x) = 2x^2 - 5$

42. $h(x) = 3x^3 + 2$

43. $F(x) = \sqrt[3]{x}$

44. $G(x) = \sqrt{x}$

45. $f(x) = x + |x|$

46. $f(x) = \sqrt[3]{2x^2 + 1}$

47. $g(x) = \dfrac{1}{x^2}$

48. $h(x) = \dfrac{x}{x^2 - 1}$

49. $h(x) = \dfrac{x^3}{3x^2 - 9}$

50. $F(x) = \dfrac{x}{|x|}$

51. How many x-intercepts can a function defined on an interval have if it is increasing on that interval? Explain.

52. How many y-intercepts can a function have? Explain.

In Problems 53–58:

(a) Find the domain of each function.

(b) Locate any intercepts.

(c) Graph each function.

(d) Based on the graph, find the range.

53. $f(x) = \begin{cases} 2x & \text{if } x \neq 0 \\ 0 & \text{if } x = 0 \end{cases}$

54. $f(x) = \begin{cases} 3x & \text{if } x \neq 0 \\ 4 & \text{if } x = 0 \end{cases}$

55. $f(x) = \begin{cases} 1 + x & \text{if } x < 0 \\ x^2 & \text{if } x \geq 0 \end{cases}$

56. $f(x) = \begin{cases} 1/x & \text{if } x < 0 \\ \sqrt[3]{x} & \text{if } x \geq 0 \end{cases}$

57. $f(x) = \begin{cases} |x| & \text{if } -2 \leq x < 0 \\ 1 & \text{if } x = 0 \\ x^3 & \text{if } x > 0 \end{cases}$

58. $f(x) = \begin{cases} 3 + x & \text{if } -3 \leq x < 0 \\ 3 & \text{if } x = 0 \\ \sqrt{x} & \text{if } x > 0 \end{cases}$

In Problems 59–66, find the following for each function:

(a) $f(-x)$ *(b) $-f(x)$*   *(c) $f(2x)$*   *(d) $f(x - 3)$*   *(e) $f(1/x)$*   *(f) $1/f(x)$*

59. $f(x) = 2x + 5$

60. $f(x) = 3 - x$

61. $f(x) = 2x^2 - 4$

62. $f(x) = x^3 + 1$

63. $f(x) = x^3 - 3x$

64. $f(x) = x^2 + x$

65. $f(x) = |x|$

66. $f(x) = \dfrac{1}{x}$

Problems 67–70 require the following definition. **Secant Line:** *The slope of the secant line containing the two points $(x, f(x))$ and $(x + h, f(x + h))$ on the graph of a function $y = f(x)$ may be given as*

$$\frac{f(x + h) - f(x)}{h}$$

In Problems 67–70, express the slope of the secant line of each function in terms of x and h. Be sure to simplify your answer.

67. $f(x) = 2x + 5$

68. $f(x) = -3x + 2$

69. $f(x) = x^2 + 2x$

70. $f(x) = 1/x$

In Problems 71–74, the graph of a piecewise-defined function is given. Write a definition for each function.

71.

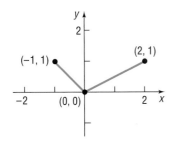

72.

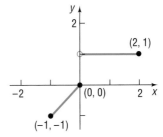

73.

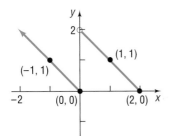

74.

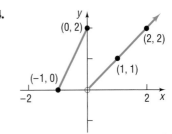

In Problems 75 and 76, decide whether each function is even. Give a reason.

75. $f(x) = \begin{cases} x^2 + 4 & \text{if } x \neq 2 \\ 6 & \text{if } x = 2 \end{cases}$

76. $f(x) = \begin{cases} x^2 + 4 & \text{if } x \neq 2 \\ 5 & \text{if } x = 2 \end{cases}$

77. Growth of Bacteria The following data represents the population of an unknown bacteria.

Time (Days)	Population
0	50
1	153
2	234
3	357
4	547
5	839
6	1280

(a) Draw a scatter diagram of the data, treating the time as the independent variable.
(b) Draw a line from the point (0, 50) to (1, 153) on the scatter diagram found in (a).
(c) Find the average rate of change of the population between 0 and 1 days.
(d) Interpret the average rate of change found in (c).
(e) Draw a line from the point (5, 839) to (6, 1280) on the scatter diagram found in (a).
(f) Find the average rate of change of the population between 5 and 6 days.

(g) Interpret the average rate of change found in (f).
(h) What is happening to the average rate of change of the population as time passes?

78. Cost of Manufacturing Bikes The following data represents the monthly cost of producing bicycles at Tunney's Bicycle Shop.

Number of Bicycles, x	Total Cost of Production, C (Dollars)
0	24,000
25	27,750
60	31,500
102	35,250
150	39,000
190	42,750
223	46,500
249	50,250

(a) Draw a scatter diagram of the data, treating the number of bicycles produced as the independent variable.

(b) Draw a line from the point (0, 24,000) to (25,27,750) on the scatter diagram found in (a).

(c) Find the average rate of change of the cost between 0 and 25 bicycles.

(d) Interpret the average rate of change found in (c).

(e) Draw a line from the point (190, 42,750) to (223, 46,500) on the scatter diagram found in (a).

(f) Find the average rate of change of the cost between 190 and 223 bicycles.

(g) Interpret the average rate of change found in (f).

79. Revenue from Selling Bikes The following data represent the total revenue that would be received from selling x bicycles at Tunney's Bicycle Shop.

Number of Bicycles, x	Total Revenue, R (Dollars)
0	0
25	28,000
60	45,000
102	53,400
150	59,160
190	62,360
223	64,835
249	66,525

(a) Draw a scatter diagram of the data, treating the number of bicycles produced as the independent variable.

(b) Draw a line from the point (0, 0) to (25, 28,000) on the scatter diagram found in (a).

(c) Find the average rate of change of revenue between 0 and 25 bicycles.

(d) Interpret the average rate of change found in (c).

(e) Draw a line from the point (190, 62,360) to (223, 64,835) on the scatter diagram found in (a).

(f) Find the average rate of change of revenue between 190 and 223 bicycles.

(g) Interpret the average rate of change found in (f).

(h) What is happening to the average rate of change of revenue as the number of bicycles sold increases?

(i) Refer to Problem 78(d). A firm's profit will increase as long as the average rate of change in revenue is greater than the average rate of change in cost. At what level

of output does Tunney's Bicycle Shop's profits stop increasing?

80. Cost of Tuition The following data represent the average cost of tuition and required fees (in dollars) at public four-year colleges.

Year	Average Cost of Tuition
1990	2,035
1991	2,159
1992	2,410
1993	2,604
1994	2,822

Source: U.S. National Center for Education Statistics.

(a) Draw a scatter diagram of the data, treating the year as the independent variable.

(b) Draw a line from the point (1990, 2035) to (1992, 2410) on the scatter diagram found in (a).

(c) Find the average rate of change of the cost of tuition and required fees between 1990 and 1992.

(d) Interpret the average rate of change found in (c).

(e) Draw a line from the point (1992, 2410) to (1994, 2822) on the scatter diagram found in (a).

(f) Find the average rate of change of the cost of tuition and required fees between 1992 and 1994.

(g) Interpret the average rate of change found in (f).

(h) What is happening to the average rate of change for the cost of tuition and required fees as time passes?

81. Cost of Natural Gas In January 1997, the Peoples Gas Company had the following rate schedule* for natural gas usage in single-family residences:

Monthly service charge	$9.00
Per therm service charge,	
1st 50 therms	$0.36375/therm
Over 50 therms	$0.11445/therm
Gas charge	$0.3256/therm

(a) What is the charge for using 50 therms in a month?

(b) What is the charge for using 500 therms in a month?

Source: The Peoples Gas Company, Chicago, Illinois.

(c) Construct a function that relates the monthly charge C for x therms of gas.

(d) Graph this function.

82. **Cost of Natural Gas** In January 1997, Northern Illinois Gas Company had the following rate schedule* for natural gas usage in single-family residences:

Monthly customer charge $6.00

Distribution charge,

1st 20 therms	$0.2012/therm
Next 30 therms	$0.1117/therm
Over 50 therms	$0.0374/therm

Gas supply charge $0.3113/therm

(a) What is the charge for using 40 therms in a month?

(b) What is the charge for using 202 therms in a month?

(c) Construct a function that gives the monthly charge C for x therms of gas.

(d) Graph this function.

83. Graph $y = x^2$. Then on the same screen graph $y = x^2 + 2$, followed by $y = x^2 + 4$, followed by $y = x^2 - 2$. What pattern do you observe? Can you predict the graph of $y = x^2 - 4$? Of $y = x^2 + 5$?

84. Graph $y = x^2$. Then on the same screen graph $y = (x - 2)^2$, followed by $y = (x - 4)^2$, followed by $y = (x + 2)^2$. What pattern do you observe? Can you predict the graph of $y = (x + 4)^2$? Of $y = (x - 5)^2$?

85. Graph $y = |x|$. Then on the same screen graph $y = 2|x|$, followed by $y = 4|x|$, followed by $y = \frac{1}{2}|x|$. What pattern do you observe? Can you predict the graph of $y = \frac{1}{4}|x|$? Of $y = 5|x|$?

86. Graph $y = x^2$. Then on the same screen graph $y = -x^2$. What pattern do you observe? Now try $y = |x|$ and $y = -|x|$. What do you conclude?

87. Graph $y = \sqrt{x}$. Then on the same screen graph $y = \sqrt{-x}$. What pattern do you observe? Now try $y = 2x + 1$ and $y = 2(-x) + 1$. What do you conclude?

88. Graph $y = x^3$. Then on the same screen graph $y = (x - 1)^3 + 2$. Could you have predicted the result?

89. Graph $y = x^2$, $y = x^4$, and $y = x^6$ on the same screen. What do you notice that is the same about each graph? What do you notice that is different?

90. Graph $y = x^3$, $y = x^5$, and $y = x^7$ on the same screen. What do you notice that is the same about each graph? What do you notice that is different?

91. Consider the equation

$$y = \begin{cases} 1 & \text{if } x \text{ is rational} \\ 0 & \text{if } x \text{ is irrational} \end{cases}$$

Is this a function? What is its domain? What is its range? What is its y-intercept, if any? What are its x-intercepts, if any? Is it even, odd, or neither? How would you describe its graph?

92. Define some functions that pass through $(0, 0)$ and $(1, 1)$ and are increasing for $x \geq 0$. Begin your list with $y = \sqrt{x}, y = x,$ and $y = x^2$. Can you propose a general result about such functions?

93. Can you think of a function that is both even and odd?

*Source: Northern Illinois Gas Company, Naperville, Illinois.

1.5   GRAPHING TECHNIQUES: TRANSFORMATIONS

   1   Graph Functions Using Horizontal and Vertical Shifts

   2   Graph Functions Using Compressions and Stretches

   3   Graph Functions Using Reflections about the x-Axis or y-Axis

At this stage, if you were asked to graph any of the functions defined by $y = x$, $y = x^2$, $y = x^3$, $y = \sqrt{x}$, $y = |x|$, or $y = 1/x$, your response should be, "Yes, I recognize these functions and know the general shapes of their graphs." (If this is not your answer, review the previous section and Figures 52 through 57.)

Sometimes, we are asked to graph a function that is "almost" like one that we already know how to graph. In this section, we look at some of these functions and develop techniques for graphing them. Collectively, these techniques are referred to as **transformations.**

1

Vertical Shifts

E X A M P L E   1

Vertical Shift Up

Use the graph of $f(x) = x^2$ to obtain the graph of $g(x) = x^2 + 3$.

Solution   We begin by obtaining some points on the graphs of f and g. For example, when $x = 0$, then $y = f(0) = 0$ and $y = g(0) = 3$. When $x = 1$, then $y = f(1) = 1$ and $y = g(1) = 4$. Table 9 lists these and a few other points on each graph. We conclude that the graph of g is identical to that of f, except that it is shifted vertically up 3 units. See Figure 62.

TABLE 9		
x	$y = f(x)$ $= x^2$	$y = g(x)$ $= x^2 + 3$
-2	4	7
-1	1	4
0	0	3
1	1	4
2	4	7

FIGURE 62

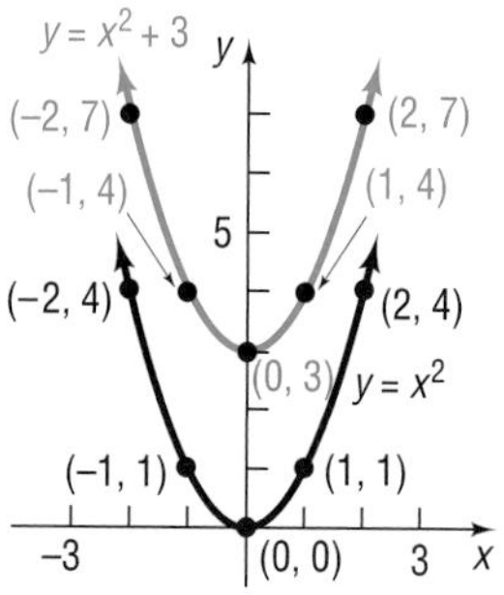

$y = x^2 + 3$
$(-2, 7)$ $(2, 7)$
$(-1, 4)$ $(1, 4)$
$(-2, 4)$ $(2, 4)$
$(0, 3)$ $y = x^2$
$(-1, 1)$ $(1, 1)$
-3 $(0, 0)$ 3 x

FIGURE 63

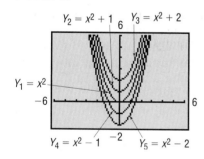

$Y_2 = x^2 + 1$ $Y_3 = x^2 + 2$
$Y_1 = x^2$
-6 6
$Y_4 = x^2 - 1$ -2 $Y_5 = x^2 - 2$

Seeing the Concept   On the same screen, graph each of the following functions:

$$Y_1 = x^2$$
$$Y_2 = x^2 + 1$$
$$Y_3 = x^2 + 2$$
$$Y_4 = x^2 - 1$$
$$Y_5 = x^2 - 2$$

Figure 63 illustrates the graphs. You should have observed a general pattern. With $Y_1 = x^2$ on the screen, the graph of $Y_2 = x^2 + 1$ is identical to that of $Y_1 = x^2$, except that it is shifted vertically up 1 unit. Similarly, $Y_3 = x^2 + 2$ is identical to that of $Y_1 = x^2$, except that it is shifted vertically up 2 units. The graph of $Y_4 = x^2 - 1$ is identical to that of $Y_1 = x^2$, except that it is shifted vertically down 1 unit.

We are led to the following conclusion:

> If a real number c is added to the right side of a function $y = f(x)$, the graph of the new function $y = f(x) + c$ is the graph of f **shifted vertically** up (if $c > 0$) or down (if $c < 0$).

Let's look at another example.

E X A M P L E   2   Vertical Shift Down

Use the graph of $f(x) = x^2$ to obtain the graph of $h(x) = x^2 - 4$.

Solution   Table 10 lists some points on the graphs of f and h. The graph of h is identical to that of f, except that is shifted down 4 units. See Figure 64.

TABLE 10		
x	$y = f(x)$ $= x^2$	$y = h(x)$ $= x^2 - 4$
-2	4	0
-1	1	-3
0	0	-4
1	1	-3
2	4	0

FIGURE 64

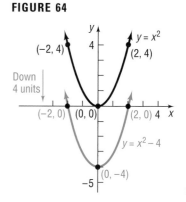

 Now work Problem 25.

Horizontal Shifts

E X A M P L E   3   Horizontal Shift to the Right

Use the graph of $f(x) = x^2$ to obtain the graph of $g(x) = (x - 2)^2$.

Solution   The function $g(x) = (x - 2)^2$ is basically a square function. Table 11 lists some points on the graphs of f and g. Note that when $f(x) = 0$ then $x = 0$, and when $g(x) = 0$, then $x = 2$. Also, when $f(x) = 4$, then $x = -2$ or 2, and when $g(x) = 4$, then $x = 0$ or 4. We conclude that the graph of g is identical to that of f, except that it is shifted 2 units to the right. See Figure 65.

TABLE 11		
x	$y = f(x)$ $= x^2$	$y = g(x)$ $= (x - 2)^2$
-2	4	16
0	0	4
2	4	0
4	16	4

FIGURE 65

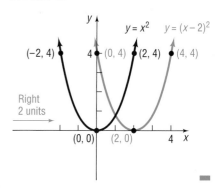

FIGURE 66

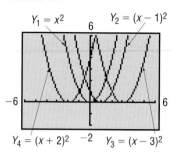

$Y_1 = x^2$ $Y_2 = (x - 1)^2$

$Y_4 = (x + 2)^2$ $Y_3 = (x - 3)^2$

Seeing the Concept   On the same screen, graph each of the following functions:

$$Y_1 = x^2$$
$$Y_2 = (x - 1)^2$$
$$Y_3 = (x - 3)^2$$
$$Y_4 = (x + 2)^2$$

Figure 66 illustrates the graphs.

You should have observed the following pattern. With the graph of $Y_1 = x^2$ on the screen, the graph of $Y_2 = (x - 1)^2$ is identical to that of $Y = x^2$, except that it is shifted horizontally to the right 1 unit. Similarly, the graph of $Y_3 = (x - 3)^2$ is identical to that of $Y_1 = x^2$, except that it is shifted horizontally to the right 3 units. Finally, the graph of $Y_4 = (x + 2)^2$ is identical to that of $Y_1 = x^2$, except that it is shifted horizontally to the left 2 units.

We are led to the following conclusion.

> If the argument x of a function f is replaced by $x - c$, c a real number, the graph of the new function $g(x) = f(x - c)$ is the graph of f **shifted horizontally** left (if $c < 0$) or right (if $c > 0$).

E X A M P L E   4   Horizontal Shift to the Left

Use the graph of $f(x) = x^2$ to obtain the graph of $h(x) = (x + 4)^2$.

Solution   Again, the function $h(x) = (x + 4)^2$ is basically a square function. Thus, its graph is the same as that of f, except that it is shifted 4 units to the left. (Do you see why? $(x + 4)^2 = [x - (-4)]^2$) See Figure 67.

FIGURE 67

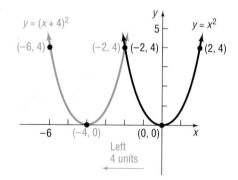

$y = (x + 4)^2$ $y = x^2$

$(-6, 4)$ $(-2, 4)$ $(-2, 4)$ $(2, 4)$

-6 $(-4, 0)$ $(0, 0)$

Left
4 units

Now work Problem 31.

Vertical and horizontal shifts are sometimes combined.

E X A M P L E   5   Combining Vertical and Horizontal Shifts

FIGURE 68

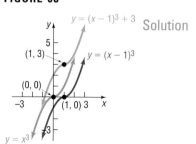

$y = (x - 1)^3 + 3$

$(1, 3)$ $y = (x - 1)^3$

$(0, 0)$

-3 $(1, 0)$ 3 x

$y = x^3$

Solution   Graph the function $f(x) = (x - 1)^3 + 3$.

We graph f in steps. First, we note that the rule for f is basically a cube function. Thus, we begin with the graph of $y = x^3$, shown in blue in Figure 68. Next, to get the graph of $y = (x - 1)^3$, we shift the graph of $y = x^3$ horizontally 1 unit to the right. See the graph shown in red in Figure 68. Finally, to get the graph of $y = (x - 1)^3 + 3$, we shift the graph of $y = (x - 1)^3$ vertically up 3 units. See the graph shown in green in Figure 68. Note the points that have been plotted on each graph. Using key points such as these can be helpful in keeping track of just what is taking place.

In Example 5, if the vertical shift had been done first, followed by the horizontal shift, the final graph would have been the same. (Try it for yourself.)

Now work Problem 43.

Compressions and Stretches

Vertical Stretch

E X A M P L E 6

Use the graph of $f(x) = |x|$ to obtain the graph of $g(x) = 2|x|$.

Solution To see the relationship between the graphs of f and g, we form Table 12, listing points on each graph. For each x, the y-coordinate of a point on the graph of g is 2 times as large as the corresponding y-coordinate on the graph of f. The graph of $f(x) = |x|$ is vertically stretched by a factor of 2 [for example, from (1, 1) to (1, 2)] to obtain the graph of $g(x) = 2|x|$. See Figure 69.

TABLE 12		
x	$y = f(x)$ $= |x|$	$y = g(x)$ $= 2|x|$
-2	2	4
-1	1	2
0	0	0
1	1	2
2	2	4

FIGURE 69

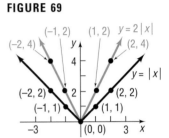

E X A M P L E 7

Vertical Compression

Use the graph of $f(x) = |x|$ to obtain the graph of $h(x) = \frac{1}{2}|x|$.

Solution For each x, the y-coordinate of a point on the graph of h is $\frac{1}{2}$ as large as the corresponding y-coordinate on the graph of f. The graph of $f(x) = |x|$ is vertically compressed by a factor of $\frac{1}{2}$ [for example, from (2, 2) to (2, 1)] to obtain the graph of $h(x) = \frac{1}{2}|x|$. See Table 13 and Figure 70.

TABLE 13		
x	$y = f(x)$ $= |x|$	$y = h(x)$ $= \frac{1}{2}|x|$
-2	2	1
-1	1	$\frac{1}{2}$
0	0	0
1	1	$\frac{1}{2}$
2	2	1

FIGURE 70

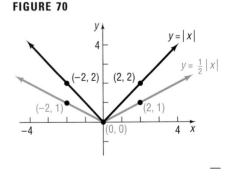

When the right side of a function $y = f(x)$ is multiplied by a positive number k, the graph of the new function $y = kf(x)$ is a **vertically compressed** (if $0 < k < 1$) or **stretched** (if $k > 1$) version of the graph of $y = f(x)$.

Now work Problem 33.

What happens if the argument x of a function $y = f(x)$ is multiplied by a positive number k, creating a new function $y = f(kx)$? To find the answer, we look at the following Exploration.

 Exploration On the same screen, graph each of the following functions:

$$Y_1 = f(x) = x^2$$
$$Y_2 = f(2x) = (2x)^2$$
$$Y_3 = f\left(\frac{1}{2}x\right) = \left(\frac{1}{2}x\right)^2$$

You should have obtained the graphs shown in Figure 71. The graph of $Y_2 = (2x)^2$ is the graph of $Y_1 = x^2$ compressed horizontally. Look at Table 14(a). Notice that $(1, 1)$, $(2, 4)$, $(4, 16)$, and $(16, 256)$ are points on the graph of $Y_1 = x^2$. Also, $(0.5, 1)$, $(1, 4)$, $(2, 16)$, and $(8, 256)$ are points on the graph of $Y_2 = (2x)^2$. Thus, the graph of $Y_2 = (2x)^2$ is obtained by multiplying the x-coordinate of each point on the graph of $Y_1 = x^2$ by $\frac{1}{2}$. The graph of $Y_3 = \left(\frac{1}{2}x\right)^2$ is the graph of $Y_1 = x^2$ stretched horizontally. Look at Table 14(b). Notice that $(0.5, 0.25)$, $(1, 1)$, $(2, 4)$, $(4, 16)$ are points on the graph of $Y_1 = x^2$. Also, $(1, 0.25)$, $(2, 1)$, $(4, 4)$, $(8, 16)$ are points on the graph of $Y_3 = \left(\frac{1}{2}x\right)^2$. Thus, the graph of $Y_3 = \left(\frac{1}{2}x\right)^2$ is obtained by multiplying the x-coordinate of each point on the graph of $Y_1 = x^2$ by a factor of 2.

FIGURE 71

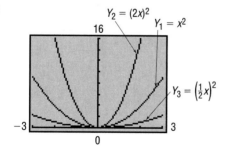

TABLE 14

X	Y1	Y2
0	0	0
.5	.25	1
1	1	4
2	4	16
4	16	64
8	64	256
16	256	1024

Y2 ☐ (2X)²

(a)

X	Y1	Y3
0	0	0
.5	.25	.0625
1	1	.25
2	4	1
4	16	4
8	64	16
16	256	64

Y3 ☐ (X/2)²

(b)

> If the argument x of a function $y = f(x)$ is multiplied by a positive number k, the graph of the new function $y = f(kx)$ is obtained by multiplying each x-coordinate of $y = f(x)$ by $1/k$. A **horizontal compression** results if $k > 1$ and a **horizontal stretch** occurs if $0 < k < 1$.

Let's look at an example.

E X A M P L E   8

Graphing Using Stretches and Compressions

The graph of $y = f(x)$ is given in Figure 72. Use this graph to find the graphs of

(a) $y = 3f(x)$ (b) $y = f(3x)$

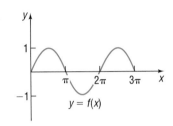

FIGURE 72

$y = f(x)$

Solution   (a) The graph of $y = 3f(x)$ is obtained by multiplying each y-coordinate of $y = f(x)$ by a factor of 3. See Figure 73(a).

(b) The graph of $y = f(3x)$ is obtained from the graph of $y = f(x)$ by multiplying each x-coordinate of $y = f(x)$ by a factor of $\frac{1}{3}$. See Figure 73(b).

FIGURE 73

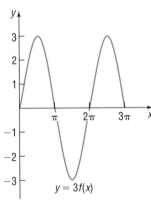

$y = 3f(x)$

(a)

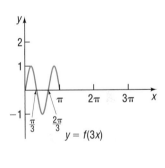

$y = f(3x)$

(b)

Now work Problems 55(d) and 55(f).

3

Reflections about the *x*-Axis and the *y*-Axis

E X A M P L E   9

Reflection about the *x*-Axis

Graph the function: $f(x) = -x^2$

Solution   We begin with the graph of $y = x^2$, as shown in Figure 74. For each point (x, y) on the graph of $y = x^2$, the point $(x, -y)$ is on the graph of $y = -x^2$, as indicated in Table 15. Thus, we can draw the graph of $y = -x^2$ by reflecting the graph of $y = x^2$ about the x-axis. See Figure 74.

TABLE 15		
x	$y = x^2$	$y = -x^2$
-2	4	-4
-1	1	-1
0	0	0
1	1	-1
2	4	-4

FIGURE 74

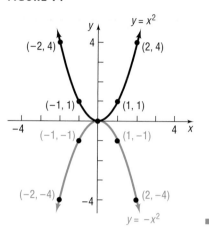

> When the right side of the equation $y = f(x)$ is multiplied by -1, the graph of the new function $y = -f(x)$ is the **reflection about the x-axis** of the graph of the function $y = f(x)$.

  Now work Problem 37.

E X A M P L E   10

Reflection about the y-Axis

Graph the function: $f(x) = \sqrt{-x}$

Solution   First, notice that the domain of f consists of all real numbers x for which $-x \geq 0$ or equivalently, $x \leq 0$. To get the graph of $f(x) = \sqrt{-x}$, we begin with the graph of $y = \sqrt{x}$, as shown in Figure 75. For each point (x, y) on the graph of $y = \sqrt{x}$, the point $(-x, y)$ is on the graph of $y = \sqrt{-x}$. Thus, we get the graph of $y = \sqrt{-x}$ by reflecting the graph of $y = \sqrt{x}$ about the y-axis. See Figure 75.

FIGURE 75

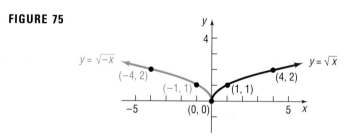

> When the graph of the function $y = f(x)$ is known, the graph of the new function $y = f(-x)$ is the **reflection about the y-axis** of the graph of the function $y = f(x)$.

SUMMARY OF GRAPHING TECHNIQUES

Table 16 summarizes the graphing procedures we have just discussed.

TABLE 16		
To Graph:	**Draw the Graph of *f* and:**	**Functional Change to *f*(x)**
Vertical shifts		
$y = f(x) + c, \quad c > 0$	Raise the graph of *f* by *c* units.	Add *c* to *f*(x).
$y = f(x) - c, \quad c > 0$	Lower the graph of *f* by *c* units.	Subtract *c* from *f*(x).
Horizontal shifts		
$y = f(x + c), \quad c > 0$	Shift the graph of *f* to the left *c* units.	Replace *x* by *x* + *c*.
$y = f(x - c), \quad c > 0$	Shift the graph of *f* to the right *c* units.	Replace *x* by *x* − *c*.
Compressing or stretching		
$y = kf(x), \quad k > 0$	Multiply each *y* coordinate of $y = f(x)$ by *k*.	Multiply *f*(x) by *k*.
$y = f(kx), \quad k > 0$	Multiply each *x* coordinate of $y = f(x)$ by $\dfrac{1}{k}$.	Replace *x* by *kx*.
Reflection about the *x*-axis		
$y = -f(x)$	Reflect the graph of *f* about the *x*-axis.	Multiply *f*(x) by −1.
Reflection about the *y*-axis		
$y = f(-x)$	Reflect the graph of *f* about the *y*-axis.	Replace *x* by −*x*.

The examples that follow combine some of the procedures outlined in this section to get the required graph.

E X A M P L E   11   Determining the Function Obtained from a Series of Transformations

Find the function that is finally graphed after the following three transformations are applied to the graph of $y = |x|$.

1. Shift left 2 units
2. Shift up 3 units
3. Reflect about the *y*-axis

Solution   1. Shift left 2 units: Replace *x* by *x* + 2 $y = |x + 2|$
2. Shift up 3 units: Add 3 $y = |x + 2| + 3$
3. Reflect about the *y*-axis: Replace *x* by −*x* $y = |-x + 2| + 3$ ▬

 Now work Problem 21.

E X A M P L E   12   Combining Graphing Procedures

Graph the function: $f(x) = \dfrac{3}{x - 2} + 1.$

Solution   We use the following steps to obtain the graph of *f*:

STEP 1: $y = \dfrac{1}{x}$ Reciprocal function.

STEP 2: $y = \dfrac{3}{x}$ Multiply by 3; vertical stretch of the graph of $y = \dfrac{1}{x}$ by a factor of 3.

STEP 3: $y = \dfrac{3}{x - 2}$ Replace x by $x - 2$; horizontal shift to the right 2 units.

STEP 4: $y = \dfrac{3}{x - 2} + 1$ Add 1; vertical shift up 1 unit.

See Figure 76.

FIGURE 76

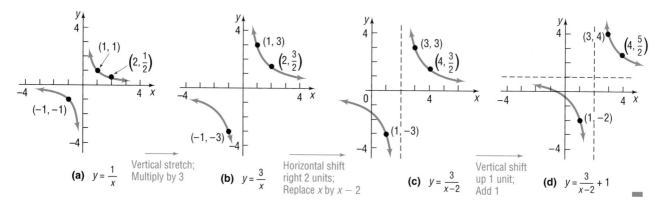

(a) $y = \dfrac{1}{x}$ Vertical stretch; Multiply by 3

(b) $y = \dfrac{3}{x}$ Horizontal shift right 2 units; Replace x by $x - 2$

(c) $y = \dfrac{3}{x-2}$ Vertical shift up 1 unit; Add 1

(d) $y = \dfrac{3}{x-2} + 1$

There are other orderings of the steps shown in Example 12 that would also result in the graph of f. For example, try this one:

STEP 1: $y = \dfrac{1}{x}$ Reciprocal function.

STEP 2: $y = \dfrac{1}{x - 2}$ Replace x by $x - 2$; horizontal shift to the right 2 units.

STEP 3: $y = \dfrac{3}{x - 2}$ Multiply by 3; vertical stretch of the graph of $y = \dfrac{1}{x - 2}$ by a factor of 3.

STEP 4: $y = \dfrac{3}{x - 2} + 1$ Add 1; vertical shift up 1 unit.

E X A M P L E   13   Combining Graphing Procedures

Graph the function: $f(x) = \sqrt{1 - x} + 2$.

Solution   We use the following steps to get the graph of $y = \sqrt{1 - x} + 2$:

STEP 1: $y = \sqrt{x}$ Square root function.

STEP 2: $y = \sqrt{x + 1}$ Replace x by $x + 1$; horizontal shift left 1 unit.

STEP 3: $y = \sqrt{-x + 1} = \sqrt{1 - x}$ Replace x by $-x$; reflect about y-axis.

STEP 4: $y = \sqrt{1 - x} + 2$ Add 2; vertical shift up 2 units.

See Figure 77.

FIGURE 77

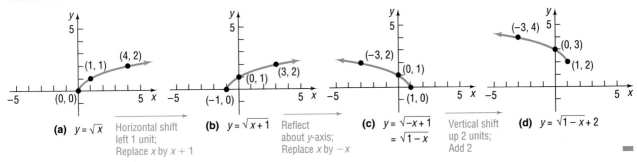

(a) $y = \sqrt{x}$ Horizontal shift
left 1 unit;
Replace x by $x + 1$

(b) $y = \sqrt{x + 1}$ Reflect
about y-axis;
Replace x by $-x$

(c) $y = \sqrt{-x + 1}$ Vertical shift
$= \sqrt{1 - x}$ up 2 units;
Add 2

(d) $y = \sqrt{1 - x} + 2$

1.5 | EXERCISES

In Problems 1–12, match each graph to one of the following functions:

A. $y = x^2 + 2$ B. $y = -x^2 + 2$ C. $y = |x| + 2$ D. $y = -|x| + 2$

E. $y = (x - 2)^2$ F. $y = -(x + 2)^2$ G. $y = |x - 2|$ H. $y = -|x + 2|$

I. $y = 2x^2$ J. $y = -2x^2$ K. $y = 2|x|$ L. $y = -2|x|$

1.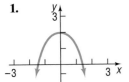

2.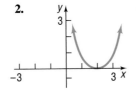

3.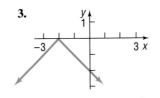

4.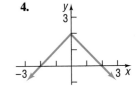

5.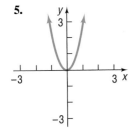

6.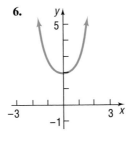

7.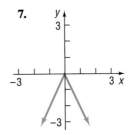

8.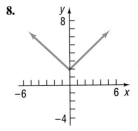

9.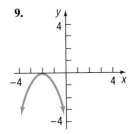

10.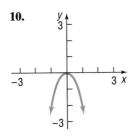

11.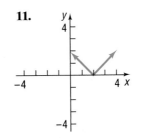

12.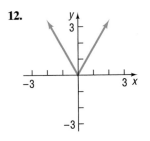

In Problems 13–20, write the function whose graph is the graph of $y = x^3$, but is

13. Shifted to the right 4 units

14. Shifted to the left 4 units

15. Shifted up 4 units

16. Shifted down 4 units

17. Reflected about the *y*-axis

18. Reflected about the *x*-axis

19. Vertically stretched by a factor of 4

20. Horizontally stretched by a factor of 4

In Problems 21–24, find the function that is finally graphed after the following transformations are applied to the graph of $y = \sqrt{x}$.

21. (1) Shift up 2 units
(2) Reflect about the *x*-axis
(3) Reflect about the *y*-axis

22. (1) Reflect about the *x*-axis
(2) Shift right 3 units
(3) Shift down 2 units

23. (1) Reflect about the *x*-axis
(2) Shift up 2 units
(3) Shift left 3 units

24. (1) Shift up 2 units
(2) Reflect about the *y*-axis
(3) Shift left 3 units

In Problems 25–54, graph each function using the techniques of shifting, compressing, stretching, and/or reflecting. Start with the graph of the basic function (for example, $y = x^2$) and show all stages.

25. $f(x) = x^2 - 1$

26. $f(x) = x^2 + 4$

27. $g(x) = x^3 + 1$

28. $g(x) = x^3 - 1$

29. $h(x) = \sqrt{x - 2}$

30. $h(x) = \sqrt{x + 1}$

31. $f(x) = (x - 1)^3$

32. $f(x) = (x + 2)^3$

33. $g(x) = 4\sqrt{x}$

34. $g(x) = \frac{1}{2}\sqrt{x}$

35. $h(x) = \dfrac{1}{2x}$

36. $h(x) = \dfrac{4}{x}$

37. $f(x) = -|x|$

38. $f(x) = -\sqrt{x}$

39. $g(x) = -\dfrac{1}{x}$

40. $g(x) = -x^3$

41. $h(x) = \text{int}(-x)$

42. $h(x) = \dfrac{1}{-x}$

43. $f(x) = (x + 1)^2 - 3$

44. $f(x) = (x - 2)^2 + 1$

45. $g(x) = \sqrt{x - 2} + 1$

46. $g(x) = |x + 1| - 3$

47. $h(x) = \sqrt{-x} - 2$

48. $h(x) = \dfrac{4}{x} + 2$

49. $f(x) = (x + 1)^3 - 1$

50. $f(x) = 4\sqrt{x - 1}$

51. $g(x) = 2|1 - x|$

52. $g(x) = 4\sqrt{2 - x}$

53. $h(x) = 2\,\text{int}(x - 1)$

54. $h(x) = -x^3 + 2$

In Problems 55–60, the graph of a function f is illustrated. Use the graph of f as the first step toward graphing each of the following functions:

(a) $F(x) = f(x) + 3$

(b) $G(x) = f(x + 2)$

(c) $P(x) = -f(x)$

(d) $Q(x) = \frac{1}{2}f(x)$

(e) $g(x) = f(-x)$

(f) $h(x) = f(2x)$

55.

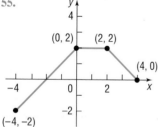

56.

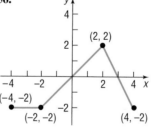

57.

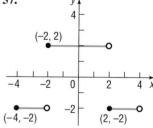

58.

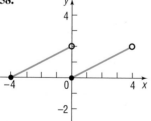

59.

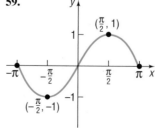

60.

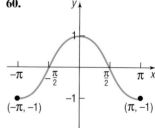

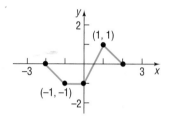

61. Exploration
(a) Use a graphing utility to graph $y = x + 1$ and $y = |x + 1|$.
(b) Graph $y = 4 - x^2$ and $y = |4 - x^2|$.
(c) Graph $y = x^3 + x$ and $y = |x^3 + x|$.
(d) What do you conclude about the relationship between the graphs of $y = f(x)$ and $y = |f(x)|$?

62. Exploration
(a) Use a graphing utility to graph $y = x + 1$ and $y = |x| + 1$.
(b) Graph $y = 4 - x^2$ and $y = 4 - |x|^2$.
(c) Graph $y = x^3 + x$ and $y = |x|^3 + |x|$.
(d) What do you conclude about the relationship between the graphs of $y = f(x)$ and $y = f(|x|)$?

63. The graph of a function f is illustrated in the figure.
(a) Draw the graph of $y = |f(x)|$.
(b) Draw the graph of $y = f(|x|)$.

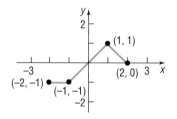

64. The graph of a function f is illustrated in the figure.
(a) Draw the graph of $y = |f(x)|$.
(b) Draw the graph of $y = f(|x|)$.

In Problems 65–70, complete the square of each quadratic expression. Then graph each function using the technique of shifting.
Hint: If $f(x) = x^2 - 2x$, then $f(x) = (x^2 - 2x + 1) - 1 = (x - 1)^2 - 1$.

65. $f(x) = x^2 + 2x$

66. $f(x) = x^2 - 6x$

67. $f(x) = x^2 - 8x + 1$

68. $f(x) = x^2 + 4x + 2$

69. $f(x) = x^2 + x + 1$

70. $f(x) = x^2 - x + 1$

71. The equation $y = (x - c)^2$ defines a *family of parabolas*, one parabola for each value of c. On one set of coordinate axes, graph the members of the family for $c = 0, c = 3, c = -2$.

72. Repeat Problem 71 for the family of parabolas $y = x^2 + c$.

73. Temperature Measurements The relationship between the Celsius, (°C) and Fahrenheit (°F) scales for measuring temperature is given by the equation

$$F = \frac{9}{5}C + 32$$

The relationship between the Celsius (°C) and Kelvin (K) scales is $K = C + 273$. Graph the equation $F = \frac{9}{5}C + 32$ using degrees Fahrenheit on the y-axis and degrees Celsius on the x-axis. Use the techniques introduced in this section to obtain the graph showing the relationship between Kelvin and Fahrenheit temperatures.

74. Period of a Pendulum The period T (in seconds) of a simple pendulum is a function of its length l (in feet) defined by the equation

$$T = 2\pi\sqrt{\frac{l}{g}}$$

where $g \approx 32.2$ feet per second per second is the acceleration of gravity.
(a) Use a graphing utility to graph the function $T = T(l)$.
(b) Now graph the functions $T = T(l + 1)$, $T = T(l + 2)$, and $T = T(l + 3)$.

(c) Discuss how adding to the length l changes the period T.
(d) Now graph the functions $T = T(2l)$, $T = T(3l)$, and $T = T(4l)$.
(e) Discuss how multiplying the length l by factors of 2, 3, and 4 changes the period T.

1.6 ONE-TO-ONE FUNCTIONS; INVERSE FUNCTIONS

> 1 Determine Whether a Function Is One-to-One
> 2 Obtain the Graph of the Inverse Function from the Graph of the Function
> 3 Find an Inverse Function

In Section 1.3, we said a function f can be thought of as a machine that receives as input a number, say x, from the domain, manipulates it, and outputs the value $f(x)$. The **inverse** of f receives as input a number $f(x)$, manipulates it, and outputs the value x. If the function f is the set of ordered pairs (x, y), then the inverse of f is the set of ordered pairs (y, x).

E X A M P L E   1

Finding the Inverse of a Function

Find the inverse of the following functions.

(a) Let the domain of the function represent the employees of Yolanda's Preowned Car Mart and let the range represent their base salaries.

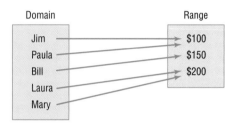

(b) Let the domain of the function represent the employees of Yolanda's Preowned Car Mart and let the range represent their spouse's names.

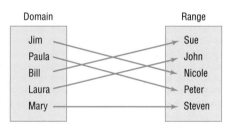

(c) $\{(-3, -27), (-2, -8), (-1, -1), (0, 0), (1, 1), (2, 8), (3, 27)\}$

(d) $\{(-3, 9), (-2, 4), (-1, 1), (0, 0), (1, 1), (2, 4), (3, 9)\}$

Solution
(a) The elements in the domain represent inputs to the function, and the elements in the range represent the outputs. To find the inverse, interchange the elements in the domain with the elements in the range. For example, the function receives as input Bill and outputs $150. So the inverse receives as input $150 and outputs Bill. The inverse of the given function takes the form:

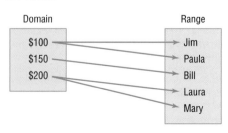

Optimal Hiring Practices at Platinum Chip, Incorporated

Your team works in the marketing department of Platinum Chip, Incorporated. You've been asked to perform some research concerning the firm's labor options. Upon consulting with the company's engineers, it has been determined that the number Q of units produced is given as a function of labor, L, in hours, by $Q = 8L^{\frac{1}{2}}$. For example, if an employee works 9 hours, then there will be $Q = 8(9)^{\frac{1}{2}} = 24$ units produced. In addition, the firm's engineers provide you with the following cost function that relates the cost per month, C, to Q, the units produced by the firm per month:

$$C(Q) = 0.002Q^3 + 0.5Q^2 - Q + 1000.$$

1. Management wishes to know cost as a function of labor.
2. A microeconomist hired by Platinum found that demand is given by $Q(P) = 5000 - 20P$, where P is the price per unit. Find revenue R as a function of units produced, Q. (Hint: Revenue equals the number of units produced times the price per unit.)
3. The Chief Financial Officer (CFO) has asked your team to find revenue as a function of labor.
4. The CFO has also asked you to determine profit as a function of labor.
5. Your team decides to provide a graphical solution to management. Graph profit as a function of labor. What is the implied domain of this function?
6. Human resources wishes to know the optimal level of labor for the company.
7. Investor relations is about to hold a conference call with institutional investors and needs to know the monthly maximum profit of the firm.
8. The production department needs to know the optimal level of units produced each month.
9. Determine the optimal price to charge for each unit.
10. The production function $Q = 8L^{\frac{1}{2}}$ is a variation of the Cobb-Douglas production function, $Q = K^\alpha L^\beta$ where K is capital (machinery, factory, etc.) and L is labor. In the short-run, capital is assumed constant. In the long-run, all inputs are variable. If $\alpha + \beta = 1$, the firm experiences constant returns to scale which means if all inputs increase $x\%$, then output will also increase $x\%$. What must α be in order for Platinum Chip, Inc. to experience constant returns to scale?
11. Graph the production function $Q = 8L^{\frac{1}{2}}$. Explain why the shape of the curve reasonably portrays what actually occurs in the production process as labor increases while all other inputs are fixed.

(b) The inverse of the given function is:

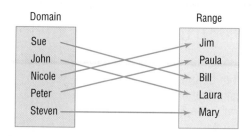

(c) The inverse of the given function is found by interchanging the entries in each ordered pair and so is given by:

$$\{(-27, -3), (-8, -2), (-1, -1), (0, 0), (1, 1), (8, 2), (27, 3)\}.$$

(d) The inverse of the given function is

$$\{(9, -3), (4, -2), (1, -1), (0, 0), (1, 1), (4, 2), (9, 3)\}.$$

We notice that the inverses found in Examples 1(b) and (c) represent functions, since each element in the domain corresponds to a unique element in the range. The inverses found in Examples 1(a) and (d) do not represent functions, since each element in the domain does not correspond to a unique element in the range. Compare the function in Example 1(c) with the function in Example 1(d). For the function in Example 1(c), to every unique x-coordinate there corresponds a unique y-coordinate; for the function in Example 1(d), every unique x-coordinate does not correspond to a unique y-coordinate. Functions for which unique x-coordinates correspond to unique y-coordinates are called *one-to-one* functions. So, in order for the inverse of a function f to be a function itself, f must be one-to-one.

> A function f is said to be **one-to-one** if, for any choice of numbers x_1 and x_2, $x_1 \neq x_2$, in the domain of f, then $f(x_1) \neq f(x_2)$.

In other words, if f is a one-to-one function, then for each x in the domain of f there is exactly one y in the range, and no y in the range is the image of more than one x in the domain. See Figure 78.

FIGURE 78

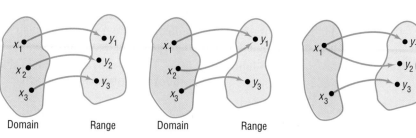

(a) One-to-one function: Each x in the domain has one and only one image in the range

(b) Not a one-to-one function: y_1 is the image of both x_1 and x_2

(c) Not a function: x_1 has two images, y_1 and y_2

Now work Problem 1.

If the graph of a function f is known, there is a simple test, called the **horizontal line test,** to determine whether f is one-to-one.

FIGURE 79
$f(x_1) = f(x_2) = h$, but $x_1 \neq x_2$; f is not a one-to-one function.

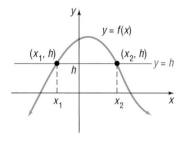

Theorem   Horizontal Line Test

If every horizontal line intersects the graph of a function f in at most one point, then f is one-to-one.

The reason this test works can be seen in Figure 79, where the horizontal line $y = h$ intersects the graph at two distinct points, (x_1, h) and (x_2, h), with the same second element. Thus, f is not one-to-one.

E X A M P L E   2   Using the Horizontal Line Test

For each given function, use the graph to determine whether the function is one-to-one.

(a) $f(x) = x^2$ (b) $g(x) = x^3$

Solution
(a) Figure 80(a) illustrates the horizontal line test for $f(x) = x^2$. The horizontal line $y = 1$ meets the graph of f twice, at $(1, 1)$ and at $(-1, 1)$, so f is not one-to-one.

(b) Figure 80(b) illustrates the horizontal line test for $g(x) = x^3$. Because each horizontal line will intersect the graph of g exactly once, it follows that g is one-to-one.

FIGURE 80

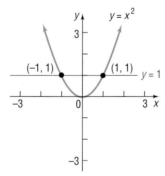

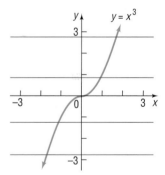

(a) A horizontal line intersects the graph twice; thus, f is not one-to-one

(b) Horizontal lines intersect the graph exactly once; thus, g is one-to-one

Now work Problem 9.

Let's look more closely at the one-to-one function $g(x) = x^3$. This function is an increasing function. Because an increasing (or decreasing) function will always have different y values for unequal x values, it follows that a function that is increasing (or decreasing) on its domain is also a one-to-one function.

Theorem

An increasing (or decreasing) function is a one-to-one function.

Inverse of a Function $y = f(x)$

For a function $y = f(x)$ to have an inverse *function*, f must be one-to-one. Then for each x in its domain there is exactly one y in its range; furthermore, to each y in the range, there corresponds exactly one x in the domain. The correspondence from the range of f onto the domain of f is, therefore, also a function. It is this function that is the *inverse of f*.

We mentioned earlier that a function $y = f(x)$ can be thought of as a rule that tells us to do something to the argument x. For example, the function $f(x) = 2x$ multiplies the argument by 2. An *inverse function* of f undoes whatever f does. Thus, the function $g(x) = \frac{1}{2}x$, which divides the argument by 2, is an inverse of $f(x) = 2x$. See Figure 81.

To put it another way, if we think of f as an input/output machine that processes an input x into $f(x)$, then the inverse function reverses this process, taking $f(x)$ back to x. A definition is given next.

FIGURE 81

$f(x) = 2x; \ g(x) = \dfrac{1}{2}x$

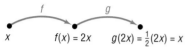

> Let f denote a one-to-one function $y = f(x)$. The **inverse of f,** denoted by f^{-1}, is a function such that $f^{-1}(f(x)) = x$ for every x in the domain of f, and $f(f^{-1}(x)) = x$ for every x in the domain of f^{-1}.

Warning Be careful! The -1 used in f^{-1} is not an exponent. Thus, f^{-1} does *not* mean the reciprocal of f; it means the inverse of f.

FIGURE 82

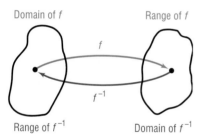

Figure 82 illustrates the definition.

Two facts are now apparent about a function f and its inverse f^{-1}.

> Domain of f = Range of f^{-1} Range of f = Domain of f^{-1}

Look again at Figure 82 to visualize the relationship. If we start with x, apply f, and then apply f^{-1}, we get x back again. If we start with x, apply f^{-1}, and then apply f, we get the number x back again. To put it simply, what f does, f^{-1} undoes, and vice versa:

$$\boxed{\text{Input } x} \xrightarrow{\text{Apply } f.} \boxed{f(x)} \xrightarrow{\text{Apply } f^{-1}.} \boxed{f^{-1}(f(x)) = x}$$

$$\boxed{\text{Input } x} \xrightarrow{\text{Apply } f^{-1}.} \boxed{f^{-1}(x)} \xrightarrow{\text{Apply } f.} \boxed{f(f^{-1}(x)) = x}$$

In other words,

> $$f^{-1}(f(x)) = x \quad \text{and} \quad f(f^{-1}(x)) = x$$

The preceding conditions can be used to verify that a function is, in fact, the inverse of f, as Example 3 demonstrates.

E X A M P L E   3

Verifying Inverse Functions

(a) We verify that the inverse of $g(x) = x^3$ is $g^{-1}(x) = \sqrt[3]{x}$ by showing that

$$g^{-1}(g(x)) = g^{-1}(x^3) = \sqrt[3]{x^3} = x$$

and

$$g(g^{-1}(x)) = g(\sqrt[3]{x}) = (\sqrt[3]{x})^3 = x$$

(b) We verify that the inverse of $h(x) = 3x$ is $h^{-1}(x) = \frac{1}{3}x$ by showing that

$$h^{-1}(h(x)) = h^{-1}(3x) = \frac{1}{3}(3x) = x$$

and

$$h(h^{-1}(x)) = h(\frac{1}{3}x) = 3(\frac{1}{3}x) = x$$

(c) We verify that the inverse of $f(x) = 2x + 3$ is $f^{-1}(x) = \frac{1}{2}(x - 3)$ by showing that

$$f^{-1}(f(x)) = f^{-1}(2x + 3) = \frac{1}{2}[(2x + 3) - 3] = \frac{1}{2}(2x) = x$$

and

$$f(f^{-1}(x)) = f(\frac{1}{2}(x - 3)) = 2[\frac{1}{2}(x - 3)] + 3 = (x - 3) + 3 = x \quad \blacksquare$$

Now work Problem 21.

Exploration: Simultaneously graph $Y_1 = x$, $Y_2 = x^3$, and $Y_3 = \sqrt[3]{x}$ on a square screen, using the viewing rectangle $-3 \le x \le 3$, $-2 \le y \le 2$. What do you observe about the graphs of $Y_2 = x^3$, its inverse $Y_3 = \sqrt[3]{x}$, and the line $Y_1 = x$?
    Repeat this experiment by simultaneously graphing $Y_1 = x$, $Y_2 = 2x + 3$, and $Y_3 = \frac{1}{2}(x - 3)$, using the viewing rectangle $-6 \le x \le 3$, $-8 \le y \le 4$. Do you see the symmetry of the graph of Y_2 and its inverse Y_3 with respect to the line $Y_1 = x$? $\quad \blacksquare$

Geometric Interpretation

FIGURE 83

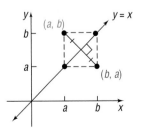

Suppose that (a, b) is a point on the graph of the one-to-one function f defined by $y = f(x)$. Then $b = f(a)$. This means that $a = f^{-1}(b)$, so (b, a) is a point on the graph of the inverse function f^{-1}. The relationship between the point (a, b) on f and the point (b, a) on f^{-1} is shown in Figure 83. The line joining (a, b) and (b, a) is perpendicular to the line $y = x$ and is bisected by the line $y = x$. (Do you see why?) It follows that the point (b, a) on f^{-1} is the reflection about the line $y = x$ of the point (a, b) on f.

Theorem

The graph of a function f and the graph of its inverse function f^{-1} are symmetric with respect to the line $y = x$.

Figure 84 illustrates this result. Notice that, once the graph of f is known, the graph of f^{-1} may be obtained by reflecting the graph of f about the line $y = x$.

FIGURE 84

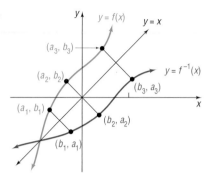

E X A M P L E   4

Graphing the Inverse Function

The graph in Figure 85(a) is that of a one-to-one function $y = f(x)$. Draw the graph of its inverse.

Solution   We begin by adding the graph of $y = x$ to Figure 85(a). Since the points $(-2, -1), (-1, 0)$, and $(2, 1)$ are on the graph of f, we know that the points $(-1, -2), (0, -1)$, and $(1, 2)$ must be on the graph of f^{-1}. Keeping in mind that the graph of f^{-1} is the reflection about the line $y = x$ of the graph of f, we can draw f^{-1}. See Figure 85(b).

FIGURE 85

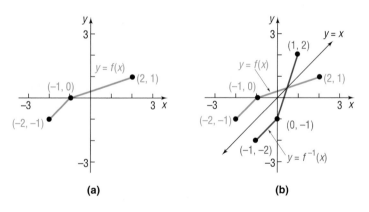

(a)   (b)

Now work Problem 15.

Finding the Inverse Function

 The fact that the graph of a one-to-one function f and its inverse are symmetric with respect to the line $y = x$ tells us more. It says that we can obtain f^{-1} by interchanging the roles of x and y in f. Look again at Figure 84. If f is defined by the equation

$$y = f(x)$$

then f^{-1} is defined by the equation

$$x = f(y)$$

The equation $x = f(y)$ defines f^{-1} *implicitly*. If we can solve this equation for y, we will have the *explicit* form of f^{-1}, that is,

$$y = f^{-1}(x)$$

Let's use this procedure to find the inverse of $f(x) = 2x + 3$. (Since f is a linear function and is increasing, we know that f is one-to-one.)

EXAMPLE 5   Finding the Inverse Function

Find the inverse of $f(x) = 2x + 3$. Also find the domain and range of f and f^{-1}. Graph f and f^{-1} on the same coordinate axes.

Solution   In the equation $y = 2x + 3$, interchange the variables x and y. The result,

$$x = 2y + 3$$

is an equation that defines the inverse f^{-1} implicitly. Solving for y, we obtain

$$2y + 3 = x$$
$$2y = x - 3$$
$$y = \tfrac{1}{2}(x - 3)$$

FIGURE 86

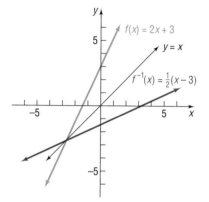

The explicit form of the inverse f^{-1} is therefore

$$f^{-1}(x) = \tfrac{1}{2}(x - 3)$$

which we verified in Example 3(c).

Then we find

$$\text{Domain } f = \text{Range } f^{-1} = (-\infty, \infty)$$
$$\text{Range } f = \text{Domain } f^{-1} = (-\infty, \infty)$$

The graphs of $f(x) = 2x + 3$ and its inverse $f^{-1}(x) = \tfrac{1}{2}(x - 3)$ are shown in Figure 86. Note the symmetry of the graphs with respect to the line $y = x$.   ▬

We outline next the steps to follow for finding the inverse of a one-to-one function.

Procedure for Finding the Inverse of a One-to-One Function

STEP 1: In $y = f(x)$, interchange the variables x and y to obtain

$$x = f(y)$$

This equation defines the inverse function f^{-1} implicitly.

STEP 2: If possible, solve the implicit equation for y in terms of x to obtain the explicit form of f^{-1}:

$$y = f^{-1}(x)$$

STEP 3: Check the result by showing that

$$f^{-1}(f(x)) = x \quad \text{and} \quad f(f^{-1}(x)) = x$$

EXAMPLE 6   Finding the Inverse Function

The function

$$f(x) = \frac{2x + 1}{x - 1} \qquad x \neq 1$$

is one-to-one. Find its inverse and check the result.

Solution   STEP 1: Interchange the variables x and y in

$$y = \frac{2x + 1}{x - 1}$$

to obtain

$$x = \frac{2y + 1}{y - 1}$$

STEP 2: Solve for y:

$$x = \frac{2y + 1}{y - 1}$$
$$x(y - 1) = 2y + 1$$
$$xy - x = 2y + 1$$
$$xy - 2y = x + 1$$
$$(x - 2)y = x + 1$$
$$y = \frac{x + 1}{x - 2}$$

The inverse is

$$f^{-1}(x) = \frac{x + 1}{x - 2} \qquad x \neq 2$$

STEP 3: *Check:*

$$f^{-1}(f(x)) = f^{-1}\left(\frac{2x + 1}{x - 1}\right) = \frac{\dfrac{2x + 1}{x - 1} + 1}{\dfrac{2x + 1}{x - 1} - 2} = \frac{2x + 1 + x - 1}{2x + 1 - 2(x - 1)} = \frac{3x}{3} = x$$

$$f(f^{-1}(x)) = f\left(\frac{x + 1}{x - 2}\right) = \frac{2\left(\dfrac{x + 1}{x - 2}\right) + 1}{\dfrac{x + 1}{x - 2} - 1} = \frac{2(x + 1) + x - 2}{x + 1 - (x - 2)} = \frac{3x}{3} = x$$

◼

Exploration   In Example 6, we found that, if $f(x) = (2x + 1)/(x - 1)$, then $f^{-1}(x) = (x + 1)/(x - 2)$. Graph $y = f(f^{-1}(x))$ on a square screen. What do you see? Are you surprised?

◼

Now work Problem 33.

If a function is not one-to-one, then it will have no inverse function. Sometimes, though, an appropriate restriction on the domain of such a function will yield a new function that is one-to-one. Let's look at an example of this common practice.

E X A M P L E   7   Finding the Inverse Function

Find the inverse of $y = f(x) = x^2$ if $x \geq 0$.

Solution   The function $f(x) = x^2$ is not one-to-one. [Refer to Example 2(a).] However, if we restrict f to only that part of its domain for which $x \geq 0$, as indicated, we have a new function that is increasing and therefore is one-to-one. As a result, the function defined by $y = x^2$, $x \geq 0$, has an inverse function, f^{-1}.

We follow the steps given previously to find f^{-1}:

STEP 1: In the equation $y = x^2$, $x \geq 0$, interchange the variables x and y. The result is

$$x = y^2 \qquad y \geq 0$$

This equation defines (implicitly) the inverse function.

STEP 2: We solve for y to get the explicit form of the inverse. Since $y \geq 0$, only one solution for y is obtained,

$$y = \sqrt{x}$$

so $f^{-1}(x) = \sqrt{x}$.

STEP 3: *Check:* $f^{-1}(f(x)) = f^{-1}(x^2) = \sqrt{x^2} = |x| = x$, since $x \geq 0$.

$$f(f^{-1}(x)) = f(\sqrt{x}) = (\sqrt{x})^2 = x$$

Figure 87 illustrates the graphs of $f(x) = x^2$, $x \geq 0$, and $f^{-1}(x) = \sqrt{x}$. ■

FIGURE 87

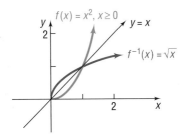

SUMMARY

1. If a function f is one-to-one, then it has an inverse function f^{-1}.
2. Domain f = Range f^{-1};   Range f = Domain f^{-1}.
3. To verify that f^{-1} is the inverse of f, show that $f^{-1}(f(x)) = x$ and $f(f^{-1}(x)) = x$.
4. The graphs of f and f^{-1} are symmetric with respect to the line $y = x$.

1.6 | EXERCISES

In Problems 1–8, (a) find the inverse and (b) determine whether the inverse represents a function.

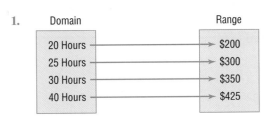

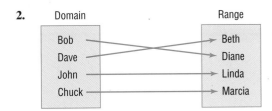

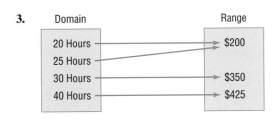

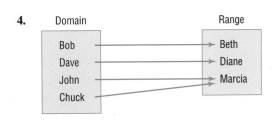

5. $\{(2, 6), (-3, 6), (4, 9), (1, 10)\}$

6. $\{(-2, 5), (-1, 3), (3, 7), (4, 12)\}$

7. $\{(0, 0), (1, 1), (2, 16), (3, 81)\}$

8. $\{(1, 2), (2, 8), (3, 18), (4, 32)\}$

In Problems 9–14, the graph of a function f is given. Use the horizontal line test to determine whether f is one-to-one.

9.
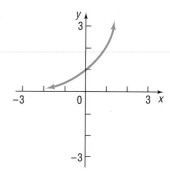

10.
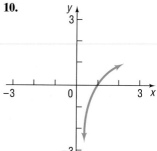

11.
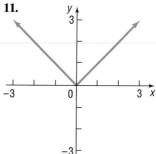

12.
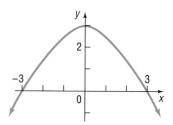

13.
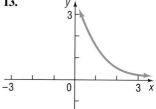

14.
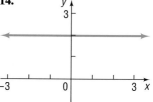

In Problems 15–20, the graph of a one-to-one function f is given. Find the graph of the inverse function f^{-1}. For convenience (and as a hint), the graph of y = x is also given.

15.
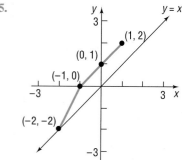

16.
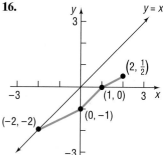

17.
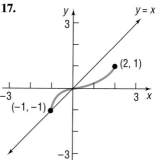

18.
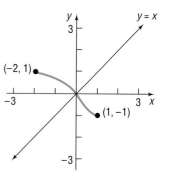

19.
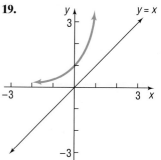

20.
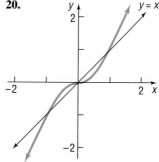

In Problems 21–30, verify that the functions f and g are inverses of each other by showing that f(g(x)) = x and g(f(x)) = x.

21. $f(x) = 3x + 4$; $g(x) = \frac{1}{3}(x - 4)$

22. $f(x) = 3 - 2x$; $g(x) = -\frac{1}{2}(x - 3)$

23. $f(x) = 4x - 8$; $g(x) = \frac{x}{4} + 2$

24. $f(x) = 2x + 6$; $g(x) = \frac{1}{2}x - 3$

25. $f(x) = x^3 - 8;\quad g(x) = \sqrt[3]{x + 8}$

26. $f(x) = (x - 2)^2,\quad x \ge 2;\quad g(x) = \sqrt{x} + 2,\quad x \ge 0$

27. $f(x) = \dfrac{1}{x};\quad g(x) = \dfrac{1}{x}$

28. $f(x) = x;\quad g(x) = x$

29. $f(x) = \dfrac{2x + 3}{x + 4};\quad g(x) = \dfrac{4x - 3}{2 - x}$

30. $f(x) = \dfrac{x - 5}{2x + 3};\quad g(x) = \dfrac{3x + 5}{1 - 2x}$

In Problems 31–42, the function f is one-to-one. Find its inverse and check your answer. State the domain and range of f and f^{-1}. Graph f, f^{-1}, and y = x on the same coordinate axes.

31. $f(x) = 3x$

32. $f(x) = -4x$

33. $f(x) = 4x + 2$

34. $f(x) = 1 - 3x$

35. $f(x) = x^3 - 1$

36. $f(x) = x^3 + 1$

37. $f(x) = x^2 + 4,\quad x \ge 0$

38. $f(x) = x^2 + 9,\quad x \ge 0$

39. $f(x) = \dfrac{4}{x}$

40. $f(x) = -\dfrac{3}{x}$

41. $f(x) = \dfrac{1}{x - 2}$

42. $f(x) = \dfrac{4}{x + 2}$

In Problems 43–54, the function f is one-to-one. Find its inverse and check your answer. State the domain and range of f and f^{-1}.

43. $f(x) = \dfrac{2}{3 + x}$

44. $f(x) = \dfrac{4}{2 - x}$

45. $f(x) = (x + 2)^2,\quad x \ge -2$

46. $f(x) = (x - 1)^2,\quad x \ge 1$

47. $f(x) = \dfrac{2x}{x - 1}$

48. $f(x) = \dfrac{3x + 1}{x}$

49. $f(x) = \dfrac{3x + 4}{2x - 3}$

50. $f(x) = \dfrac{2x - 3}{x + 4}$

51. $f(x) = \dfrac{2x + 3}{x + 2}$

52. $f(x) = \dfrac{-3x - 4}{x - 2}$

53. $f(x) = 2\sqrt[3]{x}$

54. $f(x) = \dfrac{4}{\sqrt{x}}$

55. Find the inverse of the linear function:
$f(x) = mx + b, m \ne 0$.

56. Find the inverse of the function:
$f(x) = \sqrt{r^2 - x^2}, 0 \le x \le r$.

 57. Can an even function be one-to-one? Explain.

58. Is every odd function one-to-one? Explain.

59. A function f has an inverse. If the graph of f lies in quadrant I, in which quadrant does the graph of f^{-1} lie?

60. A function f has an inverse. If the graph of f lies in quadrant II, in which quadrant does the graph of f^{-1} lie?

61. The function $f(x) = |x|$ is not one-to-one. Find a suitable restriction on the domain of f so that the new function that results is one-to-one. Then find the inverse of f.

62. The function $f(x) = x^4$ is not one-to-one. Find a suitable restriction on the domain of f so that the new function that results is one-to-one. Then find the inverse of f.

63. **Temperature Conversion** To convert from x degrees Celsius to y degrees Fahrenheit, we use the formula $y = f(x) = \frac{9}{5}x + 32$. To convert from x degrees Fahrenheit to y degrees Celsius,

we use the formula $y = g(x) = \frac{5}{9}(x - 32)$. Show that f and g are inverse functions.

64. **Demand for Corn** The demand for corn obeys the equation $p(x) = 300 - 50x$, where p is the price per bushel (in dollars) and x is the number of bushels produced, in millions. Express the production amount x as a function of the price p.

65. **Period of a Pendulum** The period T (in seconds) of a simple pendulum is a function of its length l (in feet), given by $T(l) = 2\pi\sqrt{l/g}$, where $g \approx 32.2$ feet per second per second is the acceleration of gravity. Express the length l as a function of the period T.

66. Give an example of a function whose domain is the set of real numbers and that is neither increasing nor decreasing on its domain, but is one-to-one.

[**Hint:** Use a piecewise defined function.]

67. Given
$$f(x) = \frac{ax + b}{cx + d}$$
find $f^{-1}(x)$. If $c \ne 0$, under what conditions on a, b, c, and d is $f = f^{-1}$?

68. We said earlier that finding the range of a function f is not easy. However, if f is one-to-one, we can find its range by finding the domain of the inverse function f^{-1}. Use this technique to find the range of each of the following one-to-one functions:

(a) $f(x) = \dfrac{2x + 5}{x - 3}$

(b) $g(x) = 4 - \dfrac{2}{x}$

(c) $F(x) = \dfrac{3}{4 - x}$

 69. If the graph of a function and its inverse intersect, where must this necessarily occur? Can they intersect anywhere else? Must they intersect?

70. Can a one-to-one function and its inverse be equal? What must be true about the graph of f for this to happen? Give some examples to support your conclusion.

71. Draw the graph of a one-to-one function that contains the points $(-2, -3)$, $(0, 0)$, and $(1, 5)$. Now draw the graph of its inverse. Compare your graph to those of other students. Discuss any similarities. What differences do you see?

CHAPTER REVIEW

THINGS TO KNOW

Absolute value	$|a| = a$ if $a \geq 0$, $\quad |a| = -a$ if $a < 0$
Principal nth root of a	$\sqrt[n]{a} = b$ means $a = b^n$, where $a \geq 0$ and $b \geq 0$ if $n \geq 2$ is even and a, b are any real numbers if $n \geq 3$ is odd
Pythagorean Theorem	In a right triangle, the square of the length of the hypotenuse is equal to the sum of the squares of the lengths of the legs.

Function

A relation between two sets of real numbers so that each number x in the first set, the domain, has corresponding to it exactly one number y in the second set. The range is the set of y values of the function for the x values in the domain. x is the independent variable; y is the dependent variable.

A function f may be defined implicitly by an equation involving x and y or explicitly by writing $y = f(x)$.

A function can also be characterized as a set of ordered pairs (x, y) or $(x, f(x))$ in which no two distinct pairs have the same first element.

Function notation

$y = f(x)$

f is a symbol for the function.

x is the argument, or independent variable.

y is the dependent variable.

$f(x)$ is the value of the function at x, or the image of x.

Domain

If unspecified, the domain of a function f is the largest set of real numbers for which the rule defines a real number.

Vertical line test

A set of points in the plane is the graph of a function if and only if every vertical line intersects the graph in at most one point.

Even function f

$f(-x) = f(x)$ for every x in the domain ($-x$ must also be in the domain).

Odd function f

$f(-x) = -f(x)$ for every x in the domain ($-x$ must also be in the domain).

One-to-one function f	If $x_1 \neq x_2$, then $f(x_1) \neq f(x_2)$ for any choice of x_1 and x_2 in the domain.
Horizontal-line test	If horizontal lines intersect the graph of a function f in at most one point, then f is one-to-one.
Inverse function f^{-1} of f	Domain of f = range of f^{-1}; range of f = domain of f^{-1}.
	$f^{-1}(f(x)) = x$ and $f(f^{-1}(x)) = x$.
	Graphs of f and f^{-1} are symmetric with respect to the line $y = x$.

LIBRARY OF FUNCTIONS

Linear function

$f(x) = mx + b$ Graph is a straight line with slope m and y-intercept b.

Constant function

$f(x) = b$ Graph is a horizontal line with y-intercept b (see Figure 51).

Identity function

$f(x) = x$ Graph is a straight line with slope 1 and y-intercept 0 (see Figure 52).

Square function

$f(x) = x^2$ Graph is a parabola with intercept at $(0, 0)$ (see Figure 53).

Cube function

$f(x) = x^3$ See Figure 54.

Square root function

$f(x) = \sqrt{x}$ See Figure 55.

Reciprocal function

$f(x) = 1/x$ See Figure 56.

Absolute value function

$f(x) = |x|$ See Figure 57.

FORMULAS

Distance formula

$$d = \sqrt{(x_2 - x_1)^2 + (y_2 - y_1)^2}$$

Midpoint formula

$$(x, y) = \left(\frac{x_1 + x_2}{2}, \frac{y_1 + y_2}{2} \right)$$

Standard form of an equation of a circle

$$(x - h)^2 + (y - k)^2 = r^2; \quad r \text{ is the radius of the circle, } (h, k) \text{ is the center of the circle}$$

General form of an equation of a circle

$$x^2 + y^2 + ax + by + c = 0$$

HOW TO

Use the distance formula

Graph equations by plotting points

Find the intercepts of a graph

Test an equation for symmetry

Find the center and radius of a circle, given the equation

Obtain the equation of a circle

Graph circles

Determine whether a relation represents a function

Find the domain and range of a function from its graph

Find the domain of a function given its equation

Determine algebraically whether a function is even or odd

Graph certain functions by shifting, compressing, stretching, and/or reflecting (see Table 16).

Given the graph of a function, determine where it is increasing or decreasing

Find the inverse of certain one-to-one functions (see the procedure given on page 89).

Graph f^{-1} given the graph of f

FILL-IN-THE-BLANK ITEMS

1. If, for every point (x, y) on a graph, the point $(-x, y)$ is also on the graph, then the graph is symmetric with respect to the _____.

2. The set of points in the xy-plane that is a fixed distance from a fixed point is called a(n) _____. The fixed distance is called the _____; the fixed point is called the _____.

3. If f is a function defined by the equation $y = f(x)$, then x is called the _____ variable and y is the _____ variable.

4. A set of points in the xy-plane is the graph of a function if and only if no _____ line contains more than one point of the set.

5. A(n) _____ function f is one for which $f(-x) = f(x)$ for every x in the domain of f; a(n) _____ function f is one for which $f(-x) = -f(x)$ for every x in the domain of f.

6. Suppose that the graph of a function f is known. Then the graph of $y = f(x - 2)$ may be obtained by a(n) _____ shift of the graph of f to the _____ a distance of 2 units.

7. If every horizontal line intersects the graph of a function f at no more than one point, then f is a(n) _____ function.

8. If f^{-1} denotes the inverse of a function f, then the graphs of f and f^{-1} are symmetric with respect to the line _____.

TRUE/FALSE ITEMS

T F   **1.**   The circumference of a circle of diameter d is πd.

T F   **2.**   The radius of the circle $x^2 + y^2 = 9$ is 3.

T F   **3.**   Every relation is a function.

T F   **4.**   Vertical lines intersect the graph of a function in no more than one point.

T F   **5.**   The y-intercept of the graph of the function $y = f(x)$ whose domain is all real numbers is $f(0)$.

T F   **6.**   Even functions have graphs that are symmetric with respect to the origin.

T F   **7.**   The graph of $y = f(-x)$ is the reflection about the y-axis of the graph of $y = f(x)$.

T F   **8.**   If f and g are inverse functions, then the domain of f is the same as the domain of g.

T F   **9.**   If f and g are inverse functions, then their graphs are symmetric with respect to the line $y = x$.

REVIEW EXERCISES

Blue problem numbers indicate the author's suggestions for use in a Practice Test.

In Problems 1–4, find the center and radius of each circle.

1. $x^2 + y^2 - 2x + 4y - 4 = 0$
2. $x^2 + y^2 + 4x - 4y - 1 = 0$
3. $3x^2 + 3y^2 - 6x + 12y = 0$
4. $2x^2 + 2y^2 - 4x = 0$

In Problems 5–12, list the x- and y-intercepts and test for symmetry.

5. $2x = 3y^2$
6. $y = 5x$
7. $4x^2 + y^2 = 1$
8. $x^2 - 9y^2 = 9$
9. $y = x^4 + 2x^2 + 1$
10. $y = x^3 - x$
11. $x^2 + x + y^2 + 2y = 0$
12. $x^2 + 4x + y^2 - 2y = 0$

13. Given that f is a linear function, $f(4) = -5$, and $f(0) = 3$, write the equation that defines f.

14. Given that g is a linear function with slope $= -4$ and $g(-2) = 2$, write the equation that defines g.

15. A function f is defined by

$$f(x) = \frac{Ax + 5}{6x - 2}$$

If $f(1) = 4$, find A.

16. A function g is defined by

$$g(x) = \frac{A}{x} + \frac{8}{x^2}$$

If $g(-1) = 0$, find A.

17. Tell which of the following graphs are graphs of functions.

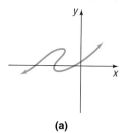

(a)

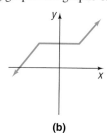

(b)

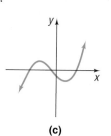

(c)

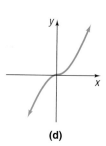

(d)

18. Use the graph of the function f shown to find
   (a) The domain and range of f
   (b) The intervals on which f is increasing
   (c) The intervals on which f is constant
   (d) The intercepts of f

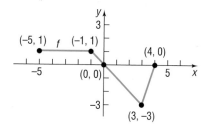

In Problems 19–24, find the following for each function:

   (a) $f(-x)$ (b) $-f(x)$ (c) $f(x + 2)$ (d) $f(x - 2)$

19. $f(x) = \dfrac{3x}{x^2 - 4}$ **20.** $f(x) = \dfrac{x^2}{x + 2}$ **21.** $f(x) = \sqrt{x^2 - 4}$ **22.** $f(x) = |x^2 - 4|$

23. $f(x) = \dfrac{x^2 - 4}{x^2}$ **24.** $f(x) = \dfrac{x^3}{x^2 - 4}$

In Problems 25–30, determine (algebraically) whether the given function is even, odd, or neither.

25. $f(x) = x^3 - 4x$ **26.** $g(x) = \dfrac{4 + x^2}{1 + x^4}$ **27.** $h(x) = \dfrac{1}{x^4} + \dfrac{1}{x^2} + 1$ **28.** $F(x) = \sqrt{1 - x^3}$

29. $G(x) = 1 - x + x^3$ **30.** $H(x) = 1 + x + x^2$

In Problems 31–42, find the domain of each function.

31. $f(x) = \dfrac{x}{x^2 - 9}$ **32.** $f(x) = \dfrac{3x^2}{x - 2}$ **33.** $f(x) = \sqrt{2 - x}$ **34.** $f(x) = \sqrt{x + 2}$

35. $h(x) = \dfrac{\sqrt{x}}{|x|}$ **36.** $g(x) = \dfrac{|x|}{x}$ **37.** $f(x) = \dfrac{x}{x^2 + 2x - 3}$ **38.** $F(x) = \dfrac{1}{x^2 - 3x - 4}$

39. $G(x) = \begin{cases} |x| & \text{if } -1 \le x \le 1 \\ 1/x & \text{if } x > 1 \end{cases}$ **40.** $H(x) = \begin{cases} 1/x & \text{if } 0 < x < 4 \\ x - 4 & \text{if } 4 \le x \le 8 \end{cases}$

41. $f(x) = \begin{cases} 1/(x - 2) & \text{if } x > 2 \\ 0 & \text{if } x = 2 \\ 3x & \text{if } 0 \le x < 2 \end{cases}$ **42.** $g(x) = \begin{cases} |1 - x| & \text{if } x < 1 \\ 3 & \text{if } x = 1 \\ x + 1 & \text{if } 1 < x \le 3 \end{cases}$

In Problems 43–46, find the average rate of change between 0 and x for each function f. Be sure to simplify.

43. $f(x) = 2 - 5x$ **44.** $f(x) = 2x^2 + 7$ **45.** $f(x) = 3x - 4x^2$ **46.** $f(x) = x^2 - 3x + 2$

In Problems 47–66, graph each function using the techniques of shifting, compressing or stretching, and reflections. Identify any intercepts on the graph. State the domain and, based on the graph, find the range.

47. $F(x) = |x| - 4$ **48.** $f(x) = |x| + 4$ **49.** $g(x) = -|x|$ **50.** $g(x) = \tfrac{1}{2}|x|$

51. $h(x) = \sqrt{x - 1}$ **52.** $h(x) = \sqrt{x} - 1$ **53.** $f(x) = \sqrt{1 - x}$ **54.** $f(x) = -\sqrt{x}$

55. $F(x) = \begin{cases} x^2 + 4 & \text{if } x < 0 \\ 4 - x^2 & \text{if } x \geq 0 \end{cases}$

56. $H(x) = \begin{cases} |1 - x| & \text{if } 0 \leq x \leq 2 \\ |x - 1| & \text{if } x > 2 \end{cases}$

57. $h(x) = (x - 1)^2 + 2$

58. $h(x) = (x + 2)^2 - 3$

59. $g(x) = (x - 1)^3 + 1$

60. $g(x) = (x + 2)^3 - 8$

61. $f(x) = \begin{cases} 2\sqrt{x} & \text{if } x \geq 4 \\ x & \text{if } 0 < x < 4 \end{cases}$

62. $f(x) = \begin{cases} 3|x| & \text{if } x < 0 \\ \sqrt{1 - x} & \text{if } 0 \leq x \leq 1 \end{cases}$

63. $g(x) = \dfrac{1}{x - 1} + 1$

64. $g(x) = \dfrac{1}{x + 2} - 2$

65. $h(x) = \text{int}(-x)$

66. $h(x) = -\text{int}(x)$

In Problems 67–72, the function f is one-to-one. Find the inverse of each function, and verify your answer. Find the domain and range of f and f^{-1}.

67. $f(x) = \dfrac{2x + 3}{x - 2}$

68. $f(x) = \dfrac{2 - x}{1 + x}$

69. $f(x) = \dfrac{1}{x - 1}$

70. $f(x) = \sqrt{x - 2}$

71. $f(x) = \dfrac{3}{x^{1/3}}$

72. $f(x) = x^{1/3} + 1$

73. For the graph of the function f shown in the figure:
(a) Draw the graph of $y = f(-x)$.
(b) Draw the graph of $y = -f(x)$.
(c) Draw the graph of $y = f(x + 2)$.
(d) Draw the graph of $y = f(x) + 2$.
(e) Draw the graph of $y = 2f(x)$.
(f) Draw the graph of $y = f(3x)$.

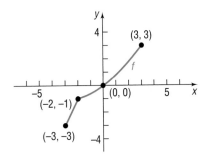

74. Repeat Problem 73 for the graph of the function g shown in the figure.

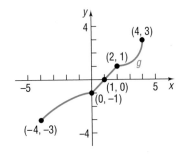

75. **Temperature Conversion**   The temperature T of the air is approximately a linear function of the altitude h for altitudes within 10,000 meters of the surface of Earth. If the surface temperature is 30°C and the temperature at 10,000 meters is 5°C, find the function $T = T(h)$.

76. **Speed as a Function of Time**   The speed v (in feet per second) of a car is a linear function of the time t (in seconds) for $10 \leq t \leq 30$. If after each second the speed of the car has increased by 5 feet per second and if after 20 seconds the speed is 80 feet per second, how fast is the car going after 30 seconds? Find the function $v = v(t)$.

77. **Strength of a Beam**   The strength of a rectangular wooden beam is proportional to the product of the width and the cube of its depth (see the figure). If the beam is to be cut from a log in the shape of a cylinder of radius 3 feet, express the strength S of the beam as a function of the width x. What is the domain of S?

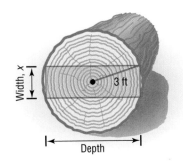

78. Productivity versus Earnings The following data represent the average hourly earnings and productivity (output per hour) of production workers for the years 1986–1995. Let productivity x be the independent variable and average hourly earnings y be the dependent variable.

Productivity	Average Hourly Earnings
94.2	8.76
94.1	8.98
94.6	9.28
95.3	9.66
96.1	10.01
96.7	10.32
100	10.57
100.2	10.83
100.7	11.12
100.8	11.44

Source: Bureau of Labor Statistics.

(a) Draw a scatter diagram of the data.
(b) Draw a line from the point (94.2, 8.76) to (96.7, 10.32) on the scatter diagram found in part (a).
(c) Find the average rate of change of hourly earnings for productivity between 94.2 and 96.7.
(d) Interpret the average rate of change found in part (c).
(e) Draw a line from the point (96.7, 10.32) to (100.8, 11.44) on the scatter diagram found in part (a).
(f) Find the average rate of change of hourly earnings for productivity between 96.7 and 100.8.
(g) Interpret the average rate of change found in part (f).
(h) What is happening to the average rate of change of hourly earnings as productivity increases?

Trigonometric Functions

Alternative energy production is being studied and put into practice throughout the world. On the following page you will find your project. Use the Sullivan website at:

www.prenhall.com/sullivan

to link to the Internet resources needed to prepare your report.

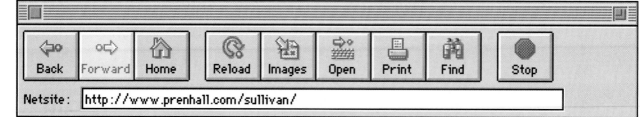

TIDAL ENERGY: MAKING ELECTRICITY FROM OCEAN TIDES

Using tidal flows to generate electrical power seems to be a viable means of alternative energy production. Suppose you are the project manager for an engineering firm. The city of Hilo, HI has contacted your company to do a feasibility study to determine whether or not tidal energy is a practical method for generating electrical power for the city. You begin your study by doing an Internet search on "tidal energy conversion." Your plan is to model Hilo Harbor tides by a mathematical function. You know by Harmonic Analysis of the tidal oscillations that a good model is a function of the form:

$$y = A \sin(bx + c) + D \sin(ex + f)$$

The constants for A, b, c, D, e, f can be calculated using data found at the *School of Ocean and Earth Science and Technology* (SOEST). NOTE: An explanation of Harmonic Analysis can be found at: *The Harmonic Analysis Tutorial;* information on a tidal energy conversion experiment can be found at: *The TOPEX/Poseidon Project.*

1. What are the advantages/disadvantages of using tidal energy to produce electrical power? (Clean energy, reduces coal burning . . . only works part time . . . etc.) What are some of the different methods available for tidal energy conversion? (Dams, floating platforms . . .)
2. From the Topex Ocean Tides graph at the SOEST website, does Hawaii seem like a good location for a tidal energy plant?
3. By looking at the SOEST Tide Chart for Hilo Harbor, can you estimate how many hours per day the power plant would produce electricity? What additional information do you need?
4. Using the data from SOEST, determine a function that models the tide at Hilo Harbor. Graph this function. How does your function compare with *the NOAA tide chart?*
5. By using a dam type of power generating plant, your plant will be able to generate 1 kW per hour if the difference between the high and low tides is 6 meters or more. How many days per month would the plant be able to generate 1 kW per hour at some time during the day? Use the function found in question 4.
6. If a dam type of power plant could produce electrical power for only 70 percent of the time between high and low tides, how many kWs could be generated in one day if the tides differ by more than 6 meters?
7. What other information would you need before you could determine the monthly electrical output for a Hilo Harbor Tidal Conversion Facility? If 70 percent of the time of the incoming and outgoing tides you would be able to generate 1 kW of power per hour, how many kilowatts of power would the plant produce in 1 month? Use the function found in question 4.
8. Before you prepare your final report for the city of Hilo you might want to take a look at the website of the *Office of Scientific and Technical Information (OSTI),* an office of the Department of Energy.

Trigonometry was developed by Greek astronomers, who regarded the sky as the inside of a sphere, so it was natural that triangles on a sphere were investigated early (by Menelaus of Alexandria about AD 100) and that triangles in the plane were studied much later. The first book containing a systematic treatment of plane and spherical trigonometry was written by the Persian astronomer Nasîr Eddîn (about AD 1250).

Regiomontanus (1436–1476) is the man most responsible for moving trigonometry from astronomy into mathematics. His work was improved by Copernicus (1473–1543) and Copernicus's student Rhaeticus (1514–1576). Rhaeticus's book was the first to define the six trigonometric functions as ratios of sides of triangles, although he did not give the functions their present names. Credit for this is due to Thomas Finck (1583), but Finck's notation was by no means universally accepted at the time. The notation was finally stabilized by the textbooks of Leonhard Euler (1707–1783).

Trigonometry has since evolved from its use by surveyors, navigators, and engineers to present applications involving ocean tides, the rise and fall of food supplies in certain ecologies, brain wave patterns, and many other phenomena.

There are two widely accepted approaches to the development of the trigonometric functions: One uses circles, especially the unit circle; the other uses right triangles. In this book, we develop the trigonometric functions using the unit circle. In Section 2.4, we shall show that right triangle trigonometry is a special case of the unit circle approach.

2.1 ANGLES AND THEIR MEASURE

1 Convert between Degrees, Minutes, Seconds, and Decimal Forms for Angles
2 Find the Arc Length of a Circle
3 Convert from Degrees to Radians
4 Convert from Radians to Degrees
5 Find the Linear Speed of an Object Traveling in Circular Motion

A **ray,** or **half-line,** is that portion of a line that starts at a point V on the line and extends indefinitely in one direction. The starting point V of a ray is called its **vertex.** See Figure 1.

FIGURE 1

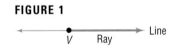

If two rays are drawn with a common vertex, they form an **angle.** We call one of the rays of an angle the **initial side** and the other the **terminal side.** The angle that is formed is identified by showing the direction and amount of rotation from the initial side to the terminal side. If the rotation is in the counterclockwise direction, the angle is **positive;** if the rotation is clockwise, the angle is **negative.** See Figure 2. Lowercase Greek letters, such as α (alpha), β (beta), γ (gamma), and θ (theta), will be used to denote angles. Notice in Figure 2(a) that the angle α is positive because the direction of the rotation from the initial side to the terminal side is counterclockwise. The angle β in Figure 2(b) is negative because the rotation is clockwise. The angle γ in Figure 2(c) is positive. Notice that the angle α in Figure 2(a) and the angle γ have the same initial side and the same terminal side. However, α and γ are unequal, because the amount of rotation required to go from the initial side to the terminal side is greater for angle γ than for angle α.

FIGURE 2

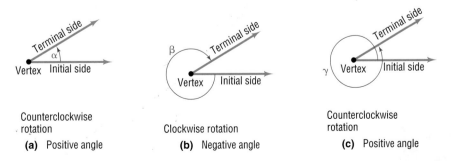

An angle θ is said to be in **standard position** if its vertex is at the origin of a rectangular coordinate system and its initial side coincides with the positive x-axis. See Figure 3.

FIGURE 3

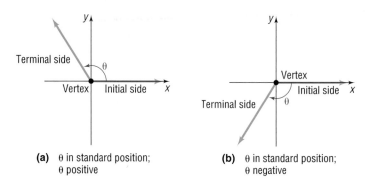

(a) θ in standard position; θ positive

(b) θ in standard position; θ negative

When an angle θ is in standard position, the terminal side either will lie in a quadrant, in which case we say that θ **lies in that quadrant,** or it will lie on the x-axis or the y-axis, in which case we say that θ is a **quadrantal angle.** For example, the angle θ in Figure 4(a) lies in quadrant II, the angle θ in Figure 4(b) lies in quadrant IV, and the angle θ in Figure 4(c) is a quadrantral angle.

FIGURE 4

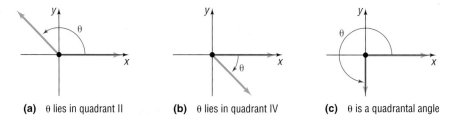

(a) θ lies in quadrant II

(b) θ lies in quadrant IV

(c) θ is a quadrantal angle

We measure angles by determining the amount of rotation needed for the initial side to become coincident with the terminal side. There are two commonly used measures for angles: *degrees* and *radians*.

Degrees

The angle formed by rotating the initial side exactly once in the counterclockwise direction until it coincides with itself (1 revolution) is said to measure 360 degrees, abbreviated 360°. Thus **one degree, 1°,** is $\frac{1}{360}$ revolution. A **right angle** is an angle that measures 90°, or $\frac{1}{4}$ revolution; a **straight angle** is an angle that measures 180°, or $\frac{1}{2}$ revolution. See Figure 5. As Figure 5(b) shows, it is customary to indicate a right angle by using the symbol ⌐.

FIGURE 5

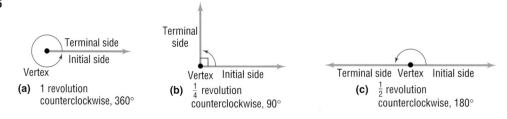

(a) 1 revolution counterclockwise, 360°

(b) $\frac{1}{4}$ revolution counterclockwise, 90°

(c) $\frac{1}{2}$ revolution counterclockwise, 180°

It is also customary to refer to an angle that measures θ degrees as an angle of θ degrees.

EXAMPLE 1

Drawing an Angle

Draw each angle.

(a) 45°   (b) −90°   (c) 225°   (d) 405°

Solution (a) An angle of 45° is $\frac{1}{2}$ of a right angle. See Figure 6.

(b) An angle of −90° is $\frac{1}{4}$ revolution in the clockwise direction. See Figure 7.

FIGURE 6

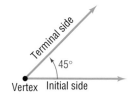

FIGURE 7

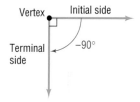

(c) An angle of 225° consists of a rotation through 180° followed by a rotation through 45°. See Figure 8.

(d) An angle of 405° consists of 1 revolution (360°) followed by a rotation through 45°. See Figure 9.

FIGURE 8

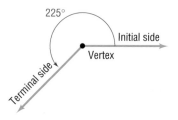

FIGURE 9

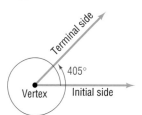

 Now work Problem 1.

 Although subdivisions of a degree may be obtained by using decimals, we also may use the notion of *minutes* and *seconds*. **One minute,** denoted by **1′,** is defined as $\frac{1}{60}$ degree. **One second,** denoted by **1″,** is defined as $\frac{1}{60}$ minute, or equivalently, $\frac{1}{3600}$ degree. An angle of, say, 30 degrees, 40 minutes, 10 seconds is written compactly as 30°40′10″. To summarize:

$$1 \text{ counterclockwise revolution} = 360°$$
$$60' = 1° \qquad 60'' = 1'$$

(1)

Because calculators use decimals, it is important to be able to convert from the degree, minute, second notation (D°M′S″) to a decimal form, and vice versa. Check your calculator; it may have a special key that does the conversion for you. If you do not have this key on your calculator, you can perform the conversion as described next.

Before getting started, though, you must set the calculator to receive degrees, because there are two common ways to measure angles. Many calculators show which mode is currently in use by displaying $\boxed{\text{DEG}}$ for degree mode or $\boxed{\text{RAD}}$ for radian mode. (We will define radians shortly.) Usually, a key is used to change from one mode to another. Check your instruction manual to find out how your particular calculator works.

Now let's see how to convert from the degree, minute, second notation (D°M′S″) to a decimal form, and vice versa, by looking at some examples: $15°30′ = 15.5°$, because $30′ = \frac{1}{2}° = 0.5°$, and $32.25° = 32°15′$, because $0.25° = \frac{1}{4}° = 15′$. For most conversions, a calculator will be helpful.

E X A M P L E   2   Converting between Degrees, Minutes, Seconds, and Decimal Forms

(a) Convert 50°6′21″ to a decimal in degrees.
(b) Convert 21.256° to the D°M′S″ form.

Solution   (a) Because $1′ = \frac{1}{60}°$ and $1″ = \frac{1}{60}′ = (\frac{1}{60} \cdot \frac{1}{60})°$, we convert as follows:

$$50°6′21′ = (50 + 6 \cdot \frac{1}{60} + 21 \cdot \frac{1}{60} \cdot \frac{1}{60})°$$
$$\approx (50 + 0.1 + 0.005833)°$$
$$= 50.105833°$$

(b) We start with the decimal part of 21.256°, that is, 0.256°:

$$0.256° = (0.256)(1°) = (0.256)(60′) = 15.36′$$
$$\underset{1° = 60′}{\uparrow}$$

Now we work with the decimal part of 15.36′, that is, 0.36′:

$$0.36′ = (0.36)(1′) = (0.36)(60″) = 21.6″ \approx 22″$$
$$\underset{1′ = 60″}{\uparrow}$$

Thus,

$$21.256° = 21° + 0.256° = 21° + 15.36′ = 21° + 15′ + 0.36′$$
$$= 21° + 15′ + 21.6″ \approx 21°15′22″$$

 Now work Problems 61 and 67.

In many applications, such as describing the exact location of a star or the precise position of a boat at sea, angles measured in degrees, minutes, and even seconds are used. For calculation purposes, these are transformed to decimal form. In many other applications, especially those in calculus, angles are measured using *radians*.

Radians

Consider a circle of radius r. Construct an angle whose vertex is at the center of this circle, called a **central angle,** and whose rays subtend an arc on the circle whose length equals r. See Figure 10(a). The measure of such an angle is **1 radian.** Thus, for a circle of radius 1, the rays of a central angle with measure 1 radian would subtend an arc of length 1. For a circle of radius 3, the rays of a central angle with measure 1 radian would subtend an arc of length 3. See Figure 10(b).

FIGURE 10

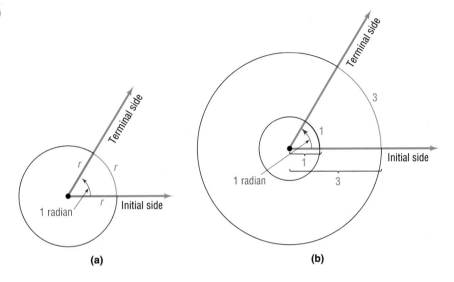

(a)   (b)

2

FIGURE 11

$$\frac{\theta}{\theta_1} = \frac{s}{s_1}$$

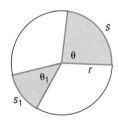

Now consider a circle and two central angles, θ and θ_1. Suppose that these central angles subtend arcs of lengths s and s_1, respectively, as shown in Figure 11. From geometry, we know that the ratio of the measures of the angles equals the ratio of the corresponding lengths of the arcs subtended by those angles; that is,

$$\frac{\theta}{\theta_1} = \frac{s}{s_1} \tag{2}$$

Suppose that θ and θ_1 are measured in radians, and suppose that $\theta_1 = 1$ radian. Refer again to Figure 11. Then the amount of arc s_1 subtended by the central angle $\theta_1 = 1$ radian equals the radius r of the circle. Thus, $s_1 = r$, so formula (2) reduces to

$$\frac{\theta}{1} = \frac{s}{r} \quad \text{or} \quad s = r\theta \tag{3}$$

Theorem   Arc Length

For a circle of radius r, a central angle of θ radians subtends an arc whose length s is

$$s = r\theta \tag{4}$$

Note: Formulas must be consistent with regard to the units used. In equation (4), we write

$$s = r\theta$$

To see the units, however, we must go back to equation (3) and write

$$\frac{\theta \text{ radians}}{1 \text{ radian}} = \frac{s \text{ length units}}{r \text{ length units}}$$

$$s \text{ length units} = (r \text{ length units})\frac{\theta \text{ radians}}{1 \text{ radian}}$$

Since the radians cancel, we are left with

$$s \text{ length units} = (r \text{ length units})\theta \quad s = r\theta$$

where θ appears to be "dimensionless" but, in fact, is measured in radians. So. in using the formula $s = r\theta$, the dimension for θ is radians, and any convenient unit of length (such as inches or meters) may be used for s and r.

E X A M P L E   3

Finding the Length of Arc of a Circle

Find the length of the arc of a circle of radius 2 meters subtended by a central angle of 0.25 radian.

Solution   We use equation (4) with $r = 2$ meters and $\theta = 0.25$. The length s of the arc is

$$s = r\theta = 2(0.25) = 0.5 \text{ meter}$$

Now work Problem 37.

Relationship between Degrees and Radians

Consider a circle of radius r. A central angle of 1 revolution will subtend an arc equal to the circumference of the circle (Figure 12). Because the circumference of a circle equals $2\pi r$, we use $s = 2\pi r$ in equation (4) to find that, for an angle θ of 1 revolution,

$$s = r\theta$$
$$2\pi r = r\theta$$
$$\theta = 2\pi \text{ radians}$$

FIGURE 12
1 revolution $= 2\pi$ radians

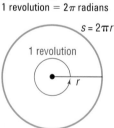

$s = 2\pi r$

1 revolution

r

Thus,

$$\boxed{1 \text{ revolution} = 2\pi \text{ radians}} \qquad (5)$$

so

$$360° = 2\pi \text{ radians}$$

or

$$180° = \pi \text{ radians} \qquad (6)$$

Divide both sides of equation (6) by 180. Then

$$1 \text{ degree} = \frac{\pi}{180} \text{ radian}$$

Divide both sides of (6) by π. Then

$$\frac{180}{\pi} \text{ degrees} = 1 \text{ radian}$$

Thus, we have the following two conversion formulas:

$$1 \text{ degree} = \frac{\pi}{180} \text{ radian} \qquad 1 \text{ radian} = \frac{180}{\pi} \text{ degrees} \qquad (7)$$

E X A M P L E   4

Converting from Degrees to Radians

3 Convert each angle in degrees to radians.

(a) 60°   (b) 150°   (c) −45°   (d) 90°

Solution   (a) $60° = 60 \cdot 1 \text{ degree} = 60 \cdot \dfrac{\pi}{180} \text{ radian} = \dfrac{\pi}{3} \text{ radians}$

(b) $150° = 150 \cdot \dfrac{\pi}{180} \text{ radian} = \dfrac{5\pi}{6} \text{ radians}$

(c) $-45° = -45 \cdot \dfrac{\pi}{180} \text{ radian} = -\dfrac{\pi}{4} \text{ radian}$

(d) $90° = 90 \cdot \dfrac{\pi}{180} \text{ radian} = \dfrac{\pi}{2} \text{ radians}$

Now work Problem 13.

Example 4 illustrates that angles that are fractions of a revolution are expressed in radian measure as fractional multiples of π, rather than as decimals. Thus, a right angle, as in Example 4(d), is left in the form $\pi/2$ radians, which is exact, rather than using the approximation $\pi/2 \approx 3.1416/2 = 1.5708$ radians.

E X A M P L E   5

Converting Radians to Degrees

4 Convert each angle in radians to degrees.

(a) $\dfrac{\pi}{6} \text{ radian}$ (b) $\dfrac{3\pi}{2} \text{ radians}$ (c) $-\dfrac{3\pi}{4} \text{ radians}$ (d) $\dfrac{7\pi}{3} \text{ radians}$

Solution   (a) $\dfrac{\pi}{6}$ radian $= \dfrac{\pi}{6} \cdot 1$ radian $= \dfrac{\pi}{6} \cdot \dfrac{180}{\pi}$ degrees $= 30°$

(b) $\dfrac{3\pi}{2}$ radians $= \dfrac{3\pi}{2} \cdot \dfrac{180}{\pi}$ degrees $= 270°$

(c) $-\dfrac{3\pi}{4}$ radians $= -\dfrac{3\pi}{4} \cdot \dfrac{180}{\pi}$ degrees $= -135°$

(d) $\dfrac{7\pi}{3}$ radians $= \dfrac{7\pi}{3} \cdot \dfrac{180}{\pi}$ degrees $= 420°$

Now work Problem 25.

Table 1 lists the degree and radian measures of some commonly encountered angles. You should learn to feel equally comfortable using degree
or radian measure for these angles.

TABLE 1

Degrees	0°	30°	45°	60°	90°	120°	135°	150°	180°
Radians	0	$\dfrac{\pi}{6}$	$\dfrac{\pi}{4}$	$\dfrac{\pi}{3}$	$\dfrac{\pi}{2}$	$\dfrac{2\pi}{3}$	$\dfrac{3\pi}{4}$	$\dfrac{5\pi}{6}$	π
Degrees	210°	225°	240°	270°	300°	315°	330°	360°	
Radians	$\dfrac{7\pi}{6}$	$\dfrac{5\pi}{4}$	$\dfrac{4\pi}{3}$	$\dfrac{3\pi}{2}$	$\dfrac{5\pi}{3}$	$\dfrac{7\pi}{4}$	$\dfrac{11\pi}{6}$	2π	

E X A M P L E   6   Finding the Length of Arc of a Circle

Find the length of the arc of a circle of radius $r = 3$ feet subtended by a central angle of 30°.

Solution   We use equation (4), $s = r\theta$, but first we must convert the central angle of
30° to radians. Since $30° = \pi/6$ radian, we use $\theta = \pi/6$ and $r = 3$ feet in equation (4). The length of the arc is

$$s = r\theta = 3 \cdot \dfrac{\pi}{6} = \dfrac{\pi}{2} \approx \dfrac{3.14}{2} = 1.57 \text{ feet}$$

When an angle is measured in degrees, the degree symbol will always be
shown. However, when an angle is measured in radians, we will follow the
usual practice and omit the word *radians*. Thus, if the measure of an angle
is given as $\pi/6$, it is understood to mean $\pi/6$ radian.

Circular Motion

5   We have already defined the average speed of an object as the distance traveled divided by the elapsed time. Suppose that an object moves along a circle of radius r at a constant speed. If s is the distance traveled in time t along
this circle, then the **linear speed** v of the object is defined as

$$v = \dfrac{s}{t} \tag{8}$$

FIGURE 13

$$v = \frac{s}{t} \quad \omega = \frac{\theta}{t}$$

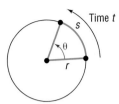

As this object travels along the circle, suppose that θ (measured in radians) is the central angle swept out in time t (see Figure 13). Then the **angular speed** ω (the Greek letter omega) of this object is the angle (measured in radians) swept out divided by the elapsed time; that is,

$$\omega = \frac{\theta}{t} \qquad (9)$$

Angular speed is the way the speed of a phonograph record is described. For example, a 45 rpm (revolutions per minute) record is one that rotates at an angular speed of

$$45\frac{\text{revolutions}}{\text{minute}} = 45\frac{\text{revolutions}}{\text{minute}} \cdot 2\pi\frac{\text{radians}}{\text{revolution}} = 90\pi\frac{\text{radians}}{\text{minute}}$$

There is an important relationship between linear speed and angular speed. In the formula $s = r\theta$, divide each side by t:

$$\frac{s}{t} = r\frac{\theta}{t}$$

Then, using equations (8) and (9), we obtain

$$v = r\omega \qquad (10)$$

When using equation (10), remember that $v = s/t$ (the linear speed) has the dimensions of length per unit of time (such as feet per second or miles per hour), r (the radius of the circular motion) has the same length dimension as s, and ω (the angular speed) has the dimensions of radians per unit of time. As noted earlier, we leave the radian dimension off the numerical value of the angular speed ω so that both sides of the equation will be dimensionally consistent (with "length per unit of time"). If the angular speed is given in terms of *revolutions* per unit of time (as is often the case), be sure to convert it to *radians* per unit of time before attempting to use equation (10).

E X A M P L E   7   Finding Linear Speed

Find the linear speed of a $33\frac{1}{3}$ rpm record at the point where the needle is 3 inches from the spindle (center of the record).

FIGURE 14

Solution   Look at Figure 14. The point P is traveling along a circle of radius $r = 3$ inches. The angular speed ω of the record is

$$\omega = 33\frac{1}{3}\frac{\text{revolutions}}{\text{minute}} = \frac{100}{3}\frac{\text{revolutions}}{\text{minutes}} \cdot 2\pi\frac{\text{radians}}{\text{revolution}}$$
$$= \frac{200\pi}{3}\frac{\text{radians}}{\text{minute}}$$

From equation (10), the linear speed v of the point P is

$$v = r\omega = 3\text{ inches} \cdot \frac{200\pi}{3}\frac{\text{radians}}{\text{minutes}} = 200\pi\frac{\text{inches}}{\text{minute}} \approx 628\frac{\text{inches}}{\text{minute}}$$

 Now work Problem 75.

2.1 | EXERCISES

In Problems 1–12, draw each angle.

1. $30°$ 2. $60°$ 3. $135°$ 4. $-120°$ 5. $450°$ 6. $540°$
7. $3\pi/4$ 8. $4\pi/3$ 9. $-\pi/6$ 10. $-2\pi/3$ 11. $16\pi/3$ 12. $21\pi/4$

In Problems 13–24, convert each angle in degrees to radians. Express your answer as a multiple of π.

13. $30°$ 14. $120°$ 15. $240°$ 16. $330°$ 17. $-60°$ 18. $-30°$
19. $180°$ 20. $270°$ 21. $-135°$ 22. $-225°$ 23. $-90°$ 24. $-180°$

In Problems 25–36, convert each angle in radians to degrees.

25. $\pi/3$ 26. $5\pi/6$ 27. $-5\pi/4$ 28. $-2\pi/3$ 29. $\pi/2$ 30. 4π
31. $\pi/12$ 32. $5\pi/12$ 33. $-\pi/2$ 34. $-\pi$ 35. $-\pi/6$ 36. $-3\pi/4$

In Problems 37–44, s denotes the length of arc of a circle of radius r subtended by the central angle θ. Find the missing quantity.

37. $r = 10$ meters, $\theta = \frac{1}{2}$ radian, $s = ?$
38. $r = 6$ feet, $\theta = 2$ radians, $s = ?$
39. $\theta = \frac{1}{3}$ radian, $s = 2$ feet, $r = ?$
40. $\theta = \frac{1}{4}$ radian, $s = 6$ centimeters, $r = ?$
41. $r = 5$ miles, $s = 3$ miles, $\theta = ?$
42. $r = 6$ meters, $s = 8$ meters, $\theta = ?$
43. $r = 2$ inches, $\theta = 30°$, $s = ?$
44. $r = 3$ meters, $\theta = 120°$, $s = ?$

In Problems 45–52, convert each angle in degrees to radians. Express your answer in decimal form, rounded to two decimal places.

45. $17°$ 46. $73°$ 47. $-40°$ 48. $-51°$ 49. $125°$ 50. $200°$ 51. $340°$ 52. $350°$

In Problems 53–60, convert each angle in radians to degrees. Express your answer in decimal form, rounded to two decimal places.

53. 3.14 54. π 55. 10.25 56. 0.75 57. 2 58. 3 59. 6.32 60. $\sqrt{2}$

In Problems 61–66, convert each angle to a decimal in degrees. Round your answer to two decimal places.

61. $40°10'25''$ 62. $61°42'21''$ 63. $1°2'3''$ 64. $73°40'40''$ 65. $9°9'9''$ 66. $98°22'45''$

In Problems 67–72, convert each angle to $D°M'S''$ form. Round your answer to the nearest second.

67. $40.32°$ 68. $61.24°$ 69. $18.255°$ 70. $29.411°$ 71. $19.99°$ 72. $44.01°$

73. **Minute Hand of a Clock** The minute hand of a clock is 6 inches long. How far does the tip of the minute hand move in 15 minutes? How far does it move in 25 minutes?

74. **Movement of a Pendulum** A pendulum swings through an angle of 20° each second. If the pendulum is 40 inches long, how far does its tip move each second?

75. An object is traveling around a circle with a radius of 5 centimeters. If in 20 seconds a central angle of $\frac{1}{3}$ radian is swept out, what is the angular speed of the object? What is its linear speed?

76. An object is traveling around a circle with a radius of 2 meters. If in 20 seconds the object travels 5 meters, what is its angular speed? What is its linear speed?

77. **Bicycle Wheels**   The diameter of each wheel of a bicycle is 26 inches. If you are traveling at a speed of 35 miles per hour on this bicycle, through how many revolutions per minute are the wheels turning?

78. **Car Wheels**   The radius of each wheel of a car is 15″. If the wheels are turning at the rate of 3 revolutions per second, how fast is the car moving? Express your answer in inches per second and in miles per hour.

79. **Windshield Wipers**   The windshield wiper of a car is 18 inches long. How many inches will the tip of the wiper trace out in $\frac{1}{3}$ revolution?

80. **Windshield Wipers**   The windshield wiper of a car is 18 inches long. If it takes 1 second to trace out $\frac{1}{3}$ revolution, how fast is the tip of the wiper moving?

81. **Speed of the Moon**   The mean distance of the Moon from Earth is 2.39×10^5 miles. Assuming that the orbit of the Moon around Earth is circular and that 1 revolution takes 27.3 days, find the linear speed of the Moon. Express your answer in miles per hour.

82. **Speed of the Earth**   The mean distance of Earth from the Sun is 9.29×10^7 miles. Assuming that the orbit of Earth around the Sun is circular and that 1 revolution takes 365 days, find the linear speed of Earth. Express your answer in miles per hour.

83. **Pulleys**   Two pulleys, one with radius 2 inches and the other with radius 8 inches, are connected by a belt. (See the figure.) If the 2-inch pulley is caused to rotate at 3 revolutions per minute, determine the revolutions per minute of the 8-inch pulley.
[**Hint:** The linear speeds of the pulley, that is, the speed of the belt, are the same.]

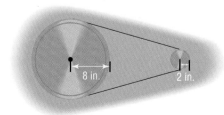

84. **Ferris Wheels**   A neighborhood carnival has a Ferris wheel whose radius is 30 feet. You measure the time it takes for one revolution to be 70 seconds. What is the speed of this Ferris wheel?

85. **Computing the Speed of a River Current**   To approximate the speed of the current of a river, a circular paddle wheel with radius 4 feet is lowered into the water. If the current causes the wheel to rotate at a speed of 10 revolutions per minute, what is the speed of the current? Express your answer in miles per hour.

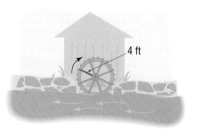

86. **Spin Balancing Tires**   A spin balancer rotates the wheel of a car at 480 revolutions per minute. If the diameter of the wheel is 26 inches, what road speed is being tested? Express your answer in miles per hour. At how many revolutions per minute should the balancer be set to test a road speed of 80 miles per hour?

87. **The Cable Cars of San Francisco**   At the Cable Car Museum you can see the four cable lines that are used to pull cable cars up and down the hills of San Francisco. Each cable travels at a speed of 9.55 miles per hour, caused by a rotating wheel whose diameter is 8.5 feet. How fast is the wheel rotating? Express your answer in revolutions per minute.

88. **Difference in Time of Sunrise**   Naples, Florida, is approximately 90 miles due west of Ft. Lauderdale. How much sooner would a person in Ft. Lauderdale first see the rising Sun than a person in Naples?

[**Hint:** Consult the figure at the top of page 114. When a person at Q sees the first rays of the Sun, a person at P is still in the dark. The person at P sees the first rays after Earth has rotated so that P is at the location Q. Now use the fact that, at the latitude of Ft. Lauderdale, in 24 hours a length of arc of $2\pi(3150)$ miles is subtended.]

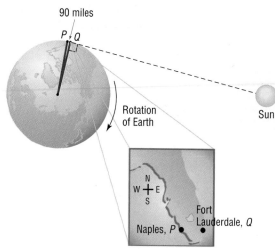

90 miles

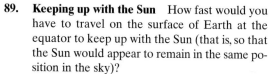

Rotation of Earth

Sun

N
W + E
S

Fort Lauderdale, Q

Naples, P

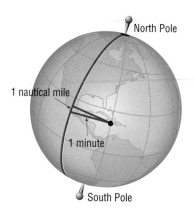

North Pole

1 nautical mile

1 minute

South Pole

89. Keeping up with the Sun How fast would you have to travel on the surface of Earth at the equator to keep up with the Sun (that is, so that the Sun would appear to remain in the same position in the sky)?

[**Hint:** The radius of Earth is 3960 miles.]

90. Nautical Miles A **nautical mile** equals the length of arc subtended by a central angle of 1 minute on a great circle* on the surface of Earth. (See the figure.) If the radius of Earth is taken as 3960 miles, express 1 nautical mile in terms of ordinary, or **statute,** miles (5280 feet).

*Any circle drawn on the surface of Earth that divides Earth into two equal hemispheres.

91. Pulleys Two pulleys, one with radius r_1 and the other with radius r_2, are connected by a belt. The pulley with radius r_1 rotates at ω_1 revolutions per minute, whereas the pulley with radius r_2 rotates at ω_2 revolutions per minute. Show that $r_1/r_2 = \omega_2/\omega_1$.

92. Do you prefer to measure angles using degrees or radians? Provide justification and a rationale for your choice.

93. Discuss why ships and airplanes use nautical miles to measure distance. Explain the difference between a nautical mile and a statute mile.

94. Investigate the way that speed bicycles work. In particular, explain the differences and similarities between 10-speed and 18-speed derailleurs. Be sure to include a discussion of linear speed and angular speed.

2.2 | TRIGONOMETRIC FUNCTIONS: UNIT CIRCLE APPROACH

1. Find the Exact Value of the Trigonometric Functions Using a Point on the Unit Circle
2. Find the Exact Value of the Trigonometric Functions of Quadrantal Angles
3. Find the Exact Value of the Trigonometric Functions of 45° Angles
4. Find the Exact Value of the Trigonometric Functions of 60° Angles
5. Find the Exact Value of the Trigonometric Functions of 30° Angles
6. Use a Calculator to Approximate the Value of the Trigonometric Functions

We are now ready to introduce trigonometric functions. As we said earlier, the approach taken uses the unit circle.

The Unit Circle

Recall that the **unit circle** is a circle whose radius is 1 and whose center is at the origin of a rectangular coordinate system. Because the radius r of the unit circle is 1, we see from the formula $s = r\theta$ that on the unit circle a central angle of θ radians subtends an arc whose length s is

$$s = \theta$$

FIGURE 15
Unit circle: $x^2 + y^2 = 1$

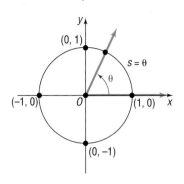

See Figure 15. Thus, on the unit circle, the length measure of the arc s equals the radian measure of the central angle θ. In other words, on the unit circle, the real number used to measure an angle θ in radians corresponds exactly with the real number used to measure the length of the arc subtended by that angle.

For example, suppose that $r = 1$ foot. Then, if $\theta = 3$ radians, $s = 3$ feet; if $\theta = 8.2$ radians, then $s = 8.2$ feet; and so on.

Now, let t be any real number and let θ be the angle, in standard position, equal to t radians. Let P be the point on the unit circle that is also on the terminal side of θ. If $t \geq 0$, this point P is reached by moving *counterclockwise* along the unit circle, starting at $(1, 0)$, for a length of arc equal to t units. See Figure 16(a). If $t < 0$, this point P is reached by moving *clockwise* along the unit circle, starting at $(1, 0)$, for a length of arc equal to $|t|$ units. See Figure 16(b).

FIGURE 16

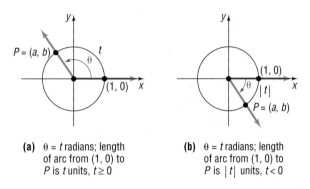

(a) $\theta = t$ radians; length of arc from $(1, 0)$ to P is t units, $t \geq 0$

(b) $\theta = t$ radians; length of arc from $(1, 0)$ to P is $|t|$ units, $t < 0$

Thus, to each real number t there corresponds a unique point $P = (a, b)$ on the unit circle. This is the important idea here. No matter what real number t is chosen, there corresponds a unique point P on the unit circle. We use the coordinates of the point $P = (a, b)$ on the unit circle corresponding to the real number t to define the **six trigonometric functions.**

Let t be a real number and let $P = (a, b)$ be the point on the unit circle that corresponds to t.

The **sine function** associates with t the y-coordinate of P and is denoted by

$$\sin t = b$$

The **cosine function** associates with t the x-coordinate of P and is denoted by

$$\cos t = a$$

If $a \neq 0$, the **tangent function** is defined as

$$\tan t = \frac{b}{a}$$

If $b \neq 0$, the **cosecant function** is defined as

$$\csc t = \frac{1}{b}$$

If $a \neq 0$, the **secant function** is defined as

$$\sec t = \frac{1}{a}$$

If $b \neq 0$, the **cotangent function** is defined as

$$\cot t = \frac{a}{b}$$

Notice in these definitions that if $a = 0$, that is, if the point $P = (0, b)$ is on the y-axis, then the tangent function and the secant function are undefined. Also, if $b = 0$, that is, if the point $P = (a, 0)$ is on the x-axis, then the cosecant function and the cotangent function are undefined.

Because we use the unit circle in these definitions of the trigonometric functions, they are also sometimes referred to as **circular functions.**

EXAMPLE 1

FIGURE 17
$\theta = t$ radians

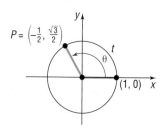

Finding the Value of the Six Trigonometric Functions

Let t be a real number and let $P = (-\frac{1}{2}, \frac{\sqrt{3}}{2})$ be the point on the unit circle that corresponds to t. See Figure 17.
Then

$$\sin t = \frac{\sqrt{3}}{2} \qquad \cos t = -\frac{1}{2} \qquad \tan t = \frac{\frac{\sqrt{3}}{2}}{-\frac{1}{2}} = -\sqrt{3}$$

$$\csc t = \frac{1}{\frac{\sqrt{3}}{2}} = \frac{2\sqrt{3}}{3} \qquad \sec t = \frac{1}{-\frac{1}{2}} = -2 \qquad \cot t = \frac{-\frac{1}{2}}{\frac{\sqrt{3}}{2}} = \frac{-\sqrt{3}}{3}$$

Trigonometric Functions of Angles

Let P be the point on the unit circle corresponding to the real number t. Then the angle θ, in standard position and measured in radians, whose terminal side is the ray from the origin through P is

$$\theta = t \text{ radians}$$

FIGURE 18

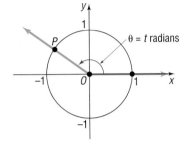

See Figure 18.

Thus, on the unit circle, the measure of the angle θ in radians equals the value of the real number t. As a result, we can say that

$$\sin t = \sin \theta$$

$\uparrow$ $\uparrow$

Real number $\theta = t$ radians

and so on. We now can define the trigonometric functions of the angle θ.

If $\theta = t$ radians, the **six trigonometric functions of the angle θ** are defined as

$$\sin \theta = \sin t \qquad \cos \theta = \cos t \qquad \tan \theta = \tan t$$
$$\csc \theta = \csc t \qquad \sec \theta = \sec t \qquad \cot \theta = \cot t$$

Even though the distinction between trigonometric functions of real numbers and trigonometric functions of angles is important, it is customary to refer to trigonometric functions of real numbers and trigonometric functions of angles collectively as *the trigonometric functions*. We shall follow this practice from now on.

If an angle θ is measured in degrees, we shall use the degree symbol when writing a trigonometric function of θ, as, for example, in $\sin 30°$ and $\tan 45°$. If an angle θ is measured in radians, then no symbol is used when writing a trigonometric function of θ, as, for example, in $\cos \pi$ and $\sec (\pi/3)$.

Finally, since the values of the trigonometric functions of an angle θ are determined by the coordinates of the point $P = (a, b)$ on the unit circle corresponding to θ, the units used to measure the angle θ are irrelevant. For example, it does not matter whether we write $\theta = \pi/2$ radians or $\theta = 90°$. The point on the unit circle corresponding to this angle is $P = (0, 1)$. Hence,

$$\sin \frac{\pi}{2} = \sin 90° = 1 \quad \text{and} \quad \cos \frac{\pi}{2} = \cos 90° = 0$$

Evaluating the Trigonometric Functions

To find the exact value of a trigonometric function of an angle θ requires that we locate the corresponding point P on the unit circle. In fact, though, any circle whose center is at the origin can be used.

Let θ be any nonquadrantal angle placed in standard position. Let $P* = (a*, b*)$ be the point where the terminal side of θ intersects the unit circle. See Figure 19.

FIGURE 19

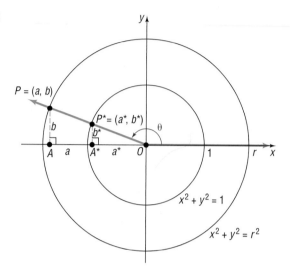

Let $P = (a, b)$ be the point on the terminal side of θ that is also on the circle $x^2 + y^2 = r^2$. Refer again to Figure 19. Notice that the triangles $OA*P*$ and OAP are similar; thus, the ratios of corresponding sides are equal:

$$\frac{b*}{1} = \frac{b}{r} \qquad \frac{a*}{1} = \frac{a}{r} \qquad \frac{b*}{a*} = \frac{b}{a}$$

$$\frac{1}{b*} = \frac{r}{b} \qquad \frac{1}{a*} = \frac{r}{a} \qquad \frac{a*}{b*} = \frac{a}{b}$$

These results lead us to formulate the following theorem:

Theorem

For an angle θ in standard position, let $P = (a, b)$ be the point on the terminal side of θ that is also on the circle $x^2 + y^2 = r^2$. Then

$$\sin \theta = \frac{b}{r} \qquad \cos \theta = \frac{a}{r} \qquad \tan \theta = \frac{b}{a}, \quad a \neq 0$$

$$\csc \theta = \frac{r}{b}, \quad b \neq 0 \qquad \sec \theta = \frac{r}{a}, \quad a \neq 0 \qquad \cot \theta = \frac{a}{b}, \quad b \neq 0$$

E X A M P L E   2   Finding the Exact Value of the Six Trigonometric Functions

Find the exact value of each of the six trigonometric functions of an angle θ if $(4, -3)$ is a point on its terminal side.

FIGURE 20

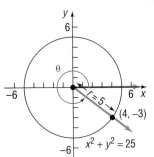

Solution Figure 20 illustrates the situation for θ a positive angle. For the point $(a, b) = (4, -3)$, we have $a = 4$ and $b = -3$. Since $r = \sqrt{a^2 + b^2} = \sqrt{16 + 9} = 5$, the point $(4, -3)$ is also on the circle $x^2 + y^2 = 25$. Thus,

$$\sin\theta = \frac{b}{r} = -\frac{3}{5} \qquad \cos\theta = \frac{a}{r} = \frac{4}{5} \qquad \tan\theta = \frac{b}{a} = -\frac{3}{4}$$

$$\csc\theta = \frac{r}{b} = -\frac{5}{3} \qquad \sec\theta = \frac{r}{a} = \frac{5}{4} \qquad \cot\theta = \frac{a}{b} = -\frac{4}{3}$$

∎

Now work Problem 1.

To evaluate the trigonometric functions of a given angle θ in standard position, we need to find the coordinates of any point on the terminal side of that angle. This is not always easy to do. In the examples that follow, we will evaluate the trigonometric functions of certain angles for which this process is relatively easy. A calculator will be used to evaluate the trigonometric functions of most angles.

E X A M P L E   3

Finding the Exact Value of the Six Trigonometric Functions of Quadrantal Angles

2

Find the exact value of the six trigonometric functions at:

(a) $\theta = 0 = 0°$ (b) $\theta = \pi/2 = 90°$
(c) $\theta = \pi = 180°$ (d) $\theta = 3\pi/2 = 270°$

FIGURE 21(a,b,c)

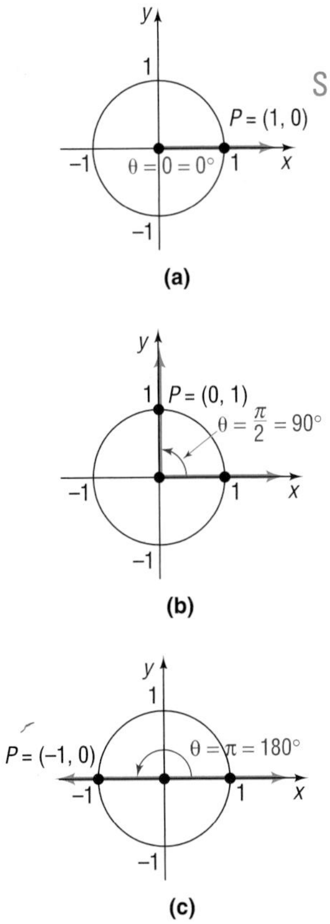

Solution (a) The point $P = (1, 0)$ is on the terminal side of $\theta = 0 = 0°$ and is on the unit circle. See Figure 21(a). Thus,

$$\sin 0 = \sin 0° = 0 \qquad \cos 0 = \cos 0° = 1$$
$$\tan 0 = \tan 0° = 0 \qquad \sec 0 = \sec 0° = 1$$

Since the y-coordinate of P is 0, csc 0 and cot 0 are not defined.

(b) The point $P = (0, 1)$ is on the terminal side of $\theta = \pi/2 = 90°$ and is on the unit circle. See Figure 21(b). Thus,

$$\sin\frac{\pi}{2} = \sin 90° = 1 \qquad \cos\frac{\pi}{2} = \cos 90° = 0$$

$$\csc\frac{\pi}{2} = \csc 90° = 1 \qquad \cot\frac{\pi}{2} = \cot 90° = 0$$

Since the x-coordinate of P is 0, $\tan(\pi/2)$ and $\sec(\pi/2)$ are not defined.

(c) The point $P = (-1, 0)$ is on the terminal side of $\theta = \pi = 180°$ and is on the unit circle. See Figure 21(c). Thus,

$$\sin\pi = \sin 180° = 0 \qquad \cos\pi = \cos 180° = -1$$
$$\tan\pi = \tan 180° = 0 \qquad \sec\pi = \sec 180° = -1$$

Since the y-coordinate of P is 0, csc π and cot π are not defined.

FIGURE 21(d)

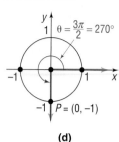

(d)

(d) The point $P = (0, -1)$ is on the terminal side of $\theta = 3\pi/2 = 270°$ and is on the unit circle. See Figure 21(d). Thus,

$$\sin \frac{3\pi}{2} = \sin 270° = -1 \qquad \cos \frac{3\pi}{2} = \cos 270° = 0$$

$$\csc \frac{3\pi}{2} = \csc 270° = -1 \qquad \cot \frac{3\pi}{2} = \cot 270° = 0$$

Since the x-coordinate of P is 0, $\tan (3\pi/2)$ and $\sec (3\pi/2)$ are not defined.

Note: The results obtained in Example 3 are the same whether we pick a point on the unit circle or a point on any circle whose center is at the origin. For example, $P = (0, 5)$ is a point on the terminal side of $\theta = \pi/2 = 90°$ and is a distance of 5 units from the origin. Using this point, $\sin \frac{\pi}{2} = \frac{5}{5} = 1$, $\cos \frac{\pi}{2} = \frac{0}{5} = 0$, and so on, as in Example 3(b).

Table 2 summarizes the values of the trigonometric functions found in Example 3.

TABLE 2	Quadrantal Angles						
θ (Radians)	θ (Degrees)	$\sin \theta$	$\cos \theta$	$\tan \theta$	$\csc \theta$	$\sec \theta$	$\cot \theta$
0	0°	0	1	0	Not defined	1	Not defined
$\pi/2$	90°	1	0	Not defined	1	Not defined	0
π	180°	0	-1	0	Not defined	-1	Not defined
$3\pi/2$	270°	-1	0	Not defined	-1	Not defined	0

Now work Problems 27 and 39.

E X A M P L E   4

Finding Exact Values of Trigonometric Functions ($\theta = \pi/4 = 45°$)

Find the exact value of the six trigonometric functions of $\dfrac{\pi}{4} = 45°$.

FIGURE 22
$\theta = \pi/4 = 45°$

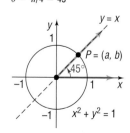

Solution   We seek the coordinates of a point $P = (a, b)$ on the terminal side of $\theta = \pi/4 = 45°$. See Figure 22. First, we observe that P lies on the line $y = x$, so $a = b$. (Do you see why? Since $\theta = 45° = \frac{1}{2} \cdot 90°$, P must lie on the line that bisects quadrant I). Suppose that $P = (a, b)$ also lies on the unit circle, $x^2 + y^2 = 1$. Then it follows that

$$a^2 + b^2 = 1 \quad \text{\small $a = b, a > 0, b > 0$}$$
$$a^2 + a^2 = 1$$
$$2a^2 = 1$$

$$a = \frac{1}{\sqrt{2}} = \frac{\sqrt{2}}{2}, \qquad b = \frac{\sqrt{2}}{2}$$

Thus,

$$\sin \frac{\pi}{4} = \sin 45° = \frac{\sqrt{2}}{2} \qquad \cos \frac{\pi}{4} = \cos 45° = \frac{\sqrt{2}}{2} \qquad \tan \frac{\pi}{4} = \tan 45° = \frac{\sqrt{2}/2}{\sqrt{2}/2} = 1$$

$$\csc \frac{\pi}{4} = \csc 45° = \frac{1}{\sqrt{2}/2} = \sqrt{2} \qquad \sec \frac{\pi}{4} = \sec 45° = \frac{1}{\sqrt{2}/2} = \sqrt{2} \qquad \cot \frac{\pi}{4} = \cot 45° = \frac{\sqrt{2}/2}{\sqrt{2}/2} = 1$$

E X A M P L E   5   Finding the Exact Value of a Trigonometric Expression

Find the exact value of each expression:

(a) $\sin 45° \cos 180°$ (b) $\tan \dfrac{\pi}{4} - \sin \dfrac{3\pi}{2}$

Solution   (a) $\sin 45° \cos 180° = \dfrac{\sqrt{2}}{2} \cdot (-1) = \dfrac{-\sqrt{2}}{2}$

From Example 4   From Table 2

(b) $\tan \dfrac{\pi}{4} - \sin \dfrac{3\pi}{2} = 1 - (-1) = 2$

From Example 4   From Table 2

 Now work Problem 15.

Trigonometric Functions of 30° and 60°

Consider a right triangle in which one of the angles is 30°. It then follows that the other angle is 60°. Figure 23(a) illustrates such a triangle with hypotenuse of length 1. Our problem is to determine a and b.

We begin by placing next to this triangle another triangle congruent to the first, as shown in Figure 23(b). Notice that we now have a triangle whose angles are each 60°. This triangle is therefore equilateral, so each side is of length 1. In particular, the base is $2a = 1$, and so $a = \frac{1}{2}$. By the Pythagorean Theorem, b satisfies the equation $a^2 + b^2 = c^2$, so we have

$$a^2 + b^2 = c^2 \quad a = \tfrac{1}{2}, c = 1$$
$$\tfrac{1}{4} + b^2 = 1$$
$$b^2 = 1 - \tfrac{1}{4} = \tfrac{3}{4}$$
$$b = \frac{\sqrt{3}}{2}$$

This results in Figure 23(c).

FIGURE 23

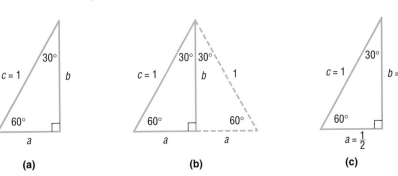

(a)   (b)   (c)

E X A M P L E   6   Finding Exact Values of the Trigonometric Functions ($\theta = \pi/3 = 60°$)

[4]   Find the exact value of the six trigonometric functions of $\pi/3 = 60°$.

FIGURE 24   Solution   Position the triangle in Figure 23(c) so that the 60° angle is in the standard position. See Figure 24. The point $P = (a, b)$ is on the terminal side of 60° and is also on the unit circle. The coordinates of P are $(1/2, \sqrt{3}/2)$. Thus,

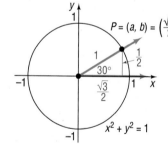

$$\sin\frac{\pi}{3} = \sin 60° = \frac{\sqrt{3}}{2} \qquad \cos\frac{\pi}{3} = \cos 60° = \frac{1}{2}$$

$$\csc\frac{\pi}{3} = \csc 60° = \frac{1}{\sqrt{3}/2} = \frac{2}{\sqrt{3}} = \frac{2\sqrt{3}}{3} \qquad \sec\frac{\pi}{3} = \sec 60° = \frac{1}{1/2} = 2$$

$$\tan\frac{\pi}{3} = \tan 60° = \frac{\sqrt{3}/2}{1/2} = \sqrt{3} \qquad \cot\frac{\pi}{3} = \cot 60° = \frac{1/2}{\sqrt{3}/2} = \frac{1}{\sqrt{3}} = \frac{\sqrt{3}}{3}$$

E X A M P L E   7   Finding Exact Values of the Trigonometric Functions ($\theta = \pi/6 = 30°$)

[5]   Find the exact value of the trigonometric functions of $\pi/6 = 30°$.

FIGURE 25   Solution   Position the triangle in Figure 23(c) so that the 30° angle is in the standard position. See Figure 25. The point $P = (a, b)$ is on the terminal side of 30° and is also on the unit circle. The coordinates of P are $(\sqrt{3}/2, 1/2)$. Thus,

$$\sin\frac{\pi}{6} = \sin 30° = \frac{1}{2} \qquad \cos\frac{\pi}{6} = \cos 30° = \frac{\sqrt{3}}{2}$$

$$\csc\frac{\pi}{6} = \csc 30° = \frac{1}{1/2} = 2 \qquad \sec\frac{\pi}{6} = \sec 30° = \frac{1}{\sqrt{3}/2} = \frac{2}{\sqrt{3}} = \frac{2\sqrt{3}}{3}$$

$$\tan\frac{\pi}{6} = \tan 30° = \frac{1/2}{\sqrt{3}/2} = \frac{1}{\sqrt{3}} = \frac{\sqrt{3}}{3} \qquad \cot\frac{\pi}{6} = \cot 30° = \frac{\sqrt{3}/2}{1/2} = \sqrt{3}$$

Table 3 summarizes the information just derived for $\pi/6$ (30°), $\pi/4$ (45°), and $\pi/3$ (60°). Until you memorize the entries in Table 3, you should draw an appropriate diagram to determine the values given in the table.

TABLE 3							
θ (Radians)	θ (Degrees)	$\sin \theta$	$\cos \theta$	$\tan \theta$	$\csc \theta$	$\sec \theta$	$\cot \theta$
$\pi/6$	30°	$1/2$	$\sqrt{3}/2$	$\sqrt{3}/3$	2	$2\sqrt{3}/3$	$\sqrt{3}$
$\pi/4$	45°	$\sqrt{2}/2$	$\sqrt{2}/2$	1	$\sqrt{2}$	$\sqrt{2}$	1
$\pi/3$	60°	$\sqrt{3}/2$	$1/2$	$\sqrt{3}$	$2\sqrt{3}/3$	2	$\sqrt{3}/3$

     Now work Problems 21 and 31.

EXAMPLE 8   Constructing a Rain Gutter

A rain gutter is to be constructed of aluminum sheets 12 inches wide. After marking off a length of 4 inches from each edge, this length is bent up at an angle θ. See Figure 26. The area A of the opening may be expressed as a function of θ as:

FIGURE 26

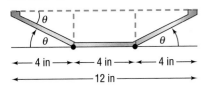

$$A(\theta) = 16 \sin \theta(\cos \theta + 1)$$

Find the area A of the opening for $\theta = 30°$, $\theta = 45°$, and $\theta = 60°$.

Solution   For $\theta = 30°$: $A(30°) = 16 \sin 30° (\cos 30° + 1)$

$$= 16\left(\frac{1}{2}\right)\left(\frac{\sqrt{3}}{2} + 1\right) = 4\sqrt{3} + 8$$

The area of the opening for $\theta = 30°$ is about 14.9 square inches.

For $\theta = 45°$: $A(45°) = 16 \sin 45° (\cos 45° + 1)$

$$= 16\left(\frac{\sqrt{2}}{2}\right)\left(\frac{\sqrt{2}}{2} + 1\right) = 8 + 8\sqrt{2}$$

The area of the opening for $\theta = 45°$ is about 19.3 square inches.

For $\theta = 60°$: $A(60°) = 16 \sin 60° (\cos 60° + 1)$

$$= 16\left(\frac{\sqrt{3}}{2}\right)\left(\frac{1}{2} + 1\right) = 12\sqrt{3}$$

The area of the opening for $\theta = 60°$ is about 20.8 square inches.   ■

Using a Calculator to Find Values of Trigonometric Functions

Before getting started, you must first decide whether to enter the angle in the calculator using radians or degrees and then set the calculator to the correct MODE. If your calculator does not display the mode, you can determine the current mode by entering 30 and then pressing the key marked $\boxed{\sin}$. If you are in the degree mode, the display will show $\boxed{0.5}$ (sin 30° = 0.5). If you are in the radian mode, the display will show $\boxed{-0.9880316}$. Most calculators have a key that allows you to change from one mode to the other. (Check your instruction manual to find out how your calculator handles degrees and radians.)

Your calculator probably only has the keys marked $\boxed{\sin}$, $\boxed{\cos}$, and $\boxed{\tan}$. To find the values of the remaining three trigonometric functions, secant, cosecant, and cotangent, we use the facts that

$$\sec \theta = \frac{1}{\cos \theta} \qquad \csc \theta = \frac{1}{\sin \theta} \qquad \cot \theta = \frac{1}{\tan \theta}$$

These facts are a direct consequence of the definition of the six trigonometric functions.

E X A M P L E   9   Using a Calculator to Approximate the Value
of Trigonometric Functions

Use a calculator to find the approximate value of:

(a) $\cos 48°$ (b) $\csc 21°$ (c) $\tan \dfrac{\pi}{12}$

Express your answer rounded to two decimal places.

Solution   (a) First, we set the mode to receive degrees.

Keystrokes:*   48   cos

Display: 48   0.6691306

Thus, rounded to two decimal places,

   $\cos 48° = 0.67$

(b) Most calculators do not have a csc key. The manufacturers assume that
the user knows some trigonometry. Thus, to find the value of $\csc 21°$, we
use the fact that $\csc 21° = 1/(\sin 21°)$ and proceed as follows:

Keystrokes: 21   sin $1/x$

Display: 21   0.3583679   2.7904281

Thus, rounded to two decimal places,

   $\csc 21° = 2.79$

(c) Set the mode to receive radians. Then, to find $\tan(\pi/12)$,

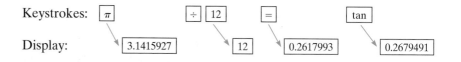

Keystrokes: π $\div$ 12   =   tan

Display: 3.1415927   12   0.2617993   0.2679491

Thus, rounded to two decimal places,

   $\tan \dfrac{\pi}{12} = 0.27$

Now work Problem 47.

*On some calculators, the function key is pressed first, followed by the angle. Consult your
manual.

SUMMARY

For the quadrantal angles and the angles 30°, 45°, and 60° and their integral multiples, we can find the exact value of each trigonometric function by using the geometric features of these angles and symmetry.

Figure 27 shows points on the unit circle that are on the terminal sides of some angles that are integral multiples of $\pi/6$ (30°), $\pi/4$ (45°), and $\pi/3$ (60°).

FIGURE 27

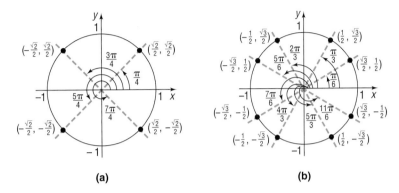

(a)   (b)

Notice the relationship between integral multiples of an angle and symmetry. For example, in Figure 27(a), the point on the unit circle corresponding to $7\pi/4$ is the reflection about the x-axis of the point corresponding to $\pi/4$. The point on the unit circle corresponding to $5\pi/4$ is the reflection about the origin of the point corresponding to $\pi/4$. Thus, using Figure 27(a), we see that

$$\sin\frac{7\pi}{4} = -\frac{\sqrt{2}}{2} \qquad \cos\left(-\frac{\pi}{4}\right) = \frac{\sqrt{2}}{2} \qquad \tan\frac{5\pi}{4} = \frac{-\sqrt{2}/2}{-\sqrt{2}/2} = 1$$

Using Figure 27(b), we see that

$$\sin\frac{11\pi}{6} = -\frac{1}{2} \qquad \cos\frac{7\pi}{6} = -\frac{\sqrt{3}}{2} \qquad \tan\frac{4\pi}{3} = \frac{-\sqrt{3}/2}{-\frac{1}{2}} = \sqrt{3}$$

For most other angles, besides the quadrantal angles and those listed in Figure 27, we can only approximate the value of each trigonometric function using a calculator.

HISTORICAL FEATURE The name *sine* for the sine function is due to a medieval confusion. The name comes from the Sanskrit word *jiva* (meaning chord), first used in India by Aryabhata the Elder (AD 510). He really meant half-chord, but abbreviated it. This was brought into Arabic as *jība*, which was meaningless. Because the proper Arabic word *jaib* would be written the same way (short vowels are not written out in Arabic), *jība* was pronounced as *jaib*, which meant bosom or hollow, and *jaib* remains as the Arabic word for sine to this day. Scholars translating the Arabic works into Latin found that the word *sinus* also meant bosom or hollow, and from *sinus* we get the word *sine*.

The name *tangent*, due to Thomas Finck (1583), can be understood by looking at Figure 28. The line segment $\overline{DC}$ is tangent to the circle at C. If $d(O, B) = d(O, C) = 1$, then the length of the line segment $\overline{DC}$ is

$$d(D, C) = \frac{d(D, C)}{1} = \frac{d(D, C)}{d(O, C)} = \tan\alpha$$

FIGURE 28

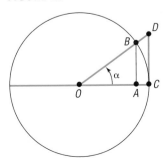

The old name for the tangent is *umbra versa* (meaning turned shadow), referring to the use of the tangent in solving height problems with shadows.

The names of the remaining functions came about as follows. If α and β are complementary angles, then cos α = sin β. Because β is the complement of α, it was natural to write the cosine of α as *sin co α*. Probably for reasons involving ease of pronunciation, the *co* migrated to the front, and then cosine received a three-letter abbreviation to match sin, sec, and tan. The two other cofunctions were similarly treated, except that the long forms *cotan* and *cosec* survive to this day in some countries.

2.2 EXERCISES

In Problems 1–10, a point on the terminal side of an angle θ is given. Find the exact value of the six trigonometric functions of θ.

1. $(-3, 4)$ **2.** $(5, -12)$ **3.** $(2, -3)$ **4.** $(-1, -2)$ **5.** $(-2, -2)$

6. $(1, -1)$ **7.** $(-3, -2)$ **8.** $(2, 2)$ **9.** $(\frac{1}{3}, -\frac{1}{4})$ **10.** $(-0.3, -0.4)$

In Problems 11–30, find the exact value of each expression. Do not use a calculator.

11. sin 45° + cos 60°   **12.** sin 30° − cos 45°   **13.** sin 90° + tan 45°

14. cos 180° − sin 180°   **15.** sin 45° cos 45°   **16.** tan 45° cos 30°

17. csc 45° tan 60°   **18.** sec 30° cot 45°   **19.** 4 sin 90° − 3 tan 180°

20. 5 cos 90° − 8 sin 270°   **21.** $2 \sin \dfrac{\pi}{3} - 3 \tan \dfrac{\pi}{6}$ **22.** $2 \sin \dfrac{\pi}{4} + 3 \tan \dfrac{\pi}{4}$

23. $\sin \dfrac{\pi}{4} - \cos \dfrac{\pi}{4}$ **24.** $\tan \dfrac{\pi}{3} + \cos \dfrac{\pi}{3}$ **25.** $2 \sec \dfrac{\pi}{4} + 4 \cot \dfrac{\pi}{3}$

26. $3 \csc \dfrac{\pi}{3} + \cot \dfrac{\pi}{4}$ **27.** $\tan \pi - \cos 0$ **28.** $\sin \dfrac{3\pi}{2} + \tan \pi$

29. $\csc \dfrac{\pi}{2} + \cot \dfrac{\pi}{2}$ **30.** $\sec \pi - \csc \dfrac{\pi}{2}$

In Problems 31–46, find the exact value of the six trigonometric functions of the given angle. If any are not defined, say "not defined." Do not use a calculator.

31. $2\pi/3$ **32.** $3\pi/4$ **33.** $150°$ **34.** $330°$

35. $-\pi/6$ **36.** $-\pi/3$ **37.** $225°$ **38.** $210°$

39. $5\pi/2$ **40.** 3π **41.** $-180°$ **42.** $-270°$

43. $3\pi/2$ **44.** $-\pi$ **45.** $450°$ **46.** $-90°$

In Problems 47–70, use a calculator to find the approximate value of each expression rounded to two decimal places.

47. sin 28°   **48.** cos 14°   **49.** tan 21°   **50.** sin 15°

51. sec 41°   **52.** csc 55°   **53.** cot 70°   **54.** tan 80°

55. $\sin \dfrac{\pi}{10}$ **56.** $\cos \dfrac{\pi}{8}$ **57.** $\tan \dfrac{5\pi}{12}$ **58.** $\sin \dfrac{3\pi}{10}$

59. $\sec \dfrac{\pi}{12}$ **60.** $\csc \dfrac{5\pi}{13}$ **61.** $\cot \dfrac{\pi}{18}$ **62.** $\sin \dfrac{\pi}{18}$

63. sin 1   **64.** tan 1   **65.** sin 1°   **66.** tan 1°

67. cos 21.5°   **68.** cos 35.2°   **69.** tan 0.3   **70.** tan 0.1

In Problems 71–82, find the exact value of each expression if $\theta = 60°$. Do not use a calculator.

71. $\sin \theta$

72. $\cos \theta$

73. $\sin \dfrac{\theta}{2}$

74. $\cos \dfrac{\theta}{2}$

75. $(\sin \theta)^2$

76. $(\cos \theta)^2$

77. $\sin 2\theta$

78. $\cos 2\theta$

79. $2 \sin \theta$

80. $2 \cos \theta$

81. $\dfrac{\sin \theta}{2}$

82. $\dfrac{\cos \theta}{2}$

83. Find the exact value of $\sin 45° + \sin 135° + \sin 225° + \sin 315°$.

84. Find the exact value of $\tan 60° + \tan 150°$.

85. If $\sin \theta = 0.1$, find $\sin(\theta + \pi)$.

86. If $\cos \theta = 0.3$, find $\cos(\theta + \pi)$.

87. If $\tan \theta = 3$, find $\tan(\theta + \pi)$.

88. If $\cot \theta = -2$, find $\cot(\theta + \pi)$.

89. If $\sin \theta = \frac{1}{5}$, find $\csc \theta$.

90. If $\cos \theta = \frac{2}{3}$, find $\sec \theta$.

Projectile Motion *The path of a projectile fired at an inclination θ to the horizontal with initial speed v_0 is a parabola (see the figure).*

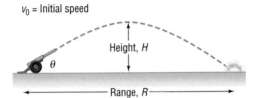

v_0 = Initial speed

Height, H

θ

Range, R

The range R of the projectile, that is, the horizontal distance that the projectile travels, is found by using the formula

$$R = \dfrac{v_0^2 \sin 2\theta}{g}$$

where $g \approx 32.2$ feet per second per second ≈ 9.8 meters per second per second is the acceleration due to gravity. The maximum height H of the projectile is

$$H = \dfrac{v_0^2 \sin^2 \theta}{2g}$$

In Problems 91–94, find the range R and maximum height H.

91. The projectile is fired at an angle of 45° to the horizontal with an initial speed of 100 feet per second.

92. The projectile is fired at an angle of 30° to the horizontal with an initial speed of 150 meters per second.

93. The projectile is fired at an angle of 25° to the horizontal with an initial speed of 500 meters per second.

94. The projectile is fired at an angle of 50° to the horizontal with an initial speed of 200 feet per second.

95. Inclined Plane If friction is ignored, the time t (in seconds) required for a block to slide down an inclined plane (see the figure) is given by the formula

$$t = \sqrt{\dfrac{2a}{g \sin \theta \cos \theta}}$$

where a is the length (in feet) of the base and $g \approx 32$ feet per second per second is the acceleration of gravity. How long does it take a block to slide down an inclined plane with base $a = 10$ feet when

(a) $\theta = 30°$?   (b) $\theta = 45°$?   (c) $\theta = 60°$?

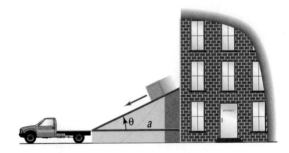

96. Piston Engines In a certain piston engine, the distance x (in meters) from the center of the drive shaft to the head of the piston is given by

$$x = \cos \theta + \sqrt{16 + 0.5 \cos 2\theta}$$

where θ is the angle between the crank and the path of the piston head (see the figure). Find x when $\theta = 30°$ and when $\theta = 45°$.

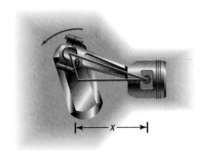

x

97. Calculating the Time of a Trip Two oceanfront homes are located 8 miles apart on a straight stretch of beach, each a distance of 1 mile from a paved road that parallels the ocean. Sally can jog 8 miles per hour along the paved road, but only 3 miles per hour in the sand on the beach. Because of a river directly between the two houses, it is necessary to jog in the sand to the road, continue on the road, and then jog directly back in the sand to get from one house to the other. See the illustration. The time T to get from one house to the other as a function of the angle θ shown in the illustration is

$$T(\theta) = 1 + \frac{2}{3 \sin \theta} - \frac{1}{4 \tan \theta}, \qquad 0° < \theta < 90°$$

(a) Calculate the time T for $\theta = 30°$. How long is Sally on the paved road?

(b) Calculate the time T for $\theta = 45°$. How long is Sally on the paved road?

(c) Calculate the time T for $\theta = 60°$. How long is Sally on the paved road?

 (d) Calculate the time T for $\theta = 90°$. Describe the path taken. Why can't the formula for T be used?

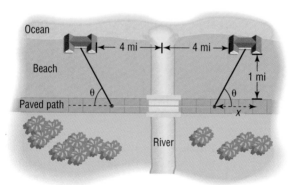

98. Designing Fine Decorative Pieces A designer of decorative art plans to market solid gold spheres encased in clear crystal cones. Each sphere is of fixed radius R and will be enclosed in a cone of height h and radius r. See the illustration. Many cones can be used to enclose the sphere, each having a different slant angle θ. The volume V of the cone can be expressed as a function of the slant angle θ of the cone:

$$V(\theta) = \frac{1}{3}\pi R^3 \frac{(1 + \sec \theta)^3}{\tan^2 \theta}, \qquad 0° < \theta < 90°$$

What volume V is required to enclose a sphere of radius 2 centimeters in a cone whose slant angle θ is 30°? 45°? 60°?

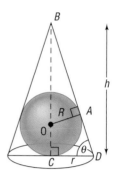

99. Projectile Motion An object is propelled upward at an angle θ, $45° < \theta < 90°$, to the horizontal with an initial velocity of v_0 feet per second from the base of a plane that makes an angle of 45° with the horizontal. See the illustration. If air resistance is ignored, the distance R it travels up the inclined plane is given by

$$R = \frac{v_0^2 \sqrt{2}}{32}(\sin 2\theta - \cos 2\theta - 1)$$

(a) Find the distance R that the object travels along the inclined plane if the initial velocity is 32 feet per second and $\theta = 60°$.

 (b) Graph $R = R(\theta)$ if the initial velocity is 32 feet per second.

(c) What value of θ makes R largest?

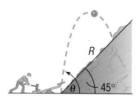

100. If θ $(0 < \theta < \pi)$ is the angle between a horizontal ray directed to the right (say, the positive x-axis) and a nonhorizontal, nonvertical line L, show that the slope m of L equals $\tan \theta$. The angle θ is called the **inclination** of L.

[**Hint:** See the illustration, where we have drawn the line L^* parallel to L and passing through the origin.

Use the fact that L^* intersects the unit circle at the point $(\cos \theta, \sin \theta)$.]

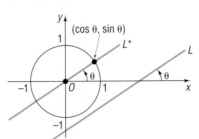

101. Write a brief paragraph that explains how to quickly compute the trigonometric functions of 30°, 45°, and 60°.

102. Write a brief paragraph that explains how to quickly compute the trigonometric functions of 0°, 90°, 180°, and 270°.

103. How would you explain the meaning of the sine function to a fellow student who has just completed college algebra?

2.3 | PROPERTIES OF THE TRIGONOMETRIC FUNCTIONS

- 1 Determine the Domain and Range of the Trigonometric Functions
- 2 Determine the Period of the Trigonometric Functions
- 3 Determine the Sign of the Trigonometric Functions
- 4 Find the Value of the Trigonometric Functions Utilizing Fundamental Identities
- 5 Use Even–Odd Properties to Find the Exact Value of the Trigonometric Functions

Domain and Range of the Trigonometric Functions

FIGURE 29

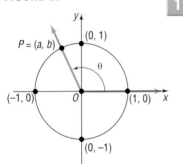

1

Let θ be an angle in standard position, and let $P = (a, b)$ be a point on the terminal side of θ. Suppose, for convenience, that P also lies on the unit circle. See Figure 29. Then, by definition,

$$\sin \theta = b \qquad \cos \theta = a \qquad \tan \theta = \frac{b}{a}, \quad a \neq 0$$

$$\csc \theta = \frac{1}{b}, \quad b \neq 0 \qquad \sec \theta = \frac{1}{a}, \quad a \neq 0 \qquad \cot \theta = \frac{a}{b}, \quad b \neq 0$$

For $\sin \theta$ and $\cos \theta$, θ can be any angle, so it follows that the domain of the sine function and cosine function is the set of all real numbers.

> The domain of the sine function is the set of all real numbers.
> The domain of the cosine function is the set of all real numbers.

If $a = 0$, then the tangent function and the secant function are not defined. Thus, for the tangent function and secant function, the x-coordinate of $P = (a, b)$ cannot be 0. On the unit circle, there are two such points $(0, 1)$ and $(0, -1)$. These two points correspond to the angles $\pi/2$ (90°) and $3\pi/2$ (270°) or, more generally, to any angle that is an odd multiple of $\pi/2$ (90°), such as $\pi/2$ (90°), $3\pi/2$ (270°), $5\pi/2$ (450°), $-\pi/2$ (−90°), $-3\pi/2$ (−270°), and so on. Such angles must therefore be excluded from the domain of the tangent function and secant function.

> The domain of the tangent function is the set of all real numbers, except odd multiples of $\pi/2$ (90°).
>
> The domain of the secant function is the set of all real numbers, except odd multiples of $\pi/2$ (90°).

If $b = 0$, then the cotangent function and the cosecant function are not defined. Thus, for the cotangent function and cosecant function the y-coordinate of $P = (a, b)$ cannot be 0. On the unit circle, there are two such points, $(1, 0)$ and $(-1, 0)$. These two points correspond to the angles 0 (0°) and π (180°) or, more generally, to any angle that is an integral multiple of π (180°), such as 0 (0°), π (180°), 2π (360°), 3π (540°), $-\pi$ (−180°), and so on. Such angles must therefore be excluded from the domain of the cotangent function and cosecant function.

> The domain of the cotangent function is the set of all real numbers, except integral multiples of π (180°).
>
> The domain of the cosecant function is the set of all real numbers, except integral multiples of π (180°).

Next, we determine the range of each of the six trigonometric functions. Refer again to Figure 29. Let $P = (a, b)$ be the point on the unit circle that corresponds to the angle θ. It follows that $-1 \le a \le 1$ and $-1 \le b \le 1$. Consequently, since $\sin \theta = b$ and $\cos \theta = a$, we have

$$-1 \le \sin \theta \le 1 \qquad -1 \le \cos \theta \le 1$$

Thus, the range of both the sine function and the cosine function consists of all real numbers between −1 and 1, inclusive. In terms of absolute value notation, we have $|\sin \theta| \le 1$ and $|\cos \theta| \le 1$.

Similarly, if θ is not a multiple of π (180°), then $\csc \theta = 1/b$. Since $b = \sin \theta$ and $|b| = |\sin \theta| \le 1$, it follows that $|\csc \theta| = 1/|\sin \theta| = 1/|b| \ge 1$. Thus, the range of the cosecant function consists of all real numbers less than or equal to −1 or greater than or equal to 1. That is,

$$\csc \theta \le -1 \quad \text{or} \quad \csc \theta \ge 1$$

If θ is not an odd multiple of $\pi/2$ (90°), then, by definition, $\sec \theta = 1/a$. Since $a = \cos \theta$ and $|a| = |\cos \theta| \le 1$, it follows that $|\sec \theta| = 1/|\cos \theta| = 1/|a| \ge 1$. Thus, the range of the secant function consists of all real numbers less than or equal to −1 or greater than or equal to 1.

$$\sec \theta \leq -1 \quad \text{or} \quad \sec \theta \geq 1$$

The range of both the tangent function and the cotangent function consists of all real numbers. You are asked to prove this in Problems 109 and 110.

$$-\infty < \tan \theta < \infty \qquad -\infty < \cot \theta < \infty$$

Table 4 summarizes these results.

TABLE 4

Function	Symbol	Domain	Range
sine	$f(\theta) = \sin \theta$	All real numbers	All real numbers from −1 to 1, inclusive
cosine	$f(\theta) = \cos \theta$	All real numbers	All real numbers from −1 to 1, inclusive
tangent	$f(\theta) = \tan \theta$	All real numbers, except odd multiples of $\pi/2$ (90°)	All real numbers
cosecant	$f(\theta) = \csc \theta$	All real numbers, except integral multiples of π (180°)	All real numbers greater than or equal to 1 or less than or equal to −1
secant	$f(\theta) = \sec \theta$	All real numbers, except odd multiples of $\pi/2$ (90°)	All real numbers greater than or equal to 1 or less than or equal to −1
cotangent	$f(\theta) = \cot \theta$	All real numbers, except integral multiples of π (180°)	All real numbers

Now work Problem 85.

Period of the Trigonometric Functions

FIGURE 30

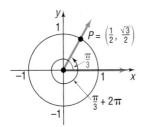

Look at Figure 30. This figure shows that for an angle of $\pi/3$ radians, the corresponding point P on the unit circle is $(1/2, \sqrt{3}/2)$. Notice that, for an angle of $\pi/3 + 2\pi$ radians, the corresponding point P on the unit circle is also $(1/2, \sqrt{3}/2)$. Thus,

$$\sin \frac{\pi}{3} = \frac{\sqrt{3}}{2} \quad \text{and} \quad \sin\left(\frac{\pi}{3} + 2\pi\right) = \frac{\sqrt{3}}{2}$$

$$\cos \frac{\pi}{3} = \frac{1}{2} \quad \text{and} \quad \cos\left(\frac{\pi}{3} + 2\pi\right) = \frac{1}{2}$$

This example illustrates a more general situation. For a given angle θ, measured in radians, suppose that we know the corresponding point $P = (a, b)$ on the unit circle. Now add 2π to θ. The point on the unit circle corresponding to $\theta + 2\pi$ is identical to the point P on the unit circle corresponding to θ. See Figure 31. Thus, the values of the trigonometric functions of $\theta + 2\pi$ are equal to the values of the corresponding trigonometric functions of θ.

FIGURE 31

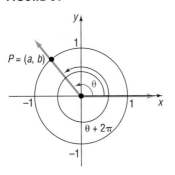

If we add (or subtract) integral multiples of 2π to θ, the trigonometric values remain unchanged. That is, for all θ,

$$\sin(\theta + 2\pi k) = \sin \theta \qquad \cos(\theta + 2\pi k) = \cos \theta \tag{1}$$

where k is any integer.

Functions that exhibit this kind of behavior are called *periodic functions.*

Seeing the Concept To see the periodic behavior of the sine function, graph $Y_1 = \sin x$, $Y_2 = \sin(x + 2\pi)$, $Y_3 = \sin(x - 2\pi)$, and $Y_4 = \sin(x + 4\pi)$ on the same screen. ∎

A function f is called **periodic** if there is a positive number p such that, whenever θ is in the domain of f, so is $\theta + p$, and

$$f(\theta + p) = f(\theta)$$

If there is a smallest such number p, this smallest value is called the **(fundamental) period** of f.

Thus, based on equation (1), the sine and cosine functions are periodic. In fact, the sine and cosine functions have period 2π. You are asked to prove this fact in Problems 111 and 112. The secant and cosecant functions are also periodic with period 2π, and the tangent and cotangent functions are periodic with period π. You are asked to prove these statements in Problems 113 through 116.

Periodic Properties

$$\sin(\theta + 2\pi k) = \sin \theta \quad \cos(\theta + 2\pi k) = \cos \theta \quad \tan(\theta + \pi k) = \tan \theta$$
$$\csc(\theta + 2\pi k) = \csc \theta \quad \sec(\theta + 2\pi k) = \sec \theta \quad \cot(\theta + \pi k) = \cot \theta$$

where k is any integer.

Because the sine, cosine, secant, and cosecant functions have period 2π, once we know their values for $0 \leq \theta < 2\pi$, we know all their values; similarly, since the tangent and cotangent functions have period π, once we know their values for $0 \leq \theta < \pi$, we know all their values.

E X A M P L E 1

FIGURE 32(a,b)

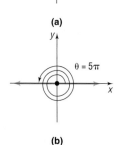

(a)

(b)

Finding Exact Values Using Periodic Properties

Find the exact value of:

(a) $\sin \dfrac{17\pi}{4}$ (b) $\cos 5\pi$ (c) $\tan \dfrac{5\pi}{4}$

Solution (a) It is best to sketch the angle first, as shown in Figure 32(a). Since the period of the sine function is 2π, each full revolution can be ignored. This leaves the angle $\pi/4$. Thus,

$$\sin \frac{17\pi}{4} = \sin\left(\frac{\pi}{4} + 4\pi\right) = \sin \frac{\pi}{4} = \frac{\sqrt{2}}{2}$$

(b) See Figure 32(b). Since the period of the cosine function is 2π, each full revolution can be ignored. This leaves the angle π. Thus,

$$\cos 5\pi = \cos(\pi + 4\pi) = \cos \pi = -1$$

FIGURE 32(c)

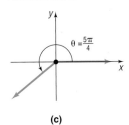

(c)

(c) See Figure 32(c). Since the period of the tangent function is π, each half-revolution can be ignored. This leaves the angle $\pi/4$. Thus,

$$\tan\frac{5\pi}{4} = \tan\left(\frac{\pi}{4} + \pi\right) = \tan\frac{\pi}{4} = 1$$

The periodic properties of the trigonometric functions will be very helpful to us when we study their graphs later in the chapter.

Now work Problem 1.

The Signs of the Trigonometric Functions

FIGURE 33

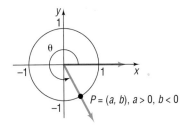

$P = (a, b), a > 0, b < 0$

Let $P = (a, b)$ be the point on the unit circle that corresponds to the angle θ. If we know in which quadrant the point P lies, then we can determine the signs of the trigonometric functions of θ. For example, if $P = (a, b)$ lies in quadrant IV, as shown in Figure 33, then we know that $a > 0$ and $b < 0$. Consequently,

$$\sin\theta = b < 0 \qquad \cos\theta = a > 0 \qquad \tan\theta = \frac{b}{a} < 0$$

$$\csc\theta = \frac{1}{b} < 0 \qquad \sec\theta = \frac{1}{a} > 0 \qquad \cot\theta = \frac{a}{b} < 0$$

Table 5 lists the signs of the six trigonometric functions for each quadrant. See also Figure 34.

TABLE 5			
Quadrant of P	$\sin\theta, \csc\theta$	$\cos\theta, \sec\theta$	$\tan\theta, \cot\theta$
I	Positive	Positive	Positive
II	Positive	Negative	Negative
III	Negative	Negative	Positive
IV	Negative	Positive	Negative

FIGURE 34

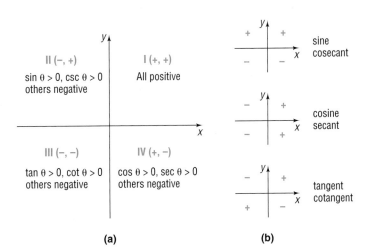

(a)   (b)

E X A M P L E   2   Finding the Quadrant in Which an Angle θ Lies

If $\sin \theta < 0$ and $\cos \theta < 0$, name the quadrant in which the angle θ lies.

Solution   Let $P = (a, b)$ be the point on the unit circle corresponding to θ. Then $\sin \theta = b < 0$ and $\cos \theta = a < 0$. Thus, $P = (a, b)$ must be in quadrant III, so θ lies in quadrant III.   ▬

 Now work Problem 17.

Fundamental Identities

If $P = (a, b)$ is the point on the unit circle corresponding to θ, then

$$\sin \theta = b \qquad\qquad \cos \theta = a \qquad\qquad \tan \theta = \frac{b}{a}, \ \text{ if } a \neq 0$$

$$\csc \theta = \frac{1}{b}, \ \text{ if } b \neq 0 \qquad \sec \theta = \frac{1}{a}, \ \text{ if } a \neq 0 \qquad \cot \theta = \frac{a}{b}, \ \text{ if } b \neq 0$$

Thus, we have the **reciprocal identities:**

Reciprocal Identities

$$\csc \theta = \frac{1}{\sin \theta} \qquad \sec \theta = \frac{1}{\cos \theta} \qquad \cot \theta = \frac{1}{\tan \theta} \qquad (2)$$

Two other fundamental identities that are easy to see are the **quotient identities.**

Quotient Identities

$$\tan \theta = \frac{\sin \theta}{\cos \theta} \qquad \cot \theta = \frac{\cos \theta}{\sin \theta} \qquad (3)$$

The proofs of formulas (2) and (3) follow from the definitions of the trigonometric functions. (See Problems 117 and 118.)

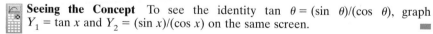 **Seeing the Concept** To see the identity $\tan \theta = (\sin \theta)/(\cos \theta)$, graph $Y_1 = \tan x$ and $Y_2 = (\sin x)/(\cos x)$ on the same screen.   ▬

 If $\sin \theta$ and $\cos \theta$ are known, formulas (2) and (3) make it easy to find the values of the remaining trigonometric functions.

E X A M P L E   3   Finding Exact Values Using Identities When Sine and Cosine Are Given

Given $\sin \theta = \sqrt{5}/5$ and $\cos \theta = 2\sqrt{5}/5$, find the exact values of the four remaining trigonometric functions of θ.

Solution   Based on a quotient identity from formula (3), we have

$$\tan \theta = \frac{\sin \theta}{\cos \theta} = \frac{\sqrt{5}/5}{2\sqrt{5}/5} = \frac{1}{2}$$

Then we use the reciprocal identities from formula (2) to get

$$\csc \theta = \frac{1}{\sin \theta} = \frac{1}{\sqrt{5}/5} = \frac{5}{\sqrt{5}} = \sqrt{5} \qquad \sec \theta = \frac{1}{\cos \theta} = \frac{1}{2\sqrt{5}/5} = \frac{5}{2\sqrt{5}} = \frac{\sqrt{5}}{2}$$

$$\cot \theta = \frac{1}{\tan \theta} = \frac{1}{\frac{1}{2}} = 2$$

 Now work Problem 25.

The equation of the unit circle is $x^2 + y^2 = 1$. Thus, if $P = (a, b)$ is the point on the terminal side of an angle θ and if P lies on the unit circle, then

$$b^2 + a^2 = 1$$

But $b = \sin \theta$ and $a = \cos \theta$. Thus,

$$(\sin \theta)^2 + (\cos \theta)^2 = 1 \qquad\qquad (4)$$

It is customary to write $\sin^2 \theta$ instead of $(\sin \theta)^2$, $\cos^2 \theta$ instead of $(\cos \theta)^2$, and so on. With this notation, we can rewrite equation (4) as

$$\sin^2 \theta + \cos^2 \theta = 1 \qquad\qquad (5)$$

If $\cos \theta \neq 0$, we can divide each side of equation (5) by $\cos^2 \theta$.

$$\frac{\sin^2 \theta}{\cos^2 \theta} + 1 = \frac{1}{\cos^2 \theta}$$

$$\left(\frac{\sin \theta}{\cos \theta}\right)^2 + 1 = \left(\frac{1}{\cos \theta}\right)^2$$

Now use formulas (2) and (3) to get

$$\tan^2 \theta + 1 = \sec^2 \theta \qquad\qquad (6)$$

Similarly, if $\sin \theta \neq 0$, we can divide equation (5) by $\sin^2 \theta$ and use formulas (2) and (3) to get the result.

$$1 + \cot^2 \theta = \csc^2 \theta \tag{7}$$

Collectively, the identities in equations (5), (6), and (7) are referred to as the **Pythagorean identities.**

Let's pause here to summarize the fundamental identities.

Fundamental Identities

$$\tan \theta = \frac{\sin \theta}{\cos \theta} \qquad \cot \theta = \frac{\cos \theta}{\sin \theta}$$

$$\cot \theta = \frac{1}{\tan \theta} \qquad \sec \theta = \frac{1}{\cos \theta} \qquad \csc \theta = \frac{1}{\sin \theta}$$

$$\sin^2 \theta + \cos^2 \theta = 1 \qquad \tan^2 \theta + 1 = \sec^2 \theta \qquad 1 + \cot^2 \theta = \csc^2 \theta$$

The Pythagorean identity

$$\sin^2 \theta + \cos^2 \theta = 1$$

can be solved for $\sin \theta$ in terms of $\cos \theta$ (or vice versa) as follows:

$$\sin^2 \theta = 1 - \cos^2 \theta$$
$$\sin \theta = \pm \sqrt{1 - \cos^2 \theta}$$

where the $+$ sign is used if $\sin \theta > 0$ and the $-$ sign is used if $\sin \theta < 0$.

E X A M P L E 4

Finding Exact Values Given One Value and the Sign of Another

Given that $\sin \theta = \frac{1}{3}$ and $\cos \theta < 0$, find the exact value of each of the remaining five trigonometric functions.

Solution   We solve this problem in two ways: the first way uses the definition of the trigonometric functions; the second method uses the fundamental identities.

Solution 1   Using the Definition

Suppose that $P = (a, b)$ is a point on the terminal side of θ that is on the circle $x^2 + y^2 = 9$. See Figure 35. (Do you see why we chose the circle with radius $r = 3$? Notice that $\sin \theta = \frac{1}{3} = \frac{b}{r}$. The choice $r = 3$ will make our calculations easy.) With this choice, $b = 1$ and $r = 3$. Since $\cos \theta = a/r < 0$, it follows that $a < 0$. Thus,

$$a^2 + b^2 = r^2 \qquad b = 1, r = 3, a < 0$$
$$a^2 + 1 = 9$$
$$a^2 = 8$$
$$a = -2\sqrt{2}$$

FIGURE 35

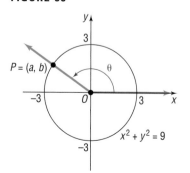

Thus,

$$\cos\theta = \frac{a}{r} = \frac{-2\sqrt{2}}{3} \qquad \tan\theta = \frac{b}{a} = \frac{1}{-2\sqrt{2}} = \frac{-\sqrt{2}}{4}$$

$$\csc\theta = \frac{r}{b} = \frac{3}{1} = 3 \qquad \sec\theta = \frac{r}{a} = \frac{3}{-2\sqrt{2}} = \frac{-3\sqrt{2}}{4} \qquad \cot\theta = \frac{a}{b} = \frac{-2\sqrt{2}}{1} = -2\sqrt{2}$$

Solution 2   Using Identities

First, we solve equation (5) for $\cos\theta$.

$$\sin^2\theta + \cos^2\theta = 1$$
$$\cos^2\theta = 1 - \sin^2\theta$$
$$\cos\theta = \pm\sqrt{1 - \sin^2\theta}$$

Because $\cos\theta < 0$, we choose the minus sign.

$$\cos\theta = -\sqrt{1 - \sin^2\theta} = -\sqrt{1 - \frac{1}{9}} = -\sqrt{\frac{8}{9}} = -\frac{2\sqrt{2}}{3}$$
$$\uparrow$$
$$\scriptstyle \sin\theta = \frac{1}{3}$$

Now we know the values of $\sin\theta$ and $\cos\theta$, so we can use formulas (2) and (3) to get

$$\tan\theta = \frac{\sin\theta}{\cos\theta} = \frac{1/3}{-2\sqrt{2}/3} = \frac{1}{-2\sqrt{2}} = \frac{-\sqrt{2}}{4} \qquad \cot\theta = \frac{1}{\tan\theta} = -2\sqrt{2}$$

$$\sec\theta = \frac{1}{\cos\theta} = \frac{1}{-2\sqrt{2}/3} = \frac{-3}{2\sqrt{2}} = \frac{-3\sqrt{2}}{4} \qquad \csc\theta = \frac{1}{\sin\theta} = \frac{1}{1/3} = 3$$

 Now work Problem 33.

Even–Odd Properties

Recall that a function f is even if $f(-\theta) = f(\theta)$ for all θ in the domain of f; a function f is odd if $f(-\theta) = -f(\theta)$ for all θ in the domain of f. We will now show that the trigonometric functions sine, tangent, cotangent, and cosecant are odd functions, whereas the functions cosine and secant are even functions.

Theorem   Even–Odd Properties

$$\sin(-\theta) = -\sin\theta \qquad \cos(-\theta) = \cos\theta \qquad \tan(-\theta) = -\tan\theta$$
$$\csc(-\theta) = -\csc\theta \qquad \sec(-\theta) = \sec\theta \qquad \cot(-\theta) = -\cot\theta$$

FIGURE 36

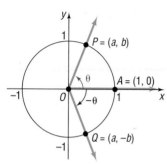

Proof Let $P = (a, b)$ be the point on the terminal side of the angle θ that is on the unit circle. (See Figure 36.) The point Q on the terminal side of the angle $-\theta$ that is on the unit circle will have coordinates $(a, -b)$. Using the definition for the trigonometric functions, we have

$$\sin \theta = b \qquad \cos \theta = a \qquad \sin(-\theta) = -b \qquad \cos(-\theta) = a$$

so

$$\sin(-\theta) = -b = -\sin \theta \qquad \cos(-\theta) = a = \cos \theta$$

Now, using these results and some of the fundamental identities, we have

$$\tan(-\theta) = \frac{\sin(-\theta)}{\cos(-\theta)} = \frac{-\sin \theta}{\cos \theta} = -\tan \theta \qquad \cot(-\theta) = \frac{1}{\tan(-\theta)} = \frac{1}{-\tan \theta} = -\cot \theta$$

$$\sec(-\theta) = \frac{1}{\cos(-\theta)} = \frac{1}{\cos \theta} = \sec \theta \qquad \csc(-\theta) = \frac{1}{\sin(-\theta)} = \frac{1}{-\sin \theta} = -\csc \theta$$

    ■

Seeing the Concept To see that the cosine function is even, graph $Y_1 = \cos x$ and then $Y_2 = \cos(-x)$. Clear the screen. To see that the sin function is odd, graph $Y_1 = -\sin x$ and $Y_2 = \sin(-x)$.   ■

EXAMPLE 5   Finding Exact Values Using Even–Odd Properties

[5] Find the exact value of:

    (a) $\sin(-45°)$ (b) $\cos(-\pi)$ (c) $\cot(-3\pi/2)$ (d) $\tan(-37\pi/4)$

Solution   (a) $\sin(-45°) \underset{\uparrow}{=} -\sin 45° = -\dfrac{\sqrt{2}}{2}$ (b) $\cos(-\pi) \underset{\uparrow}{=} \cos \pi = -1$
                      Odd function   Even function

    (c) $\cot\left(-\dfrac{3\pi}{2}\right) \underset{\uparrow}{=} -\cot \dfrac{3\pi}{2} = 0$
                  Odd function

    (d) $\tan\left(-\dfrac{37\pi}{4}\right) \underset{\uparrow}{=} -\tan \dfrac{37\pi}{4} = -\tan\left(\dfrac{\pi}{4} + 9\pi\right) \underset{\uparrow}{=} -\tan \dfrac{\pi}{4} = -1$
                      Odd function   Period is π ■

 Now work Problem 49.

2.3 EXERCISES

In Problems 1–16, use the fact that the trigonometric functions are periodic to find the exact value of each expression. Do no use a calculator.

    1. $\sin 405°$ 2. $\cos 420°$ 3. $\tan 405°$ 4. $\sin 390°$ 5. $\csc 450°$ 6. $\sec 540°$

    7. $\cot 390°$ 8. $\sec 420°$ 9. $\cos \dfrac{33\pi}{4}$ 10. $\sin \dfrac{9\pi}{4}$ 11. $\tan 21\pi$ 12. $\csc \dfrac{9\pi}{2}$

    13. $\sec \dfrac{17\pi}{4}$ 14. $\cot \dfrac{17\pi}{4}$ 15. $\tan \dfrac{19\pi}{6}$ 16. $\sec \dfrac{25\pi}{6}$

In Problems 17–24, name the quadrant in which the angle θ lies.

17. $\sin\theta > 0$, $\cos\theta < 0$

18. $\sin\theta < 0$, $\cos\theta > 0$

19. $\sin\theta < 0$, $\tan\theta < 0$

20. $\cos\theta > 0$, $\tan\theta > 0$

21. $\cos\theta > 0$, $\tan\theta < 0$

22. $\cos\theta < 0$, $\tan\theta > 0$

23. $\sec\theta < 0$, $\sin\theta > 0$

24. $\csc\theta > 0$, $\cos\theta < 0$

In Problems 25–32, sin θ and cos θ are given. Find the exact value of each of the four remaining trigonometric functions. In Problems 31–32, round your answer to four decimal places.

25. $\sin\theta = 2\sqrt{5}/5$, $\cos\theta = \sqrt{5}/5$

26. $\sin\theta = -\sqrt{5}/5$, $\cos\theta = -2\sqrt{5}/5$

27. $\sin\theta = \frac{1}{2}$, $\cos\theta = \sqrt{3}/2$

28. $\sin\theta = \sqrt{3}/2$, $\cos\theta = \frac{1}{2}$

29. $\sin\theta = -\frac{1}{3}$, $\cos\theta = 2\sqrt{2}/3$

30. $\sin\theta = 2\sqrt{2}/3$, $\cos\theta = -\frac{1}{3}$

31. $\sin\theta = 0.2588$, $\cos\theta = 0.9659$

32. $\sin\theta = 0.6428$, $\cos\theta = 0.7660$

In Problems 33–48, find the exact value of each of the remaining trigonometric functions of θ.

33. $\sin\theta = \frac{12}{13}$, θ in quadrant II

34. $\cos\theta = \frac{3}{5}$, θ in quadrant IV

35. $\cos\theta = -\frac{4}{5}$, θ in quadrant III

36. $\sin\theta = -\frac{5}{13}$, θ in quadrant III

37. $\sin\theta = \frac{5}{13}$, $90° < \theta < 180°$

38. $\cos\theta = \frac{4}{5}$, $270° < \theta < 360°$

39. $\cos\theta = -\frac{1}{3}$, $\frac{\pi}{2} < \theta < \pi$

40. $\sin\theta = -\frac{2}{3}$, $\pi < \theta < 3\pi/2$

41. $\sin\theta = \frac{2}{3}$, $\tan\theta < 0$

42. $\cos\theta = -\frac{1}{4}$, $\tan\theta > 0$

43. $\sec\theta = 2$, $\sin\theta < 0$

44. $\csc\theta = 3$, $\cot\theta < 0$

45. $\tan\theta = \frac{3}{4}$, $\sin\theta < 0$

46. $\cot\theta = \frac{4}{3}$, $\cos\theta < 0$

47. $\tan\theta = -\frac{1}{3}$, $\sin\theta > 0$

48. $\sec\theta = -2$, $\tan\theta > 0$

In Problems 49–66, use the even–odd properties to find the exact value of each expression. Do not use a calculator.

49. $\sin(-60°)$

50. $\cos(-30°)$

51. $\tan(-30°)$

52. $\sin(-135°)$

53. $\sec(-60°)$

54. $\csc(-30°)$

55. $\sin(-90°)$

56. $\cos(-270°)$

57. $\tan\left(-\dfrac{\pi}{4}\right)$

58. $\sin(-\pi)$

59. $\cos\left(-\dfrac{\pi}{4}\right)$

60. $\sin\left(-\dfrac{\pi}{3}\right)$

61. $\tan(-\pi)$

62. $\sin\left(-\dfrac{3\pi}{2}\right)$

63. $\csc\left(-\dfrac{\pi}{4}\right)$

64. $\sec(-\pi)$

65. $\sec\left(-\dfrac{\pi}{6}\right)$

66. $\csc\left(-\dfrac{\pi}{3}\right)$

In Problems 67–78, find the exact value of each expression. Do not use a calculator.

67. $\sin(-\pi) + \cos 5\pi$

68. $\tan\left(-\dfrac{5\pi}{6}\right) - \cot\dfrac{7\pi}{2}$

69. $\sec(-\pi) + \csc\left(-\dfrac{\pi}{2}\right)$

70. $\tan(-6\pi) + \cos\dfrac{9\pi}{4}$

71. $\sin\left(-\dfrac{9\pi}{4}\right) - \tan\left(-\dfrac{9\pi}{4}\right)$

72. $\cos\left(-\dfrac{17\pi}{4}\right) - \sin\left(-\dfrac{3\pi}{2}\right)$

73. $\sin^2 40° + \cos^2 40°$

74. $\sec^2 18° - \tan^2 18°$

75. $\sin 80° \csc 80°$

76. $\tan 10° \cot 10°$

77. $\tan 40° - \dfrac{\sin 40°}{\cos 40°}$

78. $\cot 20° - \dfrac{\cos 20°}{\sin 20°}$

79. If $\sin\theta = 0.3$, find the value of $\sin\theta + \sin(\theta + 2\pi) + \sin(\theta + 4\pi)$.

80. If $\cos\theta = 0.2$, find the value of $\cos\theta + \cos(\theta + 2\pi) + \cos(\theta + 4\pi)$.

81. If $\tan\theta = 3$, find the value of $\tan\theta + \tan(\theta + \pi) + \tan(\theta + 2\pi)$.

82. If $\cot\theta = -2$, find the value of $\cot\theta + \cot(\theta - \pi) + \cot(\theta - 2\pi)$.

83. What is the domain of the sine function?

84. What is the domain of the cosine function?

85. For what numbers θ is $f(\theta) = \tan \theta$ not defined?

86. For what numbers θ is $f(\theta) = \cot \theta$ not defined?

87. For what numbers θ is $f(\theta) = \sec \theta$ not defined?

88. For what numbers θ is $f(\theta) = \csc \theta$ not defined?

89. What is the range of the sine function?

90. What is the range of the cosine function?

91. What is the range of the tangent function?

92. What is the range of the cotangent function?

93. What is the range of the secant function?

94. What is the range of the cosecant function?

95. Is the sine function even, odd, or neither? Is its graph symmetric? With respect to what?

96. Is the cosine function even, odd, or neither? Is its graph symmetric? With respect to what?

97. Is the tangent function even, odd, or neither? Is its graph symmetric? With respect to what?

98. Is the cotangent function even, odd, or neither? Is its graph symmetric? With respect to what?

99. Is the secant function even, odd, or neither? Is its graph symmetric? With respect to what?

100. Is the cosecant function even, odd, or neither? Is its graph symmetric? With respect to what?

In Problems 101–106, use the periodic and even–odd properties.

101. If $f(x) = \sin x$ and $f(a) = 1/3$, find the exact value of:
    (a) $f(-a)$ (b) $f(a) + f(a + 2\pi) + f(a + 4\pi)$

102. If $f(x) = \cos x$ and $f(a) = 1/4$, find the exact value of:
    (a) $f(-a)$ (b) $f(a) + f(a + 2\pi) + f(a - 2\pi)$

103. If $f(x) = \tan x$ and $f(a) = 2$, find the exact value of:
    (a) $f(-a)$ (b) $f(a) + f(a + \pi) + f(a + 2\pi)$

104. If $f(x) = \cot x$ and $f(a) = -3$, find the exact value of:
    (a) $f(-a)$ (b) $f(a) + f(a + \pi) + f(a + 4\pi)$

105. If $f(x) = \sec x$ and $f(a) = -4$, find the exact value of:
    (a) $f(-a)$ (b) $f(a) + f(a + 2\pi) + f(a + 4\pi)$

106. If $f(x) = \csc x$ and $f(a) = 2$, find the exact value of:
    (a) $f(-a)$ (b) $f(a) + f(a + 2\pi) + f(a + 4\pi)$

107. **Calculating the Time of a Trip** From a parking lot, you want to walk to a house on the ocean. The house is located 1500 feet down a paved path that parallels the ocean, which is 500 feet away. See the illustration. Along the path you can walk 300 feet per minute, but in the sand on the beach you can only walk 100 feet per minute.

    The time T to get from the parking lot to the beachhouse can be expressed as a function of the angle θ shown in the illustration and is

$$T(\theta) = 5 - \frac{5}{3 \tan \theta} + \frac{5}{\sin \theta}, \qquad 0 < \theta < \frac{\pi}{2}$$

Calculate the time T if you walk directly from the parking lot to the house.
[**Hint:** $\tan \theta = 500/1500$.]

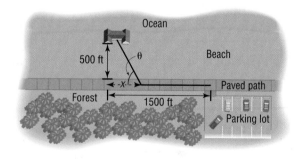

108. **Calculating the Time of a Trip** Two oceanfront homes are located 8 miles apart on a straight stretch of beach, each a distance of 1 mile from a paved road that parallels the ocean. Sally can jog 8 miles per hour along the paved road, but only 3 miles per hour in the sand on the beach. Because of a river directly between the two houses, it is necessary to jog in the sand to the road, continue on the road, and then jog directly back in the sand to get from one house to the other. See the illustration. The time T to get from one house to the other as a function of the angle θ shown in the illustration is

$$T(\theta) = 1 + \frac{2}{3 \sin \theta} - \frac{1}{4 \tan \theta}, \qquad 0 < \theta < \frac{\pi}{2}$$

(a) Calculate the time T for $\tan \theta = 1/4$.
(b) Describe the path taken.
(c) Explain why θ must be larger than 14°.

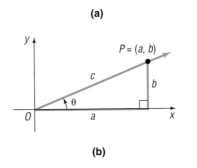

109. Show that the range of the tangent function is the set of all real numbers.

110. Show that the range of the cotangent function is the set of all real numbers.

111. Show that the period of $f(\theta) = \sin \theta$ is 2π.
[**Hint:** Assume that $0 < p < 2\pi$ exists so that $\sin(\theta + p) = \sin \theta$ for all θ. Let $\theta = 0$ to find p. Then let $\theta = \pi/2$ to obtain a contradiction.]

112. Show that the period of $f(\theta) = \cos \theta$ is 2π.

113. Show that the period of $f(\theta) = \sec \theta$ is 2π.

114. Show that the period of $f(\theta) = \csc \theta$ is 2π.

115. Show that the period of $f(\theta) = \tan \theta$ is π.

116. Show that the period of $f(\theta) = \cot \theta$ is π.

117. Prove the reciprocal identities given in formula (2).

118. Prove the quotient identities given in formula (3).

119. Establish the identity:

$$(\sin \theta \cos \phi)^2 + (\sin \theta \sin \phi)^2 + \cos^2 \theta = 1$$

120. Write down five characteristics of the tangent function. Explain the meaning of each.

121. Describe your understanding of the meaning of a periodic function.

2.4 | RIGHT TRIANGLE TRIGONOMETRY

1 Find the Value of Trigonometric Functions of Acute Angles
2 Use the Complementary Angle Theorem
3 Find the Reference Angle

FIGURE 37

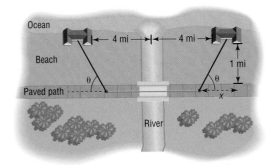

(a)

(b)

1 A triangle in which one angle is a right angle (90°) is called a **right triangle.** Recall that the side opposite the right angle is called the **hypotenuse,** and the remaining two sides are called the **legs** of the triangle. In Figure 37(a), we have labeled the hypotenuse as c, to indicate that its length is c units, and, in a like manner, we have labeled the legs as a and b. Because the triangle is a right triangle, the Pythagorean Theorem tells us that

$$a^2 + b^2 = c^2$$

Now, suppose that θ is an **acute angle;** that is, $0° < \theta < 90°$ (if θ is measured in degrees) or $0 < \theta < \pi/2$ (if θ is measured in radians). Place θ in standard position, and let $P = (a, b)$ be any point except the origin O on the terminal side of θ. Form a right triangle by dropping the perpendicular from P to the x-axis, as shown in Figure 37(b).

By referring to the lengths of the sides of the triangle by the names hypotenuse (c), opposite (b), and adjacent (a), as indicated in Figure 38, we can express the trigonometric functions of θ as ratios of the sides of a right triangle:

FIGURE 38

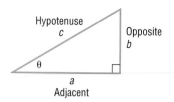

$$\sin \theta = \frac{\text{Opposite}}{\text{Hypotenuse}} = \frac{b}{c} \qquad \cos \theta = \frac{\text{Adjacent}}{\text{Hypotenuse}} = \frac{a}{c}$$

$$\tan \theta = \frac{\text{Opposite}}{\text{Adjacent}} = \frac{b}{a} \qquad \csc \theta = \frac{\text{Hypotenuse}}{\text{Opposite}} = \frac{c}{b} \qquad (1)$$

$$\sec \theta = \frac{\text{Hypotenuse}}{\text{Adjacent}} = \frac{c}{a} \qquad \cot \theta = \frac{\text{Adjacent}}{\text{Opposite}} = \frac{a}{b}$$

Notice that each of the trigonometric functions of the acute angle θ is positive.

E X A M P L E   1

Finding the Value of Trigonometric Functions from a Right Triangle

Find the exact value of the six trigonometric functions of the angle θ in Figure 39.

FIGURE 39

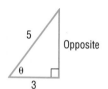

Solution   We see in Figure 39 that the two given sides of the triangle are

$$c = \text{Hypotenuse} = 5 \qquad a = \text{Adjacent} = 3$$

To find the length of the opposite side, we use the Pythagorean Theorem:

$$(\text{Adjacent})^2 + (\text{Opposite})^2 = (\text{Hypotenuse})^2$$
$$3^2 + (\text{Opposite})^2 = 5^2$$
$$(\text{Opposite})^2 = 25 - 9 = 16$$
$$\text{Opposite} = 4$$

Now that we know the lengths of the three sides, we use the ratios in equations (1) to find the value of each of the six trigonometric functions:

$$\sin \theta = \frac{\text{Opposite}}{\text{Hypotenuse}} = \frac{4}{5} \qquad \cos \theta = \frac{\text{Adjacent}}{\text{Hypotenuse}} = \frac{3}{5} \qquad \tan \theta = \frac{\text{Opposite}}{\text{Adjacent}} = \frac{4}{3}$$

$$\csc \theta = \frac{\text{Hypotenuse}}{\text{Opposite}} = \frac{5}{4} \qquad \sec \theta = \frac{\text{Hypotenuse}}{\text{Adjacent}} = \frac{5}{3} \qquad \cot \theta = \frac{\text{Adjacent}}{\text{Opposite}} = \frac{3}{4}$$

Now work Problem 1.

Thus, the values of the trigonometric functions of an acute angle are ratios of the lengths of the sides of a right triangle. This way of viewing the trigonometric functions leads to many applications and, in fact, was the point of view used by early mathematicians (before calculus) in studying the subject of trigonometry.

Complementary Angles: Cofunctions

2   Two acute angles are called **complementary** if their sum is a right angle. Because the sum of the angles of any triangle is 180°, it follows that, for a right triangle, the two acute angles are complementary.

Identifying the Mountains of the Hawaiian Islands Seen from Oahu

A wealthy tourist with a strong desire for the perfect scrapbook has called on your consulting firm for help in labeling a photograph he took on a clear day from the southeast shore of Oahu. He shows you the photograph in which you see mostly sea and sky. But on the horizon are three mountain peaks, equally spaced and apparently all of the same height. The tourist tells you that the T-shirt vendor at the beach informed him that the three mountain peaks are the volcanoes of Lanai, Maui, and the Big Island (Hawaii). The vendor even added, "You almost never see the Big Island from here."

The tourist thinks his photograph may have some special value. He has some misgivings, however. He is wondering if, in fact, it is ever really possible to see the Big Island from Oahu. He wants accurate labels on his photos, so he asks you for help.

You realize that some trigonometry is needed, as well as a good atlas and encyclopedia. After doing some research, you discover the following facts:

The heights of the peaks vary. Lanaihale, the tallest peak on Lanai, is only about 3370 feet above sea level. Maui's Haleakala is 10,023 feet above sea level. Mauna Kea on the Big Island is the highest of all, 13,796 feet above sea level. (If measured from its base on the ocean floor, it would qualify as the tallest mountain on Earth.)

Measuring from the southeast corner of Oahu, the distance to Lanai is about 65 miles, to Maui it is about 110 miles, and to the Big Island it is about 190 miles. These distances, measured along the surface of Earth, represent the length of arc that originates at sea level on Oahu and terminates in an imaginary location directly below the peak of each mountain at what would be sea level.

1. These volcanic peaks are all different heights. If your wealthy client took a photo of three mountain peaks that all look the same height, how could they possibly represent these three volcanoes?

2. The radius of Earth is approximately 3960 miles. Based on this figure, what is the circumference of Earth? How is the distance between islands related to the circumference of Earth?

3. To determine which of these mountain peaks would actually be visible from Oahu, you need to consider that the tourist standing on the shore and looking "straight out" would have a line of sight tangent to the surface of Earth at that point. Make a rough sketch of the right triangle formed by the tourist's line of sight, the radius from the center of Earth to the tourist, and a line from the center of Earth that passes through Mauna Kea. Can you determine the angle formed at the center of Earth?

4. What would the length of the hypotenuse of the triangle be? Can you tell from that whether Mauna Kea would be visible from Oahu?

5. Repeat this procedure with the information about Maui and Lanai. Will they be visible from Oahu?

6. There is another island off Oahu that the T-shirt vendor did not mention. Molokai is about 40 miles from Oahu, and its highest peak, Kamakou, is 4961 feet above sea level. Would Kamakou be visible from Oahu?

7. Your team should now be prepared to name the three volcanic peaks in the photograph. How do you decide which one is which?

FIGURE 40

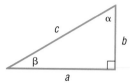

Refer now to Figure 40; we have labeled the angle opposite side b as β and the angle opposite side a as α. Notice that side b is adjacent to angle α and side a is adjacent to angle β. As a result,

$$\sin \beta = \frac{b}{c} = \cos \alpha \qquad \cos \beta = \frac{a}{c} = \sin \alpha \qquad \tan \beta = \frac{b}{a} = \cot \alpha$$

$$\csc \beta = \frac{c}{b} = \sec \alpha \qquad \sec \beta = \frac{c}{a} = \csc \alpha \qquad \cot \beta = \frac{a}{b} = \tan \alpha$$

(2)

Because of these relationships, the functions sine and cosine, tangent and cotangent, and secant and cosecant are called **cofunctions** of each other. The identities (2) may be expressed in words as follows:

Complementary Angle Theorem

Cofunctions of complementary angles are equal.

Examples of this theorem are given next:

Complementary angles   Complementary angles   Complementary angles

$$\sin 30° = \cos 60° \qquad \tan 40° = \cot 50° \qquad \sec 80° = \csc 10°$$

Cofunctions   Cofunctions   Cofunctions

If θ is an acute angle measured in degrees, the angle $90° - \theta$ (or $\pi/2 - \theta$, if θ is in radians) is the angle complementary to θ. Table 6 restates the preceding theorem on cofunctions.

TABLE 6	
θ **(Degrees)**	θ **(Radians)**
$\sin \theta = \cos(90° - \theta)$	$\sin \theta = \cos(\pi/2 - \theta)$
$\cos \theta = \sin(90° - \theta)$	$\cos \theta = \sin(\pi/2 - \theta)$
$\tan \theta = \cot(90° - \theta)$	$\tan \theta = \cot(\pi/2 - \theta)$
$\csc \theta = \sec(90° - \theta)$	$\csc \theta = \sec(\pi/2 - \theta)$
$\sec \theta = \csc(90° - \theta)$	$\sec \theta = \csc(\pi/2 - \theta)$
$\cot \theta = \tan(90° - \theta)$	$\cot \theta = \tan(\pi/2 - \theta)$

Although the angle θ in Table 6 is acute, we will see later that these results are valid for any angle θ.

Seeing the Concept   Graph $Y_1 = \sin x$ and $Y_2 = \cos(90° - x)$. Be sure that the mode is set to degrees.

E X A M P L E   2   Using the Complementary Angle Theorem

(a) $\sin 62° = \cos(90° - 62°) = \cos 28°$

(b) $\tan \dfrac{\pi}{12} = \cot\left(\dfrac{\pi}{2} - \dfrac{\pi}{12}\right) = \cot \dfrac{5\pi}{12}$

(c) $\cos \dfrac{\pi}{4} = \sin\left(\dfrac{\pi}{2} - \dfrac{\pi}{4}\right) = \sin \dfrac{\pi}{4}$

(d) $\csc \dfrac{\pi}{6} = \sec\left(\dfrac{\pi}{2} - \dfrac{\pi}{6}\right) = \sec \dfrac{\pi}{3}$

 Now work Problem 57(a).

Reference Angle

Next, we concentrate on angles that lie in a quadrant. Once we know in which quadrant an angle lies, we know the sign of each value of the trigonometric functions of that angle. The use of a certain reference angle may help us evaluate the trigonometric functions of such an angle.

> Let θ denote a nonacute angle that lies in a quadrant. The acute angle formed by the terminal side of θ and either the positive x-axis or the negative x-axis is called the **reference angle** for θ.

Figure 41 illustrates the reference angle for some general angles θ. Note that a reference angle is always an acute angle, that is, an angle whose measure is between 0° and 90°.

Although formulas can be given for calculating reference angles, usually it is easier to find the reference angle for a given angle by making a quick sketch of the angle.

FIGURE 41

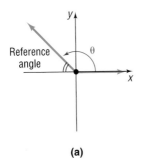 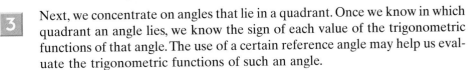 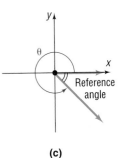 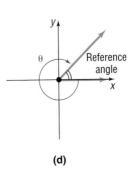

(a)   (b)   (c)   (d)

E X A M P L E   3   Finding Reference Angles

Find the reference angle for each of the following angles:

(a) 150°   (b) −45°   (c) $9\pi/4$ (d) $-5\pi/6$

Solution (a) Refer to Figure 42. The reference angle for 150° is 30°.

(b) Refer to Figure 43. The reference angle for −45° is 45°.

FIGURE 42

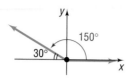

FIGURE 43

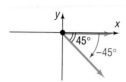

(c) Refer to Figure 44. The reference angle for $9\pi/4$ is $\pi/4$.

(d) Refer to Figure 45. The reference angle for $-5\pi/6$ is $\pi/6$.

FIGURE 44

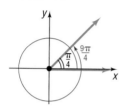

FIGURE 45

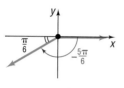

Now work **Problem 11.**

The advantage of using reference angles is that, except for the correct sign, the values of the trigonometric functions of a general angle θ equal the values of the trigonometric functions of its reference angle.

Theorem   Reference Angles

If θ is an angle that lies in a quadrant and if α is its reference angle, then

$$\sin \theta = \pm \sin \alpha \qquad \cos \theta = \pm \cos \alpha \qquad \tan \theta = \pm \tan \alpha$$
$$\csc \theta = \pm \csc \alpha \qquad \sec \theta = \pm \sec \alpha \qquad \cot \theta = \pm \cot \alpha \tag{3}$$

where the + or − sign depends on the quadrant in which θ lies.

FIGURE 46

$\sin \theta = b/c$, $\sin \alpha = b/c$;
$\cos \theta = a/c$, $\cos \alpha = |a|/c$

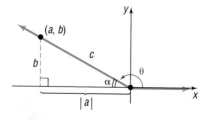

For example, suppose that θ lies in quadrant II and α is its reference angle. See Figure 46. If (a, b) is a point on the terminal side of θ and if $c = \sqrt{a^2 + b^2}$, we have

$$\sin \theta = \frac{b}{c} = \sin \alpha \qquad \cos \theta = \frac{a}{c} = \frac{-|a|}{c} = -\cos \alpha$$

and so on.

The next example illustrates how the theorem on reference angles is used.

E X A M P L E   4

Using Reference Angles to Find the Value of Trigonometric Functions

Find the exact value of each of the following trigonometric functions using reference angles:

(a) $\sin 135°$　　(b) $\cos 240°$　　(c) $\cos \dfrac{5\pi}{6}$　　(d) $\tan\left(-\dfrac{\pi}{3}\right)$

Solution

(a) Refer to Figure 47. The angle 135° is in quadrant II, where the sine function is positive. The reference angle for 135° is 45°. Thus,

$$\sin 135° = \sin 45° = \frac{\sqrt{2}}{2}$$

FIGURE 47

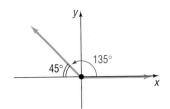

(b) Refer to Figure 48. The angle 240° is in quadrant III, where the cosine function is negative. The reference angle for 240° is 60°. Thus,

$$\cos 240° = -\cos 60° = -\tfrac{1}{2}$$

FIGURE 48

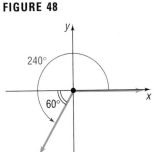

(c) Refer to Figure 49. The angle $5\pi/6$ is in quadrant II, where the cosine function is negative. The reference angle for $5\pi/6$ is $\pi/6$. Thus,

$$\cos \frac{5\pi}{6} = -\cos \frac{\pi}{6} = -\frac{\sqrt{3}}{2}$$

FIGURE 49

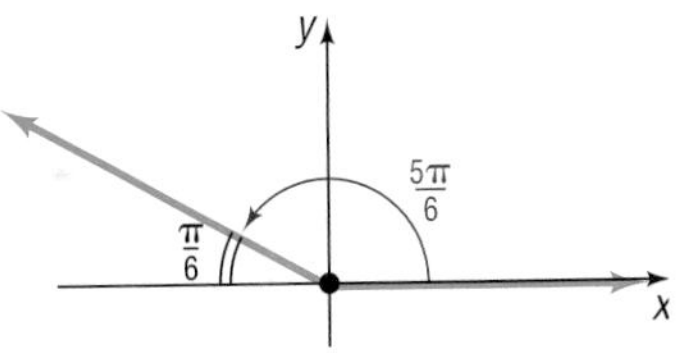

(d) Refer to Figure 50. The angle $-\pi/3$ is in quadrant IV, where the tangent function is negative. The reference angle for $-\pi/3$ is $\pi/3$. Thus,

$$\tan\left(-\frac{\pi}{3}\right) = -\tan \frac{\pi}{3} = -\sqrt{3}$$

FIGURE 50

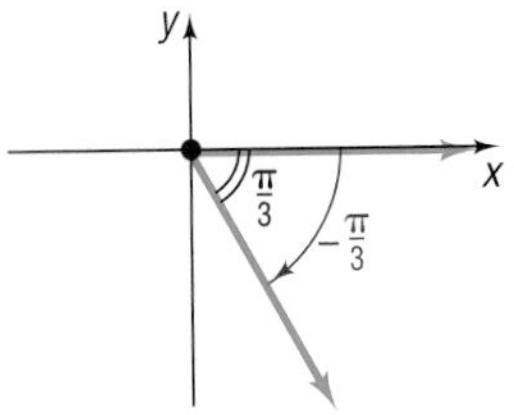

Now work Problems 27 and 45.

E X A M P L E   5

Using Reference Angles to Find Values When One Value Is Known

Given that $\cos \theta = -\tfrac{2}{3}$, $\pi/2 < \theta < \pi$, find the exact value of each of the remaining trigonometric functions.

Solution

Using Reference Angles

We have already discussed two ways of solving this type of problem: using the definition of the trigonometric functions and using identities. (Refer to Example 4 in Section 2.3). Here we present a third method: using reference angles.

The angle θ lies in quadrant II, so we know that $\sin \theta$ and $\csc \theta$ are positive, whereas the other trigonometric functions are negative. If α is the reference angle for θ, then $\cos \alpha = \frac{2}{3}$. The values of the remaining trigonometric functions of the angle α can be found by drawing the appropriate triangle. We use Figure 51 to obtain

FIGURE 51

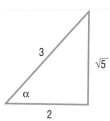

$$\sin \alpha = \frac{\sqrt{5}}{3} \qquad \cos \alpha = \frac{2}{3} \qquad \tan \alpha = \frac{\sqrt{5}}{2}$$

$$\csc \alpha = \frac{3}{\sqrt{5}} = \frac{3\sqrt{5}}{5} \qquad \sec \alpha = \frac{3}{2} \qquad \cot \alpha = \frac{2}{\sqrt{5}} = \frac{2\sqrt{5}}{5}$$

Now, we assign the appropriate sign to each of these values to find the values of the trigonometric functions of θ.

$$\sin \theta = \frac{\sqrt{5}}{3} \qquad \cos \theta = -\frac{2}{3} \qquad \tan \theta = -\frac{\sqrt{5}}{2}$$

$$\csc \theta = \frac{3\sqrt{5}}{5} \qquad \sec \theta = -\frac{3}{2} \qquad \cot \theta = -\frac{2\sqrt{5}}{5}$$

EXAMPLE 6

Constructing a Rain Gutter

FIGURE 52

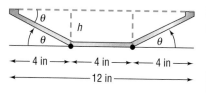

A rain gutter is to be constructed of aluminum sheets 12 inches wide. After marking off a length of 4 inches from each edge, this length is bent up at an angle θ. See Figure 52.

(a) Express the area A of the opening as a function of θ.

(b) Graph $A = A(\theta)$. Find the angle θ that makes A largest. (This bend will allow the most water to flow through the gutter.)

Solution

(a) The area A of the opening is the sum of the areas of two right triangles and one rectangle. Each triangle has legs h and $\sqrt{16 - h^2}$ and hypotenuse 4. The rectangle has length 4 and height h. Thus,

$$A = 2 \cdot \frac{1}{2} \cdot h\sqrt{16 - h^2} + 4h = (4 \sin \theta)(4 \cos \theta) + 4(4 \sin \theta)$$

$$\underset{\sin \theta = \frac{h}{4}, \, \cos \theta = \frac{\sqrt{16 - h^2}}{4}}{\uparrow}$$

$$A(\theta) = 16 \sin \theta(\cos \theta + 1)$$

(b) See Figure 53. The angle θ that makes A largest is $60°$.

FIGURE 53

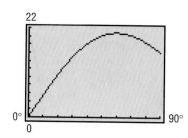

2.4 | EXERCISES

In Problems 1–10, find the exact value of the six trigonometric functions of the angle θ in each figure.

1.
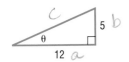

2.

3.

4.

5.
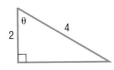

6.

7.

8.

9.
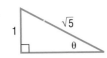

10.

In Problems 11–26, find the reference angle of each angle.

11. $-30°$ **12.** $60°$ **13.** $120°$ **14.** $300°$ **15.** $210°$

16. $330°$ **17.** $5\pi/4$ **18.** $5\pi/6$ **19.** $8\pi/3$ **20.** $7\pi/4$

21. $-135°$ **22.** $-240°$ **23.** $-2\pi/3$ **24.** $-7\pi/6$ **25.** $420°$

26. $480°$

In Problems 27–56, find the exact value of each expression. Do not use a calculator.

27. $\sin 150°$ **28.** $\cos 210°$ **29.** $\cos 315°$ **30.** $\sin 120°$

31. $\sec 240°$ **32.** $\csc 300°$ **33.** $\cot 330°$ **34.** $\tan 225°$

35. $\sin \dfrac{3\pi}{4}$ **36.** $\cos \dfrac{2\pi}{3}$ **37.** $\cot \dfrac{7\pi}{6}$ **38.** $\csc \dfrac{7\pi}{4}$

39. $\cos(-60°)$ **40.** $\tan(-120°)$ **41.** $\sin\left(-\dfrac{2\pi}{3}\right)$ **42.** $\cot\left(-\dfrac{\pi}{6}\right)$

43. $\tan \dfrac{14\pi}{3}$ **44.** $\sec \dfrac{11\pi}{4}$ **45.** $\csc(-315°)$ **46.** $\sec(-225°)$

47. $\sin 38° - \cos 52°$ **48.** $\tan 12° - \cot 78°$ **49.** $\dfrac{\cos 10°}{\sin 80°}$ **50.** $\dfrac{\cos 40°}{\sin 50°}$

51. $1 - \cos^2 20° - \cos^2 70°$ **52.** $1 + \tan^2 5° - \csc^2 85°$ **53.** $\tan 20° - \dfrac{\cos 70°}{\cos 20°}$

54. $\cot 40° - \dfrac{\sin 50°}{\sin 40°}$ **55.** $\cos 35° \sin 55° + \sin 35° \cos 55°$ **56.** $\sec 35° \csc 55° - \tan 35° \cot 55°$

57. If $\sin \theta = \frac{1}{3}$, find the exact value of: (a) $\cos(90° - \theta)$ (b) $\cos^2 \theta$ (c) $\csc \theta$ (d) $\sec\left(\frac{\pi}{2} - \theta\right)$

58. If $\sin \theta = 0.2$, find the exact value of: (a) $\cos\left(\frac{\pi}{2} - \theta\right)$ (b) $\cos^2 \theta$ (c) $\sec(90° - \theta)$ (d) $\csc \theta$

59. If $\tan \theta = 4$, find the exact value of: (a) $\sec^2 \theta$ (b) $\cot \theta$ (c) $\cot\left(\frac{\pi}{2} - \theta\right)$ (d) $\csc^2 \theta$

60. If $\sec \theta = 3$, find the exact value of: (a) $\cos \theta$ (b) $\tan^2 \theta$ (c) $\csc(90° - \theta)$ (d) $\sin^2 \theta$

61. If $\csc \theta = 4$, find the exact value of: (a) $\sin \theta$ (b) $\cot^2 \theta$ (c) $\sec(90° - \theta)$ (d) $\sec^2 \theta$

62. If $\cot \theta = 2$, find the exact value of: (a) $\tan \theta$ (b) $\csc^2 \theta$ (c) $\tan\left(\frac{\pi}{2} - \theta\right)$ (d) $\sec^2 \theta$

63. If $\sin \theta = 0.3$, find the exact value of $\sin \theta + \cos\left(\frac{\pi}{2} - \theta\right)$.

64. If $\tan \theta = 4$, find the exact value of $\tan \theta + \tan\left(\frac{\pi}{2} - \theta\right)$.

65. Find the exact value of $\sin 1° + \sin 2° + \sin 3° + \cdots + \sin 358° + \sin 359°$.

66. Find the exact value of $\cos 1° + \cos 2° + \cos 3° + \cdots + \cos 358° + \cos 359°$.

67. Find the acute angle θ that satisfies the equation $\sin \theta = \cos(2\theta + 30°)$.

68. Find the acute angle θ that satisfies the equation $\tan \theta = \cot(\theta + 45°)$.

69. **Calculating the Time of a Trip** Two oceanfront homes are located 8 miles apart on a straight stretch of beach, each a distance of 1 mile from a paved road that parallels the ocean. Sally can jog 8 miles per hour along the paved road, but only 3 miles per hour in the sand on the beach. Because of a river directly between the two houses, it is necessary to jog in the sand to the road, continue on the road, and then jog directly back in the sand to get from one house to the other. See the illustration.
    (a) Express the time T to get from one house to the other as a function of the angle θ shown in the illustration.
     (b) Graph $T = T(\theta)$. What angle θ results in the least time? What is the least time? How long is Sally on the paved road?

70. **Designing Fine Decorative Pieces** A designer of decorative art plans to market solid gold spheres enclosed in clear crystal cones. Each sphere is of fixed radius R and will be enclosed in a cone of height h and radius r. See the illustration. Many cones can be used to enclose the sphere, each having a different slant angle θ.
    (a) Express the volume V of the cone as a function of the slant angle θ of the cone.
      [**Hint:** The volume V of a cone of height h and radius r is $V = \frac{1}{3}\pi r^2 h$.]
    (b) What slant angle θ should be used for the volume V of the cone to be a minimum? (This choice minimizes the amount of crystal required and gives maximum emphasis to the gold sphere).

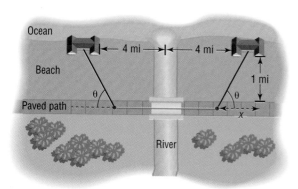

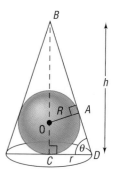

71. **Calculating the Time of a Trip** From a parking lot, you want to walk to a house on the

ocean. The house is located 1500 feet down a paved path that parallels the ocean, which is 500 feet away. See the illustration. Along the path you can walk 300 feet per minute, but in the sand on the beach you can only walk 100 feet per minute.

(a) Calculate the time T if you walk 1500 feet along the paved path and then 500 feet in the sand to the house.

(b) Calculate the time T if you walk in the sand first for 500 feet and then walk along the beach for 1500 feet to the house.

(c) Express the time T to get from the parking lot to the beachhouse as a function of the angle θ shown in the illustration.

(d) Calculate the time T if you walk 1000 feet along the paved path and then walk directly to the house.

 (e) Graph $T = T(\theta)$. For what angle θ is T least? What is the least time? What is x for this angle?

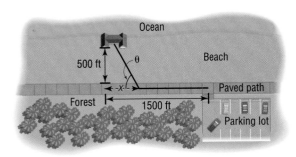

72. **Carrying a Ladder around a Corner** A ladder of length L is carried horizontally around a corner from a hall 3 feet wide into a hall 4 feet wide. See the illustration. Find the length L of the ladder as a function of the angle θ shown in the illustration.

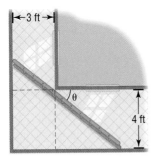

73. Give three examples that will show a fellow student how to use the Theorem on Reference Angles. Give them to the fellow student and ask for a critique.

74. Refer to Example 4, Section 2.3, and Example 5, Section 2.4. Which of the three methods of solution do you prefer? Which is your least favorite? Are there situations in which one method is sometimes best and others in which another method is best? Give reasons.

75. Use a calculator in radian mode to complete the following table. What can you conclude about the ratio $(\sin \theta)/\theta$ as θ approaches 0?

θ	0.5	0.4	0.2	0.1	0.01	0.001	0.0001	0.00001
$\sin \theta$								
$\dfrac{\sin \theta}{\theta}$								

76. Use a calculator in radian mode to complete the following table. What can you conclude about the ratio $(\cos \theta - 1)/\theta$ as θ approaches 0?

θ	0.5	0.4	0.2	0.1	0.01	0.001	0.0001	0.00001
$\cos \theta - 1$								
$\dfrac{\cos \theta - 1}{\theta}$								

77. Suppose that the angle θ is a central angle of a circle of radius 1 (see the figure). Show that

(a) Angle $OAC = \dfrac{\theta}{2}$

(b) $|CD| = \sin\theta$ and $|OD| = \cos\theta$

(c) $\tan\dfrac{\theta}{2} = \dfrac{\sin\theta}{1 + \cos\theta}$

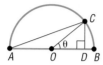

78. Show that the area of an isosceles triangle is $A = a^2 \sin\theta\cos\theta$, where a is the length of one of the two equal sides and θ is the measure of one of the two equal angles (see the figure).

79. Let $n > 0$ be any real number, and let θ be any angle for which $0 < \theta < \pi/(1 + n)$. Then we can construct a triangle with the angles θ and $n\theta$ and included side of length 1 (do you see why?) and place it on the unit circle as illustrated. Now, drop the perpendicular from C to $D = (x, 0)$ and show that

$$x = \frac{\tan n\theta}{\tan\theta + \tan n\theta}$$

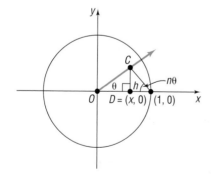

80. Refer to the given figure. The smaller circle, whose radius is a, is tangent to the larger circle, whose radius is b. The ray OA contains a diameter of each circle, and the ray OB is tangent to each circle. Show that

$$\cos\theta = \frac{\sqrt{ab}}{\dfrac{a + b}{2}}$$

(That is, $\cos\theta$ equals the ratio of the geometric mean of a and b to the arithmetic mean of a and b.)

[**Hint:** First show that $\sin\theta = (b - a)/(b + a)$.]

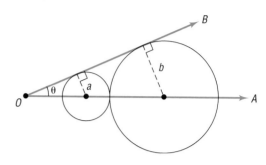

81. Refer to the figure. If $|OA| = 1$, show that

(a) Area $\triangle OAC = \frac{1}{2}\sin\alpha\cos\alpha$

(b) Area $\triangle OCB = \frac{1}{2}|OB|^2 \sin\beta\cos\beta$

(c) Area $\triangle OAB = \frac{1}{2}|OB|\sin(\alpha + \beta)$

(d) $|OB| = \dfrac{\cos\alpha}{\cos\beta}$

(e) $\sin(\alpha + \beta) = \sin\alpha\cos\beta + \cos\alpha\sin\beta$

[**Hint:** Area $\triangle OAB =$ area $\triangle OAC +$ area $\triangle OCB$.]

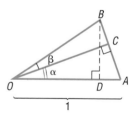

82. Refer to the figure on page 153, where a unit circle is drawn. The line DB is tangent to the circle.

(a) Express the area of $\triangle OBC$ in terms of $\sin\theta$ and $\cos\theta$.

(b) Express the area of $\triangle OBD$ in terms of $\sin\theta$ and $\cos\theta$.

(c) The area of the sector $\overset{\frown}{OBC}$ of the circle is $\frac{1}{2}\theta$, where θ is measured in radians. Use the results of parts (a) and (b) and the fact that

Area $\triangle OBC <$ area $\overset{\frown}{OBC} <$ area $\triangle OBD$

to show that

$$1 < \frac{\theta}{\sin\theta} < \frac{1}{\cos\theta}$$

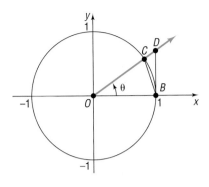

83. If $\cos \alpha = \tan \beta$ and $\cos \beta = \tan \alpha$, where α and β are acute angles, show that

$$\sin \alpha = \sin \beta = \sqrt{\dfrac{3 - \sqrt{5}}{2}}$$

 84. If θ is an acute angle, explain why $\sec \theta > 1$.

85. If θ is an acute angle, explain why $0 < \sin \theta < 1$.

2.5 | GRAPHS OF THE TRIGONOMETRIC FUNCTIONS

1 Graph Transformations of the Sine Function
2 Graph Transformations of the Cosine Function
3 Graph Transformations of the Tangent Function
4 Graph Transformations of the Cosecant, Secant, and Cotangent Functions

We have discussed properties of the trigonometric functions $f(\theta) = \sin \theta$, $f(\theta) = \cos \theta$, and so on. In this section, we shall use the traditional symbols x to represent the independent variable (or argument) and y for the dependent variable (or value at x) for each function. Thus, we write the six trigonometric functions as

$y = \sin x$	$y = \cos x$	$y = \tan x$
$y = \csc x$	$y = \sec x$	$y = \cot x$

Our purpose in this section is to graph each of these functions. Unless indicated otherwise, we shall use radian measure throughout for the independent variable x.

The Graph of $y = \sin x$

Since the sine function has period 2π, we need to graph $y = \sin x$ only on the interval $[0, 2\pi]$. The remainder of the graph will consist of repetitions of this portion of the graph.

We begin by constructing Table 7, which lists some points on the graph of $y = \sin x$, $0 \le x \le 2\pi$. As the table shows, the graph of $y = \sin x$, $0 \le x \le 2\pi$, begins at the origin. As x increases from 0 to $\pi/2$, the value of $y = \sin x$ increases from 0 to 1; as x increases from $\pi/2$ to π to $3\pi/2$, the value of y decreases from 1 to 0 to -1; as x increases from $3\pi/2$ to 2π, the value of y

TABLE 7

x	$y = \sin x$	(x, y)
0	0	$(0, 0)$
$\pi/6$	$\frac{1}{2}$	$(\pi/6, \frac{1}{2})$
$\pi/2$	1	$(\pi/2, 1)$
$5\pi/6$	$\frac{1}{2}$	$(5\pi/6, \frac{1}{2})$
π	0	$(\pi, 0)$
$7\pi/6$	$-\frac{1}{2}$	$(7\pi/6, -\frac{1}{2})$
$3\pi/2$	-1	$(3\pi/2, -1)$
$11\pi/6$	$-\frac{1}{2}$	$(11\pi/6, -\frac{1}{2})$
2π	0	$(2\pi, 0)$

increases from -1 to 0. If we plot the points listed in Table 7 and connect them with a smooth curve, we obtain the graph shown in Figure 54.

FIGURE 54
$y = \sin x,$
$0 \leq x \leq 2\pi$

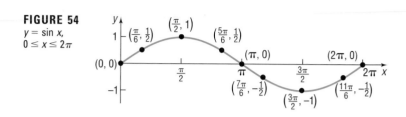

The graph in Figure 54 is one period of the graph of $y = \sin x$. To obtain a complete graph of $y = \sin x$, we repeat this period in each direction, as shown in Figure 55.

FIGURE 55
$y = \sin x,$
$-\infty < x < \infty$

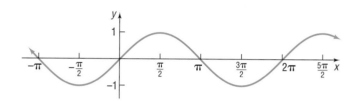

The graph of $y = \sin x$ illustrates some of the facts that we already know about the sine function.

Characteristics of the Sine Function

1. The domain is the set of all real numbers.
2. The range consists of all real numbers from -1 to 1, inclusive.
3. The sine function is an odd function, as the symmetry of the graph with respect to the origin indicates.
4. The sine function is periodic, with period 2π.
5. The x-intercepts are $\ldots, -2\pi, -\pi, 0, \pi, 2\pi, 3\pi, \ldots$; the y-intercept is 0.
6. The maximum value is 1 and occurs at $x = \ldots, -3\pi/2, \pi/2, 5\pi/2, 9\pi/2, \ldots$; the minimum value is -1 and occurs at $x = \ldots, -\pi/2, 3\pi/2, 7\pi/2, 11\pi/2, \ldots$.

Now work Problems 1, 3, and 5.

The graphing techniques introduced in Chapter 1 may be used to graph functions that are variations of the sine function.

E X A M P L E   1   Graphing Variations of $y = \sin x$ Using Transformations

Use the graph of $y = \sin x$ to graph $y = -\sin x + 2$.

Solution   Figure 56 illustrates the steps.

FIGURE 56

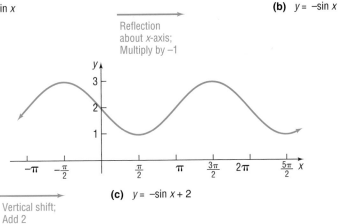

(a) $y = \sin x$

Reflection
about x-axis;
Multiply by -1

(b) $y = -\sin x$

(c) $y = -\sin x + 2$

Vertical shift;
Add 2

Check: Graph $Y_1 = -\sin x + 2$ and compare the result with Figure 56(c).

E X A M P L E   2   Graphing Variations of $y = \sin x$ Using Transformations

Use the graph of $y = \sin x$ to graph $y = \sin\left(x - \dfrac{\pi}{4}\right)$.

Solution   Figure 57 illustrates the steps.

FIGURE 57

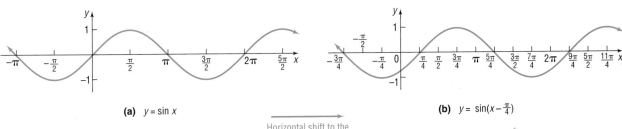

(a) $y = \sin x$

Horizontal shift to the
right $\frac{\pi}{4}$ units;
Replace x by $x - \frac{\pi}{4}$

(b) $y = \sin\left(x - \frac{\pi}{4}\right)$

Check: Graph $Y_1 = \sin\left(x - \dfrac{\pi}{4}\right)$ and compare the result with Figure 57(b).

Now work Problem 35.

The Graph of $y = \cos x$

The cosine function also has period 2π. Thus, we proceed as we did with the sine function by constructing Table 8, which lists some points on the graph of $y = \cos x, 0 \leq x \leq 2\pi$. As the table shows, the graph of $y = \cos x, 0 \leq x \leq 2\pi$, begins at the point $(0, 1)$. As x increases from 0 to $\pi/2$ to π, the value of y decreases from 1 to 0 to -1; as x increases from π to $3\pi/2$ to 2π, the value of y increases from -1 to 0 to 1. As before, we plot the points in Table 8 to get one period of the graph of $y = \cos x$. See Figure 58.

TABLE 8		
x	$y = \cos x$	(x, y)
0	1	$(0, 1)$
$\pi/3$	$\frac{1}{2}$	$(\pi/3, \frac{1}{2})$
$\pi/2$	0	$(\pi/2, 0)$
$2\pi/3$	$-\frac{1}{2}$	$(2\pi/3, -\frac{1}{2})$
π	-1	$(\pi, -1)$
$4\pi/3$	$-\frac{1}{2}$	$(4\pi/3, -\frac{1}{2})$
$3\pi/2$	0	$(3\pi/2, 0)$
$5\pi/3$	$\frac{1}{2}$	$(5\pi/3, \frac{1}{2})$
2π	1	$(2\pi, 1)$

FIGURE 58
$y = \cos x,$
$0 \leq x \leq 2\pi$

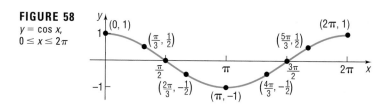

A complete graph of $y = \cos x$ is obtained by repeating this period in each direction, as shown in Figure 59.

FIGURE 59
$y = \cos x, -\infty < x < \infty$

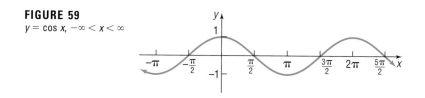

The graph of $y = \cos x$ illustrates some of the facts that we already know about the cosine function.

Characteristics
of the Cosine Function

1. The domain is the set of all real numbers.
2. The range consists of all real numbers from -1 to 1, inclusive.
3. The cosine function is an even function, as the symmetry of the graph with respect to the y-axis indicates.
4. The cosine function is periodic, with period 2π.
5. The x-intercepts are $\dots, -3\pi/2, -\pi/2, \pi/2, 3\pi/2, 5\pi/2, \dots$; the y-intercept is 1.
6. The maximum value is 1 and occurs at $x = \dots, -2\pi, 0, 2\pi, 4\pi, 6\pi, \dots$; the minimum value is -1 and occurs at $x = \dots, -\pi, \pi, 3\pi, 5\pi, \dots$.

Again, the graphing techniques from Chapter 1 may be used to graph variations of the cosine function.

E X A M P L E   3   Graphing Variations of $y = \cos x$ Using Transformations

Use the graph of $y = \cos x$ to graph $y = 2 \cos x$

Solution   Figure 60 illustrates the graph, which is a vertical stretch of the graph of $y = \cos x$.

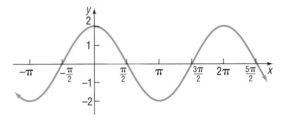

FIGURE 60
$y = 2 \cos x$

Check: Graph $Y_1 = 2 \cos x$ and compare the result with Figure 60.

E X A M P L E   4   Graphing Variations of $y = \cos x$ Using Transformations

Use the graph of $y = \cos x$ to graph $y = \cos 3x$.

Solution   Figure 61 illustrates the graph, which is a horizontal compression by a factor of $\frac{1}{3}$ of the graph of $y = \cos x$. Notice that, due to this compression, the period of $y = \cos 3x$ is $2\pi/3$, whereas the period of $y = \cos x$ is 2π. We shall comment more about this in the next section.

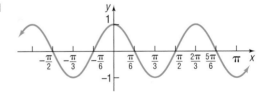

FIGURE 61
$y = \cos 3x$

Check: Graph $Y_1 = \cos 3x$. Use TRACE to verify that the period is $2\pi/3$.

Now work Problem 43.

The Graph of $y = \tan x$

Because the tangent function has period π, we only need to determine the graph over some interval of length π. The rest of the graph will consist of repetitions of that graph. Because the tangent function is not defined at . . . , $-3\pi/2, -\pi/2, \pi/2, 3\pi/2, \ldots$, we will concentrate on the interval $(-\pi/2, \pi/2)$, of length π, and construct Table 9, which lists some points on the graph of $y = \tan x, -\pi/2 < x < \pi/2$. We plot the points in the table and connect them with a smooth curve. See Figure 62 for a partial graph of $y = \tan x$, where $-\pi/3 \leq x \leq \pi/3$.

TABLE 9

x	y = tan x	(x, y)
$-\pi/3$	$-\sqrt{3} \approx -1.73$	$(-\pi/3, -\sqrt{3})$
$-\pi/4$	-1	$(-\pi/4, -1)$
$-\pi/6$	$-\sqrt{3}/3 \approx -0.58$	$(-\pi/6, -\sqrt{3}/3)$
0	0	$(0, 0)$
$\pi/6$	$\sqrt{3}/3 \approx 0.58$	$(\pi/6, \sqrt{3}/3)$
$\pi/4$	1	$(\pi/4, 1)$
$\pi/3$	$\sqrt{3} \approx 1.73$	$(\pi/3, \sqrt{3})$

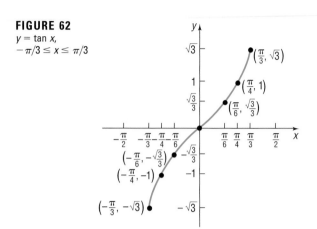

FIGURE 62
$y = \tan x$,
$-\pi/3 \leq x \leq \pi/3$

To complete one period of the graph of $y = \tan x$, we need to investigate the behavior of the function as x approaches $-\pi/2$ and $\pi/2$. We must be careful, though, because $y = \tan x$ is not defined at these numbers. To determine this behavior, we use the identity

$$\tan x = \frac{\sin x}{\cos x}$$

See Table 10. If x is close to $\pi/2 \approx 1.5708$, but remains less than $\pi/2$, then $\sin x$ will be close to 1 and $\cos x$ will be positive and close to 0. (Refer back to the graphs of the sine function and the cosine function.) Hence, the ratio $(\sin x)/(\cos x)$ will be positive and large. In fact, the closer x gets to $\pi/2$, the closer $\sin x$ gets to 1 and $\cos x$ gets to 0, so $\tan x$ approaches ∞. In other words, the vertical line $x = \pi/2$ is a vertical asymptote to the graph of $y = \tan x$.

TABLE 10

x	sin x	cos x	y = tan x
$\frac{\pi}{3} \approx 1.05$	$\frac{\sqrt{3}}{2}$	$\frac{1}{2}$	$\sqrt{3} \approx 1.73$
1.5	0.9975	0.0707	14.1
1.57	0.9999	$7.96E^{-4}$	1255.8
1.5707	0.9999	$9.6E^{-5}$	10381
$\frac{\pi}{2} \approx 1.5708$	1	0	Undefined

If x is close to $-\pi/2$, but remains greater than $-\pi/2$, then $\sin x$ will be close to -1 and $\cos x$ will be positive and close to 0. Hence, the ratio $(\sin x)/(\cos x)$ approaches $-\infty$. In other words, the vertical line $x = -\pi/2$ is also a vertical asymptote to the graph.

With these observations, we can complete one period of the graph. We obtain the complete graph of $y = \tan x$ by repeating this period, as shown in Figure 63.

FIGURE 63
$y = \tan x, \ -\infty < x < \infty,$
x not equal to odd multiples of $\pi/2$

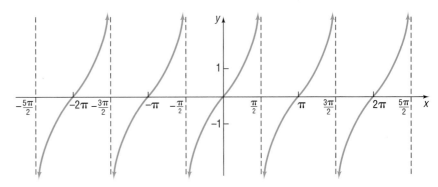

 Check: Graph $y = \tan x$ and compare the result with Figure 63. Use TRACE to see what happens as x gets close to $\pi/2$, but is less than $\pi/2$. Be sure to set the WINDOW accordingly.

The graph of $y = \tan x$ illustrates some of the facts that we already know about the tangent function.

Characteristics of the Tangent Function

1. The domain is the set of all real numbers, except odd multiples of $\pi/2$.
2. The range consists of all real numbers.
3. The tangent function is an odd function, as the symmetry of the graph with respect to the origin indicates.
4. The tangent function is periodic, with period π.
5. The x-intercepts are $\ldots, -2\pi, -\pi, 0, \pi, 2\pi, 3\pi, \ldots$; the y-intercept is 0.
6. Vertical asymptotes occur at $x = \ldots, -3\pi/2, -\pi/2, \pi/2, 3\pi/2, \ldots$.

 Now work Problems 11 and 19.

EXAMPLE 5   Graphing Variations of $y = \tan x$ Using Transformations

 Graph: $y = \tan\left(x + \dfrac{\pi}{4}\right)$

Solution   We start with the graph of $y = \tan x$ and shift it horizontally to the left $\pi/4$ unit to obtain the graph of $y = \tan(x + \pi/4)$. See Figure 64.

FIGURE 64
$y = \tan\left(x + \dfrac{\pi}{4}\right)$

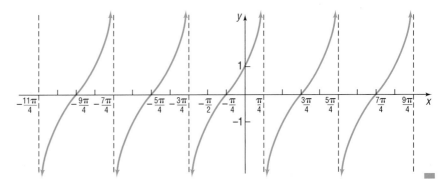

 Now work Problem 51.

The Graphs of $y = \csc x$, $y = \sec x$, and $y = \cot x$

The cosecant and secant functions, sometimes referred to as **reciprocal functions,** are graphed by making use of the reciprocal identities

$$\csc x = \frac{1}{\sin x} \quad \text{and} \quad \sec x = \frac{1}{\cos x}$$

For example, the value of the cosecant function $y = \csc x$ at a given number x equals the reciprocal of the corresponding value of the sine function, provided the value of the sine function is not 0. If the value of $\sin x$ is 0, then, at such numbers x, the cosecant function is not defined. In fact, the graph of the cosecant function has vertical asymptotes at integral multiples of π. Figure 65 shows the graph.

FIGURE 65
$y = \csc x$, $-\infty < x < \infty$, x not equal to integral multiples of π, $|y| \geq 1$

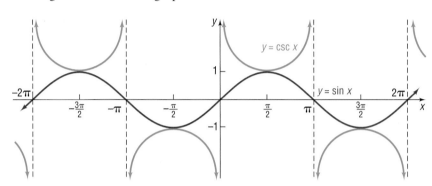

Check: Graph $Y_1 = \csc x$ and compare the result with Figure 65. Use TRACE to see what happens when x is close to 0.

E X A M P L E   6

Graphing Variations of $y = \csc x$ Using Transformations

Graph: $y = 2 \csc(x - \pi/2)$, $-\pi \leq x \leq \pi$

Solution   Figure 66 shows the required steps.

FIGURE 66

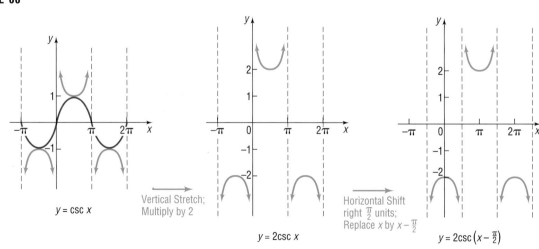

Check: Graph $Y_1 = 2 \csc (x - \pi/2)$ and compare the result with Figure 66.

Using the idea of reciprocals, we can similarly obtain the graph of $y = \sec x$. See Figure 67.

FIGURE 67
$y = \sec x$, $-\infty < x < \infty$, x not equal to odd multiples of $\pi/2$, $|y| \geq 1$

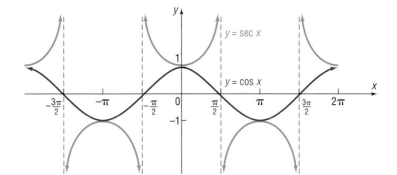

TABLE 11

x	$y = \cot x$	(x, y)
$\pi/6$	$\sqrt{3}$	$(\pi/6, \sqrt{3})$
$\pi/4$	1	$(\pi/4, 1)$
$\pi/3$	$\sqrt{3}/3$	$(\pi/3, \sqrt{3}/3)$
$\pi/2$	0	$(\pi/2, 0)$
$2\pi/3$	$-\sqrt{3}/3$	$(2\pi/3, -\sqrt{3}/3)$
$3\pi/4$	-1	$(3\pi/4, -1)$
$5\pi/6$	$-\sqrt{3}$	$(5\pi/6, -\sqrt{3})$

We obtain the graph of $y = \cot x$ as we did the graph of $y = \tan x$. The period of $y = \cot x$ is π. Because the cotangent function is not defined for integral multiples of π, we will concentrate on the interval $(0, \pi)$. Table 11 lists some points on the graph of $y = \cot x$, $0 < x < \pi$. As x approaches 0, but remains greater than 0, the value of $\cos x$ will be close to 1 and the value of $\sin x$ will be positive and close to 0. Hence, the ratio $(\cos x)/(\sin x) = \cot x$ will be positive and large, so as x approaches 0, $\cot x$ approaches ∞. Similarly, as x approaches π, but remains less than π, the value of $\cos x$ will be close to -1 and the value of $\sin x$ will be positive and close to 0. Hence, the ratio $(\cos x)/(\sin x) = \cot x$ will be negative and will approach $-\infty$ as x approaches π. Figure 68 shows the graph.

FIGURE 68
$y = \cot x$, $-\infty < x < \infty$, x not equal to integral multiples of π, $-\infty < y < \infty$

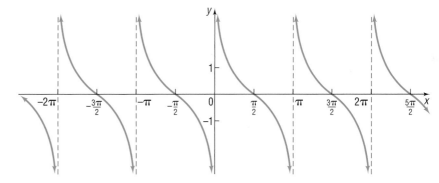

Check: Graph $Y_1 = \cot x$ and compute the result with Figure 68. Use TRACE to see what happens when x is close to 0. ∎

2.5 | EXERCISES

In Problems 1–20, refer to the graphs to answer each question, if necessary.

1. What is the y-intercept of $y = \sin x$?

2. What is the y-intercept of $y = \cos x$?

3. For what numbers x, $-\pi \leq x \leq \pi$ is the graph of $y = \sin x$ increasing?

4. For what numbers x, $-\pi \leq x \leq \pi$, is the graph of $y = \cos x$ decreasing?

5. What is the largest value of $y = \sin x$?

6. What is the smallest value of $y = \cos x$?

7. For what numbers x, $0 \le x \le 2\pi$, does $\sin x = 0$?

8. For what numbers x, $0 \le x \le 2\pi$, does $\cos x = 0$?

9. For what numbers x, $-2\pi \le x \le 2\pi$, does $\sin x = 1$? What about $\sin x = -1$?

10. For what numbers x, $-2\pi \le x \le 2\pi$, does $\cos x = 1$? What about $\cos x = -1$?

11. What is the y-intercept of $y = \tan x$?

12. What is the y-intercept of $y = \cot x$?

13. What is the y-intercept of $y = \sec x$?

14. What is the y-intercept of $y = \csc x$?

15. For what numbers x, $-2\pi \le x \le 2\pi$, does $\sec x = 1$? What about $\sec x = -1$?

16. For what numbers x, $-2\pi \le x \le 2\pi$, does $\csc x = 1$? What about $\csc x = -1$?

17. For what numbers x, $-2\pi \le x \le 2\pi$, does the graph of $y = \sec x$ have vertical asymptotes?

18. For what numbers x, $-2\pi \le x \le 2\pi$, does the graph of $y = \csc x$ have vertical asymptotes?

19. For what numbers x, $-2\pi \le x \le 2\pi$, does the graph of $y = \tan x$ have vertical asymptotes?

20. For what numbers x, $-2\pi \le x \le 2\pi$, does the graph of $y = \cot x$ have vertical asymptotes?

In Problems 21 and 22, match the graph to a function. Three answers are possible.

A. $y = -\sin x$

B. $y = -\cos x$

C. $y = \sin\left(x - \dfrac{\pi}{2}\right)$

D. $y = -\cos\left(x - \dfrac{\pi}{2}\right)$

E. $y = \sin(x + \pi)$

F. $y = \cos(x + \pi)$

21.

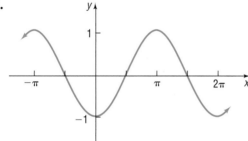

22.

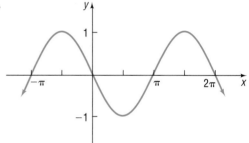

In Problems 23–26, match each function to its graph.

A. $y = \sin 2x$ B. $y = \sin 4x$ C. $y = \cos 2x$ D. $y = \cos 4x$

23.

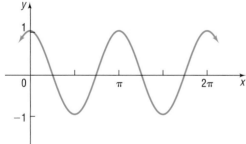

24.

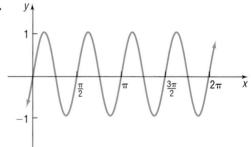

25.

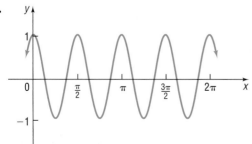

26.

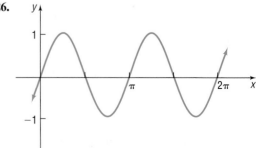

In Problems 27–30, match each graph to a function.

A. $y = -\tan x$ B. $y = \tan\left(x + \dfrac{\pi}{2}\right)$ C. $y = \tan(x + \pi)$ D. $y = -\tan\left(x - \dfrac{\pi}{2}\right)$

27.

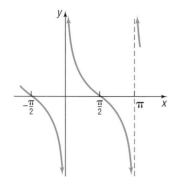

28.

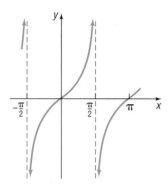

29.

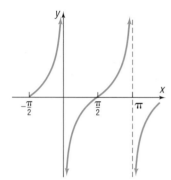

30.

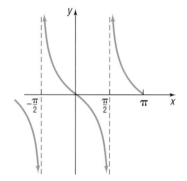

In Problems 31–62, use transformations to graph each function.

31. $y = 3 \sin x$

32. $y = 4 \cos x$

33. $y = \cos\left(x + \dfrac{\pi}{4}\right)$

34. $y = \sin(x - \pi)$

35. $y = \sin x - 1$

36. $y = \cos x + 1$

37. $y = -2 \sin x$

38. $y = -3 \cos x$

39. $y = \sin \pi x$

40. $y = \cos\dfrac{\pi}{2}x$

41. $y = 2 \sin x + 2$

42. $y = 3 \cos x + 3$

43. $y = -2 \cos\left(x - \dfrac{\pi}{2}\right)$

44. $y = -3 \sin\left(x + \dfrac{\pi}{2}\right)$

45. $y = 3 \sin (\pi - x)$

46. $y = 2 \cos (\pi - x)$

47. $y = -\sec x$

48. $y = -\cot x$

49. $y = \sec\left(x - \dfrac{\pi}{2}\right)$

50. $y = \csc(x - \pi)$

51. $y = \tan(x - \pi)$

52. $y = \cot(x - \pi)$

53. $y = 3 \tan 2x$

54. $y = 4 \tan \frac{1}{2}x$

55. $y = \sec 2x$

56. $y = \csc \frac{1}{2}x$

57. $y = \cot \pi x$

58. $y = \cot 2x$

59. $y = -3 \tan 4x$

60. $y = -3 \tan 2x$

61. $y = 2 \sec \frac{1}{2}x$

62. $y = 2 \sec 3x$

63. On the same coordinate axes, graph $y = 2 \sin x$ and $y = \sin 2x, 0 \le x \le 2\pi$. Compare each graph's maximum and minimum value. Compare each graph's period.

64. Repeat Problem 63 for $y = 2 \cos x$ and $y = \cos 2x, 0 \le x \le 2\pi$.

65. Repeat Problem 63 for $y = 4 \sin x$ and $y = \sin 4x, 0 \le x \le 2\pi$.

66. Repeat Problem 63 for $y = 4 \cos x$ and $y = \cos 4x, 0 \le x \le 2\pi$.

67. Graph: $y = |\sin x|, 0 \le x \le 2\pi$.

68. Graph: $y = |\cos x|, 0 \le x \le 2\pi$.

69. Carrying a Ladder around a Corner A ladder of length L is carried horizontally around a corner from a hall 3 feet wide and into a hall 4 feet wide. See the illustration.

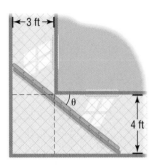

(a) Show that the length L of the ladder as a function of the angle θ is

$$L = 3 \sec \theta + 4 \csc \theta$$

 (b) Graph $L, 0 < \theta < \dfrac{\pi}{2}$.

 (c) Where is L the least?

(d) What is the length of the largest ladder that can be carried around the corner? Why is this also the least value of L?

70. Graph $y = \sin x$, $y = 2 \sin x$, $y = \frac{1}{2} \sin x$, and $y = 8 \sin x$. What do you conclude about the graph of $y = A \sin x$, $A > 0$?

71. Graph $y = \sin x$, $y = \sin 2x$, $y = \sin 4x$, and $y = \sin \frac{1}{2}x$. What do you conclude about the graph of $y = \sin \omega x$?

72. Graph $y = \sin x$, $y = \sin[x - (\pi/3)]$, $y = \sin[x - (\pi/4)]$, and $y = \sin[x - (\pi/6)]$. What do you conclude about the graph of $y = \sin(x - \phi)$, $\phi > 0$?

73. Graph:

$$y = \sin x \quad \text{and} \quad y = \cos\left(x - \frac{\pi}{2}\right)$$

Do you think that $\sin x = \cos\left(x - \dfrac{\pi}{2}\right)$?

74. Graph:

$$y = \tan x \quad \text{and} \quad y = -\cot\left(x + \frac{\pi}{2}\right)$$

Do you think that $\tan x = -\cot\left(x + \dfrac{\pi}{2}\right)$?

2.6 SINUSOIDAL GRAPHS; SINUSOIDAL CURVE FITTING

1 Determine the Amplitude and Period of Sinusoidal Functions
2 Find an Equation for a Sinusoidal Graph
3 Determine the Phase Shift of a Sinusoidal Function
4 Graph Sinusoidal Functions
5 Find a Sinusoidal Function from Data

The graph of $y = \cos x$, when compared to the graph of $y = \sin x$, suggests that the graph of $y = \sin x$ is the same as the graph of $y = \cos x$ after a horizontal shift of $\pi/2$ units to the right. See Figure 69.

FIGURE 69

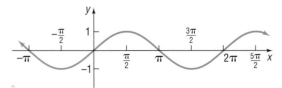

(a) $y = \sin x$

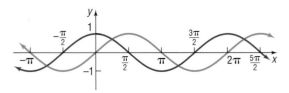

(b) $y = \cos x$ $\quad y = \cos\left(x - \frac{\pi}{2}\right)$

Based on Figure 69, we conjecture that

$$\sin x = \cos\left(x - \frac{\pi}{2}\right)$$

(We shall prove this fact in Chapter 3.) Because of this similarity, the graphs of sine functions and cosine functions are referred to as **sinusoidal graphs.**

Seeing the Concept   Graph $Y_1 = \sin x$ and $Y_2 = \cos\left(x - \frac{\pi}{2}\right)$. How many graphs do you see?

Let's look at some general characteristics of sinusoidal graphs.

In Example 3 of Section 2.5, we obtained the graph of $y = 2 \cos x$, which we reproduce in Figure 70. Notice that the values of $y = 2 \cos x$ lie between -2 and 2, inclusive.

FIGURE 70
$y = 2 \cos x$

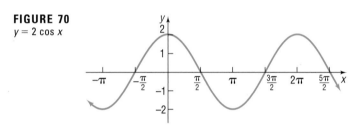

In general, the values of the functions $y = A \sin x$ and $y = A \cos x$, where $A \neq 0$, will always satisfy the inequalities.

$$-|A| \leq A \sin x \leq |A| \quad \text{and} \quad -|A| \leq A \cos x \leq |A|$$

respectively. The number $|A|$ is called the **amplitude** of $y = A \sin x$ or $y = A \cos x$. See Figure 71.

FIGURE 71

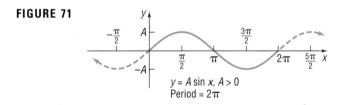

$y = A \sin x, A > 0$
Period $= 2\pi$

In Example 4 of Section 2.5, we obtained the graph of $y = \cos 3x$, which we reproduce in Figure 72. Notice that the period of this function is $2\pi/3$.

FIGURE 72
$y = \cos 3x$

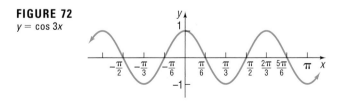

In general, if $\omega > 0$, the functions $y = \sin \omega x$ and $y = \cos \omega x$ will have period $T = 2\pi/\omega$. To see why, recall that the graph of $y = \sin \omega x$ is obtained from the graph of $y = \sin x$ by performing a horizontal compression or stretch by a factor of $1/\omega$. This horizontal compression or stretch replaces the interval $[0, 2\pi]$, which contains one period of the graph of $y = \sin x$, by the interval $[0, 2\pi/\omega]$, which contains one period of the graph of $y = \sin \omega x$. Thus, the period of the functions $y = \sin \omega x$ and $y = \cos \omega x$, $\omega > 0$, is $2\pi/\omega$.

For example, for the function $y = \cos 3x$, graphed in Figure 72, $\omega = 3$, so the period is $2\pi/3$.

Figure 73 illustrates the general situation.

FIGURE 73

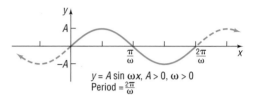

$y = A \sin \omega x, A > 0, \omega > 0$
Period $= \frac{2\pi}{\omega}$

If $\omega < 0$ in $y = \sin \omega x$ or $y = \cos \omega x$, we use the Even–Odd Properties of the sine and cosine functions as follows:

$$\sin \omega x = -\sin(-\omega x) \quad \text{and} \quad \cos \omega x = \cos(-\omega x)$$

This gives us an equivalent form in which the coefficient of x is positive.

Theorem

If $\omega > 0$, the amplitude and period of $y = A \sin \omega x$ and $y = A \cos \omega x$ are given by

$$\text{Amplitude} = |A| \qquad \text{Period} = T = \frac{2\pi}{\omega} \qquad (1)$$

E X A M P L E   1   Finding the Amplitude and Period of a Sinusoidal Function

Determine the amplitude and period of $y = 3 \sin 4x$, and graph the function.

Solution   Comparing $y = 3 \sin 4x$ to $y = A \sin \omega x$, we find that $A = 3$ and $\omega = 4$. Thus, from equation (1),

$$\text{Amplitude} = |A| = 3 \qquad \text{Period} = T = 2\pi/\omega = 2\pi/4 = \pi/2$$

We use this information to graph $y = 3 \sin 4x$ by beginning as shown in Figure 74(a). Notice that the amplitude is used to scale the y-axis, and the period is used to scale the x-axis. (This will usually result in a different scale for each axis.) Now we fill in the graph of the sine function. See Figure 74(b).

FIGURE 74

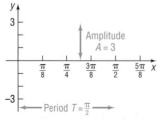

Amplitude
$A = 3$

Period $T = \frac{\pi}{2}$

(a)

(b) $y = 3 \sin 4x$

Check: Graph $Y_1 = 3 \sin 4x$ and compare the result with Figure 74(b). ▬

Now work Problem 5.

E X A M P L E   2   Finding the Amplitude and Period of a Sinusoidal Function

Determine the amplitude and period of $y = -4 \cos \pi x$, and graph the function.

Solution   Comparing $y = -4 \cos \pi x$ with $y = A \cos \omega x$, we find that $A = -4$ and $\omega = \pi$. Thus, the amplitude is $|A| = |-4| = 4$, and the period is $T = 2\pi/\omega = 2\pi/\pi = 2$.

We use the amplitude to scale the y-axis and the period to scale the x-axis and fill in the graph of the cosine function, thus obtaining the graph of $y = 4 \cos \pi x$ shown in Figure 75(a). Now, since we want the graph of $y = -4 \cos \pi x$, we reflect the graph of $y = 4 \cos \pi x$ about the x-axis, as shown in Figure 75(b).

FIGURE 75

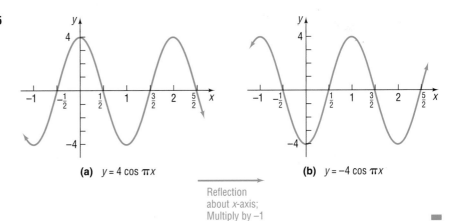

(a) $y = 4 \cos \pi x$

(b) $y = -4 \cos \pi x$

Reflection
about x-axis;
Multiply by -1

▬

Check: Graph $Y_1 = -4 \cos \pi x$ and compare the result with Figure 75(b).

▬

E X A M P L E   3   Finding the Amplitude and Period of a Sinusoidal Function

Determine the amplitude and period of $y = 2 \sin(-\pi x)$, and graph the function.

Solution   Since the sine function is odd, we use the equivalent form

$$y = -2 \sin \pi x$$

Comparing $y = -2 \sin \pi x$ to $y = A \sin \omega x$, we find that $A = -2$ and $\omega = \pi$. Thus, the amplitude is $|A| = 2$, and the period is $T = 2\pi/\omega = 2\pi/\pi = 2$.

To graph $y = -2 \sin \pi x$, we use the amplitude to scale the y-axis, and the period to scale the x-axis and fill in the graph of the sine function. Figure 76(a) shows the resulting graph of $y = 2 \sin \pi x$. To obtain the graph of

$y = -2 \sin \pi x$, we reflect the graph of $y = 2 \sin \pi x$ about the x-axis, as shown in Figure 76(b).

FIGURE 76

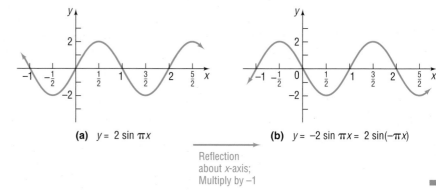

(a) $y = 2 \sin \pi x$ **(b)** $y = -2 \sin \pi x = 2 \sin(-\pi x)$

Reflection
about x-axis;
Multiply by –1

Check: Graph $Y_1 = 2 \sin(-\pi x)$ and compare the result with Figure 76(b).

Now work Problem 29.

We also can use the ideas of amplitude and period to identify a sinusoidal function when its graph is given.

E X A M P L E   4   Finding an Equation for a Sinusoidal Graph

Find an equation for the graph shown in Figure 77.

FIGURE 77

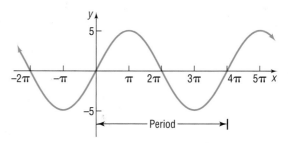

Solution   This graph can be viewed as the graph of a sine function* with amplitude $A = 5$. The period T is observed to be 4π. Thus, by equation (1),

$$T = \frac{2\pi}{\omega}$$

$$4\pi = \frac{2\pi}{\omega}$$

$$\omega = \frac{2\pi}{4\pi} = \frac{1}{2}$$

A sine function whose graph is given in Figure 77 is

$$y = A \sin \omega x = 5 \sin \frac{x}{2}$$

Check: Graph $Y_1 = 5 \sin \frac{x}{2}$ and compare the result with Figure 77.

*The equation could also be viewed as a cosine function with a horizontal shift, but viewing it as a sine function is easier.

E X A M P L E   5   Finding an Equation for a Sinusoidal Graph

Find an equation for the graph shown in Figure 78.

FIGURE 78

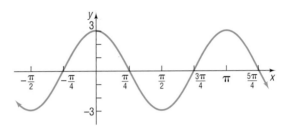

Solution   From the graph we conclude that it is easier to view the equation as a cosine function with amplitude $A = 3$ and period $T = \pi$. Thus, $2\pi/\omega = \pi$, so $\omega = 2$. A cosine function whose graph is given in Figure 78 is

$$y = A \cos \omega x = 3 \cos 2x$$

 Check: Graph $Y_1 = 3 \cos 2x$ and compare the result with Figure 78.

E X A M P L E   6   Finding an Equation for a Sinusoidal Graph

FIGURE 79

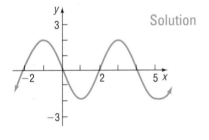

Find an equation for the graph shown in Figure 79.

Solution   The graph is sinusoidal, with amplitude $A = 2$. The period is 4, so $2\pi/\omega = 4$ or $\omega = \pi/2$. Since the graph passes through the origin, it is easiest to view the equation as a sine function, but notice that the graph is actually the reflection of a sine function about the x-axis (since the graph is decreasing near the origin). Thus, we have

$$y = -A \sin \omega x = -2 \sin \frac{\pi}{2} x$$

 Check: Graph $Y_1 = -2 \sin \frac{\pi}{2} x$ and compare the result with Figure 79.

 Now work Problem 37.

Phase Shift

We have seen that the graph of $y = A \sin \omega x$, $\omega > 0$, has amplitude $|A|$ and period $T = 2\pi/\omega$. Thus, one period can be drawn as x varies from 0 to $2\pi/\omega$ or, equivalently, as ωx varies from 0 to 2π. See Figure 80.

We now want to discuss the graph of

FIGURE 80
One period of $y = A \sin \omega x$, $A > 0$, $\omega > 0$

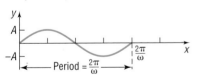

$$y = A \sin(\omega x - \phi) = A \sin \omega \left(x - \frac{\phi}{\omega} \right)$$

where $\omega > 0$ and ϕ (the Greek letter phi) are real numbers. The graph will be a sine curve of amplitude $|A|$. As $\omega x - \phi$ varies from 0 to 2π, one period will be traced out. This period will begin with

$$\omega x - \phi = 0 \quad \text{or} \quad x = \frac{\phi}{\omega}$$

and will end when

$$\omega x - \phi = 2\pi \quad \text{or} \quad x = \frac{2\pi}{\omega} + \frac{\phi}{\omega}$$

FIGURE 81
One period of $y = A \sin(\omega x - \phi)$, $A > 0$, $\omega > 0$, $\phi > 0$

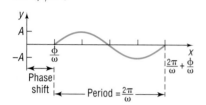

See Figure 81.

Thus, we see that the graph of $y = A \sin(\omega x - \phi) = A \sin \omega\left(x - \dfrac{\phi}{\omega}\right)$ is the same as the graph of $y = A \sin \omega x$, except that it has been shifted ϕ/ω units (to the right if $\phi > 0$ and to the left if $\phi < 0$). This number ϕ/ω is called the **phase shift** of the graph of $y = A \sin(\omega x - \phi)$.

For the graphs of $y = A \sin(\omega x - \phi)$ or $y = A \cos(\omega x - \phi)$, $\omega > 0$,

$$\text{Amplitude} = |A| \qquad \text{Period} = T = \frac{2\pi}{\omega} \qquad \text{Phase shift} = \frac{\phi}{\omega}$$

The phase shift is to the left if $\phi < 0$ and to the right if $\phi > 0$.

E X A M P L E   7   Finding the Amplitude, Period, and Phase Shift of a Sinusoidal Function

Find the amplitude, period, and phase shift of $y = 3 \sin(2x - \pi)$, and graph the function.

Solution   Comparing $y = 3 \sin(2x - \pi) = 3 \sin 2(x - \pi/2)$ to $y = A \sin(\omega x - \phi) = A \sin \omega\left(x - \dfrac{\phi}{\omega}\right)$, we find that $A = 3$, $\omega = 2$, and $\phi = \pi$. The graph is a sine curve with amplitude $A = 3$, period $T = 2\pi/\omega = 2\pi/2 = \pi$, and phase shift $= \dfrac{\phi}{\omega} = \dfrac{\pi}{2}$.

The graph of $y = 3 \sin(2x - \pi) = 3 \sin 2\left(x - \dfrac{\pi}{2}\right)$ is obtained by using transformations. See Figure 82.

FIGURE 82

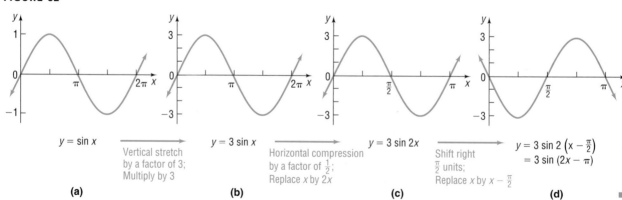

Check: Graph $y = 3 \sin(2x - \pi)$ and compare the result with Figure 82(d).

E X A M P L E   8   Finding the Amplitude, Period, and Phase Shift
                         of a Sinusoidal Function

Find the amplitude, period, and phase shift of $y = 2 \cos(4x + 3\pi)$, and graph
the function.

Solution   Comparing $y = 2 \cos(4x + 3\pi) = 2 \cos 4\left(x + \dfrac{3\pi}{4}\right)$ to $y = A \cos(\omega x - \phi) =$

$A \cos \omega\left(x - \dfrac{\phi}{\omega}\right)$, we see that $A = 2$, $\omega = 4$, and $\phi = -3\pi$. The graph is a
cosine curve with amplitude $A = 2$, period $T = 2\pi/\omega = 2\pi/4 = \pi/2$, and

phase   shift $= \dfrac{\phi}{\omega} = \dfrac{-3\pi}{4}$. The stages to graph $y = 2 \cos(4x + 3\pi) =$

$2 \cos 4\left(x + \dfrac{3\pi}{4}\right)$ are shown in Figure 83.

FIGURE 83

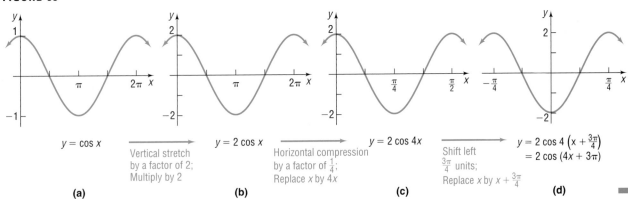

(a) $y = \cos x$

Vertical stretch by a factor of 2; Multiply by 2

(b) $y = 2 \cos x$

Horizontal compression by a factor of $\frac{1}{4}$; Replace x by $4x$

(c) $y = 2 \cos 4x$

Shift left $\frac{3\pi}{4}$ units; Replace x by $x + \frac{3\pi}{4}$

(d) $y = 2 \cos 4\left(x + \frac{3\pi}{4}\right)$ $= 2 \cos(4x + 3\pi)$

Check: Graph $y = 2 \cos(4x + 3\pi)$ and compare the result with Figure 83(d).

Now work Problem 51.

Finding Sinusoidal Functions from Data

Scatter diagrams of data sometimes take the form of a sinusoidal function.
Let's look at an example.

The data given in Table 11 represent the average monthly temperatures
in Denver, Colorado. Since the data represent *average* monthly temperatures
collected over many years, the data will not vary much from year to year and
so will essentially repeat each year. In other words, the data are periodic.
Figure 84 shows the scatter diagram of these data repeated over two years,
where $x = 1$ represents January, $x = 2$ represents February, and so on.

TABLE 11

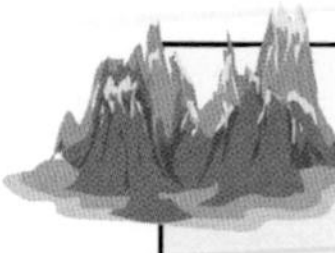

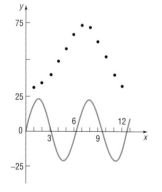

Month, x	Average Monthly Temperature, °F
January, 1	29.7
February, 2	33.4
March, 3	39.0
April, 4	48.2
May, 5	57.2
June, 6	66.9
July, 7	73.5
August, 8	71.4
September, 9	62.3
October, 10	51.4
November, 11	39.0
December, 12	31.0

Source: U.S. National Oceanic and Atmospheric Administration

FIGURE 84

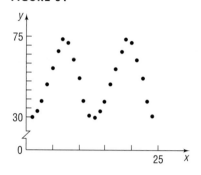

Notice that the scatter diagram looks like the graph of a sinusoidal function. We choose to fit the data to a sine function of the form

$$y = A\sin(\omega x - \phi) + B$$

where A, B, ω, and ϕ are constants.

E X A M P L E   9   Finding a Sinusoidal Function from Temperature Data

Fit a sine function to the data in Table 11.

FIGURE 85

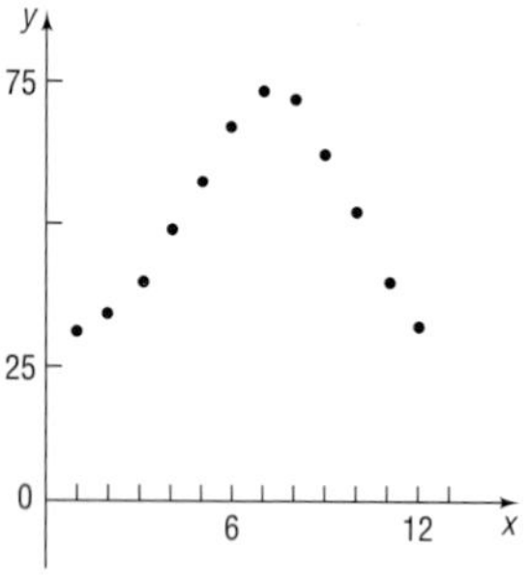

Solution   We begin with a scatter diagram of the data for one year. See Figure 85. The data will be fitted to a sine function of the form

$$y = A\sin(\omega x - \phi) + B$$

STEP 1: To find the amplitude A, we compute

Amplitude = (largest data value − smallest data value)/2
$$= (73.5 - 29.7)/2 = 21.9$$

To see the remaining steps in this process, we superimpose the graph of the function $y = 21.9\sin x$, where x represents months, on the scatter diagram. Figure 86 shows the two graphs.

To fit the data, the graph needs to be shifted vertically, shifted horizontally, and stretched horizontally.

FIGURE 86

STEP 2: We determine the vertical shift by finding the average of the highest and lowest data value.

vertical shift = $(73.5 + 29.7)/2 = 51.6$

Now we superimpose the graph of $y = 21.9\sin x + 51.6$ on the scatter diagram. See Figure 87.

FIGURE 87

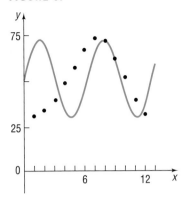

FIGURE 88

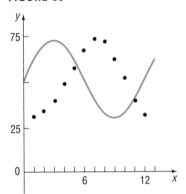

FIGURE 89

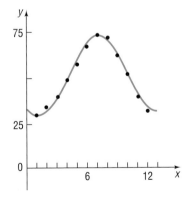

We see that the graph needs to be shifted horizontally and stretched horizontally.

STEP 3: It is easier to find the horizontal stretch factor first. Since the temperatures repeat every 12 months, the period of the function is $T = 12$. Since $T = \dfrac{2\pi}{\omega} = 12$,

$$\omega = \frac{2\pi}{12} = \frac{\pi}{6}$$

Now we superimpose the graph of $y = 21.9 \sin \dfrac{\pi x}{6} + 51.6$ on the scatter diagram. See Figure 88.

We see that the graph still needs to be shifted horizontally.

STEP 4: To determine the horizontal shift, we solve the equation

$$y = 21.9 \sin\left(\frac{\pi}{6} x - \phi\right) + 51.6$$

for ϕ by letting $y = 29.7$ and $x = 1$ (the average temperature in Denver in January).*

$$29.7 = 21.9 \sin\left(\frac{\pi}{6} \cdot 1 - \phi\right) + 51.6$$

$$-21.9 = 21.9 \sin\left(\frac{\pi}{6} - \phi\right) \qquad \text{Subtract 51.6 from both sides of the equation.}$$

$$-1 = \sin\left(\frac{\pi}{6} - \phi\right) \qquad \text{Divide both sides of the equation by 21.9.}$$

$$\frac{\pi}{6} - \phi = -\frac{\pi}{2} \qquad \text{sin } \theta = -1 \text{ when } \theta = -\pi/2.$$

$$\phi = \frac{2\pi}{3} \qquad \text{Solve for } \phi.$$

Thus,

$$y = 21.9 \sin\left(\frac{\pi}{6}x - \frac{2\pi}{3}\right) + 51.6$$

The graph of $y = 21.9 \sin (\pi/6\, x - 2\pi/3) + 51.6$, and the scatter diagram of the data is shown in Figure 89. ▬

The steps to fit a sine function

$$y = A \sin(\omega x - \phi) + B$$

to sinusoidal data follow:

Steps for Fitting Data to a Sine Function
$y = A \sin(\omega x - \phi) + B$

> STEP 1: Determine A, the amplitude of the function.
>
> $$\text{Amplitude} = \frac{\text{largest data value} - \text{smallest data value}}{2}$$
>
> STEP 2: Determine B, the vertical shift of the function.

*The data point selected here to find ϕ is arbitrary. Selecting a different data point will usually result in a different value for ϕ. To maintain consistency, we will always choose the data point for which y is smallest (in this case, January gives the lowest temperature).

> $$\text{Vertical shift} = \frac{\text{largest data value} + \text{smallest data value}}{2}$$
>
> STEP 3: Determine ω. Since the period T, the time it takes for the data to repeat, is $T = \dfrac{2\pi}{\omega}$, we have
>
> $$\omega = \frac{2\pi}{T}$$
>
> STEP 4: Determine the horizontal shift of the function by solving the equation
>
> $$y = A\sin(\omega x - \phi) + B$$
>
> for ϕ by choosing an ordered pair (x, y) from the data. Since answers will vary depending on the ordered pair selected, we will always choose the ordered pair for which y is smallest in order to maintain consistency.

Certain graphing utilities (such as a TI-83 and TI-86) have the capability of finding the sine function of best fit for sinusoidal data. At least four data points are required for this process.

E X A M P L E   10   Finding the Sine Function of Best Fit

Use a graphing utility to find the sine function of best fit for the data in Table 11. Graph this function with the scatter diagram of the data.

Solution   Enter the data from Table 11 and execute the SINe REGression program. The result is shown in Figure 90.

FIGURE 90

The output that the utility provides shows us the equation

$$y = a\sin(bx + c) + d.$$

The sinusoidal function of best fit is

$$y = 21.15\sin(0.55x - 2.35) + 51.19$$

where x represents the month and y represents the average temperature.

Figure 91 shows the graph of the sinusoidal function of best fit on the scatter diagram.   ▬

FIGURE 91

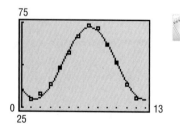

Now work Problem 79.

Since the number of hours of sunlight in a day cycles annually, the number of hours of sunlight in a day for a given location can be modeled by a sinusoidal function.

The longest day of the year (in terms of hours of sunlight) occurs on the day of the summer solstice. The summer solstice is the time when the sun is farthest north (for locations in the northern hemisphere). In 1997, the summer solstice occurs on June 21 (the 172nd day of the year) at 8:21 AM Greenwich mean time (GMT). The shortest day of the year occurs on the day of the winter solstice. The winter solstice is the time when the sun is farthest south (again, for locations in the northern hemisphere). In 1997, the winter solstice occurred on December 21 (the 355th day of the year) at 8:09 PM (GMT).

E X A M P L E   11   Finding a Sinusoidal Function for Hours of Daylight

According to the *Old Farmer's Almanac*, the number of hours of sunlight in Boston on the day of the summer solstice is 15.283 and the number of hours of sunlight on the day of the winter solstice is 9.067.

(a) Find a sinusoidal function of the form $y = A \sin(\omega x - \phi) + B$ that fits the data.*

(b) Use the function found in (a) to predict the number of hours of sunlight on April 1, 1997, the 91st day of the year.

(c) Draw a graph of the function found in (a).

(d) Look up the number of hours of sunlight on April 1, 1997, in the *Old Farmer's Almanac* and compare the actual hours of daylight to the results found in (b).

Solution   (a) STEP 1: Amplitude = (largest data value − smallest data value)/2
$$= (15.283 - 9.067)/2 = 3.108$$

STEP 2: Vertical shift = (largest data value + smallest data value)/2
$$= (15.283 + 9.067)/2 = 12.175$$

STEP 3: The data repeat every 365 days. Since $T = \dfrac{2\pi}{\omega} = 365$, we find

$$\omega = \frac{2\pi}{365}$$

STEP 4: To determine the horizontal shift, we solve the equation

$$y = 3.108 \sin\left(\frac{2\pi}{365}x - \phi\right) + 12.175$$

for ϕ by letting $y = 9.067$ and $x = 355$ (the number of hours of daylight in Boston on December 21).

$$9.067 = 3.108 \sin\left(\frac{2\pi}{365} \cdot 355 - \phi\right) + 12.175$$

$$-3.108 = 3.108 \sin\left(\frac{2\pi}{365} \cdot 355 - \phi\right) \qquad \text{Subtract 12.175 from both sides of the equation.}$$

$$-1 = \sin\left(\frac{2\pi}{365} \cdot 355 - \phi\right) \qquad \text{Divide both sides of the equation by 3.108.}$$

$$\frac{2\pi}{365} \cdot 355 - \phi = -\frac{\pi}{2} \qquad \text{sin } \theta = -1 \text{ when } \theta = -\pi/2.$$

$$\phi \approx 2.45\pi \qquad \text{Solve for } \phi.$$

Thus, the function that provides the number of hours of daylight in Boston for any day, x, is given by

$$y = 3.108 \sin\left(\frac{2\pi}{365}x - 2.45\pi\right) + 12.175$$

*Notice that only two data points are given, so a graphing utility cannot be used to find the sine function of best fit.

FIGURE 92

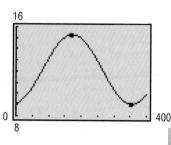

(b) To predict the number of hours of daylight on April 1, we let $x = 91$ in the function found in (a) and obtain

$$y = 3.108 \sin\left(\frac{2\pi}{365} \cdot 91 - 2.45\pi\right) + 12.175$$

$$= 3.108 \sin(-1.95\pi) + 12.175$$

$$\approx 12.69$$

So we predict that there will be about 12.69 hours of sunlight on April 1, 1997, in Boston.

(c) The graph of the function found in (a) is given in Figure 92.

(d) According to the *Old Farmer's Almanac*, there will be 12 hours 43 minutes of sunlight on April 1, 1997, in Boston. Our prediction of 12.69 hours converts to 12 hours 42 minutes.

2.6 EXERCISES

In Problems 1–10, determine the amplitude and period of each function without graphing.

1. $y = 2 \sin x$ **2.** $y = 3 \cos x$ **3.** $y = -4 \cos 2x$ **4.** $y = -\sin \frac{1}{2}x$

5. $y = 6 \sin \pi x$ **6.** $y = -3 \cos 3x$ **7.** $y = -\frac{1}{2} \cos \frac{3}{2}x$ **8.** $y = \frac{4}{3} \sin \frac{2}{3}x$

9. $y = \frac{5}{3} \sin(-\frac{2\pi}{3}x)$ **10.** $y = \frac{9}{5} \cos(-\frac{3\pi}{2}x)$

In Problems 11–20, match the given function to one of the graphs (A)–(J).

11. $y = 2 \sin \frac{\pi}{2}x$ **12.** $y = 2 \cos \frac{\pi}{2}x$ **13.** $y = 2 \cos \frac{1}{2}x$

14. $y = 3 \cos 2x$ **15.** $y = -3 \sin 2x$ **16.** $y = 2 \sin \frac{1}{2}x$

17. $y = -2 \cos \frac{1}{2}x$ **18.** $y = -2 \cos \frac{\pi}{2}x$ **19.** $y = 3 \sin 2x$

20. $y = -2 \sin \frac{1}{2}x$

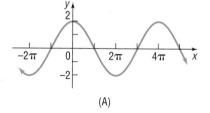

(A)

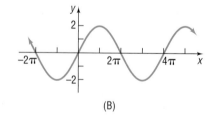

(B)

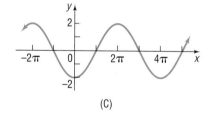

(C)

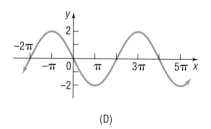

(D)

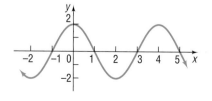

(E)

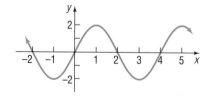

(F)

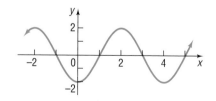

(G)

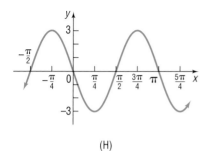

(H)

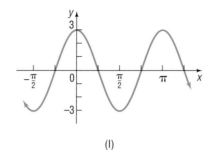

(I)

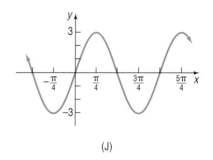

(J)

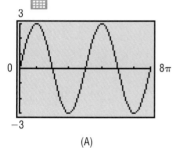

In Problems 21–26, match the given function to one of the graphs (A)–(F).

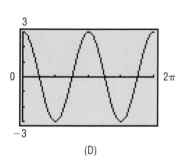

(A)

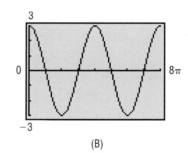

(B)

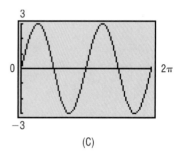

(C)

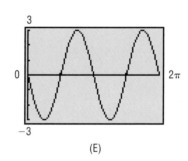

(D)

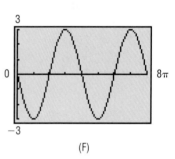

(E)

(F)

21. $y = 3 \sin \frac{1}{2}x$ **22.** $y = -3 \sin 2x$ **23.** $y = 3 \cos 2x$

24. $y = 3 \sin 2x$ **25.** $y = 3 \cos \frac{1}{2}x$ **26.** $y = -3 \sin \frac{1}{2}x$

In Problems 27–36, graph each function.

27. $y = 5 \sin 4x$ **28.** $y = 4 \cos 6x$ **29.** $y = 5 \cos \pi x$ **30.** $y = 2 \sin \pi x$

31. $y = -2 \cos 2\pi x$ **32.** $y = -5 \cos 2\pi x$ **33.** $y = -4 \sin \frac{1}{2}x$ **34.** $y = -2 \cos \frac{1}{2}x$

35. $y = \frac{3}{2} \sin(-\frac{2}{3}x)$ **36.** $y = \frac{4}{3} \cos(-\frac{1}{3}x)$

In Problems 37–50, find an equation for each graph.

37.

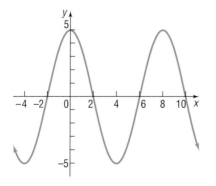

38.

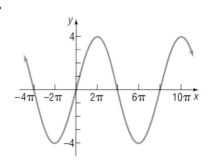

39.

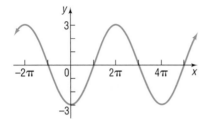

40.

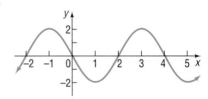

41.

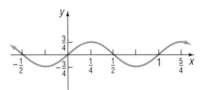

42.

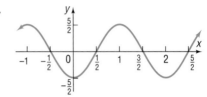

43.

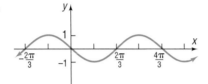

44.

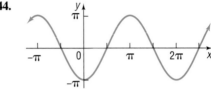

45.

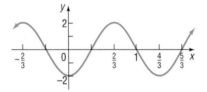

46.

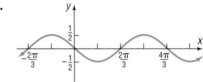

47.

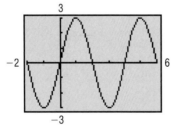

48.

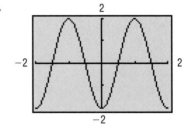

49.

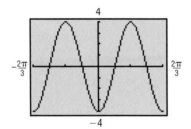

50.

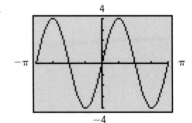

In Problems 51–62, find the amplitude, period, and phase shift of each function. Use transformations to graph the function. Show at least one period.

51. $y = 4 \sin(2x - \pi)$

52. $y = 3 \sin(3x - \pi)$

53. $y = 2 \cos\left(3x + \dfrac{\pi}{2}\right)$

54. $y = 3 \cos(2x + \pi)$

55. $y = -3 \sin\left(2x + \dfrac{\pi}{2}\right)$

56. $y = -2 \cos\left(2x - \dfrac{\pi}{2}\right)$

57. $y = 4 \sin(\pi x + 2)$ **58.** $y = 2 \cos(2\pi x + 4)$ **59.** $y = 3 \cos(\pi x - 2)$

60. $y = 2 \cos(2\pi x - 4)$ **61.** $y = 3 \sin\left(-2x + \dfrac{\pi}{2}\right)$ **62.** $y = 3 \cos\left(-2x + \dfrac{\pi}{2}\right)$

In Problems 63–70, write the equation of a sine function that has the given characteristics.

63. Amplitude: 3
Period: π

64. Amplitude: 2
Period: 4π

65. Amplitude: 3
Period: 2

66. Amplitude: 4
Period: 1

67. Amplitude: 2
Period: π
Phase shift: $\frac{1}{2}$

68. Amplitude: 3
Period: $\pi/2$
Phase shift: 2

69. Amplitude: 3
Period: 3π
Phase shift: $-\frac{1}{3}$

70. Amplitude: 2
Period: π
Phase shift: -2

71. **Alternating Current (ac) Circuits** The current I, in amperes, flowing through an ac (alternating current) circuit at time t is

$$I = 220 \sin 60\pi t, \qquad t \geq 0$$

What is the period? What is the amplitude? Graph this function over two periods.

72. **Alternating Current (ac) Circuits** The current I, in amperes, flowing through an ac (alternating current) circuit at time t is

$$I = 120 \sin 30\pi t, \qquad t \geq 0$$

What is the period? What is the amplitude? Graph this function over two periods.

73. **Alternating Current (ac) Circuits** The current I, in amperes, flowing through an ac (alternating current) circuit at time t is

$$I = 120 \sin\left(30\pi t - \dfrac{\pi}{3}\right), \qquad t \geq 0$$

What is the period? What is the amplitude? What is the phase shift? Graph this function over two periods.

74. **Alternating Current (ac) Circuits** The current I, in amperes, flowing through an ac (alternating current) circuit at time t is

$$I = 220 \sin\left(60\pi t - \dfrac{\pi}{6}\right), \qquad t \geq 0$$

What is the period? What is the amplitude? What is the phase shift? Graph this function over two periods.

75. **Alternating Current (ac) Generators** The voltage V produced by an ac generator is

$$V = 220 \sin 120\pi t$$

(a) What is the amplitude? What is the period?
(b) Graph V over two periods, beginning at $t = 0$.
(c) If a resistance of $R = 10$ ohms is present, what is the current I?
 [**Hint:** Use Ohm's Law, $V = IR$.]
(d) What is the amplitude and period of the current I?
(e) Graph I over two periods, beginning at $t = 0$.

76. **Alternating Current (ac) Generators** The voltage V produced by an ac generator is

$$V = 120 \sin 120\pi t$$

(a) What is the amplitude? What is the period?
(b) Graph V over two periods, beginning at $t = 0$.
(c) If a resistance of $R = 20$ ohms is present, what is the current I?
 [**Hint:** Use Ohm's Law, $V = IR$.]
(d) What is the amplitude and period of the current I?
(e) Graph I over two periods, beginning at $t = 0$.

77. **Alternating Current (ac) Generators** The voltage V produced by an ac generator is sinusoidal. As a function of time, the voltage V is

$$V = V_0 \sin 2\pi f t$$

where f is the **frequency,** the number of complete oscillations (cycles) per second. [In the United States and Canada, f is 60 hertz (Hz).] The **power** P delivered to a resistance R at any time t is defined as

$$P = \dfrac{V^2}{R}$$

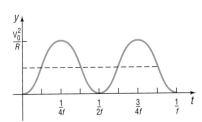

Power in an ac generator

(a) Show that $P = \dfrac{V_0^2}{R} \sin^2 2\pi f t$.
(b) The graph of P is shown in the figure. Express P as a sinusoidal function.
(c) Deduce that

$$\sin^2 2\pi f t = \dfrac{1}{2}(1 - \cos 4\pi f t)$$

78. Biorhythms In the theory of biorhythms, a sine function of the form

$$P = 50 \sin \omega t + 50$$

is used to measure the percent P of a person's potential at time t, where t is measured in days and $t = 0$ is the person's birthday. Three characteristics are commonly measured:

Physical potential: period of 23 days
Emotional potential: period of 28 days
Intellectual potential: period of 33 days

(a) Find ω for each characteristic.

(b) Graph all three functions.
(c) Is there a time t when all three characteristics have 100% potential? When is it?
(d) Suppose that you are 20 years old today ($t = 7305$ days). Describe your physical, emotional, and intellectual potential for the next 30 days.

79. Monthly Temperature The following data represent the average monthly temperatures for Juneau, Alaska.

Month, x	Average Monthly Temperature, °F
January, 1	24.2
February, 2	28.4
March, 3	32.7
April, 4	39.7
May, 5	47.0
June, 6	53.0
July, 7	56.0
August, 8	55.0
September, 9	49.4
October, 10	42.2
November, 11	32.0
December, 12	27.1

Source: U.S. National Oceanic and Atmospheric Administration.

(a) Draw a scatter diagram of the data for one period.
(b) Find a sinusoidal function of the form $y = A \sin(\omega x - \phi) + B$ that fits the data.
(c) Draw the sinusoidal function found in (b) on the scatter diagram.
(d) Use a graphing utility to find the sinusoidal function of best fit.
(e) Draw the sinusoidal function of best fit on the scatter diagram.

80. Monthly Temperature The following data represent the average monthly temperatures for Washington, D.C.

Month, x	Average Monthly Temperature, °F
January, 1	34.6
February, 2	37.5
March, 3	47.2
April, 4	56.5
May, 5	66.4
June, 6	75.6
July, 7	80.0
August, 8	78.5
September, 9	71.3
October, 10	59.7
November, 11	49.8
December, 12	39.4

Source: U.S. National Oceanic and Atmospheric Administration.

(a) Draw a scatter diagram of the data for one period.
(b) Find a sinusoidal function of the form $y = A \sin(\omega x - \phi) + B$ that fits the data.
(c) Draw the sinusoidal function found in (b) on the scatter diagram.
(d) Use a graphing utility to find the sinusoidal function of best fit.
(e) Draw the sinusoidal function of best fit on the scatter diagram.

81. Monthly Temperature The following data represent the average monthly temperatures for Indianapolis, Indiana.

Month, x	Average Monthly Temperature, °F
January, 1	25.5
February, 2	29.6
March, 3	41.4
April, 4	52.4
May, 5	62.8
June, 6	71.9
July, 7	75.4
August, 8	73.2
September, 9	66.6
October, 10	54.7
November, 11	43.0
December, 12	30.9

Source: U.S. National Oceanic and Atmospheric Administration.

(a) Draw a scatter diagram of the data for one period.
(b) Find a sinusoidal function of the form $y = A \sin(\omega x - \phi) + B$ that fits the data.
(c) Draw the sinusoidal function found in (b) on the scatter diagram.
 (d) Use a graphing utility to find the sinusoidal function of best fit.
(e) Draw the sinusoidal function of best fit on the scatter diagram.

82. Monthly Temperature The data to the right represent the average monthly temperatures for Baltimore, Maryland.
(a) Draw a scatter diagram of the data for one period.
(b) Find a sinusoidal function of the form $y = A \sin(\omega x - \phi) + B$ that fits the data.
(c) Draw the sinusoidal function found in (b) on the scatter diagram.
(d) Use a graphing utility to find the sinusoidal function of best fit.
(e) Draw the sinusoidal function of best fit on the scatter diagram.

Month, x	Average Monthly Temperature, °F
January, 1	31.8
February, 2	34.8
March, 3	44.1
April, 4	53.4
May, 5	63.4
June, 6	72.5
July, 7	77.0
August, 8	75.6
September, 9	68.5
October, 10	56.6
November, 11	46.8
December, 12	36.7

Source: U.S. National Oceanic and Atmospheric Administration.

83. Tides Suppose that the length of time between consecutive high tides is approximately 12.5 hours. According to the National Oceanic and Atmospheric Administration, on Saturday, June 28, 1997, in Savannah, Georgia, high tide occurred at 3:38 AM (03.6333 hours) and low tide occurred at 10:08 AM (10.1333 hours). Water heights are measured as the amounts above or below the mean lower low water. The height of the water at high tide was 8.2 feet and the height of the water at low tide was −0.6 foot.
(a) Approximately when will the next high tide occur?
(b) Find a sinusoidal function of the form $y = A \sin(\omega x - \phi) + B$ that fits the data.
(c) Draw a graph of the function found in (b).
(d) Use the function found in (b) to predict the height of the water at the next high tide.

84. Tides Suppose that the length of time between consecutive high tides is approximately 12.5 hours. According to the National Oceanic and Atmospheric Administration, on Saturday, June 28, 1997, in Juneau, Alaska, high tide occurred at 8:11 AM (08.1833 hours) and low tide occurred at 2:14 PM (14.2333 hours). Water heights are measured as the amounts above or below the mean lower low water. The height of the water at high tide was 13.2 feet and the height of the water at low tide was 2.2 feet.
(a) Approximately when will the next high tide occur?
(b) Find a sinusoidal function of the form $y = A \sin(\omega x - \phi) + B$ that fits the data.
(c) Draw a graph of the function found in (b).
(d) Use the function found in (b) to predict the height of the water at the next high tide.

85. **Hours of Daylight** According to the *Old Farmer's Almanac,* in Miami, Florida, the number of hours of sunlight on the day of the summer solstice is 12.75 and the number of hours of sunlight on the day of the winter solstice is 10.583.
    (a) Find a sinusoidal function of the form $y = A \sin(\omega x - \phi) + B$ that fits the data.
    (b) Draw a graph of the function found in (a).
    (c) Use the function found in (a) to predict the number of hours of sunlight on April 1, the 91st day of the year.
     (d) Look up the number of hours of sunlight on April 1, 1997, in the *Old Farmer's Almanac,* and compare the actual hours of daylight to the results found in (c).

86. **Hours of Daylight** According to the *Old Farmer's Almanac,* in Detroit, Michigan, the number of hours of sunlight on the day of the summer solstice is 13.65 and the number of hours of sunlight on the day of the winter solstice is 9.067.
    (a) Find a sinusoidal function of the form $y = A \sin(\omega x - \phi) + B$ that fits the data.
    (b) Draw a graph of the function found in (a).
    (c) Use the function found in (a) to predict the number of hours of sunlight on April 1, the 91st day of the year.
    (d) Look up the number of hours of sunlight on April 1, 1997, in the *Old Farmer's Almanac,* and compare the actual hours of daylight to the results found in (c).

87. **Hours of Daylight** According to the *Old Farmer's Almanac,* in Anchorage, Alaska, the number of hours of sunlight on the day of the

summer solstice is 16.233 and the number of hours of sunlight on the day of the winter solstice is 5.45.
    (a) Find a sinusoidal function of the form $y = A \sin(\omega x - \phi) + B$ that fits the data.
    (b) Draw a graph of the function found in (a).
    (c) Use the function found in (a) to predict the number of hours of sunlight on April 1, the 91st day of the year.
    (d) Look up the number of hours of sunlight on April 1, 1997, in the *Old Farmer's Almanac,* and compare the actual hours of daylight to the results found in (c).

88. **Hours of Daylight** According to the *Old Farmer's Almanac,* in Honolulu, Hawaii, the number of hours of sunlight on the day of the summer solstice is 12.767 and the number of hours of sunlight on the day of the winter solstice is 10.783.
    (a) Find a sinusoidal function of the form $y = A \sin(\omega x - \phi) + B$ that fits the data.
    (b) Draw a graph of the function found in (a).
    (c) Use the function found in (a) to predict the number of hours of sunlight on April 1, the 91st day of the year.
    (d) Look up the number of hours of sunlight on April 1, 1997, in the *Old Farmer's Almanac,* and compare the actual hours of daylight to the results found in (c).

89. Explain how the amplitude and period of a sinusoidal graph are used to establish the scale on each coordinate axis.

90. Find an application in your major field that leads to a sinusoidal graph. Write a paper about your findings.

CHAPTER REVIEW

THINGS TO KNOW

Definitions

Angle in standard position	Vertex is at the origin; initial side is along the positive x-axis
Degree (1°)	$1° = \frac{1}{360}$ revolution
Radian (1)	The measure of a central angle whose rays subtend an arc whose length is the radius of the circle
Unit circle	Center at origin; radius = 1 Equation: $x^2 + y^2 = 1$
Trigonometric functions	$P = (a, b)$ is the point on the unit circle corresponding to $\theta = t$ radians:

$$\sin t = \sin \theta = b \qquad\qquad \cos t = \cos \theta = a$$

$$\tan t = \tan \theta = \frac{b}{a}, \quad a \neq 0 \qquad \cot t = \cot \theta = \frac{a}{b}, \quad b \neq 0$$

$$\csc t = \csc \theta = \frac{1}{b}, \quad b \neq 0 \qquad \sec t = \sec \theta = \frac{1}{a}, \quad a \neq 0$$

Periodic function	$f(\theta + p) = f(\theta)$, for all θ, $p > 0$, where the smallest such p is the fundamental period
Acute angle	An angle θ whose measure is $0° < \theta < 90°$ (or $0 < \theta < \pi/2$)
Complementary angles	Two acute angles whose sum is $90°$ ($\pi/2$)

Cofunction	The following pairs of functions are cofunctions of each other: sine and cosine; tangent and cotangent; secant and cosecant
Reference angle of θ	The acute angle formed by the terminal side of θ and either the positive or negative x-axis

Formulas

1 revolution = $360°$
            = 2π radians

$s = r\theta$ θ is measured in radians; s is the length of arc subtended by the central angle θ of the circle of radius r

$v = r\omega$ v is the linear speed along the circle of radius r; ω is the angular speed (measured in radians per unit time)

TABLE OF VALUES

θ (Radians)	θ (Degrees)	$\sin\theta$	$\cos\theta$	$\tan\theta$	$\csc\theta$	$\sec\theta$	$\cot\theta$
0	0°	0	1	0	Not defined	1	Not defined
$\pi/6$	30°	$\frac{1}{2}$	$\sqrt{3}/2$	$\sqrt{3}/3$	2	$2\sqrt{3}/3$	$\sqrt{3}$
$\pi/4$	45°	$\sqrt{2}/2$	$\sqrt{2}/2$	1	$\sqrt{2}$	$\sqrt{2}$	1
$\pi/3$	60°	$\sqrt{3}/2$	$\frac{1}{2}$	$\sqrt{3}$	$2\sqrt{3}/3$	2	$\sqrt{3}/3$
$\pi/2$	90°	1	0	Not defined	1	Not defined	0
π	180°	0	-1	0	Not defined	-1	Not defined
$3\pi/2$	270°	-1	0	Not defined	-1	Not defined	0

Sinusoidal graphs	$y = A \sin \omega x,\quad \omega > 0$	Period = $2\pi/\omega$
	$y = A \cos \omega x,\quad \omega > 0$	Amplitude = $|A|$
	$y = A \sin(\omega x - \phi) = A \sin \omega(x - \phi/\omega)$	Phase shift = ϕ/ω
	$y = A \cos(\omega x - \phi) = A \cos \omega(x - \phi/\omega)$	

Identities

$$\tan\theta = \frac{\sin\theta}{\cos\theta}, \quad \cot\theta = \frac{\cos\theta}{\sin\theta}$$

$$\cot\theta = \frac{1}{\tan\theta}, \quad \sec\theta = \frac{1}{\cos\theta}, \quad \csc\theta = \frac{1}{\sin\theta}$$

$$\sin^2\theta + \cos^2\theta = 1, \quad \tan^2\theta + 1 = \sec^2\theta, \quad 1 + \cot^2\theta = \csc^2\theta$$

Properties of the Trigonometric Functions

$y = \sin x$ Domain: $-\infty < x < \infty$
                  Range: $-1 \le y \le 1$
                  Periodic: period = 2π (360°)
                  Odd function

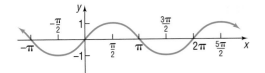

$y = \cos x$ Domain: $-\infty < x < \infty$
                  Range: $-1 \le y \le 1$
                  Periodic: period = 2π (360°)
                  Even function

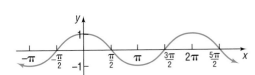

$y = \tan x$ Domain: $-\infty < x < \infty$, except odd multiples of $\pi/2$ (90°)
                  Range: $-\infty < y < \infty$
                  Periodic: period = π (180°)
                  Odd function

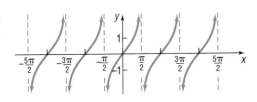

$y = \csc x$ Domain: $-\infty < x < \infty$, except integral multiples of π (180°)
Range: $|y| \geq 1$
Periodic: period $= 2\pi$ (360°)
Odd function

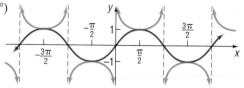

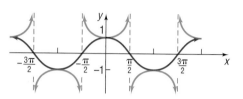

$y = \sec x$ Domain: $-\infty < x < \infty$, except odd multiples of $\pi/2$ (90°)
Range: $|y| \geq 1$
Periodic: period $= 2\pi$ (360°)
Even function

$y = \cot x$ Domain: $-\infty < x < \infty$, except integral multiples of π (180°)
Range: $-\infty < y < \infty$
Periodic: period $= \pi$ (180°)
Odd function

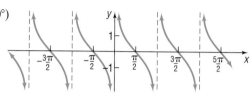

HOW TO

Convert an angle from radian measure to degree measure

Convert an angle from degree measure to radian measure

Solve applied problems involving circular motion

Find the value of each of the remaining trigonometric functions if the value of one function and the quadrant of the angle are given

Use the theorem on cofunctions of complementary angles

Use reference angles to find the value of a trigonometric function

Use a calculator to find the value of a trigonometric function

Graph the trigonometric functions, including transformations

Find the period and amplitude of a sinusoidal function and use them to graph the function

Find a function whose sinusoidal graph is given

Find a sinusoidal function from data

FILL-IN-THE-BLANK ITEMS

1. Two rays drawn with a common vertex form a(n) _____. One of the rays is called the _____ _____; the other is called the _____ _____.

2. In the formula $s = r\theta$ for measuring the length s of arc along a circle of radius r, the angle θ must be measured in _____.

3. 180 degrees = _____ radians.

4. Two acute angles whose sum is a right angle are called _____.

5. The sine and _____ functions are cofunctions.

6. An angle is in _____ _____ if its vertex is at the origin and its initial side coincides with the positive x-axis.

7. The reference angle of 135° is _____.

8. The sine, cosine, cosecant, and secant functions have period _____; the tangent and cotangent functions have period _____.

9. Which of the trigonometric functions have graphs that are symmetric with respect to the y-axis?

10. Which of the trigonometric functions have graphs that are symmetric with respect to the origin?

11. The following function has amplitude 3 and period 2:
    $y = $ _____ sin _____ x

12. The function $y = 3 \sin 6x$ has amplitude _____ and period _____.

TRUE/FALSE ITEMS

T F **1.** In the formula $s = r\theta$, r is the radius of a circle and s is the length of arc subtended by a central angle θ, where θ is measured in degrees.

T F **2.** $|\sin \theta| \leq 1$

T F **3.** $1 + \tan^2 \theta = \csc^2 \theta$

T F **4.** The only even trigonometric functions are the cosine and secant functions.

T F **5.** $\tan 62° = \cot 38°$

T F **6.** $\sin 182° = \cos 2°$

T F **7.** The graphs of $y = \tan x$, $y = \cot x$, $y = \sec x$, and $y = \csc x$ have infinitely many vertical asymptotes.

T F **8.** The graphs of $y = \sin x$ and $y = \cos x$ are identical except for a horizontal shift.

T F **9.** For $y = 2 \sin \pi x$, the amplitude is 2 and the period is $\pi/2$.

REVIEW EXERCISES

Blue problem numbers indicate the author's suggestions for use in a Practice Test.

In Problems 1–4, convert each angle in degrees to radians. Express your answer as a multiple of π.

1. 135°   **2.** 210°   **3.** 18°   **4.** 15°

In Problems 5–8, convert each angle in radians to degrees.

5. $3\pi/4$ **6.** $2\pi/3$ **7.** $-5\pi/2$ **8.** $-3\pi/2$

In Problems 9–30, find the exact value of each expression. Do not use a calculator.

9. $\tan \dfrac{\pi}{4} - \sin \dfrac{\pi}{6}$

10. $\cos \dfrac{\pi}{3} + \sin \dfrac{\pi}{2}$

11. $3 \sin 45° - 4 \tan \dfrac{\pi}{6}$

12. $4 \cos 60° + 3 \tan \dfrac{\pi}{3}$

13. $6 \cos \dfrac{3\pi}{4} + 2 \tan\left(-\dfrac{\pi}{3}\right)$

14. $3 \sin \dfrac{2\pi}{3} - 4 \cos \dfrac{5\pi}{2}$

15. $\sec\left(-\dfrac{\pi}{3}\right) - \cot\left(-\dfrac{5\pi}{4}\right)$

16. $4 \csc \dfrac{3\pi}{4} - \cot\left(-\dfrac{\pi}{4}\right)$

17. $\tan \pi + \sin \pi$

18. $\cos \dfrac{\pi}{2} - \csc\left(-\dfrac{\pi}{2}\right)$

19. $\cos 180° - \tan(-45°)$

20. $\sin 270° + \cos(-180°)$

21. $\sin^2 20° + \dfrac{1}{\sec^2 20°}$

22. $\dfrac{1}{\cos^2 40°} - \dfrac{1}{\cot^2 40°}$

23. $\sec 50° \cos 50°$

24. $\tan 10° \cot 10°$

25. $\dfrac{\sin 50°}{\cos 40°}$

26. $\dfrac{\tan 20°}{\cot 70°}$

27. $\dfrac{\sin(-40°)}{\cos 50°}$

28. $\tan(-20°) \cot 20°$

29. $\sin 400° \sec(-50°)$

30. $\cot 200° \cot(-70°)$

In Problems 31–46, find the exact value of each of the remaining trigonometric functions.

31. $\sin \theta = -\frac{4}{5}$, $\cos \theta > 0$ **32.** $\cos \theta = -\frac{3}{5}$, $\sin \theta < 0$ **33.** $\tan \theta = \frac{12}{5}$, $\sin \theta < 0$

34. $\cot \theta = \frac{12}{5}$, $\cos \theta < 0$ **35.** $\sec \theta = -\frac{5}{4}$, $\tan \theta < 0$ **36.** $\csc \theta = -\frac{5}{3}$, $\cot \theta < 0$

37. $\sin \theta = \frac{12}{13}$, θ in quadrant II   **38.** $\cos \theta = -\frac{3}{5}$, θ in quadrant III   **39.** $\sin \theta = -\frac{5}{13}$, $3\pi/2 < \theta < 2\pi$

40. $\cos \theta = \frac{12}{13}$, $3\pi/2 < \theta < 2\pi$ **41.** $\tan \theta = \frac{1}{3}$, $180° < \theta < 270°$ **42.** $\tan \theta = -\frac{2}{3}$, $90° < \theta < 180°$

43. $\sec \theta = 3$, $3\pi/2 < \theta < 2\pi$ **44.** $\csc \theta = -4$, $\pi < \theta < 3\pi/2$ **45.** $\cot \theta = -2$, $\pi/2 < \theta < \pi$

46. $\tan \theta = -2$, $3\pi/2 < \theta < 2\pi$

In Problems 47–54, graph each function. Each graph should contain at least one period.

47. $y = 2 \sin 4x$ **48.** $y = -3 \cos 2x$ **49.** $y = -2 \cos\left(x + \dfrac{\pi}{2}\right)$ **50.** $y = 3 \sin(x - \pi)$

51. $y = \tan(x + \pi)$ **52.** $y = -\tan\left(x - \dfrac{\pi}{2}\right)$ **53.** $y = -2 \tan 3x$ **54.** $y = 4 \tan 2x$

In Problems 55–58, determine the amplitude and period of each function without graphing.

55. $y = 4 \cos x$ **56.** $y = \sin 2x$ **57.** $y = -8 \sin \dfrac{\pi}{2}x$ **58.** $y = -2 \cos 3\pi x$

In Problems 59–66, find the amplitude, period, and phase shift of each function. Graph each function. Show at least one period.

59. $y = 4 \sin 3x$ **60.** $y = 2 \cos \frac{1}{3}x$ **61.** $y = -2 \sin\left(\dfrac{\pi}{2}x + \dfrac{1}{2}\right)$ **62.** $y = -6 \sin(2\pi x - 2)$

63. $y = \frac{1}{2} \sin(\frac{3}{2}x - \pi)$ **64.** $y = \frac{3}{2} \cos(6x + 3\pi)$ **65.** $y = -\frac{2}{3} \cos(\pi x - 6)$ **66.** $y = -7 \sin\left(\dfrac{\pi}{3}x + \dfrac{4}{3}\right)$

In Problems 67–70, find a function whose graph is given.

67.

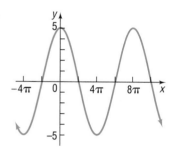

68.

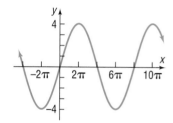

69.

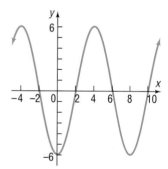

70.

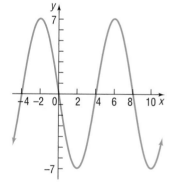

71. Find the length of arc subtended by a central angle of 30° on a circle of radius 2 feet.

72. The minute hand of a clock is 8 inches long. How far does the tip of the minute hand move in 30 minutes? How far does it move in 20 minutes?

73. **Angular Speed of a Race Car** A race car is driven around a circular track at a constant speed of 180 miles per hour. If the diameter of the track is $\frac{1}{2}$ mile, what is the angular speed of the car? Express your answer in revolutions per hour (which is equivalent to laps per hour).

74. **Merry-Go-Rounds** A neighborhood carnival has a merry-go-round whose radius is 25 feet. If the time for one revolution is 30 seconds, how fast is the merry-go-round going?

75. **Lighthouse Beacons** The Montauk Point Lighthouse on Long Island has dual beams (two light sources opposite each other). Ships at sea observe a blinking light every 5 seconds. What rotation speed is required to do this?

76. Spin Balancing Tires The radius of each wheel of a car is 16 inches. At how many revolutions per minute should a spin balancer be set to balance the tires at a speed of 90 miles per hour? Is the setting different for a wheel of radius 14 inches? What is this setting?

77. Alternating Voltage The electromotive force E, in volts, in a certain ac circuit obeys the equation

$$E = 120 \sin 120\pi t, \qquad t \geq 0$$

where t is measured in seconds.
(a) What is the maximum value of E?
(b) What is the period?
(c) Graph this function over two periods.

78. Alternating Current The current I, in amperes, flowing through an ac (alternating current) circuit at time t is

$$I = 220 \sin\left(30\pi t + \frac{\pi}{6}\right), \qquad t \geq 0$$

(a) What is the period?
(b) What is the amplitude?
(c) What is the phase shift?
(d) Graph this function over two periods.

79. Monthly Temperature The following data represent the average monthly temperatures for Phoenix, Arizona.

Month, x	Average Monthly Temperature, °F
January, 1	51
February, 2	55
March, 3	63
April, 4	67
May, 5	77
June, 6	86
July, 7	90
August, 8	90
September, 9	84
October, 10	71
November, 11	59
December, 12	52

Source: U.S. National Oceanic and Atmospheric Administration.

(a) Draw a scatter diagram of the data for one period.
(b) Find a sinusoidal function of the form $y = A \sin(\omega x - \phi) + B$ that fits the data.
(c) Draw the sinusoidal function found in (b) on the scatter diagram.
 (d) Use a graphing utility to find the sinusoidal function of best fit.

(e) Draw the sinusoidal function of best fit on the scatter diagram.

80. Monthly Temperature The following data represent the average monthly temperatures for Chicago, Illinois.

Month, x	Average Monthly Temperature, °F
January, 1	25
February, 2	28
March, 3	36
April, 4	48
May, 5	61
June, 6	72
July, 7	74
August, 8	75
September, 9	66
October, 10	55
November, 11	39
December, 12	28

Source: U.S. National Oceanic and Atmospheric Administration.

(a) Draw a scatter diagram of the data for one period.
(b) Find a sinusoidal function of the form $y = A \sin(\omega x - \phi) + B$ that fits the data.
(c) Draw the sinusoidal function found in (b) on the scatter diagram.
(d) Use a graphing utility to find the sinusoidal function of best fit.
(e) Draw the sinusoidal function of best fit on the scatter diagram.

81. Hours of Daylight According to the *Old Farmer's Almanac*, in Las Vegas, Nevada, the number of hours of sunlight on the day of the summer solstice is 13.367 and the number of hours of sunlight on the day of the winter solstice is 9.667.
(a) Find a sinusoidal function of the form $y = A \sin(\omega x - \phi) + B$ that fits the data.
(b) Draw a graph of the function found in (a).
(c) Use the function found in (a) to predict the number of hours of sunlight on April 1, the 91st day of the year.
 (d) Look up the number of hours of sunlight on April 1, 1997, in the *Old Farmer's Almanac* and compare the actual hours of daylight to the results found in (c).

82. **Hours of Daylight** According to the *Old Farmer's Almanac,* in Seattle, Washington, the number of hours of sunlight on the day of the summer solstice is 13.967 and the number of hours of sunlight on the day of the winter solstice is 8.417.

   (a) Find a sinusoidal function of the form
$$y = A \sin(\omega x - \phi) + B$$ that fits the data.

   (b) Draw a graph of the function found in (a).

   (c) Use the function found in (a) to predict the number of hours of sunlight on April 1, the 91st day of the year.

    (d) Look up the number of hours of sunlight on April 1, 1997, in the *Old Farmer's Almanac* and compare the actual hours of daylight to the results found in (c).

CHAPTER 3

Analytic Trigonometry

Extra credit is always welcome in school, and this time your physics professor has given you a golden opportunity to shine! All you need to do is to create a display on rainbows for the college open house. You will want to use the Sullivan website at

www.prenhall.com/sullivan

to find the information needed to complete your display.

PREPARING FOR THIS CHAPTER

Before getting started on this chapter, review the following concepts:

Fundamental Identities *(p. 136)*

Values of the Trigonometric Functions of Certain Angles *(pp. 120 and 122)*

Even–Odd Properties *(p. 137)*

Inverse Functions *(pp. 86–91)*

Graphs of the Trigonometric Functions *(Section 2.5)*

OUTLINE

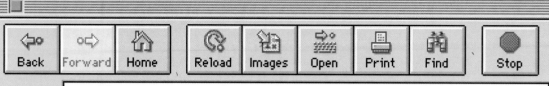

THE MATHEMATICS OF RAINBOWS

Rainbows, those incredible displays of light and water, will always be a source of wonder and inspiration. From pots-o-gold to bridges for the gods, countless myths and legends have been told to explain rainbows.

In 1621, the basic law of refraction was proposed by the Dutch scientist Willebrord Snell. The law of refraction and the law of reflection provide the keys to the modern explanation for the formation of rainbows. Refraction is the bending of the light's path as it passes from one medium into another. When this happens, the different wavelengths of light separate. The algebraic expression for Snell's Law looks similar to the Law of Sines (discussed in Chapter 4), except that the constants no longer refer to the lengths of the sides of a triangle, but instead refer to a *refractive index*.

You have been asked by your physics professor to make a display on rainbows for the college open house.

1. You begin working on the project by brainstorming. Since this is the age of the World Wide Web, you decide to create an interactive display! This gets the adrenaline going. You make a web search to collect some ideas. Your web search generates thousands of sites! You select some of your favorites. At this point you want to make a list of questions that your display might answer. For example, did you ever notice that you generally will see rainbows only in the morning or afternoon? If you were to make a display of a rainbow, how would you do it?
2. Since your display is for the physics department, you want to be sure to find some examples that reveal the physical properties of rainbows. What would you add? For some examples, access *keys*.
3. To get the visitors involved in your booth, you decide to perform the following experiment: Draw a circle on a piece of paper. This will represent a water droplet. Now, with a ruler draw a line, a light ray, from the edge of your paper to your "water drop". Make a sketch of your guess of the path of the light ray as it passes through the water drop. You try this out on some of your friends to see if it will work. Do you want to use it? Why or why not?
4. Next imagine a curtain of water droplets and a light source. Try to make an illustration as to where the light source, the water curtain, the rainbow, and an observer will be located with respect to one another. For a solution check out *Descartes and Rainbows*.
5. For the graphic of your display you decide to make a large-scale illustration of the experiment of question 3. (For this exercise you will want to review *Snell's Law*.) You use colored felt-tip pens, a compass, a ruler, a calculator, and a protractor to make an *exact drawing*. You want this to be right, since visitors will ask you to explain your drawing to them. Make the drawing and then try to explain it to a friend. How did you do?

In this chapter we continue the derivation of identities involving the trigonometric functions. These identities play an important role in calculus, the physical and life sciences, and economics, where they are used to simplify complicated expressions.

The last two sections of this chapter deal with the inverse trigonometric functions and with equations that contain trigonometric functions.

3.1 | TRIGONOMETRIC IDENTITIES

1 Establish Identities

We saw in the previous chapter that the trigonometric functions lend themselves to a wide variety of identities. Before establishing some additional identities, let's review the definition of an *identity*.

Two functions f and g are said to be **identically equal** if

$$f(x) = g(x)$$

for every value of x for which both functions are defined. Such an equation is referred to as an **identity**. An equation that is not an identity is called a **conditional equation.**

For example, the following are identities:

$$(x + 1)^2 = x^2 + 2x + 1 \qquad \sin^2 x + \cos^2 x = 1 \qquad \csc x = \frac{1}{\sin x}$$

The following are conditional equations:

$$2x + 5 = 0 \qquad \text{True only if } x = -\frac{5}{2}.$$

$$\sin x = 0 \qquad \text{True only if } x = k\pi, \ k \text{ an integer.}$$

$$\sin x = \cos x \qquad \text{True only if } x = \frac{\pi}{4} + 2k\pi \text{ or } x = \frac{5\pi}{4} + 2k\pi, \ k \text{ an integer.}$$

The following boxes summarize the trigonometric identities that we have established thus far.

Quotient Identities

$$\tan \theta = \frac{\sin \theta}{\cos \theta} \qquad \cot \theta = \frac{\cos \theta}{\sin \theta}$$

Reciprocal Identities

$$\csc \theta = \frac{1}{\sin \theta} \qquad \sec \theta = \frac{1}{\cos \theta} \qquad \cot \theta = \frac{1}{\tan \theta}$$

Pythagorean Identities

$$\sin^2 \theta + \cos^2 \theta = 1 \qquad \tan^2 \theta + 1 = \sec^2 \theta$$
$$1 + \cot^2 \theta = \csc^2 \theta$$

Even–Odd Identities

$$\sin(-\theta) = -\sin \theta \qquad \cos(-\theta) = \cos \theta \qquad \tan(-\theta) = -\tan \theta$$
$$\csc(-\theta) = -\csc \theta \qquad \sec(-\theta) = \sec \theta \qquad \cot(-\theta) = -\cot \theta$$

This list of identities comprises what we shall refer to as the **basic trigonometric identities.** These identities should not merely be memorized, but should be *known* (just as you know your name rather than have it memorized). In fact, minor variations of a basic identity are often used. For example, we might want to use $\sin^2 \theta = 1 - \cos^2 \theta$ or $\cos^2 \theta = 1 - \sin^2 \theta$ instead of $\sin^2 \theta + \cos^2 \theta = 1$. For this reason, among others, you need to know these relationships and be quite comfortable with variations of them.

In the examples that follow, the directions will read "Establish the identity . . . ." As you will see, this is accomplished by starting with one side of the given equation (usually the one containing the more complicated expression) and, using appropriate basic identities and algebraic manipulations, arriving at the other side. The selection of an appropriate basic identity to obtain the desired result is learned only through experience and lots of practice.

EXAMPLE 1

Establishing an Identity

Establish the identity: $\csc \theta \cdot \tan \theta = \sec \theta$

Solution

We start with the left side, because it contains the more complicated expression, and apply a reciprocal identity and a quotient identity.

$$\csc \theta \cdot \tan \theta = \frac{1}{\sin \theta} \cdot \frac{\sin \theta}{\cos \theta} = \frac{1}{\cos \theta} = \sec \theta$$

Having arrived at the right side, the identity is established.

Comment A graphing utility can be used to provide evidence of an identity. For example, if we graph $Y_1 = \csc \theta \cdot \tan \theta$ and $Y_2 = \sec \theta$, the graphs appear to be the same. This provides evidence that $Y_1 = Y_2$. However, it does not prove their equality. A graphing utility *cannot be used to establish an identity*—identities must be established algebraically.

Now work Problem 1.

E X A M P L E 2   Establishing an Identity

Establish the identity: $\sin^2(-\theta) + \cos^2(-\theta) = 1$

Solution   We begin with the left side and apply even–odd identities.

$$
\begin{aligned}
\sin^2(-\theta) + \cos^2(-\theta) &= [\sin(-\theta)]^2 + [\cos(-\theta)]^2 \\
&= (-\sin\theta)^2 + (\cos\theta)^2 && \text{Even–odd identities.} \\
&= (\sin\theta)^2 + (\cos\theta)^2 && \text{Pythagorean identity.} \\
&= 1
\end{aligned}
$$
■

E X A M P L E 3   Establishing an Identity

Establish the identity: $\dfrac{\sin^2(-\theta) - \cos^2(-\theta)}{\sin(-\theta) - \cos(-\theta)} = \cos\theta - \sin\theta$

Solution   We begin with two observations: The left side appears to contain the more complicated expression. Also, the left side contains expressions with the argument $-\theta$, whereas the right side contains expressions with the argument θ. We decide, therefore, to start with the left side and apply even–odd identities.

$$
\begin{aligned}
\frac{\sin^2(-\theta) - \cos^2(-\theta)}{\sin(-\theta) - \cos(-\theta)} &= \frac{[\sin(-\theta)]^2 - [\cos(-\theta)]^2}{\sin(-\theta) - \cos(-\theta)} \\
&= \frac{(-\sin\theta)^2 - (\cos\theta)^2}{-\sin\theta - \cos\theta} && \text{Even–odd identities.} \\
&= \frac{(\sin\theta)^2 - (\cos\theta)^2}{-\sin\theta - \cos\theta} && \text{Simplify.} \\
&= \frac{(\sin\theta - \cos\theta)(\sin\theta + \cos\theta)}{-(\sin\theta + \cos\theta)} && \text{Factor.} \\
&= \cos\theta - \sin\theta && \text{Cancel and simplify.}
\end{aligned}
$$
■

E X A M P L E 4   Establishing an Identity

Establish the identity: $\dfrac{1 + \tan\theta}{1 + \cot\theta} = \tan\theta$

Solution $\dfrac{1 + \tan\theta}{1 + \cot\theta} = \dfrac{1 + \tan\theta}{1 + \dfrac{1}{\tan\theta}} = \dfrac{1 + \tan\theta}{\dfrac{\tan\theta + 1}{\tan\theta}} = \dfrac{\tan\theta(1 + \tan\theta)}{\tan\theta + 1} = \tan\theta$
■

Now work Problem 9.

When sums or differences of quotients appear, it is usually best to rewrite them as a single quotient, especially if the other side of the identity consists of only one term.

E X A M P L E 5   Establishing an Identity

Establish the identity: $\dfrac{\sin\theta}{1+\cos\theta} + \dfrac{1+\cos\theta}{\sin\theta} = 2\csc\theta$

Solution   The left side is more complicated, so we start with it and proceed to add.

$$\dfrac{\sin\theta}{1+\cos\theta} + \dfrac{1+\cos\theta}{\sin\theta} = \dfrac{\sin^2\theta + (1+\cos\theta)^2}{(1+\cos\theta)(\sin\theta)}$$ Add the quotients.

$$= \dfrac{\sin^2\theta + 1 + 2\cos\theta + \cos^2\theta}{(1+\cos\theta)(\sin\theta)}$$ Remove parentheses in numerator.

$$= \dfrac{(\sin^2\theta + \cos^2\theta) + 1 + 2\cos\theta}{(1+\cos\theta)(\sin\theta)}$$ Regroup.

$$= \dfrac{2 + 2\cos\theta}{(1+\cos\theta)(\sin\theta)}$$ Pythagorean Identity.

$$= \dfrac{2(1+\cos\theta)}{(1+\cos\theta)(\sin\theta)}$$ Factor and cancel.

$$= \dfrac{2}{\sin\theta}$$ Reciprocal identity.

$$= 2\csc\theta$$

Sometimes it helps to write one side in terms of sines and cosines only.

E X A M P L E 6   Establishing an Identity

Establish the identity: $\dfrac{\tan\theta + \cot\theta}{\sec\theta\csc\theta} = 1$

Solution

$$\dfrac{\tan\theta + \cot\theta}{\sec\theta\csc\theta} = \dfrac{\dfrac{\sin\theta}{\cos\theta} + \dfrac{\cos\theta}{\sin\theta}}{\dfrac{1}{\cos\theta}\cdot\dfrac{1}{\sin\theta}} = \dfrac{\dfrac{\sin^2\theta + \cos^2\theta}{\cos\theta\sin\theta}}{\dfrac{1}{\cos\theta\sin\theta}}$$

Change to sines and cosines.   Add the quotients in the numerator.

$$= \dfrac{1}{\cos\theta\sin\theta} \cdot \dfrac{\cos\theta\sin\theta}{1} = 1$$

Divide quotients;
$\sin^2\theta + \cos^2\theta = 1$

 Now work Problem 51.

Sometimes, multiplying the numerator and denominator by an appropriate factor will result in a simplification.

EXAMPLE 7   Establishing an Identity

Establish the identity: $\dfrac{1 - \sin \theta}{\cos \theta} = \dfrac{\cos \theta}{1 + \sin \theta}$

Solution   We start with the left side and multiply the numerator and the denominator by $1 + \sin \theta$. (Alternatively, we could multiply the numerator and denominator of the right side by $1 - \sin \theta$).

$$\frac{1 - \sin \theta}{\cos \theta} = \frac{1 - \sin \theta}{\cos \theta} \cdot \frac{1 + \sin \theta}{1 + \sin \theta} \qquad \text{Multiply numerator and denominator by } 1 + \sin \theta.$$

$$= \frac{1 - \sin^2 \theta}{\cos \theta (1 + \sin \theta)} \qquad 1 - \sin^2 \theta = \cos^2 \theta$$

$$= \frac{\cos^2 \theta}{\cos \theta (1 + \sin \theta)} \qquad \text{Cancel.}$$

$$= \frac{\cos \theta}{1 + \sin \theta}$$

■

Now work Problem 35.

Although a lot of practice is the only real way to learn how to establish identities, the following guidelines should prove helpful.

Guidelines
for Establishing Identities

1. It is almost always preferable to start with the side containing the more complicated expression.
2. Rewrite sums or differences of quotients as a single quotient.
3. Sometimes rewriting one side in terms of sines and cosines only will help.
4. Always keep your goal in mind. As you manipulate one side of the expression, you must keep in mind the form of the expression on the other side.

Warning   Be careful not to handle identities to be established as if they were equations. You *cannot* establish an identity by such methods as adding the same expression to each side and obtaining a true statement. This practice is not allowed, because the original statement is precisely the one that you are trying to establish. You do not know until it has been established that it is, in fact, true.   ■

3.1 | EXERCISES

In Problems 1–80, establish each identity.

1. $\csc \theta \cdot \cos \theta = \cot \theta$
2. $\sec \theta \cdot \sin \theta = \tan \theta$
3. $1 + \tan^2(-\theta) = \sec^2 \theta$
4. $1 + \cot^2(-\theta) = \csc^2 \theta$
5. $\cos \theta(\tan \theta + \cot \theta) = \csc \theta$
6. $\sin \theta(\cot \theta + \tan \theta) = \sec \theta$
7. $\tan \theta \cot \theta - \cos^2 \theta = \sin^2 \theta$
8. $\sin \theta \csc \theta - \cos^2 \theta = \sin^2 \theta$
9. $(\sec \theta - 1)(\sec \theta + 1) = \tan^2 \theta$
10. $(\csc \theta - 1)(\csc \theta + 1) = \cot^2 \theta$
11. $(\sec \theta + \tan \theta)(\sec \theta - \tan \theta) = 1$
12. $(\csc \theta + \cot \theta)(\csc \theta - \cot \theta) = 1$
13. $\cos^2 \theta(1 + \tan^2 \theta) = 1$
14. $(1 - \cos^2 \theta)(1 + \cot^2 \theta) = 1$
15. $(\sin \theta + \cos \theta)^2 + (\sin \theta - \cos \theta)^2 = 2$
16. $\tan^2 \theta \cos^2 \theta + \cot^2 \theta \sin^2 \theta = 1$

17. $\sec^4 \theta - \sec^2 \theta = \tan^4 \theta + \tan^2 \theta$

18. $\csc^4 \theta - \csc^2 \theta = \cot^4 \theta + \cot^2 \theta$

19. $\sec \theta - \tan \theta = \dfrac{\cos \theta}{1 + \sin \theta}$

20. $\csc \theta - \cot \theta = \dfrac{\sin \theta}{1 + \cos \theta}$

21. $3 \sin^2 \theta + 4 \cos^2 \theta = 3 + \cos^2 \theta$

22. $9 \sec^2 \theta - 5 \tan^2 \theta = 5 + 4 \sec^2 \theta$

23. $1 - \dfrac{\cos^2 \theta}{1 + \sin \theta} = \sin \theta$

24. $1 - \dfrac{\sin^2 \theta}{1 - \cos \theta} = -\cos \theta$

25. $\dfrac{1 + \tan \theta}{1 - \tan \theta} = \dfrac{\cot \theta + 1}{\cot \theta - 1}$

26. $\dfrac{\csc \theta - 1}{\csc \theta + 1} = \dfrac{1 - \sin \theta}{1 + \sin \theta}$

27. $\dfrac{\sec \theta}{\csc \theta} + \dfrac{\sin \theta}{\cos \theta} = 2 \tan \theta$

28. $\dfrac{\csc \theta - 1}{\cot \theta} = \dfrac{\cot \theta}{\csc \theta + 1}$

29. $\dfrac{1 + \sin \theta}{1 - \sin \theta} = \dfrac{\csc \theta + 1}{\csc \theta - 1}$

30. $\dfrac{\cos \theta + 1}{\cos \theta - 1} = \dfrac{1 + \sec \theta}{1 - \sec \theta}$

31. $\dfrac{1 - \sin \theta}{\cos \theta} + \dfrac{\cos \theta}{1 - \sin \theta} = 2 \sec \theta$

32. $\dfrac{\cos \theta}{1 + \sin \theta} + \dfrac{1 + \sin \theta}{\cos \theta} = 2 \sec \theta$

33. $\dfrac{\sin \theta}{\sin \theta - \cos \theta} = \dfrac{1}{1 - \cot \theta}$

34. $1 - \dfrac{\sin^2 \theta}{1 + \cos \theta} = \cos \theta$

35. $\dfrac{1 - \sin \theta}{1 + \sin \theta} = (\sec \theta - \tan \theta)^2$

36. $\dfrac{1 - \cos \theta}{1 + \cos \theta} = (\csc \theta - \cot \theta)^2$

37. $\dfrac{\cos \theta}{1 - \tan \theta} + \dfrac{\sin \theta}{1 - \cot \theta} = \sin \theta + \cos \theta$

38. $\dfrac{\cot \theta}{1 - \tan \theta} + \dfrac{\tan \theta}{1 - \cot \theta} = 1 + \tan \theta + \cot \theta$

39. $\tan \theta + \dfrac{\cos \theta}{1 + \sin \theta} = \sec \theta$

40. $\dfrac{\sin \theta \cos \theta}{\cos^2 \theta - \sin^2 \theta} = \dfrac{\tan \theta}{1 - \tan^2 \theta}$

41. $\dfrac{\tan \theta + \sec \theta - 1}{\tan \theta - \sec \theta + 1} = \tan \theta + \sec \theta$

42. $\dfrac{\sin \theta - \cos \theta + 1}{\sin \theta + \cos \theta - 1} = \dfrac{\sin \theta + 1}{\cos \theta}$

43. $\dfrac{\tan \theta - \cot \theta}{\tan \theta + \cot \theta} = \sin^2 \theta - \cos^2 \theta$

44. $\dfrac{\sec \theta - \cos \theta}{\sec \theta + \cos \theta} = \dfrac{\sin^2 \theta}{1 + \cos^2 \theta}$

45. $\dfrac{\tan \theta - \cot \theta}{\tan \theta + \cot \theta} + 1 = 2 \sin^2 \theta$

46. $\dfrac{\tan \theta - \cot \theta}{\tan \theta + \cot \theta} + 2 \cos^2 \theta = 1$

47. $\dfrac{\sec \theta + \tan \theta}{\cot \theta + \cos \theta} = \tan \theta \sec \theta$

48. $\dfrac{\sec \theta}{1 + \sec \theta} = \dfrac{1 - \cos \theta}{\sin^2 \theta}$

49. $\dfrac{1 - \tan^2 \theta}{1 + \tan^2 \theta} + 1 = 2 \cos^2 \theta$

50. $\dfrac{1 - \cot^2 \theta}{1 + \cot^2 \theta} + 2 \cos^2 \theta = 1$

51. $\dfrac{\sec \theta - \csc \theta}{\sec \theta \csc \theta} = \sin \theta - \cos \theta$

52. $\dfrac{\sin^2 \theta - \tan \theta}{\cos^2 \theta - \cot \theta} = \tan^2 \theta$

53. $\sec \theta - \cos \theta - \sin \theta \tan \theta = 0$

54. $\tan \theta + \cot \theta - \sec \theta \csc \theta = 0$

55. $\dfrac{1}{1 - \sin \theta} + \dfrac{1}{1 + \sin \theta} = 2 \sec^2 \theta$

56. $\dfrac{1 + \sin \theta}{1 - \sin \theta} - \dfrac{1 - \sin \theta}{1 + \sin \theta} = 4 \tan \theta \sec \theta$

57. $\dfrac{\sec \theta}{1 - \sin \theta} = \dfrac{1 + \sin \theta}{\cos^3 \theta}$

58. $\dfrac{1 - \sin \theta}{1 + \sin \theta} = (\sec \theta - \tan \theta)^2$

59. $\dfrac{(\sec \theta - \tan \theta)^2 + 1}{\csc \theta (\sec \theta - \tan \theta)} = 2 \tan \theta$

60. $\dfrac{\sec^2 \theta - \tan^2 \theta + \tan \theta}{\sec \theta} = \sin \theta + \cos \theta$

61. $\dfrac{\sin \theta + \cos \theta}{\cos \theta} - \dfrac{\sin \theta - \cos \theta}{\sin \theta} = \sec \theta \csc \theta$

62. $\dfrac{\sin \theta + \cos \theta}{\sin \theta} - \dfrac{\cos \theta - \sin \theta}{\cos \theta} = \sec \theta \csc \theta$

63. $\dfrac{\sin^3 \theta + \cos^3 \theta}{\sin \theta + \cos \theta} = 1 - \sin \theta \cos \theta$

64. $\dfrac{\sin^3 \theta + \cos^3 \theta}{1 - 2 \cos^2 \theta} = \dfrac{\sec \theta - \sin \theta}{\tan \theta - 1}$

65. $\dfrac{\cos^2 \theta - \sin^2 \theta}{1 - \tan^2 \theta} = \cos^2 \theta$

66. $\dfrac{\cos \theta + \sin \theta - \sin^3 \theta}{\sin \theta} = \cot \theta + \cos^2 \theta$

67. $\dfrac{(2 \cos^2 \theta - 1)^2}{\cos^4 \theta - \sin^4 \theta} = 1 - 2 \sin^2 \theta$

68. $\dfrac{1 - 2 \cos^2 \theta}{\sin \theta \cos \theta} = \tan \theta - \cot \theta$

69. $\dfrac{1 + \sin \theta + \cos \theta}{1 + \sin \theta - \cos \theta} = \dfrac{1 + \cos \theta}{\sin \theta}$

70. $\dfrac{1 + \cos \theta + \sin \theta}{1 + \cos \theta - \sin \theta} = \sec \theta + \tan \theta$

71. $(a \sin \theta + b \cos \theta)^2 + (a \cos \theta - b \sin \theta)^2 = a^2 + b^2$

72. $(2a \sin \theta \cos \theta)^2 + a^2(\cos^2 \theta - \sin^2 \theta)^2 = a^2$

73. $\dfrac{\tan \alpha + \tan \beta}{\cot \alpha + \cot \beta} = \tan \alpha \tan \beta$

74. $(\tan \alpha + \tan \beta)(1 - \cot \alpha \cot \beta) + (\cot \alpha + \cot \beta)(1 - \tan \alpha \tan \beta) = 0$

75. $(\sin \alpha + \cos \beta)^2 + (\cos \beta + \sin \alpha)(\cos \beta - \sin \alpha) = 2 \cos \beta(\sin \alpha + \cos \beta)$

76. $(\sin \alpha - \cos \beta)^2 + (\cos \beta + \sin \alpha)(\cos \beta - \sin \alpha) = -2 \cos \beta(\sin \alpha - \cos \beta)$

77.* $\ln|\sec \theta| = -\ln|\cos \theta|$

78. $\ln|\tan \theta| = \ln|\sin \theta| - \ln|\cos \theta|$

79. $\ln|1 + \cos \theta| + \ln|1 - \cos \theta| = 2 \ln|\sin \theta|$

80. $\ln|\sec \theta + \tan \theta| + \ln|\sec \theta - \tan \theta| = 0$

 81. Write a few paragraphs outlining your strategy for establishing identities.

3.2 | SUM AND DIFFERENCE FORMULAS

1 Use Sum and Difference Formulas to Find Exact Values
2 Use Sum and Difference Formulas to Establish Identities

In this section, we continue our derivation of trigonometric identities by obtaining formulas that involve the sum or difference of two angles, such as $\cos(\alpha + \beta)$, $\cos(\alpha - \beta)$, or $\sin(\alpha + \beta)$. These formulas are referred to as the **sum and difference formulas.** We begin with the formulas for $\cos(\alpha + \beta)$ and $\cos(\alpha - \beta)$.

Theorem   Sum and Difference Formulas for Cosines

$$\cos(\alpha + \beta) = \cos \alpha \cos \beta - \sin \alpha \sin \beta \qquad (1)$$
$$\cos(\alpha - \beta) = \cos \alpha \cos \beta + \sin \alpha \sin \beta \qquad (2)$$

In words, formula (1) states that the cosine of the sum of two angles equals the cosine of the first times the cosine of the second minus the sine of the first times the sine of the second.

*See Chapter 7 for a discussion of logarithm functions.

FIGURE 1

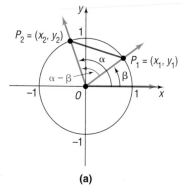

(a)

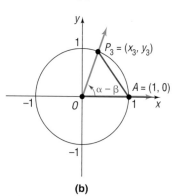

(b)

Proof   We will prove formula (2) first. Although this formula is true for all numbers α and β, we shall assume in our proof that $0 < \beta < \alpha < 2\pi$. We begin with a circle with center at the origin $(0, 0)$ and radius of 1 unit (the unit circle), and we place the angles α and β in standard position, as shown in Figure 1(a). The point $P_1 = (x_1, y_1)$ lies on the terminal side of β, and the point $P_2 = (x_2, y_2)$ lies on the terminal side of α.

Now, place the angle $\alpha - \beta$ in standard position, as shown in Figure 1(b), where the point A has coordinates $(1, 0)$ and the point $P_3 = (x_3, y_3)$ is on the terminal side of the angle $\alpha - \beta$.

Looking at triangle OP_1P_2 in Figure 1(a) and triangle OAP_3 in Figure 1(b), we see that these triangles are congruent. (Do you see why? Two sides and the included angle, $\alpha - \beta$, are equal.) Hence, the unknown side of each triangle must be equal; that is,

$$d(A, P_3) = d(P_1, P_2)$$

Using the distance formula, we find that

$$\sqrt{(x_3 - 1)^2 + y_3^2} = \sqrt{(x_2 - x_1)^2 + (y_2 - y_1)^2}$$
$$(x_3 - 1)^2 + y_3^2 = (x_2 - x_1)^2 + (y_2 - y_1)^2 \quad \text{Square each side.}$$
$$x_3^2 - 2x_3 + 1 + y_3^2 = x_2^2 - 2x_1x_2 + x_1^2 + y_2^2 - 2y_1y_2 + y_1^2 \quad (3)$$

Since $P_1 = (x_1, y_1)$, $P_2 = (x_2, y_2)$, and $P_3 = (x_3, y_3)$ are points on the unit circle $x^2 + y^2 = 1$, it follows that

$$x_1^2 + y_1^2 = 1 \qquad x_2^2 + y_2^2 = 1 \qquad x_3^2 + y_3^2 = 1$$

Consequently, equation (3) simplifies to

$$x_3^2 + y_3^2 - 2x_3 + 1 = (x_2^2 + y_2^2) + (x_1^2 + y_1^2) - 2x_1x_2 - 2y_1y_2$$
$$2 - 2x_3 = 2 - 2x_1x_2 - 2y_1y_2$$
$$x_3 = x_1x_2 + y_1y_2 \quad (4)$$

But $P_1 = (x_1, y_1)$ is on the terminal side of angle β and is on the unit circle. Thus,

$$\sin \beta = y_1 \qquad \cos \beta = x_1 \quad (5)$$

Similarly,

$$\sin \alpha = y_2 \qquad \cos \alpha = x_2 \qquad \cos(\alpha - \beta) = x_3 \quad (6)$$

Using equations (5) and (6) in equation (4), we get

$$\cos(\alpha - \beta) = \cos \alpha \cos \beta + \sin \alpha \sin \beta$$

which is formula (2).

The proof of formula (1) follows from formula (2). We use the fact that $\alpha + \beta = \alpha - (-\beta)$. Then

$$\cos(\alpha + \beta) = \cos[\alpha - (-\beta)]$$
$$= \cos \alpha \cos(-\beta) + \sin \alpha \sin(-\beta) \quad \text{Use formula (2).}$$
$$= \cos \alpha \cos \beta - \sin \alpha \sin \beta \quad \text{Even–odd identities.} \quad \blacksquare$$

1 One use of formulas (1) and (2) is to obtain the exact value of the cosine of an angle that can be expressed as the sum or difference of angles whose sine and cosine are known exactly.

E X A M P L E   1

Using the Sum Formula to Find Exact Values

Find the exact value of cos 75°.

Solution   Since $75° = 45° + 30°$, we use formula (1) to obtain

$$\cos 75° = \cos(45° + 30°) = \underset{\substack{\uparrow \\ \text{Formula (1)}}}{\cos 45° \cos 30°} - \sin 45° \sin 30°$$

$$= \frac{\sqrt{2}}{2} \cdot \frac{\sqrt{3}}{2} - \frac{\sqrt{2}}{2} \cdot \frac{1}{2} = \frac{1}{4}(\sqrt{6} - \sqrt{2})$$

∎

E X A M P L E   2

Using the Difference Formula to Find Exact Values

Find the exact value of $\cos(\pi/12)$.

Solution

$$\cos \frac{\pi}{12} = \cos\left(\frac{3\pi}{12} - \frac{2\pi}{12}\right) = \cos\left(\frac{\pi}{4} - \frac{\pi}{6}\right)$$

$$= \cos \frac{\pi}{4} \cos \frac{\pi}{6} + \sin \frac{\pi}{4} \sin \frac{\pi}{6} \qquad \text{Use formula (2).}$$

$$= \frac{\sqrt{2}}{2} \cdot \frac{\sqrt{3}}{2} + \frac{\sqrt{2}}{2} \cdot \frac{1}{2} = \frac{1}{4}(\sqrt{6} + \sqrt{2})$$

∎

 Now work Problem 3.

Another use of formulas (1) and (2) is to establish other identities. One important pair of identities is given next.

$$\cos\left(\frac{\pi}{2} - \theta\right) = \sin \theta \qquad\qquad (7a)$$

$$\sin\left(\frac{\pi}{2} - \theta\right) = \cos \theta \qquad\qquad (7b)$$

Seeing the Concept   Graph $Y_1 = \cos(\pi/2 - \theta)$ and $Y_2 = \sin \theta$ on the same screen. Does this demonstrate the result 7(a)? How would you demonstrate the result 7(b)?

∎

Proof   To prove formula (7a), we use the formula for $\cos(\alpha - \beta)$ with $\alpha = \pi/2$ and $\beta = \theta$.

$$\cos\left(\frac{\pi}{2} - \theta\right) = \cos \frac{\pi}{2} \cos \theta + \sin \frac{\pi}{2} \sin \theta$$

$$= 0 \cdot \cos \theta + 1 \cdot \sin \theta$$

$$= \sin \theta$$

To prove formula (7b), we make use of the identity (7a) just established.

$$\sin\left(\frac{\pi}{2} - \theta\right) = \underset{\substack{\uparrow \\ \text{Use (7a).}}}{\cos}\left[\frac{\pi}{2} - \left(\frac{\pi}{2} - \theta\right)\right] = \cos \theta$$

∎

Formulas (7a) and (7b) should look familiar. They are the basis for the theorem stated in Chapter 2: Cofunctions of complementary angles are equal.

Also, since

$$\cos\left(\frac{\pi}{2} - \theta\right) = \cos\left[-\left(\theta - \frac{\pi}{2}\right)\right] = \cos\left(\theta - \frac{\pi}{2}\right)$$
<small>↑
Even Property
of Cosine</small>

and

$$\cos\left(\frac{\pi}{2} - \theta\right) = \sin\theta$$
<small>↑
7(a)</small>

it follows that $\cos(\theta - \pi/2) = \sin\theta$. Thus, the graphs of $y = \cos(\theta - \pi/2)$ and $y = \sin\theta$ are identical, a fact that we conjectured earlier in Section 2.6.

 Now work Problem 31.

Formulas for $\sin(\alpha + \beta)$ and $\sin(\alpha - \beta)$

Having established the identities in formulas (7a) and (7b), we now can derive the sum and difference formulas for $\sin(\alpha + \beta)$ and $\sin(\alpha - \beta)$.

Proof

$$\begin{aligned}
\sin(\alpha + \beta) &= \cos\left[\frac{\pi}{2} - (\alpha + \beta)\right] && \text{Formula (7a).} \\
&= \cos\left[\left(\frac{\pi}{2} - \alpha\right) - \beta\right] \\
&= \cos\left(\frac{\pi}{2} - \alpha\right)\cos\beta + \sin\left(\frac{\pi}{2} - \alpha\right)\sin\beta && \text{Formula (2).} \\
&= \sin\alpha\cos\beta + \cos\alpha\sin\beta && \text{Formulas (7a) and (7b).} \\
\sin(\alpha - \beta) &= \sin[\alpha + (-\beta)] \\
&= \sin\alpha\cos(-\beta) + \cos\alpha\sin(-\beta) \\
&= \sin\alpha\cos\beta + \cos\alpha(-\sin\beta) && \text{Even–odd identities.} \\
&= \sin\alpha\cos\beta - \cos\alpha\sin\beta
\end{aligned}$$

Thus,

Theorem Sum and Difference Formulas for Sines

$$\sin(\alpha + \beta) = \sin\alpha\cos\beta + \cos\alpha\sin\beta \tag{8}$$
$$\sin(\alpha - \beta) = \sin\alpha\cos\beta - \cos\alpha\sin\beta \tag{9}$$

In words, formula (8) states that the sine of the sum of two angles equals the sine of the first times the cosine of the second plus the cosine of the first times the sine of the second.

E X A M P L E   3

Using the Sum Formula to Find Exact Values

Find the exact value of $\sin(7\pi/12)$.

Solution

$$\sin\frac{7\pi}{12} = \sin\left(\frac{3\pi}{12} + \frac{4\pi}{12}\right) = \sin\left(\frac{\pi}{4} + \frac{\pi}{3}\right)$$

$$= \sin\frac{\pi}{4}\cos\frac{\pi}{3} + \cos\frac{\pi}{4}\sin\frac{\pi}{3} \quad \text{Formula (8).}$$

$$= \frac{\sqrt{2}}{2}\cdot\frac{1}{2} + \frac{\sqrt{2}}{2}\cdot\frac{\sqrt{3}}{2} = \frac{1}{4}(\sqrt{2} + \sqrt{6})$$ ∎

 Now work Problem 9.

E X A M P L E   4

Using the Difference Formula to Find Exact Values

Find the exact value of $\sin 80° \cos 20° - \cos 80° \sin 20°$.

Solution

The form of the expression $\sin 80° \cos 20° - \cos 80° \sin 20°$ is that of the right side of the formula for $\sin(\alpha - \beta)$ with $\alpha = 80°$ and $\beta = 20°$. Thus,

$$\sin 80° \cos 20° - \cos 80° \sin 20° = \sin(80° - 20°) = \sin 60° = \frac{\sqrt{3}}{2}$$ ∎

 Now work Problem 19.

E X A M P L E   5

Finding Exact Values

If it is known that $\sin\alpha = \frac{4}{5}$, $\pi/2 < \alpha < \pi$, and that $\sin\beta = -2/\sqrt{5} = -2\sqrt{5}/5$, $\pi < \beta < 3\pi/2$, find the exact value of

(a) $\cos\alpha$ (b) $\cos\beta$ (c) $\cos(\alpha + \beta)$ (d) $\sin(\alpha + \beta)$

Solution

(a) Since $\sin\alpha = \frac{4}{5} = \frac{b}{r}$ and $\frac{\pi}{2} < \alpha < \pi$, we let $b = 4$ and $r = 5$. See Figure 2. Also, since $(a, 4)$ is on the circle $x^2 + y^2 = 25$ and is in Quadrant II, we have

$$a^2 + 4^2 = 25, \quad a < 0$$
$$a^2 = 25 - 16 = 9$$
$$a = -3$$

Thus,

$$\cos\alpha = \frac{a}{r} = -\frac{3}{5}$$

Alternatively, we can find $\cos\alpha$ using identities, as follows:

$$\cos\alpha = -\sqrt{1 - \sin^2\alpha} = -\sqrt{1 - \frac{16}{25}} = -\sqrt{\frac{9}{25}} = -\frac{3}{5}$$

↑
α in quadrant II
$\cos\alpha < 0$

FIGURE 2
Given $\sin\alpha = \frac{4}{5}$, $\pi/2 < \alpha < \pi$

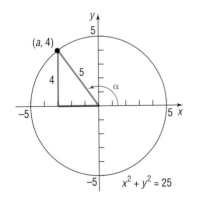

FIGURE 3
Given $\sin \beta = -2/\sqrt{5}$, $\pi < \beta < 3\pi/2$

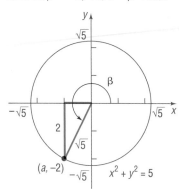

(b) Since $\sin \beta = \dfrac{-2}{\sqrt{5}} = \dfrac{b}{r}$ and $\pi < \beta < \dfrac{3\pi}{2}$, we let $b = -2$ and $r = \sqrt{5}$. See Figure 3. Since $(a, -2)$ is on the circle $x^2 + y^2 = 5$ and is in Quadrant III, we have

$$a^2 + (-2)^2 = 5, \qquad a < 0$$
$$a^2 = 5 - 4 = 1$$
$$a = -1$$

Thus,

$$\cos \beta = \frac{a}{r} = -\frac{1}{\sqrt{5}} = -\frac{\sqrt{5}}{5}$$

Alternatively, we can find $\cos \beta$ using identities, as follows:

$$\cos \beta = -\sqrt{1 - \sin^2 \beta} = -\sqrt{1 - \frac{4}{5}} = -\sqrt{\frac{1}{5}} = -\frac{\sqrt{5}}{5}$$

(c) Using the results found in parts (a) and (b) and formula (1), we have

$$\cos(\alpha + \beta) = \cos \alpha \cos \beta - \sin \alpha \sin \beta$$
$$= -\frac{3}{5}\left(-\frac{\sqrt{5}}{5}\right) - \frac{4}{5}\left(-\frac{2\sqrt{5}}{5}\right) = \frac{11\sqrt{5}}{25}$$

(d) $$\sin(\alpha + \beta) = \sin \alpha \cos \beta + \cos \alpha \sin \beta$$
$$= \frac{4}{5}\left(-\frac{\sqrt{5}}{5}\right) + \left(-\frac{3}{5}\right)\left(-\frac{2\sqrt{5}}{5}\right) = \frac{2\sqrt{5}}{25}$$

Now work Problem 23, parts (a), (b), and (c).

E X A M P L E   6

Establishing an Identity

Establish the identity: $\dfrac{\cos(\alpha - \beta)}{\sin \alpha \sin \beta} = \cot \alpha \cot \beta + 1$

Solution

$$\frac{\cos(\alpha - \beta)}{\sin \alpha \sin \beta} = \frac{\cos \alpha \cos \beta + \sin \alpha \sin \beta}{\sin \alpha \sin \beta}$$
$$= \frac{\cos \alpha \cos \beta}{\sin \alpha \sin \beta} + \frac{\sin \alpha \sin \beta}{\sin \alpha \sin \beta}$$
$$= \frac{\cos \alpha}{\sin \alpha} \frac{\cos \beta}{\sin \beta} + 1$$
$$= \cot \alpha \cot \beta + 1$$

Now work Problem 43.

Formulas for tan($\alpha + \beta$) and tan($\alpha - \beta$)

We use the identity $\tan \theta = (\sin \theta)/(\cos \theta)$ and the sum formulas for $\sin(\alpha + \beta)$ and $\cos(\alpha + \beta)$ to derive a formula for $\tan(\alpha + \beta)$.

Proof

$$\tan(\alpha + \beta) = \frac{\sin(\alpha + \beta)}{\cos(\alpha + \beta)} = \frac{\sin \alpha \cos \beta + \cos \alpha \sin \beta}{\cos \alpha \cos \beta - \sin \alpha \sin \beta}$$

Now we divide the numerator and denominator by $\cos \alpha \cos \beta$.

$$\tan(\alpha + \beta) = \frac{\dfrac{\sin \alpha \cos \beta + \cos \alpha \sin \beta}{\cos \alpha \cos \beta}}{\dfrac{\cos \alpha \cos \beta - \sin \alpha \sin \beta}{\cos \alpha \cos \beta}} = \frac{\dfrac{\sin \alpha \cos \beta}{\cos \alpha \cos \beta} + \dfrac{\cos \alpha \sin \beta}{\cos \alpha \cos \beta}}{\dfrac{\cos \alpha \cos \beta}{\cos \alpha \cos \beta} - \dfrac{\sin \alpha \sin \beta}{\cos \alpha \cos \beta}}$$

$$= \frac{\dfrac{\sin \alpha}{\cos \alpha} + \dfrac{\sin \beta}{\cos \beta}}{1 - \dfrac{\sin \alpha \sin \beta}{\cos \alpha \cos \beta}} = \frac{\tan \alpha + \tan \beta}{1 - \tan \alpha \tan \beta}$$

We use the sum formula for $\tan(\alpha + \beta)$ and even–odd properties to get the difference formula.

$$\tan(\alpha - \beta) = \tan[\alpha + (-\beta)] = \frac{\tan \alpha + \tan(-\beta)}{1 - \tan \alpha \tan(-\beta)} = \frac{\tan \alpha - \tan \beta}{1 + \tan \alpha \tan \beta}$$

Thus, we have proved the following results:

Theorem   Sum and Difference Formulas for Tangents

$$\tan(\alpha + \beta) = \frac{\tan \alpha + \tan \beta}{1 - \tan \alpha \tan \beta} \qquad (10)$$

$$\tan(\alpha - \beta) = \frac{\tan \alpha - \tan \beta}{1 + \tan \alpha \tan \beta} \qquad (11)$$

In words, formula (10) states that the tangent of the sum of two angles equals the tangent of the first plus the tangent of the second divided by 1 minus their product.

EXAMPLE 7

Establishing an Identity

Prove the identity: $\tan(\theta + \pi) = \tan \theta$

Solution

$$\tan(\theta + \pi) = \frac{\tan \theta + \tan \pi}{1 - \tan \theta \tan \pi} = \frac{\tan \theta + 0}{1 - \tan \theta \cdot 0} = \tan \theta$$

The result obtained in Example 7 verifies that the tangent function is periodic with period π, a fact that we mentioned earlier.

Warning Be careful when using formulas (10) and (11). These formulas can be used only for angles α and β for which $\tan \alpha$ and $\tan \beta$ are defined, that is, all angles except odd multiples of $\pi/2$. ∎

E X A M P L E 8   Establishing an Identity

Prove the identity: $\tan\left(\theta + \dfrac{\pi}{2}\right) = -\cot \theta$

Solution   We cannot use formula (10), since $\tan(\pi/2)$ is not defined. Instead, we proceed as follows:

$$\tan\left(\theta + \frac{\pi}{2}\right) = \frac{\sin\left(\theta + \dfrac{\pi}{2}\right)}{\cos\left(\theta + \dfrac{\pi}{2}\right)} = \frac{\sin \theta \cos \dfrac{\pi}{2} + \cos \theta \sin \dfrac{\pi}{2}}{\cos \theta \cos \dfrac{\pi}{2} - \sin \theta \sin \dfrac{\pi}{2}}$$

$$= \frac{(\sin \theta)(0) + (\cos \theta)(1)}{(\cos \theta)(0) - (\sin \theta)(1)} = \frac{\cos \theta}{-\sin \theta} = -\cot \theta$$

∎

SUMMARY

The following box summarizes the sum and difference formulas.

Sum and Difference Formulas

$$\cos(\alpha + \beta) = \cos \alpha \cos \beta - \sin \alpha \sin \beta$$

$$\cos(\alpha - \beta) = \cos \alpha \cos \beta + \sin \alpha \sin \beta$$

$$\sin(\alpha + \beta) = \sin \alpha \cos \beta + \cos \alpha \sin \beta$$

$$\sin(\alpha - \beta) = \sin \alpha \cos \beta - \cos \alpha \sin \beta$$

$$\tan(\alpha + \beta) = \frac{\tan \alpha + \tan \beta}{1 - \tan \alpha \tan \beta}$$

$$\tan(\alpha - \beta) = \frac{\tan \alpha - \tan \beta}{1 + \tan \alpha \tan \beta}$$

3.2 | EXERCISES

In Problems 1–12, find the exact value of each trigonometric function.

1. $\sin \dfrac{5\pi}{12}$ **2.** $\sin \dfrac{\pi}{12}$ **3.** $\cos \dfrac{7\pi}{12}$ **4.** $\tan \dfrac{7\pi}{12}$ **5.** $\cos 165°$ **6.** $\sin 105°$

7. $\tan 15°$ **8.** $\tan 195°$ **9.** $\sin \dfrac{17\pi}{12}$ **10.** $\tan \dfrac{19\pi}{12}$ **11.** $\sec\left(-\dfrac{\pi}{12}\right)$ **12.** $\cot\left(-\dfrac{5\pi}{12}\right)$

In Problems 13–22, find the exact value of each expression.

13. $\sin 20° \cos 10° + \cos 20° \sin 10°$

14. $\sin 20° \cos 80° - \cos 20° \sin 80°$

15. $\cos 70° \cos 20° - \sin 70° \sin 20°$

16. $\cos 40° \cos 10° + \sin 40° \sin 10°$

17. $\dfrac{\tan 20° + \tan 25°}{1 - \tan 20° \tan 25°}$

18. $\dfrac{\tan 40° - \tan 10°}{1 + \tan 40° \tan 10°}$

19. $\sin \dfrac{\pi}{12} \cos \dfrac{7\pi}{12} - \cos \dfrac{\pi}{12} \sin \dfrac{7\pi}{12}$

20. $\cos \dfrac{5\pi}{12} \cos \dfrac{7\pi}{12} - \sin \dfrac{5\pi}{12} \sin \dfrac{7\pi}{12}$

21. $\cos \dfrac{\pi}{12} \cos \dfrac{5\pi}{12} + \sin \dfrac{5\pi}{12} \sin \dfrac{\pi}{12}$

22. $\sin \dfrac{\pi}{18} \cos \dfrac{5\pi}{18} + \cos \dfrac{\pi}{18} \sin \dfrac{5\pi}{18}$

In Problems 23–28, find the exact value of each of the following under the given conditions:

(a) $\sin(\alpha + \beta)$ (b) $\cos(\alpha + \beta)$ (c) $\sin(\alpha - \beta)$ (d) $\tan(\alpha - \beta)$

23. $\sin \alpha = \frac{3}{5}, \quad 0 < \alpha < \pi/2; \quad \cos \beta = 2\sqrt{5}/5, \quad -\pi/2 < \beta < 0$

24. $\cos \alpha = \sqrt{5}/5, \quad 0 < \alpha < \pi/2; \quad \sin \beta = -\frac{4}{5}, \quad -\pi/2 < \beta < 0$

25. $\tan \alpha = -\frac{4}{3}, \quad \pi/2 < \alpha < \pi; \quad \cos \beta = \frac{1}{2}, \quad 0 < \beta < \pi/2$

26. $\tan \alpha = \frac{5}{12}, \quad \pi < \alpha < 3\pi/2; \quad \sin \beta = -\frac{1}{2}, \quad \pi < \beta < 3\pi/2$

27. $\sin \alpha = \frac{5}{13}, \quad -3\pi/2 < \alpha < -\pi; \quad \tan \beta = -\sqrt{3}, \quad \pi/2 < \beta < \pi$

28. $\cos \alpha = \frac{1}{2}, \quad -\pi/2 < \alpha < 0; \quad \sin \beta = \frac{1}{3}, \quad 0 < \beta < \pi/2$

29. If $\sin \theta = \frac{1}{3}$, θ in quadrant II, find the exact value of

(a) $\cos \theta$ (b) $\sin(\theta + \pi/6)$ (c) $\cos(\theta - \pi/3)$ (d) $\tan(\theta + \pi/4)$

30. If $\cos \theta = \frac{1}{4}$, θ in quadrant IV, find the exact value of

(a) $\sin \theta$ (b) $\sin(\theta - \pi/6)$ (c) $\cos(\theta + \pi/3)$ (d) $\tan(\theta - \pi/4)$

In Problems 31–56, establish each identity.

31. $\sin\left(\dfrac{\pi}{2} + \theta\right) = \cos \theta$

32. $\cos\left(\dfrac{\pi}{2} + \theta\right) = -\sin \theta$

33. $\sin(\pi - \theta) = \sin \theta$

34. $\cos(\pi - \theta) = -\cos \theta$

35. $\sin(\pi + \theta) = -\sin \theta$

36. $\cos(\pi + \theta) = -\cos \theta$

37. $\tan(\pi - \theta) = -\tan \theta$

38. $\tan(2\pi - \theta) = -\tan \theta$

39. $\sin\left(\dfrac{3\pi}{2} + \theta\right) = -\cos \theta$

40. $\cos\left(\dfrac{3\pi}{2} + \theta\right) = \sin \theta$

41. $\sin(\alpha + \beta) + \sin(\alpha - \beta) = 2 \sin \alpha \cos \beta$

42. $\cos(\alpha + \beta) + \cos(\alpha - \beta) = 2 \cos \alpha \cos \beta$

43. $\dfrac{\sin(\alpha + \beta)}{\sin \alpha \cos \beta} = 1 + \cot \alpha \tan \beta$

44. $\dfrac{\sin(\alpha + \beta)}{\cos \alpha \cos \beta} = \tan \alpha + \tan \beta$

45. $\dfrac{\cos(\alpha + \beta)}{\cos \alpha \cos \beta} = 1 - \tan \alpha \tan \beta$

46. $\dfrac{\cos(\alpha - \beta)}{\sin \alpha \cos \beta} = \cot \alpha + \tan \beta$

47. $\dfrac{\sin(\alpha + \beta)}{\sin(\alpha - \beta)} = \dfrac{\tan \alpha + \tan \beta}{\tan \alpha - \tan \beta}$

48. $\dfrac{\cos(\alpha + \beta)}{\cos(\alpha - \beta)} = \dfrac{1 - \tan \alpha \tan \beta}{1 + \tan \alpha \tan \beta}$

49. $\cot(\alpha + \beta) = \dfrac{\cot \alpha \cot \beta - 1}{\cot \beta + \cot \alpha}$

50. $\cot(\alpha - \beta) = \dfrac{\cot \alpha \cot \beta + 1}{\cot \beta - \cot \alpha}$

51. $\sec(\alpha + \beta) = \dfrac{\csc \alpha \csc \beta}{\cot \alpha \cot \beta - 1}$

52. $\sec(\alpha - \beta) = \dfrac{\sec \alpha \sec \beta}{1 + \tan \alpha \tan \beta}$

53. $\sin(\alpha - \beta) \sin(\alpha + \beta) = \sin^2 \alpha - \sin^2 \beta$

54. $\cos(\alpha - \beta) \cos(\alpha + \beta) = \cos^2 \alpha - \sin^2 \beta$

55. $\sin(\theta + k\pi) = (-1)^k \sin \theta, \quad k$ any integer

56. $\cos(\theta + k\pi) = (-1)^k \cos \theta, \quad k$ any integer

57. Calculus Show that the difference quotient for $f(x) = \sin x$ is given by

$$\frac{f(x + h) - f(x)}{h} = \frac{\sin(x + h) - \sin x}{h}$$

$$= \cos x \cdot \frac{\sin h}{h} - \sin x \cdot \frac{1 - \cos h}{h}$$

58. Calculus Show that the difference quotient for $f(x) = \cos x$ is given by

$$\frac{f(x + h) - f(x)}{h} = \frac{\cos(x + h) - \cos x}{h}$$

$$= -\sin x \cdot \frac{\sin h}{h} - \cos x \cdot \frac{1 - \cos h}{h}$$

59. Explain why formula (11) cannot be used to show that

$$\tan\left(\frac{\pi}{2} - \theta\right) = \cot \theta$$

Establish this identity by using formulas (7a) and (7b).

60. If $\tan \alpha = x + 1$ and $\tan \beta = x - 1$, show that $2 \cot(\alpha - \beta) = x^2$.

61. Geometry: Angle between Two Lines Let L_1 and L_2 denote two nonvertical intersecting lines, and let θ denote the acute angle between L_1 and L_2 (see the figure). Show that

$$\tan \theta = \frac{m_2 - m_1}{1 + m_1 m_2}$$

where m_1 and m_2 are the slopes of L_1 and L_2, respectively.

[**Hint:** Use the facts that $\tan \theta_1 = m_1$ and $\tan \theta_2 = m_2$.]

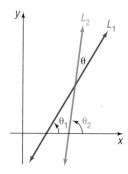

62. If $\alpha + \beta + \gamma = 180°$ and

$$\cot \theta = \cot \alpha + \cot \beta + \cot \gamma, 0 < \theta < 90°,$$

show that

$$\sin^3 \theta = \sin(\alpha - \theta) \sin(\beta - \theta) \sin(\gamma - \theta)$$

 63. Discuss the following derivation:

$$\tan\left(\theta + \frac{\pi}{2}\right) = \frac{\tan \theta + \tan(\pi/2)}{1 - \tan \theta \tan(\pi/2)}$$

$$= \frac{\dfrac{\tan \theta}{\tan(\pi/2)} + 1}{\dfrac{1}{\tan(\pi/2)} - \tan \theta} = \frac{0 + 1}{0 - \tan \theta}$$

$$= \frac{1}{-\tan \theta} = -\cot \theta$$

Can you justify each step?

3.3 DOUBLE-ANGLE AND HALF-ANGLE FORMULAS

> 1 Use Double-Angle Formulas to Find Exact Values
> 2 Use Double-Angle and Half-Angle Formulas to Establish Identities
> 3 Use Half-Angle Formulas to Find Exact Values

In this section we derive formulas for $\sin 2\theta$, $\cos 2\theta$, $\sin\left(\frac{1}{2}\theta\right)$, and $\cos\left(\frac{1}{2}\theta\right)$ in terms of $\sin \theta$ and $\cos \theta$. They are easily derived using the sum formulas.

Double-Angle Formulas

In the sum formulas for $\sin(\alpha + \beta)$ and $\cos(\alpha + \beta)$, let $\alpha = \beta = \theta$. Then

$$\sin(\alpha + \beta) = \sin \alpha \cos \beta + \cos \alpha \sin \beta$$
$$\sin(\theta + \theta) = \sin \theta \cos \theta + \cos \theta \sin \theta$$
$$\sin 2\theta = 2 \sin \theta \cos \theta \qquad\qquad (1)$$

and

$$\cos(\alpha + \beta) = \cos\alpha\cos\beta - \sin\alpha\sin\beta$$
$$\cos(\theta + \theta) = \cos\theta\cos\theta - \sin\theta\sin\theta$$
$$\cos 2\theta = \cos^2\theta - \sin^2\theta \qquad (2)$$

An application of the Pythagorean identity $\sin^2\theta + \cos^2\theta = 1$ results in two other ways to write formula (2) for $\cos 2\theta$.

$$\cos 2\theta = \cos^2\theta - \sin^2\theta = (1 - \sin^2\theta) - \sin^2\theta = 1 - 2\sin^2\theta$$

and

$$\cos 2\theta = \cos^2\theta - \sin^2\theta = \cos^2\theta - (1 - \cos^2\theta) = 2\cos^2\theta - 1$$

Thus, we have established the following **double-angle formulas:**

Theorem Double-Angle Formulas

$$\sin 2\theta = 2\sin\theta\cos\theta \qquad (3)$$
$$\cos 2\theta = \cos^2\theta - \sin^2\theta \qquad (4a)$$
$$\cos 2\theta = 1 - 2\sin^2\theta \qquad (4b)$$
$$\cos 2\theta = 2\cos^2\theta - 1 \qquad (4c)$$

E X A M P L E   1

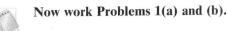

Finding Exact Values Using the Double-Angle Formula

If $\sin\theta = \frac{3}{5}$, $\pi/2 < \theta < \pi$, find the exact value of

(a) $\sin 2\theta$ (b) $\cos 2\theta$

Solution

(a) Because $\sin 2\theta = 2\sin\theta\cos\theta$ and we already know that $\sin\theta = \frac{3}{5}$, we only need to find $\cos\theta$. See Figure 4. Since $\sin\theta = \frac{3}{5} = \frac{b}{r}$, we let $b = 3$ and $r = 5$. Since $\pi/2 < \theta < \pi$, the point $(a, 3)$ is on the circle $x^2 + y^2 = 25$ and is in Quadrant II. Thus,

$$a^2 + 3^2 = 25, \qquad a < 0$$
$$a^2 = 25 - 9 = 16$$
$$a = -4$$

Thus, $\cos\theta = -\frac{4}{5}$. Now we use formula (3) to obtain

$$\sin 2\theta = 2\sin\theta\cos\theta = 2(\tfrac{3}{5})(-\tfrac{4}{5}) = -\tfrac{24}{25}$$

(b) Because we are given $\sin\theta = \frac{3}{5}$, it is easiest to use formula (4b) to get $\cos 2\theta$.

$$\cos 2\theta = 1 - 2\sin^2\theta = 1 - 2(\tfrac{9}{25}) = 1 - \tfrac{18}{25} = \tfrac{7}{25}$$

FIGURE 4

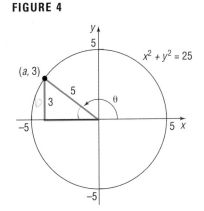

Warning In finding $\cos 2\theta$ in Example 1(b), we chose to use a version of the double-angle formula, formula (4b). Note that we are unable to use the Pythagorean identity $\cos 2\theta = \pm\sqrt{1 - \sin^2 2\theta}$, with $\sin 2\theta = -\frac{24}{25}$, because we have no way of knowing which sign to choose.

Now work Problems 1(a) and (b).

E X A M P L E 2

Establishing Identities

(a) Develop a formula for tan 2θ in terms of tan θ.

(b) Develop a formula for sin 3θ in terms of sin θ and cos θ.

Solution (a) In the sum formula for tan$(\alpha + \beta)$, let $\alpha = \beta = \theta$. Then

$$\tan(\alpha + \beta) = \frac{\tan \alpha + \tan \beta}{1 - \tan \alpha \tan \beta}$$

$$\tan(\theta + \theta) = \frac{\tan \theta + \tan \theta}{1 - \tan \theta \tan \theta}$$

$$\tan 2\theta = \frac{2 \tan \theta}{1 - \tan^2 \theta} \qquad (5)$$

(b) To get a formula for sin 3θ, we use the sum formula and write 3θ as $2\theta + \theta$.

$$\sin 3\theta = \sin(2\theta + \theta) = \sin 2\theta \cos \theta + \cos 2\theta \sin \theta$$

Now use the double-angle formulas to get

$$\sin 3\theta = (2 \sin \theta \cos \theta)(\cos \theta) + (\cos^2 \theta - \sin^2 \theta)(\sin \theta)$$
$$= 2 \sin \theta \cos^2 \theta + \sin \theta \cos^2 \theta - \sin^3 \theta$$
$$= 3 \sin \theta \cos^2 \theta - \sin^3 \theta$$

The formula obtained in Example 2(b) can also be written as

$$\sin 3\theta = 3 \sin \theta \cos^2 \theta - \sin^3 \theta = 3 \sin \theta(1 - \sin^2 \theta) - \sin^3 \theta$$
$$= 3 \sin \theta - 4 \sin^3 \theta$$

That is, sin 3θ is a third-degree polynomial in the variable sin θ. In fact, sin $n\theta$, n a positive odd integer, can always be written as a polynomial of degree n in the variable sin θ.*

Now work Problem 47.

Other Variations of the Double-Angle Formulas

By rearranging the double-angle formulas (4b) and (4c), we obtain other formulas that we will use a little later in this section.

We begin with formula (4b) and proceed to solve for $\sin^2 \theta$.

$$\cos 2\theta = 1 - 2 \sin^2 \theta$$
$$2 \sin^2 \theta = 1 - \cos 2\theta$$

*Due to the work done by P. L. Chebyshëv, these polynomials are sometimes called *Chebyshëv polynomials*.

$$\sin^2 \theta = \frac{1 - \cos 2\theta}{2} \qquad (6)$$

Similarly, using formula (4c), we proceed to solve for $\cos^2 \theta$:

$$\cos 2\theta = 2 \cos^2 \theta - 1$$
$$2 \cos^2 \theta = 1 + \cos 2\theta$$

$$\cos^2 \theta = \frac{1 + \cos 2\theta}{2} \qquad (7)$$

Formulas (6) and (7) can be used to develop a formula for $\tan^2 \theta$:

$$\tan^2 \theta = \frac{\sin^2 \theta}{\cos^2 \theta} = \frac{\dfrac{1 - \cos 2\theta}{2}}{\dfrac{1 + \cos 2\theta}{2}}$$

$$\tan^2 \theta = \frac{1 - \cos 2\theta}{1 + \cos 2\theta} \qquad (8)$$

Formulas (6) through (8) do not have to be memorized since their derivations are so straightforward.

Formulas (6) and (7) are important in calculus. The next example illustrates a problem that arises in calculus requiring the use of formula (7).

EXAMPLE 3

Establishing an Identity

Write an equivalent expression for $\cos^4 \theta$ that does not involve any powers of sine or cosine greater than 1.

Solution   The idea here is to apply formula (7) twice.

$$\cos^4 \theta = (\cos^2 \theta)^2 = \left(\frac{1 + \cos 2\theta}{2}\right)^2 \qquad \text{Formula (7).}$$

$$= \frac{1}{4}(1 + 2 \cos 2\theta + \cos^2 2\theta)$$

$$= \frac{1}{4} + \frac{1}{2} \cos 2\theta + \frac{1}{4} \cos^2 2\theta$$

$$= \frac{1}{4} + \frac{1}{2} \cos 2\theta + \frac{1}{4}\left[\frac{1 + \cos 2(2\theta)}{2}\right] \qquad \text{Formula (7).}$$

$$= \frac{1}{4} + \frac{1}{2} \cos 2\theta + \frac{1}{8}(1 + \cos 4\theta)$$

$$= \frac{3}{8} + \frac{1}{2} \cos 2\theta + \frac{1}{8} \cos 4\theta$$

Now work Problem 23.

Identities, such as the double-angle formulas, can sometimes be used to rewrite expressions in a more suitable form. Let's look at an example.

E X A M P L E   4   Projectile Motion

FIGURE 5

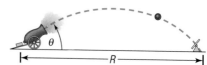

An object is propelled upward at an angle θ to the horizontal with an initial velocity of v_0 feet per second. See Figure 5. If air resistance is ignored, the **range** R, the horizontal distance R the object travels, is given by

$$R = \frac{1}{16} v_0^2 \sin \theta \cos \theta$$

(a) Show that $R = \dfrac{1}{32} v_0^2 \sin 2\theta$.

(b) Find the angle θ for which R is a maximum.

Solution   (a) We rewrite the given expression for the range using the double-angle formula $\sin 2\theta = 2 \sin \theta \cos \theta$. Then

$$R = \frac{1}{16} v_0^2 \sin \theta \cos \theta = \frac{1}{16} v_0^2 \frac{2 \sin \theta \cos \theta}{2} = \frac{1}{32} v_0^2 \sin 2\theta$$

(b) In this form, the largest value for the range R can be easily found. For a fixed initial speed v_0, the angle θ of inclination to the horizontal determines the value of R. Since the largest value of a sine function is 1, occurring when the argument 2θ is 90°, it follows that for maximum R we must have

$$2\theta = 90°$$
$$\theta = 45°$$

An inclination to the horizontal of 45° results in maximum range. ▪

Half-Angle Formulas

Another important use of formulas (6) through (8) is to prove the **half-angle formulas.** In formulas (6) through (8), let $\theta = \alpha/2$. Then

$$\sin^2 \frac{\alpha}{2} = \frac{1 - \cos \alpha}{2} \qquad \cos^2 \frac{\alpha}{2} = \frac{1 + \cos \alpha}{2}$$

$$\tan^2 \frac{\alpha}{2} = \frac{1 - \cos \alpha}{1 + \cos \alpha} \qquad (9)$$

If we solve for the trigonometric functions on the left sides of equations (9), we obtain the half-angle formulas.

Theorem   Half-Angle Formulas

$$\sin\frac{\alpha}{2} = \pm\sqrt{\frac{1-\cos\alpha}{2}} \tag{10a}$$

$$\cos\frac{\alpha}{2} = \pm\sqrt{\frac{1+\cos\alpha}{2}} \tag{10b}$$

$$\tan\frac{\alpha}{2} = \pm\sqrt{\frac{1-\cos\alpha}{1+\cos\alpha}} \tag{10c}$$

where the $+$ or $-$ sign is determined by the quadrant of the angle $\alpha/2$.

We use the half-angle formulas in the next example.

E X A M P L E   5   Finding Exact Values Using Half-Angle Formulas

3   Find the exact value of

(a) $\cos 15°$ (b) $\sin(-15°)$

Solution   (a) Because $15° = 30°/2$, we can use the half-angle formula for $\cos(\alpha/2)$ with $\alpha = 30°$. Also, because $15°$ is in quadrant I, $\cos 15° > 0$, and we choose the $+$ sign in using formula (10b):

$$\cos 15° = \cos\frac{30°}{2} = \sqrt{\frac{1+\cos 30°}{2}}$$

$$= \sqrt{\frac{1+\sqrt{3}/2}{2}} = \sqrt{\frac{2+\sqrt{3}}{4}} = \frac{\sqrt{2+\sqrt{3}}}{2}$$

(b) We use the fact that $\sin(-15°) = -\sin 15°$ and then apply formula (10a).

$$\sin(-15°) = -\sin\frac{30°}{2} = -\sqrt{\frac{1-\cos 30°}{2}}$$

$$= -\sqrt{\frac{1-\sqrt{3}/2}{2}} = -\sqrt{\frac{2-\sqrt{3}}{4}} = -\frac{\sqrt{2-\sqrt{3}}}{2}$$

It is interesting to compare the answer found in Example 5(a) with the answer to Example 2 of Section 3.2. There we calculated

$$\cos\frac{\pi}{12} = \cos 15° = \frac{1}{4}(\sqrt{6}+\sqrt{2})$$

Based on these results, we conclude that

$$\frac{1}{4}(\sqrt{6}+\sqrt{2}) \quad\text{and}\quad \frac{\sqrt{2+\sqrt{3}}}{2}$$

are equal. (Since each expression is positive, you can verify this equality by squaring each expression.) Thus, two very different looking, yet correct, answers can be obtained, depending on the approach taken to solve a problem.

  Now work Problem 13.

E X A M P L E   6   Finding Exact Values Using Half-Angle Formulas

If $\cos \alpha = -\frac{3}{5}$, $\pi < \alpha < 3\pi/2$, find the exact value of

(a) $\sin \dfrac{\alpha}{2}$ (b) $\cos \dfrac{\alpha}{2}$ (c) $\tan \dfrac{\alpha}{2}$

Solution   First, we observe that if $\pi < \alpha < 3\pi/2$, then $\pi/2 < \alpha/2 < 3\pi/4$. As a result, $\alpha/2$ lies in quadrant II.

(a) Because $\alpha/2$ lies in quadrant II, $\sin(\alpha/2) > 0$. Thus, we use the $+$ sign in formula (10a) to get

$$\sin \frac{\alpha}{2} = \sqrt{\frac{1 - \cos \alpha}{2}} = \sqrt{\frac{1 - (-\frac{3}{5})}{2}}$$

$$= \sqrt{\frac{\frac{8}{5}}{2}} = \sqrt{\frac{4}{5}} = \frac{2}{\sqrt{5}} = \frac{2\sqrt{5}}{5}$$

(b) Because $\alpha/2$ lies in quadrant II, $\cos(\alpha/2) < 0$. Thus, we use the $-$ sign in formula (10b) to get

$$\cos \frac{\alpha}{2} = -\sqrt{\frac{1 + \cos \alpha}{2}} = -\sqrt{\frac{1 + (-\frac{3}{5})}{2}}$$

$$= -\sqrt{\frac{\frac{2}{5}}{2}} = -\frac{1}{\sqrt{5}} = -\frac{\sqrt{5}}{5}$$

(c) Because $\alpha/2$ lies in quadrant II, $\tan(\alpha/2) < 0$. Thus, we use the $-$ sign in formula (10c) to get

$$\tan \frac{\alpha}{2} = -\sqrt{\frac{1 - \cos \alpha}{1 + \cos \alpha}} = -\sqrt{\frac{1 - (-\frac{3}{5})}{1 + (-\frac{3}{5})}} = -\sqrt{\frac{\frac{8}{5}}{\frac{2}{5}}} = -2$$

Another way to solve Example 6(c) is to use the solutions found in parts (a) and (b).

$$\tan \frac{\alpha}{2} = \frac{\sin(\alpha/2)}{\cos(\alpha/2)} = \frac{2\sqrt{5}/5}{-\sqrt{5}/5} = -2$$

    Now work Problems 1(c) and (d).

There is a formula for $\tan(\alpha/2)$ that does not contain $+$ and $-$ signs, making it more useful than Formula 10(c). Because

$$1 - \cos \alpha = 2 \sin^2\left(\frac{\alpha}{2}\right) \qquad \text{Formula 9.}$$

and

$$\sin \alpha = \sin 2\left(\frac{\alpha}{2}\right) = 2 \sin\left(\frac{\alpha}{2}\right) \cos\left(\frac{\alpha}{2}\right) \qquad \text{Double-angle formula.}$$

we have

$$\frac{1 - \cos \alpha}{\sin \alpha} = \frac{2 \sin^2\left(\dfrac{\alpha}{2}\right)}{2 \sin\left(\dfrac{\alpha}{2}\right) \cos\left(\dfrac{\alpha}{2}\right)} = \frac{\sin\left(\dfrac{\alpha}{2}\right)}{\cos\left(\dfrac{\alpha}{2}\right)} = \tan\left(\dfrac{\alpha}{2}\right)$$

Since it also can be shown that

$$\frac{1 - \cos \alpha}{\sin \alpha} = \frac{\sin \alpha}{1 + \cos \alpha}$$

we have the following two half-angle formulas:

Half-Angle Formulas for tan $\alpha/2$

$$\tan\left(\frac{\alpha}{2}\right) = \frac{1 - \cos \alpha}{\sin \alpha} = \frac{\sin \alpha}{1 + \cos \alpha} \qquad (11)$$

With this formula, the solution to Example 6(c) can be given as

$$\cos \alpha = -\frac{3}{5}$$

$$\sin \alpha = -\sqrt{1 - \cos^2 \alpha} = -\sqrt{1 - \frac{9}{25}} = -\sqrt{\frac{16}{25}} = -\frac{4}{5}$$

Then, by equation (11),

$$\tan \frac{\alpha}{2} = \frac{1 - \cos \alpha}{\sin \alpha} = \frac{1 - (-\frac{3}{5})}{-\frac{4}{5}} = \frac{\frac{8}{5}}{-\frac{4}{5}} = -2$$

The next example illustrates a problem that arises in calculus.

EXAMPLE 7   **Using Half-Angle Formulas in Calculus**

If $z = \tan(\alpha/2)$, show that

(a) $\sin \alpha = \dfrac{2z}{1 + z^2}$ (b) $\cos \alpha = \dfrac{1 - z^2}{1 + z^2}$

Solution   (a) $\dfrac{2z}{1 + z^2} = \dfrac{2 \tan(\alpha/2)}{1 + \tan^2(\alpha/2)} = \dfrac{2 \tan(\alpha/2)}{\sec^2(\alpha/2)} = \dfrac{2 \cdot \dfrac{\sin(\alpha/2)}{\cos(\alpha/2)}}{\dfrac{1}{\cos^2(\alpha/2)}}$

$$= \frac{2 \sin(\alpha/2)}{\cos(\alpha/2)} \cdot \cos^2 \frac{\alpha}{2} = 2 \sin \frac{\alpha}{2} \cos \frac{\alpha}{2}$$

$$= \underset{\underset{\text{Double-angle formula (3)}}{\uparrow}}{\sin 2\left(\frac{\alpha}{2}\right)} = \sin \alpha$$

$$(b) \quad \frac{1 - z^2}{1 + z^2} = \frac{1 - \tan^2(\alpha/2)}{1 + \tan^2(\alpha/2)} = \frac{1 - \dfrac{\sin^2(\alpha/2)}{\cos^2(\alpha/2)}}{1 + \dfrac{\sin^2(\alpha/2)}{\cos^2(\alpha/2)}}$$

$$= \frac{\cos^2(\alpha/2) - \sin^2(\alpha/2)}{\cos^2(\alpha/2) + \sin^2(\alpha/2)} = \underset{\uparrow}{\frac{\cos 2(\alpha/2)}{1}} = \cos \alpha$$

Double-angle formula (4a)
and Pythagorean identity

3.3 | EXERCISES

In Problems 1–12, use the information given about the angle θ, $0 \leq \theta \leq 2\pi$, to find the exact value of

(a) $\sin 2\theta$ (b) $\cos 2\theta$ (c) $\sin \dfrac{\theta}{2}$ (d) $\cos \dfrac{\theta}{2}$

1. $\sin \theta = \frac{3}{5}$, $0 < \theta < \pi/2$ 2. $\cos \theta = \frac{3}{5}$, $0 < \theta < \pi/2$ 3. $\tan \theta = \frac{4}{3}$, $\pi < \theta < 3\pi/2$

4. $\tan \theta = \frac{1}{2}$, $\pi < \theta < 3\pi/2$ 5. $\cos \theta = -\sqrt{6}/3$, $\pi/2 < \theta < \pi$ 6. $\sin \theta = -\sqrt{3}/3$, $3\pi/2 < \theta < 2\pi$

7. $\sec \theta = 3$, $\sin \theta > 0$ 8. $\csc \theta = -\sqrt{5}$, $\cos \theta < 0$ 9. $\cot \theta = -2$, $\sec \theta < 0$

10. $\sec \theta = 2$, $\csc \theta < 0$ 11. $\tan \theta = -3$, $\sin \theta < 0$ 12. $\cot \theta = 3$, $\cos \theta < 0$

In Problems 13–22, use the half-angle formulas to find the exact value of each trigonometric function.

13. $\sin 22.5°$ 14. $\cos 22.5°$ 15. $\tan \dfrac{7\pi}{8}$ 16. $\tan \dfrac{9\pi}{8}$ 17. $\cos 165°$

18. $\sin 195°$ 19. $\sec \dfrac{15\pi}{8}$ 20. $\csc \dfrac{7\pi}{8}$ 21. $\sin\left(-\dfrac{\pi}{8}\right)$ 22. $\cos\left(-\dfrac{3\pi}{8}\right)$

23. Show that $\sin^4 \theta = \frac{3}{8} - \frac{1}{2}\cos 2\theta + \frac{1}{8}\cos 4\theta$.

24. Develop a formula for $\cos 3\theta$ as a third-degree polynomial in the variable $\cos \theta$.

25. Show that $\sin 4\theta = (\cos \theta)(4 \sin \theta - 8 \sin^3 \theta)$.

26. Develop a formula for $\cos 4\theta$ as a fourth-degree polynomial in the variable $\cos \theta$.

27. Find an expression for $\sin 5\theta$ as a fifth-degree polynomial in the variable $\sin \theta$.

28. Find an expression for $\cos 5\theta$ as a fifth-degree polynomial in the variable $\cos \theta$.

In Problems 29–48, establish each identity.

29. $\cos^4 \theta - \sin^4 \theta = \cos 2\theta$

30. $\dfrac{\cot \theta - \tan \theta}{\cot \theta + \tan \theta} = \cos 2\theta$

31. $\cot 2\theta = \dfrac{\cot^2 \theta - 1}{2 \cot \theta}$

32. $\cot 2\theta = \frac{1}{2}(\cot \theta - \tan \theta)$

33. $\sec 2\theta = \dfrac{\sec^2 \theta}{2 - \sec^2 \theta}$

34. $\csc 2\theta = \frac{1}{2} \sec \theta \csc \theta$

35. $\cos^2 2\theta - \sin^2 2\theta = \cos 4\theta$

36. $(4 \sin \theta \cos \theta)(1 - 2 \sin^2 \theta) = \sin 4\theta$

37. $\dfrac{\cos 2\theta}{1 + \sin 2\theta} = \dfrac{\cot \theta - 1}{\cot \theta + 1}$

38. $\sin^2 \theta \cos^2 \theta = \frac{1}{8}(1 - \cos 4\theta)$

39. $\sec^2 \dfrac{\theta}{2} = \dfrac{2}{1 + \cos \theta}$

40. $\csc^2 \dfrac{\theta}{2} = \dfrac{2}{1 - \cos \theta}$

41. $\cot^2 \dfrac{\theta}{2} = \dfrac{\sec \theta + 1}{\sec \theta - 1}$

42. $\tan \dfrac{\theta}{2} = \csc \theta - \cot \theta$

43. $\cos \theta = \dfrac{1 - \tan^2(\theta/2)}{1 + \tan^2(\theta/2)}$

44. $1 - \frac{1}{2} \sin 2\theta = \dfrac{\sin^3 \theta + \cos^3 \theta}{\sin \theta + \cos \theta}$

45. $\dfrac{\sin 3\theta}{\sin \theta} - \dfrac{\cos 3\theta}{\cos \theta} = 2$

46. $\dfrac{\cos \theta + \sin \theta}{\cos \theta - \sin \theta} - \dfrac{\cos \theta - \sin \theta}{\cos \theta + \sin \theta} = 2 \tan 2\theta$

47. $\tan 3\theta = \dfrac{3 \tan \theta - \tan^3 \theta}{1 - 3 \tan^2 \theta}$

48. $\tan \theta + \tan(\theta + 120°) + \tan(\theta + 240°) = 3 \tan 3\theta$

49. If $x = 2 \tan \theta$, express $\sin 2\theta$ as a function of x.

50. If $x = 2 \tan \theta$, express $\cos 2\theta$ as a function of x.

51. Find the value of the number C:

$\frac{1}{2} \sin^2 x + C = -\frac{1}{4} \cos 2x$

52. Find the value of the number C:

$\frac{1}{2} \cos^2 x + C = \frac{1}{4} \cos 2x$

53. **Area of an Isosceles Triangle** Show that the area A of an isosceles triangle whose equal sides are of length s and the angle between them is θ is

$A = \frac{1}{2}s^2 \sin \theta$

[**Hint:** See the illustration. The height h bisects the angle θ and is the perpendicular bisector of the base.]

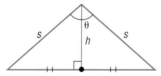

54. **Geometry** A rectangle is inscribed in a semi-circle of radius 1. See the illustration.

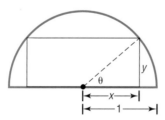

(a) Express the area A of the rectangle as a function of the angle θ shown in the illustration.
(b) Show that $A = \sin 2\theta$.
(c) Find the angle θ that results in the largest area A.
(d) Find the dimensions of this largest rectangle.

55. Graph $f(x) = \sin^2 x = (1 - \cos 2x)/2$ for $0 \le x \le 2\pi$ by using transformations.

56. Repeat Problem 55 for $g(x) = \cos^2 x$.

57. Use the fact that

$$\cos \frac{\pi}{12} = \frac{1}{4}(\sqrt{6} + \sqrt{2})$$

to find $\sin(\pi/24)$ and $\cos(\pi/24)$.

58. Show that

$$\cos \frac{\pi}{8} = \frac{\sqrt{2 - \sqrt{2}}}{2}$$

and use it to find $\sin(\pi/16)$ and $\cos(\pi/16)$.

59. Show that

$$\sin^3 \theta + \sin^3(\theta + 120°) + \sin^3(\theta + 240°) = -\frac{3}{4} \sin 3\theta$$

60. If $\tan \theta = a \tan (\theta/3)$, express $\tan(\theta/3)$ in terms of a.

In Problems 61 and 62, establish each identity.

61. $\ln|\sin \theta| = \frac{1}{2}(\ln|1 - \cos 2\theta| - \ln 2)$

62. $\ln|\cos \theta| = \frac{1}{2}(\ln|1 + \cos 2\theta| - \ln 2)$

*See Chapter 7 for a discussion of logarithm functions.

63. Projectile Motion An object is propelled upward at an angle θ, $45° < \theta < 90°$, to the horizontal with an initial velocity of v_0 feet per second from the base of a plane that makes an angle of 45° with the horizontal. See the illustration. If air resistance is ignored, the distance R that it travels up the inclined plane is given by

$$R = \frac{v_0^2 \sqrt{2}}{16} \cos \theta (\sin \theta - \cos \theta)$$

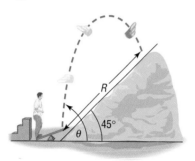

 (a) Show that

$$R = \frac{v_0^2 \sqrt{2}}{32} (\sin 2\theta - \cos 2\theta - 1)$$

(b) Graph $R = R(\theta)$. (Use $v_0 = 32$ feet per second.)

(c) What value of θ makes R the largest? (Use $v_0 = 32$ feet per second.)

64. Sawtooth Curve An oscilloscope often displays a sawtooth curve. This curve can be approximated by sinusoidal curves of varying periods and amplitudes. A first approximation to the sawtooth curve is given by

$$y = \frac{1}{2} \sin 2\pi x + \frac{1}{4} \sin 4\pi x$$

Show that $y = \sin 2\pi x \cos^2 \pi x$.

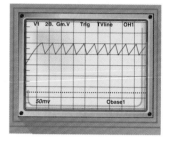

 65. Go to the library and research Chebyshëv polynomials. Write a report on your findings.

3.4 PRODUCT-TO-SUM AND SUM-TO-PRODUCT FORMULAS

1 Express Products as Sums

2 Express Sums as Products

1 The sum and difference formulas can be used to derive formulas for writing the products of sines and/or cosines as sums or differences. These identities are usually called the **Product-to-Sum Formulas.**

Theorem Product-to-Sum Formulas

$$\sin \alpha \sin \beta = \tfrac{1}{2}[\cos(\alpha - \beta) - \cos(\alpha + \beta)] \qquad (1)$$
$$\cos \alpha \cos \beta = \tfrac{1}{2}[\cos(\alpha - \beta) + \cos(\alpha + \beta)] \qquad (2)$$
$$\sin \alpha \cos \beta = \tfrac{1}{2}[\sin(\alpha + \beta) + \sin(\alpha - \beta)] \qquad (3)$$

These formulas do not have to be memorized. Instead, you should remember how they are derived. Then, when you want to use them, either look them up or derive them, as needed.

M I S S I O N   P O S S I B L E

How Far and How High Does a Baseball Need to Go for an Out-of-the-Park Home Run?

The sportscasters for the Cleveland Indians decided that they had better be prepared for any eventuality when the new ballpark, Jacob's Field, opened in 1994. One eventuality they worried about was Jim Thome hitting a home run that went so high and so far that it left the ball park. What would they tell their listeners about the height and distance that the ball went?

   So they called in the Mission Possible team. The sportscaster wanted to know the distance from home plate to the highest point of the stadium and the distance from the base of the outfield fence to the highest point of the stadium for every 5° from the third-base line around to the first-base line. The problem was complicated by the fact that the distance from home plate to the outfield fence varied from 325 feet to 410 feet. Furthermore, the height that would have to be cleared also varied, depending on where the ball was hit.

1. First, how many distances and heights do the sportscasters want?

2. To see one solution, use the distance from home plate across second base to the outfield fence, 410 feet. Using a transit, you find the angle of elevation from home plate to the highest point of the stadium is 10° and the angle of elevation from the base of the outfield fence to the highest point of the stadium is 32.5°. Use this information to find the minimum distance from home plate to the highest point of the stadium and the distance from ground level to the highest point of the stadium.

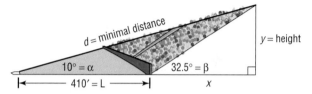

3. You tell the sportscasters that if they provide you the two angles of elevation for each 5° movement around the stadium you can solve their problem. After receiving the two angles of elevation for each 5° movement around the stadium, you decide that it would be faster to develop a general formula so that by inputting the two angles and the distance from home plate to the fence, the distance from home plate to the top of the roof and the height of the roof would be computed. Let α and β denote the angles of elevation to the top of the roof from home plate and to the top of the roof from the base of the outfield fence, respectively. Let L be the distance from home plate to the outfield fence. What are the correct formulas?

4. Compare your formulas with those of other groups. Are they all the same? Are they all equivalent? Which ones are simplest?

5. Now write each correct formula in the following forms:

$$\text{height} = \frac{L \sin \alpha \sin \beta}{\sin(\beta - \alpha)} \qquad \text{distance} = \frac{L \sin \beta}{\sin(\beta - \alpha)}$$

6. What is the probable trajectory of a hit baseball on a windless day? How might the wind and other factors affect the path of the baseball?

7. If a player actually did hit an out-of-the-park home run, how would the distance the ball traveled compare to the figures that you have been developing for the sportscasters? What other factors will add to the distance?

8. Could there be a longest home run? How would you measure it?

To derive formulas (1) and (2), write down the sum and difference formulas for the cosine:

$$\cos(\alpha - \beta) = \cos \alpha \cos \beta + \sin \alpha \sin \beta \qquad (4)$$
$$\cos(\alpha + \beta) = \cos \alpha \cos \beta - \sin \alpha \sin \beta \qquad (5)$$

Subtract equation (5) from equation (4) to get

$$\cos(\alpha - \beta) - \cos(\alpha + \beta) = 2 \sin \alpha \sin \beta$$

from which

$$\sin \alpha \sin \beta = \tfrac{1}{2}[\cos(\alpha - \beta) - \cos(\alpha + \beta)]$$

Now, add equations (4) and (5) to get

$$\cos(\alpha - \beta) + \cos(\alpha + \beta) = 2 \cos \alpha \cos \beta$$

from which

$$\cos \alpha \cos \beta = \tfrac{1}{2}[\cos(\alpha - \beta) + \cos(\alpha + \beta)]$$

To derive Product-to-Sum Formula (3), use the sum and difference formulas for sine in a similar way. (You are asked to do this in Problem 41.)

EXAMPLE 1

Expressing Products as Sums

Express each of the following products as a sum containing only sines or cosines.

(a) $\sin 6\theta \sin 4\theta$ (b) $\cos 3\theta \cos \theta$ (c) $\sin 3\theta \cos 5\theta$

Solution

(a) We use formula (1) to get

$$\sin 6\theta \sin 4\theta = \tfrac{1}{2}[\cos(6\theta - 4\theta) - \cos(6\theta + 4\theta)]$$
$$= \tfrac{1}{2}(\cos 2\theta - \cos 10\theta)$$

(b) We use formula (2) to get

$$\cos 3\theta \cos \theta = \tfrac{1}{2}[\cos(3\theta - \theta) + \cos(3\theta + \theta)]$$
$$= \tfrac{1}{2}(\cos 2\theta + \cos 4\theta)$$

(c) We use formula (3) to get

$$\sin 3\theta \cos 5\theta = \tfrac{1}{2}[\sin(3\theta + 5\theta) + \sin(3\theta - 5\theta)]$$
$$= \tfrac{1}{2}[\sin 8\theta + \sin(-2\theta)] = \tfrac{1}{2}(\sin 8\theta - \sin 2\theta)$$

 Now work Problem 1.

2   The **Sum-to-Product Formulas** are given next.

Theorem   Sum-to-Product Formulas

$$\sin \alpha + \sin \beta = 2 \sin \frac{\alpha + \beta}{2} \cos \frac{\alpha - \beta}{2} \qquad (6)$$

$$\sin \alpha - \sin \beta = 2 \sin \frac{\alpha - \beta}{2} \cos \frac{\alpha + \beta}{2} \qquad (7)$$

$$\cos \alpha + \cos \beta = 2 \cos \frac{\alpha + \beta}{2} \cos \frac{\alpha - \beta}{2} \qquad (8)$$

$$\cos \alpha - \cos \beta = -2 \sin \frac{\alpha + \beta}{2} \sin \frac{\alpha - \beta}{2} \qquad (9)$$

We will derive formula (6) and leave the derivations of formulas (7) through (9) as exercises (see Problems 42 through 44).

Proof

$$2 \sin \frac{\alpha + \beta}{2} \cos \frac{\alpha - \beta}{2} = 2 \cdot \frac{1}{2} \left[\sin\left(\frac{\alpha + \beta}{2} + \frac{\alpha - \beta}{2} \right) + \sin\left(\frac{\alpha + \beta}{2} - \frac{\alpha - \beta}{2} \right) \right]$$

↑
Product-to-Sum Formula (3)

$$= \sin \frac{2\alpha}{2} + \sin \frac{2\beta}{2} = \sin \alpha + \sin \beta \qquad ∎$$

E X A M P L E   2

Expressing Sums (or Differences) as a Product

Express each sum or difference as a product of sines and/or cosines.

(a) $\sin 5\theta - \sin 3\theta$ (b) $\cos 3\theta + \cos 2\theta$

Solution   (a) We use formula (7) to get

$$\sin 5\theta - \sin 3\theta = 2 \sin \frac{5\theta - 3\theta}{2} \cos \frac{5\theta + 3\theta}{2}$$

$$= 2 \sin \theta \cos 4\theta$$

(b) $\cos 3\theta + \cos 2\theta = 2 \cos \dfrac{3\theta + 2\theta}{2} \cos \dfrac{3\theta - 2\theta}{2}$ Formula (8)

$$= 2 \cos \frac{5\theta}{2} \cos \frac{\theta}{2} \qquad ∎$$

 Now work Problem 11.

3.4 │ EXERCISES

In Problems 1–10, express each product as a sum containing only sines or cosines.

1. $\sin 4\theta \sin 2\theta$

2. $\cos 4\theta \cos 2\theta$

3. $\sin 4\theta \cos 2\theta$

4. $\sin 3\theta \sin 5\theta$

5. $\cos 3\theta \cos 5\theta$

6. $\sin 4\theta \cos 6\theta$

7. $\sin \theta \sin 2\theta$

8. $\cos 3\theta \cos 4\theta$

9. $\sin \dfrac{3\theta}{2} \cos \dfrac{\theta}{2}$

10. $\sin \dfrac{\theta}{2} \cos \dfrac{5\theta}{2}$

In Problems 11–18, express each sum or difference as a product of sines and/or cosines.

11. $\sin 4\theta - \sin 2\theta$

12. $\sin 4\theta + \sin 2\theta$

13. $\cos 2\theta + \cos 4\theta$

14. $\cos 5\theta - \cos 3\theta$

15. $\sin \theta + \sin 3\theta$

16. $\cos \theta + \cos 3\theta$

17. $\cos \dfrac{\theta}{2} - \cos \dfrac{3\theta}{2}$

18. $\sin \dfrac{\theta}{2} - \sin \dfrac{3\theta}{2}$

In Problems 19–36, establish each identity.

19. $\dfrac{\sin \theta + \sin 3\theta}{2 \sin 2\theta} = \cos \theta$

20. $\dfrac{\cos \theta + \cos 3\theta}{2 \cos 2\theta} = \cos \theta$

21. $\dfrac{\sin 4\theta + \sin 2\theta}{\cos 4\theta + \cos 2\theta} = \tan 3\theta$

22. $\dfrac{\cos \theta - \cos 3\theta}{\sin 3\theta - \sin \theta} = \tan 2\theta$

23. $\dfrac{\cos \theta - \cos 3\theta}{\sin \theta + \sin 3\theta} = \tan \theta$

24. $\dfrac{\cos \theta - \cos 5\theta}{\sin \theta + \sin 5\theta} = \tan 2\theta$

25. $\sin\theta(\sin\theta + \sin 3\theta) = \cos\theta(\cos\theta - \cos 3\theta)$

26. $\sin\theta(\sin 3\theta + \sin 5\theta) = \cos\theta(\cos 3\theta - \cos 5\theta)$

27. $\dfrac{\sin 4\theta + \sin 8\theta}{\cos 4\theta + \cos 8\theta} = \tan 6\theta$

28. $\dfrac{\sin 4\theta - \sin 8\theta}{\cos 4\theta - \cos 8\theta} = -\cot 6\theta$

29. $\dfrac{\sin 4\theta + \sin 8\theta}{\sin 4\theta - \sin 8\theta} = -\dfrac{\tan 6\theta}{\tan 2\theta}$

30. $\dfrac{\cos 4\theta - \cos 8\theta}{\cos 4\theta + \cos 8\theta} = \tan 2\theta \tan 6\theta$

31. $\dfrac{\sin\alpha + \sin\beta}{\sin\alpha - \sin\beta} = \tan\dfrac{\alpha+\beta}{2}\cot\dfrac{\alpha-\beta}{2}$

32. $\dfrac{\cos\alpha + \cos\beta}{\cos\alpha - \cos\beta} = -\cot\dfrac{\alpha+\beta}{2}\cot\dfrac{\alpha-\beta}{2}$

33. $\dfrac{\sin\alpha + \sin\beta}{\cos\alpha + \cos\beta} = \tan\dfrac{\alpha+\beta}{2}$

34. $\dfrac{\sin\alpha - \sin\beta}{\cos\alpha - \cos\beta} = -\cot\dfrac{\alpha+\beta}{2}$

35. $1 + \cos 2\theta + \cos 4\theta + \cos 6\theta = 4\cos\theta\cos 2\theta\cos 3\theta$

36. $1 - \cos 2\theta + \cos 4\theta - \cos 6\theta = 4\sin\theta\cos 2\theta\sin 3\theta$

37. **Touch-Tone Phones** On a Touch-Tone phone, each button produces a unique sound. The sound produced is the sum of two tones, given by

$$y = \sin 2\pi lt \quad \text{and} \quad y = \sin 2\pi ht$$

where l and h are the low and high frequencies (cycles per second) shown on the illustration. For example, if you touch 7, the low frequency is $l = 852$ cycles per second and the high frequency is $h = 1209$ cycles per second. The sound emitted by touching 7 is

$$y = \sin 2\pi(852)t + \sin 2\pi(1209)t$$

Touch-Tone phone

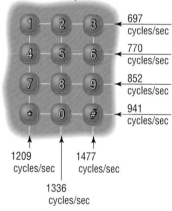

697 cycles/sec
770 cycles/sec
852 cycles/sec
941 cycles/sec

1209 cycles/sec   1477 cycles/sec

1336 cycles/sec

(a) Write this sound as a product of sines and/or cosines.

(b) Determine the maximum value of y.

(c) Graph the sound emitted by touching 7.

38. **Touch-Tone Phones**

(a) Write the sound emitted by touching the # key as a product of sines and/or cosines.

(b) Determine the maximum value of y.

(c) Graph the sound emitted by touching the # key.

39. If $\alpha + \beta + \gamma = \pi$, show that

$$\sin 2\alpha + \sin 2\beta + \sin 2\gamma = 4\sin\alpha\sin\beta\sin\gamma$$

40. If $\alpha + \beta + \gamma = \pi$, show that

$$\tan\alpha + \tan\beta + \tan\gamma = \tan\alpha\tan\beta\tan\gamma$$

41. Derive formula (3).

42. Derive formula (7).

43. Derive formula (8).

44. Derive formula (9).

3.5 THE INVERSE TRIGONOMETRIC FUNCTIONS

1 Find the Exact Value of an Inverse Trigonometric Function

2 Find the Approximate Value of an Inverse Trigonometric Function

In Section 1.6 we discussed inverse functions, and we noted that if a function is one-to-one it will have an inverse function. We also observed that if a function is not one-to-one it may be possible to restrict its domain in some suitable manner such that the restricted function is one-to-one. In this section, we use these ideas to define inverse trigonometric functions. (You may wish to review Section 1.6 at this time.) We begin with the inverse of the sine function.

The Inverse Sine Function

In Figure 6, we reproduce the graph of $y = \sin x$. Because every horizontal line $y = b$, where b is between -1 and 1, intersects the graph of $y = \sin x$ infinitely many times, it follows from the horizontal-line test that the function $y = \sin x$ is not one-to-one.

FIGURE 6
$y = \sin x, -\infty < x < \infty,$
$-1 \le y \le 1$

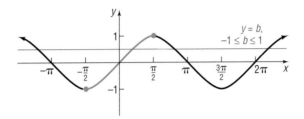

However, if we restrict the domain of $y = \sin x$ to the interval $[-\pi/2, \pi/2]$, the restricted function

$$y = \sin x, \qquad -\frac{\pi}{2} \le x \le \frac{\pi}{2}$$

is one-to-one and, hence, will have an inverse function.* See Figure 7.

FIGURE 7
$y = \sin x,$
$-\pi/2 \le x \le \pi/2, -1 \le y \le 1$

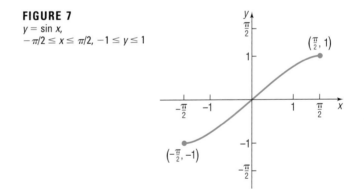

An equation for the inverse is obtained by interchanging x and y. The implicit form of the inverse function is: $x = \sin y, -\frac{\pi}{2} \le y \le \frac{\pi}{2}$. The explicit form is called the **inverse sine** of x and is symbolized by $y = \sin^{-1} x$. Thus,

$$y = \sin^{-1} x \quad \text{means} \quad x = \sin y$$
$$\text{where} \quad -\frac{\pi}{2} \le y \le \frac{\pi}{2} \quad \text{and} \quad -1 \le x \le 1 \tag{1}$$

---

*Although there are many other ways to restrict the domain and obtain a one-to-one function, mathematicians have agreed on a consistent use of the interval $[-\pi/2, \pi/2]$ in order to define the inverse of $y = \sin x$.

Because $y = \sin^{-1} x$ means $x = \sin y$, we read $y = \sin^{-1} x$ as "y is the angle whose sine equals x." Alternatively, we can say that "y is the inverse sine of x." Be careful about the notation used. The superscript -1 that appears in $y = \sin^{-1} x$ is not an exponent, but is reminiscent of the symbolism f^{-1} used to denote the inverse of a function f. (To avoid this notation, some books use the notation $y = \arcsin x$ instead of $y = \sin^{-1} x$.)

When we discussed functions and their inverses in Section 1.6, we found that $f^{-1}(f(x)) = x$ and $f(f^{-1}(x)) = x$. In terms of the sine function and its inverse, these properties are of the form

$$\sin^{-1}(\sin u) = u, \qquad \text{where } -\frac{\pi}{2} \le u \le \frac{\pi}{2}$$
$$\sin(\sin^{-1} v) = v, \qquad \text{where } -1 \le v \le 1 \tag{2}$$

Let's examine the function $y = \sin^{-1} x$. Its domain is $-1 \le x \le 1$, and its range is $-\pi/2 \le y \le \pi/2$. Its graph can be obtained by reflecting the restricted portion of the graph of $y = \sin x$ about the line $y = x$, as shown in Figure 8.

FIGURE 8
$y = \sin^{-1} x, -1 \le x \le 1,$
$-\pi/2 \le y \le \pi/2$

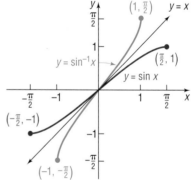

 Check: Graph $Y = \sin^{-1} x$ and compare the result with Figure 8.

 For some numbers x it is possible to find the exact value of $y = \sin^{-1} x$.

E X A M P L E   1   Finding the Exact Value of an Inverse Sine Function

Find the exact value of: $\sin^{-1} 1$

Solution   Let $\theta = \sin^{-1} 1$. We seek the angle θ, $-\pi/2 \le \theta \le \pi/2$, whose sine equals 1.

$$\theta = \sin^{-1} 1, \qquad -\frac{\pi}{2} \le \theta \le \frac{\pi}{2}$$

$$\sin \theta = 1, \qquad -\frac{\pi}{2} \le \theta \le \frac{\pi}{2} \quad \text{By definition of } y = \sin^{-1} x$$

Now look at Table 1 and Figure 9.

TABLE 1

θ	sin θ
$-\pi/2$	-1
$-\pi/3$	$-\sqrt{3}/2$
$-\pi/4$	$-\sqrt{2}/2$
$-\pi/6$	$-1/2$
0	0
$\pi/6$	$1/2$
$\pi/4$	$\sqrt{2}/2$
$\pi/3$	$\sqrt{3}/2$
$\pi/2$	1

FIGURE 9

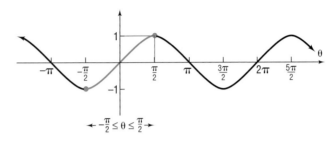

We see that the only angle θ within the interval $[-\pi/2, \pi/2]$ whose sine is 1 is $\pi/2$. (Note that $\sin(5\pi/2)$ also equals 1, but $5\pi/2$ lies outside the interval $[-\pi/2, \pi/2]$ and hence is not admissible.) So, since $\sin \pi/2 = 1$ and $\pi/2$ is in $[-\pi/2, \pi/2]$, we conclude that

$$\sin^{-1} 1 = \frac{\pi}{2}$$

 Now work Problem 1.

E X A M P L E   2

Finding the Exact Value of an Inverse Sine Function

Find the exact value of: $\sin^{-1}(\sqrt{3}/2)$

Solution   Let $\theta = \sin^{-1}(\sqrt{3}/2)$. We seek the angle θ, $-\pi/2 \le \theta \le \pi/2$, whose sine equals $\sqrt{3}/2$.

$$\theta = \sin^{-1}\frac{\sqrt{3}}{2}, \qquad -\frac{\pi}{2} \le \theta \le \frac{\pi}{2}$$

$$\sin \theta = \frac{\sqrt{3}}{2}, \qquad -\frac{\pi}{2} \le \theta \le \frac{\pi}{2} \qquad \text{By definition of } y = \sin^{-1} x$$

Look at Table 1 and Figure 10.

FIGURE 10

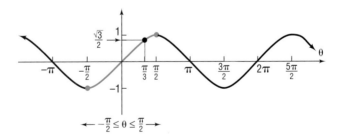

We see that the only angle θ within the interval $[-\pi/2, \pi/2]$ whose sine is $\sqrt{3}/2$ is $\pi/3$. (Note that $\sin(2\pi/3)$ also equals $\sqrt{3}/2$, but $2\pi/3$ lies outside the interval $[-\pi/2, \pi/2]$ and hence is not admissible.) So, since $\sin \pi/3 = \sqrt{3}/2$ and $\pi/3$ is in $[-\pi/2, \pi/2]$, we conclude that

$$\sin^{-1}\frac{\sqrt{3}}{2} = \frac{\pi}{3}$$

E X A M P L E   3

Finding the Exact Value of an Inverse Sine Function

Find the exact value of: $\sin^{-1}\left(-\frac{1}{2}\right)$

Solution   Let $\theta = \sin^{-1}\left(-\frac{1}{2}\right)$. We seek the angle θ, $-\pi/2 \le \theta \le \pi/2$, whose sine equals $-\frac{1}{2}$.

$$\theta = \sin^{-1}\left(-\frac{1}{2}\right), \qquad -\frac{\pi}{2} \le \theta \le \frac{\pi}{2}$$

$$\sin\theta = -\frac{1}{2}, \qquad -\frac{\pi}{2} \le \theta \le \frac{\pi}{2}$$

(Refer to Table 1 and Figure 9 if necessary.) The only angle within the interval $[-\pi/2, \pi/2]$ whose sine is $-\frac{1}{2}$ is $-\pi/6$. So, since $\sin(-\pi/6) = -1/2$ and $-\pi/6$ is in $[-\pi/2, \pi/2]$, we conclude that

$$\sin^{-1}\left(-\frac{1}{2}\right) = -\frac{\pi}{6}$$

 Now work Problem 3.

For most numbers x, the value $y = \sin^{-1}x$ must be approximated.

E X A M P L E   4   Finding an Approximate Value of an Inverse Sine Function

Find the approximate value of

(a) $\sin^{-1}\dfrac{1}{3}$ (b) $\sin^{-1}(-0.25)$

Express the answer in radians rounded to two decimal places.

Solution   Because we want the angle measured in radians, we first set the mode to radians.

(a) Keystrokes:*   [1]   [÷]   [3]   [=]   [SHIFT]   [sin]

Display: [1]   [3]   [0.3333333]   [0.3398369]

Thus, $\sin^{-1}\frac{1}{3} = 0.34$, rounded to two decimal places.

(b) Keystrokes: [0.25]   [+/−]   [SHIFT]   [sin]

Display: [0.25]   [−0.25]   [−0.2526802]

Thus, $\sin^{-1}(-0.25) = -0.25$, rounded to two decimal places.

Now work Problem 13.

*On some calculators, [sin^{-1}] is pressed first; then [1/3] is entered. Consult your owner's manual for the correct sequence.

The Inverse Cosine Function

In Figure 11 we reproduce the graph of $y = \cos x$. Because every horizontal line $y = b$, where b is between -1 and 1, intersects the graph of $y = \cos x$ infinitely many times, it follows that the cosine function is not one-to-one.

FIGURE 11
$y = \cos x,$
$-\infty < x < \infty, -1 \leq y \leq 1$

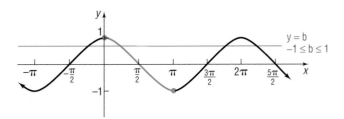

However, if we restrict the domain of $y = \cos x$ to the interval $[0, \pi]$, the restricted function

$$y = \cos x, \qquad 0 \leq x \leq \pi$$

is one-to-one and hence will have an inverse function.* See Figure 12.

FIGURE 12
$y = \cos x,$
$0 \leq x \leq \pi, -1 \leq y \leq 1$

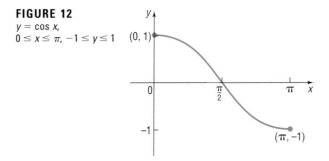

An equation for the inverse is obtained by interchanging x and y. The implicit form of the inverse function is: $x = \cos y, 0 \leq y \leq \pi$. The explicit form is called the **inverse cosine** of x and is symbolized by $y = \cos^{-1} x$ (or by $y = \arccos x$). Thus,

$$y = \cos^{-1} x \quad \text{means} \quad x = \cos y \qquad (3)$$
$$\text{where} \quad 0 \leq y \leq \pi \quad \text{and} \quad -1 \leq x \leq 1$$

*This is the generally accepted restriction to define the inverse.

Here, y is the angle whose cosine is x. The domain of the function $y = \cos^{-1} x$ is $-1 \le x \le 1$, and its range is $0 \le y \le \pi$. The graph of $y = \cos^{-1} x$ can be obtained by reflecting the restricted portion of the graph of $y = \cos x$ about the line $y = x$, as shown in Figure 13.

FIGURE 13
$y = \cos^{-1} x,$
$-1 \le x \le 1, 0 \le y \le \pi$

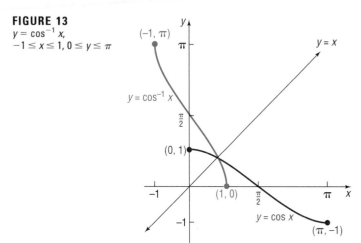

Check: Graph $Y = \cos^{-1} x$ and compare the result with Figure 13.

For the cosine function and its inverse, the following properties hold:

$$\cos^{-1}(\cos u) = u, \qquad \text{where } 0 \le u \le \pi$$
$$\cos(\cos^{-1} v) = v, \qquad \text{where } -1 \le v \le 1$$

E X A M P L E   5

Finding the Exact Value of an Inverse Cosine Function

Find the exact value of: $\cos^{-1} 0$

Solution   Let $\theta = \cos^{-1} 0$. We seek the angle $\theta, 0 \le \theta \le \pi$, whose cosine equals 0.

$$\theta = \cos^{-1} 0, \qquad 0 \le \theta \le \pi$$
$$\cos \theta = 0, \qquad\quad 0 \le \theta \le \pi$$

Look at Table 2 and Figure 14.

TABLE 2	
θ	$\cos \theta$
0	1
$\pi/6$	$\sqrt{3}/2$
$\pi/4$	$\sqrt{2}/2$
$\pi/3$	$1/2$
$\pi/2$	0
$2\pi/3$	$-1/2$
$3\pi/4$	$-\sqrt{2}/2$
$5\pi/6$	$-\sqrt{3}/2$
π	-1

FIGURE 14

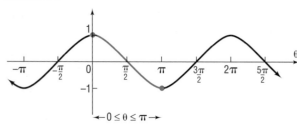

We see that the only angle θ within the interval $[0, \pi]$ whose cosine is 0 is $\pi/2$. (Note that $\cos 3\pi/2$ also equals 0, but $3\pi/2$ lies outside the interval $[0, \pi]$ and hence is not admissible.) So, since $\cos \pi/2 = 0$ and $\pi/2$ is in $[0, \pi]$, we conclude that

$$\cos^{-1} 0 = \frac{\pi}{2}$$

E X A M P L E 6

Finding the Exact Value of an Inverse Cosine Function

Find the exact value of: $\cos^{-1}(\sqrt{2}/2)$

Solution   Let $\theta = \cos^{-1}(\sqrt{2}/2)$. We seek the angle $\theta, 0 \le \theta \le \pi$, whose cosine equals $\sqrt{2}/2$.

$$\theta = \cos^{-1} \frac{\sqrt{2}}{2}, \qquad 0 \le \theta \le \pi$$

$$\cos \theta = \frac{\sqrt{2}}{2}, \qquad 0 \le \theta \le \pi$$

Look at Table 2 and Figure 15.

FIGURE 15

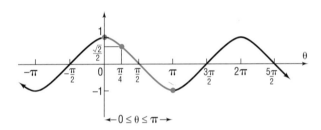

We see that the only angle θ within the interval $[0, \pi]$ whose cosine is $\sqrt{2}/2$ is $\pi/4$. So, since $\cos \pi/4 = \sqrt{2}/2$ and $\pi/4$ is in $[0, \pi]$, we conclude that

$$\cos^{-1} \frac{\sqrt{2}}{2} = \frac{\pi}{4}$$

E X A M P L E 7

Finding the Exact Value of an Inverse Cosine Function

Find the exact value of: $\cos^{-1}(-\frac{1}{2})$

Solution   Let $\theta = \cos^{-1}(-\frac{1}{2})$. We seek the angle $\theta, 0 \le \theta \le \pi$, whose cosine equals $-\frac{1}{2}$.

$$\theta = \cos^{-1}(-\tfrac{1}{2}), \qquad 0 \le \theta \le \pi$$
$$\cos \theta = -\tfrac{1}{2}, \qquad 0 \le \theta \le \pi$$

(Refer to Table 2 and Figure 14 if necessary.) The only angle within the interval $[0, \pi]$ whose cosine is $-\frac{1}{2}$ is $2\pi/3$. So, since $\cos 2\pi/3 = -1/2$ and $2\pi/3$ is in $[0, \pi]$, we conclude that

$$\cos^{-1}\left(-\frac{1}{2}\right) = \frac{2\pi}{3}$$

 Now work Problem 11.

The Inverse Tangent Function

In Figure 16 we reproduce the graph of $y = \tan x$. Because every horizontal line intersects the graph infinitely many times, it follows that the tangent function is not one-to-one.

FIGURE 16
$y = \tan x$, $-\infty < x < \infty$, x not equal to odd multiples of $\pi/2$, $-\infty < y < \infty$

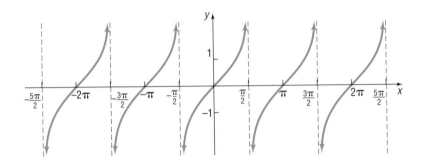

However, if we restrict the domain of $y = \tan x$ to the interval $(-\pi/2, \pi/2)$,* the restricted function

$$y = \tan x, \qquad -\frac{\pi}{2} < x < \frac{\pi}{2}$$

is one-to-one and hence has an inverse function. See Figure 17.

FIGURE 17
$y = \tan x$,
$-\pi/2 < x < \pi/2$, $-\infty < y < \infty$

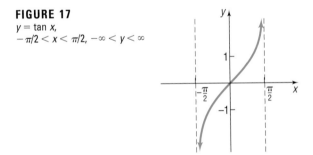

An equation for the inverse is obtained by interchanging x and y. The implicit form of the inverse function is: $x = \tan y$, $-\frac{\pi}{2} < y < \frac{\pi}{2}$. The explicit form is called the **inverse tangent** of x and is symbolized by $y = \tan^{-1} x$ (or by $y = \arctan x$). Thus,

$$y = \tan^{-1} x \quad \text{means} \quad x = \tan y$$
$$\text{where} \quad -\frac{\pi}{2} < y < \frac{\pi}{2} \quad \text{and} \quad -\infty < x < \infty \tag{4}$$

*This is the generally accepted restriction.

Here, y is the angle whose tangent is x. The domain of the function $y = \tan^{-1} x$ is $-\infty < x < \infty$, and its range is $-\pi/2 < y < \pi/2$. The graph of $y = \tan^{-1} x$ can be obtained by reflecting the restricted portion of the graph of $y = \tan x$ about the line $y = x$, as shown in Figure 18.

FIGURE 18
$y = \tan^{-1} x, -\infty < x < \infty,$
$-\pi/2 < y < \pi/2$

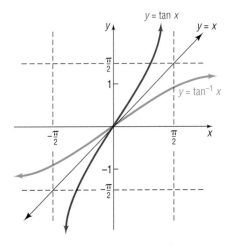

Check: Graph $Y = \tan^{-1} x$ and compare the results with Figure 18.

For the tangent function and its inverse, the following properties hold:

$$\tan^{-1}(\tan u) = u, \quad \text{where} \quad -\frac{\pi}{2} < u < \frac{\pi}{2}$$

$$\tan(\tan^{-1} v) = v, \quad \text{where} \quad -\infty < v < \infty$$

E X A M P L E 8

Finding the Exact Value of an Inverse Tangent Function

Find the exact value of: $\tan^{-1} 1$

Solution   Let $\theta = \tan^{-1} 1$. We seek the angle θ, $-\pi/2 < \theta < \pi/2$, whose tangent equals 1.

$$\theta = \tan^{-1} 1, \qquad -\frac{\pi}{2} < \theta < \frac{\pi}{2}$$

$$\tan \theta = 1, \qquad -\frac{\pi}{2} < \theta < \frac{\pi}{2}$$

Look at Table 3 and Figure 17. The only angle θ within the interval $(-\pi/2, \pi/2)$ whose tangent is 1 is $\pi/4$. So, since $\tan \pi/4 = 1$ and $\pi/4$ is in $(-\pi/2, \pi/2)$, we conclude that

$$\tan^{-1} 1 = \frac{\pi}{4}$$

TABLE 3

θ	$\tan \theta$
$-\pi/2$	Undefined
$-\pi/3$	$-\sqrt{3}$
$-\pi/4$	-1
$-\pi/6$	$-\sqrt{3}/3$
0	0
$\pi/6$	$\sqrt{3}/3$
$\pi/4$	1
$\pi/3$	$\sqrt{3}$
$\pi/2$	Undefined

E X A M P L E   9

Finding the Exact Value of an Inverse Tangent Function

Find the exact value of: $\tan^{-1}(-\sqrt{3})$

Solution   Let $\theta = \tan^{-1}(-\sqrt{3})$. We seek the angle θ, $-\pi/2 < \theta < \pi/2$, whose tangent equals $-\sqrt{3}$.

$$\theta = \tan^{-1}(-\sqrt{3}), \qquad -\frac{\pi}{2} < \theta < \frac{\pi}{2}$$

$$\tan \theta = -\sqrt{3} \qquad\qquad -\frac{\pi}{2} < \theta < \frac{\pi}{2}$$

Look at Table 3 and Figure 17 if necessary. The only angle θ within the interval $(-\pi/2, \pi/2)$ whose tangent is $-\sqrt{3}$ is $-\pi/3$. So, since $\tan(-\pi/3) = -\sqrt{3}$ and $-\pi/3$ is in $(-\pi/2, \pi/2)$, we conclude that

$$\tan^{-1}(-\sqrt{3}) = -\frac{\pi}{3}$$

◼

Now work Problem 5.

E X A M P L E   10

Finding the Exact Value of Expressions Involving Inverse Trigonometric Functions

Find the exact value of: $\sin^{-1}(\sin 5\pi/4)$

Solution $$\sin^{-1}\left(\sin\frac{5\pi}{4}\right) = \sin^{-1}\left(-\frac{\sqrt{2}}{2}\right) = -\frac{\pi}{4}$$

◼

How do you reconcile this solution with equation (2) on page 222?   ◼

Now work Problem 37.

E X A M P L E   11

Finding the Exact Value of Expressions Involving Inverse Trigonometric Functions

Find the exact value of: $\cos[\tan^{-1}(-1)]$

Solution   Let $\theta = \tan^{-1}(-1)$. We seek the angle θ, $-\pi/2 < \theta < \pi/2$, whose tangent equals -1.

$$\tan \theta = -1, \qquad -\frac{\pi}{2} < \theta < \frac{\pi}{2}$$

$$\theta = -\frac{\pi}{4}$$

Now

$$\cos[\tan^{-1}(-1)] = \cos \theta = \cos\left(-\frac{\pi}{4}\right) = \frac{\sqrt{2}}{2}$$

◼

E X A M P L E   12

Finding the Exact Value of Expressions Involving Inverse Trigonometric Functions

Find the exact value of: $\sec(\sin^{-1}\frac{1}{2})$

Solution   Let $\theta = \sin^{-1}\frac{1}{2}$. We seek the angle θ, $-\pi/2 \le \theta \le \pi/2$, whose sine equals $\frac{1}{2}$.

$$\sin \theta = \frac{1}{2}, \qquad -\frac{\pi}{2} \le \theta \le \frac{\pi}{2}$$

$$\theta = \frac{\pi}{6}$$

Now

$$\sec\left(\sin^{-1}\frac{1}{2}\right) = \sec \theta = \sec \frac{\pi}{6} = \frac{2}{\sqrt{3}} = \frac{2\sqrt{3}}{3}$$

 Now work Problem 25.

It is not necessary to be able to find the angle in order to solve problems like those given in Examples 11 and 12.

E X A M P L E   13

Finding the Exact Value of Expressions Involving Inverse Trigonometric Functions

Find the exact value of: $\sin(\tan^{-1}\frac{1}{2})$

FIGURE 19
$\tan \theta = \frac{1}{2}$

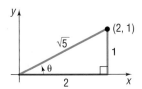

Solution   Let $\theta = \tan^{-1}\frac{1}{2}$. Then $\tan \theta = \frac{1}{2}$, where $-\pi/2 < \theta < \pi/2$. Because $\tan \theta > 0$, it follows that $0 < \theta < \pi/2$. Now, in Figure 19, we draw a triangle in the appropriate quadrant depicting $\tan \theta = \frac{1}{2}$. The hypotenuse of this triangle is easily found to be $\sqrt{5}$. Hence, the sine of θ is $1/\sqrt{5}$, so

$$\sin\left(\tan^{-1}\frac{1}{2}\right) = \sin \theta = \frac{1}{\sqrt{5}} = \frac{\sqrt{5}}{5}$$

E X A M P L E   14

Finding the Exact Value of Expressions Involving Inverse Trigonometric Functions

FIGURE 20
$\sin \theta = -\frac{1}{3}$

Find the exact value of: $\cos[\sin^{-1}(-\frac{1}{3})]$

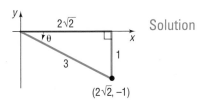

Solution   Let $\theta = \sin^{-1}(-\frac{1}{3})$. Then $\sin \theta = -\frac{1}{3}$ and $-\pi/2 \le \theta \le \pi/2$. Because $\sin \theta < 0$, it follows that $-\pi/2 \le \theta < 0$. Based on Figure 20, we conclude that

$$\cos\left[\sin^{-1}\left(-\frac{1}{3}\right)\right] = \cos \theta = \frac{2\sqrt{2}}{3}$$

E X A M P L E   15

Finding the Exact Value of Expressions Involving Inverse Trigonometric Functions

FIGURE 21
$\cos \theta = -\frac{1}{3}$

Find the exact value of: $\tan[\cos^{-1}(-\frac{1}{3})]$

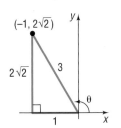

Solution   Let $\theta = \cos^{-1}(-\frac{1}{3})$. Then $\cos \theta = -\frac{1}{3}$ and $0 \le \theta \le \pi$. Because $\cos \theta < 0$, it follows that $\pi/2 < \theta \le \pi$. Based on Figure 21, we conclude that

$$\tan\left[\cos^{-1}\left(-\frac{1}{3}\right)\right] = \tan \theta = \frac{2\sqrt{2}}{-1} = -2\sqrt{2}$$

 Now work Problem 43.

E X A M P L E 16   Finding the Exact Value of Expressions Involving Inverse
Trigonometric Functions

Find the exact value of: $\sin(\cos^{-1} \frac{1}{2} + \sin^{-1} \frac{3}{5})$

Solution   Let $\alpha = \cos^{-1} \frac{1}{2}$ and $\beta = \sin^{-1} \frac{3}{5}$. Then

$$\cos \alpha = \tfrac{1}{2}, \quad 0 \le \alpha \le \pi, \quad \text{and} \quad \sin \beta = \tfrac{3}{5}, \quad -\frac{\pi}{2} \le \beta \le \frac{\pi}{2}$$

Based on Figure 22, we find $\sin \alpha = \sqrt{3}/2$ and $\cos \beta = \tfrac{4}{5}$. Thus,

$$\sin\left(\cos^{-1} \frac{1}{2} + \sin^{-1} \frac{3}{5}\right) = \sin(\alpha + \beta) = \sin \alpha \cos \beta + \cos \alpha \sin \beta$$

$$= \frac{\sqrt{3}}{2} \cdot \frac{4}{5} + \frac{1}{2} \cdot \frac{3}{5} = \frac{4\sqrt{3} + 3}{10}$$

FIGURE 22

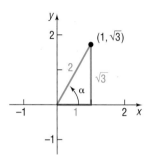 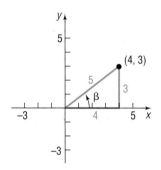

(a) $\cos \alpha = \tfrac{1}{2}, 0 \le \alpha \le \pi$

(b) $\sin \beta = \tfrac{3}{5}, -\pi/2 \le \beta \le \pi/2$

Now work Problem 55.

E X A M P L E 17   Writing a Trigonometric Expression as an Algebraic Expression

Write $\sin(\sin^{-1} u + \cos^{-1} v)$ as an algebraic expression containing u and v
(that is, without any trigonometric functions).

Solution   Let $\alpha = \sin^{-1} u$ and $\beta = \cos^{-1} v$. Then

$$\sin \alpha = u, \quad -\frac{\pi}{2} \le \alpha \le \frac{\pi}{2}, \quad \text{and} \quad \cos \beta = v, \quad 0 \le \beta \le \pi$$

Since $-\pi/2 \le \alpha \le \pi/2$, we know that $\cos \alpha \ge 0$. As a result,

$$\cos \alpha = \sqrt{1 - \sin^2 \alpha} = \sqrt{1 - u^2}$$

Similarly, since $0 \le \beta \le \pi$, we know that $\sin \beta \ge 0$. Thus,

$$\sin \beta = \sqrt{1 - \cos^2 \beta} = \sqrt{1 - v^2}$$

Now,

$$\sin(\sin^{-1} u + \cos^{-1} v) = \sin(\alpha + \beta) = \sin \alpha \cos \beta + \cos \alpha \sin \beta$$

$$= uv + \sqrt{1 - u^2}\sqrt{1 - v^2}$$

EXAMPLE 18   Establishing an Identity Involving Inverse Trigonometric Functions

Show that $\sin(\tan^{-1} v) = \dfrac{v}{\sqrt{1 + v^2}}$.

Solution   Let $\theta = \tan^{-1} v$, so that $\tan \theta = v$, $-\pi/2 < \theta < \pi/2$. As a result, we know that $\sec \theta > 0$. Thus,

$$\sin(\tan^{-1} v) = \sin \theta = \sin \theta \cdot \frac{\cos \theta}{\cos \theta} = \tan \theta \cos \theta = \frac{\tan \theta}{\sec \theta} = \frac{\tan \theta}{\sqrt{1 + \tan^2 \theta}} = \frac{v}{\sqrt{1 + v^2}}$$

$$\frac{\sin \theta}{\cos \theta} = \tan \theta \qquad \begin{array}{c} \sec^2 \theta = 1 + \tan^2 \theta \\ \sec \theta > 0 \end{array}$$ ∎

Now work Problem 83.

The Remaining Inverse Trigonometric Functions

The inverse cotangent, inverse secant, and inverse cosecant functions are defined as follows:

$$y = \cot^{-1} x \quad \text{means} \quad x = \cot y \qquad (5)$$
$$\text{where} \quad -\infty < x < \infty \quad \text{and} \quad 0 < y < \pi$$

$$y = \sec^{-1} x \quad \text{means} \quad x = \sec y \qquad (6)$$
$$\text{where} \quad |x| \geq 1 \quad \text{and} \quad 0 \leq y \leq \pi, \quad y \neq \frac{\pi}{2}$$

$$y = \csc^{-1} x \quad \text{means} \quad x = \csc y \qquad (7)$$
$$\text{where} \quad |x| \geq 1 \quad \text{and} \quad -\frac{\pi}{2} \leq y \leq \frac{\pi}{2}, \quad y \neq 0$$

Most calculators do not have keys for evaluating these inverse trigonometric functions. The easiest way to evaluate them is to convert to an inverse trigonometric function whose range is the same as the one to be evaluated. In this regard, notice that $y = \cot^{-1} x$ and $y = \sec^{-1} x$ (except where undefined) each have the same range as $y = \cos^{-1} x$, while $y = \csc^{-1} x$ (except where undefined) has the same range as $y = \sin^{-1} x$.

EXAMPLE 19   Approximating the Value of Inverse Trigonometric Functions

Use a calculator to approximate each expression in radians rounded to two decimal places.

(a) $\sec^{-1} 3$ (b) $\csc^{-1}(-4)$ (c) $\cot^{-1} \frac{1}{2}$ (d) $\cot^{-1}(-2)$

Solution   First, set your calculator to radian mode.

(a) Let $\theta = \sec^{-1} 3$. Then $\sec \theta = 3$ and $0 \leq \theta \leq \pi$, $\theta \neq \pi/2$. Thus, $\cos \theta = \frac{1}{3}$ and $\theta = \cos^{-1} \frac{1}{3}$. Then

$$\sec^{-1} 3 = \theta = \cos^{-1} \frac{1}{3} \approx 1.23$$

Use a calculator.

FIGURE 23
$\cot \theta = \frac{1}{2}, 0 < \theta < \pi$

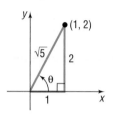

FIGURE 24
$\cot \theta = -2, 0 < \theta < \pi$

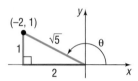

(b) Let $\theta = \csc^{-1}(-4)$. Then $\csc \theta = -4$, $-\pi/2 \leq \theta \leq \pi/2$, $\theta \neq 0$. Thus, $\sin \theta = -\frac{1}{4}$ and $\theta = \sin^{-1}(-\frac{1}{4})$. Then

$$\csc^{-1}(-4) = \theta = \sin^{-1}(-\tfrac{1}{4}) \approx -0.25$$

(c) Let $\theta = \cot^{-1}\frac{1}{2}$. Then $\cot \theta = \frac{1}{2}, 0 < \theta < \pi$. From these facts we know that θ is in quadrant I. We draw Figure 23 to help us to find $\cos \theta$. Thus, $\cos \theta = 1/\sqrt{5}, 0 < \theta < \pi/2, \theta = \cos^{-1}(\frac{1}{\sqrt{5}})$, and

$$\cot^{-1}\frac{1}{2} = \theta = \cos^{-1}\left(\frac{1}{\sqrt{5}}\right) \approx 1.11$$

(d) Let $\theta = \cot^{-1}(-2)$. Then $\cot \theta = -2, 0 < \theta < \pi$. From these facts we know that θ lies in quadrant II. We draw Figure 24 to help us to find $\cos \theta$. Thus, $\cos \theta = -2/\sqrt{5}, \pi/2 < \theta < \pi, \theta = \cos^{-1}(-\frac{2}{\sqrt{5}})$, and

$$\cot^{-1}(-2) = \theta = \cos^{-1}\left(-\frac{2}{\sqrt{5}}\right) \approx 2.68$$

Now work Problem 95.

3.5 | EXERCISES

In Problems 1–12, find the exact value of each expression.

1. $\sin^{-1} 0$ **2.** $\cos^{-1} 1$ **3.** $\sin^{-1}(-1)$ **4.** $\cos^{-1}(-1)$

5. $\tan^{-1} 0$ **6.** $\tan^{-1}(-1)$ **7.** $\sin^{-1}\dfrac{\sqrt{2}}{2}$ **8.** $\tan^{-1}\dfrac{\sqrt{3}}{3}$

9. $\tan^{-1}\sqrt{3}$ **10.** $\sin^{-1}\left(-\dfrac{\sqrt{3}}{2}\right)$ **11.** $\cos^{-1}\left(-\dfrac{\sqrt{3}}{2}\right)$ **12.** $\sin^{-1}\left(-\dfrac{\sqrt{2}}{2}\right)$

In Problems 13–24, use a calculator to find the value of each expression rounded to two decimal places.

13. $\sin^{-1} 0.1$ **14.** $\cos^{-1} 0.6$ **15.** $\tan^{-1} 5$ **16.** $\tan^{-1} 0.2$

17. $\cos^{-1}\frac{7}{8}$ **18.** $\sin^{-1}\frac{1}{8}$ **19.** $\tan^{-1}(-0.4)$ **20.** $\tan^{-1}(-3)$

21. $\sin^{-1}(-0.12)$ **22.** $\cos^{-1}(-0.44)$ **23.** $\cos^{-1}\dfrac{\sqrt{2}}{3}$ **24.** $\sin^{-1}\dfrac{\sqrt{3}}{5}$

In Problems 25–52, find the exact value of each expression.

25. $\cos(\sin^{-1}\frac{\sqrt{2}}{2})$ **26.** $\sin(\cos^{-1}\frac{1}{2})$ **27.** $\tan[\cos^{-1}(-\frac{\sqrt{3}}{2})]$ **28.** $\tan[\sin^{-1}(-\frac{1}{2})]$

29. $\sec(\cos^{-1}\frac{1}{2})$ **30.** $\cot[\sin^{-1}(-\frac{1}{2})]$ **31.** $\csc(\tan^{-1} 1)$ **32.** $\sec(\tan^{-1}\sqrt{3})$

33. $\sin[\tan^{-1}(-1)]$ **34.** $\cos[\sin^{-1}(-\frac{\sqrt{3}}{2})]$ **35.** $\sec[\sin^{-1}(-\frac{1}{2})]$ **36.** $\csc[\cos^{-1}(-\frac{\sqrt{3}}{2})]$

37. $\cos^{-1}(\cos\frac{5\pi}{4})$ **38.** $\tan^{-1}(\tan\frac{2\pi}{3})$ **39.** $\sin^{-1}[\sin(-\frac{7\pi}{6})]$ **40.** $\cos^{-1}[\cos(-\frac{\pi}{3})]$

41. $\tan(\sin^{-1}\frac{1}{3})$ **42.** $\tan(\cos^{-1}\frac{1}{3})$ **43.** $\sec(\tan^{-1}\frac{1}{2})$ **44.** $\cos(\sin^{-1}\frac{\sqrt{2}}{3})$

45. $\cot[\sin^{-1}(-\frac{\sqrt{2}}{3})]$ **46.** $\csc[\tan^{-1}(-2)]$ **47.** $\sin[\tan^{-1}(-3)]$ **48.** $\cot[\cos^{-1}(-\frac{\sqrt{3}}{3})]$

49. $\sec(\sin^{-1}\frac{2\sqrt{5}}{5})$ **50.** $\csc(\tan^{-1}\frac{1}{2})$ **51.** $\sin^{-1}(\cos\frac{3\pi}{4})$ **52.** $\cos^{-1}(\sin\frac{7\pi}{6})$

In Problems 53–76, find the exact value of each expression.

53. $\sin(\sin^{-1}\frac{1}{2} + \cos^{-1}0)$

54. $\sin(\sin^{-1}\frac{\sqrt{3}}{2} + \cos^{-1}1)$

55. $\sin[\sin^{-1}\frac{3}{5} - \cos^{-1}(-\frac{4}{5})]$

56. $\sin[\sin^{-1}(-\frac{4}{5}) - \tan^{-1}\frac{3}{4}]$

57. $\cos(\tan^{-1}\frac{4}{3} + \cos^{-1}\frac{5}{13})$

58. $\sin[\tan^{-1}\frac{5}{12} - \sin^{-1}(-\frac{3}{5})]$

59. $\sec(\sin^{-1}\frac{5}{13} - \tan^{-1}\frac{3}{4})$

60. $\sec(\tan^{-1}\frac{4}{3} + \cot^{-1}\frac{5}{12})$

61. $\cot(\sec^{-1}\frac{5}{3} + \frac{\pi}{6})$

62. $\cos(\frac{\pi}{4} - \csc^{-1}\frac{5}{3})$

63. $\sin(2\sin^{-1}\frac{1}{2})$

64. $\sin[2\sin^{-1}\frac{\sqrt{3}}{2}]$

65. $\cos(2\sin^{-1}\frac{3}{5})$

66. $\cos(2\cos^{-1}\frac{4}{5})$

67. $\tan[2\cos^{-1}(-\frac{3}{5})]$

68. $\tan(2\tan^{-1}\frac{3}{4})$

69. $\sin(2\cos^{-1}\frac{4}{5})$

70. $\cos[2\tan^{-1}(-\frac{4}{3})]$

71. $\sin^2(\frac{1}{2}\cos^{-1}\frac{3}{5})$

72. $\cos^2(\frac{1}{2}\sin^{-1}\frac{3}{5})$

73. $\sec(2\tan^{-1}\frac{3}{4})$

74. $\csc[2\sin^{-1}(-\frac{3}{5})]$

75. $\cot^2(\frac{1}{2}\tan^{-1}\frac{4}{3})$

76. $\cot^2(\frac{1}{2}\cos^{-1}\frac{5}{13})$

In Problems 77–82, write each trigonometric expression as an algebraic expression containing u and v.

77. $\cos(\cos^{-1}u + \sin^{-1}v)$

78. $\sin(\sin^{-1}u - \cos^{-1}v)$

79. $\sin(\tan^{-1}u - \sin^{-1}v)$

80. $\cos(\tan^{-1}u + \tan^{-1}v)$

81. $\tan(\sin^{-1}u - \cos^{-1}v)$

82. $\sec(\tan^{-1}u + \cos^{-1}v)$

In Problems 83–94, establish each identity.

83. Show that $\sec(\tan^{-1}v) = \sqrt{1 + v^2}$.

84. Show that $\tan(\sin^{-1}v) = v/\sqrt{1 - v^2}$.

85. Show that $\tan(\cos^{-1}v) = \sqrt{1 - v^2}/v$.

86. Show that $\sin(\cos^{-1}v) = \sqrt{1 - v^2}$.

87. Show that $\cos(\sin^{-1}v) = \sqrt{1 - v^2}$.

88. Show that $\cos(\tan^{-1}v) = 1/\sqrt{1 + v^2}$.

89. Show that $\sin^{-1}v + \cos^{-1}v = \pi/2$.

90. Show that $\tan^{-1}v + \cot^{-1}v = \pi/2$.

91. Show that $\tan^{-1}(1/v) = \pi/2 - \tan^{-1}v$, if $v > 0$.

92. Show that $\cot^{-1}e^v = \tan^{-1}e^{-v}$.

93. Show that $\sin(\sin^{-1}v + \cos^{-1}v) = 1$.

94. Show that $\cos(\sin^{-1}v + \cos^{-1}v) = 0$.

In Problems 95–106, use a calculator to find the value of each expression rounded to two decimal places.

95. $\sec^{-1}4$

96. $\csc^{-1}5$

97. $\cot^{-1}2$

98. $\sec^{-1}(-3)$

99. $\csc^{-1}(-3)$

100. $\cot^{-1}(-\frac{1}{2})$

101. $\cot^{-1}(-\sqrt{5})$

102. $\cot^{-1}(-8.1)$

103. $\csc^{-1}(-\frac{3}{2})$

104. $\sec^{-1}(-\frac{4}{3})$

105. $\cot^{-1}(-\frac{3}{2})$

106. $\cot^{-1}(-\sqrt{10})$

107. **Being the First to See the Rising Sun** Cadillac Mountain, elevation 1530 feet, is located in Acadia National Park, Maine, and is the highest peak on the east coast of the United States. It is said that a person standing on the summit will be the first person in the United States to see the rays of the rising Sun. How much sooner would a person atop Cadillac Mountain see the first rays than a person standing below, at sea level?

[**Hint:** Consult the figure on page 236. When the person at D sees the first rays of the Sun, the person at P does not. The person at P sees the first rays of the Sun only after Earth has rotated so that

P is at location *Q*. Compute the length of arc *s* subtended by the central angle θ. Then use the fact that, at the latitude of Cadillac Mountain, in 24 hours a length of $2\pi(2170)$ miles is subtended, and find the time it takes to subtend the length *s*.]

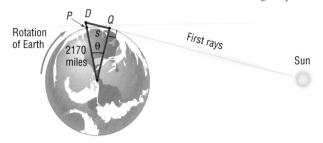

108. For what numbers *x* does $\sin(\sin^{-1} x) = x$?

109. For what numbers *x* does $\cos(\cos^{-1} x) = x$?

110. For what numbers *x* does $\sin^{-1}(\sin x) = x$?

111. For what numbers *x* does $\cos^{-1}(\cos x) = x$?

 112. Graph $y = \cot^{-1} x$.

113. Graph $y = \sec^{-1} x$.

114. Graph $y = \csc^{-1} x$.

115. Explain in your own words how you would use your calculator to find the value of $\cot^{-1} 10$.

116. Consult three books on calculus and write down the definition in each of $y = \sec^{-1} x$ and $y = \csc^{-1} x$. Compare these with the definitions given in this book.

3.6 | TRIGONOMETRIC EQUATIONS

1 Solve Trigonometric Equations

1 The first four sections of this chapter were devoted to trigonometric identities, that is, equations involving trigonometric functions that are satisfied by every value in the domain of the variable. In this section, we discuss **trigonometric equations,** that is, equations involving trigonometric functions that are satisfied only by some values of the variable (or, possibly, are not satisfied by any values of the variable). The values that satisfy the equation are called **solutions** of the equation.

E X A M P L E   1   **Checking Whether a Given Number Is a Solution of a Trigonometric Equation**

Determine whether $\theta = \pi/4$ is a solution of the equation $\sin\theta = \frac{1}{2}$. Is $\theta = \pi/6$ a solution?

Solution   Replace θ by $\pi/4$ in the given equation. The result is

$$\sin\frac{\pi}{4} = \frac{\sqrt{2}}{2} \neq \frac{1}{2}$$

We conclude that $\pi/4$ is not a solution.

Next, replace θ by $\pi/6$ in the equation. The result is

$$\sin\frac{\pi}{6} = \frac{1}{2}$$

Thus, $\pi/6$ is a solution of the given equation. ▬

The equation given in Example 1 has other solutions besides $\theta = \pi/6$. For example, $\theta = 5\pi/6$ is also a solution, as is $\theta = 13\pi/6$. (You should check this for yourself.) In fact, the equation has an infinite number of solutions due to the periodicity of the sine function, as can be seen in Figure 25.

FIGURE 25

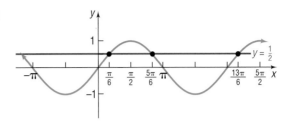

Unless the domain of the variable is restricted, we need to find *all* the solutions of a trigonometric equation. As the next example illustrates, finding all the solutions can be accomplished by first finding solutions over an interval whose length equals the period of the function and then adding multiples of that period to the solutions found. Let's look at some examples.

E X A M P L E   2

Finding All the Solutions of a Trigonometric Equation

FIGURE 26
$\cos \theta = \frac{1}{2}$

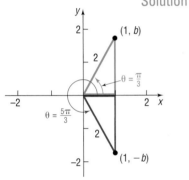

Solve the equation: $\cos \theta = \frac{1}{2}$

Solution

The period of the cosine function is 2π. In the interval $[0, 2\pi)$, there are two angles θ for which $\cos \theta = \frac{1}{2}$: $\theta = \pi/3$ and $\theta = 5\pi/3$. See Figure 26. Because the cosine function has period 2π, all the solutions of $\cos \theta = \frac{1}{2}$ may be given by

$$\theta = \frac{\pi}{3} + 2k\pi \quad \text{or} \quad \theta = \frac{5\pi}{3} + 2k\pi \quad \text{\small{k any integer.}}$$

Some of the solutions are

$$\underbrace{\frac{\pi}{3}, \frac{5\pi}{3}}_{k=0}, \underbrace{\frac{7\pi}{3}, \frac{11\pi}{3}}_{k=1}, \underbrace{\frac{13\pi}{3}, \frac{17\pi}{3}}_{k=2}, \quad \text{and so on}$$

FIGURE 27

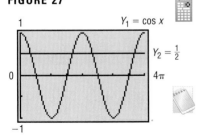

Check: We can verify the solution by graphing $Y_1 = \cos x$ and $Y_2 = \frac{1}{2}$ to determine where the graphs intersect. (Be sure to graph in radian mode.) See Figure 27. The graph of Y_1 intersects the graph of Y_2 at $x = 1.04 \ (\approx \pi/3)$, 5.23 $(\approx 5\pi/3)$, $7.33 \ (\approx 7\pi/3)$, and $11.51 \ (\approx 11\pi/3)$, correct to two decimal places.

Now work Problem 1.

In most of our work, we shall be interested only in finding solutions of trigonometric equations for $0 \le \theta < 2\pi$.

E X A M P L E   3

Solving a Trigonometric Equation

Solve the equation: $\sin 2\theta = 1, \quad 0 \le \theta < 2\pi$

Solution

The period of the sine function is 2π. In the interval $[0, 2\pi)$, the sine function has the value 1 only at $\pi/2$. Because the argument is 2θ in the given equation, we have

$$2\theta = \frac{\pi}{2} + 2k\pi \quad \text{\small{k any integer.}}$$

To find θ, we divide each side by 2.

$$\theta = \frac{\pi}{4} + k\pi$$

In the interval $[0, 2\pi)$, the solutions of $\sin 2\theta = 1$ are $\pi/4$ $(k = 0)$ and $\pi/4 + \pi = 5\pi/4$ $(k = 1)$. Note that $k = -1$ gives $\theta = -3\pi/4$ and $k = 2$ gives $\theta = 9\pi/4$, both of which are outside $[0, 2\pi)$. ∎

 Check: Verify that $\pi/4$ and $5\pi/4$ are solutions by graphing $Y_1 = \sin(2x)$ and $Y_2 = 1$ for $0 \leq x \leq 2\pi$. ∎

Warning In solving a trigonometric equation for θ, $0 \leq \theta < 2\pi$, in which the argument is not θ (as in Example 3), you must write down all the solutions, first, and then list those that are in the interval $[0, 2\pi)$. Otherwise, solutions may be lost. For example, in solving $\sin 2\theta = 1$, if you merely write the solution $2\theta = \pi/2$, you will find only $\theta = \pi/4$ and miss the solution $\theta = 5\pi/4$. ∎

 Now work Problem 7.

E X A M P L E   4

Solving a Trigonometric Equation

Solve the equation: $\sin 2\theta = \frac{1}{2}, \quad 0 \leq \theta < 2\pi$

FIGURE 28
$\sin 2\theta = \frac{1}{2}, 0 \leq \theta < 2\pi$

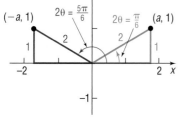

Solution   The period of the sine function is 2π. In the interval $[0, 2\pi)$, the sine function has a value $\frac{1}{2}$ at $\pi/6$ and $5\pi/6$. See Figure 28. Consequently, because the argument is 2θ in the equation $\sin 2\theta = \frac{1}{2}$, we have

$$2\theta = \frac{\pi}{6} + 2k\pi \quad \text{or} \quad 2\theta = \frac{5\pi}{6} + 2k\pi \qquad k \text{ any integer.}$$

$$\theta = \frac{\pi}{12} + k\pi \qquad\qquad \theta = \frac{5\pi}{12} + k\pi \qquad \text{Divide by 2.}$$

In the interval $[0, 2\pi)$, the solutions of $\sin 2\theta = \frac{1}{2}$ are

$$\pi/12, \quad \pi/12 + \pi = 13\pi/12, \quad 5\pi/12, \quad 5\pi/12 + \pi = 17\pi/12 \qquad ∎$$

E X A M P L E   5

Solving a Trigonometric Equation

Solve the equation: $\tan \dfrac{\theta}{4} = 1, \quad 0 \leq \theta < 2\pi$

Solution   The period of the tangent function is π. In the interval $[0, \pi)$, the tangent function has the value 1 only at $\pi/4$. Because the argument is $\theta/4$ in the given equation, we have

$$\frac{\theta}{4} = \frac{\pi}{4} + k\pi \qquad k \text{ any integer.}$$

$$\theta = \pi + 4k\pi$$

In the interval $[0, 2\pi)$, $\theta = \pi$ is the only solution. ∎

The next example illustrates how to solve trigonometric equations using a calculator. Remember that the function keys on a calculator will only give values consistent with the definition of the function.

E X A M P L E 6

Solving a Trigonometric Equation with a Calculator

Use a calculator to solve the equation: $\sin \theta = 0.3$, $0 \le \theta < 2\pi$
Express any solutions in radians, rounded to two decimal places.

Solution

To solve $\sin \theta = 0.3$ on a calculator, we first set the mode to radians. Then we use the $\boxed{\sin^{-1}}$ key as follows:

$$\sin^{-1}$$

Keystrokes: $\boxed{0.3}$ $\boxed{\text{SHIFT}}$ $\boxed{\sin}$

Display: $\searrow \boxed{0.3}$ $\searrow \boxed{0.3046927}$

FIGURE 29
$\sin \theta = 0.3$

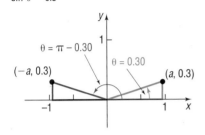

Because of the definition of $y = \sin^{-1} x$, the angle we obtain, 0.30 radian, rounded to two decimal places, is the angle $-\pi/2 \le \theta \le \pi/2$ for which $\sin \theta = 0.3$. Another angle for which $\sin \theta = 0.3$ is $\pi - 0.30$. See Figure 29. The angle $\pi - 0.30$ is the angle in quadrant II, where $\sin \theta = 0.3$. Thus, the solutions for $\sin \theta = 0.3, 0 \le \theta < 2\pi$, are

$$\theta = 0.30 \text{ radian} \quad \text{or} \quad \theta = \pi - 0.30 \approx 2.84 \text{ radians} \qquad \blacksquare$$

Warning Example 6 illustrates that caution must be exercised when solving trigonometric equations on a calculator. Remember that the calculator supplies an angle only within the restrictions of the definition of the inverse trigonometric function. To find the remaining solutions, you must identify other quadrants, if any, in which the angle may be located. $\blacksquare$

Now work Problem 17.

Many trigonometric equations can be solved by applying techniques that we already know from algebra, such as applying the quadratic formula (if the equation is a second-degree polynomial) or factoring. (Refer to Section A.4 in Appendix A, The Quadratic Formula).

E X A M P L E 7

Solving a Trigonometric Equation Quadratic in Form

Solve the equation: $2 \sin^2 \theta - 3 \sin \theta + 1 = 0$, $0 \le \theta < 2\pi$

Solution

The equation that we wish to solve is a quadratic equation (in $\sin \theta$) that can be factored.

$$2 \sin^2\theta - 3 \sin \theta + 1 = 0 \quad \text{\small $2x^2 - 3x + 1 = 0$, $x = \sin \theta$}$$
$$(2 \sin \theta - 1)(\sin \theta - 1) = 0 \quad \text{\small $(2x - 1)(x - 1) = 0$}$$
$$2 \sin \theta - 1 = 0 \quad \text{or} \quad \sin \theta - 1 = 0$$
$$\sin \theta = \tfrac{1}{2} \qquad\qquad \sin \theta = 1$$

Thus,

$$\theta = \frac{\pi}{6} \qquad \theta = \frac{5\pi}{6} \qquad \theta = \frac{\pi}{2} \qquad\qquad \blacksquare$$

Now work Problem 27.

When a trigonometric equation contains more than one trigonometric function, identities sometimes can be used to obtain an equivalent equation that contains only one trigonometric function.

E X A M P L E   8

Solving a Trigonometric Equation Using Identities

Solve the equation: $3 \cos \theta + 3 = 2 \sin^2 \theta, \quad 0 \le \theta < 2\pi$

Solution   The equation in its present form contains sines and cosines. However, a form of the Pythagorean Identity can be used to transform the equation into an equivalent expression containing only cosines:

$$3 \cos \theta + 3 = 2 \sin^2 \theta$$
$$3 \cos \theta + 3 = 2(1 - \cos^2 \theta) \quad \text{sin}^2 \theta = 1 - \cos^2 \theta$$
$$3 \cos \theta + 3 = 2 - 2 \cos^2 \theta$$
$$2 \cos^2 \theta + 3 \cos \theta + 1 = 0 \quad \text{Quadratic in } \cos \theta$$
$$(2 \cos \theta + 1)(\cos \theta + 1) = 0 \quad \text{Factor.}$$
$$2 \cos \theta + 1 = 0 \quad \text{or} \quad \cos \theta + 1 = 0$$
$$\cos \theta = -\tfrac{1}{2} \qquad\qquad \cos \theta = -1$$

Thus,

$$\theta = \frac{2\pi}{3} \qquad \theta = \frac{4\pi}{3} \qquad \theta = \pi$$

Check: Graph $Y_1 = 3 \cos x + 3$ and $Y_2 = 2 \sin^2 x, 0 \le x \le 2\pi$, and find the points of intersection. How close are your approximate solutions to the exact ones found in this example?

E X A M P L E   9

Solving a Trigonometric Equation Using Identities

Solve the equation: $\cos 2\theta + 3 = 5 \cos \theta, \quad 0 \le \theta < 2\pi$

Solution   First, we observe that the given equation contains two cosine functions, but with different arguments, θ and 2θ. We use the double-angle formula $\cos 2\theta = 2 \cos^2 \theta - 1$ to obtain an equivalent equation containing only $\cos \theta$.

$$\cos 2\theta + 3 = 5 \cos \theta$$
$$(2 \cos^2 \theta - 1) + 3 = 5 \cos \theta$$
$$2 \cos^2 \theta - 5 \cos \theta + 2 = 0$$
$$(\cos \theta - 2)(2 \cos \theta - 1) = 0$$
$$\cos \theta = 2 \quad \text{or} \quad \cos \theta = \tfrac{1}{2}$$

For any angle θ, $-1 \le \cos \theta \le 1$; thus, the equation $\cos \theta = 2$ has no solution. The solutions of $\cos \theta = \frac{1}{2}, 0 \le \theta < 2\pi$, are

$$\theta = \frac{\pi}{3} \qquad \theta = \frac{5\pi}{3}$$

Check: Graph $Y_1 = \cos(2x) + 3$ and $Y_2 = 5 \cos x, 0 \le x \le 2\pi$, and find the points of intersection. Compare your results with those of Example 9.

Now work Problem 37.

E X A M P L E 10   Solving a Trigonometric Equation Using Identities

Solve the equation: $\cos^2 \theta + \sin \theta = 2, \quad 0 \le \theta < 2\pi$

Solution   We use a form of the Pythagorean Identity:

$$\cos^2 \theta + \sin \theta = 2$$
$$(1 - \sin^2 \theta) + \sin \theta = 2 \quad \cos^2 \theta = 1 - \sin^2 \theta$$
$$\sin^2 \theta - \sin \theta + 1 = 0$$

This is a quadratic equation in $\sin \theta$. The discriminant is $b^2 - 4ac = 1 - 4 = -3 < 0$. Therefore, the equation has no real solution. ▬

Check: Graph $Y_1 = \cos^2 x + \sin x$ and $Y_2 = 2$ to see that the two graphs do not intersect anywhere. ▬

E X A M P L E 11   Solving a Trigonometric Equation Using Identities

Solve the equation: $\sin \theta \cos \theta = -\frac{1}{2}, \quad 0 \le \theta < 2\pi$

Solution   The left side of the given equation is in the form of the double-angle formula $2 \sin \theta \cos \theta = \sin 2\theta$, except for a factor of 2. Thus, we multiply each side by 2.

$$\sin \theta \cos \theta = -\frac{1}{2}$$
$$2 \sin \theta \cos \theta = -1$$
$$\sin 2\theta = -1$$

The argument here is 2θ. Thus, we need to write all the solutions of this equation and then list those that are in the interval $[0, 2\pi)$.

$$2\theta = \frac{3\pi}{2} + 2k\pi \quad k \text{ any integer.}$$

$$\theta = \frac{3\pi}{4} + k\pi$$

The solutions in the interval $[0, 2\pi)$ are

$$\theta = \frac{3\pi}{4} \qquad \theta = \frac{7\pi}{4}$$ ▬

Sometimes it is necessary to square both sides of an equation in order to obtain expressions that allow the use of identities. Remember, however, that when squaring both sides extraneous solutions may be introduced. As a result, apparent solutions must be checked.

E X A M P L E   12   Other Methods for Solving a Trigonometric Equation

Solve the equation: $\sin \theta + \cos \theta = 1, \quad 0 \le \theta < 2\pi$

Solution A   Attempts to use available identities do not lead to equations that are easy to solve. (Try it yourself.) Given the form of this equation, we decide to square each side.

$$\sin \theta + \cos \theta = 1$$
$$(\sin \theta + \cos \theta)^2 = 1$$
$$\sin^2 \theta + 2 \sin \theta \cos \theta + \cos^2 \theta = 1$$
$$2 \sin \theta \cos \theta = 0 \qquad \sin^2 \theta + \cos^2 \theta = 1$$
$$\sin \theta \cos \theta = 0$$

Thus,

$$\sin \theta = 0 \quad \text{or} \quad \cos \theta = 0$$

and the apparent solutions are

$$\theta = 0 \qquad \theta = \pi \qquad \theta = \frac{\pi}{2} \qquad \theta = \frac{3\pi}{2}$$

Because we squared both sides of the original equation, we must check these apparent solutions to see if any are extraneous.

$\theta = 0$: $\sin 0 + \cos 0 = 0 + 1 = 1$ A solution

$\theta = \pi$: $\sin \pi + \cos \pi = 0 + (-1) = -1$ Not a solution

$\theta = \dfrac{\pi}{2}$: $\sin \dfrac{\pi}{2} + \cos \dfrac{\pi}{2} = 1 + 0 = 1$ A solution

$\theta = \dfrac{3\pi}{2}$: $\sin \dfrac{3\pi}{2} + \cos \dfrac{3\pi}{2} = -1 + 0 = -1$ Not a solution

Thus, $\theta = 3\pi/2$ and $\theta = \pi$ are extraneous. The actual solutions are $\theta = 0$ and $\theta = \pi/2$.   ∎

We can solve the equation given in Example 12 another way.

Solution B   We start with the equation

$$\sin \theta + \cos \theta = 1$$

and divide each side by $\sqrt{2}$. (The reason for this choice will become apparent shortly.) Then

$$\frac{1}{\sqrt{2}} \sin \theta + \frac{1}{\sqrt{2}} \cos \theta = \frac{1}{\sqrt{2}}$$

The left side now resembles the formula for the sine of the sum of two angles, one of which is θ. The other angle is unknown (call it ϕ.) Then

$$\sin(\theta + \phi) = \sin \theta \cos \phi + \cos \theta \sin \phi = \frac{1}{\sqrt{2}} \tag{1}$$

where

$$\cos \phi = \frac{1}{\sqrt{2}} \qquad \sin \phi = \frac{1}{\sqrt{2}}, \qquad 0 \le \phi < 2\pi$$

The angle ϕ is therefore $\pi/4$. As a result, equation (1) becomes

$$\sin\left(\theta + \frac{\pi}{4}\right) = \frac{1}{\sqrt{2}}$$

We solve this equation to get

$$\theta + \frac{\pi}{4} = \frac{\pi}{4} \quad \text{or} \quad \theta + \frac{\pi}{4} = \frac{3\pi}{4}$$

$$\theta = 0 \qquad\qquad \theta = \frac{\pi}{2}$$

These solutions agree with the solutions found earlier. ■

This second method of solution can be used to solve any linear equation in the variables $\sin\theta$ and $\cos\theta$.

E X A M P L E   13

Solving a Trigonometric Equation Linear in $\sin\theta$ and $\cos\theta$

Solve: $a \sin\theta + b \cos\theta = c, \quad 0 \le \theta < 2\pi$ (2)

where a, b, and c are constants and either $a \ne 0$ or $b \ne 0$.

Solution   We divide each side of equation (2) by $\sqrt{a^2 + b^2}$. Then

$$\frac{a}{\sqrt{a^2 + b^2}} \sin\theta + \frac{b}{\sqrt{a^2 + b^2}} \cos\theta = \frac{c}{\sqrt{a^2 + b^2}} \tag{3}$$

There is a unique angle $\phi, 0 \le \phi < 2\pi$, for which

$$\cos\phi = \frac{a}{\sqrt{a^2 + b^2}} \quad \text{and} \quad \sin\phi = \frac{b}{\sqrt{a^2 + b^2}} \tag{4}$$

FIGURE 30

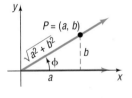

(see Figure 30). Thus, equation (3) may be written as

$$\sin\theta\cos\phi + \cos\theta\sin\phi = \frac{c}{\sqrt{a^2 + b^2}}$$

or, equivalently,

$$\sin(\theta + \phi) = \frac{c}{\sqrt{a^2 + b^2}} \tag{5}$$

where ϕ satisfies equations (4).

If $|c| > \sqrt{a^2 + b^2}$, then $\sin(\theta + \phi) > 1$ or $\sin(\theta + \phi) < -1$, and equation (5) has no solution.

If $|c| \le \sqrt{a^2 + b^2}$, then the solutions of equation (5) are

$$\theta + \phi = \sin^{-1}\frac{c}{\sqrt{a^2 + b^2}} \quad \text{or} \quad \theta + \phi = \pi - \sin^{-1}\frac{c}{\sqrt{a^2 + b^2}}$$

Because the angle ϕ is determined by equation (4), these are the solutions to equation (2). ■

Now work Problem 51.

Graphing Utility Solutions

The techniques introduced in this section apply only to certain types of trigonometric equations. Solutions for other types are usually studied in calculus, using numerical methods. In the next example, we show how a graphing utility may be used to obtain solutions.

E X A M P L E   14   Solving Trigonometric Equations Using a Graphing Utility

Solve: $5 \sin x + x = 3$

Express the solution(s) correct to two decimal places.

Solution   This type of trigonometric equation cannot be solved by previous methods. A graphing utility, though, can be used here. The solution(s) of this equation are the same as the points of intersection of the graphs of $Y_1 = 5 \sin x + x$ and $Y_2 = 3$. See Figure 31.

FIGURE 31

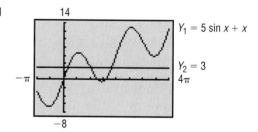

There are three points of intersection; the x-coordinates are the solutions that we seek. Using INTERSECT, we find

$$x = 0.51 \qquad x = 3.17 \qquad x = 5.71$$

correct to two decimal places.

3.6 | EXERCISES

In Problems 1–16, solve each equation on the interval $0 \le \theta < 2\pi$.

1. $\sin \theta = \dfrac{1}{2}$

2. $\tan \theta = 1$

3. $\tan \theta = -\dfrac{\sqrt{3}}{3}$

4. $\cos \theta = -\dfrac{\sqrt{3}}{2}$

5. $\cos \theta = 0$

6. $\sin \theta = \dfrac{\sqrt{2}}{2}$

7. $\sin 3\theta = -1$

8. $\tan \dfrac{\theta}{2} = \sqrt{3}$

9. $\cos 2\theta = -\dfrac{1}{2}$

10. $\tan 2\theta = -1$

11. $\sec \dfrac{3\theta}{2} = -2$

12. $\cot \dfrac{2\theta}{3} = -\sqrt{3}$

13. $\cos\left(2\theta - \dfrac{\pi}{2}\right) = -1$

14. $\sin\left(3\theta + \dfrac{\pi}{18}\right) = 1$

15. $\tan\left(\dfrac{\theta}{2} + \dfrac{\pi}{3}\right) = 1$

16. $\cos\left(\dfrac{\theta}{3} - \dfrac{\pi}{4}\right) = \dfrac{1}{2}$

In Problems 17–24, use a calculator to solve each equation on the interval $0 \leq \theta < 2\pi$. Round answers to two decimal places.

17. $\sin \theta = 0.4$ **18.** $\cos \theta = 0.6$ **19.** $\tan \theta = 5$ **20.** $\cot \theta = 2$

21. $\cos \theta = -0.9$ **22.** $\sin \theta = -0.2$ **23.** $\sec \theta = -4$ **24.** $\csc \theta = -3$

In Problems 25–56, solve each equation on the interval $0 \leq \theta < 2\pi$.

25. $2\cos^2 \theta + \cos \theta = 0$ **26.** $\sin^2 \theta - 1 = 0$ **27.** $2\sin^2 \theta - \sin \theta - 1 = 0$

28. $2\cos^2 \theta + \cos \theta - 1 = 0$ **29.** $(\tan \theta - 1)(\sec \theta - 1) = 0$ **30.** $(\cot \theta + 1)(\csc \theta - \frac{1}{2}) = 0$

31. $\cos \theta = \sin \theta$ **32.** $\cos \theta + \sin \theta = 0$ **33.** $\tan \theta = 2 \sin \theta$

34. $\sin 2\theta = \cos \theta$ **35.** $\sin \theta = \csc \theta$ **36.** $\tan \theta = \cot \theta$

37. $\cos 2\theta = \cos \theta$ **38.** $\sin 2\theta \sin \theta = \cos \theta$ **39.** $\sin 2\theta + \sin 4\theta = 0$

40. $\cos 2\theta + \cos 4\theta = 0$ **41.** $\cos 4\theta - \cos 6\theta = 0$ **42.** $\sin 4\theta - \sin 6\theta = 0$

43. $1 + \sin \theta = 2\cos^2 \theta$ **44.** $\sin^2 \theta = 2\cos \theta + 2$ **45.** $\tan^2 \theta = \frac{3}{2}\sec \theta$

46. $\csc^2 \theta = \cot \theta + 1$ **47.** $3 - \sin \theta = \cos 2\theta$ **48.** $\cos 2\theta + 5\cos \theta + 3 = 0$

49. $\sec^2 \theta + \tan \theta = 0$ **50.** $\sec \theta = \tan \theta + \cot \theta$ **51.** $\sin \theta - \sqrt{3}\cos \theta = 1$

52. $\sqrt{3}\sin \theta + \cos \theta = 1$ **53.** $\tan 2\theta + 2\sin \theta = 0$ **54.** $\tan 2\theta + 2\cos \theta = 0$

55. $\sin \theta + \cos \theta = \sqrt{2}$ **56.** $\sin \theta + \cos \theta = -\sqrt{2}$

 In Problems 57–62, solve each equation for $-\pi \leq x \leq \pi$. Express the solution(s) correct to two decimal places.

57. Solve the equation $\cos x = e^x$ by graphing $Y_1 = \cos x$ and $Y_2 = e^x$ and finding their point(s) of intersection.

58. Solve the equation $\cos x = e^x$ by graphing $Y_1 = \cos x - e^x$ and finding the x-intercept(s).

59. Solve the equation $2\sin x = 0.7x$ by graphing $Y_1 = 2\sin x$ and $Y_2 = 0.7x$ and finding their point(s) of intersection.

60. Solve the equation $2\sin x = 0.7x$ by graphing $Y_1 = 2\sin x - 0.7x$ and finding the x-intercept(s).

61. Solve the equation $\cos x = x^2$ by graphing $Y_1 = \cos x$ and $Y_2 = x^2$ and finding their point(s) of intersection.

62. Solve the equation $\cos x = x^2$ by graphing $Y_1 = \cos x - x^2$ and finding the x-intercept(s).

In Problems 63–74, use a graphing utility to solve each equation. Express the solution(s) correct to two decimal places.

63. $x + 5\cos x = 0$ **64.** $x - 4\sin x = 0$ **65.** $22x - 17\sin x = 3$

66. $19x + 8\cos x = 2$ **67.** $\sin x + \cos x = x$ **68.** $\sin x - \cos x = x$

69. $x^2 - 2\cos x = 0$ **70.** $x^2 + 3\sin x = 0$ **71.** $x^2 - 2\sin 2x = 3x$

72. $x^2 = x + 3\cos 2x$ **73.** $6\sin x - e^x = 2$, $x > 0$ **74.** $4\cos 3x - e^x = 1$, $x > 0$

75. Constructing a Rain Gutter   A rain gutter is to be constructed of aluminum sheets 12 inches wide. After marking off a length of 4 inches from each edge, this length is bent up at an angle θ. See the illustration. The area A of the opening as a function of θ is given by

$$A = 16 \sin \theta(\cos \theta + 1), \quad 0° < \theta < 90°$$

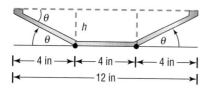

(a) In calculus, you will be asked to find the angle θ that maximizes A by solving the equation

$$\cos 2\theta + \cos \theta = 0, \quad 0° < \theta < 90°$$

Solve this equation for θ by using the double-angle formula.
(b) Solve the equation for θ by writing the sum of the two cosines as a product.
(c) What is the maximum area A of the opening?
 (d) Graph A, $0° \le \theta \le 90°$, and find the angle θ that maximizes the area A. Also find the maximum area. Compare the results to the answers found earlier.

76. Projectile Motion   An object is propelled upward at an angle θ, $45° < \theta < 90°$, to the horizontal with an initial velocity of v_0 feet per second from the base of a plane that makes an angle of $45°$ with the horizontal. See the illustration. If air resistance is ignored, the distance R that it travels up the inclined plane is given by

$$R = \frac{v_0^2 \sqrt{2}}{32}(\sin 2\theta - \cos 2\theta - 1)$$

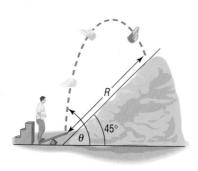

(a) In calculus, you will be asked to find the angle θ that maximizes R by solving the equation

$$\sin 2\theta + \cos 2\theta = 0$$

Solve this equation for θ using the method of Example 13.
(b) Solve this equation for θ by dividing each side by $\cos 2\theta$.
(c) What is the maximum distance R if $v_0 = 32$ feet per second?
(d) Graph R, $45° \le \theta \le 90°$, and find the angle θ that maximizes the distance R. Also find the maximum distance. Use $v_0 = 32$ feet per second. Compare the results with the answers found earlier.

77. Heat Transfer   In the study of heat transfer, the equation $x + \tan x = 0$ occurs. Graph $Y_1 = -x$ and $Y_2 = \tan x$ for $x \ge 0$. Conclude that there are an infinite number of points of intersection of these two graphs. Now find the first two positive solutions of $x + \tan x = 0$ correct to two decimal places.

78. Carrying a Ladder around a Corner   A ladder of length L is carried horizontally around a corner from a hall 3 feet wide into a hall 4 feet wide. See the illustration.

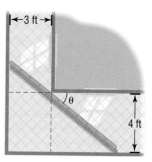

(a) Express L as a function of θ.
(b) In calculus, you will be asked to find the length of the longest ladder that can turn the corner by solving the equation

$$3 \sec \theta \tan \theta - 4 \csc \theta \cot \theta = 0, \quad 0° < \theta < 90°$$

Solve this equation for θ.
(c) What is the length of the longest ladder that can be carried around the corner?
(d) Graph L, $0° \le \theta \le 90°$, and find the angle θ that maximizes the length L. Also find the maximum length. Compare the results with the ones found in parts (b) and (c).

79. Projectile Motion The horizontal distance that a projectile will travel in the air is given by the equation

$$R = \frac{v_0^2 \sin 2\theta}{g}$$

where v_0 is the initial velocity of the projectile, θ is the angle of elevation, and g is acceleration due to gravity (9.8 meters per second squared).

(a) If you can throw a baseball with an initial speed of 34.8 meters per second, at what angle of elevation θ should you direct the throw so that the ball travels a distance of 107 meters before striking the ground?

(b) Determine the maximum distance that you can throw the ball.

(c) Graph R, with $v_0 = 34.8$ meters per second.

(d) Verify the results obtained in parts (a) and (b) using ZERO or ROOT.

80. Projectile Motion Refer to Problem 79.

(a) If you can throw a baseball with an initial speed of 40 meters per second, at what angle of elevation θ should you direct the throw so that the ball travels a distance of 110 meters before striking the ground?

(b) Determine the maximum distance that you can throw the ball.

(c) Graph R, with $v_0 = 40$ meters per second.

(d) Verify the results obtained in parts (a) and (b) using ZERO or ROOT.

*The following discussion of **Snell's Law of Refraction** (named after Willebrod Snell, 1591–1626) is needed for Problems 81–87. Light, sound, and other waves travel at different speeds, depending on the media (air, water, wood, and so on) through which they pass. Suppose that light travels from a point A in one medium, where its speed is v_1, to a point B in another medium, where its speed is v_2. Refer to the figure, where the angle θ_1 is called the **angle of incidence** and the angle θ_2 is the **angle of refraction**. Snell's Law,* which can be proved using calculus, states that*

$$\frac{\sin \theta_1}{\sin \theta_2} = \frac{v_1}{v_2}$$

*The ratio v_1/v_2 is called the **index of refraction**. Some values are given in the following table.*

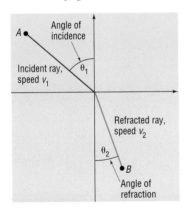

Some Indexes of Refraction	
Medium	**Index of Refraction**
Water	1.33
Ethyl alcohol	1.36
Carbon bisulfide	1.63
Air (1 atm and 20°C)	1.0003
Methylene iodide	1.74
Fused quartz	1.46
Glass, crown	1.52
Glass, dense flint	1.66
Sodium chloride	1.53

For light of wavelength 589 nanometers, measured with respect to a vacuum. The index with respect to air is negligibly different in most cases.

81. The index of refraction of light in passing from a vacuum into water is 1.33. If the angle of incidence is 40°, determine the angle of refraction.

82. The index of refraction of light in passing from a vacuum into dense glass is 1.66. If the angle of incidence is 50°, determine the angle of refraction.

83. Ptolemy, who lived in the city of Alexandria in Egypt during the second century A.D., gave the measured values in the table to the right for the angle of incidence θ_1 and the angle of refraction θ_2 for a light beam passing from air into water. Do these values agree with Snell's Law? If so, what index of refraction results? (These data are interesting as the oldest recorded physical measurements.)†

θ_1	θ_2
10°	7°45′
20°	15°30′
30°	22°30′
40°	29°0′
50°	35°0′
60°	40°30′
70°	45°30′
80°	50°0′

*Because this law was also deduced by René Descartes in France, it is also known as Descartes' Law.

†Adapted from Halliday and Resnick, *Physics, Parts 1 & 2,* 3rd ed. New York: Wiley, 1978, p. 953.

84. The speed of yellow sodium light (wavelength of 589 nanometers) in a certain liquid is measured to be 1.92×10^8 meters per second. What is the index of refraction of this liquid, with respect to air, for sodium light?*

[**Hint:** The speed of light in air is approximately 2.99×10^8 meters per second.]

85. A beam of light with a wavelength of 589 nanometers traveling in air makes an angle of incidence of 40° upon a slab of transparent material, and the refracted beam makes an angle of refraction of 26°. Find the index of refraction of the material.*

*Adapted from Serway, *Physics*, 3rd ed. Philadelphia: W. B. Saunders, p. 805.

86. A light ray with a wavelength of 589 nanometers (produced by a sodium lamp) traveling through air makes an angle of incidence of 30° on a smooth, flat slab of crown glass. Find the angle of refraction.*

87. A light beam passes through a thick slab of material whose index of refraction is n_2. Show that the emerging beam is parallel to the incident beam.*

 88. Explain in your own words how you would use your calculator to solve the equation $\sin x = 0.3, 0 \le x < 2\pi$. How would you modify your approach in order to solve the equation $\cot x = 5, 0 < x < 2\pi$?

CHAPTER REVIEW

THINGS TO KNOW

Formulas

Sum and Difference Formulas

$$\cos(\alpha + \beta) = \cos \alpha \cos \beta - \sin \alpha \sin \beta$$
$$\cos(\alpha - \beta) = \cos \alpha \cos \beta + \sin \alpha \sin \beta$$
$$\sin(\alpha + \beta) = \sin \alpha \cos \beta + \cos \alpha \sin \beta$$
$$\sin(\alpha - \beta) = \sin \alpha \cos \beta - \cos \alpha \sin \beta$$
$$\tan(\alpha + \beta) = \frac{\tan \alpha + \tan \beta}{1 - \tan \alpha \tan \beta}$$
$$\tan(\alpha - \beta) = \frac{\tan \alpha - \tan \beta}{1 + \tan \alpha \tan \beta}$$

Double-Angle Formulas

$$\sin 2\theta = 2 \sin \theta \cos \theta$$
$$\cos 2\theta = \cos^2 \theta - \sin^2 \theta$$
$$\cos 2\theta = 1 - 2 \sin^2 \theta$$
$$\cos 2\theta = 2 \cos^2 \theta - 1$$
$$\tan 2\theta = \frac{2 \tan \theta}{1 - \tan^2 \theta}$$

Half-Angle Formulas

$$\sin^2 \frac{\alpha}{2} = \frac{1 - \cos \alpha}{2}$$

$$\cos^2 \frac{\alpha}{2} = \frac{1 + \cos \alpha}{2}$$

$$\tan^2 \frac{\alpha}{2} = \frac{1 - \cos \alpha}{1 + \cos \alpha}$$

$$\sin \frac{\alpha}{2} = \pm \sqrt{\frac{1 - \cos \alpha}{2}}$$

$$\cos \frac{\alpha}{2} = \pm \sqrt{\frac{1 + \cos \alpha}{2}}$$

Where the + or − sign is determined by the quadrant of the angle $\alpha/2$.

$$\tan \frac{\alpha}{2} = \pm \sqrt{\frac{1 - \cos \alpha}{1 + \cos \alpha}} = \frac{1 - \cos \alpha}{\sin \alpha} = \frac{\sin \alpha}{1 + \cos \alpha}$$

Product-to-Sum Formulas

$$\sin \alpha \sin \beta = \tfrac{1}{2}[\cos(\alpha - \beta) - \cos(\alpha + \beta)]$$
$$\cos \alpha \cos \beta = \tfrac{1}{2}[\cos(\alpha - \beta) + \cos(\alpha + \beta)]$$
$$\sin \alpha \cos \beta = \tfrac{1}{2}[\sin(\alpha + \beta) + \sin(\alpha - \beta)]$$

Sum-to-Product Formulas

$$\sin \alpha + \sin \beta = 2 \sin \frac{\alpha + \beta}{2} \cos \frac{\alpha - \beta}{2}$$
$$\sin \alpha - \sin \beta = 2 \sin \frac{\alpha - \beta}{2} \cos \frac{\alpha + \beta}{2}$$
$$\cos \alpha + \cos \beta = 2 \cos \frac{\alpha + \beta}{2} \cos \frac{\alpha - \beta}{2}$$
$$\cos \alpha - \cos \beta = -2 \sin \frac{\alpha + \beta}{2} \sin \frac{\alpha - \beta}{2}$$

Definitions of the six inverse trigonometric functions

$y = \sin^{-1} x$ means $x = \sin y$ where $-1 \le x \le 1, \; -\pi/2 \le y \le \pi/2$
$y = \cos^{-1} x$ means $x = \cos y$ where $-1 \le x \le 1, \; 0 \le y \le \pi$
$y = \tan^{-1} x$ means $x = \tan y$ where $-\infty < x < \infty, \; -\pi/2 < y < \pi/2$
$y = \cot^{-1} x$ means $x = \cot y$ where $-\infty < x < \infty, \; 0 < y < \pi$
$y = \sec^{-1} x$ means $x = \sec y$ where $|x| \ge 1, \; 0 \le y \le \pi, \; y \ne \pi/2$
$y = \csc^{-1} x$ means $x = \csc y$ where $|x| \ge 1, \; -\pi/2 \le y \le \pi/2, \; y \ne 0$

HOW TO

Establish identities

Find the exact value of certain inverse trigonometric functions

Use a calculator to find the approximate values of inverse trigonometric functions

Solve a trigonometric equation

FILL-IN-THE-BLANK ITEMS

1. Suppose that f and g are two functions with the same domain. If $f(x) = g(x)$ for every x in the domain, the equation is called a(n) _____. Otherwise, it is called a(n) _____ equation.

2. $\cos(\alpha + \beta) = \cos \alpha \cos \beta$ _____ $\sin \alpha \sin \beta$

3. $\sin(\alpha + \beta) = \sin \alpha \cos \beta$ _____ $\cos \alpha \sin \beta$

4. $\cos 2\theta = \cos^2 \theta -$ _____ $=$ _____ $- 1 = 1 -$ _____

5. $\sin^2 \dfrac{\alpha}{2} = \dfrac{}{2}$

6. The function $y = \sin^{-1} x$ has domain _____ and range _____.

7. The value of $\sin^{-1}[\cos(\pi/2)]$ is _____.

TRUE/FALSE ITEMS

T F **1.** $\sin(-\theta) + \sin \theta = 0$ for all θ.

T F **2.** $\sin(\alpha + \beta) = \sin \alpha + \sin \beta + 2 \sin \alpha \sin \beta$.

T F **3.** $\cos 2\theta$ has three equivalent forms: $\cos^2 \theta - \sin^2 \theta, 1 - 2 \sin^2 \theta$, and $2 \cos^2 \theta - 1$.

T F **4.** $\cos \dfrac{\alpha}{2} = \pm \dfrac{\sqrt{1 + \cos \alpha}}{2}$, where the $+$ or $-$ sign depends on the angle $\alpha/2$.

T F **5.** The domain of $y = \sin^{-1} x$ is $-\pi/2 \le x \le \pi/2$.

T F **6.** $\cos(\sin^{-1} 0) = 1$ and $\sin(\cos^{-1} 0) = 1$.

T F **7.** Most trigonometric equations have unique solutions.

T F **8.** The equation $\tan \theta = \pi/2$ has no solution.

REVIEW EXERCISES

Blue problem numbers indicate the author's suggestions for use in a Practice Test.

In Problems 1–32, establish each identity.

1. $\tan \theta \cot \theta - \sin^2 \theta = \cos^2 \theta$

2. $\sin \theta \csc \theta - \sin^2 \theta = \cos^2 \theta$

3. $\cos^2 \theta(1 + \tan^2 \theta) = 1$

4. $(1 - \cos^2 \theta)(1 + \cot^2 \theta) = 1$

5. $4 \cos^2 \theta + 3 \sin^2 \theta = 3 + \cos^2 \theta$

6. $4 \sin^2 \theta + 2 \cos^2 \theta = 4 - 2 \cos^2 \theta$

7. $\dfrac{1 - \cos \theta}{\sin \theta} + \dfrac{\sin \theta}{1 - \cos \theta} = 2 \csc \theta$

8. $\dfrac{\sin \theta}{1 + \cos \theta} + \dfrac{1 + \cos \theta}{\sin \theta} = 2 \csc \theta$

9. $\dfrac{\cos \theta}{\cos \theta - \sin \theta} = \dfrac{1}{1 - \tan \theta}$

10. $1 - \dfrac{\cos^2 \theta}{1 + \sin \theta} = \sin \theta$

11. $\dfrac{\csc \theta}{1 + \csc \theta} = \dfrac{1 - \sin \theta}{\cos^2 \theta}$

12. $\dfrac{1 + \sec \theta}{\sec \theta} = \dfrac{\sin^2 \theta}{1 - \cos \theta}$

13. $\csc \theta - \sin \theta = \cos \theta \cot \theta$

14. $\dfrac{\csc \theta}{1 - \cos \theta} = \dfrac{1 + \cos \theta}{\sin^3 \theta}$

15. $\dfrac{1 - \sin \theta}{\sec \theta} = \dfrac{\cos^3 \theta}{1 + \sin \theta}$

16. $\dfrac{1 - \cos \theta}{1 + \cos \theta} = (\csc \theta - \cot \theta)^2$

17. $\dfrac{1 - 2 \sin^2 \theta}{\sin \theta \cos \theta} = \cot \theta - \tan \theta$

18. $\dfrac{(2 \sin^2 \theta - 1)^2}{\sin^4 \theta - \cos^4 \theta} = 1 - 2 \cos^2 \theta$

19. $\dfrac{\cos(\alpha + \beta)}{\cos \alpha \sin \beta} = \cot \beta - \tan \alpha$

20. $\dfrac{\sin(\alpha - \beta)}{\sin \alpha \cos \beta} = 1 - \cot \alpha \tan \beta$

21. $\dfrac{\cos(\alpha - \beta)}{\cos \alpha \cos \beta} = 1 + \tan \alpha \tan \beta$

22. $\dfrac{\cos(\alpha + \beta)}{\sin \alpha \cos \beta} = \cot \alpha - \tan \beta$

23. $(1 + \cos \theta)\left(\tan \dfrac{\theta}{2}\right) = \sin \theta$

24. $\sin \theta \tan \dfrac{\theta}{2} = 1 - \cos \theta$

25. $2 \cot \theta \cot 2\theta = \cot^2 \theta - 1$

26. $2 \sin 2\theta(1 - 2 \sin^2 \theta) = \sin 4\theta$

27. $1 - 8 \sin^2 \theta \cos^2 \theta = \cos 4\theta$

28. $\dfrac{\sin 3\theta \cos \theta - \sin \theta \cos 3\theta}{\sin 2\theta} = 1$

29. $\dfrac{\sin 2\theta + \sin 4\theta}{\cos 2\theta + \cos 4\theta} = \tan 3\theta$

30. $\dfrac{\sin 2\theta + \sin 4\theta}{\sin 2\theta - \sin 4\theta} + \dfrac{\tan 3\theta}{\tan \theta} = 0$

31. $\dfrac{\cos 2\theta - \cos 4\theta}{\cos 2\theta + \cos 4\theta} - \tan \theta \tan 3\theta = 0$

32. $\cos 2\theta - \cos 10\theta = (\tan 4\theta)(\sin 2\theta + \sin 10\theta)$

In Problems 33–40, find the exact value of each expression.

33. $\sin 165°$

34. $\tan 105°$

35. $\cos \dfrac{5\pi}{12}$

36. $\sin\left(-\dfrac{\pi}{12}\right)$

37. $\cos 80° \cos 20° + \sin 80° \sin 20°$

38. $\sin 70° \cos 40° - \cos 70° \sin 40°$

39. $\tan \dfrac{\pi}{8}$

40. $\sin \dfrac{5\pi}{8}$

In Problems 41–50, use the information given about the angles α and β to find the exact value of:

(a) $\sin(\alpha + \beta)$ (b) $\cos(\alpha + \beta)$ (c) $\sin(\alpha - \beta)$ (d) $\tan(\alpha + \beta)$

(e) $\sin 2\alpha$ (f) $\cos 2\beta$ (g) $\sin \dfrac{\beta}{2}$ (h) $\cos \dfrac{\alpha}{2}$

41. $\sin \alpha = \frac{4}{5}$, $\quad 0 < \alpha < \pi/2$; $\quad \sin \beta = \frac{5}{13}$, $\quad \pi/2 < \beta < \pi$

42. $\cos \alpha = \frac{4}{5}$, $\quad 0 < \alpha < \pi/2$; $\quad \cos \beta = \frac{5}{13}$, $\quad -\pi/2 < \beta < 0$

43. $\sin \alpha = -\frac{3}{5}$, $\quad \pi < \alpha < 3\pi/2$; $\quad \cos \beta = \frac{12}{13}$, $\quad 3\pi/2 < \beta < 2\pi$

44. $\sin \alpha = -\frac{4}{5}, \quad -\pi/2 < \alpha < 0; \quad \cos \beta = -\frac{5}{13}, \quad \pi/2 < \beta < \pi$

45. $\tan \alpha = \frac{3}{4}, \quad \pi < \alpha < 3\pi/2; \quad \tan \beta = \frac{12}{5}, \quad 0 < \beta < \pi/2$

46. $\tan \alpha = -\frac{4}{3}, \quad \pi/2 < \alpha < \pi; \quad \cot \beta = \frac{12}{5}, \quad \pi < \beta < 3\pi/2$

47. $\sec \alpha = 2, \quad -\pi/2 < \alpha < 0; \quad \sec \beta = 3, \quad 3\pi/2 < \beta < 2\pi$

48. $\csc \alpha = 2, \quad \pi/2 < \alpha < \pi; \quad \sec \beta = -3, \quad \pi/2 < \beta < \pi$

49. $\sin \alpha = -\frac{2}{3}, \quad \pi < \alpha < 3\pi/2; \quad \cos \beta = -\frac{2}{3}, \quad \pi < \beta < 3\pi/2$

50. $\tan \alpha = -2, \quad \pi/2 < \alpha < \pi; \quad \cot \beta = -2, \quad \pi/2 < \beta < \pi$

In Problems 51–70, find the exact value of each expression.

51. $\sin^{-1} 1$

52. $\cos^{-1} 0$

53. $\tan^{-1} 1$

54. $\sin^{-1}\left(-\frac{1}{2}\right)$

55. $\cos^{-1}\left(-\frac{\sqrt{3}}{2}\right)$

56. $\tan^{-1}(-\sqrt{3})$

57. $\sin\left(\cos^{-1}\frac{\sqrt{2}}{2}\right)$

58. $\cos(\sin^{-1} 0)$

59. $\tan\left[\sin^{-1}\left(-\frac{\sqrt{3}}{2}\right)\right]$

60. $\tan\left[\cos^{-1}\left(-\frac{1}{2}\right)\right]$

61. $\sec\left(\tan^{-1}\frac{\sqrt{3}}{3}\right)$

62. $\csc\left(\sin^{-1}\frac{\sqrt{3}}{2}\right)$

63. $\sin\left(\tan^{-1}\frac{3}{4}\right)$

64. $\cos\left(\sin^{-1}\frac{3}{5}\right)$

65. $\tan\left[\sin^{-1}\left(-\frac{4}{5}\right)\right]$

66. $\tan\left[\cos^{-1}\left(-\frac{3}{5}\right)\right]$

67. $\sin^{-1}\left(\cos\frac{2\pi}{3}\right)$

68. $\cos^{-1}\left(\tan\frac{3\pi}{4}\right)$

69. $\tan^{-1}\left(\tan\frac{7\pi}{4}\right)$

70. $\cos^{-1}\left(\cos\frac{7\pi}{6}\right)$

In Problems 71–76, find the exact value of each expression.

71. $\cos(\sin^{-1}\frac{3}{5} - \cos^{-1}\frac{1}{2})$

72. $\sin(\cos^{-1}\frac{5}{13} - \cos^{-1}\frac{4}{5})$

73. $\tan[\sin^{-1}\left(-\frac{1}{2}\right) - \tan^{-1}\frac{3}{4}]$

74. $\cos[\tan^{-1}(-1) + \cos^{-1}\left(-\frac{4}{5}\right)]$

75. $\sin[2\cos^{-1}\left(-\frac{3}{5}\right)]$

76. $\cos(2\tan^{-1}\frac{4}{3})$

In Problems 77–96, solve each equation on the interval $0 \le \theta < 2\pi$.

77. $\cos \theta = \frac{1}{2}$

78. $\sin \theta = -\sqrt{3}/2$

79. $\cos \theta = -\sqrt{2}/2$

80. $\tan \theta = -\sqrt{3}$

81. $\sin 2\theta = -1$

82. $\cos 2\theta = 0$

83. $\tan 2\theta = 0$

84. $\sin 3\theta = 1$

85. $\sin \theta = 0.9$

86. $\tan \theta = 25$

87. $\sin \theta = \tan \theta$

88. $\cos \theta = \sec \theta$

89. $\sin \theta + \sin 2\theta = 0$

90. $\cos 2\theta = \sin \theta$

91. $\sin 2\theta - \cos \theta - 2 \sin \theta + 1 = 0$

92. $\sin 2\theta - \sin \theta - 2 \cos \theta + 1 = 0$

93. $2 \sin^2 \theta - 3 \sin \theta + 1 = 0$

94. $2 \cos^2 \theta + \cos \theta - 1 = 0$

95. $\sin \theta - \cos \theta = 1$

96. $\sin \theta + 2 \cos \theta = 1$

In Problems 97–102, use a graphing utility to solve each equation on the interval $0 \le x \le 2\pi$. Approximate any solutions correct to two decimal places.

97. $2x = 5 \cos x$

98. $2x = 5 \sin x$

99. $2 \sin x + 3 \cos x = 4x$

100. $3 \cos x + x = \sin x$

101. $\sin x = \ln x$

102. $\sin x = e^{-x}$

Applications of Trigonometric Functions

Many of the everyday items around us are based upon mathematics. On the following page, you accept a commission to create a stained-glass lamp with a mathematical shape. The information you will need to build this lamp can be found on the Sullivan website at:

www.prenhall.com/sullivan

PREPARING FOR THIS CHAPTER

Before getting started on this chapter, review the following concepts:

Pythagorean Theorem *(p. 10)*

Right Triangle Trigonometry *(Section 2.4)*

OUTLINE

POLYHEDRA STAINED-GLASS LAMPS

Polyhedra, or many sided 3-D objects, have been known and studied for a long time. The Greek philosopher Plato studied the five regular sided polyhedra, now known as *platonic solids*. Many objects found in nature, such as crystals, some types of viruses, and buckyballs, have the shape of polyhedra. Ever since *Buckminster Fuller* popularized the geodesic dome there has been a resurgence of interest in polyhedra.

Suppose that you are the owner and operator of a stained-glass business and a customer who lives in a geodesic dome wants to commission you to make her a stained-glass lamp in the shape of a polyhedron. As a master craftsman, building such a lamp will not pose much of a problem, but you have no ready-made designs drawn up for the customer to choose from, so you ask the customer if she would like to choose a pattern from the *World Wide Web Library of Polyhedra*. The virtual reality displays make a big impression on your client. She selects a *5cube* as her choice for a lamp.

1. The stained-glass lamp that you are about to build at first appears to be very complicated, but as you continue to look at the 5cube you begin to see an underlying structure. Can you see the underlying structure? Do you think you could build a 5cube?
2. Since the cost of a stained-glass lamp runs about $5 per piece of glass, (and each triangle will be a piece), what do you estimate is the cost of the lamp?
3. You will be able to build the lamp by making six square sections. Each of these sections is made up only of triangles. How many different types of triangles are there? How many pieces will you need to cut for each of the different triangles?
4. Next you have the challenge of figuring out how to make the patterns for cutting the glass. You plan to make a *paper model*. You decide to make the lamp with an overall diameter of 20 inches. Can you use your calculator and a formula to determine the size of the cubes?
5. Once you have the side lengths of the cubes that compose the 5cube, what do you need to know in order to calculate the triangles? Can you determine these lengths by computation?
6. If you spend 20 minutes per piece making the lamp, how much would you make per hour if you charge the price that you estimated in question 2?

In this chapter we use the trigonometric functions to solve applied problems. The first four sections deal with applications involving right triangles and *oblique triangles,* triangles that do not have a right angle. To solve problems involving oblique triangles, we will develop the Law of Sines and the Law of Cosines. We will also develop formulas for finding the area of a triangle.

The final section deals with applications involving simple harmonic motion and damped motion.

4.1 | SOLVING RIGHT TRIANGLES

 Solve Right Triangles

 Solve Applied Problems

FIGURE 1

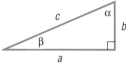

 In the discussion that follows, we will always label a right triangle so that side a is opposite angle α, side b is opposite angle β, and side c is the hypotenuse, as shown in Figure 1. **To solve a right triangle** means to find the missing lengths of its sides and the measurements of its angles. We shall follow the practice of expressing the lengths of the sides rounded to two decimal places and of expressing angles in degrees rounded to one decimal place.

To solve a right triangle, we need to know either an angle (besides the 90° one) and a side or else two sides. Then we make use of the Pythagorean Theorem and the fact that the sum of the angles of a triangle is 180°. The sum of the unknown angles in a right triangle is therefore 90°.

For the right triangle shown in Figure 1, we have

$$c^2 = a^2 + b^2 \qquad \alpha + \beta = 90°$$

E X A M P L E   1 — Solving a Right Triangle

FIGURE 2

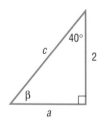

Use Figure 2. If $b = 2$ and $\alpha = 40°$, find a, c, and β.

Solution Since $\alpha = 40°$ and $\alpha + \beta = 90°$, we find that $\beta = 50°$. To find the sides a and c, we use the facts that

$$\tan 40° = \frac{a}{2} \quad \text{and} \quad \cos 40° = \frac{2}{c}$$

Now solve for a and c.

$$a = 2 \tan 40° \approx 1.68 \quad \text{and} \quad c = \frac{2}{\cos 40°} \approx 2.61$$

 Now work Problem 1.

E X A M P L E   2 — Solving a Right Triangle

FIGURE 3

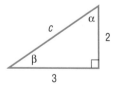

Use Figure 3. If $a = 3$ and $b = 2$, find c, α, and β.

Solution Since $a = 3$ and $b = 2$, then, by the Pythagorean Theorem, we have

$$c^2 = a^2 + b^2 = 9 + 4 = 13$$
$$c = \sqrt{13} \approx 3.61$$

To find angle α, we use the fact that

$$\tan \alpha = \frac{3}{2} \quad \text{so} \quad \alpha = \tan^{-1}\frac{3}{2}$$

Set the mode on your calculator to degrees. Then, rounded to one decimal place, we find $\alpha = 56.3°$. Since $\alpha + \beta = 90°$, we find that $\beta = 33.7°$.

Note   In subsequent examples and in the exercises that follow, unless otherwise indicated, we will measure angles in degrees and round off to one decimal place; we will round off all sides to two decimal places. To avoid round-off errors when using a calculator, we will store unrounded values in memory for use in subsequent calculations.

Now work Problem 11.

One common use for trigonometry is to measure heights and distances that are either awkward or impossible to measure by ordinary means.

E X A M P L E   3   Finding the Width of a River

A surveyor can measure the width of a river by setting up a transit* at a point C on one side of the river and taking a sighting of a point A on the other side. Refer to Figure 4. After turning through an angle of 90° at C, the surveyor walks a distance of 200 meters to point B. Using the transit at B, the angle β is measured and found to be 20°. What is the width of the river?

FIGURE 4

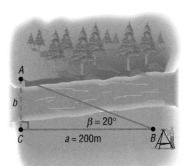

Solution   We seek the length of side b. We know a and β, so we use the fact that

$$\tan \beta = \frac{b}{a}$$

to get

$$\tan 20° = \frac{b}{200}$$
$$b = 200 \tan 20° \approx 72.79 \text{ meters}$$

Thus, the width of the river is approximately 73 meters, rounded to the nearest meter.   ▬

Now work Problem 21.

E X A M P L E   4   Finding the Inclination of a Mountain Trail

A straight trail with a uniform inclination leads from the Alpine Hotel, elevation 8000 feet, to a scenic overlook, elevation 11,100 feet. The length of the trail is 14,100 feet. What is the inclination (grade) of the trail? That is, what is the angle β in Figure 5?

FIGURE 5

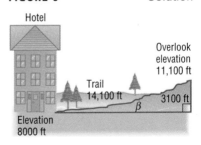

Solution   As Figure 5 illustrates, the angle β obeys the equation

$$\sin \beta = \frac{3100}{14,100}$$

Using a calculator,

$$\beta = \sin^{-1} \frac{3100}{14,100} \approx 12.7°$$

The inclination (grade) of the trail is approximately 12.7°.   ▬

*An instrument used in surveying to measure angles.

Vertical heights can sometimes be measured using either the angle of elevation or the angle of depression. If a person is looking up at an object, the acute angle measured from the horizontal to a line-of-sight observation of the object is called the **angle of elevation.** See Figure 6(a).

If a person is standing on a cliff looking down at an object, the acute angle made by the line of sight observation of the object and the horizontal is called the **angle of depression.** See Figure 6(b).

FIGURE 6

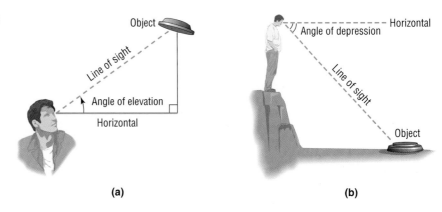

(a)   (b)

E X A M P L E   5   Finding Heights Using the Angle of Elevation

To determine the height of a radio transmission tower, a surveyor walks off a distance of 300 meters from the base of the tower. The angle of elevation is then measured and found to be 40°. If the transit is 2 meters off the ground when the sighting is taken, how high is the radio tower? See Figure 7(a).

FIGURE 7

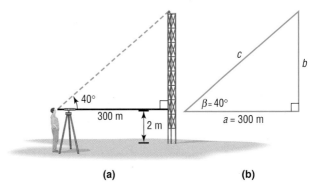

(a)   (b)

Solution   Figure 7(b) shows a triangle that replicates the illustration in Figure 7(a). To find the length b, we use the fact that $\tan \beta = b/a$. Then

$$b = a \tan \beta = 300 \tan 40° \approx 251.73 \text{ meters}$$

Because the transit is 2 meters high, the actual height of the tower is approximately 254 meters, rounded to the nearest meter.

Now work Problem 23.

The idea behind Example 5 can also be used to find the height of an object with a base that is not accessible to the horizontal.

EXAMPLE 6   Finding the Height of a Statue on a Building

Adorning the top of the Board of Trade building in Chicago is a statue of Ceres, the Greek goddess of wheat. From street level, two observations are taken 400 feet from the center of the building. The angle of elevation to the base of the statue is found to be 45°; the angle of elevation to the top of the statue is 47.2°. See Figure 8(a). What is the height of the statue?

FIGURE 8

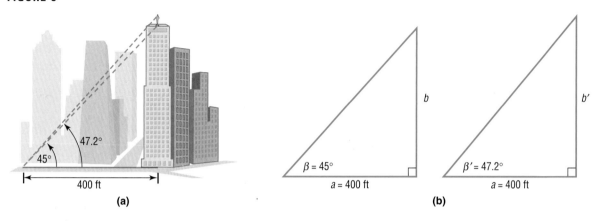

(a)   (b)

Solution   Figure 8(b) shows two triangles that replicate Figure 8(a). The height of the statue of Ceres will be $b' - b$. To find b and b', we refer to Figure 8(b).

$$\tan 45° = \frac{b}{400} \qquad\qquad \tan 47.2° = \frac{b'}{400}$$

$$b = 400 \tan 45° = 400 \qquad\qquad b' = 400 \tan 47.2° \approx 432$$

The height of the statue is approximately $432 - 400 = 32$ feet. ▬

When it is not possible to walk off a distance from the base of the object whose height we seek, a more imaginative solution is required.

EXAMPLE 7   Finding the Height of a Mountain

To measure the height of a mountain, a surveyor takes two sightings of the peak at a distance 900 meters apart on a direct line to the mountain.* See Figure 9(a). The first observation results in an angle of elevation of 47°, whereas the second results in an angle of elevation of 35°. If the transit is 2 meters high, what is the height h of the mountain?

FIGURE 9

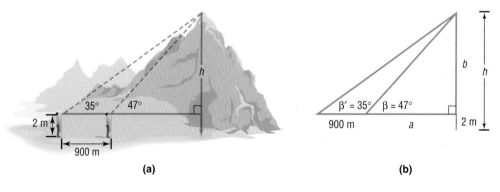

(a)   (b)

*For simplicity, we assume that these sightings are at the same level.

Solution   Figure 9(b) shows two triangles that replicate the illustration in Figure 9(a). From the two right triangles shown, we find

$$\tan \beta' = \frac{b}{a + 900} \qquad \tan \beta = \frac{b}{a}$$

$$\tan 35° = \frac{b}{a + 900} \qquad \tan 47° = \frac{b}{a}$$

This is a system of two equations involving two variables, a and b. Because we seek b, we choose to solve the right-hand equation for a and substitute the result, $a = b/\tan 47° = b \cot 47°$, in the left-hand equation. The result is

$$\tan 35° = \frac{b}{b \cot 47° + 900}$$

$$b = (b \cot 47° + 900) \tan 35° \qquad \text{Clear fraction.}$$

$$b = b \cot 47° \tan 35° + 900 \tan 35° \qquad \text{Remove parentheses.}$$

$$b(1 - \cot 47° \tan 35°) = 900 \tan 35° \qquad \text{Isolate } b \text{ on left side.}$$

$$b = \frac{900 \tan 35°}{1 - \cot 47° \tan 35°} = \frac{900 \tan 35°}{1 - \dfrac{\tan 35°}{\tan 47°}} \approx 1816 \qquad \text{Solve for } b.$$

$$\text{Use a calculator.}$$

The height of the peak from ground level is therefore approximately $1816 + 2 = 1818$ meters.   ▄

   Now work Problem 29.

E X A M P L E   8   The Gibb's Hill Lighthouse, Southampton, Bermuda

In operation since 1846, the Gibb's Hill Lighthouse stands 117 feet high on a hill 245 feet high, so its beam of light is 362 feet above sea level. A brochure states that the light can be seen on the horizon about 26 miles distant. Verify the accuracy of this statement.

Solution   Figure 10 illustrates the situation. The central angle θ obeys the equation

FIGURE 10

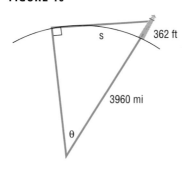

362 ft

3960 mi

θ

s

$$\cos \theta = \frac{3960}{3960 + \dfrac{362}{5280}} \approx 0.999982687 \qquad \text{1 mile = 5280 feet}$$

so

$$\theta \approx 0.00588 \text{ radian} \approx 0.33715° \approx 20.23'$$

The brochure does not indicate whether the distance is measured in nautical miles or statute miles. The distance s in nautical miles is 20.23, the measurement of θ in minutes. (Refer to Problem 90, Section 2.1). The distance s in statute miles is

$$s = r\theta = 3960(0.00588) = 23.3 \text{ miles}$$

In either case, it would seem that the brochure overstated the distance somewhat.   ▄

In navigation and in surveying, the **direction** or **bearing** from a point O to a point P equals the acute angle θ between the ray OP and the vertical line through O, the north–south line.

Figure 11 illustrates some bearings. Notice that the bearing from O to P_1 is denoted by the symbolism N30°E, indicating that the bearing is 30° east of north. In writing the bearing from O to P, the direction north or south always appears first, followed by an acute angle, followed by east or west. Thus, in Figure 11, the bearing from O to P_2 is S50°W, and from O to P_3 it is N70°W.

FIGURE 11

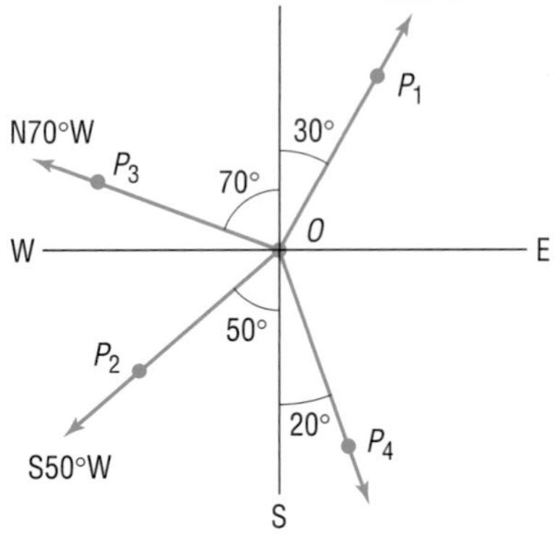

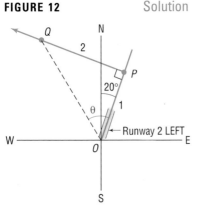

E X A M P L E   9

Finding the Bearing of an Object

In Figure 11, what is the bearing from O to an object at P_4?

Solution   The acute angle between the ray OP_4 and the north–south line through O is given as 20°. The bearing from O to P is S20°E.   ∎

E X A M P L E   10

Finding the Bearing of an Airplane

A Boeing 777 aircraft takes off from O'Hare Airport on runway 2 LEFT, which has a bearing of N20°E.* After flying for 1 mile, the pilot of the aircraft requests permission to turn 90° and head toward the northwest. The request is granted. After the plane goes 2 miles in this direction, what bearing should the control tower use to locate the aircraft?

FIGURE 12

Solution   Figure 12 illustrates the situation. After flying 1 mile from the airport O (the control tower), the aircraft is at P. After turning 90° toward the northwest and flying 2 miles, the aircraft is at the point Q. In triangle OPQ, the angle θ obeys the equation

$$\tan \theta = \frac{2}{1} = 2 \quad \text{so} \quad \theta = \tan^{-1} 2 \approx 63.4°$$

The acute angle between north and the ray OQ is $63.4° - 20° = 43.4°$. The bearing of the aircraft from O to Q is N43.4°W.   ∎

*In air navigation, the term **azimuth** is employed to denote the positive angle measured clockwise from the north (N) to a ray OP. Thus, in Figure 11 the azimuth from O to P_1 is 30°; the azimuth from O to P_2 is 230°; the azimuth from O to P_3 is 290°. Runways are designated using azimuth. Thus, runway 2 LEFT means the left runway with a direction of azimuth 20° (bearing N20°E). Runway 23 is the runway with azimuth 230° and bearing S50°W.

4.1 | EXERCISES

In Problems 1–14, use the right triangle shown in the margin. Then, using the given information, solve the triangle.

1. $b = 5$, $\beta = 20°$; find a, c, and α

2. $b = 4$, $\beta = 10°$; find a, c, and α

3. $a = 6$, $\beta = 40°$; find b, c, and α

4. $a = 7$, $\beta = 50°$; find b, c, and α

5. $b = 4$, $\alpha = 10°$; find a, c, and β

6. $b = 6$, $\alpha = 20°$; find a, c, and β

7. $a = 5$, $\alpha = 25°$; find b, c, and β

8. $a = 6$, $\alpha = 40°$; find b, c, and β

9. $c = 9$, $\beta = 20°$; find b, a, and α

10. $c = 10$, $\alpha = 40°$; find b, a, and β

11. $a = 5$, $b = 3$; find c, α, and β

12. $a = 2$, $b = 8$; find c, α, and β

13. $a = 2$, $c = 5$; find b, α, and β

14. $b = 4$, $c = 6$; find a, α, and β

15. A right triangle has a hypotenuse of length 8 inches. If one angle is 35°, find the length of each leg.

16. A right triangle has a hypotenuse of length 10 centimeters. If one angle is 40°, find the length of each leg.

17. A right triangle contains a 25° angle. If one leg is of length 5 inches, what is the length of the hypotenuse?

   [**Hint:** Two answers are possible.]

18. A right triangle contains an angle of $\pi/8$ radian. If one leg is of length 3 meters, what is the length of the hypotenuse?

   [**Hint:** Two answers are possible.]

19. The hypotenuse of a right triangle is 5 inches. If one leg is 2 inches, find the degree measure of each angle.

20. The hypotenuse of a right triangle is 3 feet. If one leg is 1 foot, find the degree measure of each angle.

21. **Finding the Width of a Gorge** Find the distance from A to C across the gorge illustrated in the figure.

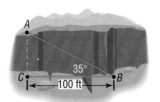

22. **Finding the Distance across a Pond** Find the distance from A to C across the pond illustrated in the figure.

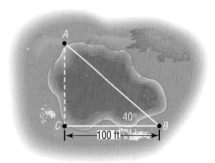

23. **The Eiffel Tower** The tallest tower built before the era of television masts, the Eiffel Tower was completed on March 31, 1889. Find the height of the Eiffel Tower (before a television mast was added to the top) using the information given in the illustration.

24. **Finding the Distance of a Ship from Shore** A ship, offshore from a vertical cliff known to be 100 feet in height, takes a sighting of the top of the cliff. If the angle of elevation is found to be 25°, how far offshore is the ship?

25. Suppose that you are headed toward a plateau 50 meters high. If the angle of elevation to the top of the plateau is 20°, how far are you from the base of the plateau?

26. **Statue of Liberty** A ship is just offshore of New York City. A sighting is taken of the Statue of Liberty, which is about 305 feet tall. If the angle of elevation to the top of the statue is 20°, how far is the ship from the base of the statue?

27. A 22 foot extension ladder leaning against a building makes a 70° angle with the ground. How far up the building does the ladder touch?

28. **Finding the Height of a Building** To measure the height of a building, two sightings are taken a distance of 50 feet apart. If the first angle of elevation is 40° and the second is 32°, what is the height of the building?

29. **Great Pyramid of Cheops** One of the original Seven Wonders of the World, the Great Pyramid of Cheops was built about 2580 BC. Its original height was 480 feet 11 inches, but due to the loss of its topmost stones, it is now shorter.* Find the current height of the Great Pyramid, using the information given in the illustration.

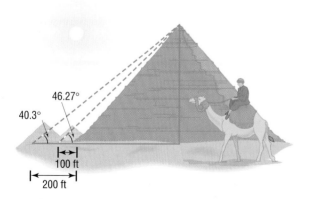

Source: Guinness Book of World Records.

30. A laser beam is to be directed through a small hole in the center of a circle of radius 10 feet. The origin of the beam is 35 feet from the circle (see the figure). At what angle of elevation should the beam be aimed to ensure that it goes through the hole?

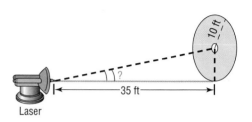

31. **Finding the Angle of Elevation of the Sun** At 10 AM on April 26, 1998, a building 300 feet high casts a shadow 50 feet long. What is the angle of elevation of the Sun?

32. **Mt. Rushmore** To measure the height of Lincoln's caricature on Mt. Rushmore, two sightings 800 feet from the base of the mountain are taken. If the angle of elevation to the bottom of Lincoln's face is 32° and the angle of elevation to the top is 35°, what is the height of Lincoln's face?

33. **Finding the Height of a Helicopter** Two observers simultaneously measure the angle of elevation of a helicopter. One angle is measured as 25°, the other as 40° (see the figure). If the observers are 100 feet apart and the helicopter lies over the line joining them, how high is the helicopter?

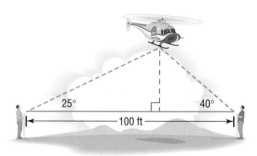

34. **Finding the Distance between Two Objects** A blimp, suspended in the air at a height of 500 feet, lies directly over a line from Soldier Field to the Adler Planetarium on Lake Michigan (see the figure). If the angle of depression from the blimp to the stadium is 32° and from the blimp to the planetarium is 23°, find the dis-

tance between Soldier Field and the Adler Planetarium.

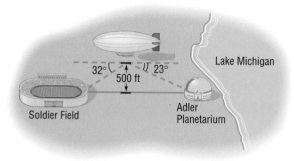

35. **Finding the Length of a Guy Wire**   A radio transmission tower is 200 feet high. How long should a guy wire be if it is to be attached to the tower 10 feet from the top and is to make an angle of 21° with the ground?

36. **Finding the Height of a Tower**   A guy wire 80 feet long is attached to the top of a radio transmission tower, making an angle of 25° with the ground. How high is the tower?

37. **Washington Monument**   The angle of elevation of the Washington Monument is 35.1° at the instant it casts a shadow 789 feet long. Use this information to calculate the height of the monument.

38. **Finding the Length of a Mountain Trail**   A straight trail with a uniform inclination of 17° leads from a hotel at an elevation of 9000 feet to a mountain lake at an elevation of 11,200 feet. What is the length of the trail?

39. **Finding the Speed of a Truck**   A state trooper is hidden 30 feet from a highway. One second after a truck passes, the angle θ between the highway and the line of observation from the patrol car to the truck is measured. See the illustration.

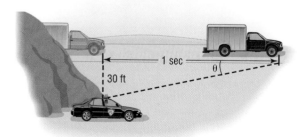

(a) If the angle measures 15°, how fast is the truck traveling? Express the answer in feet per second and in miles per hour.

(b) If the angle measures 20°, how fast is the truck traveling? Express the answer in feet per second and in miles per hour.

(c) If the speed limit is 55 miles per hour and a speeding ticket is issued for speeds of 5 miles per hour or more over the limit, for what angles should the trooper issue a ticket?

40. **Security**   A security camera in a neighborhood bank is mounted on a wall 9 feet above the floor. What angle of depression should be used if the camera is to be directed to a spot 6 feet above the floor and 12 feet from the wall?

41. **Finding the Bearing of an Aircraft**   A DC-9 aircraft leaves Midway Airport from runway 4 RIGHT, whose bearing is N40°E. After flying for 1/2 mile, the pilot requests permission to turn 90° and head toward the southeast. The permission is granted. After the airplane goes 1 mile in this direction, what bearing should the control tower use to locate the aircraft?

42. **Finding the Bearing of a Ship**   A ship leaves the port of Miami with a bearing of S80°E and a speed of 15 knots. After 1 hour, the ship turns 90° toward the south. After 2 hours, maintaining the same speed, what is the bearing to the ship from port?

43. **Shooting Free Throws in Basketball**   The eyes of a basketball player are 6 feet above the floor. The player is at the free-throw line, which is 15 feet from the center of the basket rim (see the figure). What is the angle of elevation from the player's eyes to the center of the rim?

[**Hint:** The rim is 10 feet above the floor.]

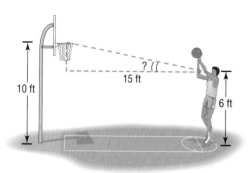

44. **Finding the Pitch of a Roof**   A carpenter is preparing to put a roof on a garage that is 20 feet by 40 feet by 20 feet. A steel support beam 46 feet in length is positioned in the center of the garage. To support the roof, another beam will be attached to the top of the center

beam (see the figure). At what angle of elevation is the new beam? In other words, what is the pitch of the roof?

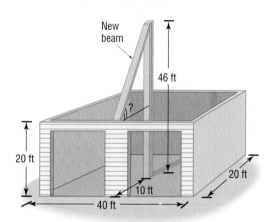

45. **Determining Distances at Sea** The navigator of a ship at sea spots two lighthouses that she recognizes as ones 3 miles apart along a straight seashore. She determines that the angles formed between two line-of-sight observations of the lighthouses and the line from the ship directly to shore are 15° and 35°. See the illustration.
    (a) How far is the ship from shore?
    (b) How far is the ship from lighthouse A?
    (c) How far is the ship from lighthouse B?

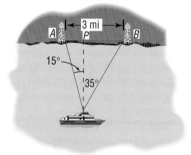

46. **Constructing a Highway** A highway whose primary directions are north–south is being constructed along the west coast of Florida. Near Naples, a bay obstructs the straight path of the road. Since the cost of a bridge is prohibitive, engineers decide to go around the bay. The illustration shows the path that they decide

on and the measurements taken. What is the length of highway needed to go around the bay?

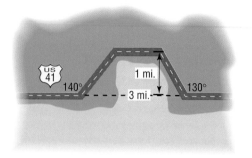

47. **Surveillance Satellites** A surveillance satellite circles Earth at a height of h miles above the surface. Suppose that d is the distance, in miles, on the surface of Earth that can be observed from the satellite. See the illustration.
    (a) Find an equation that relates the central angle θ to the height h.
    (b) Find an equation that relates the observable distance d and θ.
    (c) Find an equation that relates d and h.
    (d) If d is to be 2500 miles, how high must the satellite orbit above Earth?
    (e) If the satellite orbits at a height of 300 miles, what distance d on the surface can be observed?

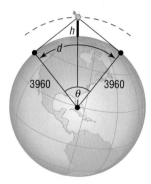

48. **Determining the Height of an Aircraft** Two sensors are spaced 700 feet apart along the approach to a small airport. When an aircraft is 2 miles from the airport, the angle of elevation from the first sensor to the aircraft is 20° and from the second sensor to the aircraft it is 15°. How high is the aircraft at this time?

49. **Photography** A camera is mounted on a tripod 4 feet high at a distance of 10 feet from George, who is 6 feet tall. See the illustration.

If the camera lens has angles of depression and elevation of 20°, will George's feet and head be seen by the lens? If not, how far back will the camera need to be moved to include George's feet and head?

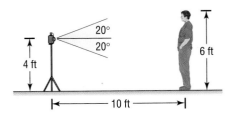

50. **Construction** A ramp for wheel chair accessibility is to be constructed with an angle of elevation of 15° and a final height of 5 feet. How long is the ramp?

 51. **The Gibb's Hill Lighthouse, Southampton, Bermuda** In operation since 1846, the Gibb's Hill Lighthouse stands 117 feet high on a hill 245 feet high, so its beam of light is 362 feet above sea level. A brochure states that ships 40 miles away can see the light and planes flying at 10,000 feet can see it 120 miles away. Verify the accuracy of these statements. What assumption did the brochure make about the height of the ship? (See the illustration for Problem 115 in Section 1.1.)

52. Compare Problem 115 in Section 1.1 with Example 8 and Problem 51. One uses the Pythagorean Theorem in its solution, while the others use trigonometry. Write a short paper that contrasts the two methods of solution. Be sure to point out the benefits as well as the shortcomings of each method. Then decide which solution you prefer. Be sure to have reasons.

4.2 | THE LAW OF SINES

1 Solve SAA or ASA Triangles
2 Solve SSA Triangles
3 Solve Applied Problems

If none of the angles of a triangle is a right angle, the triangle is called **oblique.** Thus, an oblique triangle will have either three acute angles or two acute angles and one obtuse angle (an angle between 90° and 180°). See Figure 13.

FIGURE 13
Oblique triangles

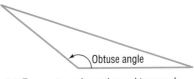

(a) All angles are acute   **(b)** Two acute angles and one obtuse angle

In the discussion that follows, we will always label an oblique triangle so that side a is opposite angle α, side b is opposite angle β, and side c is opposite angle γ, as shown in Figure 14.

FIGURE 14

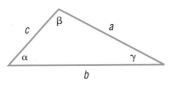

To **solve an oblique triangle** means to find the lengths of its sides and the measurements of its angles. To do this, we shall need to know the length of one side along with two other facts: either two angles, or the other two sides, or one angle and one other side.* Thus, there are four possibilities to consider:

---

*Recall from plane geometry the fact that knowing three angles of a triangle determines a family of *similar triangles*, that is, triangles that have the same shape but different sizes.

CASE 1: One side and two angles are known (SAA or ASA).
CASE 2: Two sides and the angle opposite one of them are known (SSA).
CASE 3: Two sides and the included angle are known (SAS).
CASE 4: Three sides are known (SSS).

Figure 15 illustrates the four cases.

FIGURE 15

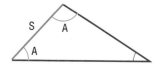

Case 1: SAA or ASA

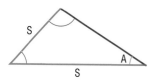

Case 2: SSA

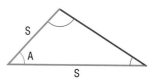

Case 3: SAS

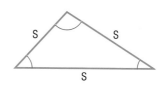

Case 4: SSS

The **Law of Sines** is used to solve triangles for which Case 1 or 2 holds.

Theorem   Law of Sines

For a triangle with sides a, b, c and opposite angles α, β, γ, respectively,

$$\frac{\sin \alpha}{a} = \frac{\sin \beta}{b} = \frac{\sin \gamma}{c} \tag{1}$$

FIGURE 16

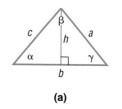

(a)

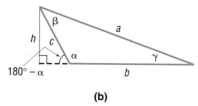

(b)

Proof   To prove the Law of Sines, we construct an altitude of length h from one of the vertices of such a triangle. Figure 16(a) shows h for a triangle with three acute angles, and Figure 16(b) shows h for a triangle with an obtuse angle. In each case, the altitude is drawn from the vertex at β. Using either illustration, we have

$$\sin \gamma = \frac{h}{a}$$

from which

$$h = a \sin \gamma \tag{2}$$

From Figure 16(a), it also follows that

$$\sin \alpha = \frac{h}{c}$$

from which

$$h = c \sin \alpha \tag{3}$$

From Figure 16(b), it follows that

$$\sin \alpha = \sin(180° - \alpha) = \frac{h}{c}$$

which again gives

$$h = c \sin \alpha$$

Thus, whether the triangle has three acute angles or has two acute angles and one obtuse angle, equations (2) and (3) hold. As a result, we may equate the expressions for h in equations (2) and (3) to get

$$a \sin \gamma = c \sin \alpha$$

from which

$$\frac{\sin \alpha}{a} = \frac{\sin \gamma}{c} \tag{4}$$

In a similar manner, by constructing the altitude h' from the vertex of angle α as shown in Figure 17, we can show that

$$\sin \beta = \frac{h'}{c} \quad \text{and} \quad \sin \gamma = \frac{h'}{b}$$

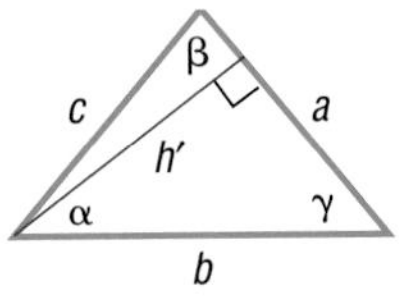

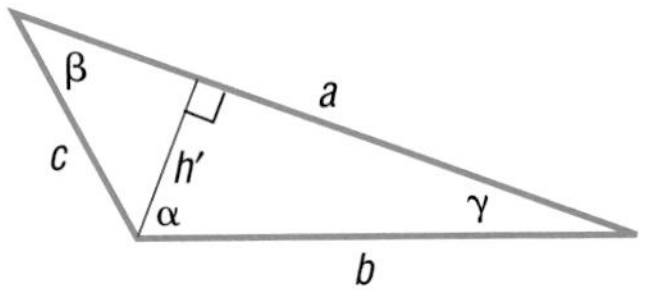

Thus,

$$h' = c \sin \beta = b \sin \gamma$$

and

$$\frac{\sin \beta}{b} = \frac{\sin \gamma}{c} \tag{5}$$

When equations (4) and (5) are combined, we have equation (1), the Law of Sines. ∎

In applying the Law of Sines to solve triangles, we use the fact that the sum of the angles of any triangle equals 180°; that is,

$$\alpha + \beta + \gamma = 180° \tag{6}$$

1   Our first two examples show how to solve a triangle when one side and two angles are known (Case 1: SAA or ASA).

FIGURE 17

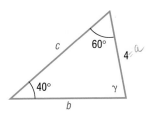

(a)

(b)

E X A M P L E   1   Using the Law of Sines to Solve a SAA Triangle

Solve the triangle: $\quad \alpha = 40°, \beta = 60°, a = 4$

Solution   Figure 18 shows the triangle that we want to solve. The third angle γ is easily found using equation (6).

$$\alpha + \beta + \gamma = 180°$$
$$40° + 60° + \gamma = 180°$$
$$\gamma = 80°$$

Now we use the Law of Sines (twice) to find the unknown sides b and c.

$$\frac{\sin \alpha}{a} = \frac{\sin \beta}{b} \qquad \frac{\sin \alpha}{a} = \frac{\sin \gamma}{c}$$

Because $a = 4$, $\alpha = 40°$, $\beta = 60°$, and $\gamma = 80°$, we have

$$\frac{\sin 40°}{4} = \frac{\sin 60°}{b} \qquad \frac{\sin 40°}{4} = \frac{\sin 80°}{c}$$

FIGURE 18

Thus,

$$b = \frac{4 \sin 60°}{\sin 40°} \approx 5.39 \qquad c = \frac{4 \sin 80°}{\sin 40°} \approx 6.13$$

From a calculator.   From a calculator.

Notice that in Example 1 we found b and c by working with the given side a. This is better than finding b first and working with a rounded value of b to find c.

 Now work Problem 1.

E X A M P L E   2   Using the Law of Sines to Solve an ASA Triangle

Solve the triangle: $\alpha = 35°, \beta = 15°, c = 5$

FIGURE 19

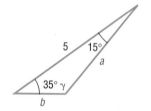

Solution   Figure 19 illustrates the triangle that we want to solve. Because we know two angles ($\alpha = 35°$ and $\beta = 15°$), it is easy to find the third angle using equation (6).

$$\alpha + \beta + \gamma = 180°$$
$$35° + 15° + \gamma = 180°$$
$$\gamma = 130°$$

Now we know the three angles and one side ($c = 5$) of the triangle. To find the remaining two sides a and b, we use the Law of Sines (twice).

$$\frac{\sin \alpha}{a} = \frac{\sin \gamma}{c} \qquad\qquad \frac{\sin \beta}{b} = \frac{\sin \gamma}{c}$$

$$\frac{\sin 35°}{a} = \frac{\sin 130°}{5} \qquad\qquad \frac{\sin 15°}{b} = \frac{\sin 130°}{5}$$

$$a = \frac{5 \sin 35°}{\sin 130°} \approx 3.74 \qquad\qquad b = \frac{5 \sin 15°}{\sin 130°} \approx 1.69$$

 Now work Problem 15.

FIGURE 20

The Ambiguous Case

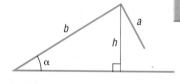

Case 2 (SSA), which applies to triangles for which two sides and the angle opposite one of them are known, is referred to as the **ambiguous case,** because the known information may result in one triangle, two triangles, or no triangle at all. Suppose that we are given sides a and b and angle α, as illustrated in Figure 20. The key to determining the possible triangles, if any, that

may be formed from the given information, lies primarily with the height h and the fact that $h = b \sin \alpha$.

No Triangle: If $a < h = b \sin \alpha$, then clearly side a is not sufficiently long to form a triangle. See Figure 21.

One Right Triangle: If $a = h = b \sin \alpha$, then side a is just long enough to form a right triangle. See Figure 22.

FIGURE 21
$a < h = b \sin \alpha$

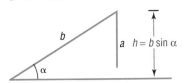

FIGURE 22
$a = b \sin \alpha$

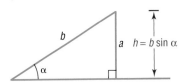

Two Triangles: If $a < b$ and $h = b \sin \alpha < a$, then two distinct triangles can be formed from the given information. See Figure 23.

One Triangle: If $a \geq b$, then only one triangle can be formed. See Figure 24.

FIGURE 23
$b \sin \alpha < a$ and $a < b$

FIGURE 24
$a \geq b$

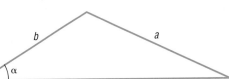

Fortunately, we do not have to rely on an illustration to draw the correct conclusion in the ambiguous case. The Law of Sines will lead us to the correct determination. Let's see how.

E X A M P L E   3

Using the Law of Sines to Solve a SSA Triangle (One Solution)

Solve the triangle: $a = 3, b = 2, \alpha = 40°$

Solution

FIGURE 25(a)

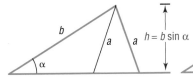

See Figure 25(a). Because $a = 3$, $b = 2$, and $\alpha = 40°$ are known, we use the Law of Sines to find the angle β.

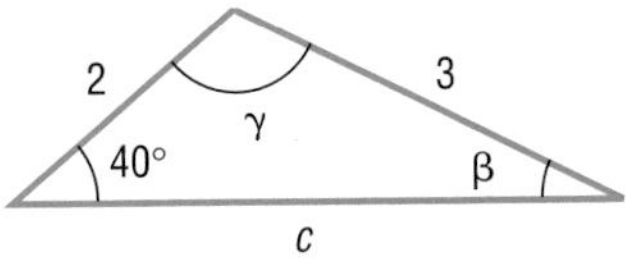

$$\frac{\sin \alpha}{a} = \frac{\sin \beta}{b}$$

Then

$$\frac{\sin 40°}{3} = \frac{\sin \beta}{2}$$

$$\sin \beta = \frac{2 \sin 40°}{3} \approx 0.43$$

There are two angles β, $0° < \beta < 180°$, for which $\sin \beta \approx 0.43$.

$$\beta_1 \approx 25.4° \quad \text{and} \quad \beta_2 \approx 154.6°$$

[**Note:** Here we computed β using the stored value of $\sin \beta$. If you use the rounded value, $\sin \beta \approx 0.43$, you will obtain slightly different results.]

The second possibility, $\beta_2 \approx 154.6°$, is ruled out, because $\alpha = 40°$, making $\alpha + \beta_2 \approx 194.6° > 180°$. Now, using $\beta_1 \approx 25.4°$, we find

$$\gamma = 180° - \alpha - \beta_1 \approx 180° - 40° - 25.4° = 114.6°$$

The third side c may now be determined using the Law of Sines.

$$\frac{\sin \gamma}{c} = \frac{\sin \alpha}{a}$$

$$\frac{\sin 114.6°}{c} = \frac{\sin 40°}{3}$$

FIGURE 25(b)

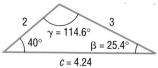

$c = 4.24$

$$c = \frac{3 \sin 114.6°}{\sin 40°} \approx 4.24$$

Figure 25(b) illustrates the solved triangle.

E X A M P L E   4   Using the Law of Sines to Solve a SSA Triangle (Two Solutions)

Solve the triangle: $a = 6, b = 8, \alpha = 35°$

Solution   See Figure 26(a). Because $a = 6$, $b = 8$, and $\alpha = 35°$ are known, we use the Law of Sines to find the angle β.

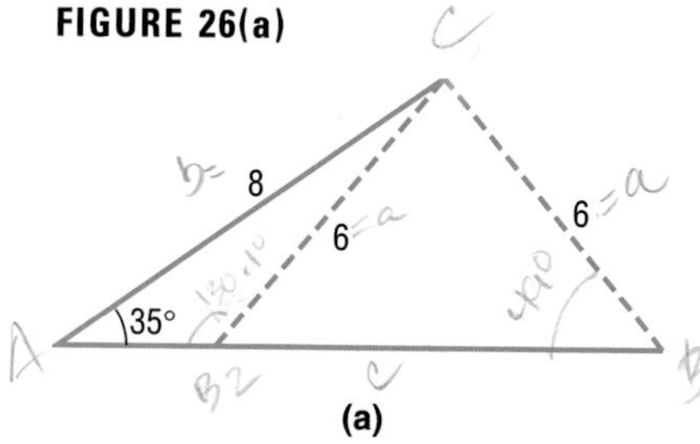

FIGURE 26(a)

$$\frac{\sin \alpha}{a} = \frac{\sin \beta}{b}$$

Then

$$\frac{\sin 35°}{6} = \frac{\sin \beta}{8}$$

$$\sin \beta = \frac{8 \sin 35°}{6} \approx 0.76$$

$$\beta_1 \approx 49.9° \quad \text{or} \quad \beta_2 \approx 130.1°$$

For both choices of β, we have $\alpha + \beta < 180°$. Hence, there are two triangles, one containing the angle $\beta_1 \approx 49.9°$ and the other containing the angle $\beta_2 \approx 130.1°$. The third angle γ is either

$$\gamma_1 = 180° - \alpha - \beta_1 \approx 95.1° \quad \text{or} \quad \gamma_2 = 180° - \alpha - \beta_2 \approx 14.9°$$

$$\begin{array}{cc} \uparrow & \uparrow \\ \alpha = 35° & \alpha = 35° \\ \beta_1 \approx 49.9° & \beta_2 \approx 130.1° \end{array}$$

FIGURE 26(b)

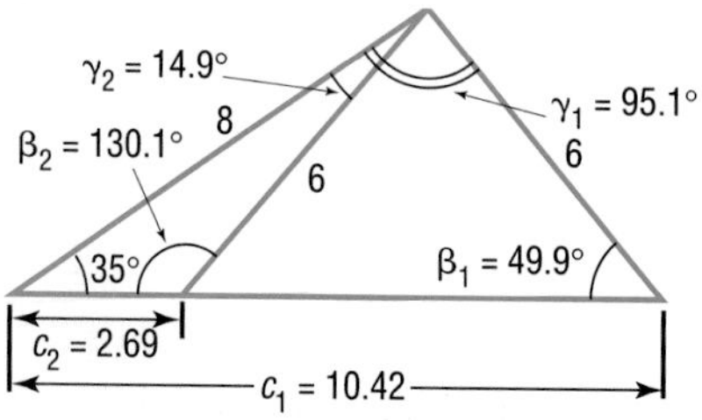

The third side c obeys the Law of Sines, so we have

$$\frac{\sin \gamma}{c} = \frac{\sin \alpha}{a}$$

$$\frac{\sin 95.1°}{c_1} = \frac{\sin 35°}{6} \quad \text{or} \quad \frac{\sin 14.9°}{c_2} = \frac{\sin 35°}{6}$$

$$c_1 = \frac{6 \sin 95.1°}{\sin 35°} \approx 10.42 \qquad c_2 = \frac{6 \sin 14.9°}{\sin 35°} \approx 2.69$$

The two solved triangles are illustrated in Figure 26(b).

E X A M P L E   5   Using the Law of Sines to Solve a SSA Triangle (No Solution)

Solve the triangle: $a = 2, c = 1, \gamma = 50°$

Solution   Because $a = 2$, $c = 1$, and $\gamma = 50°$ are known, we use the Law of Sines to find the angle α.

$$\frac{\sin \alpha}{a} = \frac{\sin \gamma}{c}$$

FIGURE 27

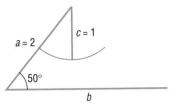

$$\frac{\sin \alpha}{2} = \frac{\sin 50°}{1}$$

$$\sin \alpha = 2 \sin 50° \approx 1.53$$

There is no angle α for which $\sin \alpha > 1$. Hence, there can be no triangle with the given measurements. Figure 27 illustrates the measurements given. Notice that, no matter how we attempt to position side c, it will never intersect side b to form a triangle.   ▬

Now work **Problems 17 and 23.**

Applied Problems

3 The Law of Sines is particularly useful for solving certain applied problems.

E X A M P L E   6   Rescue at Sea

Coast Guard Station Zulu is located 120 miles due west of Station X-ray. A ship at sea sends an SOS call that is received by each station. The call to Station Zulu indicates that the bearing of the ship from Zulu is N40°E (40° east of north). The call to Station X-ray indicates that the bearing of the ship from X-ray is N30°W (30° west of north).

(a) How far is each station from the ship?

(b) If a helicopter capable of flying 200 miles per hour is dispatched from the nearest station to the ship, how long will it take to reach the ship?

Solution   (a) Figure 28 illustrates the situation. The angle γ is found to be

FIGURE 28

$$\gamma = 180° - 50° - 60° = 70°$$

The Law of Sines can now be used to find the two distances a and b that we seek.

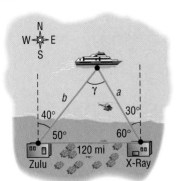

$$\frac{\sin 50°}{a} = \frac{\sin 70°}{120}$$

$$a = \frac{120 \sin 50°}{\sin 70°} \approx 97.82 \text{ miles}$$

$$\frac{\sin 60°}{b} = \frac{\sin 70°}{120}$$

$$b = \frac{120 \sin 60°}{\sin 70°} \approx 110.59 \text{ miles}$$

Thus, Station Zulu is about 111 miles from the ship, and Station X-ray is about 98 miles from the ship.

(b) The time t needed for the helicopter to reach the ship from Station X-ray is found by using the formula

$$(\text{Velocity, } v)(\text{time, } t) = \text{distance, } a$$

Then

$$t = \frac{a}{v} = \frac{97.82}{200} \approx 0.49 \text{ hour} \approx 29 \text{ minutes}$$

It will take about 29 minutes for the helicopter to reach the ship. ▬

 Now work Problem 29.

4.2 EXERCISES

In Problems 1–8, solve each triangle.

1.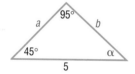

2.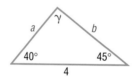

3.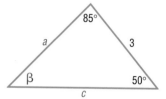

4.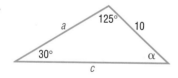

5.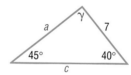

6.

7.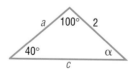

8.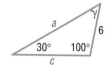

In Problems 9–16, solve each triangle.

9. $\alpha = 40°$, $\beta = 20°$, $a = 2$ **10.** $\alpha = 50°$, $\gamma = 20°$, $a = 3$ **11.** $\beta = 70°$, $\gamma = 10°$, $b = 5$

12. $\alpha = 70°$, $\beta = 60°$, $c = 4$ **13.** $\alpha = 110°$, $\gamma = 30°$, $c = 3$ **14.** $\beta = 10°$, $\gamma = 100°$, $b = 2$

15. $\alpha = 40°$, $\beta = 40°$, $c = 2$ **16.** $\beta = 20°$, $\gamma = 70°$, $a = 1$

In Problems 17–28, two sides and an angle are given. Determine whether the given information results in one triangle, two triangles, or no triangle at all. Solve any triangle(s) that results.

17. $a = 3$, $b = 2$, $\alpha = 50°$ **18.** $b = 4$, $c = 3$, $\beta = 40°$ **19.** $b = 5$, $c = 3$, $\beta = 100°$

20. $a = 2$, $c = 1$, $\alpha = 120°$ **21.** $a = 4$, $b = 5$, $\alpha = 60°$ **22.** $b = 2$, $c = 3$, $\beta = 40°$

23. $b = 4$, $c = 6$, $\beta = 20°$ **24.** $a = 3$, $b = 7$, $\alpha = 70°$ **25.** $a = 2$, $c = 1$, $\gamma = 100°$

26. $b = 4$, $c = 5$, $\beta = 95°$ **27.** $a = 2$, $c = 1$, $\gamma = 25°$ **28.** $b = 4$, $c = 5$, $\beta = 40°$

29. **Rescue at Sea** Coast Guard Station Able is located 150 miles due south of Station Baker. A ship at sea sends an SOS call that is received by each station. The call to Station Able indicates that the ship is located N55°E; the call to Station Baker indicates that the ship is located S60°E.
    (a) How far is each station from the ship?
    (b) If a helicopter capable of flying 200 miles per hour is dispatched from the nearest station to the ship, how long will it take to reach the ship?

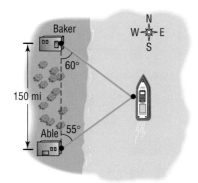

30. **Surveying** Consult the figure. To find the distance from the house at A to the house at B, a surveyor measures the angle BAC to be 40° and then walks off a distance of 100 feet to C and measures the angle ACB to be 50°. What is the distance from A to B?

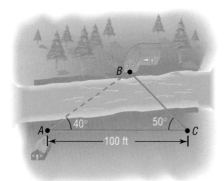

31. **Finding the Length of a Ski Lift** Consult the figure. To find the length of the span of a proposed ski lift from A to B, a surveyor measures the angle DAB to be 25° and then walks off a distance of 1000 feet to C and measures the an-

gle ACB to be 15°. What is the distance from A to B?

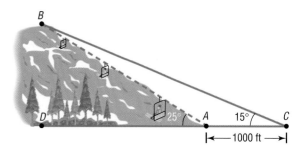

32. **Finding the Height of a Mountain** Use the illustration in Problem 31 to find the height BD of the mountain at B.

33. **Finding the Height of an Airplane** An aircraft is spotted by two observers who are 1000 feet apart. As the airplane passes over the line joining them, each observer takes a sighting of the angle of elevation to the plane, as indicated in the figure. How high is the airplane?

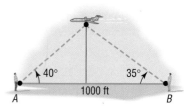

34. **Finding the Height of the Bridge over the Royal Gorge** The highest bridge in the world is the bridge over the Royal Gorge of the Arkansas River in Colorado.* Sightings to the same point at water level directly under the bridge are taken from each side of the 880-foot-long bridge, as indicated in the figure. How high is the bridge?

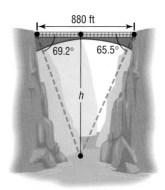

Source: Guinness Book of World Records.

35. Navigation An airplane flies from city A to city B, a distance of 150 miles, and then turns through an angle of 40° and heads toward city C, as shown in the figure.
  (a) If the distance between cities A and C is 300 miles, how far is it from city B to city C?
  (b) Through what angle should the pilot turn at city C to return to city A?

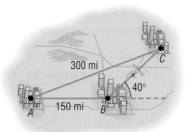

36. Time Lost due to a Navigation Error In attempting to fly from city A to city B, an aircraft followed a course that was 10° in error, as indicated in the figure. After flying a distance of 50 miles, the pilot corrected the course by turning at point C and flying 70 miles farther. If the constant speed of the aircraft was 250 miles per hour, how much time was lost due to the error?

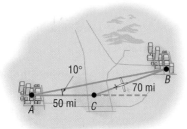

37. Finding the Lean of the Leaning Tower of Pisa The famous Leaning Tower of Pisa was originally 184.5 feet high.* At a distance of

*In their 1986 report on the fragile seven-century-old bell tower, scientists in Pisa, Italy, said the Leaning Tower of Pisa had increased its famous lean by 1 millimeter, or 0.04 inch. That is about the annual average, although the tilting had slowed to about half that much in the previous 2 years. (*Source:* United Press International, June 29, 1986.)
    PISA, ITALY. September 1995. The Leaning Tower of Pisa has suddenly shifted, jeopardizing years of preservation work to stabilize it, Italian newspapers said Sunday. The tower, built on shifting subsoil between 1174 and 1350 as a belfry for the nearby cathedral, recently moved 0.07 inches in one night. The tower has been closed to tourists since 1990.

123 feet from the base of the tower, the angle of elevation to the top of the tower is found to be 60°. Find the angle CAB indicated in the figure. Also, find the perpendicular distance from C to AB.

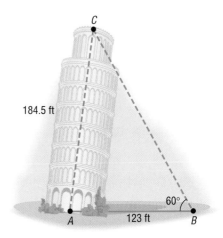

38. Crankshafts on Cars On a certain automobile, the crankshaft is 3 inches long and the connecting rod is 9 inches long (see the figure). At the time when the angle OPA is 15°, how far is the piston (P) from the center (O) of the crankshaft?

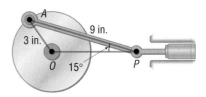

39. Constructing a Highway U.S. 41, a highway whose primary directions are north–south, is being constructed along the west coast of Florida. Near Naples, a bay obstructs the straight path of the road. Since the cost of a bridge is prohibitive, engineers decide to go around the bay. The illustrations show the path that they decide on and the measurements taken. What is the length of highway needed to go around the bay?

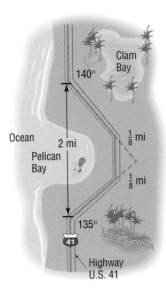

40. **Determining Distances at Sea** The navigator of a ship at sea spots two lighthouses that she knows to be 3 miles apart along a straight seashore. She determines that the angles formed between two line-of-sight observations of the lighthouses and the line from the ship directly to shore are 15° and 35°. See the illustration.
    (a) How far is the ship from lighthouse A?
    (b) How far is the ship from lighthouse B?
    (c) How far is the ship from shore?
    (d) Compare your solutions using the Law of Sines with the solutions to Problem 45 in Section 4.1. Which method do you prefer? Why?

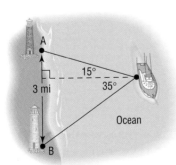

41. **Calculating Distances at Sea** The navigator of a ship at sea has the harbor in sight at which the ship is to dock. She spots a lighthouse that she knows is 1 mile up the coast from the mouth of the harbor, and she measures the angle between the line-of-sight observations of the harbor and lighthouse to be 20°. With the ship heading directly toward the harbor, she repeats this measurement after 5 minutes of traveling at 12 miles per hour. If the new angle is 30°, how far is the ship from the harbor?

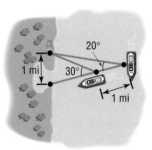

42. **Finding Distances** A forest ranger is walking on a path inclined at 5° to the horizontal directly toward a 100-foot-tall fire observation tower. The angle of elevation of the top of the tower is 40°. How far is the ranger from the tower at this time?

43. **Finding the Height of a Mountain**
    (a) Rework Example 7 in Section 4.1 using the Law of Sines.
    (b) Compare the solution using the Law of Sines with the one given in Section 4.1. Which solution do you prefer? Why?

44. **Determining the Height of an Aircraft** Two sensors are spaced 700 feet apart along the approach to a small airport. When an aircraft is 2 miles from the airport, the angle of elevation from the first sensor to the aircraft is 20°, and from the second sensor to the aircraft it is 15°.
    (a) Use the Law of Sines to determine how high the aircraft is at this time.
    (b) Compare the solution in (a) to the solution to Problem 48 in Exercise 4.1. Which solution do you prefer? Why?

45. **Landscaping** Pat needs to determine the height of a tree before cutting it down to be sure that it will not fall on a nearby fence. The angle of elevation of the tree from one position on a

flat path from the tree is 30°, and from a second position 40 feet farther along this path it is 20°. What is the height of the tree?

46. Construction A loading ramp 10 feet long that makes an angle of 18° with the horizontal is to be replaced by one that makes an angle of 12° with the horizontal. How long is the new ramp?

47. Mollweide's Formula For any triangle, Mollweide's Formula (named after Karl Mollweide, 1774–1825) states that

$$\frac{a + b}{c} = \frac{\cos \frac{1}{2}(\alpha - \beta)}{\sin \frac{1}{2} \gamma}$$

Derive it.

[Hint: Use the Law of Sines and then a sum-to-product formula. Notice that this formula involves all six parts of a triangle. As a result, it is sometimes used to check the solution of a triangle.]

48. Mollweide's Formula Another form of Mollweide's Formula is

$$\frac{a - b}{c} = \frac{\sin \frac{1}{2}(\alpha - \beta)}{\cos \frac{1}{2} \gamma}$$

Derive it.

49. For any triangle, derive the formula

$$a = b \cos \gamma + c \cos \beta$$

[Hint: Use the fact that $\sin \alpha = \sin(180° - \beta - \gamma)$.]

50. Law of Tangents For any triangle, derive the Law of Tangents:

$$\frac{a - b}{a + b} = \frac{\tan \frac{1}{2}(\alpha - \beta)}{\tan \frac{1}{2}(\alpha + \beta)}$$

[Hint: Use Mollweide's Formula.]

51. Circumscribing a Triangle Show that

$$\frac{\sin \alpha}{a} = \frac{\sin \beta}{b} = \frac{\sin \gamma}{c} = \frac{1}{2r}$$

where r is the radius of the circle circumscribing the triangle ABC whose sides are a, b, c, as shown in the figure.

[Hint: Draw the diameter AB'. Then $\beta =$ angle $ABC =$ angle $AB'C$ and angle $ACB' = 90°$.]

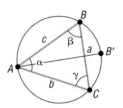

52. Make up three problems involving oblique triangles. One should result in one triangle, the second in two triangles, and the third in no triangle.

4.3 THE LAW OF COSINES

1. Solve SAS Triangles
2. Solve SSS Triangles
3. Solve Applied Problems

In Section 4.2, we used the Law of Sines to solve Case 1 (SAA or ASA) and Case 2 (SSA) of an oblique triangle. In this section, we derive the Law of Cosines and use it to solve the remaining cases, 3 and 4.

CASE 3: Two sides and the included angle are known (SAS).
CASE 4: Three sides are known (SSS).

Theorem   Law of Cosines

For a triangle with sides a, b, c and opposite angles α, β, γ, respectively,

$$c^2 = a^2 + b^2 - 2ab \cos \gamma \tag{1}$$

$$b^2 = a^2 + c^2 - 2ac \cos \beta \tag{2}$$

$$a^2 = b^2 + c^2 - 2bc \cos \alpha \tag{3}$$

FIGURE 29

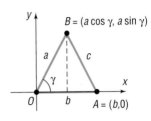

(a) Angle γ is acute

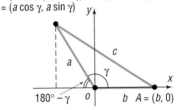

(b) Angle γ is obtuse

Proof   We will prove only formula (1) here. Formulas (2) and (3) may be proved using the same argument.

We begin by strategically placing a triangle on a rectangular coordinate system so that the vertex of angle γ is at the origin and side b lies along the positive x-axis. Regardless of whether γ is acute, as in Figure 29(a), or obtuse, as in Figure 29(b), the vertex B has coordinates $(a \cos \gamma, a \sin \gamma)$. Vertex A has coordinates $(b, 0)$.

We can now use the distance formula to compute c^2.

$$\begin{aligned}
c^2 &= (b - a \cos \gamma)^2 + (0 - a \sin \gamma)^2 \\
&= b^2 - 2ab \cos \gamma + a^2 \cos^2 \gamma + a^2 \sin^2 \gamma \\
&= b^2 - 2ab \cos \gamma + a^2(\cos^2 \gamma + \sin^2 \gamma) \\
&= a^2 + b^2 - 2ab \cos \gamma
\end{aligned}$$

Each of formulas (1), (2), and (3) may be stated in words as follows:

Theorem   Law of Cosines

The square of one side of a triangle equals the sum of the squares of the other two sides minus twice their product times the cosine of their included angle.

Observe that if the triangle is a right triangle (so that, say, $\gamma = 90°$), then formula (1) becomes the familiar Pythagorean Theorem: $c^2 = a^2 + b^2$. Thus, the Pythagorean Theorem is a special case of the Law of Cosines.

Let's see how to use the Law of Cosines to solve Case 3 (SAS), which applies to triangles for which two sides and the included angle are known.

E X A M P L E   1   Using the Law of Cosines to Solve a SAS Triangle

Solve the triangle: $a = 2$, $b = 3$, $\gamma = 60°$

FIGURE 30

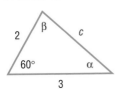

Solution   See Figure 30. The Law of Cosines makes it easy to find the third side, c.

$$\begin{aligned}
c^2 &= a^2 + b^2 - 2ab \cos \gamma \\
&= 4 + 9 - 2 \cdot 2 \cdot 3 \cdot \cos 60° \\
&= 13 - (12 \cdot \tfrac{1}{2}) = 7 \\
c &= \sqrt{7}
\end{aligned}$$

Side c is of length $\sqrt{7}$. To find the angles α and β, we may use either the Law of Sines or the Law of Cosines. It is preferable to use the Law of Cosines, since it will lead to an equation with one solution. Using the Law of Sines would lead to an equation with two solutions that would need to be checked to determine which solution fits the given data. Thus, we choose to use formulas (2) and (3) of the Law of Cosines.

For α:

$$a^2 = b^2 + c^2 - 2bc \cos \alpha$$
$$2bc \cos \alpha = b^2 + c^2 - a^2$$
$$\cos \alpha = \frac{b^2 + c^2 - a^2}{2bc} = \frac{9 + 7 - 4}{2 \cdot 3\sqrt{7}} = \frac{12}{6\sqrt{7}} = \frac{2\sqrt{7}}{7}$$
$$\alpha \approx 40.9°$$

For β:

$$b^2 = a^2 + c^2 - 2ac \cos \beta$$
$$\cos \beta = \frac{a^2 + c^2 - b^2}{2ac} = \frac{4 + 7 - 9}{4\sqrt{7}} = \frac{1}{2\sqrt{7}} = \frac{\sqrt{7}}{14}$$
$$\beta \approx 79.1°$$

Notice that $\alpha + \beta + \gamma = 40.9° + 79.1° + 60° = 180°$, as required.   ▬

Now work Problem 1.

The next example illustrates how the Law of Cosines is used when three sides of a triangle are known, Case 4 (SSS).

E X A M P L E   2

Using the Law of Cosines to Solve a SSS Triangle

Solve the triangle: $a = 4, b = 3, c = 6$

Solution   See Figure 31. To find the angles α, β, and γ, we proceed as we did in the latter part of the solution to Example 1.

FIGURE 31

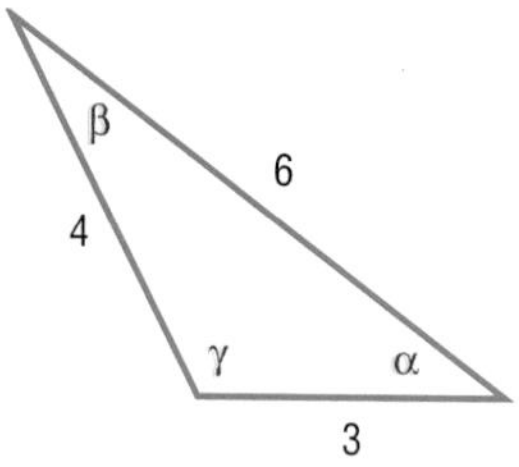

For α:

$$\cos \alpha = \frac{b^2 + c^2 - a^2}{2bc} = \frac{9 + 36 - 16}{2 \cdot 3 \cdot 6} = \frac{29}{36}$$
$$\alpha \approx 36.3°$$

For β:

$$\cos \beta = \frac{a^2 + c^2 - b^2}{2ac} = \frac{16 + 36 - 9}{2 \cdot 4 \cdot 6} = \frac{43}{48}$$
$$\beta \approx 26.4°$$

Since we know α and β,

$$\gamma = 180° - \alpha - \beta \approx 180° - 36.3° - 26.4° = 117.3°$$ ▬

Now work Problem 7.

E X A M P L E   3   Correcting a Navigational Error

A motorized sailboat leaves Naples, Florida, bound for Key West, 150 miles away. Maintaining a constant speed of 15 miles per hour, but encountering heavy crosswinds and strong currents, the crew finds, after 4 hours, that the sailboat is off course by 20°.

(a) How far is the sailboat from Key West at this time?
(b) Through what angle should the sailboat turn to correct its course?
(c) How much time has been added to the trip because of this? (Assume that the speed remains at 15 miles per hour.)

Solution   See Figure 32. With a speed of 15 miles per hour, the sailboat has gone 60 miles after 4 hours. We seek the distance x of the sailboat from Key West. We also seek the angle θ that the sailboat should turn through to correct its course.

FIGURE 32

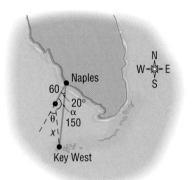

(a) To find x, we use the Law of Cosines, since we know two sides and the included angle.

$$x^2 = 150^2 + 60^2 - 2(150)(60)\cos 20° = 9186$$
$$x \approx 95.8$$

The sailboat is about 96 miles from Key West.

(b) We now know three sides of the triangle, so we can use the Law of Cosines again to find the angle α opposite the side of length 150 miles.

$$150^2 = 96^2 + 60^2 - 2(96)(60)\cos \alpha$$
$$9684 = -11{,}520 \cos \alpha$$
$$\cos \alpha \approx -0.8406$$
$$\alpha \approx 147.2°$$

The sailboat should turn through an angle of

$$\theta = 180° - \alpha \approx 180° - 147.2° = 32.8°$$

The sailboat should turn through an angle of about 33° to correct its course.

(c) The total length of the trip is now $60 + 96 = 156$ miles. The extra 6 miles will only require about 0.4 hour or 24 minutes more if the speed of 15 miles per hour is maintained.   ▬

Now work Problem 27.

HISTORICAL FEATURE   The Law of Sines was known vaguely, long before it was explicitly stated by Nasîr Eddîn (about AD 1250). Ptolemy (about AD 150) was aware of it in a form using a chord function instead of the sine function. But it was first clearly stated in Europe by Regiomontanus, writing in 1464.

The Law of Cosines appears first in Euclid's *Elements* (Book II), but in a well-disguised form in which squares built on the sides of triangles are added and a rectangle representing the cosine term is subtracted. It was thus known to all mathematicians because of their familiarity with Euclid's work. An early modern form of the Law of Cosines, that for finding the angle when the sides are known, was stated by François Vièta (in 1593).

The Law of Tangents (see Problem 50 of Exercise 4.2) has become obsolete. In the past it was used in place of the Law of Cosines, because the Law of Cosines was very inconvenient for calculation with logarithms or slide rules. Mixing of addition and multiplication is now quite easy on a calculator, however, and the Law of Tangents has been shelved along with the slide rule.

4.3 | EXERCISES

In Problems 1–8, solve each triangle.

1.
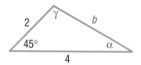

2.
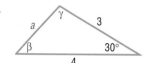

3.
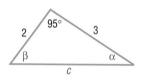

4.
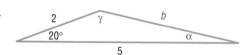

5.
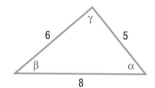

6.
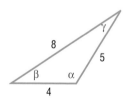

7.
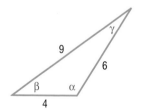

8.
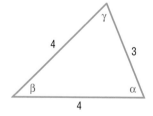

In Problems 9–24, solve each triangle.

9. $a = 3,\ b = 4,\ \gamma = 40°$ **10.** $a = 2,\ c = 1,\ \beta = 10°$ **11.** $b = 1,\ c = 3,\ \alpha = 80°$

12. $a = 6,\ b = 4,\ \gamma = 60°$ **13.** $a = 3,\ c = 2,\ \beta = 110°$ **14.** $b = 4,\ c = 1,\ \alpha = 120°$

15. $a = 2,\ b = 2,\ \gamma = 50°$ **16.** $a = 3,\ c = 2,\ \beta = 90°$ **17.** $a = 12,\ b = 13,\ c = 5$

18. $a = 4,\ b = 5,\ c = 3$ **19.** $a = 2,\ b = 2,\ c = 2$ **20.** $a = 3,\ b = 3,\ c = 2$

21. $a = 5,\ b = 8,\ c = 9$ **22.** $a = 4,\ b = 3,\ c = 6$ **23.** $a = 10,\ b = 8,\ c = 5$

24. $a = 9,\ b = 7,\ c = 10$

25. Surveying Consult the figure. To find the distance from the house at A to the house at B, a surveyor measures the angle ACB, which is found to be 70°, and then walks off the distance to each house, 50 feet and 70 feet, respectively. How far apart are the houses?

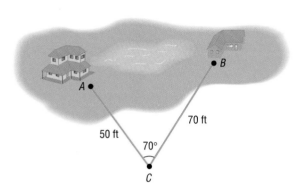

26. Navigation An airplane flies from Ft. Myers to Sarasota, a distance of 150 miles and then turns through an angle of 50° and flies to Orlando, a distance of 100 miles (see the figure).
(a) How far is it from Ft. Myers to Orlando?
(b) Through what angle should the pilot turn at Orlando to return to Ft. Myers?

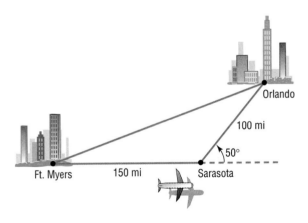

27. Revising a Flight Plan In attempting to fly from Chicago to Louisville, a distance of 330 miles, a pilot inadvertently took a course that was 10° in error, as indicated in the figure.
(a) If the aircraft maintains an average speed of 220 miles per hour and if the error in direction is discovered after 15 minutes, through what angle should the pilot turn to head toward Louisville?

(b) What new average speed should the pilot maintain so that the total time of the trip is 90 minutes?

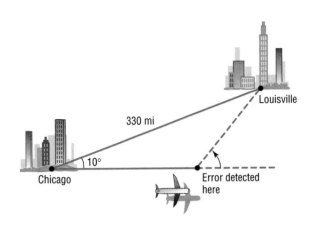

28. Avoiding a Tropical Storm A cruise ship maintains an average speed of 15 knots in going from San Juan, Puerto Rico, to Barbados, West Indies, a distance of 600 nautical miles. To avoid a tropical storm, the captain heads out of San Juan in a direction of 20° off a direct heading to Barbados. The captain maintains the 15 knot speed for 10 hours, after which time the path to Barbados becomes clear of storms.
(a) Through what angle should the captain turn to head directly to Barbados?
(b) How long will it be before the ship reaches Barbados if the same 15 knot speed is maintained?

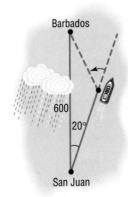

29. **Major League Baseball Field** A Major League baseball diamond is actually a square 90 feet on a side. The pitching rubber is located 60.5 feet from home plate on a line joining home plate and second base.
    (a) How far is it from the pitching rubber to first base?
    (b) How far is it from the pitching rubber to second base?
    (c) If a pitcher faces home plate, through what angle does he need to turn to face first base?

30. **Little League Baseball Field** According to Little League baseball official regulations, the diamond is a square 60 feet on a side. The pitching rubber is located 46 feet from home plate on a line joining home plate and second base.
    (a) How far is it from the pitching rubber to first base?
    (b) How far is it from the pitching rubber to second base?
    (c) If a pitcher faces home plate, through what angle does he need to turn to face first base?

31. **Finding the Length of a Guy Wire** The height of a radio tower is 500 feet, and the ground on one side of the tower slopes upward at an angle of 10° (see the figure).
    (a) How long should a guy wire be if it is to connect to the top of the tower and be secured at a point on the sloped side 100 feet from the base of the tower?
    (b) How long should a second guy wire be if it is to connect to the middle of the tower and be secured at a point 100 feet from the base on the flat side?

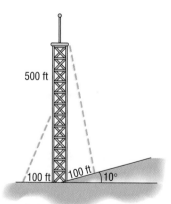

32. **Finding the Length of a Guy Wire** A radio tower 500 feet high is located on the side of a hill with an inclination to the horizontal of 5°

(see the figure). How long should two guy wires be if they are to connect to the top of the tower and be secured at two points 100 feet directly above and directly below the base of the tower?

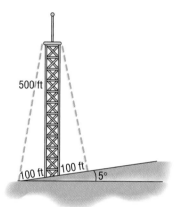

33. **Wrigley Field, Home of the Chicago Cubs** The distance from home plate to dead center of Wrigley Field is 400 feet (see the figure). How far is it from dead center to third base?

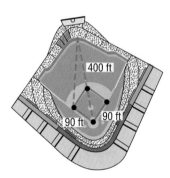

34. **Little League Baseball** The distance from home plate to dead center at the Oak Lawn Little League field is 280 feet. How far is it from dead center to third base?

    [**Hint:** The distance between the bases in Little League is 60 feet.]

35. **Rods and Pistons** Rod OA (see the figure on page 283) rotates about the fixed point O so that point A travels on a circle of radius r. Connected to point A is another rod AB of length $L > r$, and point B is connected to a piston. Show that the distance x between point O and point B is given by

$$x = r \cos \theta + \sqrt{r^2 \cos^2 \theta + L^2 - r^2}$$

where θ is the angle of rotation of rod OA.

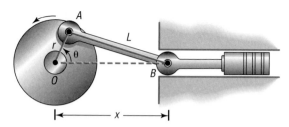

36. **Geometry** Show that the length d of a chord of a circle of radius r is given by the formula

$$d = 2r \sin \frac{\theta}{2}$$

where θ is the central angle formed by the radii to the ends of the chord (see the figure). Use this result to derive the fact that $\sin \theta < \theta$, where $\theta > 0$ is measured in radians.

37. For any triangle, show that

$$\cos \frac{\gamma}{2} = \sqrt{\frac{s(s - c)}{ab}}$$

where $s = \frac{1}{2}(a + b + c)$.
[**Hint:** Use a half-angle formula and the Law of Cosines.]

38. For any triangle show that

$$\sin \frac{\gamma}{2} = \sqrt{\frac{(s - a)(s - b)}{ab}}$$

where $s = \frac{1}{2}(a + b + c)$.

39. Use the Law of Cosines to prove the identity

$$\frac{\cos \alpha}{a} + \frac{\cos \beta}{b} + \frac{\cos \gamma}{c} = \frac{a^2 + b^2 + c^2}{2abc}$$

 40. Write down your strategy for solving an oblique triangle.

4.4 | THE AREA OF A TRIANGLE

  1   Find the Area of SAS Triangles
  2   Find the Area of SSS Triangles

In this section, we will derive several formulas for calculating the area A of a triangle. The most familiar of these is the following:

> **Theorem**
>
> The area A of a triangle is
>
> $$A = \frac{1}{2}bh \qquad (1)$$
>
> where b is the base and h is an altitude drawn to that base.

FIGURE 33
$A = \frac{1}{2}bh$

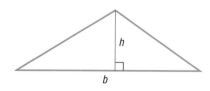

FIGURE 34

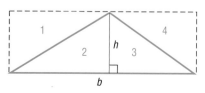

Proof The derivation of this formula is rather easy once a rectangle of base b and height h is constructed around the triangle. See Figures 33 and 34.

Triangles 1 and 2 in Figure 34 are equal in area, as are triangles 3 and 4. Consequently, the area of the triangle with base b and altitude h is exactly half the area of the rectangle, which is bh. ∎

If the base b and altitude h to that base are known, then we can easily find the area of such a triangle using formula (1). Usually, though, the information required to use formula (1) is not given. Suppose, for example,

FIGURE 35
$h = a \sin \gamma$

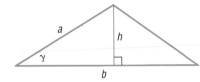

that we know two sides a and b and the included angle γ (see Figure 35). Then the altitude h can be found by noting that

$$\frac{h}{a} = \sin \gamma$$

so that

$$h = a \sin \gamma$$

Using this fact in formula (1) produces

$$A = \tfrac{1}{2}bh = \tfrac{1}{2}b(a \sin \gamma) = \tfrac{1}{2}ab \sin \gamma$$

Thus, we have the formula

$$A = \tfrac{1}{2}ab \sin \gamma \qquad (2)$$

By dropping altitudes from the other two vertices of the triangle, we obtain the following corresponding formulas:

$$A = \tfrac{1}{2}bc \sin \alpha \qquad (3)$$

$$A = \tfrac{1}{2}ac \sin \beta \qquad (4)$$

It is easiest to remember these formulas using the following wording:

Theorem

The area A of a triangle equals one-half the product of two of its sides times the sine of their included angle.

E X A M P L E   1

Finding the Area of a SAS Triangle

Find the area A of the triangle for which $a = 8$, $b = 6$, and $\gamma = 30°$.

Solution   We use formula (2) to get

$$A = \tfrac{1}{2}ab \sin \gamma = \tfrac{1}{2} \cdot 8 \cdot 6 \sin 30° = 12$$

Now work Problem 1.

If the three sides of a triangle are known, another formula, called **Heron's Formula** (named after Heron of Alexandria), can be used to find the area of a triangle.

Theorem   Heron's Formula

The area A of a triangle with sides a, b, and c is

$$A = \sqrt{s(s - a)(s - b)(s - c)} \qquad (5)$$

where $s = \tfrac{1}{2}(a + b + c)$.

Proof The proof that we shall give uses the Law of Cosines and is quite different from the proof given by Heron.

From the Law of Cosines,

$$c^2 = a^2 + b^2 - 2ab \cos \gamma$$

and the two half-angle formulas,

$$\cos^2 \frac{\gamma}{2} = \frac{1 + \cos \gamma}{2} \qquad \sin^2 \frac{\gamma}{2} = \frac{1 - \cos \gamma}{2}$$

and using $2s = a + b + c$, we find

$$\cos^2 \frac{\gamma}{2} = \frac{1 + \cos \gamma}{2} = \frac{1 + \dfrac{a^2 + b^2 - c^2}{2ab}}{2}$$

$$= \frac{a^2 + 2ab + b^2 - c^2}{4ab} = \frac{(a + b)^2 - c^2}{4ab}$$

$$= \frac{(a + b - c)(a + b + c)}{4ab} = \frac{2(s - c) \cdot 2s}{4ab} = \frac{s(s - c)}{ab} \qquad (6)$$

$$\uparrow$$

$$a + b - c = a + b + c - 2c$$
$$= 2s - 2c$$

Similarly,

$$\sin^2 \frac{\gamma}{2} = \frac{(s - a)(s - b)}{ab} \qquad (7)$$

Now we use formula (2) for the area.

$$A = \frac{1}{2}ab \sin \gamma$$

$$= \frac{1}{2}ab \cdot 2 \sin \frac{\gamma}{2} \cos \frac{\gamma}{2} \qquad \sin \gamma = \sin 2\left(\frac{\gamma}{2}\right) = 2 \sin \frac{\gamma}{2} \cos \frac{\gamma}{2}$$

$$= ab \sqrt{\frac{(s - a)(s - b)}{ab}} \sqrt{\frac{s(s - c)}{ab}} \qquad \text{Use equations (6) and (7).}$$

$$= \sqrt{s(s - a)(s - b)(s - c)} \qquad \blacksquare$$

E X A M P L E   2   Finding the Area of a SSS Triangle

Find the area of a triangle whose sides are 4, 5, and 7.

⬥ Solution   We let $a = 4$, $b = 5$, and $c = 7$. Then

$$s = \tfrac{1}{2}(a + b + c) = \tfrac{1}{2}(4 + 5 + 7) = 8$$

Heron's Formula then gives the area A as

$$A = \sqrt{s(s - a)(s - b)(s - c)} = \sqrt{8 \cdot 4 \cdot 3 \cdot 1} = \sqrt{96} = 4\sqrt{6} \qquad \blacksquare$$

  Now work Problem 7.

HISTORICAL FEATURE   Heron's formula (also known as *Hero's Formula*) is due to Heron of Alexandria (about AD 75), who had, besides his mathematical talents, a good deal of engineering skills. In various temples his mechanical devices produced effects that seemed supernatural, and visitors presumably were thus influenced to generosity. Heron's book *Metrica,* on making such devices, has survived and was discovered in 1896 in the city of Constantinople.

Heron's Formula for the area of a triangle caused some mild discomfort in Greek mathematics, because a product with two factors was an area while one with three factors was a volume, but four factors seemed contradictory in Heron's time.

4.4 EXERCISES

In Problems 1–8, find the area of each triangle. Round answers to two decimal places.

1.
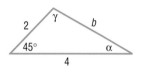

2.
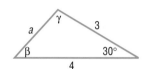

3.
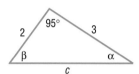

4.
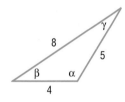

5.
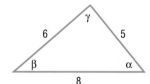

6.
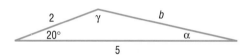

7.
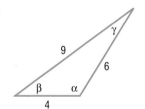

8.
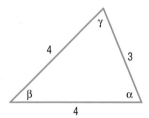

In Problems 9–24, find the area of each triangle. Round answers to two decimal places.

9. $a = 3, \quad b = 4, \quad \gamma = 40°$

10. $a = 2, \quad c = 1, \quad \beta = 10°$

11. $b = 1, \quad c = 3, \quad \alpha = 80°$

12. $a = 6, \quad b = 4, \quad \gamma = 60°$

13. $a = 3, \quad c = 2, \quad \beta = 110°$

14. $b = 4, \quad c = 1, \quad \alpha = 120°$

15. $a = 2, \quad b = 2, \quad \gamma = 50°$

16. $a = 3, \quad c = 2, \quad \beta = 90°$

17. $a = 12, \quad b = 13, \quad c = 5$

18. $a = 4, \quad b = 5, \quad c = 3$

19. $a = 2, \quad b = 2, \quad c = 2$

20. $a = 3, \quad b = 3, \quad c = 2$

21. $a = 5, \quad b = 8, \quad c = 9$

22. $a = 4, \quad b = 3, \quad c = 6$

23. $a = 10, \quad b = 8, \quad c = 5$

24. $a = 9, \quad b = 7, \quad c = 10$

25. Area of a Segment Find the area of the segment (shaded in blue in the figure) of a circle whose radius is 8 feet, formed by a central angle of 70°.

[**Hint:** The area of the sector (enclosed in red) of a circle of radius r formed by a central angle of θ radians is $1/2\, r^2\theta$. Now subtract the area of the triangle from the area of the sector to obtain the area of the segment.]

26. Area of a Segment Find the area of the segment of a circle, whose radius is 5 inches, formed by a central angle of 40°.

27. Cost of a Triangular Lot The dimensions of a triangular lot are 100 feet by 50 feet by 75 feet. If the price of such land is $3 per square foot, how much does the lot cost?

28. Amount of Material to Make a Tent A cone-shaped tent is made from a circular piece of canvas 24 feet in diameter by removing a sector with central angle 100° and connecting the ends. What is the surface area of the tent?

29. Computing Areas Find the area of the shaded region enclosed in a semicircle of diameter 8 centimeters. The length of the chord AB is 6 centimeters.

[**Hint:** Triangle ABC is a right triangle.]

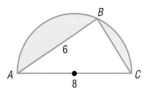

30. Computing Areas Find the area of the shaded region enclosed in a semicircle of diameter 10 inches. The length of the chord AB is 8 inches.

[**Hint:** Triangle ABC is a right triangle.]

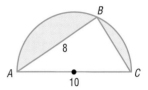

31. Area of a Triangle Prove that the area A of a triangle is given by the formula

$$A = \frac{a^2 \sin \beta \sin \gamma}{2 \sin \alpha}$$

32. Area of a Triangle Prove the two other forms of the formula given in Problem 31.

$$A = \frac{b^2 \sin \alpha \sin \gamma}{2 \sin \beta} \quad \text{and} \quad A = \frac{c^2 \sin \alpha \sin \beta}{2 \sin \gamma}$$

In Problems 33–38, use the results of Problem 31 or 32 to find the area of each triangle. Round answers to two decimal places.

33. $\alpha = 40°$, $\beta = 20°$, $a = 2$

34. $\alpha = 50°$, $\gamma = 20°$, $a = 3$

35. $\beta = 70°$, $\gamma = 10°$, $b = 5$

36. $\alpha = 70°$, $\beta = 60°$, $c = 4$

37. $\alpha = 110°$, $\gamma = 30°$, $c = 3$

38. $\beta = 10°$, $\gamma = 100°$, $b = 2$

39. Geometry Consult the figure, which shows a circle of radius r with center at O. Find the area A of the shaded region as a function of the central angle θ.

40. Approximating the Area of a Lake To approximate the area of a lake, a surveyor walks around the perimeter of the lake, taking the measurements shown in the illustration. Using this technique, what is the approximate area of the lake?

[**Hint:** Use the Law of Cosines on the three triangles shown and then find the sum of their areas.]

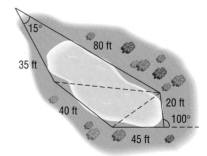

41. **The Cow Problem*** A cow is tethered to one corner of a square barn, 10 feet by 10 feet, with a rope 100 feet long. What is the maximum grazing area of the cow?

    [**Hint:** See the illustration.]

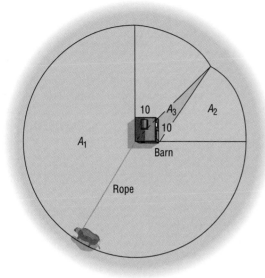

42. **Another Cow Problem** If the barn in Problem 41 is rectangular, 10 feet by 20 feet, what is the maximum grazing area for the cow?

43. If h_1, h_2, and h_3 are the altitudes dropped from A, B, and C, respectively, in a triangle (see the figure), show that

    $$\frac{1}{h_1} + \frac{1}{h_2} + \frac{1}{h_3} = \frac{s}{K}$$

    where K is the area of the triangle and $s = \frac{1}{2}(a + b + c)$.

    [**Hint:** $h_1 = 2K/a$.]

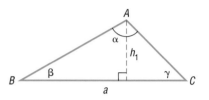

*Suggested by Professor Teddy Koukounas of SUNY at Old Westbury, who learned of it from an old farmer in Virginia. Solution provided by Professor Kathleen Miranda of SUNY at Old Westbury.

44. Show that a formula for the altitude h from a vertex to the opposite side a of a triangle is

    $$h = \frac{a \sin \beta \sin \gamma}{\sin \alpha}$$

Inscribed Circle *For Problems 45–49, the lines that bisect each angle of a triangle meet in a single point O, and the perpendicular distance r from O to each side of the triangle is the same. The circle with center at O and radius r is called the **inscribed circle** of the triangle (see the figure).*

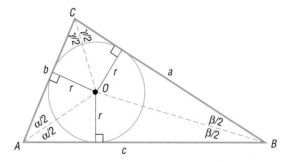

45. Apply Problem 44 to triangle OAB to show that

    $$r = \frac{c \sin(\alpha/2) \sin(\beta/2)}{\cos(\gamma/2)}$$

46. Use the results of Problem 45 and Problem 38 in Section 4.3 to show that

    $$\cot \frac{\gamma}{2} = \frac{s - c}{r}$$

47. Show that

    $$\cot \frac{\alpha}{2} + \cot \frac{\beta}{2} + \cot \frac{\gamma}{2} = \frac{s}{r}$$

48. Show that the area K of triangle ABC is $K = rs$. Then show that

    $$r = \sqrt{\frac{(s - a)(s - b)(s - c)}{s}}$$

    where $s = \frac{1}{2}(a + b + c)$.

Locating Lost Treasure

While scuba diving off Wreck Hill in Bermuda, a group of five entrepreneurs discovered a treasure map in a small watertight cask on a pirate schooner that had sunk in 1747. The map directed them to an area of Bermuda now known as The Flatts, but when they got there, they realized that the most important landmark on the map was gone. They called in the Mission Possible team to help them to re-create the map. They promised you 25% of whatever treasure was found.

The directions on the map read as follows:

1. From the tallest palm tree, sight the highest hill. Drop your eyes vertically until you sight the base of the hill.
2. Turn 40° clockwise from that line and walk 70 paces to the big red rock.
3. From the red rock, walk 50 paces back to the sight line between the palm tree and the hill. Dig there.

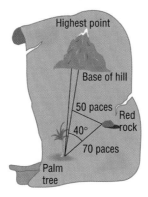

The five entrepreneurs said that they believed that they had found the red rock and the highest hill in the vicinity, but the "tallest palm tree" had long since fallen and disintegrated. It had occurred to them that the treasure must be located on a circle with radius 50 "paces" centered around the red rock, but they had decided against digging a trench 942 feet in circumference, especially since they had no assurance that the treasure was still there. (They had decided that a "pace" must be about a yard.)

1. Determine a plan to locate the position of the lost palm tree, and write out an explanation of your procedure for the entrepreneurs.
2. Unfortunately, it turns out that the entrepreneurs had more in common with the eighteenth-century pirates than you had bargained for. Once you told them the location of the lost palm tree, they tied you all to the red rock, saying that they could take it from there. From the location of the palm tree, they sighted 40° counterclockwise from the rock to the hill and then ran about 50 yards to the circle that they had traced about the rock and began to dig frantically. Nothing. After about an hour, they drove off shouting back at you, "25% of nothing is nothing!"

   Fortunately, the entrepreneurs had left the shovels. After you managed to untie yourselves, you went to the correct location and found the treasure. Where was it? How far from the palm tree? Explain.
3. People who scuba dive for sunken treasure have certain legal obligations. What are they? Should you share the treasure with a lawyer, just to make sure that you get to keep the rest?

4.5 | SIMPLE HARMONIC MOTION; DAMPED MOTION

1. Find an Equation for an Object in Simple Harmonic Motion
2. Analyze Simple Harmonic Motion
3. Analyze an Object in Damped Motion

Simple Harmonic Motion

Many physical phenomena can be described as simple harmonic motion. Radio and television waves, light waves, sound waves, and water waves exhibit motion that is simple harmonic.

The swinging of a pendulum, the vibrations of a tuning fork, and the bobbing of a weight attached to a coiled spring are examples of vibrational motion. In this type of motion, an object swings back and forth over the same path. In each illustration in Figure 36, the point B is the **equilibrium (rest) position** of the vibrating object. The **amplitude** of vibration is the distance from the object's rest position to its point of greatest displacement (either point A or point C in Figure 36). The **period** of a vibrating object is the time required to complete one vibration, that is, the time it takes to go from, say, point A through B to C and back to A.

FIGURE 36

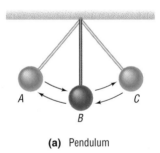

(a) Pendulum

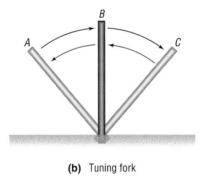

(b) Tuning fork

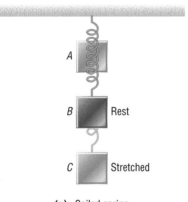

(c) Coiled spring

Simple harmonic motion is a special kind of vibrational motion in which the acceleration a of the object is directly proportional to the negative of its displacement d from its rest position. That is, $a = -kd$, $k > 0$.

For example, when the mass hanging from the spring in Figure 36(c) is pulled down from its rest position B to the point C, the force of the spring tries to restore the mass to its rest position. Assuming that there is no frictional force* to retard the motion, the amplitude will remain constant. The

---

*If friction is present, the amplitude will decrease with time to 0. This type of motion is an example of **damped motion,** which is discussed later in this section.

force increases in direct proportion to the distance that the mass is pulled from its rest position. Since the force increases directly, the acceleration of the mass of the object must do likewise, because (by Newton's Second Law of Motion) force is directly proportional to acceleration. Thus, the acceleration of the object varies directly with its displacement, and the motion is an example of simple harmonic motion.

Simple harmonic motion is related to circular motion. To see this relationship, consider a circle of radius a, with center at $(0, 0)$. See Figure 37. Suppose that an object initially placed at $(a, 0)$ moves counterclockwise around the circle at constant angular speed ω. Suppose further that after time t has elapsed the object is at the point $P = (x, y)$ on the circle. The angle θ, in radians, swept out by the ray $\overrightarrow{OP}$ in this time t is

$$\theta = \omega t$$

The coordinates of the point P at time t are

$$x = a \cos \theta = a \cos \omega t$$
$$y = a \sin \theta = a \sin \omega t$$

FIGURE 37

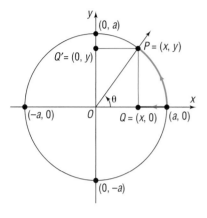

Corresponding to each position $P = (x, y)$ of the object moving about the circle, there is the point $Q = (x, 0)$, called the **projection of P on the x-axis.** As P moves around the circle at a constant rate, the point Q moves back and forth between the points $(a, 0)$ and $(-a, 0)$ along the x-axis with a motion that is simple harmonic. Similarly, for each point P there is a point $Q' = (0, y)$, called the **projection of P on the y-axis.** As P moves around the circle, the point Q' moves back and forth between the points $(0, a)$ and $(0, -a)$ on the y-axis with a motion that is simple harmonic. Thus, simple harmonic motion can be described as the projection of constant circular motion on a coordinate axis.

To put it another way, again consider a mass hanging from a spring, where the mass is pulled down from its rest position to the point C and then released. See Figure 38(a). The graph shown in Figure 38(b) describes the distance that the object is from its rest position, d, as a function of time, t, assuming that no frictional force is present.

FIGURE 38

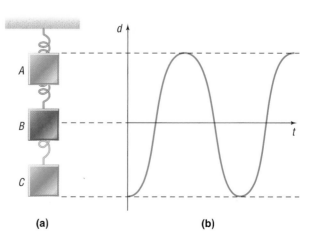

(a)   (b)

Theorem Simple Harmonic Motion

An object that moves on a coordinate axis so that its distance d from the origin at time t is given by either

$$d = a \cos \omega t \quad \text{or} \quad d = a \sin \omega t$$

where a and $\omega > 0$ are constants, moves with simple harmonic motion. The motion has amplitude $|a|$ and period $2\pi/\omega$.

The **frequency** f of an object in simple harmonic motion is the number of oscillations per unit time. Since the period is the time required for one oscillation, it follows that frequency is the reciprocal of the period; that is,

$$f = \frac{\omega}{2\pi} \qquad \omega > 0$$

E X A M P L E 1

Finding an Equation for an Object in Harmonic Motion

Suppose that the object attached to the coiled spring in Figure 36(c) is pulled down a distance of 5 inches from its rest position and then released. If the time for one oscillation is 3 seconds, write an equation that relates the distance d of the object from its rest position after time t (in seconds). Assume no friction.

Solution The motion of the object is simple harmonic. When the object is released ($t = 0$), the distance of the object from the rest position is -5 units (since the object was pulled down). Because $d = -5$ when $t = 0$, it is easier* to use the cosine function

$$d = a \cos \omega t$$

to describe the motion. Now the amplitude is $|-5| = 5$ and the period is 3. Thus,

$$a = -5 \quad \text{and} \quad \frac{2\pi}{\omega} = \text{period} = 3 \quad \text{or} \quad \omega = \frac{2\pi}{3}$$

An equation for the motion of the object is

$$d = -5 \cos \frac{2\pi}{3} t$$

Note In the solution to Example 1, we let $a = -5$, since the initial motion is down [the object was pulled down and then released (as in Figure 38a)]. If the initial direction was up, we would let $a = 5$.

Now work Problem 1.

*No phase shift is required if a cosine function is used.

E X A M P L E 2   Analyzing the Simple Harmonic Motion of an Object

Suppose that the distance d (in meters) that an object travels in time t (in seconds) satisfies the equation

$$d = 10 \sin 5t$$

(a) Describe the motion of the object.
(b) What is the maximum displacement from its resting position?
(c) What is the time required for one oscillation?
(d) What is the frequency?

Solution   We observe that the given equation is of the form

$$d = a \sin \omega t \qquad d = 10 \sin 5t$$

where $a = 10$ and $\omega = 5$.

(a) The motion is simple harmonic.
(b) The maximum displacement of the object from its resting position is the amplitude: $a = 10$ meters.
(c) The time required for one oscillation is the period.

$$\text{Period} = \frac{2\pi}{\omega} = \frac{2\pi}{5} \text{ seconds}$$

(d) The frequency is the reciprocal of the period. Thus,

$$\text{Frequency} = f = \frac{5}{2\pi} \text{ oscillations per second}$$

Now work Problem 9.

Damped Motion

Most physical phenomena are affected by friction or other resistive forces. These forces remove energy from a moving system and thereby damp its motion. For example, when a mass hanging from a spring is pulled down a distance a and released, the friction in the spring causes the distance that the mass moves from its at rest position to decrease over time. Thus, the amplitude of any real oscillating spring or swinging pendulum decreases with time due to air resistance, friction, and so forth. See Figure 39.

FIGURE 39

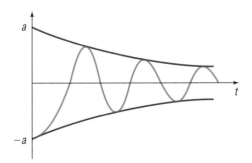

A function that describes this phenomenon maintains a sinusoidal component, but the amplitude of this component will decrease with time in order to account for the damping effect. In addition, the period of the oscillating component will be affected by the damping. The general form of a damped motion curve is

$$y = d(x) \sin \omega x \quad \text{or} \quad y = d(x) \cos \omega x$$

The function $d(x)$ is called the bounding curve for the damped motion because the graph lies between $y = d(x)$ and $y = -d(x)$.

E X A M P L E   3   Analyzing a Damped Vibration Curve

 Analyze the damped vibration curve

$$f(x) = e^{-x} \sin x, \qquad x \geq 0$$

Solution   The function f is the product of $y = e^{-x}$ and $y = \sin x$. Using properties of absolute value and the fact that $|\sin x| \leq 1$, we find that

$$|f(x)| = |e^{-x} \sin x| = |e^{-x}||\sin x| \leq |e^{-x}| = e^{-x}$$
$$\uparrow$$
$$e^{-x} > 0 \text{ for all } x$$

Thus,

$$-e^{-x} \leq f(x) \leq e^{-x}$$

and the graph of f will lie between the graphs of $y = e^{-x}$ and $y = -e^{-x}$, the bounding curves.

Also, the graph of f will touch these graphs when $|\sin x| = 1$, that is, when $x = -\pi/2, \pi/2, 3\pi/2$, and so on. The x-intercepts of the graph of f occur at $x = -\pi, 0, \pi, 2\pi$, and so on. See Table 1. See Figure 40 for the graph.

TABLE 1					
x	**0**	**$\pi/2$**	**π**	**$3\pi/2$**	**2π**
$y = e^{-x}$	1	$e^{-\pi/2}$	$e^{-\pi}$	$e^{-3\pi/2}$	$e^{-2\pi}$
$y = \sin x$	0	1	0	-1	0
$f(x) = e^{-x} \sin x$	0	$e^{-\pi/2}$	0	$-e^{-3\pi/2}$	0
Point on Graph of f	(0, 0)	$(\pi/2, e^{-\pi/2})$	$(\pi, 0)$	$(3\pi/2, -e^{-3\pi/2})$	$(2\pi, 0)$

FIGURE 40
$f(x) = e^{-x} \sin x, x \geq 0$

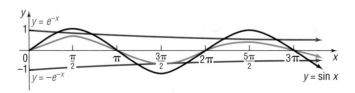

Exploration Graph $Y_1 = e^{-x} \sin x$, along with $Y_2 = e^{-x}$ and $Y_3 = -e^{-x}$ for $0 \le x \le \pi$. Determine where $Y_1 = e^{-x} \sin x$ has its first turning point (local maximum). Compare this to where Y_1 intersects Y_2.

Solution Figure 41 shows the graphs of $Y_1 = e^{-x} \sin x$, $Y_2 = e^{-x}$, and $Y_3 = -e^{-x}$. Using MAXIMUM, the first turning point occurs at $x = 0.78$; Y_1 intersects Y_2 at $x = \pi/2 \approx 1.57$.

FIGURE 41

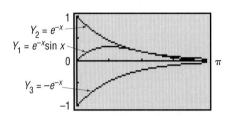

4.5 EXERCISES

In Problems 1–4, an object attached to a coiled spring is pulled down a distance a from its rest position and then released. Assuming that the motion is simple harmonic with period T, write an equation that relates the distance d of the object from its rest position after t seconds. Also assume that the positive direction of the motion is up.

1. $a = 5$; $T = 2$ seconds
2. $a = 10$; $T = 3$ seconds
3. $a = 6$; $T = \pi$ seconds
4. $a = 4$; $T = \pi/2$ seconds

5. Rework Problem 1 under the same conditions except that, at time $t = 0$, the object is at its resting position and moving down.

6. Rework Problem 2 under the same conditions except that, at time $t = 0$, the object is at its resting position and moving down.

7. Rework Problem 3 under the same conditions except that, at time $t = 0$, the object is at its resting position and moving down.

8. Rework Problem 4 under the same conditions except that, at time $t = 0$, the object is at its resting position and moving down.

In Problems 9–16, the distance d (in meters) that an object travels in time t (in seconds) is given.
(a) Describe the motion of the object.
(b) What is the maximum displacement from its resting position?
(c) What is the time required for one oscillation?
(d) What is the frequency?

9. $d = 5 \sin 3t$
10. $d = 4 \sin 2t$
11. $d = 6 \cos \pi t$
12. $d = 5 \cos \dfrac{\pi}{2} t$
13. $d = -3 \sin \frac{1}{2} t$
14. $d = -2 \cos 2t$
15. $d = 6 + 2 \cos 2\pi t$
16. $d = 4 + 3 \sin \pi t$

In Problems 17–20, graph each damped vibration curve for $0 \le x \le 2\pi$.
17. $y = e^{-x/\pi} \sin 2x$
18. $y = e^{-x/\pi} \cos x$
19. $y = e^{-x/\pi} \cos 2x$
20. $y = e^{-2x/\pi} \sin x$

21. **Charging a Capacitor** If a charged capacitor is connected to a coil by closing a switch (see the figure), energy is transferred to the coil and then back to the capacitor in an oscillatory motion. The voltage V (in volts) across the capacitor will gradually diminish to 0 with time t (in seconds). (*Continued on page 296.*)

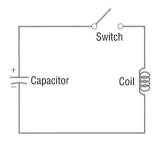

(a) Graph the equation relating V and t:

$$V(t) = e^{-t/3} \cos \pi t \qquad 0 \le t \le 3$$

(b) At what times t will the graph of V touch the graph of $y = e^{-t/3}$? When does V touch the graph of $y = -e^{-t/3}$?

(c) When will the voltage V be between -0.4 and 0.4 volt?

22. Graph the function $f(x) = (\sin\ x)/x, \ x > 0$. Based on the graph, what do you conjecture about the value of $(\sin x)/x$ for x close to 0?

23. Graph $y = x \sin x$, $y = x^2 \sin x$, and $y = x^3 \sin x$ for $x > 0$. What patterns do you observe?

24. Graph $y = \dfrac{1}{x} \sin x$, $y = \dfrac{1}{x^2} \sin x$, and

$y = \dfrac{1}{x^3} \sin x$ for $x > 0$. What patterns do you observe?

25. How would you explain to a friend what simple harmonic motion is? Damped motion?

CHAPTER REVIEW

THINGS TO KNOW

Formulas

Law of Sines	$\dfrac{\sin \alpha}{a} = \dfrac{\sin \beta}{b} = \dfrac{\sin \gamma}{c}$

Law of Cosines

$$c^2 = a^2 + b^2 - 2ab \cos \gamma$$
$$b^2 = a^2 + c^2 - 2ac \cos \beta$$
$$a^2 = b^2 + c^2 - 2bc \cos \alpha$$

Area of a triangle

$$A = \tfrac{1}{2}bh$$
$$A = \tfrac{1}{2}ab \sin \gamma$$
$$A = \tfrac{1}{2}bc \sin \alpha$$
$$A = \tfrac{1}{2}ac \sin \beta$$
$$A = \sqrt{s(s - a)(s - b)(s - c)}, \quad \text{where } s = \tfrac{1}{2}(a + b + c)$$

HOW TO

Solve right triangles

Use the Law of Sines to solve a SAA, ASA, or SSA triangle

Use the Law of Cosines to solve a SAS or SSS triangle

Find the area of a triangle

Solve applied problems involving triangles

Analyze simple harmonic motion

Analyze damped motion

FILL-IN-THE-BLANK ITEMS

1. If two sides and the angle opposite one of them are known, the Law of _____ is used to determine whether the known information results in no triangle, one triangle, or two triangles.

2. If three sides of a triangle are given, the Law of _____ is used to solve the triangle.

3. If three sides of a triangle are given, _____ Formula is used to find the area of the triangle.

4. The motion of an object obeys the equation $d = 4 \cos 6t$. Such motion is described as _____ _____ _____.

TRUE/FALSE ITEMS

T F **1.** An oblique triangle in which two sides and an angle are given always results in at least one triangle.

T F **2.** Given three sides of a triangle, there is a formula for finding its area.

T F **3.** In a right triangle, if two sides are known, we can solve the triangle.

T F **4.** The ambiguous case refers to the fact that when two sides and the angle opposite one of them is known, sometimes the Law of Sines cannot be used.

REVIEW EXERCISES

Blue problem numbers indicate the author's suggestions for use in a Practice Test.

In Problems 1–4, solve each triangle.

1.   **2.**   **3.** 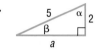 **4.**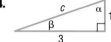

In Problems 5–24, find the remaining angle(s) and side(s) of each triangle, if it (they) exists. If no triangle exists, say "No triangle."

5. $\alpha = 50°$, $\beta = 30°$, $a = 1$ **6.** $\alpha = 10°$, $\gamma = 40°$, $c = 2$ **7.** $\alpha = 100°$, $a = 5$, $c = 2$

8. $a = 2$, $c = 5$, $\alpha = 60°$ **9.** $a = 3$, $c = 1$, $\gamma = 110°$ **10.** $a = 3$, $c = 1$, $\gamma = 20°$

11. $a = 3$, $c = 1$, $\beta = 100°$ **12.** $a = 3$, $b = 5$, $\beta = 80°$ **13.** $a = 2$, $b = 3$, $c = 1$

14. $a = 10$, $b = 7$, $c = 8$ **15.** $a = 1$, $b = 3$, $\gamma = 40°$ **16.** $a = 4$, $b = 1$, $\gamma = 100°$

17. $a = 5$, $b = 3$, $\alpha = 80°$ **18.** $a = 2$, $b = 3$, $\alpha = 20°$ **19.** $a = 1$, $b = \frac{1}{2}$, $c = \frac{4}{3}$

20. $a = 3$, $b = 2$, $c = 2$ **21.** $a = 3$, $\alpha = 10°$, $b = 4$ **22.** $a = 4$, $\alpha = 20°$, $\beta = 100°$

23. $c = 5$, $b = 4$, $\alpha = 70°$ **24.** $a = 1$, $b = 2$, $\gamma = 60°$

In Problems 25–34, find the area of each triangle.

25. $a = 2$, $b = 3$, $\gamma = 40°$ **26.** $b = 5$, $c = 4$, $\alpha = 20°$ **27.** $b = 4$, $c = 10$, $\alpha = 70°$

28. $a = 2$, $b = 1$, $\gamma = 100°$ **29.** $a = 4$, $b = 3$, $c = 5$ **30.** $a = 10$, $b = 7$, $c = 8$

31. $a = 4$, $b = 2$, $c = 5$ **32.** $a = 3$, $b = 2$, $c = 2$ **33.** $\alpha = 50°$, $\beta = 30°$, $a = 1$

34. $\alpha = 10°$, $\gamma = 40°$, $c = 3$

35. **Measuring the Length of a Lake** From a stationary hot-air balloon 500 feet above the ground, two sightings of a lake are made (see the figure). How long is the lake?

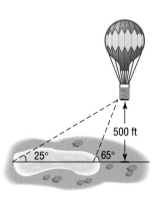

36. Finding the Speed of a Glider From a glider 200 feet above the ground, two sightings of a stationary object directly in front are taken 1 minute apart (see the figure). What is the speed of the glider?

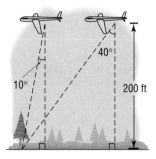

37. Finding the Width of a River Find the distance from A to C across the river illustrated in the figure.

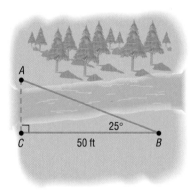

38. Finding the Height of a Building Find the height of the building shown in this figure.

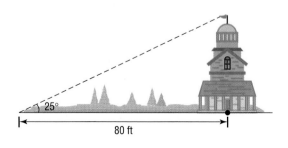

39. Finding the Distance to Shore The Sears Tower in Chicago is 1454 feet tall and is situated about 1 mile inland from the shore of Lake Michigan, as indicated in the figure. An observer in a pleasure boat on the lake directly in front of the Sears Tower looks at the top of the tower and measures the angle of elevation as 5°. How far offshore is the boat?

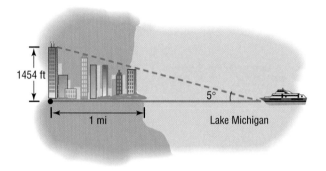

40. Finding the Grade of a Mountain Trail A straight trail with a uniform inclination leads from a hotel, elevation 5000 feet, to a lake in a valley, elevation 4100 feet. The length of the trail is 4100 feet. What is the inclination (grade) of the trail?

41. Navigation An airplane flies from city A to city B, a distance of 100 miles, and then turns through an angle of 20° and heads toward city C, as indicated in the figure. If the distance from A to C is 300 miles, how far is it from city B to city C?

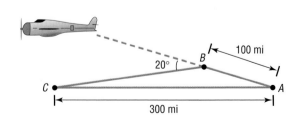

42. **Correcting a Navigation Error** Two cities A and B are 300 miles apart. In flying from city A to city B, a pilot inadvertently took a course that was 5° in error.
    (a) If the error was discovered after flying 10 minutes at a constant speed of 420 miles per hour, through what angle should the pilot turn to correct the course? (Consult the figure.)
    (b) What new constant speed should be maintained so that no time is lost due to the error? (Assume that the speed would have been a constant 420 miles per hour if no error had occurred.)

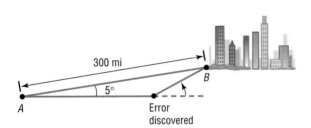

44. **Constructing a Highway** A highway whose primary directions are north–south is being constructed along the west coast of Florida. Near Naples, a bay obstructs the straight path of the road. Since the cost of a bridge is prohibitive, engineers decide to go around the bay. The illustration shows the path that they decide on and the measurements taken. What is the length of highway needed to go around the bay?

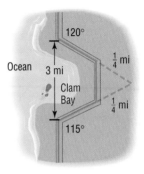

43. **Determining Distances at Sea** Rebecca, the navigator of a ship at sea, spots two lighthouses that she knows to be 2 miles apart along a straight seashore. She determines that the angles formed between two line-of-sight observations of the lighthouses and the line from the ship directly to shore are 12° and 30°. See the illustration.
    (a) How far is the ship from lighthouse A?
    (b) How far is the ship from lighthouse B?
    (c) How far is the ship from shore?

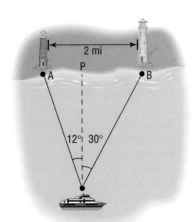

45. **Correcting a Navigational Error** A yacht leaves St. Thomas bound for an island in the British West Indies, 200 miles away. Maintaining a constant speed of 18 miles per hour, but encountering heavy crosswinds and strong currents, the crew finds after 4 hours that the sailboat is off course by 15°.
    (a) How far is the sailboat from the island at this time?
    (b) Through what angle should the sailboat turn to correct its course?
    (c) How much time has been added to the trip because of this? (Assume that the speed remains at 18 miles per hour.)

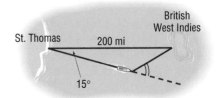

46. Surveying Two homes are located on opposite sides of a small hill. See the illustration. To measure the distance between them, a surveyor walks a distance of 50 feet from house A to point C, uses a transit to measure the angle ACB, which is found to be 80°, and then walks to house B, a distance of 60 feet. How far apart are the houses?

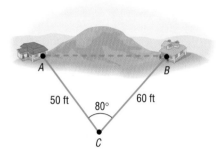

47. Approximating the Area of a Lake To approximate the area of a lake, Cindy walks around the perimeter of the lake, taking the measurements shown in the illustration. Using this technique, what is the approximate area of the lake?

[**Hint:** Use the Law of Cosines on the three triangles shown and then find the sum of their areas.]

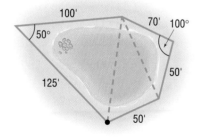

48. Calculating the Cost of Land The irregular parcel of land shown in the figure is being sold for $100 per square foot. What is the cost of this parcel?

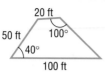

49. Area of a Segment Find the area of the segment of a circle whose radius is 6 inches formed by a central angle of 50°.

50. Finding the Bearing of a Ship The *Majesty* leaves the Port at Boston for Bermuda with a bearing of S80°E at an average speed of 10 knots. After 1 hour, the ship turns 90° toward the southwest. After 2 hours at an average speed of 20 knots, what is the bearing of the ship from Boston?

51. The drive wheel of an engine is 13 inches in diameter, and the pulley on the rotary pump is 5 inches in diameter. If the shafts of the drive wheel and the pulley are 2 feet apart, what length of belt is required to join them as shown in the figure?

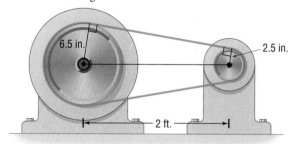

52. Rework Problem 51 if the belt is crossed, as shown in the figure.

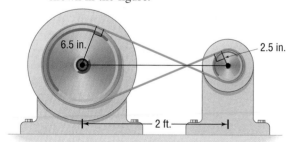

In Problems 53–56, the distance d (in feet) that an object travels in time t (in seconds) is given.

(a) Describe the motion of the object.
(b) What is the maximum displacement from its resting position?
(c) What is the time required for one oscillation?
(d) What is the frequency?

53. $d = 6 \sin 2t$

54. $d = 2 \cos 4t$

55. $d = -2 \cos \pi t$

56. $d = -3 \sin \dfrac{\pi}{2} t$

57. Graph $y = e^{-x/2\pi} \sin 2x, 0 \le x \le 2\pi$

58. Graph $y = e^{-x/3\pi} \cos 4x, 0 \le x \le 2\pi$

Polar Coordinates; Vectors

Using computer models is one way structural engineers, architects, and others try out various designs without the costs of building materials and unforeseen construction issues. On the following page you will try your hand at becoming a master bridge builder through computer modeling. The programs and other information you will need are found at:

www.prenhall.com/sullivan

PREPARING FOR THIS CHAPTER

Before getting started on this chapter, review the following concepts:

For Sections 5.1–5.3: Rectangular Coordinates
(pp. 16–17)

Graphs of Equations
(pp. 21–27)

For Sections 5.4–5.6: Rectangular Coordinates
(pp. 16–17)

Law of Cosines (Section 4.3)

OUTLINE

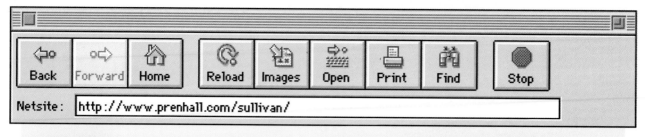

Netsite: http://www.prenhall.com/sullivan/

THE GREAT BALSA BRIDGE CONTEST

Your goal is high—you intend to become The Master Balsa Bridge Builder. Over the last two years you have performed countless hours stress-testing balsa struts to determine the strongest and lightest grains. You have performed wood-joint experiments with a host of ultrastrong superglues. You have read and reread the literature on the previous contests, analyzing those bridge designs that have failed, and those that were superior to the *competition*. You kept careful notes and have entered the data into your *bridge modeling software*. For the last few months you tested your designs on your computer. Only two survived.

You are now ready to enter into the next phase. Yours will be no *bridge disaster*. You are ready to see if your computer-generated designs match reality.

You have created a set of microtools so that your joinery will be perfect. Slowly and meticulously you build your first set of two test bridges. You need to determine how accurately your calculations match your craftsmanship. You have laminated the balsa struts, and, at last, your designs come to life! Now you place the two bridges across two chairs and begin to place weights on them . . .

1. If you think of the weights as *force vectors* acting along the struts, does it make sense that we can model the structural design for a bridge?
2. Try the *Design a Bridge* website. How did you do? Did you just overbuild everything?
3. Do you think that engineers, when they are designing a bridge, might rely a little too heavily on the computer? Do you think that might be a mistake?
4. Get out a piece of paper and make a very simple bridge design. Estimate how much weight x and y can hold. Then do a simple vector analysis using your assumptions. Are you surprised at how much your simple bridge can support?
5. Get some pieces of balsa and redo the calculations using your assumptions about the strength of the balsa. How good are you at calculating the actual values?
6. When you look carefully at a complicated bridge like the Golden Gate Bridge, how long do you think that it would take you to do all those calculations?

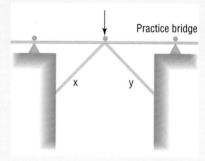

This chapter is in two parts: Polar Coordinates, Sections 5.1–5.4, and Vectors, Sections 5.5–5.7. They are independent of each other and may be covered in any order.

Sections 5.1–5.4 deal with polar coordinates (an alternative to rectangular coordinates for plotting points),

graphing in polar coordinates, complex numbers, and finding roots of complex numbers (De Moivre's Theorem).

Sections 5.5–5.7 provide an introduction to the idea of a vector, an extremely important topic in engineering and physics.

5.1 POLAR COORDINATES

1. Plot Points Using Polar Coordinates
2. Convert from Polar Coordinates to Rectangular Coordinates
3. Convert from Rectangular Coordinates to Polar Coordinates

FIGURE 1

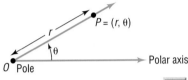

FIGURE 2

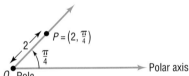

FIGURE 3

So far, we have always used a system of rectangular coordinates to plot points in the plane. Now we are ready to describe another system called *polar coordinates*. As we shall soon see, in many instances polar coordinates offer certain advantages over rectangular coordinates.

In a rectangular coordinate system, you will recall, a point in the plane is represented by an ordered pair of numbers (x, y), where x and y equal the signed distance of the point from the y-axis and x-axis, respectively. In a polar coordinate system, we select a point, called the **pole,** and then a ray with vertex at the pole, called the **polar axis.** Comparing the rectangular and polar coordinate systems, we see (in Figure 1) that the origin in rectangular coordinates coincides with the pole in polar coordinates, and the positive x-axis in rectangular coordinates coincides with the polar axis in polar coordinates.

A point P in a polar coordinate system is represented by an ordered pair of numbers (r, θ). The number r is the distance of the point from the pole, and θ is an angle (in degrees or radians) formed by the polar axis and a ray from the pole through the point. We call the ordered pair (r, θ) the **polar coordinates** of the point. See Figure 2.

As an example, suppose that the polar coordinates of a point P are $(2, \pi/4)$. We locate P by first drawing an angle of $\pi/4$ radian, placing its vertex at the pole and its initial side along the polar axis. Then we go out a distance of 2 units along the terminal side of the angle to reach the point P. See Figure 3.

Now work Problem 9.

Recall that an angle measured counterclockwise is positive, whereas one measured clockwise is negative. This convention has some interesting consequences relating to polar coordinates. Let's see what these consequences are.

E X A M P L E   1   Finding Several Polar Coordinates of a Single Point

Consider again a point P with polar coordinates $(2, \pi/4)$, as shown in Figure 4(a). Because $\pi/4$, $9\pi/4$, and $-7\pi/4$ all have the same terminal side, we also could have located this point P by using the polar coordinates $(2, 9\pi/4)$ or $(2, -7\pi/4)$, as shown in Figures 4(b) and (c).

FIGURE 4

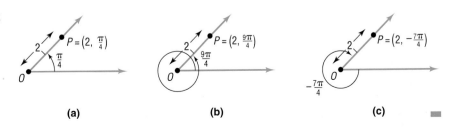

(a)   (b)   (c)

In using polar coordinates (r, θ), it is possible for the first entry r to be negative. When this happens, we follow the convention that the location of the point, instead of being on the terminal side of θ, is on the ray from the pole extending in the direction *opposite* the terminal side of θ at a distance $|r|$ from the pole. See Figure 5 for an illustration.

FIGURE 5

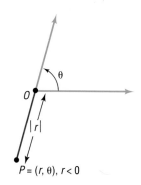

$P = (r, \theta), r < 0$

EXAMPLE 2

Polar Coordinates (r, θ), $r < 0$

Consider again the point P with polar coordinates $(2, \pi/4)$, as shown in Figure 6(a). This same point P can be assigned the polar coordinates $(-2, 5\pi/4)$, as indicated in Figure 6(b). To locate the point $(-2, 5\pi/4)$, we use the ray in

FIGURE 6

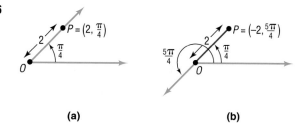

(a)   (b)

the opposite direction of $5\pi/4$ and go out 2 units along that ray to find the point P.

These examples show a major difference between rectangular coordinates and polar coordinates. In the former, each point has exactly one pair

SUMMARY

A point with polar coordinates (r, θ) also can be represented by any of the following:

$$(r, \theta + 2k\pi) \quad \text{or} \quad (-r, \theta + \pi + 2k\pi), \quad k \text{ any integer}$$

The polar coordinates of the pole are $(0, \theta)$, where θ can be any angle.

E X A M P L E   3

Plotting Points Using Polar Coordinates

Plot the points with the following polar coordinates:

(a) $(3, 5\pi/3)$ (b) $(2, -\pi/4)$ (c) $(3, 0)$ (d) $(-2, \pi/4)$

Solution   Figure 7 shows the points.

FIGURE 7

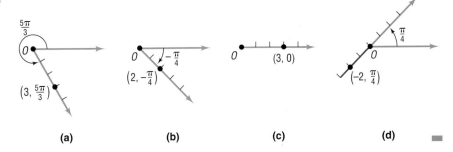

(a)   (b)   (c)   (d)

Now work Problems 1 and 17.

E X A M P L E   4

Finding Other Polar Coordinates of a Given Point

Plot the point P with polar coordinates $(3, \pi/6)$, and find other polar coordinates (r, θ) of this same point for which

(a) $r > 0$, $2\pi \le \theta < 4\pi$ (b) $r < 0$, $0 \le \theta < 2\pi$
(c) $r > 0$, $-2\pi \le \theta < 0$

FIGURE 8

$P = \left(3, \frac{\pi}{6}\right)$

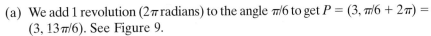

Solution   The point $(3, \pi/6)$ is plotted in Figure 8.

(a) We add 1 revolution (2π radians) to the angle $\pi/6$ to get $P = (3, \pi/6 + 2\pi) = (3, 13\pi/6)$. See Figure 9.
(b) We add $\frac{1}{2}$ revolution (π radians) to the angle $\pi/6$ and replace 3 by -3 to get $P = (-3, \pi/6 + \pi) = (-3, 7\pi/6)$. See Figure 10.
(c) We subtract 2π from the angle $\pi/6$ to get $P = (3, \pi/6 - 2\pi) = (3, -11\pi/6)$. See Figure 11.

FIGURE 9   **FIGURE 10**   **FIGURE 11**

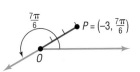   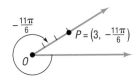

Now work Problem 21.

Conversion from Polar Coordinates to Rectangular Coordinates, and Vice Versa

It is sometimes convenient and, indeed, necessary to be able to convert coordinates or equations in rectangular form to polar form, and vice versa. To do this, we recall that the origin in rectangular coordinates is the pole in polar coordinates and that the positive x-axis in rectangular coordinates is the polar axis in polar coordinates.

Theorem   Conversion from Polar Coordinates to Rectangular
           Coordinates

If P is a point with polar coordinates (r, θ), the rectangular coordinates
(x, y) of P are given by

$$x = r \cos \theta \qquad y = r \sin \theta \qquad (1)$$

FIGURE 12

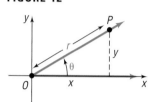

Proof   Suppose that P has the polar coordinates (r, θ). We seek the rectangular coordinates (x, y) of P. Refer to Figure 12.

If $r = 0$, then, regardless of θ, the point P is the pole, for which the rectangular coordinates are $(0, 0)$. Thus, formula (1) is valid for $r = 0$.

If $r > 0$, the point P is on the terminal side of θ and $r = d(O, P) = \sqrt{x^2 + y^2}$. Thus

$$\cos \theta = \frac{x}{r} \qquad \sin \theta = \frac{y}{r}$$

so

$$x = r \cos \theta \qquad y = r \sin \theta$$

If $r < 0$, then the point $P = (r, \theta)$ can be represented as $(-r, \pi + \theta)$, where $-r > 0$. Thus,

$$\cos(\pi + \theta) = -\cos \theta = \frac{x}{-r} \qquad \sin(\pi + \theta) = -\sin \theta = \frac{y}{-r}$$

so

$$x = r \cos \theta \qquad y = r \sin \theta \qquad \blacksquare$$

E X A M P L E   5

Converting from Polar Coordinates to Rectangular Coordinates

Find the rectangular coordinates of the points with the following polar coordinates:

(a) $(6, \pi/6)$ (b) $(-2, 5\pi/4)$ (c) $(-4, -\pi/4)$

FIGURE 13(a,b)

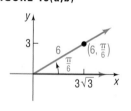

(a)

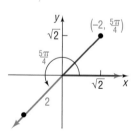

(b)

Solution   We use formula (1): $x = r \cos \theta$ and $y = r \sin \theta$.

(a) See Figure 13(a).

$$x = r \cos \theta = 6 \cos \frac{\pi}{6} = 6 \cdot \frac{\sqrt{3}}{2} = 3\sqrt{3}$$

$$y = r \sin \theta = 6 \sin \frac{\pi}{6} = 6 \cdot \frac{1}{2} = 3$$

The rectangular coordinates of the point $(6, \pi/6)$ are $(3\sqrt{3}, 3)$.

(b) See Figure 13(b).

$$x = r \cos \theta = -2 \cos \frac{5\pi}{4} = -2\left(-\frac{\sqrt{2}}{2}\right) = \sqrt{2}$$

$$y = r \sin \theta = -2 \sin \frac{5\pi}{4} = -2\left(-\frac{\sqrt{2}}{2}\right) = \sqrt{2}$$

The rectangular coordinates of the point $(-2, 5\pi/4)$ are $(\sqrt{2}, \sqrt{2})$.

FIGURE 13(c)

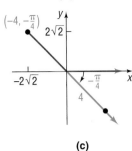

(c)

(c) See Figure 13(c).

$$x = r \cos \theta = -4 \cos\left(-\frac{\pi}{4}\right) = -4 \cdot \frac{\sqrt{2}}{2} = -2\sqrt{2}$$

$$y = r \sin \theta = -4 \sin\left(-\frac{\pi}{4}\right) = -4\left(-\frac{\sqrt{2}}{2}\right) = 2\sqrt{2}$$

The rectangular coordinates of the point $(-4, -\pi/4)$ are $(-2\sqrt{2}, 2\sqrt{2})$.

Note: Most calculators have the capability of converting from polar coordinates to rectangular coordinates. Consult your owner's manual to learn the proper key strokes. Since in most cases this procedure is tedious, you will find that using formula (1) is faster.

Now work Problems 29 and 41.

To convert from rectangular coordinates (x, y) to polar coordinates (r, θ) is a little more complicated. Notice that we begin each example by plotting the given rectangular coordinates.

E X A M P L E   6

Converting from Rectangular Coordinates to Polar Coordinates

Find polar coordinates of a point whose rectangular coordinates are $(0, 3)$.

FIGURE 14

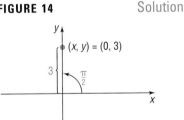

Solution   See Figure 14. The point $(0, 3)$ lies on the y-axis a distance of 3 units from the origin (pole), so $r = 3$. A ray with vertex at the pole through $(0, 3)$ forms an angle $\theta = \dfrac{\pi}{2}$ with the polar axis. Polar coordinates for this point can be given by $(3, \pi/2)$.

Figure 15 shows polar coordinates of points that lie on either the x-axis or the y-axis. In each illustration, $a > 0$.

FIGURE 15

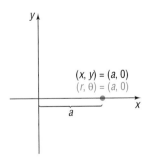

(a) $(x, y) = (a, 0)$, $a > 0$

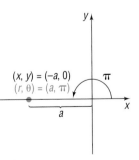

(b) $(x, y) = (0, a)$, $a > 0$

(c) $(x, y) = (-a, 0)$, $a > 0$

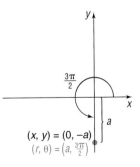

(d) $(x, y) = (0, -a)$, $a > 0$

Now work Problem 45.

E X A M P L E   7   Converting from Rectangular Coordinates to Polar Coordinates

Find polar coordinates of a point whose rectangular coordinates are $(2, -2)$.

FIGURE 16

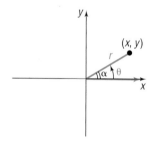

$(x, y) = (2, -2)$

Solution   See Figure 16. The distance r from the origin to the point $(2, -2)$ is

$$r = \sqrt{x^2 + y^2} = \sqrt{(2)^2 + (-2)^2} = \sqrt{8} = 2\sqrt{2}$$

To find θ, we use the reference angle α. Then

$$\alpha = \tan^{-1}\left|\frac{y}{x}\right| = \tan^{-1}\left|\frac{-2}{2}\right| = \tan^{-1} 1 = \frac{\pi}{4}$$

Thus, $\theta = -\pi/4$, and a set of polar coordinates for this point is $(2\sqrt{2}, -\pi/4)$. Other possible representations include $(2\sqrt{2}, 7\pi/4)$ and $(-2\sqrt{2}, 3\pi/4)$. ▬

E X A M P L E   8   Converting from Rectangular Coordinates to Polar Coordinates

Find polar coordinates of a point whose rectangular coordinates are $(-1, -\sqrt{3})$.

FIGURE 17

$(x, y) = (-1, -\sqrt{3})$

Solution   See Figure 17. The distance r from the origin to the point $(-1, -\sqrt{3})$ is

$$r = \sqrt{x^2 + y^2} = \sqrt{(-1)^2 + (-\sqrt{3})^2} = \sqrt{4} = 2$$

To find θ, we use the reference angle α. Then

$$\alpha = \tan^{-1}\left|\frac{y}{x}\right| = \tan^{-1}\left|\frac{-\sqrt{3}}{-1}\right| = \tan^{-1}\sqrt{3} = \frac{\pi}{3}$$

Thus,

$$\theta = \pi + \alpha = \pi + \frac{\pi}{3} = \frac{4\pi}{3}$$

and a set of polar coordinates is $(2, 4\pi/3)$. Other possible representations include $(-2, \pi/3)$ and $(2, -2\pi/3)$. ▬

Figure 18 shows how to find polar coordinates of a point that lies in a quadrant when its rectangular coordinates (x, y) are given.

FIGURE 18

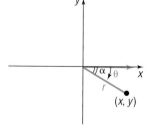
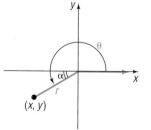

(a) $r = \sqrt{x^2 + y^2}$
$\theta = \alpha = \tan^{-1}\left|\frac{y}{x}\right|$

(b) $r = \sqrt{x^2 + y^2}$
$\theta = \pi - \alpha = \pi - \tan^{-1}\left|\frac{y}{x}\right|$

(c) $r = \sqrt{x^2 + y^2}$
$\theta = \pi + \alpha = \pi + \tan^{-1}\left|\frac{y}{x}\right|$

(d) $r = \sqrt{x^2 + y^2}$
$\theta = -\alpha = -\tan^{-1}\left|\frac{y}{x}\right|$

Based on the preceding discussion, we have the formulas

$$r^2 = x^2 + y^2 \qquad \tan \theta = \frac{y}{x} \qquad \text{if } x \neq 0 \tag{2}$$

To use Formula (2) effectively, follow these steps:

STEPS for Converting from Rectangular to Polar Coordinates

STEP 1: Always plot the point (x, y) first, as we did in Examples 7 and 8.

STEP 2: To find r, compute the distance from the origin to (x, y).

STEP 3: To find θ, it is best to compute the reference angle α of θ, $\tan \alpha = \left|\dfrac{y}{x}\right|$, if $x \neq 0$, and then use your illustration to find θ.

Look again at Figure 18 and Examples 7 and 8.

Note: Most calculators have the capability of converting from rectangular coordinates to polar coordinates. Consult your owner's manual to learn the proper key strokes. Since in most cases this procedure is tedious, you will find that using formula (2) is faster.

 Now work Problem 49.

Formulas (1) and (2) may also be used to transform equations.

E X A M P L E   9 Transforming an Equation from Polar to Rectangular Form

Transform the equation $r = 4 \sin \theta$ from polar coordinates to rectangular coordinates, and identify the graph.

Solution If we multiply each side by r, it will be easier to apply formulas (1) and (2):

$$r = 4 \sin \theta$$
$$r^2 = 4r \sin \theta \qquad \text{Multiply each side by } r.$$
$$x^2 + y^2 = 4y \qquad \text{Apply formulas (1) and (2).}$$

This is the equation of a circle; we proceed to complete the square to obtain the standard form of the equation.

$$x^2 + (y^2 - 4y) = 0 \qquad \text{General form.}$$
$$x^2 + (y^2 - 4y + 4) = 4 \qquad \text{Complete the square in } y.$$
$$x^2 + (y - 2)^2 = 4 \qquad \text{Standard form.}$$

The center of the circle is at $(0, 2)$, and its radius is 2. ▬

 Now work Problem 65.

E X A M P L E   10 Transforming an Equation from Rectangular to Polar Form

Transform the equation $4xy = 9$ from rectangular coordinates to polar coordinates.

Solution We use formula (1).

$$4xy = 9$$
$$4(r \cos \theta)(r \sin \theta) = 9 \qquad \text{Formula (1).}$$
$$4r^2 \cos \theta \sin \theta = 9$$
$$2r^2 \sin 2\theta = 9 \qquad \text{Double-angle formula.}$$

▬

5.1 EXERCISES

In Problems 1–8, match each point in polar coordinates with either A, B, C, or D on the graph.

1. $\left(2, \dfrac{-11\pi}{6}\right)$ 2. $\left(-2, \dfrac{-\pi}{6}\right)$ 3. $\left(-2, \dfrac{\pi}{6}\right)$ 4. $\left(2, \dfrac{7\pi}{6}\right)$

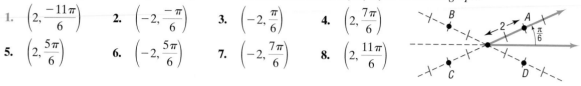

5. $\left(2, \dfrac{5\pi}{6}\right)$ 6. $\left(-2, \dfrac{5\pi}{6}\right)$ 7. $\left(-2, \dfrac{7\pi}{6}\right)$ 8. $\left(2, \dfrac{11\pi}{6}\right)$

In Problems 9–20, plot each point given in polar coordinates.

9. $(3, 90°)$ 　　　　　10. $(4, 270°)$ 　　　　　11. $(-2, 0)$ 　　　　　12. $(-3, \pi)$

13. $(6, \pi/6)$ 　　　　14. $(5, 5\pi/3)$ 　　　　15. $(-2, 135°)$ 　　　16. $(-3, 120°)$

17. $(-1, -\pi/3)$ 　　18. $(-3, -3\pi/4)$ 　　19. $(-2, -\pi)$ 　　　20. $(-3, -\pi/2)$

In Problems 21–28, plot each point given in polar coordinates, and find other polar coordinates (r, θ) of the point for which

(a) $r > 0$, $-2\pi \le \theta < 0$　　(b) $r < 0$, $0 \le \theta < 2\pi$　　(c) $r > 0$, $2\pi \le \theta < 4\pi$

21. $(5, 2\pi/3)$ 　　　　22. $(4, 3\pi/4)$ 　　　　23. $(-2, 3\pi)$ 　　　　24. $(-3, 4\pi)$

25. $(1, \pi/2)$ 　　　　26. $(2, \pi)$ 　　　　　27. $(-3, -\pi/4)$ 　　　28. $(-2, -2\pi/3)$

In Problems 29–44, polar coordinates of a point are given. Find the rectangular coordinates of each point.

29. $(3, \pi/2)$ 　　　　30. $(4, 3\pi/2)$ 　　　　31. $(-2, 0)$ 　　　　　32. $(-3, \pi)$

33. $(6, 150°)$ 　　　　34. $(5, 300°)$ 　　　　35. $(-2, 3\pi/4)$ 　　　36. $(-3, 2\pi/3)$

37. $(-1, -\pi/3)$ 　　38. $(-3, -3\pi/4)$ 　　39. $(-2, -180°)$ 　　40. $(-3, -90°)$

41. $(7.5, 110°)$ 　　　42. $(-3.1, 182°)$ 　　43. $(6.3, 3.8)$ 　　　44. $(8.1, 5.2)$

In Problems 45–56, the rectangular coordinates of a point are given. Find polar coordinates for each point.

45. $(3, 0)$ 　　　　　46. $(0, 2)$ 　　　　　47. $(-1, 0)$ 　　　　　48. $(0, -2)$

49. $(1, -1)$ 　　　　50. $(-3, 3)$ 　　　　51. $(\sqrt{3}, 1)$ 　　　　52. $(-2, -2\sqrt{3})$

53. $(1.3, -2.1)$ 　　54. $(-0.8, -2.1)$ 　　55. $(8.3, 4.2)$ 　　　56. $(-2.3, 0.2)$

In Problems 57–64, the letters x and y represent rectangular coordinates. Write each equation using polar coordinates (r, θ).

57. $2x^2 + 2y^2 = 3$ 　　58. $x^2 + y^2 = x$ 　　59. $x^2 = 4y$ 　　　60. $y^2 = 2x$

61. $2xy = 1$ 　　　　62. $4x^2y = 1$ 　　　63. $x = 4$ 　　　　64. $y = -3$

In Problems 65–72, the letters r and θ represent polar coordinates. Write each equation using rectangular coordinates (x, y).

65. $r = \cos\theta$ 　　66. $r = \sin\theta + 1$ 　　67. $r^2 = \cos\theta$ 　　68. $r = \sin\theta - \cos\theta$

69. $r = 2$ 　　　　70. $r = 4$ 　　　　71. $r = \dfrac{4}{1 - \cos\theta}$ 　　72. $r = \dfrac{3}{3 - \cos\theta}$

73. Show that the formula for the distance d between two points $P_1 = (r_1, \theta_1)$ and $P_2 = (r_2, \theta_2)$ is
$$d = \sqrt{r_1^2 + r_2^2 - 2r_1r_2\cos(\theta_2 - \theta_1)}$$

5.2 | POLAR EQUATIONS AND GRAPHS

1 Graph and Identify Polar Equations by Converting to Rectangular Equations
2 Test Polar Equations for Symmetry
3 Graph Polar Equations by Plotting Points

> An equation whose variables are polar coordinates is called a **polar equation.** The **graph of a polar equation** consists of all points whose polar coordinates satisfy the equation.

Just as a rectangular grid may be used to plot points given by rectangular coordinates, as in Figure 19(a), we can use a grid consisting of concentric circles (with centers at the pole) and rays (with vertices at the pole) to plot points given by polar coordinates, as shown in Figure 19(b). We shall use such **polar grids** to graph polar equations.

FIGURE 19

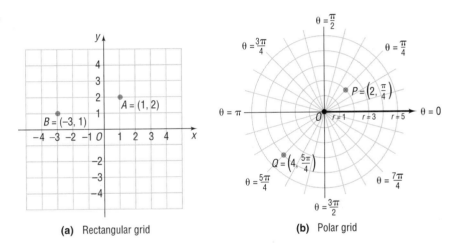

(a) Rectangular grid

(b) Polar grid

1 One method we can use to graph a polar equation is to convert the equation to rectangular coordinates. In the discussion that follows, (x, y) represent the rectangular coordinates of a point P and (r, θ) represent polar coordinates of the point P.

EXAMPLE 1   Identifying and Graphing a Polar Equation (Circle)

Identify and graph the equation: $r = 3$

Solution   We convert the polar equation to a rectangular equation.

$$r = 3$$
$$r^2 = 9$$
$$x^2 + y^2 = 9 \quad \text{Formula (2), Section 5.1, p. 309}$$

Thus, the graph of $r = 3$ is a circle, with center at the pole and radius 3. See Figure 20.

FIGURE 20
$r = 3$ or $x^2 + y^2 = 9$

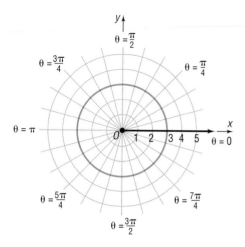

 Now work Problem 1.

E X A M P L E   2   Identifying and Graphing a Polar Equation (Line)

Identify and graph the equation: $\theta = \pi/4$

Solution   We convert the polar equation to a rectangular equation.

$$\theta = \frac{\pi}{4}$$

$$\tan \theta = \tan \frac{\pi}{4} = 1$$

$$\frac{y}{x} = 1 \qquad \text{Formula (2), Section 5.1, p. 309.}$$

$$y = x$$

The graph of $\theta = \pi/4$ is a line passing through the pole making an angle of $\pi/4$ with the polar axis. See Figure 21.

FIGURE 21
$\theta = \frac{\pi}{4}$ or $y = x$

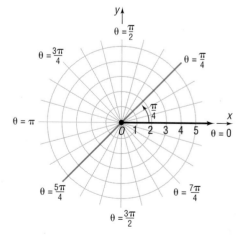

E X A M P L E   3   Identifying and Graphing a Polar Equation (Line)

Identify and graph the equation: $\theta = -\pi/6$

Solution   We convert the polar equation to a rectangular equation.

$$\theta = -\frac{\pi}{6}$$

$$\tan \theta = \tan\left(-\frac{\pi}{6}\right) = -\frac{\sqrt{3}}{3}$$

$$\frac{y}{x} = -\frac{\sqrt{3}}{3}$$

$$y = -\frac{\sqrt{3}}{3}x$$

The graph of $\theta = -\pi/6$ is a line passing through the pole making an angle of $-\pi/6$ with the polar axis. See Figure 22.

FIGURE 22
$$\theta = -\frac{\pi}{6} \text{ or } y = -\frac{\sqrt{3}}{3}x$$

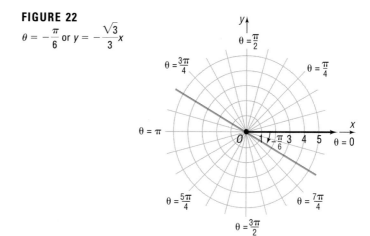

 Now work Problem 3.

E X A M P L E   4   Identifying and Graphing a Polar Equation (Horizontal Line)

Identify and graph the equation: $r \sin \theta = 2$

Solution   Since $y = r \sin \theta$ [Formula (1), p. 306], we can simply write the equation as

$$y = 2$$

We conclude that the graph of $r \sin \theta = 2$ is a horizontal line 2 units above the pole. See Figure 23.

FIGURE 23
$r \sin \theta = 2$ or $y = 2$

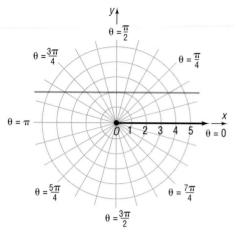

 Comment A graphing utility can be used to graph polar equations. Read Using a Graphing Utility to Graph Polar Equations, Appendix B, page 562.

E X A M P L E   5   Identifying and Graphing a Polar Equation (Vertical Line)

Identify and graph the equation: $r \cos \theta = -3$

Solution   Since $x = r \cos \theta$ [Formula (1), p. 306], we can simply write the equation as

$$x = -3$$

We conclude that the graph of $r \cos \theta = -3$ is a vertical line 3 units to the left of the pole. Figure 24 shows the graph.

FIGURE 24
$r \cos \theta = -3$ or $x = -3$

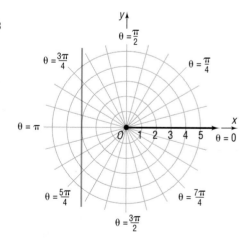

 Check   Graph $r = -3/\cos \theta$ using $\theta \min = 0$, $\theta \max = 2\pi$, and θ step $= \pi/24$. Compare the result to Figure 24.

Based on Examples 4 and 5, we are led to the following results. (The proofs are left as exercises.)

Theorem

Let a be a nonzero real number. Then the graph of the equation

$$r \sin \theta = a$$

is a horizontal line a units above the pole if $a > 0$ and $|a|$ units below the pole if $a < 0$.

The graph of the equation

$$r \cos \theta = a$$

is a vertical line a units to the right of the pole if $a > 0$ and $|a|$ units to the left of the pole if $a < 0$.

 Now work Problem 7.

E X A M P L E 6   Identifying and Graphing a Polar Equation (Circle)

Identify and graph the equation: $r = 4 \sin \theta$

Solution   To transform the equation to rectangular coordinates, we multiply each side by r.

$$r^2 = 4r \sin \theta$$

Now we use the facts that $r^2 = x^2 + y^2$ and $y = r \sin \theta$. Then

$$
\begin{aligned}
x^2 + y^2 &= 4y \\
x^2 + (y^2 - 4y) &= 0 \\
x^2 + (y^2 - 4y + 4) &= 4 \qquad \text{Complete the square in } y. \\
x^2 + (y - 2)^2 &= 4 \qquad \text{Standard equation of a circle.}
\end{aligned}
$$

This is the equation of a circle with center at $(0, 2)$ in rectangular coordinates and radius 2. Figure 25 shows the graph.

FIGURE 25
$r = 4 \sin \theta$ or $x^2 + (y - 2)^2 = 4$

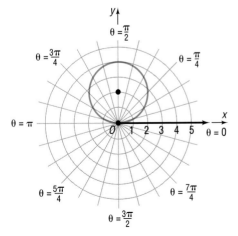

E X A M P L E   7   Identifying and Graphing a Polar Equation (Circle)

Identify and graph the equation: $r = -2 \cos \theta$

Solution   We proceed as in Example 6.

$$r^2 = -2r \cos \theta \quad \text{Multiply both sides by } r.$$
$$x^2 + y^2 = -2x$$
$$x^2 + 2x + y^2 = 0$$
$$(x^2 + 2x + 1) + y^2 = 1 \qquad \text{Complete the square in } x.$$
$$(x + 1)^2 + y^2 = 1 \qquad \text{Standard equation of a circle.}$$

This is the equation of a circle with center at $(-1, 0)$ in rectangular coordinates and radius 1. Figure 26 shows the graph.

FIGURE 26
$r = -2 \cos \theta$ or $(x + 1)^2 + y^2 = 1$

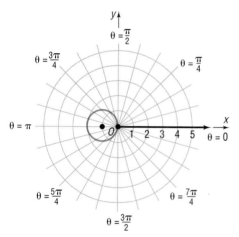

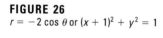 **Check:**   Graph $r = 4 \sin \theta$ and compare the result with Figure 25. Clear the screen and do the same for $r = -2 \cos \theta$ and compare with Figure 26. Be sure to use a square screen.

Exploration   Using a square screen, graph $r_1 = \sin \theta$, $r_2 = 2 \sin \theta$, and $r_3 = 3 \sin \theta$. Do you see the pattern? Clear the screen and graph $r_1 = -\sin \theta$, $r_2 = -2 \sin \theta$, and $r_3 = -3 \sin \theta$. Do you see the pattern? Clear the screen and graph $r_1 = \cos \theta$, $r_2 = 2 \cos \theta$, and $r_3 = 3 \cos \theta$. Do you see the pattern? Clear the screen and graph $r_1 = -\cos \theta$, $r_2 = -2 \cos \theta$, and $r_3 = -3 \cos \theta$. Do you see the pattern?

Based on Examples 6 and 7, we are led to the following results. (The proofs are left as exercises.)

Theorem

Let a be a positive real number. Then,

Equation	Description
(a) $r = 2a \sin \theta$	Circle: radius a; center at $(0, a)$ in rectangular coordinates
(b) $r = -2a \sin \theta$	Circle: radius a; center at $(0, -a)$ in rectangular coordinates
(c) $r = 2a \cos \theta$	Circle: radius a; center at $(a, 0)$ in rectangular coordinates
(d) $r = -2a \cos \theta$	Circle: radius a; center at $(-a, 0)$ in rectangular coordinates

Each circle passes through the pole.

Now work Problem 9.

The method of converting a polar equation to an identifiable rectangular equation in order to graph it is not always helpful, nor is it always necessary. Usually, we set up a table that lists several points on the graph. By checking for symmetry, it may be possible to reduce the number of points needed to draw the graph.

Symmetry

In polar coordinates the points (r, θ) and $(r, -\theta)$ are symmetric with respect to the polar axis (and to the x-axis). See Figure 27(a). The points (r, θ) and $(r, \pi - \theta)$ are symmetric with respect to the line $\theta = \pi/2$ (y-axis). See Figure 27(b). The points (r, θ) and $(-r, \theta)$ are symmetric with respect to the pole (origin). See Figure 27(c) on page 318.

FIGURE 27

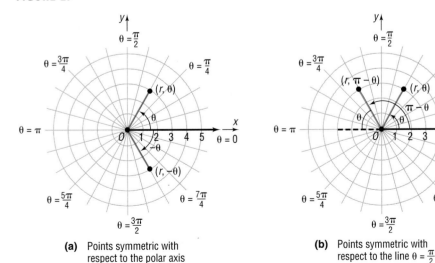

(a) Points symmetric with respect to the polar axis

(b) Points symmetric with respect to the line $\theta = \frac{\pi}{2}$

FIGURE 27
(continued)

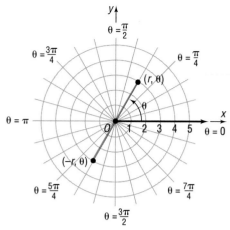

(c) Points symmetric with respect to the pole

The following tests are a consequence of these observations.

Theorem Tests for Symmetry

Symmetry with Respect to the Polar Axis (*x*-Axis):
In a polar equation, replace θ by $-\theta$. If an equivalent equation results, the graph is symmetric with respect to the polar axis.

Symmetry with Respect to the Line $\theta = \pi/2$ (*y*-Axis):
In a polar equation, replace θ by $\pi - \theta$. If an equivalent equation results, the graph is symmetric with respect to the line $\theta = \pi/2$.

Symmetry with Respect to the Pole (Origin):
In a polar equation, replace r by $-r$. If an equivalent equation results, the graph is symmetric with respect to the pole.

The three tests for symmetry given here are *sufficient* conditions for symmetry, but they are not *necessary* conditions. That is, an equation may fail these tests and still have a graph that is symmetric with respect to the polar axis, the line $\theta = \pi/2$, or the pole. For example, the graph of $r = \sin 2\theta$ turns out to be symmetric with respect to the polar axis, the line $\theta = \pi/2$, and the pole, but all three tests given here fail. See also Problems 65, 66, and 67 in Exercise 5.2.

E X A M P L E   8   Graphing a Polar Equation (Cardioid)

3   Graph the equation: $r = 1 - \sin \theta$

Solution   We check for symmetry first.

Polar Axis: Replace θ by $-\theta$. The result is

$$r = 1 - \sin(-\theta) = 1 + \sin\theta$$

The test fails, so the graph may or may not be symmetric with respect to the polar axis.

The Line $\theta = \pi/2$: Replace θ by $\pi - \theta$. The result is

$$\begin{aligned} r = 1 - \sin(\pi - \theta) &= 1 - (\sin\pi\cos\theta - \cos\pi\sin\theta) \\ &= 1 - [0 \cdot \cos\theta - (-1)\sin\theta] = 1 - \sin\theta \end{aligned}$$

Thus, the graph is symmetric with respect to the line $\theta = \pi/2$.

The Pole: Replace r by $-r$. Then the result is $-r = 1 - \sin\theta$, so $r = -1 + \sin\theta$. The test fails, so the graph may or may not be symmetric with respect to the pole.

Next, we identify points on the graph by assigning values to the angle θ and calculating the corresponding values of r. Due to the symmetry with respect to the line $\theta = \pi/2$, we only need to assign values to θ from $-\pi/2$ to $\pi/2$, as given in Table 1.

Now we plot the points (r, θ) from Table 1 and trace out the graph, beginning at the point $(2, -\pi/2)$ and ending at the point $(0, \pi/2)$. Then we reflect this portion of the graph about the line $\theta = \pi/2$ (y-axis) to obtain the complete graph. Figure 28 shows the graph.

TABLE 1	
θ	$r = 1 - \sin\theta$
$-\pi/2$	$1 + 1 = 2$
$-\pi/3$	$1 + \sqrt{3}/2 \approx 1.87$
$-\pi/6$	$1 + \frac{1}{2} = \frac{3}{2}$
0	1
$\pi/6$	$1 - \frac{1}{2} = \frac{1}{2}$
$\pi/3$	$1 - \sqrt{3}/2 \approx 0.13$
$\pi/2$	0

FIGURE 28
$r = 1 - \sin\theta$

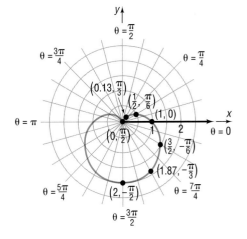

Exploration   Graph $r_1 = 1 + \sin\theta$. Clear the screen and graph $r_1 = 1 - \cos\theta$. Clear the screen and graph $r_1 = 1 + \cos\theta$. Do you see a pattern? ∎

The curve in Figure 28 is an example of a *cardioid* (a heart-shaped curve).

Cardioids are characterized by equations of the form

$r = a(1 + \cos \theta)$	$r = a(1 + \sin \theta)$
$r = a(1 - \cos \theta)$	$r = a(1 - \sin \theta)$

where $a > 0$. The graph of a cardioid passes through the pole.

 Now work Problem 25.

E X A M P L E 9   Graphing a Polar Equation (Limaçon without Inner Loop)

Graph the equation: $r = 3 + 2 \cos \theta$

Solution   We check for symmetry first.

Polar Axis: Replace θ by $-\theta$. The result is

$$r = 3 + 2 \cos(-\theta) = 3 + 2 \cos \theta$$

Thus, the graph is symmetric with respect to the polar axis.

The Line $\theta = \pi/2$: Replace θ by $\pi - \theta$. The result is

$$r = 3 + 2 \cos(\pi - \theta) = 3 + 2(\cos \pi \cos \theta + \sin \pi \sin \theta)$$
$$= 3 - 2 \cos \theta$$

The test fails, so the graph may or may not be symmetric with respect to the line $\theta = \pi/2$.

The Pole: Replace r by $-r$. The test fails, so the graph may or may not be symmetric with respect to the pole.

Next, we identify points on the graph by assigning values to the angle θ and calculating the corresponding values of r. Due to the symmetry with respect to the polar axis, we only need to assign values to θ from 0 to π, as given in Table 2.

Now we plot the points (r, θ) from Table 2 and trace out the graph, beginning at the point $(5, 0)$ and ending at the point $(1, \pi)$. Then we reflect this portion of the graph about the polar axis (x-axis) to obtain the complete graph. Figure 29 shows the graph.

TABLE 2	
θ	$r = 3 + 2 \cos \theta$
0	5
$\pi/6$	$3 + \sqrt{3} \approx 4.73$
$\pi/3$	4
$\pi/2$	3
$2\pi/3$	2
$5\pi/6$	$3 - \sqrt{3} \approx 1.27$
π	1

FIGURE 29
$r = 3 + 2 \cos \theta$

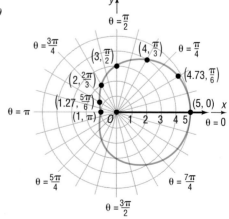

Exploration Graph $r_1 = 3 - 2 \cos \theta$. Clear the screen and graph $r_1 = 3 + 2 \sin \theta$. Clear the screen and graph $r_1 = 3 - 2 \sin \theta$. Do you see a pattern? ▬

The curve in Figure 29 is an example of a *limaçon* (the French word for *snail*) without an inner loop.

Limaçons without an inner loop are characterized by equations of the form

$$r = a + b \cos \theta \qquad r = a + b \sin \theta$$
$$r = a - b \cos \theta \qquad r = a - b \sin \theta$$

where $a > 0$, $b > 0$, and $a > b$. The graph of a limaçon without an inner loop does not pass through the pole.

Now work Problem 31.

E X A M P L E   10

Graphing a Polar Equation (Limaçon with Inner Loop)

Graph the equation: $r = 1 + 2 \cos \theta$

Solution   First, we check for symmetry.

Polar Axis: Replace θ by $-\theta$. The result is

$$r = 1 + 2 \cos(-\theta) = 1 + 2 \cos \theta$$

Thus, the graph is symmetric with respect to the polar axis.

The Line $\theta = \pi/2$: Replace θ by $\pi - \theta$. The result is

$$r = 1 + 2 \cos(\pi - \theta) = 1 + 2(\cos \pi \cos \theta + \sin \pi \sin \theta)$$
$$= 1 - 2 \cos \theta$$

The test fails, so the graph may or may not be symmetric with respect to the line $\theta = \pi/2$.

The Pole: Replace r by $-r$. The test fails, so the graph may or may not be symmetric with respect to the pole.

Next, we identify points on the graph of $r = 1 + 2 \cos \theta$ by assigning values to the angle θ and calculating the corresponding values of r. Due to the symmetry with respect to the polar axis, we only need to assign values to θ from 0 to π, as given in Table 3.

TABLE 3

θ	$r = 1 + 2 \cos \theta$
0	3
$\pi/6$	$1 + \sqrt{3} \approx 2.73$
$\pi/3$	2
$\pi/2$	1
$2\pi/3$	0
$5\pi/6$	$1 - \sqrt{3} \approx -0.73$
π	-1

Now we plot the points (r, θ) from Table 3, beginning at $(3, 0)$ and ending at $(-1, \pi)$. See Figure 30(a). Finally, we reflect this portion of the graph about the polar axis (x-axis) to obtain the complete graph. See Figure 30(b).

FIGURE 30

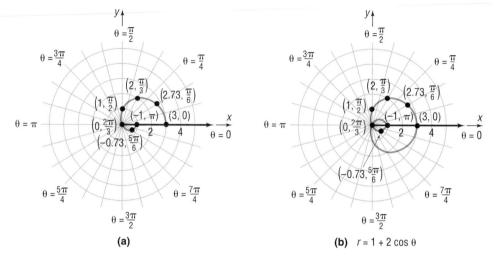

(a)   (b) $r = 1 + 2\cos\theta$

 Exploration Graph $r_1 = 1 - 2\cos\theta$. Clear the screen and graph $r_1 = 1 + 2\sin\theta$. Clear the screen and graph $r_1 = 1 - 2\sin\theta$. Do you see a pattern?

The curve in Figure 30(b) is an example of a limaçon with an inner loop.

Limaçons with an inner loop are characterized by equations of the form

$$r = a + b\cos\theta \qquad r = a + b\sin\theta$$
$$r = a - b\cos\theta \qquad r = a - b\sin\theta$$

where $a > 0$, $b > 0$, and $a < b$. The graph of a limaçon with an inner loop will pass through the pole twice.

 Now work Problem 33.

E X A M P L E   11   Graphing a Polar Equation (Rose)

Graph the equation: $r = 2\cos 2\theta$

Solution   We check for symmetry.

Polar Axis: If we replace θ by $-\theta$, the result is

$$r = 2\cos 2(-\theta) = 2\cos 2\theta$$

Thus, the graph is symmetric with respect to the polar axis.

The Line $\theta = \pi/2$: If we replace θ by $\pi - \theta$, we obtain

$$r = 2\cos 2(\pi - \theta) = 2\cos(2\pi - 2\theta) = 2\cos(-2\theta) = 2\cos 2\theta$$

Thus, the graph is symmetric with respect to the line $\theta = \pi/2$.

The Pole: Since the graph is symmetric with respect to both the polar axis and the line $\theta = \pi/2$, it must be symmetric with respect to the pole.

TABLE 4

θ	$r = 2 \cos 2\theta$
0	2
$\pi/6$	1
$\pi/4$	0
$\pi/3$	−1
$\pi/2$	−2

Next, we construct Table 4. Due to the symmetry with respect to the polar axis, the line $\theta = \pi/2$, and the pole, we consider only values of θ from 0 to $\pi/2$.

We plot and connect these points in Figure 31(a). Finally, because of symmetry, we reflect this portion of the graph first about the polar axis (x-axis) and then about the line $\theta = \pi/2$ (y-axis) to obtain the complete graph. See Figure 31(b).

FIGURE 31

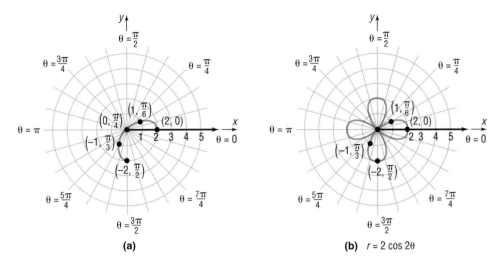

(a)

(b) $r = 2 \cos 2\theta$

 Exploration Graph $r = 2 \cos 2\theta$, and compare the result with Figure 31(b). Notice that there are four petals. Clear the screen and graph $r = 2 \cos 4\theta$; clear the screen and graph $r = 2 \cos 6\theta$. How many petals did each of these graphs have?
Clear the screen and graph, in order, each on a clear screen, $r = 2 \cos 3\theta$, $r = 2 \cos 5\theta$, and $r = 2 \cos 7\theta$. What do you notice about the number of petals?

The curve in Figure 31(b) is called a *rose* with four petals.

Rose curves are characterized by equations of the form

$$r = a \cos n\theta \qquad r = a \sin n\theta, \qquad a \neq 0$$

and have graphs that are rose shaped. If $n \neq 0$ is even, the rose has $2n$ petals; if n is odd, the rose has n petals.

Now work Problem 37.

E X A M P L E 12

Graphing a Polar Equation (Lemniscate)

Graph the equation: $r^2 = 4 \sin 2\theta$

Solution We leave it to you to verify that the graph is symmetric with respect to the pole. Table 5 lists points on the graph for values of $\theta = 0$ through $\theta = \pi/2$. Note that there are no points on the graph for $\pi/2 < \theta < \pi$ (quadrant II), since $\sin 2\theta < 0$ for such values. The points from Table 5 where $r \geq 0$ are plotted in Figure 32(a). The remaining points on the graph may be obtained by using symmetry. Figure 32(b) shows the final graph.

θ	$r^2 = 4\sin 2\theta$	r
0	0	0
$\pi/6$	$2\sqrt{3}$	± 1.9
$\pi/4$	4	± 2
$\pi/3$	$2\sqrt{3}$	± 1.9
$\pi/2$	0	0

TABLE 5

FIGURE 32

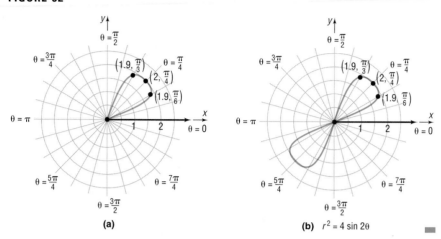

(a)

(b) $r^2 = 4\sin 2\theta$

The curve in Figure 32(b) is an example of a *lemniscate*.

Lemniscates are characterized by equations of the form

$$r^2 = a^2 \sin 2\theta \qquad r^2 = a^2 \cos 2\theta$$

where $a \neq 0$, and have graphs that are propeller shaped.

 Now work Problem 41.

E X A M P L E   13

Graphing a Polar Equation (Spiral)

Graph the equation: $r = e^{\theta/5}$

Solution The tests for symmetry with respect to the pole, the polar axis, and the line $\theta = \pi/2$ fail. Furthermore, there is no number θ for which $r = 0$. Hence, the graph does not pass through the pole. We observe that r is positive for all θ, r increases as θ increases, $r \to 0$ as $\theta \to -\infty$, and $r \to \infty$ as $\theta \to \infty$. With the help of a calculator, we obtain the values in Table 6. See Figure 33 for the graph.

The curve in Figure 33 is called a **logarithmic spiral,** since its equation may be written as $\theta = 5 \ln r$ and it spirals infinitely both toward the pole and away from it.

TABLE 6	
θ	$r = e^{\theta/5}$
$-3\pi/2$	0.39
$-\pi$	0.53
$-\pi/2$	0.73
$-\pi/4$	0.85
0	1
$\pi/4$	1.17
$\pi/2$	1.37
π	1.87
$3\pi/2$	2.57
2π	3.51

FIGURE 33
$r = e^{\theta/5}$

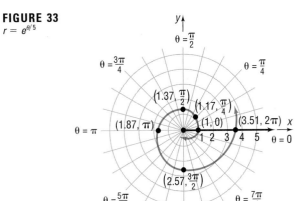

Classification of Polar Equations

The equations of some lines and circles in polar coordinates and their corresponding equations in rectangular coordinates are given in Table 7 on page 326. Also included are the names and the graphs of a few of the more frequently encountered polar equations.

Calculus Comment

For those of you who are planning to study calculus, a comment about one important role of polar equations is in order.

In rectangular coordinates, the equation $x^2 + y^2 = 1$, whose graph is the unit circle, is not the graph of a function. In fact, it requires two functions to obtain the graph of the unit circle:

$$y_1 = \sqrt{1 - x^2} \quad \text{Upper semicircle} \qquad y_2 = -\sqrt{1 - x^2} \quad \text{Lower semicircle}$$

In polar coordinates, the equation $r = 1$, whose graph is also the unit circle, does define a function. That is, for each choice of θ there is only one corresponding value of r, that is, $r = 1$. Since many uses of calculus require that functions be used, the opportunity to express nonfunctions in rectangular coordinates as functions in polar coordinates becomes extremely useful.

Note also that the vertical line test for functions is valid only for equations in rectangular coordinates.

HISTORICAL FEATURE Polar coordinates seem to have been invented by Jacob Bernoulli (1654–1705) about 1691, although, as with most such ideas, earlier traces of the notion exist. Early users of calculus remained committed to rectangular coordinates, and polar coordinates did not become widely used until the early 1800s. Even then, it was mostly geometers who used them for describing odd curves. Finally, about the mid-1800s, applied mathematicians realized the tremendous simplification that polar coordinates make possible in the description of objects with circular or cylindrical symmetry. From then on their use became widespread.

TABLE 7
Lines

Description	Line passing through the pole making an angle α with the polar axis	Vertical line	Horizontal line
Rectangular equation	$y = (\tan \alpha)x$	$x = a$	$y = b$
Polar equation	$\theta = \alpha$	$r \cos \theta = a$	$r \sin \theta = b$
Typical graph	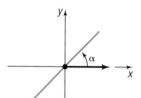		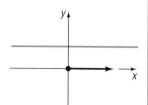

Circles

Description	Center at the pole, radius a	Passing through the pole, tangent to the line $\theta = \pi/2$, center on the polar axis, radius a	Passing through the pole, tangent to the polar axis, center on the line $\theta = \pi/2$, radius a
Rectangular equation	$x^2 + y^2 = a^2, a > 0$	$x^2 + y^2 = \pm 2ax, a > 0$	$x^2 + y^2 = \pm 2ay, a > 0$
Polar equation	$r = a, a > 0$	$r = \pm 2a \cos \theta, a > 0$	$r = \pm 2a \sin \theta, a > 0$
Typical graph	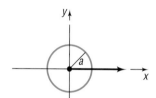	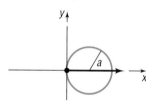	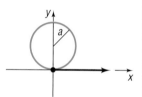

Other Equations

Name	Cardioid	Limaçon without inner loop	Limaçon with inner loop
Polar equations	$r = a \pm a \cos \theta, a > 0$ $r = a \pm a \sin \theta, a > 0$	$r = a \pm b \cos \theta, 0 < b < a$ $r = a \pm b \sin \theta, 0 < b < a$	$r = a \pm b \cos \theta, 0 < a < b$ $r = a \pm b \sin \theta, 0 < a < b$
Typical graph	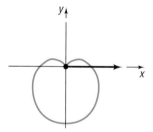	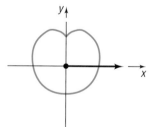	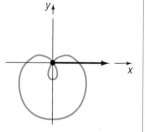

Name	Lemniscate	Rose with three petals	Rose with four petals
Polar equations	$r^2 = a^2 \cos 2\theta, a > 0$ $r^2 = a^2 \sin 2\theta, a > 0$	$r = a \sin 3\theta, a > 0$ $r = a \cos 3\theta, a > 0$	$r = a \sin 2\theta, a > 0$ $r = a \cos 2\theta, a > 0$
Typical graph	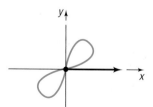	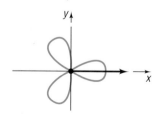	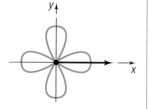

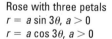

5.2 | EXERCISES

In Problems 1–16, identify each polar equation by transforming the equation to rectangular coordinates. Graph each polar equation.

1. $r = 4$
2. $r = 2$
3. $\theta = \pi/3$
4. $\theta = -\pi/4$

5. $r \sin \theta = 4$
6. $r \cos \theta = 4$
7. $r \cos \theta = -2$
8. $r \sin \theta = -2$

9. $r = 2 \cos \theta$
10. $r = 2 \sin \theta$
11. $r = -4 \sin \theta$
12. $r = -4 \cos \theta$

13. $r \sec \theta = 4$
14. $r \csc \theta = 8$
15. $r \csc \theta = -2$
16. $r \sec \theta = -4$

In Problems 17–24, match each of the graphs (A) through (H) to one of the following polar equations.

17. $r = 2$
18. $\theta = \pi/4$
19. $r = 2 \cos \theta$
20. $r \cos \theta = 2$

21. $r = 1 + \cos \theta$
22. $r = 2 \sin \theta$
23. $\theta = 3\pi/4$
24. $r \sin \theta = 2$

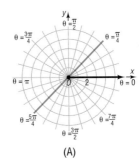

(A)

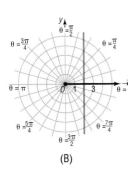

(B)

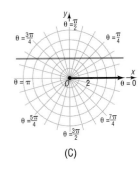

(C)

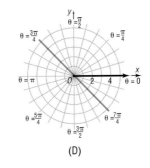

(D)

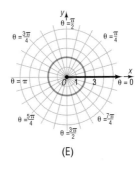

(E)

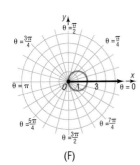

(F)

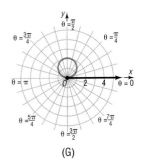

(G)

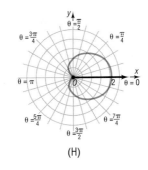

(H)

In Problems 25–48, identify and graph each polar equation. Be sure to test for symmetry.

25. $r = 2 + 2 \cos \theta$
26. $r = 1 + \sin \theta$
27. $r = 3 - 3 \sin \theta$
28. $r = 2 - 2 \cos \theta$

29. $r = 2 + \sin \theta$
30. $r = 2 - \cos \theta$
31. $r = 4 - 2 \cos \theta$
32. $r = 4 + 2 \sin \theta$

33. $r = 1 + 2 \sin \theta$
34. $r = 1 - 2 \sin \theta$
35. $r = 2 - 3 \cos \theta$
36. $r = 2 + 4 \cos \theta$

37. $r = 3 \cos 2\theta$
38. $r = 2 \sin 2\theta$
39. $r = 4 \sin 3\theta$
40. $r = 3 \cos 4\theta$

41. $r^2 = 9 \cos 2\theta$
42. $r^2 = \sin 2\theta$
43. $r = 2^\theta$
44. $r = 3^\theta$

45. $r = 1 - \cos \theta$
46. $r = 3 + \cos \theta$
47. $r = 1 - 3 \cos \theta$
48. $r = 4 \cos 3\theta$

In Problems 49–58, graph each polar equation.

49. $r = \dfrac{2}{1 - \cos\theta}$ (parabola)

50. $r = \dfrac{2}{1 - 2\cos\theta}$ (hyperbola)

51. $r = \dfrac{1}{3 - 2\cos\theta}$ (ellipse)

52. $r = \dfrac{1}{1 - \cos\theta}$ (parabola)

53. $r = \theta$, $\theta \geq 0$ (spiral of Archimedes)

54. $r = \dfrac{3}{\theta}$ (reciprocal spiral)

55. $r = \csc\theta - 2$, $0 < \theta < \pi$ (conchoid)

56. $r = \sin\theta\tan\theta$ (cissoid)

57. $r = \tan\theta$ (kappa curve), $-\dfrac{\pi}{2} < \theta < \dfrac{\pi}{2}$

58. $r = \cos\dfrac{\theta}{2}$

59. Show that the graph of the equation $r\sin\theta = a$ is a horizontal line a units above the pole, if $a > 0$ and $|a|$ units below the pole if $a < 0$.

60. Show that the graph of the equation $r\cos\theta = a$ is a vertical line a units to the right of the pole if $a > 0$ and $|a|$ units to the left of the pole if $a < 0$.

61. Show that the graph of the equation $r = 2a\sin\theta$, $a > 0$, is a circle of radius a with center at $(0, a)$ in rectangular coordinates.

62. Show that the graph of the equation $r = -2a\sin\theta$, $a > 0$, is a circle of radius a with center at $(0, -a)$ in rectangular coordinates.

63. Show that the graph of the equation $r = 2a\cos\theta$, $a > 0$, is a circle of radius a with center at $(a, 0)$ in rectangular coordinates.

64. Show that the graph of the equation $r = -2a\cos\theta$, $a > 0$, is a circle of radius a with center at $(-a, 0)$ in rectangular coordinates.

65. Explain why the following test for symmetry is valid: Replace r by $-r$ and θ by $-\theta$ in a polar equation. If an equivalent equation results, the graph is symmetric with respect to the line $\theta = \pi/2$ (y-axis).
  (a) Show that the test on page 318 fails for $r^2 = \cos\theta$, but this new test works.
  (b) Show that the test on page 318 works for $r^2 = \sin\theta$, yet this new test fails.

66. Develop a new test for symmetry with respect to the pole.
  (a) Find a polar equation for which this new test fails, yet the test on page 318 works.
  (b) Find a polar equation for which the test on page 318 fails, yet the new test works.

67. Write down two different tests for symmetry with respect to the polar axis. Find examples in which one works and the other fails. Which test do you prefer to use? Justify your answer.

5.3 COMPLEX NUMBERS

  1 Add, Subtract, Multiply, and Divide Complex Numbers
  2 Find the Square Root of a Negative Number

One property of a real number is that its square is nonnegative. For example, there is no real number x for which

$$x^2 = -1$$

To remedy this situation, we introduce a number called the **imaginary unit,** which we denote by i and whose square is -1. Thus,

$$i^2 = -1$$

This should not surprise you. If our universe were to consist only of integers, there would be no number x for which $2x = 1$. This unfortunate circumstance was remedied by introducing numbers such as $\frac{1}{2}$ and $\frac{2}{3}$, the *rational numbers.* If our universe were to consist only of rational numbers, there would be no number x whose square equals 2. That is, there would be no number x for which $x^2 = 2$. To remedy this, we introduced numbers such as $\sqrt{2}$ and $\sqrt[3]{5}$, the *irrational numbers.* The *real numbers,* you will recall, consist of the rational numbers and the irrational numbers. Now, if our universe were to consist only of real numbers, then there would be no number x whose square is -1. To remedy this, we introduce a number i, whose square is -1.

In the progression outlined, each time that we encountered a situation that was unsuitable, we introduced a new number system to remedy this situation. And, each new number system contained the earlier number system as a subset. The number system that results from introducing the number i is called the **complex number system.**

> **Complex numbers** are numbers of the form *a + bi,* where *a* and *b* are real numbers. The real number *a* is called the **real part** of the number *a + bi*; the real number *b* is called the **imaginary part** of *a + bi.*

For example, the complex number $-5 + 6i$ has the real part -5 and the imaginary part 6.

When a complex number is written in the form $a + bi$, where a and b are real numbers, we say it is in **standard form.** However, if the imaginary part of a complex number is negative, such as in the complex number $3 + (-2)i$, we agree to write it instead in the form $3 - 2i$.

Also, the complex number $a + 0i$ is usually written merely as a. This serves to remind us that the real numbers are a subset of the complex numbers. The complex number $0 + bi$ is usually written as bi. Sometimes the complex number bi is called a **pure imaginary number.**

1 Equality, addition, subtraction, and multiplication of complex numbers are defined so as to preserve the familiar rules of algebra for real numbers. Thus, two complex numbers are equal if and only if their real parts are equal and their imaginary parts are equal. That is,

Equality of Complex Numbers

$$a + bi = c + di \quad \text{if and only if} \quad a = c \text{ and } b = d \qquad (1)$$

Two complex numbers are added by forming the complex number whose real part is the sum of the real parts and whose imaginary part is the sum of the imaginary parts. That is,

Sum of Complex Numbers

$$(a + bi) + (c + di) = (a + c) + (b + d)i \qquad (2)$$

To subtract two complex numbers, we follow the rule

Difference of Complex Numbers

$$(a + bi) - (c + di) = (a - c) + (b - d)i \qquad (3)$$

E X A M P L E   1

Adding and Subtracting Complex Numbers

(a) $(3 + 5i) + (-2 + 3i) = [3 + (-2)] + (5 + 3)i = 1 + 8i$

(b) $(6 + 4i) - (3 + 6i) = (6 - 3) + (4 - 6)i = 3 + (-2)i = 3 - 2i$ ∎

Now work Problem 5.

Products of complex numbers are calculated as illustrated in Example 2.

E X A M P L E   2

Multiplying Complex Numbers

$$(5 + 3i) \cdot (2 + 7i) = 5 \cdot (2 + 7i) + 3i(2 + 7i) = 10 + 35i + 6i + 21i^2$$
$$\qquad\qquad\qquad\uparrow\qquad\qquad\qquad\qquad\uparrow$$
$$\qquad\qquad\text{Distributive property}\qquad\qquad\text{Distributive property}$$

$$= 10 + 41i + 21(-1)$$
$$\uparrow$$
$$i^2 = -1$$
$$= -11 + 41i \qquad ∎$$

Based on the procedure of Example 2, we define the **product** of two complex numbers by the formula

Product of Complex Numbers

$$(a + bi) \cdot (c + di) = (ac - bd) + (ad + bc)i \qquad (4)$$

Do not bother to memorize formula (4). Instead, whenever it is necessary to multiply two complex numbers, follow the usual rules for multiplying two binomials, as in Example 2, remembering that $i^2 = -1$. For example,

$$(2i)(2i) = 4i^2 = -4$$
$$(2 + i)(1 - i) = 2 - 2i + i - i^2 = 3 - i$$

Now work Problem 11.

Algebraic properties for addition and multiplication, such as the commutative, associative, and distributive properties, hold for complex numbers. Of these, the property that every nonzero complex number has a multiplicative inverse, or reciprocal, requires a closer look.

Conjugates

If $z = a + bi$ is a complex number, then its **conjugate**, denoted by $\bar{z}$ is defined as

$$\bar{z} = \overline{a + bi} = a - bi$$

For example, $\overline{2 + 3i} = 2 - 3i$ and $\overline{-6 - 2i} = -6 + 2i$.

E X A M P L E   3   Multiplying a Complex Number by Its Conjugate

Find the product of the complex number $z = 3 + 4i$ and its conjugate $\bar{z}$.

Solution   Since $\bar{z} = 3 - 4i$, we have

$$z\bar{z} = (3 + 4i)(3 - 4i) = 9 + 12i - 12i - 16i^2 = 9 + 16 = 25$$ ▪

The result obtained in Example 3 has an important generalization:

> **Theorem**
>
> The product of a complex number and its conjugate is a nonnegative real number. Thus, if $z = a + bi$, then
>
> $$z\bar{z} = a^2 + b^2 \qquad (5)$$

▪

Proof   If $z = a + bi$, then

$$z\bar{z} = (a + bi)(a - bi) = a^2 - (bi)^2 = a^2 - b^2i^2 = a^2 + b^2$$ ▪

To express the reciprocal of a nonzero complex number z in standard form, multiply the numerator and denominator by its conjugate $\bar{z}$. Thus, if $z = a + bi$ is a nonzero complex number, then

$$\frac{1}{a + bi} = \frac{1}{z} = \frac{1}{z} \cdot \frac{\bar{z}}{\bar{z}} = \frac{\bar{z}}{z\bar{z}} = \frac{a - bi}{(a + bi)(a - bi)}$$

$$= \frac{a - bi}{a^2 + b^2}$$

↑
Use (5).

$$= \frac{a}{a^2 + b^2} - \frac{b}{a^2 + b^2}i$$

E X A M P L E   4   Writing the Reciprocal of a Complex Number in Standard Form

Write $\dfrac{1}{3 + 4i}$ in standard form $a + bi$; that is, find the reciprocal of $3 + 4i$.

Solution   The idea is to multiply the numerator and denominator by the conjugate of $3 + 4i$, that is, the complex number $3 - 4i$. The result is

$$\frac{1}{3 + 4i} = \frac{1}{3 + 4i} \cdot \frac{3 - 4i}{3 - 4i} = \frac{3 - 4i}{9 + 16} = \frac{3}{25} - \frac{4}{25}i$$ ▪

To express the quotient of two complex numbers in standard form, we multiply the numerator and denominator of the quotient by the conjugate of the denominator.

EXAMPLE 5

Writing the Quotient of Complex Numbers in Standard Form

Write each of the following in standard form:

(a) $\dfrac{1 + 4i}{5 - 12i}$ (b) $\dfrac{2 - 3i}{4 - 3i}$

Solution (a) $\dfrac{1 + 4i}{5 - 12i} = \dfrac{1 + 4i}{5 - 12i} \cdot \dfrac{5 + 12i}{5 + 12i} = \dfrac{5 + 20i + 12i + 48i^2}{25 + 144}$

$= \dfrac{-43 + 32i}{169} = \dfrac{-43}{169} + \dfrac{32}{169}i$

(b) $\dfrac{2 - 3i}{4 - 3i} = \dfrac{2 - 3i}{4 - 3i} \cdot \dfrac{4 + 3i}{4 + 3i} = \dfrac{8 - 12i + 6i - 9i^2}{16 + 9} = \dfrac{17 - 6i}{25} = \dfrac{17}{25} - \dfrac{6}{25}i$

∎

Now work Problem 19.

EXAMPLE 6

Writing Other Expressions in Standard Form

If $z = 2 - 3i$ and $w = 5 + 2i$, write each of the following expressions in standard form:

(a) $\dfrac{z}{w}$ (b) $\overline{z + w}$ (c) $z + \overline{z}$

Solution (a) $\dfrac{z}{w} = \dfrac{z \cdot \overline{w}}{w \cdot \overline{w}} = \dfrac{(2 - 3i)(5 - 2i)}{(5 + 2i)(5 - 2i)} = \dfrac{10 - 15i - 4i + 6i^2}{25 + 4}$

$= \dfrac{4 - 19i}{29} = \dfrac{4}{29} - \dfrac{19}{29}i$

(b) $\overline{z + w} = \overline{(2 - 3i) + (5 + 2i)} = \overline{7 - i} = 7 + i$

(c) $z + \overline{z} = (2 - 3i) + (2 + 3i) = 4$

∎

The conjugate of a complex number has certain general properties that we shall find useful later.

For a real number $a = a + 0i$, the conjugate is $\overline{a} = \overline{a + 0i} = a - 0i = a$. That is,

Theorem

The conjugate of a real number is the real number itself.

∎

Other properties of the conjugate that are direct consequences of the definition are given next. In each statement, z and w represent complex numbers.

Theorem

The conjugate of the conjugate of a complex number is the complex number itself:

$$\overline{\overline{z}} = z \qquad (6)$$

The conjugate of the sum of two complex numbers equals the sum of their conjugates:

$$\overline{z + w} = \overline{z} + \overline{w} \qquad (7)$$

The conjugate of the product of two complex numbers equals the product of their conjugates:

$$\overline{z \cdot w} = \overline{z} \cdot \overline{w} \qquad (8)$$

We leave the proofs of equations (6), (7), and (8) as exercises.

Powers of i

The **powers of i** follow a pattern that is useful to know:

$$i^1 = i \qquad\qquad i^5 = i^4 \cdot i = 1 \cdot i = i$$
$$i^2 = -1 \qquad\qquad i^6 = i^4 \cdot i^2 = -1$$
$$i^3 = i^2 \cdot i = -i \qquad\qquad i^7 = i^4 \cdot i^3 = -i$$
$$i^4 = i^2 \cdot i^2 = (-1)(-1) = 1 \qquad i^8 = i^4 \cdot i^4 = 1$$

And so on. Thus, the powers of i repeat with every fourth power.

E X A M P L E   7   Evaluating Powers of i

(a) $i^{27} = i^{24} \cdot i^3 = (i^4)^6 \cdot i^3 = 1^6 \cdot i^3 = -i$

(b) $i^{101} = i^{100} \cdot i^1 = (i^4)^{25} \cdot i = 1^{25} \cdot i = i$

E X A M P L E   8   Writing the Power of a Complex Number in Standard Form

Write $(2 + i)^3$ in standard form.

Solution   We use the special product formula for $(x + a)^3$:

$$(x + a)^3 = x^3 + 3ax^2 + 3a^2x + a^3$$

Thus,

$$(2 + i)^3 = 2^3 + 3 \cdot i \cdot 2^2 + 3 \cdot i^2 \cdot 2 + i^3$$
$$= 8 + 12i + 6(-1) + (-i)$$
$$= 2 + 11i$$

Now work Problem 33.

The Square Root of a Negative Number

If N is a positive real number, we define the **principal square root of** $-N$, denoted by $\sqrt{-N}$, as

$$\sqrt{-N} = \sqrt{N}\,i$$

where i is the imaginary unit and $i^2 = -1$.

E X A M P L E   9   **2**   Evaluating the Square Root of a Negative Number

(a) $\sqrt{-1} = \sqrt{1}\,i = i$ (b) $\sqrt{-4} = \sqrt{4}\,i = 2i$
(c) $\sqrt{-8} = \sqrt{8}\,i = 2\sqrt{2}\,i$ ▬

E X A M P L E   10   Solving Equations

Solve each equation in the complex number system.

(a) $x^2 = 4$ (b) $x^2 = -9$

Solution   (a) $x^2 = 4$
$$x = \pm\sqrt{4} = \pm 2$$

The equation has two solutions, -2 and 2.

(b) $x^2 = -9$
$$x = \pm\sqrt{-9} = \pm\sqrt{9}\,i = \pm 3i$$

The equation has two solutions, $-3i$ and $3i$.   ▬

Now work Problem 45.

Warning   When working with square roots of negative numbers, do not set the square root of a product equal to the product of the square roots (which can be done with positive numbers). To see why, look at this calculation: We know that $\sqrt{100} = 10$. However, it is also true that $100 = (-25)(-4)$, so

$$10 = \sqrt{100} = \sqrt{(-25)(-4)} = \underset{\uparrow}{\sqrt{-25}}\sqrt{-4} = (\sqrt{25}i)(\sqrt{4}i) = (5i)(2i) = 10i^2 = -10$$

Here is the error.

5.3 | EXERCISES

In Problems 1–38, write each expression in the standard form a + bi.

1. $(2 - 3i) + (6 + 8i)$ **2.** $(4 + 5i) + (-8 + 2i)$ **3.** $(-3 + 2i) - (4 - 4i)$ **4.** $(3 - 4i) - (-3 - 4i)$

5. $(2 - 5i) - (8 + 6i)$ **6.** $(-8 + 4i) - (2 - 2i)$ **7.** $3(2 - 6i)$ **8.** $-4(2 + 8i)$

9. $2i(2 - 3i)$ **10.** $3i(-3 + 4i)$ **11.** $(3 - 4i)(2 + i)$ **12.** $(5 + 3i)(2 - i)$

13. $(-6 + i)(-6 - i)$ **14.** $(-3 + i)(3 + i)$ **15.** $\dfrac{10}{3 - 4i}$ **16.** $\dfrac{13}{5 - 12i}$

17. $\dfrac{2 + i}{i}$ **18.** $\dfrac{2 - i}{-2i}$ **19.** $\dfrac{6 - i}{1 + i}$ **20.** $\dfrac{2 + 3i}{1 - i}$

21. $\left(\dfrac{1}{2} + \dfrac{\sqrt{3}}{2}i\right)^2$ **22.** $\left(\dfrac{\sqrt{3}}{2} - \dfrac{1}{2}i\right)^2$ **23.** $(1 + i)^2$ **24.** $(1 - i)^2$

25. i^{23} **26.** i^{14} **27.** i^{-15} **28.** i^{-23}

29. $i^6 - 5$ **30.** $4 + i^3$ **31.** $6i^3 - 4i^5$ **32.** $4i^3 - 2i^2 + 1$

33. $(1 + i)^3$ **34.** $(3i)^4 + 1$ **35.** $i^7(1 + i^2)$ **36.** $2i^4(1 + i^2)$

37. $i^6 + i^4 + i^2 + 1$ **38.** $i^7 + i^5 + i^3 + i$

In Problems 39–44, perform the indicated operations and express your answer in the form a + bi.

39. $\sqrt{-4}$ **40.** $\sqrt{-9}$ **41.** $\sqrt{-25}$

42. $\sqrt{-64}$ **43.** $\sqrt{(3 + 4i)(4i - 3)}$ **44.** $\sqrt{(4 + 3i)(3i - 4)}$

In Problems 45–48, solve each equation in the complex number system.

45. $x^2 + 4 = 0$ **46.** $x^2 - 4 = 0$ **47.** $x^2 - 16 = 0$ **48.** $x^2 + 25 = 0$

In Problems 49–52, z = 3 − 4i and w = 8 + 3i. Write each expression in the standard form a + bi.

49. $z + \bar{z}$ **50.** $w - \bar{w}$ **51.** $z\bar{z}$ **52.** $\overline{z - w}$

53. Use $z = a + bi$ to show that $z + \bar{z} = 2a$ and that $z - \bar{z} = 2bi$.

54. Use $z = a + bi$ to show that $\bar{\bar{z}} = z$.

55. Use $z = a + bi$ and $w = c + di$ to show that $\overline{z + w} = \bar{z} + \bar{w}$.

56. Use $z = a + bi$ and $w = c + di$ to show that $\overline{z \cdot w} = \bar{z} \cdot \bar{w}$.

 57. Explain to a friend how you would add two complex numbers and how you would multiply two complex numbers. Explain any differences in the two explanations.

58. Write a brief paragraph that compares the method used to rationalize the denominator of a rational expression and the method used to write a complex number in standard form.

5.4 | THE COMPLEX PLANE; DE MOIVRE'S THEOREM

1 Convert a Complex Number from Rectangular Form to Polar Form
2 Plot Points in the Complex Plane
3 Find Products and Quotients of Complex Numbers in Polar Form
4 Use De Moivre's Theorem
5 Find Complex Roots

A complex number $z = x + yi$ can be interpreted geometrically as the point (x, y) in the xy-plane. Thus, each point in the plane corresponds to a

FIGURE 34
Complex plane

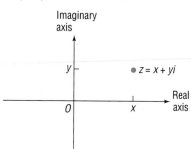

complex number and, conversely, each complex number corresponds to a point in the plane. We shall refer to the collection of such points as the **complex plane.** The x-axis will be referred to as the **real axis,** because any point that lies on the real axis is of the form $z = x + 0i = x$, a real number. The y-axis is called the **imaginary axis,** because any point that lies on it is of the form $z = 0 + yi = yi$, a pure imaginary number. See Figure 34.

Let $z = x + yi$ be a complex number. The **magnitude** or **modulus** of z, denoted by $|z|$, is defined as the distance from the origin to the point (x, y). Thus,

$$|z| = \sqrt{x^2 + y^2} \tag{1}$$

FIGURE 35

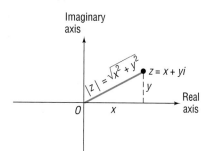

See Figure 35 for an illustration.

This definition for $|z|$ is consistent with the definition for the absolute value of a real number: If $z = x + yi$ is real, then $z = x + 0i$ and

$$|z| = \sqrt{x^2 + 0^2} = \sqrt{x^2} = |x|$$

Recall that if $z = x + yi$ then its conjugate, denoted by $\bar{z}$, is $\bar{z} = x - yi$. Because $z\bar{z} = x^2 + y^2$, it follows from equation (1) that the magnitude of z can be written as

$$|z| = \sqrt{z\bar{z}} \tag{2}$$

Polar Form of a Complex Number

1

When a complex number is written in the standard form $z = x + yi$, we say that it is in **rectangular,** or **Cartesian, form,** because (x, y) are the rectangular coordinates of the corresponding point in the complex plane. Suppose that (r, θ) are the polar coordinates of this point. Then

$$x = r \cos \theta \qquad y = r \sin \theta \tag{3}$$

If $r \geq 0$ and $0 \leq \theta < 2\pi$, the complex number $z = x + yi$ may be written in **polar form** as

FIGURE 36

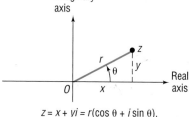

$$z = x + yi = (r \cos \theta) + (r \sin \theta)i = r(\cos \theta + i \sin \theta) \tag{4}$$

See Figure 36.

If $z = r(\cos \theta + i \sin \theta)$ is the polar form of a complex number, the angle θ, $0 \leq \theta < 2\pi$, is called the **argument of z.** Also, because $r \geq 0$, from equation (3) we have $r = \sqrt{x^2 + y^2}$. Thus, from equation (1) it follows that the magnitude of $z = r(\cos \theta + i \sin \theta)$ is

Mapping Indianapolis

Supplies needed: Graph paper with squares 1/4 inch or 1 cm. or larger per side; compass, protractor, ruler; graphing calculator; map of Indianapolis (from any atlas)

Unlike many major American cities which grew up next to large bodies of water and hence were limited in their ability to spread out in all directions, Indianapolis is surrounded by land on all sides. Major roads radiate out from the city center and intersect the ring road (Rt. 465). A map of the city looks somewhat like a rectangular grid overlaying a polar coordinate system.

The Mission Possible Team has been called to Indianapolis by a car rental company that wishes to create a computer system of interactive maps to be installed in their vehicles. The customers renting their cars would be able to indicate to the computer where they would like to go and the computer screen would then show them a map giving them the quickest route from their location to their chosen destination. Today's project represents the beginning of a study to create such a computer system.

As a first step, your team needs to set up a polar coordinate system. Using a simplified model, we designate the intersection of Meridian St. and Tenth St. as the pole. The ring road is (roughly) a circle 7 miles out from the center.

1. On graph paper sketch a large polar model of the streets of Indianapolis, using 1 square of the graph paper to represent 1 mile. The ring road would be represented by a circle with radius 7 miles. Rt. 40 is the road that intersects the ring road due east of the pole, so we will designate Rt. 40 as 0°. The other roads and their approximate angle from Rt. 40 are as follows:

Rt. 36: 30°	Rt. 65: 135°	Mann Rd.: 240°
Rt. 37: 60°	Rt. 74: 165°	Rt. 31 (Meridian St.): 270°
Rt. 31 (Meridian St.): 90°	Tenth St.: 180°	Rt. 65: 300°
Rt. 421: 105°	Rt. 70: 210°	Rt. 421: 330°

2. All but 3 of these angles are multiples of 30°. What is the greatest common factor of *all* the angles?

3. In order to establish a mathematical structure for the computer map, your team decides to create a polar equation of the petalled rose variety that would pass through all these points of intersection with the ring road and connect each to the center. What would the equation be?

4. Sketch the graph of your rose on the polar map you created for question 1. By using your graphing calculator, you can determine the order in which the petals are drawn. Number the petals on your sketch to indicate that order.

5. Find the *xy* equation determined by taking your rose equation and replacing the *r* by *y* and the θ by *x*. (Example: If your polar equation were $r = 5 \cos 7\theta$, your corresponding rectangular equation would be $y = 5 \cos 7x$. Warning: These two equations are not equivalent in the sense that they will not give the same graph.) Sketch the resulting rectangular equation in the *xy*-plane, using the domain $0 \le x \le 2\pi/3$. Label the maximum and minimum points with the numbers that correspond to the rose petals of your polar graph.

6. If your original sketch for question 1 was done on graph paper, you can now see the lines of the graph paper as representing a rectangular grid overlaying the polar graph. If not, you will need to sketch vertical and horizontal lines on your polar graph to represent the major city streets. Again, for purposes of simplifying the problem, the major streets are assumed to be occurring at 1 mile intervals.

7. Suppose that travel on the Interstate (Rt. 465, the ring road) has an average speed of 55 mph and travel on city streets has an average speed of 20 mph. Estimate the times it would take to go from the intersection of Rt. 74 and Rt. 465 to the intersection of Rt. 36 and Rt. 465 by each route. (Recall that traveling on city streets limits you to vertical and horizontal grid lines.)

8. What would be the longest time it would take to go from any intersection to another on Rt. 465, assuming you chose the shortest distance along the circle? How long would it take to go between those same two points using the city streets?

$$|z| = r$$

E X A M P L E   1

Plotting a Point in the Complex Plane and Writing a Complex Number in Polar Form

Plot the point corresponding to $z = \sqrt{3} - i$ in the complex plane, and write an expression for z in polar form.

Solution

FIGURE 37

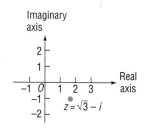

The point corresponding to $z = \sqrt{3} - i$ has the rectangular coordinates $(\sqrt{3}, -1)$. The point, located in quadrant IV, is plotted in Figure 37. Because $x = \sqrt{3}$ and $y = -1$, it follows that

$$r = \sqrt{x^2 + y^2} = \sqrt{(\sqrt{3})^2 + (-1)^2} = \sqrt{4} = 2$$

and

$$\sin \theta = \frac{y}{r} = \frac{-1}{2} \qquad \cos \theta = \frac{x}{r} = \frac{\sqrt{3}}{2}, \qquad 0 \le \theta < 2\pi$$

Thus, $\theta = 11\pi/6$ and $r = 2$; so the polar form of $z = \sqrt{3} - i$ is

$$z = r(\cos \theta + i \sin \theta) = 2\left(\cos \frac{11\pi}{6} + i \sin \frac{11\pi}{6}\right)$$

Now work Problem 1.

E X A M P L E   2

Plotting a Point in the Complex Plane and Converting from Polar to Rectangular Form

Plot the point corresponding to $z = 2(\cos 30° + i \sin 30°)$ in the complex plane, and write an expression for z in rectangular form.

Solution

FIGURE 38

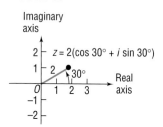

To plot the complex number $z = 2(\cos 30° + i \sin 30°)$, we plot the point whose polar coordinates are $(r, \theta) = (2, 30°)$, as shown in Figure 38. In rectangular form,

$$z = 2(\cos 30° + i \sin 30°) = 2\left(\frac{\sqrt{3}}{2} + \frac{1}{2}i\right) = \sqrt{3} + i$$

Now work Problem 13.

The polar form of a complex number provides an alternative for finding products and quotients of complex numbers.

> **Theorem**
>
> Let $z_1 = r_1(\cos \theta_1 + i \sin \theta_1)$ and $z_2 = r_2(\cos \theta_2 + i \sin \theta_2)$ be two complex numbers. Then
>
> $$z_1 z_2 = r_1 r_2 [\cos(\theta_1 + \theta_2) + i \sin(\theta_1 + \theta_2)] \tag{5}$$
>
> If $z_2 \neq 0$, then
>
> $$\frac{z_1}{z_2} = \frac{r_1}{r_2} [\cos(\theta_1 - \theta_2) + i \sin(\theta_1 - \theta_2)] \tag{6}$$

Proof We will prove formula (5). The proof of formula (6) is left as an exercise (see Problem 56).

$$
\begin{aligned}
z_1 z_2 &= [r_1(\cos \theta_1 + i \sin \theta_1)][r_2(\cos \theta_2 + i \sin \theta_2)] \\
&= r_1 r_2 [(\cos \theta_1 + i \sin \theta_1)(\cos \theta_2 + i \sin \theta_2)] \\
&= r_1 r_2 [(\cos \theta_1 \cos \theta_2 - \sin \theta_1 \sin \theta_2) + i(\sin \theta_1 \cos \theta_2 + \cos \theta_1 \sin \theta_2)] \\
&= r_1 r_2 [(\cos \theta_1 + \theta_2) + i \sin(\theta_1 + \theta_2)] \quad\blacksquare
\end{aligned}
$$

Because the magnitude of a complex number z is r and its argument is θ, when $z = r(\cos \theta + i \sin \theta)$ we can restate this theorem as follows:

> **Theorem**
>
> The magnitude of the product (quotient) of two complex numbers equals the product (quotient) of their magnitudes; the argument of the product (quotient) of two complex numbers is determined by the sum (difference) of their arguments.

Let's look at an example of how this theorem can be used.

E X A M P L E   3   Finding Products and Quotients of Complex Numbers in Polar Form

If $z = 3(\cos 20° + i \sin 20°)$ and $w = 5(\cos 100° + i \sin 100°)$, find the following (leave your answers in polar form):

(a) zw (b) $\dfrac{z}{w}$

Solution (a) $zw = [3(\cos 20° + i \sin 20°)][5(\cos 100° + i \sin 100°)]$
$$= (3 \cdot 5)[\cos(20° + 100°) + i \sin(20° + 100°)]$$
$$= 15(\cos 120° + i \sin 120°)$$

(b) $\dfrac{z}{w} = \dfrac{3(\cos 20° + i \sin 20°)}{5(\cos 100° + i \sin 100°)}$

$$= \tfrac{3}{5}[\cos(20° - 100°) + i \sin(20° - 100°)]$$
$$= \tfrac{3}{5}[\cos(-80°) + i \sin(-80°)]$$
$$= \tfrac{3}{5}(\cos 280° + i \sin 280°) \quad \text{Argument must lie}$$
$$\text{between } 0° \text{ and } 360°.$$

Now work Problem 23.

De Moivre's Theorem

De Moivre's Theorem, stated by Abraham De Moivre (1667–1754) in 1730, but already known to many people by 1710, is important for the following reason: The fundamental processes of algebra are the four operations of addition, subtraction, multiplication, and division, together with powers and the extraction of roots. De Moivre's Theorem allows these latter fundamental algebraic operations to be applied to complex numbers.

De Moivre's Theorem, in its most basic form, is a formula for raising a complex number z to the power n, where $n \ge 1$ is a positive integer. Let's see if we can guess the form of the result.

Let $z = r(\cos \theta + i \sin \theta)$ be a complex number. Then, based on equation (5), we have

$n = 2$: $z^2 = r^2(\cos 2\theta + i \sin 2\theta)$
$n = 3$: $z^3 = z^2 \cdot z$
$\qquad\qquad = [r^2(\cos 2\theta + i \sin 2\theta)][r(\cos \theta + i \sin \theta)]$
$\qquad\qquad = r^3(\cos 3\theta + i \sin 3\theta) \quad$ Equation (5).
$n = 4$: $z^4 = z^3 \cdot z$
$\qquad\qquad = [r^3(\cos 3\theta + i \sin 3\theta)][r(\cos \theta + i \sin \theta)]$
$\qquad\qquad = r^4(\cos 4\theta + i \sin 4\theta) \quad$ Equation (5).

The pattern should now be clear.

Theorem   De Moivre's Theorem

If $z = r(\cos \theta + i \sin \theta)$ is a complex number, then

$$z^n = r^n(\cos n\theta + i \sin n\theta) \qquad (7)$$

where $n \ge 1$ is a positive integer.

We will not prove De Moivre's Theorem because the proof requires mathematical induction which is not discussed in this book.

Let's look at some examples.

E X A M P L E   4   Using De Moivre's Theorem

Write $[2(\cos 20° + i \sin 20°)]^3$ in the standard form $a + bi$.

Solution $[2(\cos 20° + i \sin 20°)]^3 = 2^3[\cos(3 \cdot 20°) + i \sin(3 \cdot 20°)]$

$$= 8(\cos 60° + i \sin 60°)$$

$$= 8\left(\frac{1}{2} + \frac{\sqrt{3}}{2}i\right) = 4 + 4\sqrt{3}i$$

Now work Problem 31.

E X A M P L E   5   Using De Moivre's Theorem

Write $(1 + i)^5$ in the standard form $a + bi$.

Solution   To apply De Moivre's Theorem, we must first write the complex number in polar form. Thus, since the magnitude of $1 + i$ is $\sqrt{1^2 + 1^2} = \sqrt{2}$, we begin by writing

$$1 + i = \sqrt{2}\left(\frac{1}{\sqrt{2}} + \frac{1}{\sqrt{2}}i\right) = \sqrt{2}\left(\cos\frac{\pi}{4} + i \sin\frac{\pi}{4}\right)$$

Now

$$(1 + i)^5 = \left[\sqrt{2}\left(\cos\frac{\pi}{4} + i \sin\frac{\pi}{4}\right)\right]^5$$

$$= (\sqrt{2})^5\left[\cos\left(5 \cdot \frac{\pi}{4}\right) + i \sin\left(5 \cdot \frac{\pi}{4}\right)\right]$$

$$= 4\sqrt{2}\left(\cos\frac{5\pi}{4} + i \sin\frac{5\pi}{4}\right)$$

$$= 4\sqrt{2}\left[-\frac{1}{\sqrt{2}} + \left(-\frac{1}{\sqrt{2}}\right)i\right] = -4 - 4i$$

E X A M P L E   6   Using a Calculator with De Moivre's Theorem

Write $(3 + 4i)^3$ in the standard form $a + bi$.

Solution   Again, we start by writing $3 + 4i$ in polar form. This time we will use degrees for the argument. The magnitude of $3 + 4i$ is $\sqrt{3^2 + 4^2} = \sqrt{25} = 5$, so we write

$$3 + 4i = 5\left(\frac{3}{5} + \frac{4}{5}i\right) \approx 5(\cos 53.1° + i \sin 53.1°)$$

Although we have written the angle rounded to one decimal place (53.1°), we keep the actual value of the angle in memory. Now

$$(3 + 4i)^3 \approx [5(\cos 53.1° + i \sin 53.1°)]^3$$

$$= 5^3[\cos(3 \cdot 53.1°) + i \sin(3 \cdot 53.1°)]$$

$$\approx 125(\cos 159.4° + i \sin 159.4°)$$

$$\approx 125[-0.936 + i(0.352)] = -117 + 44i$$

In this computation, we used the actual values STOred in memory, not the rounded values shown. The final answer, $-117 + 44i$, is exact as you can verify by cubing $3 + 4i$.

Complex Roots

5 Let w be a given complex number, and let $n \geq 2$ denote a positive integer. Any complex number z that satisfies the equation

$$z^n = w$$

is called a **complex nth root** of w. In keeping with previous usage, if $n = 2$, the solutions of the equation $z^2 = w$ are called **complex square roots** of w, and if $n = 3$, the solutions of the equation $z^3 = w$ are called **complex cube roots** of w.

Theorem   Finding Complex Roots

Let $w = r(\cos \theta + i \sin \theta)$ be a complex number. If $w \neq 0$, there are n distinct complex nth roots of w, given by the formula

$$z_k = \sqrt[n]{r}\left[\cos\left(\frac{\theta}{n} + \frac{2k\pi}{n}\right) + i \sin\left(\frac{\theta}{n} + \frac{2k\pi}{n}\right)\right] \qquad (8)$$

where $k = 0, 1, 2, \ldots, n - 1$.

Proof (Outline)   We will not prove this result in its entirety. Instead, we shall show only that each z_k in equation (8) obeys the equation $z_k^n = w$ and, hence, each z_k is a complex nth root of w.

$$z_k^n = \left\{\sqrt[n]{r}\left[\cos\left(\frac{\theta}{n} + \frac{2k\pi}{n}\right) + i \sin\left(\frac{\theta}{n} + \frac{2k\pi}{n}\right)\right]\right\}^n$$
$$= (\sqrt[n]{r})^n\left[\cos n\left(\frac{\theta}{n} + \frac{2k\pi}{n}\right) + i \sin n\left(\frac{\theta}{n} + \frac{2k\pi}{n}\right)\right] \quad \text{DeMoivre's Theorem}$$
$$= r[\cos(\theta + 2k\pi) + i \sin(\theta + 2k\pi)]$$
$$= r(\cos \theta + i \sin \theta) = w$$

Thus, each z_k, $k = 0, 1, \ldots, n - 1$, is a complex nth root of w. To complete the proof, we would need to show that each z_k, $k = 0, 1, 2, \ldots, n - 1$, is, in fact, distinct and that there are no complex nth roots of w other than those given by equation (8).

E X A M P L E   7   Finding Complex Cube Roots

Find the complex cube roots of $-1 + \sqrt{3}\,i$. Leave your answers in polar form, with θ in degrees.

Solution   First, we express $-1 + \sqrt{3}\,i$ in polar form using degrees.

$$-1 + \sqrt{3}\,i = 2\left(-\frac{1}{2} + \frac{\sqrt{3}}{2}i\right) = 2(\cos 120° + i \sin 120°)$$

The three complex cube roots of $-1 + \sqrt{3}\,i = 2(\cos 120° + i \sin 120°)$ are

$$z_k = \sqrt[3]{2}\left[\cos\left(\frac{120°}{3} + \frac{360°k}{3}\right) + i\sin\left(\frac{120°}{3} + \frac{360°k}{3}\right)\right], \qquad k = 0, 1, 2$$

Thus,

$$z_0 = \sqrt[3]{2}(\cos 40° + i \sin 40°)$$
$$z_1 = \sqrt[3]{2}(\cos 160° + i \sin 160°)$$
$$z_2 = \sqrt[3]{2}(\cos 280° + i \sin 280°)$$

Notice that each of the three complex roots of $-1 + \sqrt{3}\,i$ has the same magnitude, $\sqrt[3]{2}$. This means that the points corresponding to each cube root lie the same distance from the origin; hence, the three points lie on a circle with center at the origin and radius $\sqrt[3]{2}$. Furthermore, the arguments of these cube roots are 40°, 160°, and 280°, the difference of consecutive pairs being 120°. This means that the three points are equally spaced on the circle, as shown in Figure 39. These results are not coincidental. In fact, you are asked to show that these results hold for complex nth roots in Problems 53 through 55.

FIGURE 39

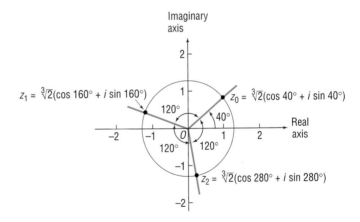

Now work Problem 43.

HISTORICAL FEATURE The Babylonians, Greeks, and Arabs considered square roots of negative quantities to be impossible and equations with complex solutions to be unsolvable. The first hint that there was some connection between real solutions of equations and complex numbers came when Girolamo Cardano (1501–1576) and Tartaglia (1499–1557) found *real* roots of cubic equations by taking cube roots of *complex* quantities. For centuries thereafter, mathematicians worked with complex numbers without much belief in their actual existence. In 1673, John Wallis appears to have been the first to suggest the graphical representation of complex numbers, a truly significant idea that was not pursued further until about 1800. Several people, including Karl Friedrich Gauss (1777–1855), then rediscovered the idea, and the graphical representation helped to establish complex numbers as equal members of the number family. In practical applications, complex numbers have found their greatest uses in the area of alternating current, where they are a commonplace tool, and in the area of subatomic physics.

HISTORICAL PROBLEMS  1. The quadratic formula will work perfectly well if the coefficients are complex numbers. Solve the following, using De Moivre's Theorem where necessary.

[**Hint:** The answers are "nice."]

(a) $z^2 - (2 + 5i)z - 3 + 5i = 0$ (b) $z^2 - (1 + i)z - 2 - i = 0$

5.4   EXERCISES

In Problems 1–12, plot each complex number in the complex plane and write it in polar form. Express the argument in degrees.

1. $1 + i$
2. $-1 + i$
3. $\sqrt{3} - i$
4. $1 - \sqrt{3}\,i$
5. $-3i$
6. -2
7. $4 - 4i$
8. $9\sqrt{3} + 9i$
9. $3 - 4i$
10. $2 + \sqrt{3}\,i$
11. $-2 + 3i$
12. $\sqrt{5} - i$

In Problems 13–22, write each complex number in rectangular form.

13. $2(\cos 120° + i \sin 120°)$
14. $3(\cos 210° + i \sin 210°)$
15. $4\left(\cos \dfrac{7\pi}{4} + i \sin \dfrac{7\pi}{4}\right)$
16. $2\left(\cos \dfrac{5\pi}{6} + i \sin \dfrac{5\pi}{6}\right)$
17. $3\left(\cos \dfrac{3\pi}{2} + i \sin \dfrac{3\pi}{2}\right)$
18. $4\left(\cos \dfrac{\pi}{2} + i \sin \dfrac{\pi}{2}\right)$
19. $0.2(\cos 100° + i \sin 100°)$
20. $0.4(\cos 200° + i \sin 200°)$
21. $2\left(\cos \dfrac{\pi}{18} + i \sin \dfrac{\pi}{18}\right)$
22. $3\left(\cos \dfrac{\pi}{10} + i \sin \dfrac{\pi}{10}\right)$

In Problems 23–30, find zw and z/w. Leave your answers in polar form.

23. $z = 2(\cos 40° + i \sin 40°)$
    $w = 4(\cos 20° + i \sin 20°)$
24. $z = \cos 120° + i \sin 120°$
    $w = \cos 100° + i \sin 100°$
25. $z = 3(\cos 130° + i \sin 130°)$
    $w = 4(\cos 270° + i \sin 270°)$
26. $z = 2(\cos 80° + i \sin 80°)$
    $w = 6(\cos 200° + i \sin 200°)$
27. $z = 2\left(\cos \dfrac{\pi}{8} + i \sin \dfrac{\pi}{8}\right)$
    $w = 2\left(\cos \dfrac{\pi}{10} + i \sin \dfrac{\pi}{10}\right)$
28. $z = 4\left(\cos \dfrac{3\pi}{8} + i \sin \dfrac{3\pi}{8}\right)$
    $w = 2\left(\cos \dfrac{9\pi}{16} + i \sin \dfrac{9\pi}{16}\right)$
29. $z = 2 + 2i$
    $w = \sqrt{3} - i$
30. $z = 1 - i$
    $w = 1 - \sqrt{3}\,i$

In Problems 31–42, write each expression in the standard form a + bi.

31. $[4(\cos 40° + i \sin 40°)]^3$
32. $[3(\cos 80° + i \sin 80°)]^3$
33. $\left[2\left(\cos \dfrac{\pi}{10} + i \sin \dfrac{\pi}{10}\right)\right]^5$
34. $\left[\sqrt{2}\left(\cos \dfrac{5\pi}{16} + i \sin \dfrac{5\pi}{16}\right)\right]^4$
35. $[\sqrt{3}(\cos 10° + i \sin 10°)]^6$
36. $[\tfrac{1}{2}(\cos 72° + i \sin 72°)]^5$
37. $\left[\sqrt{5}\left(\cos \dfrac{3\pi}{16} + i \sin \dfrac{3\pi}{16}\right)\right]^4$
38. $\left[\sqrt{3}\left(\cos \dfrac{5\pi}{18} + i \sin \dfrac{5\pi}{18}\right)\right]^6$
39. $(1 - i)^5$
40. $(\sqrt{3} - i)^6$
41. $(\sqrt{2} - i)^6$
42. $(1 - \sqrt{5}\,i)^8$

In Problems 43–50, find all the complex roots. Leave your answers in polar form with the argument in degrees.

43. The complex cube roots of $1 + i$
44. The complex fourth roots of $\sqrt{3} - i$
45. The complex fourth roots of $4 - 4\sqrt{3}\,i$
46. The complex cube roots of $-8 - 8i$

47. The complex fourth roots of $-16i$

49. The complex fifth roots of i

48. The complex cube roots of -8

50. The complex fifth roots of $-i$

51. Find the four complex fourth roots of unity (1). Plot each one.

52. Find the six complex sixth roots of unity (1). Plot each one.

53. Show that each complex nth root of a nonzero complex number w has the same magnitude.

54. Use the result of Problem 53 to draw the conclusion that each complex nth root lies on a circle with center at the origin. What is the radius of this circle?

55. Refer to Problem 54. Show that the complex nth roots of a nonzero complex number w are equally spaced on the circle.

56. Prove formula (6).

5.5 | VECTORS

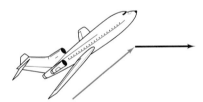

1. Graph Vectors
2. Find a Position Vector
3. Add and Subtract Vectors
4. Find a Scalar Product and the Magnitude of a Vector
5. Find a Unit Vector

In simple terms, a **vector** (derived from the Latin *vehere,* meaning "to carry") is a quantity that has both magnitude and direction. It is convenient to represent a vector by using an arrow. The length of the arrow represents the **magnitude** of the vector, and the arrowhead indicates the **direction** of the vector.

FIGURE 40

Many quantities in physics can be represented by vectors. For example, the velocity of an aircraft can be represented by an arrow that points in the direction of movement; the length of the arrow represents speed. Thus, if the aircraft speeds up, we lengthen the arrow; if the aircraft changes direction, we introduce an arrow in the new direction. See Figure 40. Based on this representation, it is not surprising that vectors and directed line segments are somehow related.

Directed Line Segments

If P and Q are two distinct points in the xy-plane, there is exactly one line containing both P and Q (Figure 41(a)). The points on that part of the line that joins P to Q, including P and Q, form what is called the **line segment** $\overline{PQ}$ (Figure 41(b)). If we order the points so that they proceed from P to Q, we have a **directed line segment** from P to Q, which we denote by $\overrightarrow{PQ}$. In a directed line segment $\overrightarrow{PQ}$, we call P the **initial point** and Q the **terminal point,** as indicated in Figure 41(c).

FIGURE 41

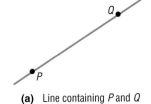

(a) Line containing P and Q

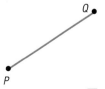

(b) Line segment $\overline{PQ}$

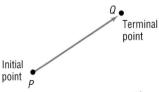

(c) Directed line segment $\overrightarrow{PQ}$

The magnitude of the directed line segment $\overrightarrow{PQ}$ is the distance from the point P to the point Q; that is, it is the length of the line segment. The direction of $\overrightarrow{PQ}$ is from P to Q. If a vector $\mathbf{v}$* has the same magnitude and the same direction as the directed line segment $\overrightarrow{PQ}$, then we write

$$\mathbf{v} = \overrightarrow{PQ}$$

The vector $\mathbf{v}$ whose magnitude is 0 is called the **zero vector, 0.** The zero vector is assigned no direction.

Two vectors $\mathbf{v}$ and $\mathbf{w}$ are **equal,** written

$$\mathbf{v} = \mathbf{w}$$

FIGURE 42

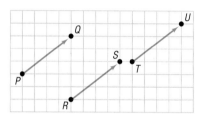

if they have the same magnitude and the same direction.

For example, the vectors shown in Figure 42 have the same magnitude and the same direction, so they are equal, even though they have different initial points and different terminal points. As a result, we find it useful to think of a vector simply as an arrow, keeping in mind that two arrows (vectors) are equal if they have the same direction and the same magnitude (length).

FIGURE 43

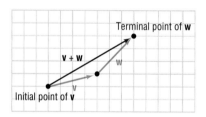

Adding Vectors

The **sum $\mathbf{v} + \mathbf{w}$** of two vectors is defined as follows: We position the vectors $\mathbf{v}$ and $\mathbf{w}$ so that the terminal point of $\mathbf{v}$ coincides with the initial point of $\mathbf{w}$, as shown in Figure 43. The vector $\mathbf{v} + \mathbf{w}$ is then the unique vector whose initial point coincides with the initial point of $\mathbf{v}$ and whose terminal point coincides with the terminal point of $\mathbf{w}$.

Vector addition is **commutative.** That is, if $\mathbf{v}$ and $\mathbf{w}$ are any two vectors, then

FIGURE 44

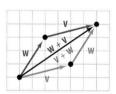

$$\mathbf{v} + \mathbf{w} = \mathbf{w} + \mathbf{v}$$

Figure 44 illustrates this fact. (Observe that the commutative property is another way of saying that opposite sides of a parallelogram are equal and parallel.)

Vector addition is also **associative.** That is, if $\mathbf{u}$, $\mathbf{v}$, and $\mathbf{w}$ are vectors, then

$$\mathbf{u} + (\mathbf{v} + \mathbf{w}) = (\mathbf{u} + \mathbf{v}) + \mathbf{w}$$

Figure 45 illustrates the associative property for vectors.

The zero vector has the property that

FIGURE 45
$(\mathbf{u} + \mathbf{v}) + \mathbf{w} = \mathbf{u} + (\mathbf{v} + \mathbf{w})$

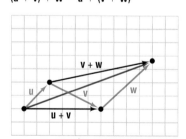

$$\mathbf{v} + \mathbf{0} = \mathbf{0} + \mathbf{v} = \mathbf{v}$$

for any vector $\mathbf{v}$.

---

*Boldface letters will be used to denote vectors, in order to distinguish them from numbers. For handwritten work, an arrow is placed over the letter to signify a vector.

FIGURE 46

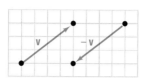

If **v** is a vector, then −**v** is the vector having the same magnitude as **v**, but whose direction is opposite to **v**, as shown in Figure 46.
Furthermore,

$$\mathbf{v} + (-\mathbf{v}) = \mathbf{0}$$

If **v** and **w** are two vectors, we define the **difference v − w** as

$$\mathbf{v} - \mathbf{w} = \mathbf{v} + (-\mathbf{w})$$

Figure 47 illustrates the relationships among **v, w, v + w,** and **v − w.**

FIGURE 47

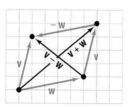

Multiplying Vectors by Numbers

When dealing with vectors, we refer to real numbers as **scalars.** Scalars are quantities that have only magnitude. Examples from physics of scalar quantities are temperature, speed, and time. We now define how to multiply a vector by a scalar.

If α is a scalar and **v** is a vector, the **scalar product** $\alpha\mathbf{v}$ is defined as

1. If $\alpha > 0$, the product $\alpha\mathbf{v}$ is the vector whose magnitude is α times the magnitude of **v** and whose direction is the same as **v.**
2. If $\alpha < 0$, the product $\alpha\mathbf{v}$ is the vector whose magnitude is $|\alpha|$ times the magnitude of **v** and whose direction is opposite that of **v.**
3. If $\alpha = 0$ or if **v = 0,** then $\alpha\mathbf{v} = \mathbf{0}$.

FIGURE 48

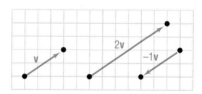

See Figure 48 for some illustrations.

For example, if **a** is the acceleration of an object of mass m due to a force **F** being exerted on it, then, by Newton's second law of motion, **F** = m**a.** Here, m**a** is the product of the scalar m and the vector **a.**

Scalar products have the following properties:

$$0\mathbf{v} = \mathbf{0} \qquad 1\mathbf{v} = \mathbf{v} \qquad -1\mathbf{v} = -\mathbf{v}$$
$$(\alpha + \beta)\mathbf{v} = \alpha\mathbf{v} + \beta\mathbf{v} \qquad \alpha(\mathbf{v} + \mathbf{w}) = \alpha\mathbf{v} + \alpha\mathbf{w}$$
$$\alpha(\beta\mathbf{v}) = (\alpha\beta)\mathbf{v}$$

E X A M P L E   1

Graphing Vectors

1

Use the vectors illustrated in Figure 49 to graph each of the following vectors.

(a) **v − w**   (b) **2v + 3w**   (c) **2v − w + 3u**

FIGURE 49

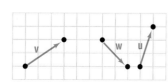

Solution   Figure 50 illustrates each graph.

FIGURE 50

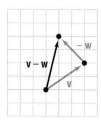

(a) v − w

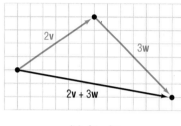

(b) 2v + 3w

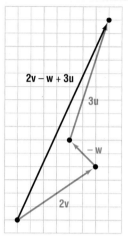

(c) 2v − w + 3u

 Now work Problems 1 and 3.

Magnitudes of Vectors

If **v** is a vector, we use the symbol $\|\mathbf{v}\|$ to represent the **magnitude** of **v.** Since $\|\mathbf{v}\|$ equals the length of a directed line segment, it follows that $\|\mathbf{v}\|$ has the following properties:

> Theorem   Properties of $\|\mathbf{v}\|$
>
> If **v** is a vector and if α is a scalar, then
>
> (a) $\|\mathbf{v}\| \geq 0$
> (b) $\|\mathbf{v}\| = 0$ if and only if $\mathbf{v} = \mathbf{0}$
> (c) $\|-\mathbf{v}\| = \|\mathbf{v}\|$
> (d) $\|\alpha\mathbf{v}\| = |\alpha|\,\|\mathbf{v}\|$

Property (a) is a consequence of the fact that distance is a nonnegative number. Property (b) follows, because the length of the directed line segment $\overrightarrow{PQ}$ is positive unless P and Q are the same point, in which case the length is 0. Property (c) follows, because the length of the line segment $\overline{PQ}$ equals the length of the line segment $\overline{QP}$. Property (d) is a direct consequence of the definition of a scalar product.

A vector **u** for which $\|\mathbf{u}\| = 1$ is called a **unit vector.**

To compute the magnitude and direction of a vector, we need an algebraic way of representing vectors.

Representing Vectors in the Plane

We use a rectangular coordinate system to represent vectors in the plane. Let **i** denote the unit vector whose direction is along the positive x-axis; let **j** denote the unit vector whose direction is along the positive y-axis. If **v** is a vector with initial point at the origin O and terminal point at $P = (a, b)$, then we can represent **v** in terms of the vectors **i** and **j** as

$$\mathbf{v} = a\mathbf{i} + b\mathbf{j}$$

FIGURE 51

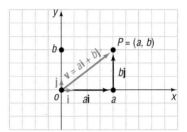

See Figure 51. The scalars a and b are called the **components** of the vector $\mathbf{v} = a\mathbf{i} + b\mathbf{j}$, with a being the component in the direction **i** and b being the component in the direction **j**.

A vector whose initial point is at the origin is called a **position vector.** The next result states that any vector whose initial point is not at the origin is equal to a unique position vector.

Theorem

Suppose that **v** is a vector with initial point $P_1 = (x_1, y_1)$, not necessarily the origin, and terminal point $P_2 = (x_2, y_2)$. If $\mathbf{v} = \overrightarrow{P_1P_2}$, then **v** is equal to the position vector

$$\mathbf{v} = (x_2 - x_1)\mathbf{i} + (y_2 - y_1)\mathbf{j} \qquad (1)$$

To see why this is true, look at Figure 52. Triangle OPA and triangle P_1P_2Q are congruent. (Do you see why? The line segments have the same magnitude, so $d(O, P) = d(P_1, P_2)$; and they have the same direction, so $\angle POA = \angle P_2P_1Q$. Since the triangles are right triangles, we have angle–side–angle.) Thus, it follows that corresponding sides are equal. As a result, $x_2 - x_1 = a$ and $y_2 - y_1 = b$, so **v** may be written as

$$\mathbf{v} = a\mathbf{i} + b\mathbf{j} = (x_2 - x_1)\mathbf{i} + (y_2 - y_1)\mathbf{j}$$

FIGURE 52
$$\mathbf{v} = a\mathbf{i} + b\mathbf{j} = (x_2 - x_1)\mathbf{i} + (y_2 - y_1)\mathbf{j}$$

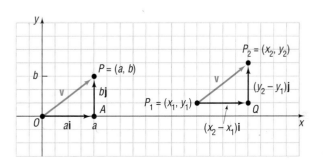

E X A M P L E   2

Finding a Position Vector

2

Find the position vector of the vector $\mathbf{v} = \overrightarrow{P_1P_2}$ if $P_1 = (-1, 2)$ and $P_2 = (4, 6)$.

FIGURE 53

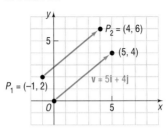

Solution

By equation (1), the position vector equal to $\mathbf{v}$ is

$$\mathbf{v} = [4 - (-1)]\mathbf{i} + (6 - 2)\mathbf{j} = 5\mathbf{i} + 4\mathbf{j}$$

See Figure 53.

Now work Problem 21.

Two position vectors $\mathbf{v}$ and $\mathbf{w}$ are equal if and only if the terminal point of $\mathbf{v}$ is the same as the terminal point of $\mathbf{w}$. This leads to the following result:

Theorem   Equality of Vectors

Two vectors $\mathbf{v}$ and $\mathbf{w}$ are equal if and only if their corresponding components are equal. That is,

If $\mathbf{v} = a_1\mathbf{i} + b_1\mathbf{j}$ and $\mathbf{w} = a_2\mathbf{i} + b_2\mathbf{j}$,
then $\mathbf{v} = \mathbf{w}$ if and only if $a_1 = a_2$ and $b_1 = b_2$.

Because of the above result, we can replace any vector (directed line segment) by a unique position vector, and vice versa. This flexibility is one of the main reasons for the wide use of vectors. Unless otherwise specified, from now on the term *vector* will mean the unique position vector equal to it.

Next, we define addition, subtraction, scalar product, and magnitude in terms of the components of a vector.

Let $\mathbf{v} = a_1\mathbf{i} + b_1\mathbf{j}$ and $\mathbf{w} = a_2\mathbf{i} + b_2\mathbf{j}$ be two vectors, and let α be a scalar. Then

$$\mathbf{v} + \mathbf{w} = (a_1 + a_2)\mathbf{i} + (b_1 + b_2)\mathbf{j} \qquad (2)$$

$$\mathbf{v} - \mathbf{w} = (a_1 - a_2)\mathbf{i} + (b_1 - b_2)\mathbf{j} \qquad (3)$$

$$\alpha\mathbf{v} = (\alpha a_1)\mathbf{i} + (\alpha b_1)\mathbf{j} \qquad (4)$$

$$\|\mathbf{v}\| = \sqrt{a_1^2 + b_1^2} \qquad (5)$$

These definitions are compatible with the geometric ones given earlier in this section. See Figure 54.

FIGURE 54

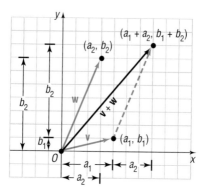

(a) Illustration of property (2)

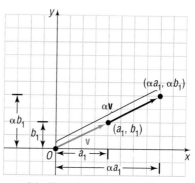

(b) Illustration of property (4)

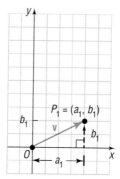

(c) Illustration of property (5):
$\| \mathbf{v} \| =$ Distance from O to P_1
$\| \mathbf{v} \| = \sqrt{a_1^2 + b_1^2}$

Thus, to add two vectors, simply add corresponding components. To subtract two vectors, subtract corresponding components.

E X A M P L E   3
3

Adding and Subtracting Vectors

If $\mathbf{v} = 2\mathbf{i} + 3\mathbf{j}$ and $\mathbf{w} = 3\mathbf{i} - 4\mathbf{j}$, find

(a) $\mathbf{v} + \mathbf{w}$ (b) $\mathbf{v} - \mathbf{w}$

Solution   (a) $\mathbf{v} + \mathbf{w} = (2\mathbf{i} + 3\mathbf{j}) + (3\mathbf{i} - 4\mathbf{j}) = (2 + 3)\mathbf{i} + (3 - 4)\mathbf{j}$
$= 5\mathbf{i} - \mathbf{j}$

(b) $\mathbf{v} - \mathbf{w} = (2\mathbf{i} + 3\mathbf{j}) - (3\mathbf{i} - 4\mathbf{j}) = (2 - 3)\mathbf{i} + [3 - (-4)]\mathbf{j}$
$= -\mathbf{i} + 7\mathbf{j}$

E X A M P L E   4
4

Finding Scalar Products and Magnitudes

If $\mathbf{v} = 2\mathbf{i} + 3\mathbf{j}$ and $\mathbf{w} = 3\mathbf{i} - 4\mathbf{j}$, find

(a) $3\mathbf{v}$ (b) $2\mathbf{v} - 3\mathbf{w}$ (c) $\|\mathbf{v}\|$

Solution   (a) $3\mathbf{v} = 3(2\mathbf{i} + 3\mathbf{j}) = 6\mathbf{i} + 9\mathbf{j}$
(b) $2\mathbf{v} - 3\mathbf{w} = 2(2\mathbf{i} + 3\mathbf{j}) - 3(3\mathbf{i} - 4\mathbf{j}) = 4\mathbf{i} + 6\mathbf{j} - 9\mathbf{i} + 12\mathbf{j}$
$= -5\mathbf{i} + 18\mathbf{j}$
(c) $\|\mathbf{v}\| = \|2\mathbf{i} + 3\mathbf{j}\| = \sqrt{2^2 + 3^2} = \sqrt{13}$

Now work Problems 27 and 41.

Recall that a unit vector $\mathbf{u}$ is one for which $\|\mathbf{u}\| = 1$. In many applications, it is useful to be able to find a unit vector $\mathbf{u}$ that has the same direction as a given vector $\mathbf{v}$.

Theorem Unit Vector in Direction of **v**

For any nonzero vector **v,** the vector

$$\mathbf{u} = \frac{\mathbf{v}}{\|\mathbf{v}\|}$$

is a unit vector that has the same direction as **v.**

Proof Let $\mathbf{v} = a\mathbf{i} + b\mathbf{j}$. Then $\|\mathbf{v}\| = \sqrt{a^2 + b^2}$ and

$$\mathbf{u} = \frac{\mathbf{v}}{\|\mathbf{v}\|} = \frac{a\mathbf{i} + b\mathbf{j}}{\sqrt{a^2 + b^2}} = \frac{a}{\sqrt{a^2 + b^2}}\mathbf{i} + \frac{b}{\sqrt{a^2 + b^2}}\mathbf{j}$$

The vector **u** is in the same direction as **v,** since $\|\mathbf{v}\| > 0$. Furthermore,

$$\|\mathbf{u}\| = \sqrt{\frac{a^2}{a^2 + b^2} + \frac{b^2}{a^2 + b^2}} = \sqrt{\frac{a^2 + b^2}{a^2 + b^2}} = 1$$

Thus, **u** is a unit vector in the direction of **v.** ∎

As a consequence of this theorem, if **u** is a unit vector in the same direction as a vector **v,** then **v** may be expressed as

$$\mathbf{v} = \|\mathbf{v}\|\,\mathbf{u} \tag{6}$$

This way of expressing a vector is useful in many applications.

EXAMPLE 5   Finding a Unit Vector

Find a unit vector in the same direction as $\mathbf{v} = 4\mathbf{i} - 3\mathbf{j}$.

Solution   We find $\|\mathbf{v}\|$ first.

$$\|\mathbf{v}\| = \|4\mathbf{i} - 3\mathbf{j}\| = \sqrt{16 + 9} = 5$$

Now we multiply **v** by the scalar $1/\|\mathbf{v}\| = \frac{1}{5}$. The result is

$$\frac{\mathbf{v}}{\|\mathbf{v}\|} = \frac{4\mathbf{i} - 3\mathbf{j}}{5} = \frac{4}{5}\mathbf{i} - \frac{3}{5}\mathbf{j}$$

Check: This vector is, in fact, a unit vector because

$$\left(\tfrac{4}{5}\right)^2 + \left(-\tfrac{3}{5}\right)^2 = \tfrac{16}{25} + \tfrac{9}{25} = \tfrac{25}{25} = 1$$

∎

 Now work Problem 43.

Applications

Forces provide an example of physical quantities that may be conveniently represented by vectors; two forces "combine" the way that vectors "add." How do we know this? Laboratory experiments bear it out. Thus, if $\mathbf{F}_1$ and $\mathbf{F}_2$ are two forces simultaneously acting on an object, the vector sum $\mathbf{F}_1 + \mathbf{F}_2$ is the force that produces the same effect on the object as that obtained when

FIGURE 55

Resultant

the forces $\mathbf{F}_1$ and $\mathbf{F}_2$ act on the object. The force $\mathbf{F}_1 + \mathbf{F}_2$ is sometimes called the **resultant** of $\mathbf{F}_1$ and $\mathbf{F}_2$. See Figure 55.

Two important applications of the resultant of two vectors occur with aircraft flying in the presence of a wind and with boats cruising across a river with a current. For example, consider the velocity of wind acting on the velocity of an airplane (see Figure 56). Suppose that $\mathbf{w}$ is a vector describing the velocity of the wind; that is, $\mathbf{w}$ represents the direction and speed of the wind. If $\mathbf{v}$ is the velocity of the airplane in the absence of wind (called its **velocity relative to the air**), then $\mathbf{v} + \mathbf{w}$ is the vector equal to the actual velocity of the airplane (called its **velocity relative to the ground**).

FIGURE 56

(a) Velocity $\mathbf{w}$ of wind relative to ground

(b) Velocity $\mathbf{v}$ of airplane relative to air

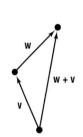

(c) Resultant $\mathbf{w} + \mathbf{v}$ equals velocity of airplane relative to the ground

Our next example illustrates a use of vectors in navigation.

EXAMPLE 6

Finding the Actual Speed of an Aircraft

A Boeing 737 aircraft maintains a constant airspeed of 500 miles per hour in the direction due south. The velocity of the jet stream is 80 miles per hour in a northeasterly direction.

(a) Find a unit vector having northeast as direction.
(b) Find a vector 80 units in magnitude having the same direction as the unit vector found in part (a).
(c) Find the actual speed of the aircraft relative to the ground.

Solution   We set up a coordinate system in which north (N) is along the positive y-axis. See Figure 57. Let

FIGURE 57

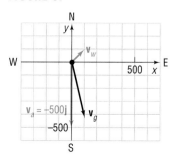

$$\mathbf{v}_a = \text{Velocity of aircraft relative to the air} = -500\mathbf{j}$$
$$\mathbf{v}_g = \text{Velocity of aircraft relative to ground}$$
$$\mathbf{v}_w = \text{Velocity of jet stream}$$

(a) A vector having northeast as direction is $\mathbf{i} + \mathbf{j}$. The unit vector in this direction is

$$\frac{\mathbf{i} + \mathbf{j}}{\|\mathbf{i} + \mathbf{j}\|} = \frac{\mathbf{i} + \mathbf{j}}{\sqrt{1 + 1}} = \frac{1}{\sqrt{2}}(\mathbf{i} + \mathbf{j})$$

(b) The velocity $\mathbf{v}_w$ of the jet stream is a vector with magnitude 80 in the direction of the unit vector $(1/\sqrt{2})(\mathbf{i} + \mathbf{j})$. Thus, from (6),

$$\mathbf{v}_w = 80\left[\frac{1}{\sqrt{2}}(\mathbf{i} + \mathbf{j})\right] = 40\sqrt{2}(\mathbf{i} + \mathbf{j})$$

(c) The velocity $\mathbf{v}_g$ of the aircraft relative to the ground is the resultant of the vectors $\mathbf{v}_a$ and $\mathbf{v}_w$. Thus,

$$\begin{aligned}\mathbf{v}_g = \mathbf{v}_a + \mathbf{v}_w &= -500\mathbf{j} + 40\sqrt{2}(\mathbf{i} + \mathbf{j}) \\ &= 40\sqrt{2}\mathbf{i} + (40\sqrt{2} - 500)\mathbf{j}\end{aligned}$$

The actual speed (speed relative to the ground) of the aircraft is

$$\|\mathbf{v}_g\| = \sqrt{(40\sqrt{2})^2 + (40\sqrt{2} - 500)^2} \approx 447 \text{ miles per hour}$$

We will find the actual direction of the aircraft in Example 4 of the next section.

HISTORICAL FEATURE   The history of vectors is surprisingly complicated for such a natural concept. In the xy-plane, complex numbers do a good job of imitating vectors. About 1840, mathematicians became interested in finding a system that would do for three dimensions what the complex numbers do for two dimensions. Hermann Grassmann (1809–1877), in Germany, and William Rowan Hamilton (1805–1865), in Ireland, both attempted to find solutions.

Hamilton's system was the *quaternions,* which are best thought of as a real number plus a vector, and do for four dimensions what complex numbers do for two dimensions. In this system the order of multiplication matters; that is, **ab** ≠ **ba.** Hamilton spent the rest of his life working out quaternion theory and trying to get it accepted in applied mathematics, but he encountered fierce resistance due to the complicated nature of quaternion multiplication. In the work with quaternions, two products of vectors emerged, the scalar (or dot) and the vector (or cross) products.

Grassmann fared even worse than Hamilton; if people did not like Hamilton's work, at least they understood it. Grassmann's abstract style, although easily read today, was almost impenetrable during the previous century, and only a few of his ideas were appreciated. Among those few were the same scalar and vector products that Hamilton had found.

About 1880, the American physicist Josiah Willard Gibbs (1839–1903) worked out an algebra involving only the simplest concepts: the vectors and the two products. He then added some calculus, and the resulting system was simple, flexible, and well adapted to expressing a large number of physical laws. This system remains in use essentially unchanged. Hamilton's and Grassmann's more extensive systems each gave birth to much interesting mathematics, but little of this mathematics is seen at elementary levels.

5.5 | EXERCISES

In Problems 1–8, use the vectors in the figure at the right to graph each of the following vectors.

1. **v + w**

2. **u + v**

3. **3v**

4. **4w**

5. **v − w**

6. **u − v**

7. **3v + u − 2w**

8. **2u − 3v + w**

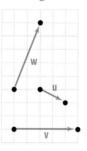

In Problems 9–16, use the figure at the right. Determine whether the given statement is true or false.

9. **A + B = F**

10. **K + G = F**

11. **C = D − E + F**

12. **G + H + E = D**

13. **E + D = G + H**

14. **H − C = G − F**

15. **A + B + K + G = 0**

16. **A + B + C + H + G = 0**

17. If $\|\mathbf{v}\| = 4$, what is $\|3\mathbf{v}\|$?

18. If $\|\mathbf{v}\| = 2$, what is $\|-4\mathbf{v}\|$?

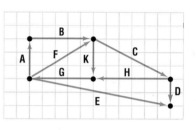

*In Problems 19–26, the vector **v** has initial point P and terminal point Q. Write **v** in the form a**i** + b**j**, that is, find its position vector.*

19. $P = (0, 0);\quad Q = (3, 4)$

20. $P = (0, 0);\quad Q = (-3, -5)$

21. $P = (3, 2);\quad Q = (5, 6)$

22. $P = (-3, 2);\quad Q = (6, 5)$

23. $P = (-2, -1);\quad Q = (6, -2)$

24. $P = (-1, 4);\quad Q = (6, 2)$

25. $P = (1, 0);\quad Q = (0, 1)$

26. $P = (1, 1);\quad Q = (2, 2)$

In Problems 27–32, find $\|\mathbf{v}\|$.

27. $\mathbf{v} = 3\mathbf{i} - 4\mathbf{j}$

28. $\mathbf{v} = -5\mathbf{i} + 12\mathbf{j}$

29. $\mathbf{v} = \mathbf{i} - \mathbf{j}$

30. $\mathbf{v} = -\mathbf{i} - \mathbf{j}$

31. $\mathbf{v} = -2\mathbf{i} + 3\mathbf{j}$

32. $\mathbf{v} = 6\mathbf{i} + 2\mathbf{j}$

In Problems 33–38, find each quantity if $\mathbf{v} = 3\mathbf{i} - 5\mathbf{j}$ and $\mathbf{w} = -2\mathbf{i} + 3\mathbf{j}$.

33. $2\mathbf{v} + 3\mathbf{w}$

34. $3\mathbf{v} - 2\mathbf{w}$

35. $\|\mathbf{v} - \mathbf{w}\|$

36. $\|\mathbf{v} + \mathbf{w}\|$

37. $\|\mathbf{v}\| - \|\mathbf{w}\|$

38. $\|\mathbf{v}\| + \|\mathbf{w}\|$

*In Problems 39–44, find the unit vector having the same direction as **v**.*

39. $\mathbf{v} = 5\mathbf{i}$

40. $\mathbf{v} = -3\mathbf{j}$

41. $\mathbf{v} = 3\mathbf{i} - 4\mathbf{j}$

42. $\mathbf{v} = -5\mathbf{i} + 12\mathbf{j}$

43. $\mathbf{v} = \mathbf{i} - \mathbf{j}$

44. $\mathbf{v} = 2\mathbf{i} - \mathbf{j}$

45. Find a vector **v** whose magnitude is 4 and whose component in the **i** direction is twice the component in the **j** direction.

46. Find a vector **v** whose magnitude is 3 and whose component in the **i** direction is equal to the component in the **j** direction.

47. If $\mathbf{v} = 2\mathbf{i} - \mathbf{j}$ and $\mathbf{w} = x\mathbf{i} + 3\mathbf{j}$, find all numbers x for which $\|\mathbf{v} + \mathbf{w}\| = 5$.

48. If $P = (-3, 1)$ and $Q = (x, 4)$, find all numbers x such that the vector represented by $\overrightarrow{PQ}$ has length 5.

49. Finding Ground Speed An airplane has an airspeed of 500 kilometers per hour in an easterly direction. If the wind velocity is 60 kilometers per hour in a northwesterly direction, find the speed of the airplane relative to the ground.

50. Finding Airspeed After 1 hour in the air, an airplane arrives at a point 200 miles due south of its departure point. If there was a steady wind of 30 miles per hour from the northwest during the entire flight, what was the average airspeed of the airplane?

51. Finding Speed without Wind An airplane travels in a northwesterly direction at a constant speed of 250 miles per hour, due to a wind from the west of 50 miles per hour. How fast would the plane have gone if there had been no wind?

52. Finding the True Speed of a Motorboat A small motorboat in still water maintains a speed of 10 miles per hour. In heading directly across

a river (that is, perpendicular to the current) whose current is 4 miles per hour, what will be the true speed of the motorboat?

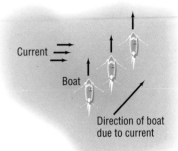

Current

Boat

Direction of boat
due to current

53. Show on the graph below the force needed to prevent an object at P from moving.

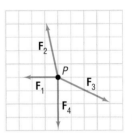

54. Explain in your own words what a vector is. Give an example of a vector.

55. Write a brief paragraph comparing the algebra of complex numbers and the algebra of vectors.

THE DOT PRODUCT

1 Find the Dot Product of Two Vectors
2 Find the Angle between Two Vectors
3 Find a Vector from Its Magnitude and Direction
4 Determine Whether Two Vectors Are Parallel
5 Determine Whether Two Vectors Are Orthogonal
6 Decompose a Vector into Two Orthogonal Vectors
7 Compute Work

The definition for a product of two vectors is somewhat unexpected. However, such a product has meaning in many geometric and physical applications.

If $\mathbf{v} = a_1\mathbf{i} + b_1\mathbf{j}$ and $\mathbf{w} = a_2\mathbf{i} + b_2\mathbf{j}$ are two vectors, the **dot product $\mathbf{v} \cdot \mathbf{w}$** is defined as

$$\mathbf{v} \cdot \mathbf{w} = a_1a_2 + b_1b_2 \qquad\qquad (1)$$

E X A M P L E 1

Finding Dot Products

If $\mathbf{v} = 2\mathbf{i} - 3\mathbf{j}$ and $\mathbf{w} = 5\mathbf{i} + 3\mathbf{j}$, find

(a) $\mathbf{v} \cdot \mathbf{w}$ (b) $\mathbf{w} \cdot \mathbf{v}$ (c) $\mathbf{v} \cdot \mathbf{v}$

(d) $\mathbf{w} \cdot \mathbf{w}$ (e) $\|\mathbf{v}\|$ (f) $\|\mathbf{w}\|$

Solution

(a) $\mathbf{v} \cdot \mathbf{w} = 2(5) + (-3)3 = 1$ (b) $\mathbf{w} \cdot \mathbf{v} = 5(2) + 3(-3) = 1$

(c) $\mathbf{v} \cdot \mathbf{v} = 2(2) + (-3)(-3) = 13$ (d) $\mathbf{w} \cdot \mathbf{w} = 5(5) + 3(3) = 34$

(e) $\|\mathbf{v}\| = \sqrt{2^2 + (-3)^2} = \sqrt{13}$ (f) $\|\mathbf{w}\| = \sqrt{5^2 + 3^2} = \sqrt{34}$ ■

Since the dot product $\mathbf{v} \cdot \mathbf{w}$ of two vectors $\mathbf{v}$ and $\mathbf{w}$ is a real number (scalar), we sometimes refer to it as the **scalar product.**

Properties

The results obtained in Example 1 suggest some general properties.

Theorem   Properties of Dot Product

If $\mathbf{u}$, $\mathbf{v}$, and $\mathbf{w}$ are vectors, then

Commutative Property

$$\mathbf{u} \cdot \mathbf{v} = \mathbf{v} \cdot \mathbf{u} \qquad\qquad (2)$$

Distributive Property

$$\mathbf{u} \cdot (\mathbf{v} + \mathbf{w}) = \mathbf{u} \cdot \mathbf{v} + \mathbf{u} \cdot \mathbf{w} \qquad\qquad (3)$$

$$\mathbf{v} \cdot \mathbf{v} = \|\mathbf{v}\|^2 \qquad\qquad (4)$$
$$\mathbf{0} \cdot \mathbf{v} = 0 \qquad\qquad (5)$$

Proof   We will prove properties (2) and (4) here and leave properties (3) and (5) as exercises (see Problems 29 and 30 at the end of this section).

To prove property (2), we let $\mathbf{u} = a_1\mathbf{i} + b_1\mathbf{j}$ and $\mathbf{v} = a_2\mathbf{i} + b_2\mathbf{j}$. Then

$$\mathbf{u} \cdot \mathbf{v} = a_1a_2 + b_1b_2 = a_2a_1 + b_2b_1 = \mathbf{v} \cdot \mathbf{u}$$

To prove property (4), we let $\mathbf{v} = a\mathbf{i} + b\mathbf{j}$. Then

$$\mathbf{v} \cdot \mathbf{v} = a^2 + b^2 = \|\mathbf{v}\|^2$$

   ■

One use of the dot product is to calculate the angle between two vectors.

Angle between Vectors

Let **u** and **v** be two vectors with the same initial point A. Then the vectors **u**, **v**, and **u** − **v** form a triangle. The angle θ at vertex A of the triangle is the **angle between the vectors u and v.** See Figure 58. We wish to find a formula for calculating the angle θ.

FIGURE 58

The sides of the triangle have lengths $\|\mathbf{v}\|$, $\|\mathbf{u}\|$, and $\|\mathbf{u} - \mathbf{v}\|$, and θ is the included angle between the sides of length $\|\mathbf{v}\|$ and $\|\mathbf{u}\|$. The Law of Cosines (Section 4.3) can be used to find the cosine of the included angle:

$$\|\mathbf{u} - \mathbf{v}\|^2 = \|\mathbf{u}\|^2 + \|\mathbf{v}\|^2 - 2\|\mathbf{u}\| \, \|\mathbf{v}\| \cos \theta$$

Now we use property (4) to rewrite this equation in terms of dot products.

$$(\mathbf{u} - \mathbf{v}) \cdot (\mathbf{u} - \mathbf{v}) = \mathbf{u} \cdot \mathbf{u} + \mathbf{v} \cdot \mathbf{v} - 2 \|\mathbf{u}\| \, \|\mathbf{v}\| \cos \theta \qquad (6)$$

Then we apply the distributive property (3) twice on the left side of (6) to obtain

$$
\begin{aligned}
(\mathbf{u} - \mathbf{v}) \cdot (\mathbf{u} - \mathbf{v}) &= \mathbf{u} \cdot (\mathbf{u} - \mathbf{v}) - \mathbf{v} \cdot (\mathbf{u} - \mathbf{v}) \\
&= \mathbf{u} \cdot \mathbf{u} - \mathbf{u} \cdot \mathbf{v} - \mathbf{v} \cdot \mathbf{u} + \mathbf{v} \cdot \mathbf{v} \\
&= \mathbf{u} \cdot \mathbf{u} + \mathbf{v} \cdot \mathbf{v} - 2 \, \mathbf{u} \cdot \mathbf{v} \qquad (7)
\end{aligned}
$$

↑
Property (2).

Combining equations (6) and (7), we have

$$\mathbf{u} \cdot \mathbf{u} + \mathbf{v} \cdot \mathbf{v} - 2 \, \mathbf{u} \cdot \mathbf{v} = \mathbf{u} \cdot \mathbf{u} + \mathbf{v} \cdot \mathbf{v} - 2 \|\mathbf{u}\| \, \|\mathbf{v}\| \cos \theta$$
$$\mathbf{u} \cdot \mathbf{v} = \|\mathbf{u}\| \, \|\mathbf{v}\| \cos \theta$$

Thus, we have proved the following result:

> **Theorem** Angle between Vectors
>
> If **u** and **v** are two nonzero vectors, the angle θ, $0 \leq \theta \leq \pi$, between **u** and **v** is determined by the formula
>
> $$\cos \theta = \frac{\mathbf{u} \cdot \mathbf{v}}{\|\mathbf{u}\| \, \|\mathbf{v}\|} \qquad (8)$$

E X A M P L E   2   Finding the Angle θ between Two Vectors

2 Find the angle θ between $\mathbf{u} = 4\mathbf{i} - 3\mathbf{j}$ and $\mathbf{v} = 2\mathbf{i} + 5\mathbf{j}$.

Solution   We compute the quantities $\mathbf{u} \cdot \mathbf{v}$, $\|\mathbf{u}\|$, and $\|\mathbf{v}\|$.

$$
\begin{aligned}
\mathbf{u} \cdot \mathbf{v} &= 4(2) + (-3)(5) = -7 \\
\|\mathbf{u}\| &= \sqrt{4^2 + (-3)^2} = 5 \\
\|\mathbf{v}\| &= \sqrt{2^2 + 5^2} = \sqrt{29}
\end{aligned}
$$

FIGURE 59

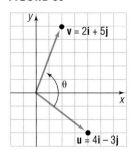

FIGURE 60

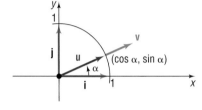

By formula (8), if θ is the angle between **u** and **v**, then

$$\cos \theta = \frac{\mathbf{u} \cdot \mathbf{v}}{\|\mathbf{u}\| \|\mathbf{v}\|} = \frac{-7}{5\sqrt{29}} \approx -0.26$$

Using a calculator, we find that $\theta \approx 105°$. See Figure 59. ■

Now work Problem 1.

Writing a Vector in Terms of Its Magnitude and Direction

Many applications describe a vector in terms of its magnitude and direction, rather than in terms of its components. Suppose that we are given the magnitude $\|\mathbf{v}\|$ of a nonzero vector **v** and the angle α between **v** and **i**. To express **v** in terms of $\|\mathbf{v}\|$ and α, we first find the unit vector **u** having the same direction as **v**.

$$\mathbf{u} = \frac{\mathbf{v}}{\|\mathbf{v}\|} \quad \text{or} \quad \mathbf{v} = \|\mathbf{v}\|\mathbf{u} \tag{9}$$

Look at Figure 60. The coordinates of the terminal point of **u** are $(\cos \alpha, \sin \alpha)$. Thus, $\mathbf{u} = \cos \alpha \, \mathbf{i} + \sin \alpha \, \mathbf{j}$ and, from (9),

$$\mathbf{v} = \|\mathbf{v}\|(\cos \alpha \, \mathbf{i} + \sin \alpha \, \mathbf{j}) \tag{10}$$

E X A M P L E   3

Writing a Vector When Its Magnitude and Direction Are Given

A force **F** of 5 pounds is applied in a direction that makes an angle of 30° with the positive x-axis. Express **F** in terms of **i** and **j**.

Solution   The magnitude of **F** is $\|\mathbf{F}\| = 5$, and the angle between the direction of **F** and **i**, the positive x-axis, is $\alpha = 30°$. Thus, by (10),

$$\mathbf{F} = \|\mathbf{F}\|(\cos \alpha \, \mathbf{i} + \sin \alpha \, \mathbf{j}) = 5(\cos 30°\mathbf{i} + \sin 30°\mathbf{j})$$

$$= 5\left(\frac{\sqrt{3}}{2}\mathbf{i} + \frac{1}{2}\mathbf{j}\right) = \frac{5}{2}(\sqrt{3}\mathbf{i} + \mathbf{j})$$ ■

E X A M P L E   4

Finding the Actual Direction of an Aircraft

A Boeing 737 aircraft maintains a constant airspeed of 500 miles per hour in the direction due south. The velocity of the jet stream is 80 miles per hour in a northeasterly direction. Find the actual direction of the aircraft relative to the ground.

Solution   This is the same information given in Example 6 of Section 5.5. We repeat the figure from that example as Figure 61.

FIGURE 61

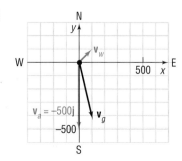

The velocity of the aircraft relative to the air is

$$\mathbf{v}_a = -500\mathbf{j}$$

The wind has magnitude 80 and direction $\alpha = 45°$. Thus, the velocity of the wind is

$$\mathbf{v}_w = 80(\cos 45°\mathbf{i} + \sin 45°\mathbf{j}) = 80\left(\frac{\sqrt{2}}{2}\mathbf{i} + \frac{\sqrt{2}}{2}\mathbf{j}\right)$$
$$= 40\sqrt{2}(\mathbf{i} + \mathbf{j})$$

The velocity of the aircraft relative to the ground is

$$\mathbf{v}_g = \mathbf{v}_a + \mathbf{v}_w = -500\mathbf{j} + 40\sqrt{2}(\mathbf{i} + \mathbf{j}) = 40\sqrt{2}\mathbf{i} + (40\sqrt{2} - 500)\mathbf{j}$$

We found in Example 6 of Section 5.5 that the speed $\|\mathbf{v}_g\|$ of the aircraft is 447 miles per hour.

The angle θ between $\mathbf{v}_g$ and the vector $\mathbf{v}_a = -500\mathbf{j}$ (the velocity of the aircraft relative to the air) is determined by the equation

$$\cos \theta = \frac{\mathbf{v}_g \cdot \mathbf{v}_a}{\|\mathbf{v}_g\| \, \|\mathbf{v}_a\|} = \frac{(40\sqrt{2} - 500)(-500)}{(447)(500)} \approx 0.9920$$
$$\theta \approx 7.3°$$

The direction of the aircraft relative to the ground is approximately S7.3° E (about 7.3° east of south). ∎

Now work Problem 19.

Parallel and Orthogonal Vectors

Two vectors $\mathbf{v}$ and $\mathbf{w}$ are said to be **parallel** if there is a nonzero scalar α so that $\mathbf{v} = \alpha\mathbf{w}$. In this case, the angle θ between $\mathbf{v}$ and $\mathbf{w}$ is 0 or π.

E X A M P L E   5   Determining Whether Vectors Are Parallel

The vectors $\mathbf{v} = 3\mathbf{i} - \mathbf{j}$ and $\mathbf{w} = 6\mathbf{i} - 2\mathbf{j}$ are parallel, since $\mathbf{v} = \frac{1}{2}\mathbf{w}$. Furthermore, since

$$\cos \theta = \frac{\mathbf{v} \cdot \mathbf{w}}{\|\mathbf{v}\| \, \|\mathbf{w}\|} = \frac{18 + 2}{\sqrt{10}\sqrt{40}} = \frac{20}{\sqrt{400}} = 1$$

the angle θ between $\mathbf{v}$ and $\mathbf{w}$ is 0. ∎

If the angle θ between two nonzero vectors $\mathbf{v}$ and $\mathbf{w}$ is $\pi/2$, the vectors $\mathbf{v}$ and $\mathbf{w}$ are called **orthogonal.***

It follows from formula (8) that if $\mathbf{v}$ and $\mathbf{w}$ are orthogonal, then $\mathbf{v} \cdot \mathbf{w} = 0$, since $\cos (\pi/2) = 0$.

On the other hand, if $\mathbf{v} \cdot \mathbf{w} = 0$, then either $\mathbf{v} = \mathbf{0}$ or $\mathbf{w} = \mathbf{0}$ or $\cos \theta = 0$. In the latter case, $\theta = \pi/2$ and $\mathbf{v}$ and $\mathbf{w}$ are orthogonal. See Figure 62. If $\mathbf{v}$ or $\mathbf{w}$ is the zero vector, then, since the zero vector has no specific direction, we adopt the convention that the zero vector is orthogonal to every vector.

FIGURE 62
$\mathbf{v} \cdot \mathbf{w} = 0$; $\mathbf{v}$ is orthogonal to $\mathbf{w}$

---

*Orthogonal, *perpendicular*, and *normal* are all terms that mean "meet at a right angle." It is customary to refer to two vectors as being *orthogonal*, two lines as being *perpendicular*, and a line and a plane or a vector and a plane as being *normal*.

> **Theorem**
>
> Two vectors **v** and **w** are orthogonal if and only if
>
> $$\mathbf{v} \cdot \mathbf{w} = 0$$

E X A M P L E   6

Determining Whether Two Vectors Are Orthogonal

FIGURE 63

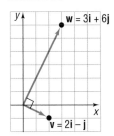

The vectors

$$\mathbf{v} = 2\mathbf{i} - \mathbf{j} \quad \text{and} \quad \mathbf{w} = 3\mathbf{i} + 6\mathbf{j}$$

are orthogonal, since

$$\mathbf{v} \cdot \mathbf{w} = 6 - 6 = 0$$

See Figure 63.

Now work Problem 11.

Projection of a Vector onto Another Vector

FIGURE 64

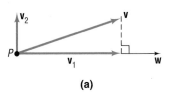

In many physical applications, it is necessary to find "how much" of a vector is applied in a given direction. Look at Figure 64. The force **F** due to gravity is pulling straight down (toward the center of Earth) on the block. To study the effect of gravity on the block, it is necessary to determine how much of **F** is actually pushing the block down the incline ($\mathbf{F}_1$) and how much is pressing the block against the incline ($\mathbf{F}_2$), at a right angle to the incline. Knowing the **decomposition** of **F** often will allow us to determine when friction is overcome and the block will slide down the incline.

Suppose that **v** and **w** are two nonzero vectors with the same initial point P. We seek to decompose **v** into two vectors: $\mathbf{v}_1$, which is parallel to **w**, and $\mathbf{v}_2$, which is orthogonal to **w**. See Figure 65(a) and (b). The vector $\mathbf{v}_1$ is called the **vector projection of v onto w** and is denoted by $\text{proj}_{\mathbf{w}}\,\mathbf{v}$.

FIGURE 65

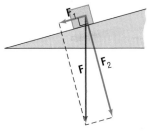

(a)

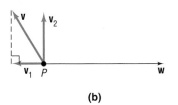

(b)

The vector $\mathbf{v}_1$ is obtained as follows: From the terminal point of **v**, drop a perpendicular to the line containing **w**. The vector $\mathbf{v}_1$ is the vector from P to the foot of this perpendicular. The vector $\mathbf{v}_2$ is given by $\mathbf{v}_2 = \mathbf{v} - \mathbf{v}_1$. Note that $\mathbf{v} = \mathbf{v}_1 + \mathbf{v}_2$, $\mathbf{v}_1$ is parallel to **w**, and $\mathbf{v}_2$ is orthogonal to **w**. This is the decomposition of **v** that we wanted.

Now we seek a formula for $\mathbf{v}_1$ that is based on a knowledge of the vectors **v** and **w**. Since $\mathbf{v} = \mathbf{v}_1 + \mathbf{v}_2$, we have

$$\mathbf{v} \cdot \mathbf{w} = (\mathbf{v}_1 + \mathbf{v}_2) \cdot \mathbf{w} = \mathbf{v}_1 \cdot \mathbf{w} + \mathbf{v}_2 \cdot \mathbf{w} \qquad (11)$$

Since $\mathbf{v}_2$ is orthogonal to **w**, we have $\mathbf{v}_2 \cdot \mathbf{w} = 0$. Since $\mathbf{v}_1$ is parallel to **w**, we have $\mathbf{v}_1 = \alpha\mathbf{w}$ for some scalar α. Thus, equation (11) can be written as

$$\mathbf{v} \cdot \mathbf{w} = \alpha\mathbf{w} \cdot \mathbf{w} = \alpha\|\mathbf{w}\|^2$$

$$\alpha = \frac{\mathbf{v} \cdot \mathbf{w}}{\|\mathbf{w}\|^2}$$

Thus,

$$\mathbf{v}_1 = \alpha\mathbf{w} = \frac{\mathbf{v} \cdot \mathbf{w}}{\|\mathbf{w}\|^2}\mathbf{w}$$

Theorem

If $\mathbf{v}$ and $\mathbf{w}$ are two nonzero vectors, the vector projection of $\mathbf{v}$ onto $\mathbf{w}$ is

$$\text{proj}_{\mathbf{w}} \mathbf{v} = \frac{\mathbf{v} \cdot \mathbf{w}}{\|\mathbf{w}\|^2} \mathbf{w}$$

The decomposition of $\mathbf{v}$ into $\mathbf{v}_1$ and $\mathbf{v}_2$, where $\mathbf{v}_1$ is parallel to $\mathbf{w}$ and $\mathbf{v}_2$ is orthogonal to $\mathbf{w}$, is

$$\mathbf{v}_1 = \text{proj}_{\mathbf{w}} \mathbf{v} = \frac{\mathbf{v} \cdot \mathbf{w}}{\|\mathbf{w}\|^2} \mathbf{w} \qquad \mathbf{v}_2 = \mathbf{v} - \mathbf{v}_1 \qquad (12)$$

E X A M P L E   7

FIGURE 66

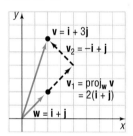

Decomposing a Vector into Two Orthogonal Vectors

Find the vector projection of $\mathbf{v} = \mathbf{i} + 3\mathbf{j}$ onto $\mathbf{w} = \mathbf{i} + \mathbf{j}$. Decompose $\mathbf{v}$ into two vectors $\mathbf{v}_1$ and $\mathbf{v}_2$, where $\mathbf{v}_1$ is parallel to $\mathbf{w}$ and $\mathbf{v}_2$ is orthogonal to $\mathbf{w}$.

Solution   We use formulas (12).

$$\mathbf{v}_1 = \text{proj}_{\mathbf{w}} \mathbf{v} = \frac{\mathbf{v} \cdot \mathbf{w}}{\|\mathbf{w}\|^2} \mathbf{w} = \frac{1 + 3}{(\sqrt{2})^2} \mathbf{w} = 2\mathbf{w} = 2(\mathbf{i} + \mathbf{j})$$

$$\mathbf{v}_2 = \mathbf{v} - \mathbf{v}_1 = (\mathbf{i} + 3\mathbf{j}) - 2(\mathbf{i} + \mathbf{j}) = -\mathbf{i} + \mathbf{j}$$

See Figure 66.

Now work Problem 13.

Work Done by a Constant Force

In elementary physics, the **work** W done by a constant force $\mathbf{F}$ in moving an object from a point A to a point B is defined as

$$W = (\text{magnitude of force})(\text{distance}) = \|\mathbf{F}\| \, \|\overrightarrow{AB}\|$$

(Work is commonly measured in foot-pounds or in newton-meters.)

In this definition, it is assumed that the force $\mathbf{F}$ is applied along the line of motion. If the constant force $\mathbf{F}$ is not along the line of motion, but, instead, is at an angle θ to the direction of motion, as illustrated in Figure 67, then the **work** W **done by** $\mathbf{F}$ in moving an object from A to B is defined as

$$W = \mathbf{F} \cdot \overrightarrow{AB} \qquad (13)$$

This definition is compatible with the force times distance definition given above, since

$$W = (\text{amount of force in the direction of } \overrightarrow{AB})(\text{distance})$$

$$= \|\text{proj}_{\overrightarrow{AB}} \mathbf{F}\| \, \|\overrightarrow{AB}\| = \frac{\mathbf{F} \cdot \overrightarrow{AB}}{\|\overrightarrow{AB}\|^2} \|\overrightarrow{AB}\| \, \|\overrightarrow{AB}\| = \mathbf{F} \cdot \overrightarrow{AB}$$

FIGURE 67

E X A M P L E   8

Computing Work

Find the work done by a force of 5 pounds acting in the direction $\mathbf{i} + \mathbf{j}$ in moving an object 1 foot from $(0, 0)$ to $(1, 0)$.

Solution   First, we must express the force **F** as a vector. The force has magnitude 5 and direction **i** + **j**. The direction of **F** therefore makes an angle of 45° with **i**. Thus, the force **F** is

$$\mathbf{F} = 5(\cos 45°\mathbf{i} + \sin 45°\mathbf{j}) = 5\left(\frac{\sqrt{2}}{2}\mathbf{i} + \frac{\sqrt{2}}{2}\mathbf{j}\right) = \frac{5\sqrt{2}}{2}(\mathbf{i} + \mathbf{j})$$

The line of motion of the object is from $A = (0, 0)$ to $B = (1, 0)$, so $\overrightarrow{AB} = \mathbf{i}$. The work W is therefore

$$W = \mathbf{F} \cdot \overrightarrow{AB} = \frac{5\sqrt{2}}{2}(\mathbf{i} + \mathbf{j}) \cdot \mathbf{i} = \frac{5\sqrt{2}}{2} \text{ foot-pounds}$$

Now work Problem 25.

E X A M P L E   9   Computing Work

Figure 68(a) shows a girl pulling a wagon with a force of 50 pounds. How much work is done in moving the wagon 100 feet if the handle makes an angle of 30° with the ground?

FIGURE 68

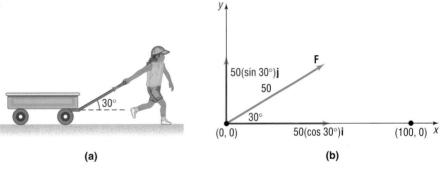

(a)   (b)

Solution   We position the vectors in a coordinate system in such a way that the wagon is moved from $(0, 0)$ to $(100, 0)$. Thus, the motion is from $A = (0, 0)$ to $B = (100, 0)$, so $\overrightarrow{AB} = 100\mathbf{i}$. The force vector **F**, as shown in Figure 68(b), is

$$\mathbf{F} = 50(\cos 30°\mathbf{i} + \sin 30°\mathbf{j}) = 50\left(\frac{\sqrt{3}}{2}\mathbf{i} + \frac{1}{2}\mathbf{j}\right) = 25\sqrt{3}\mathbf{i} + 25\mathbf{j}$$

By formula (13), the work W done is

$$W = \mathbf{F} \cdot \overrightarrow{AB} = (25\sqrt{3}\mathbf{i} + 25\mathbf{j}) \cdot 100\mathbf{i} = 2500\sqrt{3} \text{ foot-pounds}$$

HISTORICAL PROBLEM   1.   We stated in an earlier Historical Feature that complex numbers were used as vectors in the plane before the general notion of vector was clarified. Suppose that we make the correspondence

Vector ↔ Complex number
$a\mathbf{i} + b\mathbf{j} \leftrightarrow a + bi$
$c\mathbf{i} + d\mathbf{j} \leftrightarrow c + di$

Show that

$$(a\mathbf{i} + b\mathbf{j}) \cdot (c\mathbf{i} + d\mathbf{j}) = \text{real part}[\overline{(a + bi)}(c + di)]$$

This is how the dot product was found originally. The imaginary part is also interesting. It is a determinant and represents the area of the parallelogram whose edges are the vectors. This is close to some of Hermann Grassmann's ideas and is also connected with the scalar triple product of three-dimensional vectors.

5.6 | EXERCISES

In Problems 1–10, find the dot product $\mathbf{v} \cdot \mathbf{w}$ *and the angle between* $\mathbf{v}$ *and* $\mathbf{w}$.

1. $\mathbf{v} = \mathbf{i} - \mathbf{j}, \quad \mathbf{w} = \mathbf{i} + \mathbf{j}$
2. $\mathbf{v} = \mathbf{i} + \mathbf{j}, \quad \mathbf{w} = -\mathbf{i} + \mathbf{j}$
3. $\mathbf{v} = 2\mathbf{i} + \mathbf{j}, \quad \mathbf{w} = \mathbf{i} + 2\mathbf{j}$
4. $\mathbf{v} = 2\mathbf{i} + 2\mathbf{j}, \quad \mathbf{w} = \mathbf{i} + 2\mathbf{j}$
5. $\mathbf{v} = \sqrt{3}\mathbf{i} - \mathbf{j}, \quad \mathbf{w} = \mathbf{i} + \mathbf{j}$
6. $\mathbf{v} = \mathbf{i} + \sqrt{3}\mathbf{j}, \quad \mathbf{w} = \mathbf{i} - \mathbf{j}$
7. $\mathbf{v} = 3\mathbf{i} + 4\mathbf{j}, \quad \mathbf{w} = 4\mathbf{i} + 3\mathbf{j}$
8. $\mathbf{v} = 3\mathbf{i} - 4\mathbf{j}, \quad \mathbf{w} = 4\mathbf{i} - 3\mathbf{j}$
9. $\mathbf{v} = 4\mathbf{i}, \quad \mathbf{w} = \mathbf{j}$
10. $\mathbf{v} = \mathbf{i}, \quad \mathbf{w} = -3\mathbf{j}$

11. Find a such that the angle between $\mathbf{v} = a\mathbf{i} - \mathbf{j}$ and $\mathbf{w} = 2\mathbf{i} + 3\mathbf{j}$ is $\pi/2$.
12. Find b such that the angle between $\mathbf{v} = \mathbf{i} + \mathbf{j}$ and $\mathbf{w} = \mathbf{i} + b\mathbf{j}$ is $\pi/2$.

In Problems 13–18, decompose $\mathbf{v}$ *into two vectors* $\mathbf{v}_1$ *and* $\mathbf{v}_2$, *where* $\mathbf{v}_1$ *is parallel to* $\mathbf{w}$ *and* $\mathbf{v}_2$ *is orthogonal to* $\mathbf{w}$.

13. $\mathbf{v} = 2\mathbf{i} - 3\mathbf{j}, \quad \mathbf{w} = \mathbf{i} - \mathbf{j}$
14. $\mathbf{v} = -3\mathbf{i} + 2\mathbf{j}, \quad \mathbf{w} = 2\mathbf{i} + \mathbf{j}$
15. $\mathbf{v} = \mathbf{i} - \mathbf{j}, \quad \mathbf{w} = \mathbf{i} + 2\mathbf{j}$
16. $\mathbf{v} = 2\mathbf{i} - \mathbf{j}, \quad \mathbf{w} = \mathbf{i} - 2\mathbf{j}$
17. $\mathbf{v} = 3\mathbf{i} + \mathbf{j}, \quad \mathbf{w} = -2\mathbf{i} - \mathbf{j}$
18. $\mathbf{v} = \mathbf{i} - 3\mathbf{j}, \quad \mathbf{w} = 4\mathbf{i} - \mathbf{j}$

19. **Finding the Actual Speed and Distance of an Aircraft** A DC-10 jumbo jet maintains an airspeed of 550 miles per hour in a southwesterly direction. The velocity of the jet stream is a constant 80 miles per hour from the west. Find the actual speed and direction of the aircraft.

Jet stream

20. **Finding the Correct Compass Heading** The pilot of an aircraft wishes to head directly east but is faced with a wind speed of 40 miles per hour from the northwest. If the pilot maintains an airspeed of 250 miles per hour, what compass heading should be maintained? What is the actual speed of the aircraft?

21. **Correct Direction for Crossing a River** A river has a constant current of 3 kilometers per hour. At what angle to a boat dock should a motorboat, capable of maintaining a constant speed of 20 kilometers per hour, be headed in order to reach a point directly opposite the dock? If the river is $\frac{1}{2}$ kilometer wide, how long will it take to cross?

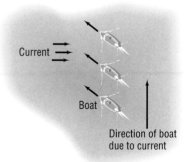

Current

Boat

Direction of boat due to current

22. **Correct Direction for Crossing a River** Repeat Problem 21 if the current is 5 kilometers per hour.

23. Correct Direction for Crossing a River A river is 500 meters wide and has a current of 1 kilometer per hour. If Tom can swim at a rate of 2 kilometers per hour, at what angle to the shore should he swim if he wishes to cross the river to a point directly opposite? How long will it take to swim across the river?

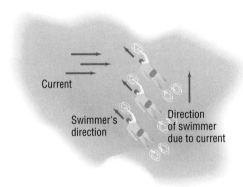

Current

Swimmer's direction

Direction of swimmer due to current

24. Finding the Position of an Airplane An airplane travels 200 miles due west and then 150 miles 60° north of west. Determine the resultant displacement.

25. Computing Work Find the work done by a force of 3 pounds acting in the direction $2\mathbf{i} + \mathbf{j}$ in moving an object 2 feet from $(0, 0)$ to $(0, 2)$.

26. Computing Work Find the work done by a force of 1 pound acting in the direction $2\mathbf{i} + 2\mathbf{j}$ in moving an object 5 feet from $(0, 0)$ to $(3, 4)$.

27. Computing Work A wagon is pulled horizontally by exerting a force of 20 pounds on the handle at an angle of 30° with the horizontal. How much work is done in moving the wagon 100 feet?

28. Finding the Angle of a Force Vector Find the acute angle that a constant unit force vector makes with the positive x-axis if the work done by the force in moving a particle from $(0, 0)$ to $(4, 0)$ equals 2.

29. Prove the distributive property.

$$\mathbf{u} \cdot (\mathbf{v} + \mathbf{w}) = \mathbf{u} \cdot \mathbf{v} + \mathbf{u} \cdot \mathbf{w}$$

30. Prove property (5), $\mathbf{0} \cdot \mathbf{v} = 0$.

31. If $\mathbf{v}$ is a unit vector and the angle between $\mathbf{v}$ and $\mathbf{i}$ is α, show that $\mathbf{v} = \cos \alpha \mathbf{i} + \sin \alpha \mathbf{j}$.

32. Suppose that $\mathbf{v}$ and $\mathbf{w}$ are unit vectors. If the angle between $\mathbf{v}$ and $\mathbf{i}$ is α and if the angle between $\mathbf{w}$ and $\mathbf{i}$ is β, use the idea of the dot product $\mathbf{v} \cdot \mathbf{w}$ to prove that

$$\cos(\alpha - \beta) = \cos \alpha \cos \beta + \sin \alpha \sin \beta$$

33. Show that the projection of $\mathbf{v}$ onto $\mathbf{i}$ is $(\mathbf{v} \cdot \mathbf{i})\mathbf{i}$. In fact, show that we can always write a vector $\mathbf{v}$ as

$$\mathbf{v} = (\mathbf{v} \cdot \mathbf{i})\mathbf{i} + (\mathbf{v} \cdot \mathbf{j})\mathbf{j}$$

34. (a) If $\mathbf{u}$ and $\mathbf{v}$ have the same magnitude, then show that $\mathbf{u} + \mathbf{v}$ and $\mathbf{u} - \mathbf{v}$ are orthogonal.
(b) Use this to prove that an angle inscribed in a semicircle is a right angle (see the figure).

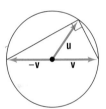

35. Let $\mathbf{v}$ and $\mathbf{w}$ denote two nonzero vectors. Show that the vector $\mathbf{v} - \alpha\mathbf{w}$ is orthogonal to $\mathbf{w}$ if $\alpha = (\mathbf{v} \cdot \mathbf{w})/\|\mathbf{w}\|^2$.

36. Let $\mathbf{v}$ and $\mathbf{w}$ denote two nonzero vectors. Show that the vectors $\|\mathbf{w}\|\mathbf{v} + \|\mathbf{v}\|\mathbf{w}$ and $\|\mathbf{w}\|\mathbf{v} - \|\mathbf{v}\|\mathbf{w}$ are orthogonal.

37. In the definition of work given in this section, what is the work done if $\mathbf{F}$ is orthogonal to $\overrightarrow{AB}$?

38. Prove the **polarization identity,**

$$\|\mathbf{u} + \mathbf{v}\|^2 - \|\mathbf{u} - \mathbf{v}\|^2 = 4(\mathbf{u} \cdot \mathbf{v})$$

39. Make up an application different from any found in the text that requires the dot product.

5.7 | VECTORS IN SPACE

1 Find the Distance between Two Points
2 Find Position Vectors
3 Perform Operations on Vectors
4 Find the Dot Product
5 Find the Angle between Two Vectors
6 Find the Direction Angles of a Vector

FIGURE 69

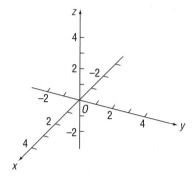

Rectangular Coordinates in Space

In the plane, each point is associated with an ordered pair of real numbers. In space, each point is associated with an ordered triple of real numbers. Through a fixed point, the *origin, O*, draw three mutually perpendicular lines, the *x-axis*, the *y-axis*, and the *z-axis*. On each of these axes, select an appropriate scale and the positive direction. See Figure 69.

The direction chosen for the positive *z*-axis in Figure 69 makes the system *right-handed*. This conforms to the *right-hand rule*, which states that, if the index finger of the right hand points in the direction of the positive *x*-axis and the middle finger points in the direction of the positive *y*-axis, then the thumb will point in the direction of the positive *z*-axis. See Figure 70.

FIGURE 70

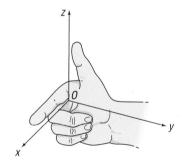

We associate with each point P an ordered triple (x, y, z) of real numbers, the *coordinates of P*. For example, the point $(2, 3, 4)$ is located by starting at the origin and moving 2 units along the positive *x*-axis, 3 units in the direction of the positive *y*-axis, and 4 units in the direction of the positive *z*-axis. See Figure 71.

FIGURE 71

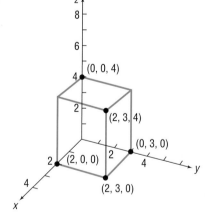

Figure 71 also shows the location of the points $(2, 0, 0)$, $(0, 3, 0)$, $(0, 0, 4)$, and $(2, 3, 0)$. Points of the form $(x, 0, 0)$ lie on the *x*-axis, while points of the form $(0, y, 0)$ and $(0, 0, z)$ lie on the *y*-axis and *z*-axis, respectively. Points of the form $(x, y, 0)$ lie in a plane, called the *xy-plane*. Its equation is $z = 0$. Similarly, points of the form $(x, 0, z)$ lie in the *xz-plane* (equation $y = 0$) and points of the form $(0, y, z)$ lie in the *yz-plane* (equation $x = 0$). See Figure 72(a). By extension of these ideas, all points obeying the equation $z = 3$ will lie in a plane parallel to and 3 units above the *xy*-plane. The equation $y = 4$

represents a plane parallel to the xz-plane and 4 units to the right of the plane $y = 0$. See Figure 72(b).

FIGURE 72

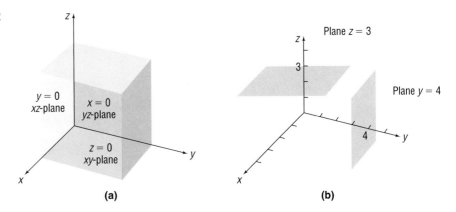

(a)   (b)

Now work Problem 3.

The formula for the distance between two points in space is an extension of the Distance Formula for points in the plane given in Chapter 1.

> **Theorem**   Distance Formula in Space
>
> If $P_1 = (x_1, y_1, z_1)$ and $P_2 = (x_2, y_2, z_2)$ are two points in space, the distance d from P_1 to P_2 is
>
> $$d = \sqrt{(x_2 - x_1)^2 + (y_2 - y_1)^2 + (z_2 - z_1)^2} \qquad (1)$$

The proof, which we omit, utilizes a double application of the Pythagorean Theorem.

E X A M P L E   1   Using the Distance Formula

Find the distance from $P_1 = (-1, 3, 2)$ to $P_2 = (4, -2, 5)$.

Solution $d = \sqrt{[4 - (-1)]^2 + [-2 - 3]^2 + [5 - 2]^2} = \sqrt{25 + 25 + 9} = \sqrt{59}$

Now work Problem 9.

Representing Vectors in Space

To represent vectors in space, we introduce the unit vectors $\mathbf{i}, \mathbf{j},$ and $\mathbf{k}$ whose directions are along the positive x-axis, positive y-axis, and positive z-axis, respectively. If $\mathbf{v}$ is a vector with initial point at the origin O and terminal point at $P = (a, b, c)$, then we can represent $\mathbf{v}$ in terms of the vectors $\mathbf{i}, \mathbf{j},$ and $\mathbf{k}$ as

$$\mathbf{v} = a\mathbf{i} + b\mathbf{j} + c\mathbf{k}$$

The scalars a, b, and c are called the **components** of the vector $\mathbf{v} = a\mathbf{i} + b\mathbf{j} + c\mathbf{k}$, with a being the component in the direction $\mathbf{i}$, b the component in the direction $\mathbf{j}$, and c the component in the direction $\mathbf{k}$. See Figure 73.

A vector whose initial point is at the origin is called a **position vector.** The next result states that any vector whose initial point is not at the origin is equal to a unique position vector.

FIGURE 73

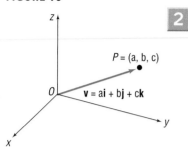

Theorem

Suppose that $\mathbf{v}$ is a vector with initial point $P_1 = (x_1, y_1, z_1)$, not necessarily the origin, and terminal point $P_2 = (x_2, y_2, z_2)$. If $\mathbf{v} = \overrightarrow{P_1 P_2}$, then $\mathbf{v}$ is equal to the position vector

$$\mathbf{v} = (x_2 - x_1)\mathbf{i} + (y_2 - y_1)\mathbf{j} + (z_2 - z_1)\mathbf{k} \qquad (2)$$

Figure 74 illustrates this result.

FIGURE 74

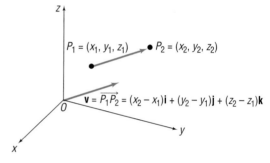

E X A M P L E   2

Finding a Position Vector

Find the position vector of the vector $\mathbf{v} = \overrightarrow{P_1 P_2}$ if $P_1 = (-1, 2, 3)$ and $P_2 = (4, 6, 2)$.

Solution   By equation (2), the position vector equal to $\mathbf{v}$ is

$$\mathbf{v} = [4 - (-1)]\mathbf{i} + (6 - 2)\mathbf{j} + (2 - 3)\mathbf{k} = 5\mathbf{i} + 4\mathbf{j} - \mathbf{k}$$

Now work Problem 23.

Next, we define equality, addition, subtraction, scalar product, and magnitude in terms of the components of a vector.

Let $\mathbf{v} = a_1\mathbf{i} + b_1\mathbf{j} + c_1\mathbf{k}$ and $\mathbf{w} = a_2\mathbf{i} + b_2\mathbf{j} + c_2\mathbf{k}$ be two vectors, and let α be a scalar. Then

$$\mathbf{v} = \mathbf{w} \quad \text{if and only if } a_1 = a_2, b_1 = b_2, \text{ and } c_1 = c_2$$
$$\mathbf{v} + \mathbf{w} = (a_1 + a_2)\mathbf{i} + (b_1 + b_2)\mathbf{j} + (c_1 + c_2)\mathbf{k}$$
$$\mathbf{v} - \mathbf{w} = (a_1 - a_2)\mathbf{i} + (b_1 - b_2)\mathbf{j} + (c_1 - c_2)\mathbf{k}$$
$$\alpha\mathbf{v} = (\alpha a_1)\mathbf{i} + (\alpha b_1)\mathbf{j} + (\alpha c_1)\mathbf{k}$$
$$\|\mathbf{v}\| = \sqrt{a_1^2 + b_1^2 + c_1^2}$$

These definitions are compatible with the geometric ones given earlier in Section 5.5.

E X A M P L E 3

Adding and Subtracting Vectors

If $\mathbf{v} = 2\mathbf{i} + 3\mathbf{j} - 2\mathbf{k}$ and $\mathbf{w} = 3\mathbf{i} - 4\mathbf{j} + 5\mathbf{k}$, find:

(a) $\mathbf{v} + \mathbf{w}$ (b) $\mathbf{v} - \mathbf{w}$

Solution (a) $\mathbf{v} + \mathbf{w} = (2\mathbf{i} + 3\mathbf{j} - 2\mathbf{k}) + (3\mathbf{i} - 4\mathbf{j} + 5\mathbf{k})$
$= (2 + 3)\mathbf{i} + (3 - 4)\mathbf{j} + (-2 + 5)\mathbf{k}$
$= 5\mathbf{i} - \mathbf{j} + 3\mathbf{k}$

(b) $\mathbf{v} - \mathbf{w} = (2\mathbf{i} + 3\mathbf{j} - 2\mathbf{k}) - (3\mathbf{i} - 4\mathbf{j} + 5\mathbf{k})$
$= (2 - 3)\mathbf{i} + [3 - (-4)]\mathbf{j} + [-2 - 5]\mathbf{k}$
$= -\mathbf{i} + 7\mathbf{j} - 7\mathbf{k}$

E X A M P L E 4

Finding Scalar Products and Magnitudes

If $\mathbf{v} = 2\mathbf{i} + 3\mathbf{j} - 2\mathbf{k}$ and $\mathbf{w} = 3\mathbf{i} - 4\mathbf{j} + 5\mathbf{k}$, find:

(a) $3\mathbf{v}$ (b) $2\mathbf{v} - 3\mathbf{w}$ (c) $\|\mathbf{v}\|$

Solution (a) $3\mathbf{v} = 3(2\mathbf{i} + 3\mathbf{j} - 2\mathbf{k}) = 6\mathbf{i} + 9\mathbf{j} - 6\mathbf{k}$
(b) $2\mathbf{v} - 3\mathbf{w} = 2(2\mathbf{i} + 3\mathbf{j} - 2\mathbf{k}) - 3(3\mathbf{i} - 4\mathbf{j} + 5\mathbf{k})$
$= 4\mathbf{i} + 6\mathbf{j} - 4\mathbf{k} - 9\mathbf{i} + 12\mathbf{j} - 15\mathbf{k} = -5\mathbf{i} + 18\mathbf{j} - 19\mathbf{k}$
(c) $\|\mathbf{v}\| = \|2\mathbf{i} + 3\mathbf{j} - 2\mathbf{k}\| = \sqrt{2^2 + 3^2 + (-2^2)} = \sqrt{17}$

 Now work Problems 27 and 33.

Recall that a unit vector $\mathbf{u}$ is one for which $\|\mathbf{u}\| = 1$. In many applications, it is useful to be able to find a unit vector $\mathbf{u}$ that has the same direction as a given vector $\mathbf{v}$.

Theorem   Unit Vector in Direction of $\mathbf{v}$

For any nonzero vector $\mathbf{v}$, the vector

$$\mathbf{u} = \frac{\mathbf{v}}{\|\mathbf{v}\|}$$

is a unit vector that has the same direction as $\mathbf{v}$.

As a consequence of this theorem, if $\mathbf{u}$ is a unit vector in the same direction as a vector $\mathbf{v}$, then $\mathbf{v}$ may be expressed as

$\mathbf{v} = \|\mathbf{v}\|\,\mathbf{u}$

This way of expressing a vector is useful in many applications.

E X A M P L E   5   Finding a Unit Vector

Find a unit vector in the same direction as $\mathbf{v} = 2\mathbf{i} - 3\mathbf{j} - 6\mathbf{k}$.

Solution   We find $\|\mathbf{v}\|$ first.

$$\|\mathbf{v}\| = \|2\mathbf{i} - 3\mathbf{j} - 6\mathbf{k}\| = \sqrt{4 + 9 + 36} = \sqrt{49} = 7$$

Now we multiply $\mathbf{v}$ by the scalar $1/\|\mathbf{v}\| = \frac{1}{7}$. The result is the unit vector

$$\mathbf{u} = \frac{\mathbf{v}}{\|\mathbf{v}\|} = \frac{2\mathbf{i} - 3\mathbf{j} - 6\mathbf{k}}{7} = \frac{2}{7}\mathbf{i} - \frac{3}{7}\mathbf{j} - \frac{6}{7}\mathbf{k}$$

  Now work Problem 41.

Dot Product

The definition of *dot product* is an extension of the definition given for vectors in the plane.

If $\mathbf{v} = a_1\mathbf{i} + b_1\mathbf{j} + c_1\mathbf{k}$ and $\mathbf{w} = a_2\mathbf{i} + b_2\mathbf{j} + c_2\mathbf{k}$ are two vectors, the **dot product $\mathbf{v} \cdot \mathbf{w}$** is defined as

$$\mathbf{v} \cdot \mathbf{w} = a_1a_2 + b_1b_2 + c_1c_2 \tag{3}$$

E X A M P L E   6   Finding Dot Products

If $\mathbf{v} = 2\mathbf{i} - 3\mathbf{j} + 6\mathbf{k}$ and $\mathbf{w} = 5\mathbf{i} + 3\mathbf{j} - \mathbf{k}$, find:

(a) $\mathbf{v} \cdot \mathbf{w}$ (b) $\mathbf{w} \cdot \mathbf{v}$ (c) $\mathbf{v} \cdot \mathbf{v}$
(d) $\mathbf{w} \cdot \mathbf{w}$ (e) $\|\mathbf{v}\|$ (f) $\|\mathbf{w}\|$

Solution   (a) $\mathbf{v} \cdot \mathbf{w} = 2(5) + (-3)3 + 6(-1) = -5$
(b) $\mathbf{w} \cdot \mathbf{v} = 5(2) + 3(-3) + (-1)(6) = -5$
(c) $\mathbf{v} \cdot \mathbf{v} = 2(2) + (-3)(-3) + 6(6) = 49$
(d) $\mathbf{w} \cdot \mathbf{w} = 5(5) + 3(3) + (-1)(-1) = 35$
(e) $\|\mathbf{v}\| = \sqrt{2^2 + (-3)^2 + 6^2} = \sqrt{49} = 7$
(f) $\|\mathbf{w}\| = \sqrt{5^2 + 3^2 + (-1)^2} = \sqrt{35}$

The dot product in space has the same properties as the dot product in the plane.

Theorem   Properties of the Dot Product

If **u, v,** and **w** are vectors, then

Commutative Property

$$\mathbf{u} \cdot \mathbf{v} = \mathbf{v} \cdot \mathbf{u}$$

Distributive Property

$$\mathbf{u} \cdot (\mathbf{v} + \mathbf{w}) = \mathbf{u} \cdot \mathbf{v} + \mathbf{u} \cdot \mathbf{w}$$

$$\mathbf{v} \cdot \mathbf{v} = \|\mathbf{v}\|^2$$
$$\mathbf{0} \cdot \mathbf{v} = 0$$

5   The angle θ between two vectors in space follows the same formula as for two vectors in the plane.

Theorem   Angle between Vectors

If **u** and **v** are two nonzero vectors, the angle θ, $0 \le \theta \le \pi$, between **u** and **v** is determined by the formula

$$\cos \theta = \frac{\mathbf{u} \cdot \mathbf{v}}{\|\mathbf{u}\| \, \|\mathbf{v}\|} \qquad (4)$$

E X A M P L E   7   Finding the Angle θ between Two Vectors

Find the angle θ between $\mathbf{u} = 2\mathbf{i} - 3\mathbf{j} + 6\mathbf{k}$ and $\mathbf{v} = 2\mathbf{i} + 5\mathbf{j} - \mathbf{k}.$

Solution   We compute the quantities $\mathbf{u} \cdot \mathbf{v}$, $\|\mathbf{u}\|$, and $\|\mathbf{v}\|$:

$$\mathbf{u} \cdot \mathbf{v} = 2(2) + (-3)(5) + 6(-1) = -17$$
$$\|\mathbf{u}\| = \sqrt{2^2 + (-3)^2 + 6^2} = \sqrt{49} = 7$$
$$\|\mathbf{v}\| = \sqrt{2^2 + 5^2 + (-1)^2} = \sqrt{30}$$

By formula (4), if θ is the angle between **u** and **v,** then

$$\cos \theta = \frac{\mathbf{u} \cdot \mathbf{v}}{\|\mathbf{u}\| \, \|\mathbf{v}\|} = \frac{-17}{7\sqrt{30}} \approx -0.443$$

Using a calculator, we find that $\theta \approx 116°.$

 Now work Problem 45.

Direction Angles of Vectors in Space

A nonzero vector **v** in space can be described by specifying its three **direction angles** α, β, and γ and its magnitude. These direction angles are defined as

α = angle between **v** and the positive x-axis, $0° \leq \alpha \leq 180°$
β = angle between **v** and the positive y-axis, $0° \leq \beta \leq 180°$
γ = angle between **v** and the positive z-axis, $0° \leq \gamma \leq 180°$

See Figure 75.

FIGURE 75

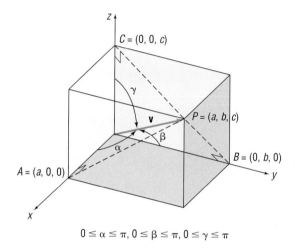

$$0 \leq \alpha \leq \pi, 0 \leq \beta \leq \pi, 0 \leq \gamma \leq \pi$$

Our first goal is to find an expression for α, β, and γ in terms of the components of a vector. Let $\mathbf{v} = a\mathbf{i} + b\mathbf{j} + c\mathbf{k}$ denote a nonzero vector. The angle α between **v** and the positive x-axis obeys

$$\cos \alpha = \frac{\mathbf{v} \cdot \mathbf{i}}{\|\mathbf{v}\| \, \|\mathbf{i}\|} = \frac{a}{\|\mathbf{v}\|}$$

Similarly,

$$\cos \beta = \frac{b}{\|\mathbf{v}\|} \qquad \cos \gamma = \frac{c}{\|\mathbf{v}\|}$$

Since $\|\mathbf{v}\| = \sqrt{a^2 + b^2 + c^2}$, we have the following result:

Theorem   Direction Angles

If $\mathbf{v} = a\mathbf{i} + b\mathbf{j} + c\mathbf{k}$ is a nonzero vector in space, the direction angles α, β, and γ obey

$$\cos \alpha = \frac{a}{\sqrt{a^2 + b^2 + c^2}}, \qquad \cos \beta = \frac{b}{\sqrt{a^2 + b^2 + c^2}}, \qquad \cos \gamma = \frac{c}{\sqrt{a^2 + b^2 + c^2}} \qquad (5)$$

The numbers $\cos \alpha$, $\cos \beta$, and $\cos \gamma$ are called the **direction cosines** of the vector **v.** They play the same role in space as slope does in the plane.

EXAMPLE 8   Finding the Direction Angles of a Vector

Find the direction angles of $\mathbf{v} = -3\mathbf{i} + 2\mathbf{j} - 6\mathbf{k}$.

Solution $\|\mathbf{v}\| = \sqrt{(-3)^2 + 2^2 + (-6)^2} = \sqrt{49} = 7$

Using the theorem on direction angles, we get

$$\cos \alpha = \frac{-3}{7} \qquad \cos \beta = \frac{2}{7} \qquad \cos \gamma = \frac{-6}{7}$$
$$\alpha = 115° \qquad \beta = 73° \qquad \gamma = 149°$$

■

Theorem   Property of Direction Cosines

If α, β, and γ are the direction angles of a nonzero vector **v** in space, then

$$\cos^2 \alpha + \cos^2 \beta + \cos^2 \gamma = 1 \qquad (6)$$

■

The proof is a direct consequence of (5).

Based on (6), when two direction cosines are known, the third is determined up to its sign. Thus, knowing two direction cosines is not sufficient to uniquely determine the direction of a vector in space.

EXAMPLE 9   Finding the Direction Angle of a Vector

The vector **v** makes an angle of $\alpha = \pi/3$ with the positive x-axis, an angle of $\beta = \pi/3$ with the positive y-axis, and an acute angle γ with the positive z-axis. Find γ.

Solution   By (6), we have

$$\cos^2 \left(\frac{\pi}{3} \right) + \cos^2 \left(\frac{\pi}{3} \right) + \cos^2 \gamma = 1$$
$$\left(\frac{1}{2} \right)^2 + \left(\frac{1}{2} \right)^2 + \cos^2 \gamma = 1$$
$$\cos^2 \gamma = \frac{1}{2}$$
$$\cos \gamma = \frac{\sqrt{2}}{2} \quad \text{or} \quad \cos \gamma = -\frac{\sqrt{2}}{2}$$
$$\gamma = \frac{\pi}{4} \quad \text{or} \quad \gamma = \frac{3\pi}{4}$$

Since we are requiring that γ be acute, the answer is $\gamma = \pi/4$.   ■

The direction cosines of a vector give information about only the direction of the vector; they provide no information about its magnitude. For example, *any* vector parallel to the *xy*-plane and making an angle of $\pi/4$ radian with the positive *x*- and *y*-axes has direction cosines

$$\cos \alpha = \frac{\sqrt{2}}{2} \qquad \cos \beta = \frac{\sqrt{2}}{2} \qquad \cos \gamma = 0$$

However, if the direction angles *and* the magnitude of a vector are known, then the vector is uniquely determined.

E X A M P L E   10   Writing a Vector in Terms of Its Magnitude and Direction Cosines

Show that any nonzero vector **v** in space can be written in terms of its magnitude and direction cosines as

$$\mathbf{v} = \|\mathbf{v}\|[(\cos \alpha)\mathbf{i} + (\cos \beta)\mathbf{j} + (\cos \gamma)\mathbf{k}] \tag{7}$$

Solution   Let $\mathbf{v} = a\mathbf{i} + b\mathbf{j} + c\mathbf{k}$. From (5), we see that

$$a = \|\mathbf{v}\| \cos \alpha \qquad b = \|\mathbf{v}\| \cos \beta \qquad c = \|\mathbf{v}\| \cos \gamma$$

Thus,

$$\mathbf{v} = a\mathbf{i} + b\mathbf{j} + c\mathbf{k} = \|\mathbf{v}\|(\cos \alpha)\mathbf{i} + \|\mathbf{v}\|(\cos \beta)\mathbf{j} + \|\mathbf{v}\|(\cos \gamma)\mathbf{k}$$
$$= \|\mathbf{v}\|[(\cos \alpha)\mathbf{i} + (\cos \beta)\mathbf{j} + (\cos \gamma)\mathbf{k}]$$

Example 10 shows that the direction cosines of a vector **v** are also the components of the unit vector in the direction of **v**.

5.7 EXERCISES

In Problems 1–8, describe the set of points (x, y, z) defined by the equation.

1. $y = 0$

2. $x = 0$

3. $z = 2$

4. $y = 3$

5. $x = -4$

6. $z = -3$

7. $x = 1$ and $y = 2$

8. $x = 3$ and $z = 1$

In Problems 9–14, find the distance from P_1 to P_2.

9. $P_1 = (0, 0, 0)$ and $P_2 = (4, 1, 2)$

10. $P_1 = (0, 0, 0)$ and $P_2 = (1, -2, 3)$

11. $P_1 = (-1, 2, -3)$ and $P_2 = (0, -2, 1)$

12. $P_1 = (-2, 2, 3)$ and $P_2 = (4, 0, -3)$

13. $P_1 = (4, -2, -2)$ and $P_2 = (3, 2, 1)$

14. $P_1 = (2, -3, -3)$ and $P_2 = (4, 1, -1)$

In Problems 15–20, opposite vertices of a rectangular box whose edges are parallel to the coordinate axes are given. List the coordinates of the other six vertices of the box.

15. $(0, 0, 0);$ $(2, 1, 3)$

16. $(0, 0, 0);$ $(4, 2, 2)$

17. $(1, 2, 3);$ $(3, 4, 5)$

18. $(5, 6, 1);$ $(3, 8, 2)$

19. $(-1, 0, 2);$ $(4, 2, 5)$

20. $(-2, -3, 0);$ $(-6, 7, 1)$

*In Problems 21–26, the vector **v** has initial point P and terminal point Q. Write **v** in the form ai + bj + ck; that is, find its position vector.*

21. $P = (0, 0, 0);$ $Q = (3, 4, -1)$

22. $P = (0, 0, 0);$ $Q = (-3, -5, 4)$

23. $P = (3, 2, -1);$ $Q = (5, 6, 0)$

24. $P = (-3, 2, 0);$ $Q = (6, 5, -1)$

25. $P = (-2, -1, 4);$ $Q = (6, -2, 4)$

26. $P = (-1, 4, -2);$ $Q = (6, 2, 2)$

In Problems 27–32, find $\|\mathbf{v}\|$.

27. $\mathbf{v} = 3\mathbf{i} - 6\mathbf{j} - 2\mathbf{k}$

28. $\mathbf{v} = -6\mathbf{i} + 12\mathbf{j} + 4\mathbf{k}$

29. $\mathbf{v} = \mathbf{i} - \mathbf{j} + \mathbf{k}$

30. $\mathbf{v} = -\mathbf{i} - \mathbf{j} + \mathbf{k}$

31. $\mathbf{v} = -2\mathbf{i} + 3\mathbf{j} - 3\mathbf{k}$

32. $\mathbf{v} = 6\mathbf{i} + 2\mathbf{j} - 2\mathbf{k}$

In Problems 33–38, find each quantity if $\mathbf{v} = 3\mathbf{i} - 5\mathbf{j} + 2\mathbf{k}$ *and* $\mathbf{w} = -2\mathbf{i} + 3\mathbf{j} - 2\mathbf{k}$.

33. $2\mathbf{v} + 3\mathbf{w}$

34. $3\mathbf{v} - 2\mathbf{w}$

35. $\|\mathbf{v} - \mathbf{w}\|$

36. $\|\mathbf{v} + \mathbf{w}\|$

37. $\|\mathbf{v}\| - \|\mathbf{w}\|$

38. $\|\mathbf{v}\| + \|\mathbf{w}\|$

In Problems 39–44, find the unit vector having the same direction as $\mathbf{v}$.

39. $\mathbf{v} = 5\mathbf{i}$

40. $\mathbf{v} = -3\mathbf{j}$

41. $\mathbf{v} = 3\mathbf{i} - 6\mathbf{j} - 2\mathbf{k}$

42. $\mathbf{v} = -6\mathbf{i} + 12\mathbf{j} + 4\mathbf{k}$

43. $\mathbf{v} = \mathbf{i} + \mathbf{j} + \mathbf{k}$

44. $\mathbf{v} = 2\mathbf{i} - \mathbf{j} + \mathbf{k}$

In Problems 45–52, find the dot product $\mathbf{v} \cdot \mathbf{w}$ *and the angle between* $\mathbf{v}$ *and* $\mathbf{w}$.

45. $\mathbf{v} = \mathbf{i} - \mathbf{j}, \quad \mathbf{w} = \mathbf{i} + \mathbf{j} + \mathbf{k}$

46. $\mathbf{v} = \mathbf{i} + \mathbf{j}, \quad \mathbf{w} = -\mathbf{i} + \mathbf{j} - \mathbf{k}$

47. $\mathbf{v} = 2\mathbf{i} + \mathbf{j} - 3\mathbf{k}, \quad \mathbf{w} = \mathbf{i} + 2\mathbf{j} + 2\mathbf{k}$

48. $\mathbf{v} = 2\mathbf{i} + 2\mathbf{j} - \mathbf{k}, \quad \mathbf{w} = \mathbf{i} + 2\mathbf{j} + 3\mathbf{k}$

49. $\mathbf{v} = 3\mathbf{i} - \mathbf{j} + 2\mathbf{k}, \quad \mathbf{w} = \mathbf{i} + \mathbf{j} - \mathbf{k}$

50. $\mathbf{v} = \mathbf{i} + 3\mathbf{j} + 2\mathbf{k}, \quad \mathbf{w} = \mathbf{i} - \mathbf{j} + \mathbf{k}$

51. $\mathbf{v} = 3\mathbf{i} + 4\mathbf{j} + \mathbf{k}, \quad \mathbf{w} = 6\mathbf{i} + 8\mathbf{j} + 2\mathbf{k}$

52. $\mathbf{v} = 3\mathbf{i} - 4\mathbf{j} + \mathbf{k}, \quad \mathbf{w} = 6\mathbf{i} - 8\mathbf{j} + 2\mathbf{k}$

In Problems 53–60, find the direction angles of each vector. Write each vector in the form of equation (7).

53. $\mathbf{v} = 3\mathbf{i} - 6\mathbf{j} - 2\mathbf{k}$

54. $\mathbf{v} = -6\mathbf{i} + 12\mathbf{j} + 4\mathbf{k}$

55. $\mathbf{v} = \mathbf{i} + \mathbf{j} + \mathbf{k}$

56. $\mathbf{v} = \mathbf{i} - \mathbf{j} - \mathbf{k}$

57. $\mathbf{v} = \mathbf{i} + \mathbf{j}$

58. $\mathbf{v} = \mathbf{j} + \mathbf{k}$

59. $\mathbf{v} = 3\mathbf{i} - 5\mathbf{j} + 2\mathbf{k}$

60. $\mathbf{v} = 2\mathbf{i} + 3\mathbf{j} - 4\mathbf{k}$

61. **The Sphere** In space, the collection of all points that are the same distance from some fixed point is called a **sphere.** See the illustration. The constant distance is called the **radius,** and the fixed point is the **center** of the sphere. Show that the equation of a sphere with center at (x_0, y_0, z_0) and radius r is

$$(x - x_0)^2 + (y - y_0)^2 + (z - z_0)^2 = r^2$$

[**Hint:** Use the Distance Formula (1).]

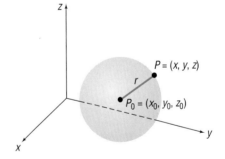

In Problems 62–64, find the equation of a sphere with radius r *and center* P_0.

62. $r = 1; \quad P_0 = (3, 1, 1)$

63. $r = 2; \quad P_0 = (1, 2, 2)$

64. $r = 3; \quad P_0 = (-1, 1, 2)$

In Problems 65–70, find the radius and center of each sphere.

65. $x^2 + y^2 + z^2 + 2x - 2y = 2$

66. $x^2 + y^2 + z^2 + 2x - 2z = -1$

67. $x^2 + y^2 + z^2 - 4x + 4y + 2z = 0$

68. $x^2 + y^2 + z^2 - 4x = 0$

69. $2x^2 + 2y^2 + 2z^2 - 8x + 4z = -1$

70. $3x^2 + 3y^2 + 3z^2 + 6x - 6y = 3$

The **work** W *done by a constant force* $\mathbf{F}$ *in moving an object from a point A in space to a point B in space is defined as* $W = \mathbf{F} \cdot \overrightarrow{AB}$. *Use this definition in Problems 71–73.*

71. **Work** Find the work done by a force of 3 newtons acting in the direction $2\mathbf{i} + \mathbf{j} + 2\mathbf{k}$ in moving an object 2 meters from $(0, 0, 0)$ to $(0, 2, 0)$.

72. **Work** Find the work done by a force of 1 newton acting in the direction $2\mathbf{i} + 2\mathbf{j} + \mathbf{k}$ in moving an object 3 meters from $(0, 0, 0)$ to $(1, 2, 2)$.

73. **Work** Find the work done in moving an object along a vector $\mathbf{u} = 3\mathbf{i} + 2\mathbf{j} - 5\mathbf{k}$ if the applied force is $\mathbf{F} = 2\mathbf{i} - \mathbf{j} - \mathbf{k}$.

CHAPTER REVIEW

THINGS TO KNOW

Relationship between polar coordinates (r, θ) and rectangular coordinates (x, y)	$x = r \cos \theta, y = r \sin \theta$ $x^2 + y^2 = r^2, \tan \theta = \dfrac{y}{x}, x \neq 0$
Standard form of a complex number	$z = x + yi, x, y$ are real numbers and i is the imaginary unit with $i^2 = -1$
Polar form of a complex number	If $z = x + yi$, then $z = r(\cos \theta + i \sin \theta)$, where $r = \|z\| = \sqrt{x^2 + y^2}, \sin \theta = \dfrac{y}{r}, \cos \theta = \dfrac{x}{r}, \quad 0 \le \theta < 2\pi$.
De Moivre's Theorem	If $z = r(\cos \theta + i \sin \theta)$, then $z^n = r^n(\cos n\theta + i \sin n\theta)$, where $n \ge 1$ is a positive integer.
nth root of a complex number	$\sqrt[n]{z} = \sqrt[n]{r}\left[\cos\left(\dfrac{\theta}{n} + \dfrac{2k\pi}{n}\right) + i \sin\left(\dfrac{\theta}{n} + \dfrac{2k\pi}{n}\right)\right], \quad k = 0, 1, \ldots, n - 1$
Vector	Quantity having magnitude and direction; equivalent to a directed line segment $\overrightarrow{PQ}$
Position vector	Vector whose initial point is at the origin
Unit vector	Vector whose magnitude is 1
Dot product	If $\mathbf{v} = a_1\mathbf{i} + b_1\mathbf{j}$ and $\mathbf{w} = a_2\mathbf{i} + b_2\mathbf{j}$, then $\mathbf{v} \cdot \mathbf{w} = a_1 a_2 + b_1 b_2$. If $\mathbf{v} = a_1\mathbf{i} + b_1\mathbf{j} + c_1\mathbf{k}$ and $\mathbf{w} = a_2\mathbf{i} + b_2\mathbf{j} + c_2\mathbf{k}$, then $\mathbf{v} \cdot \mathbf{w} = a_1 a_2 + b_1 b_2 + c_1 c_2$.
Angle θ between two nonzero vectors $\mathbf{u}$ and $\mathbf{v}$	$\cos \theta = \dfrac{\mathbf{u} \cdot \mathbf{v}}{\|\mathbf{u}\| \, \|\mathbf{v}\|}$
In space	If $\mathbf{v} = a\mathbf{i} + b\mathbf{j} + c\mathbf{k}$, then $\mathbf{v} = \|\mathbf{v}\| \, [(\cos \alpha)\mathbf{i} + (\cos \beta)\mathbf{j} + (\cos \gamma)\mathbf{k}]$, where $\cos \alpha = \dfrac{a}{\|\mathbf{v}\|}, \quad \cos \beta = \dfrac{b}{\|\mathbf{v}\|}, \quad \cos \gamma = \dfrac{c}{\|\mathbf{v}\|}$.

HOW TO

Plot polar coordinates

Convert from polar to rectangular coordinates

Convert from rectangular to polar coordinates

Graph polar equations (see Table 7, page 326)

Add, subtract, multiply, and divide complex numbers

Write a complex number in polar form, $z = r(\cos \theta + i \sin \theta), \quad 0 \le \theta < 2\pi$

Use De Moivre's Theorem to find powers of complex numbers

Find the nth roots of a complex number

Add and subtract vectors

Form scalar multiples of vectors

Find the magnitude of a vector

Solve applied problems involving vectors

Find the dot product of two vectors

Find the angle between two vectors

Determine whether two vectors in the plane are parallel

Determine whether two vectors are orthogonal

Find the vector projection of $\mathbf{v}$ onto $\mathbf{w}$

Find the distance between two points in space

Find the direction angles of a vector in space

FILL-IN-THE-BLANK ITEMS

1. In polar coordinates, the origin is called the _____, and the positive x-axis is referred to as the _____ _____.

2. Another representation in polar coordinates for the point $(2, \pi/3)$ is (_____, $4\pi/3$).

3. Using polar coordinates (r, θ), the circle $x^2 + y^2 = 2x$ takes the form _____.

4. In a polar equation, replace θ by $-\theta$. If an equivalent equation results, the graph is symmetric with respect to _____ _____.

5. In the complex number $5 + 2i$, the number 5 is called the _____ part; the number 2 is called the _____ part; the number i is called the _____ _____.

6. When a complex number z is written in the polar form $z = r(\cos \theta + i \sin \theta)$, the nonnegative number r is the _____ or _____ of z, and the angle $\theta, 0 \le \theta < 2\pi$, is the _____ of z.

7. A vector whose magnitude is 1 is called a(n) _____ vector.

8. If the angle between two vectors $\mathbf{v}$ and $\mathbf{w}$ is $\pi/2$, then the dot product $\mathbf{v} \cdot \mathbf{w}$ equals _____.

TRUE/FALSE ITEMS

T F **1.** The polar coordinates of a point are unique.

T F **2.** The rectangular coordinates of a point are unique.

T F **3.** The tests for symmetry in polar coordinates are conclusive.

T F **4.** The conjugate of $2 + \sqrt{5}i$ is $-2 - \sqrt{5}i$.

T F **5.** De Moivre's Theorem is useful for raising a complex number to a positive integer power.

T F **6.** Vectors are quantities that have magnitude and direction.

T F **7.** Force is a physical example of a vector.

T F **8.** If $\mathbf{u}$ and $\mathbf{v}$ are orthogonal vectors, then $\mathbf{u} \cdot \mathbf{v} = 0$.

T F **9.** The sum of the squares of the direction cosines of a vector in space equals 1.

REVIEW EXERCISES

Blue problem numbers indicate the author's suggestions for use in a Practice Test.

In Problems 1–6, plot each point given in polar coordinates, and find its rectangular coordinates.

1. $(3, \pi/6)$

2. $(4, 2\pi/3)$

3. $(-2, 4\pi/3)$

4. $(-1, 5\pi/4)$

5. $(-3, -\pi/2)$

6. $(-4, -\pi/4)$

In Problems 7–12, the rectangular coordinates of a point are given. Find two pairs of polar coordinates (r, θ) for each point, one with $r > 0$ and the other with $r < 0$. Express θ in radians.

7. $(-3, 3)$

8. $(1, -1)$

9. $(0, -2)$

10. $(2, 0)$

11. $(3, 4)$

12. $(-5, 12)$

In Problems 13–18, the letters x and y represent rectangular coordinates. Write each equation using polar coordinates (r, θ).

13. $3x^2 + 3y^2 = 6y$

14. $2x^2 - 2y^2 = 5y$

15. $2x^2 - y^2 = \dfrac{y}{x}$

16. $x^2 + 2y^2 = \dfrac{y}{x}$

17. $x(x^2 + y^2) = 4$

18. $y(x^2 - y^2) = 3$

In Problems 19–24, the letters r and θ represent polar coordinates. Write each polar equation as an equation in rectangular coordinates (x, y).

19. $r = 2 \sin \theta$

20. $3r = \sin \theta$

21. $r = 5$

22. $\theta = \pi/4$

23. $r \cos \theta + 3r \sin \theta = 6$

24. $r^2 \tan \theta = 1$

In Problems 25–30, sketch the graph of each polar equation. Be sure to test for symmetry.

25. $r = 4 \cos \theta$

26. $r = 3 \sin \theta$

27. $r = 3 - 3 \sin \theta$

28. $r = 2 + \cos \theta$

29. $r = 4 - \cos \theta$

30. $r = 1 - 2 \sin \theta$

In Problems 31–38, write each complex number in the standard form a + bi.

31. $(6 + 3i) - (2 - 4i)$

32. $(8 - 3i) + (-6 + 2i)$

33. $(2 - 3i)(4 + i)$

34. $(3 + 2i)(4 - 3i)$

35. $\dfrac{3}{3 + i}$

36. $\dfrac{4}{2 - i}$

37. i^{50}

38. i^{29}

In Problems 39–42, write each complex number in polar form. Express each argument in degrees.

39. $-1 - i$

40. $-\sqrt{3} + i$

41. $4 - 3i$

42. $3 - 2i$

In Problems 43–48, write each complex number in the standard form a + bi.

43. $2(\cos 150° + i \sin 150°)$

44. $3(\cos 60° + i \sin 60°)$

45. $3\left(\cos \dfrac{2\pi}{3} + i \sin \dfrac{2\pi}{3}\right)$

46. $4\left(\cos \dfrac{3\pi}{4} + i \sin \dfrac{3\pi}{4}\right)$

47. $0.1(\cos 350° + i \sin 350°)$

48. $0.5(\cos 160° + i \sin 160°)$

In Problems 49–54, find zw and z/w. Leave your answers in polar form.

49. $z = \cos 80° + i \sin 80°$
$w = \cos 50° + i \sin 50°$

50. $z = \cos 205° + i \sin 205°$
$w = \cos 85° + i \sin 85°$

51. $z = 3\left(\cos \dfrac{9\pi}{5} + i \sin \dfrac{9\pi}{5}\right)$
$w = 2\left(\cos \dfrac{\pi}{5} + i \sin \dfrac{\pi}{5}\right)$

52. $z = 2\left(\cos \dfrac{5\pi}{3} + i \sin \dfrac{5\pi}{3}\right)$
$w = 3\left(\cos \dfrac{\pi}{3} + i \sin \dfrac{\pi}{3}\right)$

53. $z = 5(\cos 10° + i \sin 10°)$
$w = \cos 355° + i \sin 355°$

54. $z = 4(\cos 50° + i \sin 50°)$
$w = \cos 340° + i \sin 340°$

In Problems 55–62, write each expression in the standard form a + bi.

55. $[3(\cos 20° + i \sin 20°)]^3$

56. $[2(\cos 50° + i \sin 50°)]^3$

57. $\left[\sqrt{2}\left(\cos \dfrac{5\pi}{8} + i \sin \dfrac{5\pi}{8}\right)\right]^4$

58. $\left[2\left(\cos \dfrac{5\pi}{16} + i \sin \dfrac{5\pi}{16}\right)\right]^4$

59. $(1 - \sqrt{3}i)^6$

60. $(2 - 2i)^8$

61. $(3 + 4i)^4$

62. $(1 - 2i)^4$

63. Find all the complex cube roots of 27.

64. Find all the complex fourth roots of -16.

*In Problems 65–72, the vector **v** is represented by the directed line segment $\overrightarrow{PQ}$. Write **v** in the form a**i** + b**j** or in the form **v** = a**i** + b**j** + c**k** and find ‖**v**‖.*

65. $P = (1, -2);\quad Q = (3, -6)$

66. $P = (-3, 1);\quad Q = (4, -2)$

67. $P = (0, -2);\quad Q = (-1, 1)$

68. $P = (3, -4);\quad Q = (-2, 0)$

69. $P = (6, 2, 1);\quad Q = (3, 0, 2)$

70. $P = (4, 7, 0);\quad Q = (0, 5, 6)$

71. $P = (-1, 0, 1);\quad Q = (2, 0, 0)$

72. $P = (6, 2, 2);\quad Q = (2, 6, 2)$

In Problems 73–80, use the vectors $\mathbf{v} = -2\mathbf{i} + \mathbf{j}$ *and* $\mathbf{w} = 4\mathbf{i} - 3\mathbf{j}$.

73. Find $4\mathbf{v} - 3\mathbf{w}$.

74. Find $-\mathbf{v} + 2\mathbf{w}$.

75. Find $\|\mathbf{v}\|$.

76. Find $\|\mathbf{v} + \mathbf{w}\|$.

77. Find $\|\mathbf{v}\| + \|\mathbf{w}\|$.

78. Find $\|2\mathbf{v}\| - 3\|\mathbf{w}\|$.

79. Find a unit vector in the same direction as $\mathbf{v}$.

80. Find a unit vector in the opposite direction of $\mathbf{w}$.

In Problems 81–88, use the vectors $\mathbf{v} = 3\mathbf{i} + \mathbf{j} - 2\mathbf{k}$ *and* $\mathbf{w} = -3\mathbf{i} + 2\mathbf{j} - \mathbf{k}$.

81. Find $4\mathbf{v} - 3\mathbf{w}$.

82. Find $-\mathbf{v} + 2\mathbf{w}$.

83. Find $\|\mathbf{v} - \mathbf{w}\|$.

84. Find $\|\mathbf{v} + \mathbf{w}\|$.

85. Find $\|\mathbf{v}\| - \|\mathbf{w}\|$.

86. Find $\|\mathbf{v}\| + \|\mathbf{w}\|$.

87. Find a unit vector in the same direction as $\mathbf{v}$ and then in the opposite direction of $\mathbf{v}$.

88. Find a unit vector in the same direction as $\mathbf{w}$ and then in the opposite direction of $\mathbf{w}$.

In Problems 89–96, find the dot product $\mathbf{v} \cdot \mathbf{w}$ *and the angle between* $\mathbf{v}$ *and* $\mathbf{w}$.

89. $\mathbf{v} = -2\mathbf{i} + \mathbf{j}$, $\mathbf{w} = 4\mathbf{i} - 3\mathbf{j}$

90. $\mathbf{v} = 3\mathbf{i} - \mathbf{j}$, $\mathbf{w} = \mathbf{i} + \mathbf{j}$

91. $\mathbf{v} = \mathbf{i} - 3\mathbf{j}$, $\mathbf{w} = -\mathbf{i} + \mathbf{j}$

92. $\mathbf{v} = \mathbf{i} + 4\mathbf{j}$, $\mathbf{w} = 3\mathbf{i} - 2\mathbf{j}$

93. $\mathbf{v} = \mathbf{i} + \mathbf{j} + \mathbf{k}$, $\mathbf{w} = \mathbf{i} - \mathbf{j} + \mathbf{k}$

94. $\mathbf{v} = \mathbf{i} - \mathbf{j} + \mathbf{k}$, $\mathbf{w} = 2\mathbf{i} + \mathbf{j} + \mathbf{k}$

95. $\mathbf{v} = 4\mathbf{i} - \mathbf{j} + 2\mathbf{k}$, $\mathbf{w} = \mathbf{i} - 2\mathbf{j} - 3\mathbf{k}$

96. $\mathbf{v} = -\mathbf{i} - 2\mathbf{j} + 3\mathbf{k}$, $\mathbf{w} = 5\mathbf{i} + \mathbf{j} + \mathbf{k}$

97. Find the vector projection of $\mathbf{v} = 2\mathbf{i} + 3\mathbf{j}$ onto $\mathbf{w} = 3\mathbf{i} + \mathbf{j}$.

98. Find the vector projection of $\mathbf{v} = -\mathbf{i} + 2\mathbf{j}$ onto $\mathbf{w} = 3\mathbf{i} - \mathbf{j}$.

99. Find the direction angles of the vector $\mathbf{v} = 3\mathbf{i} - 4\mathbf{j} + 2\mathbf{k}$.

100. Find the direction angles of the vector $\mathbf{v} = \mathbf{i} - \mathbf{j} + 2\mathbf{k}$.

101. Actual Speed and Direction of a Swimmer A swimmer can maintain a constant speed of 5 miles per hour. If the swimmer heads directly across a river that has a current moving at the rate of 2 miles per hour, what is the actual speed of the swimmer? (See the figure.) If the river is 1 mile wide, how far downstream will the swimmer end up from the point directly across the river from the starting point?

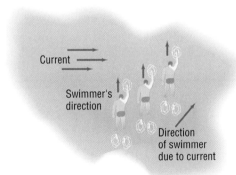

Current

Swimmer's direction

Direction of swimmer due to current

102. Actual Speed and Distance of a Motorboat A small motorboat is moving at a true speed of 11 miles per hour in a southerly direction. The current is known to be from the northeast at 3 miles per hour. What is the speed of the motorboat relative to the water? In what direction does the compass indicate that the boat is headed?

103. Correct Direction for Crossing a River A river 1 kilometer wide has a constant current of 5 kilometers per hour. At what angle to the shore should a person head a boat that is capable of maintaining a constant speed of 15 kilometers per hour in order to reach a point directly opposite?

104. Actual Speed and Direction of an Airplane An airplane has an airspeed of 500 kilometers per hour in a northerly direction. The wind velocity is 60 kilometers per hour in a southeasterly direction. Find the actual speed and direction of the plane relative to the ground.

CHAPTER 6

Analytic Geometry

Many computer users enjoy collecting and putting on their computer the latest "screen savers". As a creative entrepreneur, you have come up with an idea for a new piece of software. You will need to use the Sullivan website at

www.prenhall.com/sullivan

to find the computer programs and information to build your screen saver.

WHEELS OF FIRE

You have a great idea for a computer screen saver—an electronic spirograph! You are certain that you could sell such a hot piece of software. The great thing about making a screen saver spirograph is that you will be working with light, color-blending, and high-resolution images. You let your imagination go . . .

1. In order to produce your screen saver you first want to know if there are equations for a *spirograph*. If there are such equations, how could you tweak them to produce special effects? Can you make a spirograph on the surface of a 3-D object like a ball or a donut? Can you make 3-D spirographs like some wild celtic knot? How would you add shading?

   By changing the lengths of the arms and the gear ratios, the spirograph can produce many different patterns. For example you could generate a *trochoid*. In this lesson we will use the principles of analytic geometry to compose a collection of new curves, called **Roulettes.**

2. You need to understand how these equations work in order to adapt them to your screen saver, so you try to take them apart. Experiment by graphing the following equations for various values of R and r.

$$x = (R + r) \cos t$$
$$y = (R + r) \sin t$$

3. What kinds of designs can be made with a *spirograph?* How do the symbolic equations simulate the patterns made with the toy? Experiment by graphing the following equations for various values of R, r, and O.

$$x = (R + r) \cos t - (r + O) \cos((R + r)/r)\, t$$
$$y = (R + r) \sin t - (r + O) \sin((R + r)/r)\, t$$

4. What would happen if you were to change the equations to:

$$x = A \cos t - (r + O) \cos((R + r)/r)\, t$$
$$y = B \sin t - (r + O) \sin((R + r)/r)\, t$$

   where A is not equal to B? What does the graph look like?

5. Suppose you wanted to change the Roulette base curve to an asteroid. That is, now you want to roll the circle on a star instead of another circle. Write the equation for your "Astrograph."

6. By now you should start to see that the key idea is simply the composition of parametric equations. Once you have this idea then you are ready to play. You want to create a set of great looking designs for the advertisement of your screen saver. Start with a base curve that you like, from The MacTutor Famous Curves Index, and then try adding two motions. Write the equations and discuss the possible types of patterns that you might expect to see. Does this give you any further designs?

7. Could you extend these curves into surfaces in 3-D? Can you think of a better name than Wheels of Fire?

Historically, Apollonius (200 B.C.) was among the first to study *conics* and discover some of their interesting properties. Today, conics are still studied because of their many uses. *Paraboloids of revolution* (parabolas rotated about their axes of symmetry) are used as signal collectors (the satellite dishes used with radar and cable

TV, for example), as solar energy collectors, and as reflectors (telescopes, light projection, and so on). The planets circle the Sun in approximately elliptical orbits. Elliptical surfaces can be used to reflect signals such as light and sound from one place to another. And *hyperbolas* can be used to determine the positions of ships at sea.

The Greeks used the methods of Euclidean geometry to study conics. We shall use the more powerful methods of analytic geometry, bringing to bear both algebra and geometry, for our study of conics. Thus, we shall give a geometric description of each conic, and then, using rectangular coordinates and the distance formula, we shall find equations that represent conics. We used this same development, you may recall, when we first defined a circle in Section 1.2.

The chapter concludes with a section on equations of conics in polar coordinates, followed by a discussion of plane curves and parametric equations.

6.1 │ CONICS

> **1** Know the names of the conics

1 The word *conic* derives from the word *cone,* which is a geometric figure that can be constructed in the following way: Let *a* and *g* be two distinct lines that intersect at a point *V*. Keep the line *a* fixed. Now rotate the line *g* about *a* while maintaining the same angle between *a* and *g*. The collection of points swept out (generated) by the line *g* is called a (**right circular**) **cone.** See Figure 1. The fixed line *a* is called the **axis** of the cone; the point *V* is called its **vertex;** the lines that pass through *V* and make the same angle with *a* as *g* are called **generators** of the cone. Thus, each generator is a line that lies entirely on the cone. The cone consists of two parts, called **nappes,** that intersect at the vertex.

Conics, an abbreviation for **conic sections,** are curves that result from the intersection of a (right circular) cone and a plane. The conics we shall study arise when the plane does not contain the vertex, as shown in Figure 2. These conics are **circles** when the plane is perpendicular to the axis of the cone and intersects each generator; **ellipses** when the plane is tilted slightly so that it intersects each generator, but intersects only one nappe of the cone; **parabolas** when the plane is tilted further so that it is parallel to one (and only one) generator and intersects only one nappe of the cone; and **hyperbolas** when the plane intersects both nappes.

If the plane does contain the vertex, the intersection of the plane and the cone is a point, a line, or a pair of intersecting lines. These are usually called **degenerate conics.**

FIGURE 1

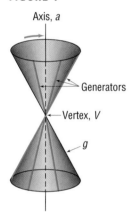

Axis, *a*

Generators

Vertex, *V*

g

FIGURE 2

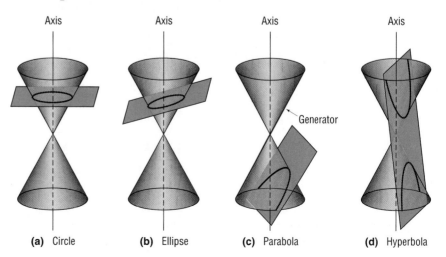

(a) Circle   **(b)** Ellipse   **(c)** Parabola   **(d)** Hyperbola

Generator

6.2 | THE PARABOLA

We begin with a geometric definition of a parabola and use it to obtain an equation.

A **parabola** is defined as the collection of all points P in the plane that are the same distance from a fixed point F as they are from a fixed line D. The point F is called the **focus** of the parabola, and the line D is its **directrix**. As a result, a parabola is the set of points P for which

$$d(F, P) = d(P, D) \tag{1}$$

FIGURE 3

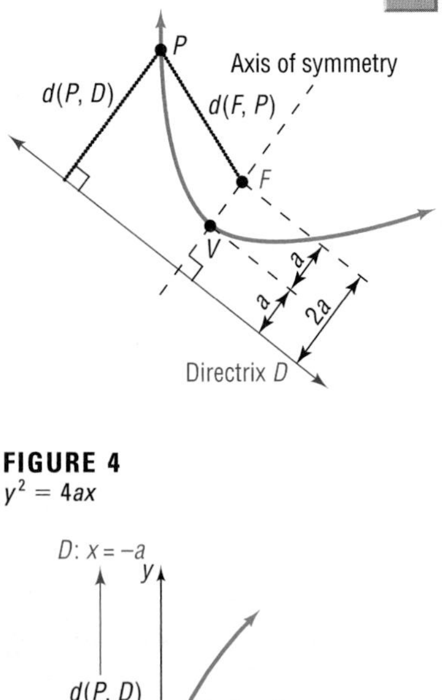

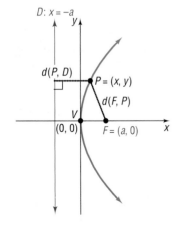

Axis of symmetry

$d(P, D)$ $d(F, P)$

Directrix D

FIGURE 4
$y^2 = 4ax$

$D: x = -a$

$d(P, D)$ $P = (x, y)$

$d(F, P)$

V
$(0, 0)$ $F = (a, 0)$

Figure 3 shows a parabola. The line through the focus F and perpendicular to the directrix D is called the **axis of symmetry** of the parabola. The point of intersection of the parabola with its axis of symmetry is called the **vertex** V.

Because the vertex V lies on the parabola, it must satisfy equation (1): $d(F, V) = d(V, D)$. Thus, the vertex is midway between the focus and the directrix. We shall let a equal the distance $d(F, V)$ from F to V. Now we are ready to derive an equation for a parabola. To do this, we use a rectangular system of coordinates positioned so that the vertex V, focus F, and directrix D of the parabola are conveniently located. If we choose to locate the vertex V at the origin $(0, 0)$, then we can conveniently position the focus F on either the x-axis or the y-axis.

First, we consider the case where the focus F is on the positive x-axis, as shown in Figure 4. Because the distance from F to V is a, the coordinates of F will be $(a, 0)$ with $a > 0$. Similarly, because the distance from V to the directrix D is also a and because D must be perpendicular to the x-axis (since the x-axis is the axis of symmetry), the equation of the directrix D must be $x = -a$. Now, if $P = (x, y)$ is any point on the parabola, then P must obey equation (1):

$$d(F, P) = d(P, D)$$

So we have

$$\sqrt{(x - a)^2 + y^2} = |x + a| \qquad \text{Use the distance formula.}$$
$$(x - a)^2 + y^2 = (x + a)^2 \qquad \text{Square both sides.}$$
$$x^2 - 2ax + a^2 + y^2 = x^2 + 2ax + a^2 \qquad \text{Simplify.}$$
$$y^2 = 4ax$$

Theorem   Equation of a Parabola; Vertex at (0, 0), Focus at (a, 0),
$a > 0$

The equation of a parabola with vertex at $(0, 0)$, focus at $(a, 0)$, and directrix $x = -a, a > 0$, is

$$y^2 = 4ax \qquad (2)$$

E X A M P L E   1

FIGURE 5
$y^2 = 12x$

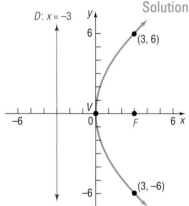

Finding the Equation of a Parabola

Find an equation of the parabola with vertex at $(0, 0)$ and focus at $(3, 0)$. Graph the equation.

Solution   The distance from the vertex $(0,0)$ to the focus $(3,0)$ is $a = 3$. Based on equation (2), the equation of this parabola is

$$y^2 = 4ax$$
$$y^2 = 12x \quad a = 3$$

To graph this parabola, it is helpful to plot the two points on the graph above and below the focus. To locate them, we let $x = 3$. Then

$$y^2 = 12x = 36$$
$$y = \pm 6$$

The points on the parabola above and below the focus are $(3, -6)$ and $(3, 6)$. These points help in graphing the parabola because they determine the "opening" of the graph. See Figure 5.

In general, the points on a parabola $y^2 = 4ax$ that lie above and below the focus $(a, 0)$ are each at a distance $2a$ from the focus. This follows from the fact that if $x = a$ then $y^2 = 4ax = 4a^2$, or $y = \pm 2a$. The line segment joining these two points is called the **latus rectum;** its length is $4a$.

Comment   To graph the parabola $y^2 = 12x$ discussed in Example 1, we need to graph the two functions $Y_1 = \sqrt{12x}$ and $Y_2 = -\sqrt{12x}$. Do this and compare what you see with Figure 5.

Now work Problem 9.

By reversing the steps we used to obtain equation (2), it follows that the graph of an equation of the form of equation (2), $y^2 = 4ax$, is a parabola; its vertex is at $(0, 0)$, its focus is at $(a, 0)$, its directrix is the line $x = -a$, and its axis of symmetry is the x-axis.

For the remainder of this section, the direction "Discuss the equation" will mean to find the vertex, focus, and directrix of the parabola and graph it.

E X A M P L E   2

Discussing the Equation of a Parabola

Discuss the equation: $y^2 = 8x$

Solution   The equation $y^2 = 8x$ is of the form $y^2 = 4ax$, where $4a = 8$. Thus, $a = 2$. Consequently, the graph of the equation is a parabola with vertex at $(0, 0)$ and

FIGURE 6
$y^2 = 8x$

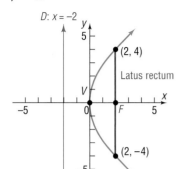

focus on the positive x-axis at $(2, 0)$. The directrix is the vertical line $x = -2$. The two points defining the latus rectum are obtained by letting $x = 2$. Then $y^2 = 16$, or $y = \pm 4$. See Figure 6. ■

Recall that we arrived at equation (2) after placing the focus on the positive x-axis. If the focus is placed on the negative x-axis, positive y-axis, or negative y-axis, a different form of the equation for the parabola results. The four forms of the equation of a parabola with vertex at $(0, 0)$ and focus on a coordinate axis a distance a from $(0, 0)$ are given in Table 1, and their graphs are given in Figure 7. Notice that each graph is symmetric with respect to its axis of symmetry.

TABLE 1 Equations of a Parabola: Vertex at (0, 0); Focus on Axis; $a > 0$

Vertex	Focus	Directrix	Equation	Description
$(0, 0)$	$(a, 0)$	$x = -a$	$y^2 = 4ax$	Parabola, axis of symmetry is the x-axis, opens to right
$(0, 0)$	$(-a, 0)$	$x = a$	$y^2 = -4ax$	Parabola, axis of symmetry is the x-axis, opens to left
$(0, 0)$	$(0, a)$	$y = -a$	$x^2 = 4ay$	Parabola, axis of symmetry is the y-axis, opens up
$(0, 0)$	$(0, -a)$	$y = a$	$x^2 = -4ay$	Parabola, axis of symmetry is the y-axis, opens down

FIGURE 7

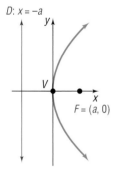

(a) $y^2 = 4ax$

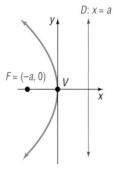

(b) $y^2 = -4ax$

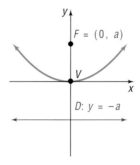

(c) $x^2 = 4ay$

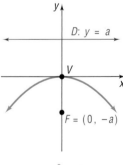

(d) $x^2 = -4ay$

E X A M P L E   3   Discussing the Equation of a Parabola

Discuss the equation: $x^2 = -12y$

Solution   The equation $x^2 = -12y$ is of the form $x^2 = -4ay$, with $a = 3$. Consequently, the graph of the equation is a parabola with vertex at $(0, 0)$, focus at $(0, -3)$, and directrix the line $y = 3$. The parabola opens down, and its axis of symmetry is the *y*-axis. To obtain the points defining the latus rectum, let $y = -3$. Then $x^2 = 36$, or $x = \pm 6$. See Figure 8.

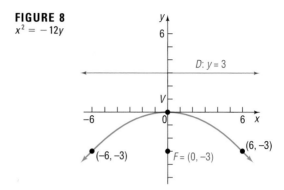

FIGURE 8
$x^2 = -12y$

Now work Problem 27.

E X A M P L E   4   Finding the Equation of a Parabola

Find the equation of the parabola with focus at $(0, 4)$ and directrix the line $y = -4$. Graph the equation.

Solution   A parabola whose focus is at $(0, 4)$ and whose directrix is the horizontal line $y = -4$ will have its vertex at $(0, 0)$. (Do you see why? The vertex is midway between the focus and the directrix.) Thus, the equation of this parabola is of the form $x^2 = 4ay$, with $a = 4$; that is,

$$x^2 = 16y$$

Figure 9 shows the graph.

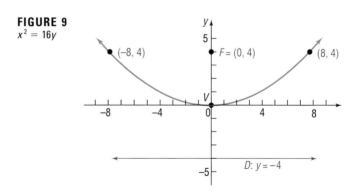

FIGURE 9
$x^2 = 16y$

E X A M P L E   5   Finding the Equation of a Parabola

Find the equation of a parabola with vertex at $(0, 0)$ if its axis of symmetry is the *x*-axis and its graph passes through the point $\left(-\frac{1}{2}, 2\right)$. Find its focus and directrix, and graph the equation.

Solution   Because the vertex is at the origin, the axis of symmetry is the x-axis, and the graph passes through a point in the second quadrant, we see from Table 1 that the form of the equation is

$$y^2 = -4ax$$

Because the point $(-\frac{1}{2}, 2)$ is on the parabola, the coordinates $x = -\frac{1}{2}, y = 2$ must satisfy the equation. Substituting $x = -\frac{1}{2}$ and $y = 2$ in the equation, we find

$$4 = -4a\left(-\frac{1}{2}\right)$$
$$a = 2$$

Thus, the equation of the parabola is

$$y^2 = -8x$$

The focus is at $(-2, 0)$ and the directrix is the line $x = 2$. Letting $x = -2$, we find $y^2 = 16$ or $y = \pm 4$. The points $(-2, 4)$ and $(-2, -4)$ define the latus rectum. See Figure 10.

FIGURE 10
$y^2 = -8x$

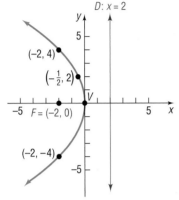

Now work Problem 19.

Vertex at (h, k)

4

If a parabola with vertex at the origin and axis of symmetry along a coordinate axis is shifted horizontally h units and then vertically k units, the result is a parabola with a vertex at (h, k) and axis of symmetry parallel to a coordinate axis. The equations of such parabolas have the same forms as those in Table 1, but with x replaced by $x - h$ (the horizontal shift) and y replaced by $y - k$ (the vertical shift). Table 2 gives the forms of the equations of such parabolas. Figure 11(a)–(d) illustrates the graphs for $h > 0, k > 0$.

TABLE 2 Parabolas with Vertex at (h, k), Axis of Symmetry Parallel to a Coordinate Axis, $a > 0$				
Vertex	**Focus**	**Directrix**	**Equation**	**Description**
(h, k)	$(h + a, k)$	$x = h - a$	$(y - k)^2 = 4a(x - h)$	Parabola, axis of symmetry parallel to x-axis, opens to right
(h, k)	$(h - a, k)$	$x = h + a$	$(y - k)^2 = -4a(x - h)$	Parabola, axis of symmetry parallel to x-axis, opens to left
(h, k)	$(h, k + a)$	$y = k - a$	$(x - h)^2 = 4a(y - k)$	Parabola, axis of symmetry parallel to y-axis, opens up
(h, k)	$(h, k - a)$	$y = k + a$	$(x - h)^2 = -4a(y - k)$	Parabola, axis of symmetry parallel to y-axis, opens down

FIGURE 11

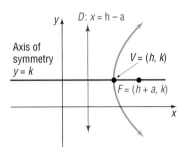

(a) $(y-k)^2 = 4a(x-h)$

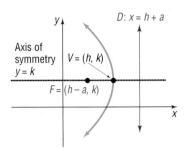

(b) $(y-k)^2 = -4a(x-h)$

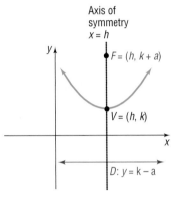

(c) $(x-h)^2 = 4a(y-k)$

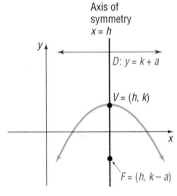

(d) $(x-h)^2 = -4a(y-k)$

E X A M P L E   6   Finding the Equation of a Parabola, Vertex Not at Origin

Find an equation of the parabola with vertex at $(-2, 3)$ and focus at $(0, 3)$. Graph the equation.

Solution   The vertex $(-2, 3)$ and focus $(0, 3)$ both lie on the horizontal line $y = 3$ (the axis of symmetry). The distance a from $(-2, 3)$ to $(0, 3)$ is $a = 2$. Also, because the focus lies to the right of the vertex, we know that the parabola opens to the right. Consequently, the form of the equation is

$$(y - k)^2 = 4a(x - h)$$

where $(h, k) = (-2, 3)$ and $a = 2$. Therefore, the equation is

$$(y - 3)^2 = 4 \cdot 2[x - (-2)]$$
$$(y - 3)^2 = 8(x + 2)$$

If $x = 0$, then $(y - 3)^2 = 16$. Thus, $y - 3 = \pm 4$ and $y = -1$ or $y = 7$. The points $(0, -1)$ and $(0, 7)$ define the latus rectum; the line $x = -4$ is the directrix. See Figure 12 (p. 390).

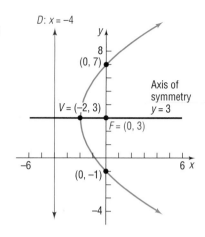

FIGURE 12
$(y - 3)^2 = 8(x + 2)$

   Now work Problem 17.

Polynomial equations define parabolas whenever they involve two variables that are quadratic in one variable and linear in the other. To discuss this type of equation, we first complete the square of the variable that is quadratic.

E X A M P L E   7   Discussing the Equation of a Parabola

Discuss the equation: $x^2 + 4x - 4y = 0$

Solution   To discuss the equation $x^2 + 4x - 4y = 0$, we complete the square involving the variable x. Thus,

FIGURE 13
$x^2 + 4x - 4y = 0$

$$x^2 + 4x - 4y = 0$$
$$x^2 + 4x = 4y \qquad \text{Isolate the terms involving } x \text{ on the left side.}$$
$$x^2 + 4x + 4 = 4y + 4 \qquad \text{Complete the square on the left side.}$$
$$(x + 2)^2 = 4(y + 1) \qquad \text{Form: } (x - h)^2 = 4a(y - k).$$

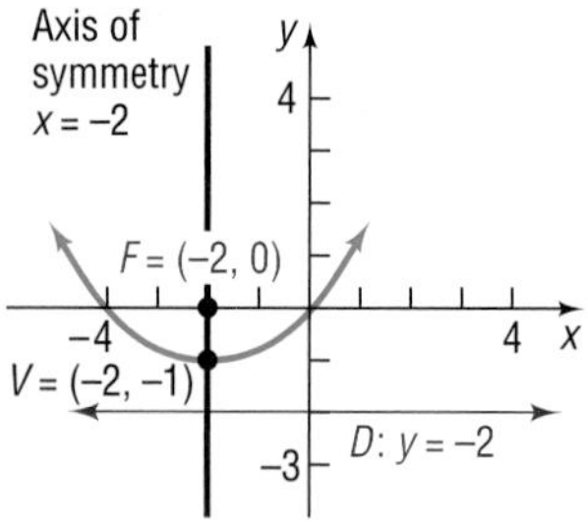

This equation is of the form $(x - h)^2 = 4a(y - k)$, with $h = -2, k = -1$, and $a = 1$. The graph is a parabola with vertex at $(h, k) = (-2, -1)$ that opens up. The focus is at $(-2, 0)$, and the directrix is the line $y = -2$. See Figure 13.

Now work Problem 35.

5   Parabolas find their way into many applications. For example, suspension bridges have cables in the shape of a parabola (See Problem 55). Another property of parabolas that is used in applications is their reflecting property.

Reflecting Property

Suppose that a mirror is shaped like a **paraboloid of revolution,** a surface formed by rotating a parabola about its axis of symmetry. If a light (or any other emitting source) is placed at the focus of the parabola, all the rays

FIGURE 14
Searchlight

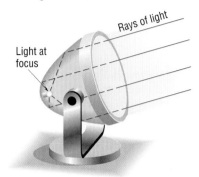

emanating from the light will reflect off the mirror in lines parallel to the axis of symmetry. This principle is used in the design of searchlights, flashlights, certain automobile headlights, and other such devices. See Figure 14.

Conversely, suppose that rays of light (or other signals) emanate from a distant source so that they are essentially parallel. When these rays strike the surface of a parabolic mirror whose axis of symmetry is parallel to these rays, they are reflected to a single point at the focus. This principle is used in the design of some solar energy devices, satellite dishes, and the mirrors used in some types of telescopes. See Figure 15.

FIGURE 15

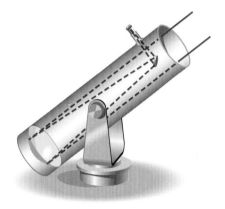

E X A M P L E   8

Satellite Dish

A satellite dish is shaped like a paraboloid of revolution. The signals that emanate from a satellite strike the surface of the dish and are reflected to a single point, where the receiver is located. If the dish is 8 feet across at its opening and is 3 feet deep at its center, at what position should the receiver be placed?

FIGURE 16

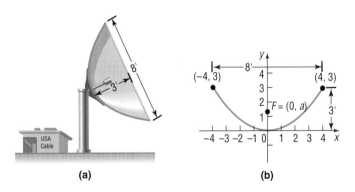

(a)   (b)

Solution   Figure 16(a) shows the satellite dish. We draw the parabola used to form the dish on a rectangular coordinate system so that the vertex of the parabola is at the origin and its focus is on the positive y-axis. See Figure 16(b). The form of the equation of the parabola is

$$x^2 = 4ay$$

and its focus is at $(0, a)$. Since $(4, 3)$ is a point on the graph, we have

$$4^2 = 4a(3)$$

$$a = \frac{4}{3}$$

The receiver should be located $1\frac{1}{3}$ feet from the base of the dish, along its axis of symmetry. ▬

 Now work Problem 51.

6.2 | EXERCISES

In Problems 1–8, the graph of a parabola is given. Match each graph to its equation.

A. $y^2 = 4x$ B. $x^2 = 4y$ C. $y^2 = -4x$

D. $x^2 = -4y$ E. $(y - 1)^2 = 4(x - 1)$ F. $(x + 1)^2 = 4(y + 1)$

G. $(y - 1)^2 = -4(x - 1)$ H. $(x + 1)^2 = -4(y + 1)$

1. 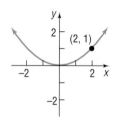   **2.** 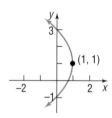   **3.** 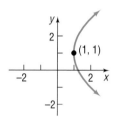   **4.**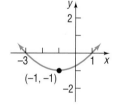

5. 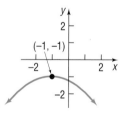   **6.** 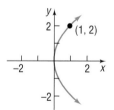   **7.** 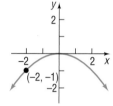   **8.**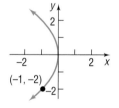

In Problems 9–24, find the equation of the parabola described. Find the two points that define the latus rectum, and graph the equation.

9. Focus at $(4, 0)$; vertex at $(0, 0)$

10. Focus at $(0, 2)$; vertex at $(0, 0)$

11. Focus at $(0, -3)$; vertex at $(0, 0)$

12. Focus at $(-4, 0)$; vertex at $(0, 0)$

13. Focus at $(-2, 0)$; directrix the line $x = 2$

14. Focus at $(0, -1)$; directrix the line $y = 1$

15. Directrix the line $y = -\frac{1}{2}$; vertex at $(0, 0)$

16. Directrix the line $x = -\frac{1}{2}$; vertex at $(0, 0)$

17. Vertex at $(2, -3)$; focus at $(2, -5)$

18. Vertex at $(4, -2)$; focus at $(6, -2)$

19. Vertex at $(0, 0)$; axis of symmetry the y-axis; passing through the point $(2, 3)$

20. Vertex at $(0, 0)$; axis of symmetry the x-axis; passing through the point $(2, 3)$

21. Focus at $(-3, 4)$; directrix the line $y = 2$

22. Focus at $(2, 4)$; directrix the line $x = -4$

23. Focus at $(-3, -2)$; directrix the line $x = 1$

24. Focus at $(-4, 4)$; directrix the line $y = -2$

In Problems 25–42, find the vertex, focus, and directrix of each parabola. Graph the equation.

25. $x^2 = 4y$ **26.** $y^2 = 8x$ **27.** $y^2 = -16x$

28. $x^2 = -4y$ **29.** $(y - 2)^2 = 8(x + 1)$ **30.** $(x + 4)^2 = 16(y + 2)$

31. $(x - 3)^2 = -(y + 1)$ **32.** $(y + 1)^2 = -4(x - 2)$ **33.** $(y + 3)^2 = 8(x - 2)$

34. $(x - 2)^2 = 4(y - 3)$ **35.** $y^2 - 4y + 4x + 4 = 0$ **36.** $x^2 + 6x - 4y + 1 = 0$

37. $x^2 + 8x = 4y - 8$ **38.** $y^2 - 2y = 8x - 1$ **39.** $y^2 + 2y - x = 0$

40. $x^2 - 4x = 2y$ **41.** $x^2 - 4x = y + 4$ **42.** $y^2 + 12y = -x + 1$

In Problems 43–50, write an equation for each parabola.

43.
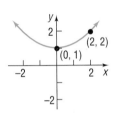

44.
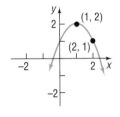

45.
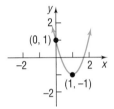

46.
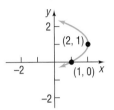

47.
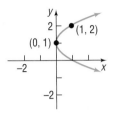

48.
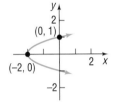

49.
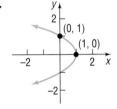

50.
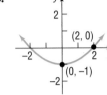

51. Satellite Dish A satellite dish is shaped like a paraboloid of revolution. The signals that emanate from a satellite strike the surface of the dish and are reflected to a single point, where the receiver is located. If the dish is 10 feet across at its opening and is 4 feet deep at its center, at what position should the receiver be placed?

52. Constructing a TV Dish A cable TV receiving dish is in the shape of a paraboloid of revolution. Find the location of the receiver, which is placed at the focus, if the dish is 6 feet across at its opening and 2 feet deep.

53. Constructing a Flashlight The reflector of a flashlight is in the shape of a paraboloid of revolution. Its diameter is 4 inches and its depth is 1 inch. How far from the vertex should the light bulb be placed so that the rays will be reflected parallel to the axis?

54. Constructing a Headlight A sealed-beam headlight is in the shape of a paraboloid of revolution. The bulb, which is placed at the focus, is 1 inch from the vertex. If the depth is to be 2 inches, what is the diameter of the headlight at its opening?

55. Suspension Bridges The cables of a suspension bridge are in the shape of a parabola, as shown in the figure. The towers supporting the cable are 600 feet apart and 80 feet high. If the cables touch the road surface midway between the towers, what is the height of the cable at a point 150 feet from the middle of the bridge?

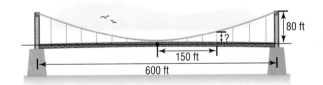

56. Suspension Bridges The cables of a suspension bridge are in the shape of a parabola. The towers supporting the cable are 400 feet apart and 100 feet high. If the cables are at a height of 10 feet midway between the towers, what is the height of the cable at a point 50 feet from a tower?

57. Searchlights A searchlight is shaped like a paraboloid of revolution. If the light source is located 2 feet from the base along the axis of symmetry and the opening is 5 feet across, how deep should the searchlight be?

58. Searchlights A searchlight is shaped like a paraboloid of revolution. If the light source is located 2 feet from the base along the axis of symmetry and the depth of the searchlight is 4 feet, what should the width of the opening be?

59. Solar Heat A mirror is shaped like a paraboloid of revolution and will be used to concentrate the rays of the sun at its focus, creating a heat source. If the mirror is 20 feet across at its opening and is 6 feet deep, where will the heat source be concentrated?

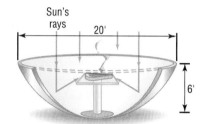

60. Reflecting Telescopes A reflecting telescope contains a mirror shaped like a paraboloid of revolution. If the mirror is 4 inches across at its opening and is 3 feet deep, where will the light collected be concentrated?

61. Parabolic Arch Bridge A bridge is to be built in the shape of a parabolic arch. The bridge has a span of 120 feet and a maximum height of 25 feet. See the illustration. Choose a suitable rectangular coordinate system and find the height of the arch at distances of 10, 30, and 50 feet from the center.

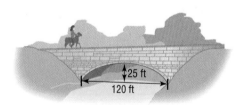

62. Parabolic Arch Bridge A bridge is to be built in the shape of a parabolic arch and is to have a span of 100 feet. The height of the arch a distance of 40 feet from the center is to be 10 feet. Find the height of the arch at its center.

63. Show that an equation of the form

$$Ax^2 + Ey = 0 \qquad A \neq 0, E \neq 0$$

is the equation of a parabola with vertex at $(0, 0)$ and axis of symmetry the y-axis. Find its focus and directrix.

64. Show that an equation of the form

$$Cy^2 + Dx = 0 \qquad C \neq 0, D \neq 0$$

is the equation of a parabola with vertex at $(0, 0)$ and axis of symmetry the x-axis. Find its focus and directrix.

65. Show that the graph of an equation of the form

$$Ax^2 + Dx + Ey + F = 0 \qquad A \neq 0$$

(a) Is a parabola if $E \neq 0$.
(b) Is a vertical line if $E = 0$ and $D^2 - 4AF = 0$.
(c) Is two vertical lines if $E = 0$ and $D^2 - 4AF > 0$.
(d) Contains no points if $E = 0$ and $D^2 - 4AF < 0$.

66. Show that the graph of an equation of the form

$$Cy^2 + Dx + Ey + F = 0 \qquad C \neq 0$$

(a) Is a parabola if $D \neq 0$.
(b) Is a horizontal line if $D = 0$ and $E^2 - 4CF = 0$.
(c) Is two horizontal lines if $D = 0$ and $E^2 - 4CF > 0$.
(d) Contains no points if $D = 0$ and $E^2 - 4CF < 0$.

6.3 | THE ELLIPSE

> 1 Find the Equation of an Ellipse
> 2 Graph Ellipses
> 3 Discuss the Equation of an Ellipse
> 4 Work with Ellipses with Center at (h, k)
> 5 Solve Applied Problems Involving Properties of Ellipses

An **ellipse** is the collection of all points in the plane the sum of whose distances from two fixed points, called the **foci**, is a constant.

FIGURE 17

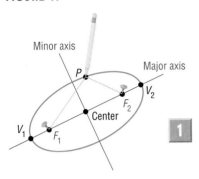

FIGURE 18
$d(F_1, P) + d(F_2, P) = 2a$

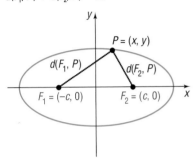

The definition actually contains within it a physical means for drawing an ellipse. Find a piece of string (the length of this string is the constant referred to in the definition). Then take two thumbtacks (the foci) and stick them on a piece of cardboard so that the distance between them is less than the length of the string. Now attach the ends of the string to the thumbtacks and, using the point of a pencil, pull the string taut. Keeping the string taut, rotate the pencil around the two thumbtacks. The pencil traces out an ellipse, as shown in Figure 17.

In Figure 17, the foci are labeled F_1 and F_2. The line containing the foci is called the **major axis.** The midpoint of the line segment joining the foci is called the **center** of the ellipse. The line through the center and perpendicular to the major axis is called the **minor axis.**

The two points of intersection of the ellipse and the major axis are the **vertices,** V_1 and V_2, of the ellipse. The distance from one vertex to the other is called the **length of the major axis.** The ellipse is symmetric with respect to its major axis and with respect to its minor axis.

With these ideas in mind, we are now ready to find the equation of an ellipse in a rectangular coordinate system. First, we place the center of the ellipse at the origin. Second, we position the ellipse so that its major axis coincides with a coordinate axis. Suppose that the major axis coincides with the x-axis, as shown in Figure 18. If c is the distance from the center to a focus, then one focus will be at $F_1 = (-c, 0)$ and the other at $F_2 = (c, 0)$. As we shall see, it is convenient to let $2a$ denote the constant distance referred to in the definition. Then, if $P = (x, y)$ is any point on the ellipse, we have

$$d(F_1, P) + d(F_2, P) = 2a \qquad \text{Sum of the distances from } P \text{ to the foci equals a constant, } 2a.$$

$$\sqrt{(x+c)^2 + y^2} + \sqrt{(x-c)^2 + y^2} = 2a \qquad \text{Use the distance formula.}$$

$$\sqrt{(x+c)^2 + y^2} = 2a - \sqrt{(x-c)^2 + y^2} \qquad \text{Isolate one radical.}$$

$$(x+c)^2 + y^2 = 4a^2 - 4a\sqrt{(x-c)^2 + y^2} \qquad \text{Square both sides.}$$
$$+ (x-c)^2 + y^2$$

$$x^2 + 2cx + c^2 + y^2 = 4a^2 - 4a\sqrt{(x-c)^2 + y^2} \qquad \text{Simplify.}$$
$$+ x^2 - 2cx + c^2 + y^2$$

$$4cx - 4a^2 = -4a\sqrt{(x-c)^2 + y^2} \qquad \text{Isolate the radical.}$$

$$cx - a^2 = -a\sqrt{(x-c)^2 + y^2} \qquad \text{Divide each side by 4.}$$

$$c^2x^2 - 2a^2cx + a^4 = a^2[(x-c)^2 + y^2] \qquad \text{Square both sides again.}$$

$$c^2x^2 - 2a^2cx + a^4 = a^2(x^2 - 2cx + c^2 + y^2)$$

$$(c^2 - a^2)x^2 - a^2y^2 = a^2c^2 - a^4$$

$$(a^2 - c^2)x^2 + a^2y^2 = a^2(a^2 - c^2) \qquad \text{Multiply each side} \qquad (1)$$
$$\text{by } -1; \text{ factor } a^2 \text{ on the right side.}$$

To obtain points on the ellipse off the x-axis, it must be that $a > c$. To see why, look again at Figure 18.

$$d(F_1, P) + d(F_2, P) > d(F_1, F_2) \qquad \text{The sum of the lengths of two sides of a triangle is greater than the length of the third side.}$$

$$2a > 2c \qquad d(F_1, P) + d(F_2, P) = 2a; \quad d(F_1, F_2) = 2c$$

$$a > c$$

Since $a > c$, we also have $a^2 > c^2$, so $a^2 - c^2 > 0$. Let $b^2 = a^2 - c^2, b > 0$. Then $a > b$ and equation (1) can be written as

$$b^2x^2 + a^2y^2 = a^2b^2$$

$$\frac{x^2}{a^2} + \frac{y^2}{b^2} = 1 \qquad \text{Divide each side by } a^2b^2.$$

Theorem   Equation of an Ellipse; Center at (0, 0); Foci at (±c, 0); Major Axis along the x-Axis

An equation of the ellipse with center at $(0, 0)$ and foci at $(-c, 0)$ and $(c, 0)$ is

$$\frac{x^2}{a^2} + \frac{y^2}{b^2} = 1 \qquad \text{where } a > b > 0 \text{ and } b^2 = a^2 - c^2 \qquad (2)$$

The major axis is the x-axis.

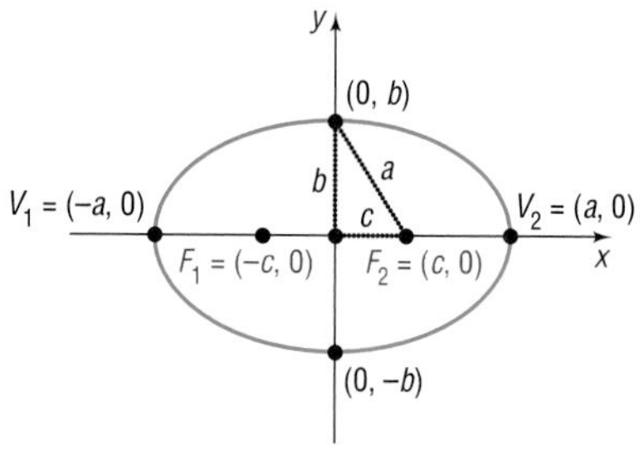

As you can verify, the ellipse defined by equation (2) is symmetric with respect to the x-axis, y-axis, and origin.

To find the vertices of the ellipse defined by equation (2), let $y = 0$. The vertices satisfy the equation $x^2/a^2 = 1$, the solutions of which are $x = \pm a$. Consequently, the vertices of the ellipse given by equation (2) are $V_1 = (-a, 0)$ and $V_2 = (a, 0)$. The y-intercepts of the ellipse, found by letting $x = 0$, have coordinates $(0, -b)$ and $(0, b)$. These four intercepts, $(a, 0)$, $(-a, 0)$, $(0, b)$, and $(0, -b)$, are used to graph the ellipse. See Figure 19.

Notice in Figure 19 the right triangle formed with the points $(0, 0)$, $(c, 0)$, and $(0, b)$. Because $b^2 = a^2 - c^2$ (or $b^2 + c^2 = a^2$), the distance from the focus at $(c, 0)$ to the point $(0, b)$ is a.

FIGURE 19

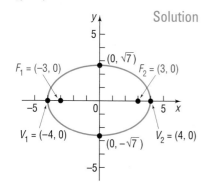

E X A M P L E   1   Finding an Equation of an Ellipse

FIGURE 20
$$\frac{x^2}{16} + \frac{y^2}{7} = 1$$

Find an equation of the ellipse with center at the origin, one focus at $(3, 0)$, and a vertex at $(-4, 0)$. Graph the equation.

Solution   The ellipse has its center at the origin, and the major axis coincides with the x-axis. One focus is at $(c, 0) = (3, 0)$, so $c = 3$. One vertex is at $(-a, 0) = (-4, 0)$, so $a = 4$. From equation (2), it follows that

$$b^2 = a^2 - c^2 = 16 - 9 = 7$$

so an equation of the ellipse is

$$\frac{x^2}{16} + \frac{y^2}{7} = 1$$

Figure 20 shows the graph.

Notice in Figure 20 how we used the intercepts of the equation to graph the ellipse. Following this practice will make it easier for you to obtain an accurate graph of an ellipse.

Comment The intercepts of the ellipse also provide information about how to set the viewing rectangle. To graph the ellipse

$$\frac{x^2}{16} + \frac{y^2}{7} = 1$$

discussed in Example 1, we would set the viewing rectangle using a square screen that includes the intercepts, perhaps $-6 \le x \le 6$, $-4 \le y \le 4$. Then we would graph the two functions

$$Y_1 = -\sqrt{7}\sqrt{1 - \frac{x^2}{16}}, \qquad Y_2 = \sqrt{7}\sqrt{1 - \frac{x^2}{16}}$$

Do this and compare what you see with Figure 20.

Now work Problem 15.

An equation of the form of equation (2), with $a > b$, is the equation of an ellipse with center at the origin, foci on the x-axis at $(-c, 0)$ and $(c, 0)$, where $c^2 = a^2 - b^2$, and major axis along the x-axis.

For the remainder of this section, the direction "Discuss the equation" will mean to find the center, major axis, foci, and vertices of the ellipse and graph it.

E X A M P L E   2   Discussing the Equation of an Ellipse

Discuss the equation: $\dfrac{x^2}{25} + \dfrac{y^2}{9} = 1$

Solution   The given equation is of the form of equation (2), with $a^2 = 25$ and $b^2 = 9$. The equation is that of an ellipse with center $(0, 0)$ and major axis along the x-axis. The vertices are at $(\pm a, 0) = (\pm 5, 0)$. Because $b^2 = a^2 - c^2$, we find

$$c^2 = a^2 - b^2 = 25 - 9 = 16$$

The foci are at $(\pm c, 0) = (\pm 4, 0)$. Figure 21 shows the graph.

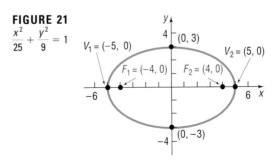

FIGURE 21
$$\frac{x^2}{25} + \frac{y^2}{9} = 1$$

Now work Problem 5.

If the major axis of an ellipse with center at $(0, 0)$ coincides with the y-axis, then the foci are at $(0, -c)$ and $(0, c)$. Using the same steps as before, the definition of an ellipse leads to the following result:

> **Theorem** Equation of an Ellipse; Center at $(0, 0)$; Foci at $(0, \pm c)$; Major Axis along the y-Axis
>
> An equation of the ellipse with center at $(0, 0)$ and foci at $(0, -c)$ and $(0, c)$ is
>
> $$\frac{x^2}{b^2} + \frac{y^2}{a^2} = 1 \qquad \text{where } a > b > 0 \text{ and } b^2 = a^2 - c^2 \qquad (3)$$
>
> The major axis is the y-axis; the vertices are at $(0, -a)$ and $(0, a)$.

FIGURE 22

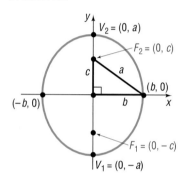

Figure 22 illustrates the graph of such an ellipse. Again, notice the right triangle with the points at $(0, 0)$, $(b, 0)$, and $(0, c)$.

Look closely at equations (2) and (3). Although they may look alike, there is a difference! In equation (2), the larger number, a^2, is in the denominator of the x^2 term, so the major axis of the ellipse is along the x-axis. In equation (3), the larger number, a^2, is in the denominator of the y^2 term, so the major axis is along the y-axis.

E X A M P L E   3

Discussing the Equation of an Ellipse

FIGURE 23

$$x^2 + \frac{y^2}{9} = 1$$

Discuss the equation: $9x^2 + y^2 = 9$

Solution

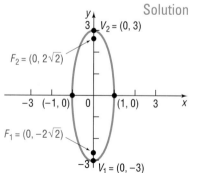

To put the equation in proper form, we divide each side by 9:

$$x^2 + \frac{y^2}{9} = 1$$

The larger number, 9, is in the denominator of the y^2 term so, based on equation (3), this is the equation of an ellipse with center at the origin and major axis along the y-axis. Also, we conclude that $a^2 = 9$, $b^2 = 1$, and $c^2 = a^2 - b^2 = 9 - 1 = 8$. The vertices are at $(0, \pm a) = (0, \pm 3)$, and the foci are at $(0, \pm c) = (0, \pm 2\sqrt{2})$. The graph is given in Figure 23.

E X A M P L E   4

Finding an Equation of an Ellipse

Find an equation of the ellipse having one focus at $(0, 2)$ and vertices at $(0, -3)$ and $(0, 3)$. Graph the equation.

FIGURE 24
$$\frac{x^2}{5} + \frac{y^2}{9} = 1$$

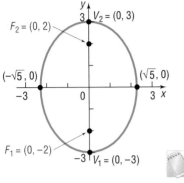

Solution   Because the vertices are at $(0, -3)$ and $(0, 3)$, the center of this ellipse is at the origin. Also, its major axis coincides with the y-axis. The given information also reveals that $c = 2$ and $a = 3$, so $b^2 = a^2 - c^2 = 9 - 4 = 5$. The form of the equation of this ellipse is given by equation (3):

$$\frac{x^2}{b^2} + \frac{y^2}{a^2} = 1$$

$$\frac{x^2}{5} + \frac{y^2}{9} = 1$$

Figure 24 shows the graph.   ■

Now work Problems 9 and 17.

The circle may be considered a special kind of ellipse. To see why, let $a = b$ in equation (2) or in equation (3). Then

$$\frac{x^2}{a^2} + \frac{y^2}{a^2} = 1$$
$$x^2 + y^2 = a^2$$

This is the equation of a circle with center at the origin and radius a. The value of c is

$$c^2 = a^2 - b^2 = 0$$

We conclude that the closer the two foci of an ellipse are the more the ellipse will look like a circle.

Center at (h, k)

4   If an ellipse with center at the origin and major axis coinciding with a coordinate axis is shifted horizontally h units and then vertically k units, the result is an ellipse with center at (h, k) and major axis parallel to a coordinate axis. The equations of such ellipses have the same forms as those given in equations (2) and (3), except that x is replaced by $x - h$ (the horizontal shift) and y is replaced by $y - k$ (the vertical shift). Table 3 gives the forms of the equations of such ellipses, and Figure 25 shows their graphs.

TABLE 3   Ellipses with Center at (h, k) and Major Axis Parallel to a Coordinate Axis

Center	Major Axis	Foci	Vertices	Equation
(h, k)	Parallel to x-axis	$(h \pm c, k)$	$(h \pm a, k)$	$\frac{(x - h)^2}{a^2} + \frac{(y - k)^2}{b^2} = 1$, $a > b$ and $b^2 = a^2 - c^2$
(h, k)	Parallel to y-axis	$(h, k \pm c)$	$(h, k \pm a)$	$\frac{(x - h)^2}{b^2} + \frac{(y - k)^2}{a^2} = 1$, $a > b$ and $b^2 = a^2 - c^2$

FIGURE 25

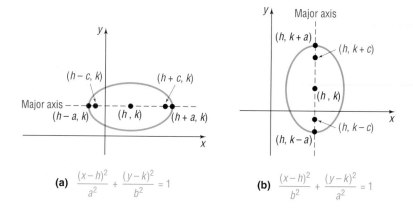

(a) $\dfrac{(x-h)^2}{a^2} + \dfrac{(y-k)^2}{b^2} = 1$

(b) $\dfrac{(x-h)^2}{b^2} + \dfrac{(y-k)^2}{a^2} = 1$

E X A M P L E   5

FIGURE 26
$$\dfrac{(x-2)^2}{9} + \dfrac{(y+3)^2}{8} = 1$$

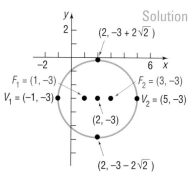

Solution

Finding an Equation of an Ellipse, Center Not at the Origin

Find an equation for the ellipse with center at $(2, -3)$, one focus at $(3, -3)$, and one vertex at $(5, -3)$. Graph the equation.

The center is at $(h, k) = (2, -3)$, so $h = 2$ and $k = -3$. The major axis is parallel to the x-axis. The distance from the center $(2, -3)$ to a focus $(3, -3)$ is $c = 1$; the distance from the center $(2, -3)$ to a vertex $(5, -3)$ is $a = 3$. Thus, $b^2 = a^2 - c^2 = 9 - 1 = 8$. The form of the equation is

$$\dfrac{(x-h)^2}{a^2} + \dfrac{(y-k)^2}{b^2} = 1 \qquad \text{where } h = 2, k = -3, a = 3, b = 2\sqrt{2}$$

$$\dfrac{(x-2)^2}{9} + \dfrac{(y+3)^2}{8} = 1$$

Figure 26 shows the graph.

Now work Problem 41.

E X A M P L E   6

Discussing the Equation of an Ellipse

Discuss the equation: $4x^2 + y^2 - 8x + 4y + 4 = 0$

FIGURE 27
$$(x-1)^2 + \dfrac{(y+2)^2}{4} = 1$$

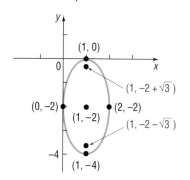

Solution

We proceed to complete the squares in x and y:

$$4x^2 + y^2 - 8x + 4y + 4 = 0$$ Group like variables; place the constant on the right.

$$4x^2 - 8x + y^2 + 4y = -4$$

$$4(x^2 - 2x) + (y^2 + 4y) = -4$$ Factor out 4.

$$4(x^2 - 2x + 1) + (y^2 + 4y + 4) = -4 + 4 + 4$$ Complete each square.

$$4(x-1)^2 + (y+2)^2 = 4$$

$$(x-1)^2 + \dfrac{(y+2)^2}{4} = 1$$ Divide each side by 4.

This is the equation of an ellipse with center at $(1, -2)$ and major axis parallel to the y-axis. Since $a^2 = 4$ and $b^2 = 1$, we have $c^2 = a^2 - b^2 = 4 - 1 = 3$. The vertices are at $(h, k \pm a) = (1, -2 \pm 2)$ or $(1, 0)$ and $(1, -4)$. The foci are at $(h, k \pm c) = (1, -2 \pm \sqrt{3})$ or $(1, -2 - \sqrt{3})$ and $(1, -2 + \sqrt{3})$. Figure 27 shows the graph.

Now work Problem 29.

Applications

5 Ellipses are found in many applications in science and engineering. For example, the orbits of the planets around the Sun are elliptical, with the Sun's position at a focus. See Figure 28.

FIGURE 28

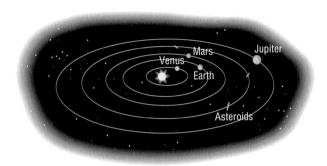

Stone and concrete bridges are often shaped as semielliptical arches. Elliptical gears are used in machinery when a variable rate of motion is required.

Ellipses also have an intersecting reflection property. If a source of light (or sound) is placed at one focus, the waves transmitted by the source will reflect off the ellipse and concentrate at the other focus. This is the principle behind "whispering galleries," which are rooms designed with elliptical ceilings. A person standing at one focus of the ellipse can whisper and be heard by a person standing at the other focus, because all the sound waves that reach the ceiling are reflected to the other person.

E X A M P L E   7   Whispering Galleries

Figure 29 shows the specifications for an elliptical ceiling in a hall designed to be a whispering gallery. In a whispering gallery, a person standing at one focus of the ellipse can whisper and be heard by another person standing at the other focus, because all the sound waves that reach the ceiling from one focus are reflected to the other focus. Where in the hall are the foci located?

FIGURE 29

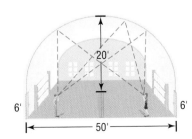

FIGURE 30

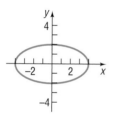

$$\frac{x^2}{a^2} + \frac{y^2}{b^2} = 1 \quad a = 25, b = 20$$

Solution   We set up a rectangular coordinate system so that the center of the ellipse is at the origin and the major axis is along the x-axis. See Figure 30. The equation of the ellipse is

$$\frac{x^2}{a^2} + \frac{y^2}{b^2} = 1$$

where $a = 25$ and $b = 20$. Since

$$c^2 = a^2 - b^2 = 25^2 - 20^2 = 625 - 400 = 225$$

we have $c = 15$. Thus, the foci are located 15 feet from the center of the ellipse along the major axis. ∎

 Now work Problem 55.

6.3 EXERCISES

In Problems 1–4, the graph of an ellipse is given. Match each graph to its equation.

A. $\dfrac{x^2}{4} + y^2 = 1$ B. $x^2 + \dfrac{y^2}{4} = 1$ C. $\dfrac{x^2}{16} + \dfrac{y^2}{4} = 1$ D. $\dfrac{x^2}{4} + \dfrac{y^2}{16} = 1$

1.

2.

3.

4.

In Problems 5–14, find the vertices and foci of each ellipse. Graph each equation.

5. $\dfrac{x^2}{25} + \dfrac{y^2}{4} = 1$ **6.** $\dfrac{x^2}{9} + \dfrac{y^2}{4} = 1$ **7.** $\dfrac{x^2}{9} + \dfrac{y^2}{25} = 1$ **8.** $x^2 + \dfrac{y^2}{16} = 1$

9. $4x^2 + y^2 = 16$ **10.** $x^2 + 9y^2 = 18$ **11.** $4y^2 + x^2 = 8$ **12.** $4y^2 + 9x^2 = 36$

13. $x^2 + y^2 = 16$ **14.** $x^2 + y^2 = 4$

In Problems 15–24, find an equation for each ellipse. Graph the equation.

15. Center at $(0, 0)$; focus at $(3, 0)$; vertex at $(5, 0)$
16. Center at $(0, 0)$; focus at $(-1, 0)$; vertex at $(3, 0)$
17. Center at $(0, 0)$; focus at $(0, -4)$; vertex at $(0, 5)$
18. Center at $(0, 0)$; focus at $(0, 1)$; vertex at $(0, -2)$
19. Foci at $(\pm 2, 0)$; length of the major axis is 6
20. Focus at $(0, -4)$; vertices at $(0, \pm 8)$
21. Foci at $(0, \pm 3)$; x-intercepts are ± 2
22. Foci at $(0, \pm 2)$; length of the major axis is 8
23. Center at $(0, 0)$; vertex at $(0, 4)$; $b = 1$
24. Vertices at $(\pm 5, 0)$; $c = 2$

In Problems 25–28, write an equation for each ellipse.

25.

26.

27.

28.

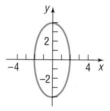
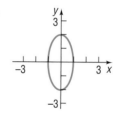
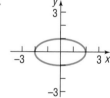
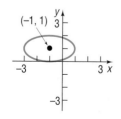
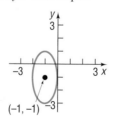
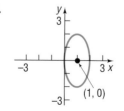
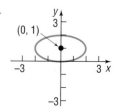

In Problems 29–40, find the center, foci, and vertices of each ellipse. Graph each equation.

29. $\dfrac{(x-3)^2}{4} + \dfrac{(y+1)^2}{9} = 1$

30. $\dfrac{(x+4)^2}{9} + \dfrac{(y+2)^2}{4} = 1$

31. $(x+5)^2 + 4(y-4)^2 = 16$

32. $9(x-3)^2 + (y+2)^2 = 18$

33. $x^2 + 4x + 4y^2 - 8y + 4 = 0$

34. $x^2 + 3y^2 - 12y + 9 = 0$

35. $2x^2 + 3y^2 - 8x + 6y + 5 = 0$

36. $4x^2 + 3y^2 + 8x - 6y = 5$

37. $9x^2 + 4y^2 - 18x + 16y - 11 = 0$

38. $x^2 + 9y^2 + 6x - 18y + 9 = 0$

39. $4x^2 + y^2 + 4y = 0$

40. $9x^2 + y^2 - 18x = 0$

In Problems 41–50, find an equation for each ellipse. Graph the equation.

41. Center at $(2, -2)$; vertex at $(7, -2)$; focus at $(4, -2)$

42. Center at $(-3, 1)$; vertex at $(-3, 3)$; focus at $(-3, 0)$

43. Vertices at $(4, 3)$ and $(4, 9)$; focus at $(4, 8)$

44. Foci at $(1, 2)$ and $(-3, 2)$; vertex at $(-4, 2)$

45. Foci at $(5, 1)$ and $(-1, 1)$; length of the major axis is 8

46. Vertices at $(2, 5)$ and $(2, -1)$; $c = 2$

47. Center at $(1, 2)$; focus at $(4, 2)$; passing through the point $(1, 3)$

48. Center at $(1, 2)$; focus at $(1, 4)$; passing through the point $(2, 2)$

49. Center at $(1, 2)$; vertex at $(4, 2)$; passing through the point $(1, 3)$

50. Center at $(1, 2)$; vertex at $(1, 4)$; passing through the point $(2, 2)$

In Problems 51–54, graph each function.

[**Hint:** Notice that each function is half an ellipse.]

51. $f(x) = \sqrt{16 - 4x^2}$

52. $f(x) = \sqrt{9 - 9x^2}$

53. $f(x) = -\sqrt{64 - 16x^2}$

54. $f(x) = -\sqrt{4 - 4x^2}$

55. **Semielliptical Arch Bridge**   An arch in the shape of the upper half of an ellipse is used to support a bridge that is to span a river 20 meters wide. The center of the arch is 6 meters above the center of the river (see the figure). Write an equation for the ellipse in which the *x*-axis coincides with the water level and the *y*-axis passes through the center of the arch.

56. **Semielliptical Arch Bridge**   The arch of a bridge is a semiellipse with a horizontal major axis. The span is 30 feet, and the top of the arch is 10 feet above the major axis. The roadway is horizontal and is 2 feet above the top of the arch. Find the vertical distance from the roadway to the arch at 5 foot intervals along the roadway.

57. **Whispering Galleries**   A hall 100 feet in length is to be designed as a whispering gallery. If the foci are located 25 feet from the center, how high will the ceiling be at the center?

58. **Whispering Galleries**   Jim, standing at one focus of a whispering gallery, is 6 feet from the nearest wall. His friend is standing at the other focus 100 feet away. What is the length of this whispering gallery? How high is its elliptical ceiling at the center?

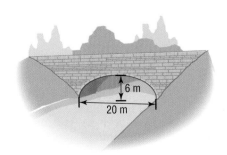

59. Semielliptical Arch Bridge A bridge is built in the shape of a semielliptical arch. The bridge has a span of 120 feet and a maximum height of 25 feet. Choose a suitable rectangular coordinate system and find the height of the arch at distances of 10, 30, and 50 feet from the center.

60. Semielliptical Arch Bridge A bridge is built in the shape of a semielliptical arch and is to have a span of 100 feet. The height of the arch at a distance of 40 feet from the center is to be 10 feet. Find the height of the arch at its center.

61. Semielliptical Arch An arch in the form of half an ellipse is 40 feet wide and 15 feet high at the center. Find the height of the arch at intervals of 10 feet along its width.

62. Semielliptical Arch Bridge An arch for a bridge over a highway is in the form of half an ellipse. The top of the arch is 20 feet above the ground level (the major axis). The highway has four lanes, each 12 feet wide; a center safety strip 8 feet wide; and two side strips, each 4 feet wide. What should the span of the bridge be (the length of its major axis) if the height 28 feet from the center is to be 13 feet?

In Problems 63–66, use the fact that the orbit of a planet about the Sun is an ellipse, with the Sun at one focus. The **aphelion** of a planet is its greatest distance from the Sun and the **perihelion** is its shortest distance. The **mean distance** of a planet from the Sun is the length of the semimajor axis of the elliptical orbit. See the illustration.

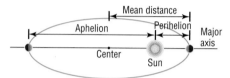

63. Earth The mean distance of Earth from the Sun is 93 million miles. If the aphelion of Earth is 94.5 million miles, what is the perihelion? Write an equation for the orbit of Earth around the Sun.

64. Mars The mean distance of Mars from the Sun is 142 million miles. If the perihelion of Mars is 128.5 million miles, what is the aphelion? Write an equation for the orbit of Mars about the Sun.

65. Jupiter The aphelion of Jupiter is 507 million miles. If the distance from the Sun to the center of its elliptical orbit is 23.2 million miles, what is the perihelion? What is the mean distance? Write an equation for the orbit of Jupiter around the Sun.

66. Pluto The perihelion of Pluto is 4551 million miles and the distance of the Sun from the center of its elliptical orbit is 897.5 million miles. Find the aphelion of Pluto. What is the mean distance of Pluto from the Sun? Write an equation for the orbit of Pluto about the Sun.

67. Racetrack Design Consult the figure. A racetrack is in the shape of an ellipse, 100 feet long and 50 feet wide. What is the width 10 feet from the side?

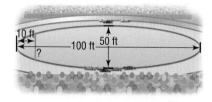

68. Racetrack Design A racetrack is in the shape of an ellipse 80 feet long and 40 feet wide. What is the width 10 feet from the side?

69. Show that an equation of the form

$$Ax^2 + Cy^2 + F = 0 \qquad A \neq 0, C \neq 0, F \neq 0$$

where A and C are of the same sign and F is of opposite sign:
(a) Is the equation of an ellipse with center at $(0, 0)$ if $A \neq C$.
(b) Is the equation of a circle with center $(0, 0)$ if $A = C$.

70. Show that the graph of an equation of the form

$$Ax^2 + Cy^2 + Dx + Ey + F = 0 \qquad A \neq 0, C \neq 0$$

where A and C are of the same sign:
(a) Is an ellipse if $(D^2/4A) + (E^2/4C) - F$ is the same sign as A.
(b) Is a point if $(D^2/4A) + (E^2/4C) - F = 0$.
(c) Contains no points if $(D^2/4A) + (E^2/4C) - F$ is of opposite sign to A.

71. The **eccentricity** e of an ellipse is defined as the number c/a, where a and c are the numbers given in equation (2). Because $a > c$, it follows that $e < 1$. Write a brief paragraph about the general shape of each of the following ellipses. Be sure to justify your conclusions.
(a) Eccentricity close to 0
(b) Eccentricity = 0.5
(c) Eccentricity close to 1

MISSION POSSIBLE

Building a Bridge over the East River

Your team is working for the transportation authority in New York City. You have been asked to study the construction plans for a new bridge over the East River in New York City. The space between supports needs to be 1050 feet; the height at the center of the arch needs to be 350 feet. One company has suggested that the support be in the shape of a parabola; another company suggests a semi-ellipse. The engineering team will determine the relative strengths of the two plans; your job is to find out if there are any differences in the channel widths.

An empty tanker needs a 280-foot clearance to pass beneath the bridge. You need to find the width of the channel for each of the two different plans.

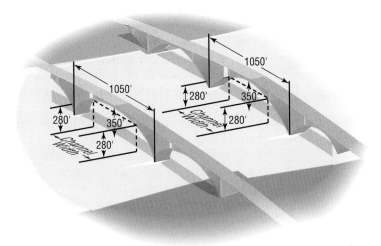

1. To determine the equation of a parabola with these characteristics, first place the parabola on coordinate axes in a convenient location and sketch it.
2. What is the equation of the parabola? (If using a decimal in the equation, you may want to carry six decimal places. If using a fraction, your answer will be more exact.)
3. How wide is the channel that the tanker can pass through if the shape of the support is parabolic?
4. To determine the equation of a semi-ellipse with these characteristics, place the semi-ellipse on coordinate axes in a convenient location and sketch it.
5. What is the equation of the ellipse? How wide is the channel that the tanker can pass through?
6. Now that you know which of the two provides the wider channel, consider some other factors. Your department is also in charge of channel depth, amount of traffic on the river, and various other factors. For example, if the river were to flood and the water level rose by 10 feet, how would the clearances be affected? Make a decision about which plan you think would be the better one as far as your department is concerned, and explain why you think so.

6.4 | THE HYPERBOLA

1. Find the Equation of a Hyperbola
2. Graph Hyperbolas
3. Discuss the Equation of a Hyperbola
4. Find the Asymptotes of a Hyperbola
5. Work with Hyperbolas with Center at (h, k)
6. Solve Applied Problems Involving Properties of Hyperbolas

> A **hyperbola** is the collection of all points in the plane the difference of whose distances from two fixed points, called the **foci**, is a constant.

FIGURE 31

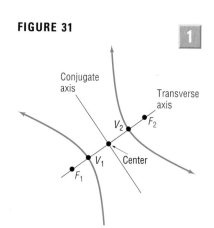

Figure 31 illustrates a hyperbola with foci F_1 and F_2. The line containing the foci is called the **transverse axis.** The midpoint of the line segment joining the foci is called the **center** of the hyperbola. The line through the center and perpendicular to the transverse axis is called the **conjugate axis.** The hyperbola consists of two separate curves, called **branches,** that are symmetric with respect to the transverse axis, conjugate axis, and center. The two points of intersection of the hyperbola and the transverse axis are the **vertices, V_1 and V_2,** of the hyperbola.

With these ideas in mind, we are now ready to find the equation of a hyperbola in a rectangular coordinate system. First, we place the center at the origin. Next, we position the hyperbola so that its transverse axis coincides with a coordinate axis. Suppose that the transverse axis coincides with the x-axis, as shown in Figure 32.

FIGURE 32
$d(F_1, P) - d(F_2, P) = \pm 2a$

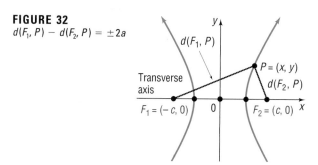

If c is the distance from the center to a focus, then one focus will be at $F_1 = (-c, 0)$ and the other at $F_2 = (c, 0)$. Now we let the constant difference of the distances from any point $P = (x, y)$ on the hyperbola to the foci F_1 and F_2 be denoted by $\pm 2a$. (If P is on the right branch, the + sign is used; if P is on the left branch, the − sign is used.) The coordinates of P must satisfy the equation

$$d(F_1, P) - d(F_2, P) = \pm 2a$$
Difference of the distances from P to the foci equals $\pm 2a$.

$$\sqrt{(x + c)^2 + y^2} - \sqrt{(x - c)^2 + y^2} = \pm 2a$$
Use the distance formula.

$$\sqrt{(x + c)^2 + y^2} = \pm 2a + \sqrt{(x - c)^2 + y^2}$$
Isolate one radical.

$$(x + c)^2 + y^2 = 4a^2 \pm 4a\sqrt{(x - c)^2 + y^2}$$
$$+ (x - c)^2 + y^2$$
Square both sides.

Next, we remove the parentheses:

$$x^2 + 2cx + c^2 + y^2 = 4a^2 \pm 4a\sqrt{(x-c)^2 + y^2} + x^2 - 2cx + c^2 + y^2$$

$$4cx - 4a^2 = \pm 4a\sqrt{(x-c)^2 + y^2} \qquad \text{Isolate the radical.}$$

$$cx - a^2 = \pm a\sqrt{(x-c)^2 + y^2}. \qquad \text{Divide each side by 4.}$$

$$(cx - a^2)^2 = a^2[(x-c)^2 + y^2] \qquad \text{Square both sides.}$$

$$c^2x^2 - 2ca^2x + a^4 = a^2(x^2 - 2cx + c^2 + y^2)$$

$$c^2x^2 + a^4 = a^2x^2 + a^2c^2 + a^2y^2 \qquad \text{Simplify.}$$

$$(c^2 - a^2)x^2 - a^2y^2 = a^2c^2 - a^4$$

$$(c^2 - a^2)x^2 - a^2y^2 = a^2(c^2 - a^2) \qquad (1)$$

To obtain points on the hyperbola off the x-axis, it must be that $a < c$. To see why, look again at Figure 32.

$$d(F_1, P) < d(F_2, P) + d(F_1, F_2) \qquad \text{Use triangle } F_1PF_2.$$

$$d(F_1, P) - d(F_2, P) < d(F_1, F_2) \qquad \begin{array}{l}\text{\small P is on the right branch,}\\ \text{\small so } d(F_1, P) - d(F_2, P) = 2a.\end{array}$$

$$2a < 2c$$

$$a < c$$

Since $a < c$, we also have $a^2 < c^2$, so $c^2 - a^2 > 0$. Let $b^2 = c^2 - a^2, b > 0$. Then equation (1) can be written as

$$b^2x^2 - a^2y^2 = a^2b^2$$

$$\frac{x^2}{a^2} - \frac{y^2}{b^2} = 1$$

To find the vertices of the hyperbola defined by this equation, let $y = 0$. The vertices satisfy the equation $x^2/a^2 = 1$, the solutions of which are $x = \pm a$. Consequently, the vertices of the hyperbola are $V_1 = (-a, 0)$ and $V_2 = (a, 0)$.

Theorem   Equation of a Hyperbola; Center at (0, 0); Foci at ($\pm c$, 0); Vertices at ($\pm a$, 0); Transverse Axis along the x-Axis

An equation of the hyperbola with center at $(0, 0)$, foci at $(-c, 0)$ and $(c, 0)$, and vertices at $(-a, 0)$ and $(a, 0)$ is

$$\frac{x^2}{a^2} - \frac{y^2}{b^2} = 1 \qquad \text{where } b^2 = c^2 - a^2 \qquad (2)$$

The transverse axis is the x-axis.

As you can verify, the hyperbola defined by equation (2) is symmetric with respect to the x-axis, y-axis, and origin. To find the y-intercepts, if any, let $x = 0$ in equation (2). This results in the equation $y^2/b^2 = -1$, which has no solution. We conclude that the hyperbola defined by equation (2) has no y-intercepts. In fact, since $x^2/a^2 - 1 = y^2/b^2 \geq 0$, it follows that $x^2/a^2 \geq 1$. Thus, there are no points on the graph for $-a < x < a$. See Figure 33.

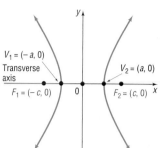

FIGURE 33

$$\frac{x^2}{a^2} - \frac{y^2}{b^2} = 1, b^2 = c^2 - a^2$$

$V_1 = (-a, 0)$
Transverse axis
$F_1 = (-c, 0)$
$V_2 = (a, 0)$
$F_2 = (c, 0)$

E X A M P L E   1

Finding an Equation of a Hyperbola

Find an equation of the hyperbola with center at the origin, one focus at $(3, 0)$, and one vertex at $(-2, 0)$. Graph the equation.

Solution   The hyperbola has its center at the origin, and the transverse axis coincides with the x-axis. One focus is at $(c, 0) = (3, 0)$, so $c = 3$. One vertex is at $(-a, 0) = (-2, 0)$, so $a = 2$. From equation (2), it follows that $b^2 = c^2 - a^2 = 9 - 4 = 5$, so an equation of the hyperbola is

$$\frac{x^2}{4} - \frac{y^2}{5} = 1$$

See Figure 34.

FIGURE 34

$$\frac{x^2}{4} - \frac{y^2}{5} = 1$$

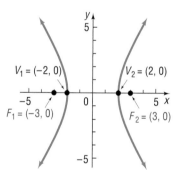

$V_1 = (-2, 0)$
$F_1 = (-3, 0)$
$V_2 = (2, 0)$
$F_2 = (3, 0)$

 Comment   To graph the hyperbola $(x^2/4) - (y^2/5) = 1$ discussed in Example 1, we need to graph the two functions $Y_1 = \sqrt{5}\sqrt{(x^2/4) - 1}$ and $Y_2 = -\sqrt{5}\sqrt{(x^2/4) - 1}$. Do this and compare what you see with Figure 34.

 Now work Problem 5.

An equation of the form of equation (2) is the equation of a hyperbola with center at the origin, foci on the x-axis at $(-c, 0)$ and $(c, 0)$, where $c^2 = a^2 + b^2$, and transverse axis along the x-axis.

 For the remainder of this section, the direction "Discuss the equation" will mean to find the center, transverse axis, vertices, and foci of the hyperbola and graph it.

E X A M P L E   2

Discussing the Equation of a Hyperbola

Discuss the equation: $\dfrac{x^2}{16} - \dfrac{y^2}{4} = 1$

FIGURE 35

$$\frac{x^2}{16} - \frac{y^2}{4} = 1$$

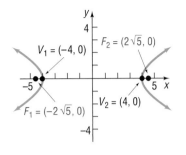

Solution The given equation is of the form of equation (2), with $a^2 = 16$ and $b^2 = 4$. Thus, the graph of the equation is a hyperbola with center at $(0, 0)$ and transverse axis along the x-axis. Also, we know that $c^2 = a^2 + b^2 = 16 + 4 = 20$. The vertices are at $(\pm a, 0) = (\pm 4, 0)$, and the foci are at $(\pm c, 0) = (\pm 2\sqrt{5}, 0)$. Figure 35 shows the graph. ∎

The next result gives the form of the equation of a hyperbola with center at the origin and transverse axis along the y-axis.

Theorem Equation of a Hyperbola; Center at (0, 0); Foci at (0, ±c); Vertices at (0, ±a); Transverse Axis along the y-Axis

An equation of the hyperbola with center at $(0, 0)$, foci at $(0, -c)$ and $(0, c)$, and vertices at $(0, -a)$ and $(0, a)$ is

$$\frac{y^2}{a^2} - \frac{x^2}{b^2} = 1 \qquad \text{where } b^2 = c^2 - a^2 \qquad (3)$$

The transverse axis is the y-axis.

FIGURE 36

$$\frac{y^2}{a^2} - \frac{x^2}{b^2} = 1, \, b^2 = c^2 - a^2$$

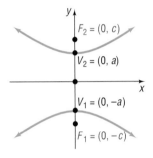

Figure 36 shows the graph of a typical hyperbola defined by equation (3). Notice the difference in the form of equations (2) and (3). When the y^2 term is subtracted from the x^2 term, the transverse axis is the x-axis. When the x^2 term is subtracted from the y^2 term, the transverse axis is the y-axis.

E X A M P L E 3 Discussing the Equation of a Hyperbola

Discuss the equation: $y^2 - 4x^2 = 4$

FIGURE 37

$$\frac{y^2}{4} - x^2 = 1$$

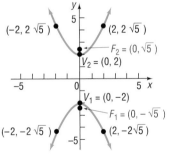

Solution To put the equation in proper form, we divide each side by 4:

$$\frac{y^2}{4} - x^2 = 1$$

Since the x^2 term is subtracted from the y^2 term, the equation is that of a hyperbola with center at the origin and transverse axis along the y-axis. Also, comparing the above equation to equation (3), we find $a^2 = 4$, $b^2 = 1$, and $c^2 = a^2 + b^2 = 5$. The vertices are at $(0, \pm a) = (0, \pm 2)$, and the foci are at $(0, \pm c) = (0, \pm \sqrt{5})$. The graph is given in Figure 37. ∎

E X A M P L E 4 Finding an Equation of a Hyperbola

Find an equation of the hyperbola having one vertex at $(0, 2)$ and foci at $(0, -3)$ and $(0, 3)$. Graph the equation.

Solution Since the foci are at $(0, -3)$ and $(0, 3)$, the center of the hyperbola is at the origin. Also, the transverse axis is along the y-axis. The given information

FIGURE 38
$$\frac{y^2}{4} - \frac{x^2}{5} = 1$$

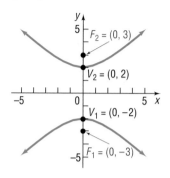

also reveals that $c = 3$, $a = 2$, and $b^2 = c^2 - a^2 = 9 - 4 = 5$. The form of the equation of the hyperbola is given by equation (3):

$$\frac{y^2}{a^2} - \frac{x^2}{b^2} = 1$$

$$\frac{y^2}{4} - \frac{x^2}{5} = 1$$

See Figure 38.

Now work Problem 7.

Look at the equations of the hyperbolas in Examples 3 and 4. For the hyperbola in Example 3, $a^2 = 4$ and $b^2 = 1$, so $a > b$; for the hyperbola in Example 4, $a^2 = 4$ and $b^2 = 5$, so $a < b$. We conclude that, for hyperbolas, there are no requirements involving the relative sizes of a and b. Contrast this situation to the case of an ellipse, in which the relative sizes of a and b dictate which axis is the major axis. Hyperbolas have another feature to distinguish them from ellipses and parabolas: Hyperbolas have asymptotes.

Asymptotes

A horizontal or oblique asymptote of a graph is a line with the property that the distance from the line to points on the graph approaches 0 as $x \to -\infty$ or as $x \to \infty$. When such asymptotes exist, they give information about the end behavior of the graph.

Theorem   Asymptotes of a Hyperbola

The hyperbola $\dfrac{x^2}{a^2} - \dfrac{y^2}{b^2} = 1$ has the two oblique asymptotes

$$y = \frac{b}{a}x \quad \text{and} \quad y = -\frac{b}{a}x$$

Proof   We begin by solving for y in the equation of the hyperbola:

$$\frac{x^2}{a^2} - \frac{y^2}{b^2} = 1$$

$$\frac{y^2}{b^2} = \frac{x^2}{a^2} - 1$$

$$y^2 = b^2\left(\frac{x^2}{a^2} - 1\right)$$

If $x \ne 0$, we can rearrange the right side in the form

$$y^2 = \frac{b^2 x^2}{a^2}\left(1 - \frac{a^2}{x^2}\right)$$

$$y = \pm\frac{bx}{a}\sqrt{1 - \frac{a^2}{x^2}}$$

Now, as $x \rightarrow -\infty$ or as $x \rightarrow \infty$, the term a^2/x^2 approaches 0, so the expression under the radical approaches 1. Thus, as $x \rightarrow -\infty$ or as $x \rightarrow \infty$, the value of y approaches $\pm bx/a$; that is, the graph of the hyperbola approaches the lines

$$y = -\frac{b}{a}x \quad \text{and} \quad y = \frac{b}{a}x$$

These lines are oblique asymptotes of the hyperbola.

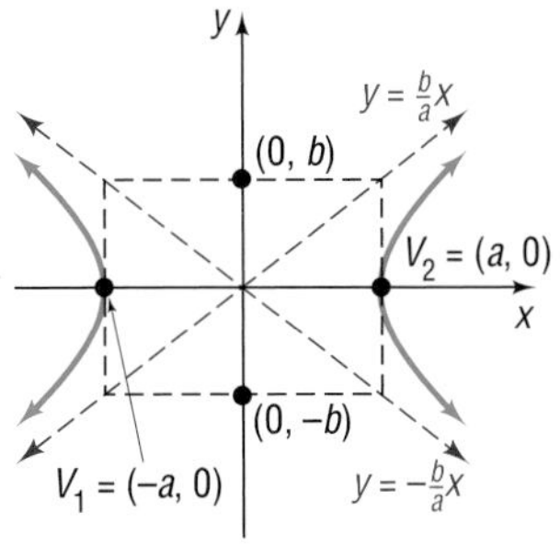

The asymptotes of a hyperbola are not part of the hyperbola, but they do serve as a guide for graphing a hyperbola. For example, suppose that we want to graph the equation

$$\frac{x^2}{a^2} - \frac{y^2}{b^2} = 1$$

We begin by plotting the vertices $(-a, 0)$ and $(a, 0)$. Then we plot the points $(0, -b)$ and $(0, b)$ and use these four points to construct a rectangle, as shown in Figure 39. The diagonals of this rectangle have slopes b/a and $-b/a$, and their extensions are the asymptotes $y = (b/a)x$ and $y = -(b/a)x$ of the hyperbola.

FIGURE 39
$$\frac{x^2}{a^2} - \frac{y^2}{b^2} = 1$$

Theorem   Asymptotes of a Hyperbola

The hyperbola $\dfrac{y^2}{a^2} - \dfrac{x^2}{b^2} = 1$ has the two oblique asymptotes

$$y = \frac{a}{b}x \quad \text{and} \quad y = -\frac{a}{b}x$$

You are asked to prove this result in Problem 60.

E X A M P L E   5   Discussing the Equation of a Hyperbola

Discuss the equation: $9x^2 - 4y^2 = 36$

FIGURE 40
$$\frac{x^2}{4} - \frac{y^2}{9} = 1$$

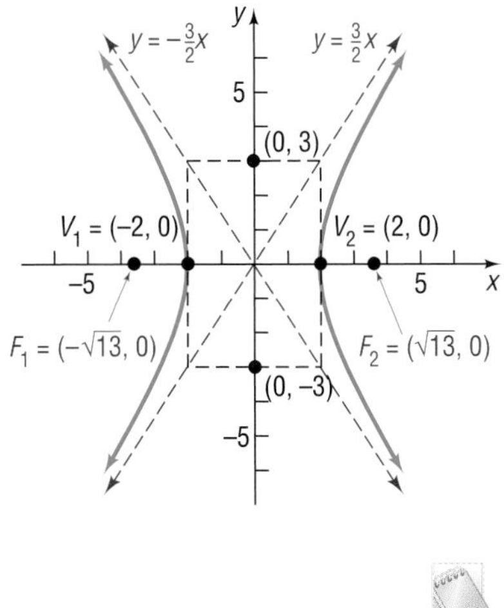

Solution   First, we divide each side by 36 to put the equation in proper form:

$$\frac{x^2}{4} - \frac{y^2}{9} = 1$$

This is the equation of a hyperbola with center at the origin and transverse axis along the x-axis. Using $a^2 = 4$ and $b^2 = 9$, we find $c^2 = a^2 + b^2 = 13$. The vertices are at $(\pm a, 0) = (\pm 2, 0)$, the foci are at $(\pm c, 0) = (\pm\sqrt{13}, 0)$, and the asymptotes have the equations

$$y = \frac{3}{2}x \quad \text{and} \quad y = -\frac{3}{2}x$$

Now form the rectangle containing the points $(\pm a, 0)$ and $(0, \pm b)$, that is, $(-2, 0), (2, 0), (0, -3)$, and $(0, 3)$. The extensions of the diagonals of this rectangle are the asymptotes. See Figure 40 for the graph.

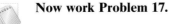

Now work Problem 17.

Exploration Graph the upper portion of the hyperbola $9x^2 - 4y^2 = 36$ discussed in Example 5 and its asymptotes $y = \frac{3}{2}x$ and $y = -\frac{3}{2}x$. Now use ZOOM and TRACE to see what happens as x becomes unbounded in the positive direction. What happens as x becomes unbounded in the negative direction?

Center at (h, k)

If a hyperbola with center at the origin and transverse axis coinciding with a coordinate axis is shifted horizontally h units and then vertically k units, the result is a hyperbola with center at (h, k) and transverse axis parallel to a coordinate axis. The equations of such hyperbolas have the same forms as those given in equations (2) and (3), except that x is replaced by $x - h$ (the horizontal shift) and y is replaced by $y - k$ (the vertical shift). Table 4 gives the forms of the equations of such hyperbolas. See Figure 41 for the graphs.

TABLE 4 Hyperbolas with Center at (h, k) and Transverse Axis Parallel to a Coordinate Axis

Center	Transverse Axis	Foci	Vertices	Equation	Asymptotes
(h, k)	Parallel to x-axis	$(h \pm c, k)$	$(h \pm a, k)$	$\dfrac{(x - h)^2}{a^2} - \dfrac{(y - k)^2}{b^2} = 1,$ $b^2 = c^2 - a^2$	$y - k = \pm\dfrac{b}{a}(x - h)$
(h, k)	Parallel to y-axis	$(h, k \pm c)$	$(h, k \pm a)$	$\dfrac{(y - k)^2}{a^2} - \dfrac{(x - h)^2}{b^2} = 1,$ $b^2 = c^2 - a^2$	$y - k = \pm\dfrac{a}{b}(x - h)$

FIGURE 41

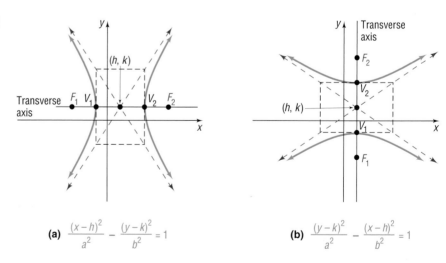

(a) $\dfrac{(x-h)^2}{a^2} - \dfrac{(y-k)^2}{b^2} = 1$

(b) $\dfrac{(y-k)^2}{a^2} - \dfrac{(x-h)^2}{b^2} = 1$

E X A M P L E   6   Finding an Equation of a Hyperbola, Center Not at the Origin

Find an equation for the hyperbola with center at $(1, -2)$, one focus at $(4, -2)$, and one vertex at $(3, -2)$. Graph the equation.

Solution   The center is at $(h, k) = (1, -2)$, so $h = 1$ and $k = -2$. The transverse axis is parallel to the x-axis. The distance from the center $(1, -2)$ to the focus

$(4, -2)$ is $c = 3$; the distance from the center $(1, -2)$ to the vertex $(3, -2)$ is $a = 2$. Thus, $b^2 = c^2 - a^2 = 9 - 4 = 5$. The equation is

$$\frac{(x - h)^2}{a^2} - \frac{(y - k)^2}{b^2} = 1$$

$$\frac{(x - 1)^2}{4} - \frac{(y + 2)^2}{5} = 1$$

See Figure 42.

FIGURE 42

$$\frac{(x - 1)^2}{4} - \frac{(y + 2)^2}{5} = 1$$

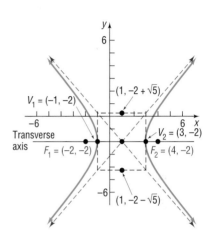

$V_1 = (-1, -2)$
$(1, -2 + \sqrt{5})$
Transverse axis
$F_1 = (-2, -2)$
$V_2 = (3, -2)$
$F_2 = (4, -2)$
$(1, -2 - \sqrt{5})$

 Now work Problem 27.

E X A M P L E   7   Discussing the Equation of a Hyperbola

Discuss the equation: $-x^2 + 4y^2 - 2x - 16y + 11 = 0$

Solution   We complete the squares in x and in y.

FIGURE 43

$(y - 2)^2 - \dfrac{(x + 1)^2}{4} = 1$

Transverse axis

$F_2 = (-1, 2 + \sqrt{5})$
$V_2 = (-1, 3)$
$(-3, 2)$
$(1, 2)$
-5
5 x
$F_1 = (-1, 2 - \sqrt{5})$
$V_1 = (-1, 1)$

$$-x^2 + 4y^2 - 2x - 16y + 11 = 0$$
$$-(x^2 + 2x) + 4(y^2 - 4y) = -11 \qquad \text{Group terms.}$$
$$-(x^2 + 2x + 1) + 4(y^2 - 4y + 4) = -1 + 16 - 11 \qquad \text{Complete each square.}$$
$$-(x + 1)^2 + 4(y - 2)^2 = 4$$
$$(y - 2)^2 - \frac{(x + 1)^2}{4} = 1 \qquad \text{Divide by 4.}$$

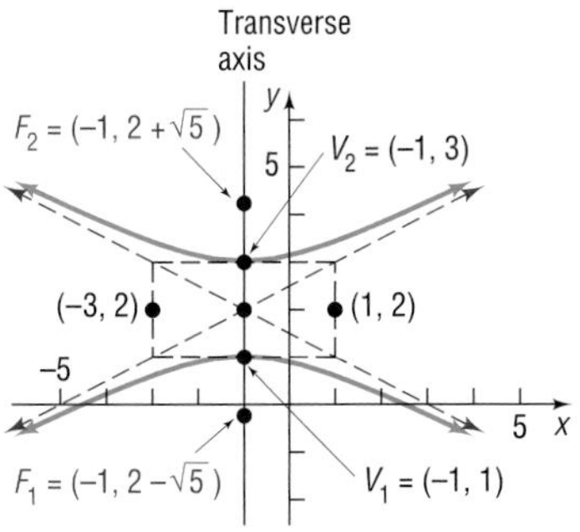

This is the equation of a hyperbola with center at $(-1, 2)$ and transverse axis parallel to the y-axis. Also, $a^2 = 1$ and $b^2 = 4$, so $c^2 = a^2 + b^2 = 5$. The vertices are at $(h, k \pm a) = (-1, 2 \pm 1)$, or $(-1, 1)$ and $(-1, 3)$. The foci are at $(h, k \pm c) = (-1, 2 \pm \sqrt{5})$. The asymptotes are $y - 2 = \frac{1}{2}(x + 1)$ and $y - 2 = -\frac{1}{2}(x + 1)$. Figure 43 shows the graph.

FIGURE 44

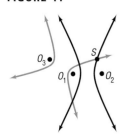

O_3
S
O_1
O_2

 Now work Problem 41.

Applications

6 See Figure 44. Suppose that a gun is fired from an unknown source S. An observer at O_1 hears the report (sound of gun shot) 1 second after another

observer at O_2. Because sound travels at about 1100 feet per second, it follows that the point S must be 1100 feet closer to O_2 than to O_1. Thus, S lies on one branch of a hyperbola with foci at O_1 and O_2. (Do you see why? The difference of the distances from S to O_1 and from S to O_2 is the constant 1100.) If a third observer at O_3 hears the same report 2 seconds after O_1 hears it, then S will lie on a branch of a second hyperbola with foci at O_1 and O_3. The intersection of the two hyperbolas will pinpoint the location of S.

LORAN

FIGURE 45

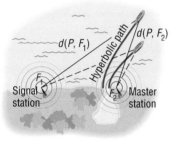

$d(P, F_1) - d(P, F_2) = \text{constant}$

In the LOng RAnge Navigation system (LORAN), a master radio sending station and a secondary sending station emit signals that can be received by a ship at sea. (See Figure 45.) Because a ship monitoring the two signals will usually be nearer to one of the two stations, there will be a difference in the distance the two signals travel, which will register as a slight time difference between the signals. As long as the time difference remains constant, the difference of the two distances will also be constant. If the ship follows a path corresponding to the fixed time difference, it will follow the path of a hyperbola whose foci are located at the positions of the two sending stations. So for each time difference a different hyperbolic path results, each bringing the ship to a different shore location. Navigation charts show the various hyperbolic paths corresponding to different time differences.

E X A M P L E   8

LORAN

Two LORAN stations are positioned 250 miles apart along a straight shore.

(a) A ship records a time difference of 0.00054 second between the LORAN signals. Set up an appropriate rectangular coordinate system to determine where the ship would reach shore if it were to follow the hyperbola corresponding to this time difference.

(b) If the ship wants to enter a harbor located between the two stations 25 miles from the master station, what time difference should it be looking for?

(c) If the ship is 80 miles off shore when the desired time difference is obtained, what is the approximate location of the ship?

[**Note:** The speed of each radio signal is 186,000 miles per second.]

Solution

FIGURE 46

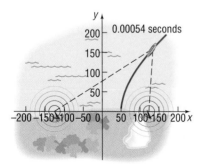

(a) We set up a rectangular coordinate system so that the two stations lie on the x-axis and the origin is midway between them. See Figure 46. The ship lies on a hyperbola whose foci are the locations of the two stations. The reason for this is that the constant time difference of the signals from each station results in a constant difference in the distance of the ship from each station. Since the time difference is 0.00054 second and the speed of the signal is 186,000 miles per second, the difference of the distances from the ship to each station (foci) is

$$\text{Distance} = \text{speed} \times \text{time} = 186,000 \times 0.00054 \approx 100 \text{ miles}$$

The difference of the distances from the ship to each station, 100, equals $2a$, so $a = 50$ and the vertex of the corresponding hyperbola is at $(50, 0)$. Since the focus is at $(125, 0)$, following this hyperbola the ship would reach shore 75 miles from the master station.

(b) To reach shore 25 miles from the master station, the ship should follow a hyperbola with vertex at $(100, 0)$. For this hyperbola, $a = 100$, so the constant difference of the distances from the ship to each station is 200. The time difference the ship should look for is

$$\text{Time} = \frac{\text{distance}}{\text{speed}} = \frac{200}{186,000} = 0.001075 \text{ second}$$

(c) To find the approximate location of the ship, we need to find the equation of the hyperbola with vertex at $(100, 0)$ and a focus at $(125, 0)$. The form of the equation of this hyperbola is

$$\frac{x^2}{a^2} - \frac{y^2}{b^2} = 1$$

where $a = 100$. Since $c = 125$, we have

$$b^2 = c^2 - a^2 = 125^2 - 100^2 = 5625$$

The equation of the hyperbola is

$$\frac{x^2}{100^2} - \frac{y^2}{5625} = 1$$

FIGURE 47

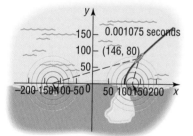

Since the ship is 80 miles from shore, we use $y = 80$ in the equation and solve for x.

$$\frac{x^2}{100^2} - \frac{80^2}{5625} = 1$$

$$\frac{x^2}{100^2} = 1 + \frac{80^2}{5625} \approx 2.14$$

$$x^2 = 100^2(2.14)$$

$$x \approx 146$$

The ship is at the position $(146, 80)$. See Figure 47.

Now work Problem 53.

6.4 EXERCISES

In Problems 1–4, the graph of a hyperbola is given. Match each graph to its equation.

A. $\dfrac{x^2}{4} - y^2 = 1$ B. $x^2 - \dfrac{y^2}{4} = 1$ C. $\dfrac{y^2}{4} - x^2 = 1$ D. $y^2 - \dfrac{x^2}{4} = 1$

1.

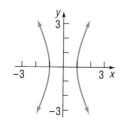

2.

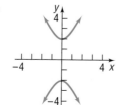

3.

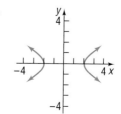

4.

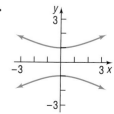

In Problems 5–14, find an equation for the hyperbola described. Graph the equation.

5. Center at $(0, 0)$; focus at $(3, 0)$; vertex at $(1, 0)$

6. Center at $(0, 0)$; focus at $(0, 5)$; vertex at $(0, 3)$

7. Center at $(0, 0)$; focus at $(0, -6)$; vertex at $(0, 4)$

8. Center at $(0, 0)$; focus at $(-3, 0)$; vertex at $(2, 0)$

9. Foci at $(-5, 0)$ and $(5, 0)$; vertex at $(3, 0)$

10. Focus at $(0, 6)$; vertices at $(0, -2)$ and $(0, 2)$

11. Vertices at $(0, -6)$ and $(0, 6)$; asymptote the line $y = 2x$

12. Vertices at $(-4, 0)$ and $(4, 0)$; asymptote the line $y = 2x$

13. Foci at $(-4, 0)$ and $(4, 0)$; asymptote the line $y = -x$

14. Foci at $(0, -2)$ and $(0, 2)$; asymptote the line $y = -x$

In Problems 15–22, find the center, transverse axis, vertices, foci, and asymptotes. Graph each equation.

15. $\dfrac{x^2}{25} - \dfrac{y^2}{9} = 1$

16. $\dfrac{y^2}{16} - \dfrac{x^2}{4} = 1$

17. $4x^2 - y^2 = 16$

18. $4y^2 - x^2 = 16$

19. $y^2 - 9x^2 = 9$

20. $x^2 - y^2 = 4$

21. $y^2 - x^2 = 25$

22. $2x^2 - y^2 = 4$

In Problems 23–26, write an equation for each hyperbola.

23.
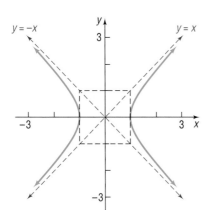

24.
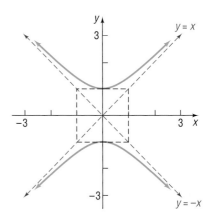

25.
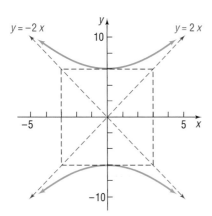

26.
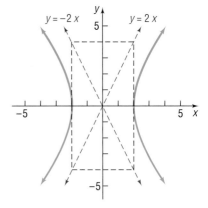

In Problems 27–34, find an equation for the hyperbola described. Graph the equation.

27. Center at $(4, -1)$; focus at $(7, -1)$; vertex at $(6, -1)$

28. Center at $(-3, 1)$; focus at $(-3, 6)$; vertex at $(-3, 4)$

29. Center at $(-3, -4)$; focus at $(-3, -8)$; vertex at $(-3, -2)$

30. Center at $(1, 4)$; focus at $(-2, 4)$; vertex at $(0, 4)$

31. Foci at $(3, 7)$ and $(7, 7)$; vertex at $(6, 7)$

32. Focus at $(-4, 0)$; vertices at $(-4, 4)$ and $(-4, 2)$

33. Vertices at $(-1, -1)$ and $(3, -1)$; asymptote the line $(x - 1)/2 = (y + 1)/3$

34. Vertices at $(1, -3)$ and $(1, 1)$; asymptote the line $(x - 1)/2 = (y + 1)/3$

In Problems 35–48, find the center, transverse axis, vertices, foci, and asymptotes. Graph each equation.

35. $\dfrac{(x-2)^2}{4} - \dfrac{(y+3)^2}{9} = 1$

36. $\dfrac{(y+3)^2}{4} - \dfrac{(x-2)^2}{9} = 1$

37. $(y-2)^2 - 4(x+2)^2 = 4$

38. $(x+4)^2 - 9(y-3)^2 = 9$

39. $(x+1)^2 - (y+2)^2 = 4$

40. $(y-3)^2 - (x+2)^2 = 4$

41. $x^2 - y^2 - 2x - 2y - 1 = 0$

42. $y^2 - x^2 - 4y + 4x - 1 = 0$

43. $y^2 - 4x^2 - 4y - 8x - 4 = 0$

44. $2x^2 - y^2 + 4x + 4y - 4 = 0$

45. $4x^2 - y^2 - 24x - 4y + 16 = 0$

46. $2y^2 - x^2 + 2x + 8y + 3 = 0$

47. $y^2 - 4x^2 - 16x - 2y - 19 = 0$

48. $x^2 - 3y^2 + 8x - 6y + 4 = 0$

In Problems 49–52, graph each function.
[**Hint:** Notice that each function is half a hyperbola.]

49. $f(x) = \sqrt{16 + 4x^2}$

50. $f(x) = -\sqrt{9 + 9x^2}$

51. $f(x) = -\sqrt{-25 + x^2}$

52. $f(x) = \sqrt{-1 + x^2}$

53. LORAN Two LORAN stations are positioned 200 miles apart along a straight shore.
(a) A ship records a time difference of 0.00038 second between the LORAN signals. Set up an appropriate rectangular coordinate system to determine where the ship would reach shore if it were to follow the hyperbola corresponding to this time difference.
(b) If the ship wants to enter a harbor located between the two stations 20 miles from the master station, what time difference should it be looking for?
(c) If the ship is 50 miles off shore when the desired time difference is obtained, what is the approximate location of the ship?
[**Note:** The speed of each radio signal is 186,000 miles per second.]

54. LORAN Two LORAN stations are positioned 100 miles apart along a straight shore.
(a) A ship records a time difference of 0.00032 second between the LORAN signals. Set up an appropriate rectangular coordinate system to determine where the ship would reach shore if it were to follow the hyperbola corresponding to this time difference.
(b) If the ship wants to enter a harbor located between the two stations 10 miles from the master station, what time difference should it be looking for?
(c) If the ship is 20 miles off shore when the desired time difference is obtained, what is the approximate location of the ship?
[**Note:** The speed of each radio signal is 186,000 miles per second.]

55. Calibrating Instruments In a test of their recording devices, a team of seismologists positioned two of the devices 2000 feet apart, with the device at point A to the west of the device at point B. At a point between the devices and 200 feet from point B, a small amount of ex-

plosive was detonated and a note made of the time at which the sound reached each device. A second explosion is to be carried out at a point directly north of point B.
(a) How far north should the site of the second explosion be chosen so that the measured time difference recorded by the devices for the second detonation is the same as that recorded for the first detonation?
 (b) Explain why this experiment can be used to calibrate the instruments.

56. Explain in your own words the LORAN system of navigation.

57. The **eccentricity** e of a hyperbola is defined as the number c/a, where a and c are the numbers given in equation (2). Because $c > a$, it follows that $e > 1$. Describe the general shape of a hyperbola whose eccentricity is close to 1. What is the shape if e is very large?

58. A hyperbola for which $a = b$ is called an **equilateral hyperbola.** Find the eccentricity e of an equilateral hyperbola.
[**Note:** The eccentricity of a hyperbola is defined in Problem 57.]

59. Two hyperbolas that have the same set of asymptotes are called **conjugate.** Show that the hyperbolas
$$\frac{x^2}{4} - y^2 = 1 \quad \text{and} \quad y^2 - \frac{x^2}{4} = 1$$
are conjugate. Graph each hyperbola on the same set of coordinate axes.

60. Prove that the hyperbola
$$\frac{y^2}{a^2} - \frac{x^2}{b^2} = 1$$
has the two oblique asymptotes
$$y = \frac{a}{b}x \quad \text{and} \quad y = -\frac{a}{b}x$$

61. Show that the graph of an equation of the form

$$Ax^2 + Cy^2 + F = 0 \qquad A \neq 0, C \neq 0, F \neq 0$$

where A and C are of opposite sign, is a hyperbola with center at $(0, 0)$.

62. Show that the graph of an equation of the form

$$Ax^2 + Cy^2 + Dx + Ey + F = 0 \quad A \neq 0, C \neq 0$$

where A and C are of opposite sign:

(a) Is a hyperbola if $(D^2/4A) + (E^2/4C) - F \neq 0$.
(b) Is two intersecting lines if

$$\left(\frac{D^2}{4A}\right) + \left(\frac{E^2}{4C}\right) - F = 0.$$

6.5 | ROTATION OF AXES; GENERAL FORM OF A CONIC

1. Identify a Conic
2. Use a Rotation of Axes to Transform Equations
3. Discuss an Equation Using a Rotation of Axes
4. Identify Conics without a Rotation of Axes

In this section, we show that the graph of a general second-degree polynomial containing two variables x and y, that is, an equation of the form

$$Ax^2 + Bxy + Cy^2 + Dx + Ey + F = 0 \tag{1}$$

where A, B, and C are not simultaneously 0, is a conic. We shall not concern ourselves here with the degenerate cases of equation (1), such as $x^2 + y^2 = 0$, whose graph is a single point $(0, 0)$; or $x^2 + 3y^2 + 3 = 0$, whose graph contains no points; or $x^2 - 4y^2 = 0$, whose graph is two lines, $x - 2y = 0$ and $x + 2y = 0$.

We begin with the case where $B = 0$. In this case, the term containing xy is not present, so equation (1) has the form

$$Ax^2 + Cy^2 + Dx + Ey + F = 0$$

where either $A \neq 0$ or $C \neq 0$.

1 We have already discussed the procedure for identifying the graph of this kind of equation; we complete the squares of the quadratic expressions in x or y, or both. Once this has been done, the conic can be identified by comparing it to one of the forms studied in Sections 6.2 through 6.4.

In fact, though, we can identify the conic directly from the equation without completing the squares.

Theorem Identifying Conics without Completing the Squares

Excluding degenerate cases, the equation

$$Ax^2 + Cy^2 + Dx + Ey + F = 0 \tag{2}$$

where either $A \neq 0$ or $C \neq 0$:

(a) Defines a parabola if $AC = 0$.
(b) Defines an ellipse (or a circle) if $AC > 0$.
(c) Defines a hyperbola if $AC < 0$.

Proof

(a) If $AC = 0$, then either $A = 0$ or $C = 0$, but not both, so the form of equation (2) is either

$$Ax^2 + Dx + Ey + F = 0, \qquad A \neq 0$$

or

$$Cy^2 + Dx + Ey + F = 0, \qquad C \neq 0$$

Using the results of Problems 65 and 66 in Exercise 6.2, it follows that, except for the degenerate cases, the equation is a parabola.

(b) If $AC > 0$, then A and C are of the same sign. Using the results of Problems 69 and 70 in Exercise 6.3, except for the degenerate cases, the equation is an ellipse if $A \neq C$ or a circle if $A = C$.

(c) If $AC < 0$, then A and C are of opposite sign. Using the results of Problems 61 and 62 in Exercise 6.4, except for the degenerate cases, the equation is a hyperbola. ▬

We will not be concerned with the degenerate cases of equation (2). However, in practice, you should be alert to the possibility of degeneracy.

E X A M P L E 1   Identifying a Conic without Completing the Squares

Identify each equation without completing the squares.

(a) $3x^2 + 6y^2 + 6x - 12y = 0$ (b) $2x^2 - 3y^2 + 6y + 4 = 0$

(c) $y^2 - 2x + 4 = 0$

Solution (a) We compare the given equation to equation (2) and conclude that $A = 3$ and $C = 6$. Since $AC = 18 > 0$, the equation is an ellipse.

(b) Here, $A = 2$ and $C = -3$, so $AC = -6 < 0$. The equation is a hyperbola.

(c) Here, $A = 0$ and $C = 1$, so $AC = 0$. The equation is a parabola. ▬

Now work Problem 1.

Although we can now identify the type of conic represented by any equation of the form of equation (2) without completing the squares, we will still need to complete the squares if we desire additional information about a conic.

Now we turn our attention to equations of the form of equation (1), where $B \neq 0$. To discuss this case, we first need to investigate a new procedure: *rotation of axes*.

FIGURE 48

Rotation of Axes

2 In a **rotation of axes,** the origin remains fixed while the x-axis and y-axis are rotated through an angle θ to a new position; the new positions of the x- and y-axes are denoted by x' and y', respectively, as shown in Figure 48(a).

Now look at Figure 48(b) on page 420. There the point P has the coordinates (x, y) relative to the xy-plane, while the same point P has coordinates (x', y') relative to the $x'y'$-plane. We seek relationships that will enable us to express x and y in terms of $x', y',$ and θ.

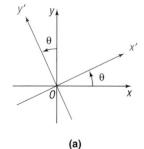

(a)

FIGURE 48 (continued)

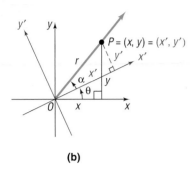

(b)

As Figure 48(b) shows, r denotes the distance from the origin O to the point P, and α denotes the angle between the positive x'-axis and the ray from O through P. Then, using the definitions of sine and cosine, we have

$$x' = r \cos \alpha \qquad\qquad y' = r \sin \alpha \qquad\qquad (3)$$
$$x = r \cos(\theta + \alpha) \qquad y = r \sin(\theta + \alpha) \qquad (4)$$

Now

$$
\begin{aligned}
x &= r \cos(\theta + \alpha) \\
&= r(\cos \theta \cos \alpha - \sin \theta \sin \alpha) &&\text{Sum formula.} \\
&= (r \cos \alpha)(\cos \theta) - (r \sin \alpha)(\sin \theta) \\
&= x' \cos \theta - y' \sin \theta &&\text{By equation (3).}
\end{aligned}
$$

Similarly,

$$
\begin{aligned}
y &= r \sin(\theta + \alpha) \\
&= r(\sin \theta \cos \alpha + \cos \theta \sin \alpha) \\
&= x' \sin \theta + y' \cos \theta
\end{aligned}
$$

Theorem   Rotation Formulas

If the x- and y-axes are rotated through an angle θ, the coordinates (x, y) of a point P relative to the xy-plane and the coordinates (x', y') of the same point relative to the new x'- and y'-axes are related by the formulas

$$x = x' \cos \theta - y' \sin \theta \qquad y = x' \sin \theta + y' \cos \theta \qquad (5)$$

E X A M P L E   2   Rotating Axes

Express the equation $xy = 1$ in terms of new $x'y'$-coordinates by rotating the axes through a $45°$ angle. Discuss the new equation.

Solution   Let $\theta = 45°$ in equation (5). Then

$$x = x' \cos 45° - y' \sin 45° = x'\frac{\sqrt{2}}{2} - y'\frac{\sqrt{2}}{2} = \frac{\sqrt{2}}{2}(x' - y')$$

$$y = x' \sin 45° + y' \cos 45° = x'\frac{\sqrt{2}}{2} + y'\frac{\sqrt{2}}{2} = \frac{\sqrt{2}}{2}(x' + y')$$

FIGURE 49

$xy = 1$ or $\dfrac{x'^2}{2} - \dfrac{y'^2}{2} = 1$

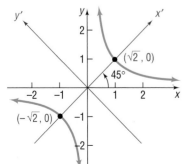

Substituting these expressions for x and y in $xy = 1$ gives

$$\left[\frac{\sqrt{2}}{2}(x' - y')\right]\left[\frac{\sqrt{2}}{2}(x' + y')\right] = 1$$

$$\frac{1}{2}(x'^2 - y'^2) = 1$$

$$\frac{x'^2}{2} - \frac{y'^2}{2} = 1$$

This is the equation of a hyperbola with center at $(0, 0)$ and transverse axis along the x'-axis. The vertices are $(\pm\sqrt{2}, 0)$ on the x'-axis; the asymptotes are $y' = x'$ and $y' = -x'$ (which correspond to the original x- and y-axes). See Figure 49 for the graph.

As Example 2 illustrates, a rotation of axes through an appropriate angle can transform a second-degree equation in x and y containing an xy-term into one in x' and y' in which no $x'y'$-term appears. In fact, we will show that a rotation of axes through an appropriate angle will transform any equation of the form of equation (1) into an equation in x' and y' without an $x'y'$-term.

To find the formula for choosing an appropriate angle θ through which to rotate the axes, we begin with equation (1),

$$Ax^2 + Bxy + Cy^2 + Dx + Ey + F = 0, \qquad B \neq 0$$

Next we rotate through an angle θ using rotation formulas (5):

$$A(x' \cos\theta - y' \sin\theta)^2 + B(x' \cos\theta - y' \sin\theta)(x' \sin\theta + y' \cos\theta)$$
$$+ C(x' \sin\theta + y' \cos\theta)^2 + D(x' \cos\theta - y' \sin\theta)$$
$$+ E(x' \sin\theta + y' \cos\theta) + F = 0$$

By expanding and collecting like terms, we obtain

$$(A \cos^2\theta + B \sin\theta \cos\theta + C \sin^2\theta)x'^2 + [B(\cos^2\theta - \sin^2\theta) + 2(C - A)(\sin\theta \cos\theta)]x'y'$$
$$+ (A \sin^2\theta - B \sin\theta \cos\theta + C \cos^2\theta)y'^2$$
$$+ (D \cos\theta + E \sin\theta)x'$$
$$+ (-D \sin\theta + E \cos\theta)y' + F = 0 \qquad (6)$$

In equation (6), the coefficient of $x'y'$ is

$$B' = 2(C - A)(\sin\theta \cos\theta) + B(\cos^2\theta - \sin^2\theta)$$

Since we want to eliminate the $x'y'$-term, we select an angle θ so that $B' = 0$. Thus,

$$2(C - A)(\sin\theta \cos\theta) + B(\cos^2\theta - \sin^2\theta) = 0$$
$$(C - A)(\sin 2\theta) + B \cos 2\theta = 0 \qquad \text{Double-angle formulas.}$$
$$B \cos 2\theta = (A - C)(\sin 2\theta)$$
$$\cot 2\theta = \frac{A - C}{B}, \qquad B \neq 0$$

Theorem

To transform the equation

$$Ax^2 + Bxy + Cy^2 + Dx + Ey + F = 0, \qquad B \neq 0$$

into an equation in x' and y' without an $x'y'$-term, rotate the axes through an angle θ that satisfies the equation

$$\cot 2\theta = \frac{A - C}{B} \qquad (7)$$

Equation (7) has an infinite number of solutions for θ. We shall adopt the convention of choosing the acute angle θ that satisfies (7). Then we have the following two possibilities:

If $\cot 2\theta \geq 0$, then $0° < 2\theta \leq 90°$ so that $0° < \theta \leq 45°$.

If $\cot 2\theta < 0$, then $90° < 2\theta < 180°$ so that $45° < \theta < 90°$.

Each of these results in a counterclockwise rotation of the axes through an acute angle θ.*

Warning: Be careful if you use a calculator to solve equation (7).

1. If $\cot 2\theta = 0$, then $2\theta = 90°$ and $\theta = 45°$.
2. If $\cot 2\theta \neq 0$, first find $\cos 2\theta$. Then use the inverse cosine function key(s) to obtain 2θ, $0° < 2\theta < 180°$. Finally, divide by 2 to obtain the correct acute angle θ.

E X A M P L E 3

Discussing an Equation Using a Rotation of Axes

Discuss the equation: $x^2 + \sqrt{3}xy + 2y^2 - 10 = 0$

Solution   Since an xy-term is present, we must rotate the axes. Using $A = 1$, $B = \sqrt{3}$, and $C = 2$ in equation (7), the appropriate acute angle θ through which to rotate the axes satisfies the equation

$$\cot 2\theta = \frac{A - C}{B} = \frac{-1}{\sqrt{3}} = \frac{-\sqrt{3}}{3}, \qquad 0° < 2\theta < 180°$$

Since $\cot 2\theta = -\sqrt{3}/3$, we find $2\theta = 120°$, so $\theta = 60°$. Using $\theta = 60°$ in rotation formulas (5), we find

$$x = \frac{1}{2}x' - \frac{\sqrt{3}}{2}y' = \frac{1}{2}(x' - \sqrt{3}y')$$

$$y = \frac{\sqrt{3}}{2}x' + \frac{1}{2}y' = \frac{1}{2}(\sqrt{3}x' + y')$$

Substituting these values into the original equation and simplifying, we have

$$x^2 + \sqrt{3}xy + 2y^2 - 10 = 0$$

FIGURE 50

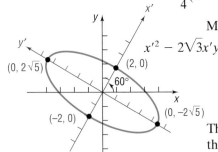

$$\frac{1}{4}(x' - \sqrt{3}y')^2 + \sqrt{3}\left[\frac{1}{2}(x' - \sqrt{3}y')\right]\left[\frac{1}{2}(\sqrt{3}x' + y')\right] + 2\left[\frac{1}{4}(\sqrt{3}x' + y')^2\right] = 10$$

Multiply both sides by 4 and expand to obtain

$$x'^2 - 2\sqrt{3}x'y' + 3y'^2 + \sqrt{3}(\sqrt{3}x'^2 - 2x'y' - \sqrt{3}y'^2) + 2(3x'^2 + 2\sqrt{3}x'y' + y'^2) = 40$$

$$10x'^2 + 2y'^2 = 40$$

$$\frac{x'^2}{4} + \frac{y'^2}{20} = 1$$

This is the equation of an ellipse with center at $(0, 0)$ and major axis along the y'-axis. The vertices are at $(0, \pm2\sqrt{5})$ on the y'-axis. See Figure 50. ■

Now work Problem 21.

In Example 3, the acute angle θ through which to rotate the axes was easy to find because of the numbers that we used in the given equation. In

---

*Any rotation (clockwise or counterclockwise) through an angle θ that satisfies $\cot 2\theta = (A - C)/B$ will eliminate the $x'y'$-term. However, the final form of the transformed equation may be different (but equivalent), depending on the angle chosen.

general, the equation $\cot 2\theta = (A - C)/B$ will not have such a "nice" solution. As the next example shows, we can still find the appropriate rotation formulas without using a calculator approximation by applying half-angle formulas.

E X A M P L E   4   Discussing an Equation Using a Rotation of Axes

Discuss the equation: $4x^2 - 4xy + y^2 + 5\sqrt{5}x + 5 = 0$

Solution   Letting $A = 4$, $B = -4$, and $C = 1$ in equation (7), the appropriate angle θ through which to rotate the axes satisfies

$$\cot 2\theta = \frac{A - C}{B} = \frac{3}{-4}$$

In order to use rotation formulas (5), we need to know the values of $\sin\theta$ and $\cos\theta$. Since we seek an acute angle θ, we know that $\sin\theta > 0$ and $\cos\theta > 0$. Thus, we use the half-angle formulas in the form

$$\sin\theta = \sqrt{\frac{1 - \cos 2\theta}{2}} \qquad \cos\theta = \sqrt{\frac{1 + \cos 2\theta}{2}}$$

Now we need to find the value of $\cos 2\theta$. Since $\cot 2\theta = -\frac{3}{4}$ and $\pi/2 < 2\theta < \pi$, it follows that $\cos 2\theta = -\frac{3}{5}$. Thus,

$$\sin\theta = \sqrt{\frac{1 - \cos 2\theta}{2}} = \sqrt{\frac{1 - (-\frac{3}{5})}{2}} = \sqrt{\frac{4}{5}} = \frac{2}{\sqrt{5}} = \frac{2\sqrt{5}}{5}$$

$$\cos\theta = \sqrt{\frac{1 + \cos 2\theta}{2}} = \sqrt{\frac{1 + (-\frac{3}{5})}{2}} = \sqrt{\frac{1}{5}} = \frac{1}{\sqrt{5}} = \frac{\sqrt{5}}{5}$$

With these values, rotation formulas (5) give us

$$x = \frac{\sqrt{5}}{5}x' - \frac{2\sqrt{5}}{5}y' = \frac{\sqrt{5}}{5}(x' - 2y')$$

$$y = \frac{2\sqrt{5}}{5}x' + \frac{\sqrt{5}}{5}y' = \frac{\sqrt{5}}{5}(2x' + y')$$

Substituting these values in the original equation and simplifying, we obtain

$$4x^2 - 4xy + y^2 + 5\sqrt{5}x + 5 = 0$$

$$4\left[\frac{\sqrt{5}}{5}(x' - 2y')\right]^2 - 4\left[\frac{\sqrt{5}}{5}(x' - 2y')\right]\left[\frac{\sqrt{5}}{5}(2x' + y')\right]$$

$$+ \left[\frac{\sqrt{5}}{5}(2x' + y')\right]^2 + 5\sqrt{5}\left[\frac{\sqrt{5}}{5}(x' - 2y')\right] = -5$$

Multiply both sides by 5 and expand to obtain

$$4(x'^2 - 4x'y' + 4y'^2) - 4(2x'^2 - 3x'y' - 2y'^2)$$
$$+ 4x'^2 + 4x'y' + y'^2 + 25(x' - 2y') = -25$$
$$25y'^2 - 50y' + 25x' = -25$$
$$y'^2 - 2y' + x' = -1$$
$$y'^2 - 2y' + 1 = -x' \qquad \text{Complete the}$$
$$(y' - 1)^2 = -x' \qquad \text{square in } y'.$$

FIGURE 51

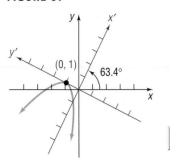

This is the equation of a parabola with vertex at $(0, 1)$ in the $x'y'$-plane. The axis of symmetry is parallel to the x'-axis. Using a calculator to solve $\sin \theta = 2\sqrt{5}/5$, we find that $\theta \approx 63.4°$. See Figure 51 for the graph. ▬

Now work Problem 27.

Identifying Conics without a Rotation of Axes

Suppose that we are required only to identify (rather than discuss) an equation of the form

$$Ax^2 + Bxy + Cy^2 + Dx + Ey + F = 0, \qquad B \neq 0 \tag{8}$$

If we apply rotation formulas (5) to this equation, we obtain an equation of the form

$$A'x'^2 + B'x'y' + C'y'^2 + D'x' + E'y' + F' = 0 \tag{9}$$

where A', B', C', D', E', and F' can be expressed in terms of A, B, C, D, E, F and the angle θ of rotation (see Problem 43). It can be shown that the value of $B^2 - 4AC$ in equation (8) and the value of $B'^2 - 4A'C'$ in equation (9) are equal no matter what angle θ of rotation is chosen (see Problem 45). In particular, if the angle θ of rotation satisfies equation (7), then $B' = 0$ in equation (9), and $B^2 - 4AC = -4A'C'$. Since equation (9) then has the form of equation (2),

$$A'x'^2 + C'y'^2 + D'x' + E'y' + F' = 0$$

we can identify it without completing the squares, as we did in the beginning of this section. In fact, now we can identify the conic described by any equation of the form of equation (8) without a rotation of axes.

> **Theorem**   Identifying Conics without a Rotation of Axes
>
> Except for degenerate cases, the equation
>
> $$Ax^2 + Bxy + Cy^2 + Dx + Ey + F = 0$$
>
> (a) Defines a parabola if $B^2 - 4AC = 0$.
> (b) Defines an ellipse (or a circle) if $B^2 - 4AC < 0$.
> (c) Defines a hyperbola if $B^2 - 4AC > 0$.

You are asked to prove this theorem in Problem 46.

E X A M P L E   5   Identifying a Conic without a Rotation of Axes

Identify the equation: $8x^2 - 12xy + 17y^2 - 4\sqrt{5}x - 2\sqrt{5}y - 15 = 0$

Solution   Here, $A = 8$, $B = -12$, and $C = 17$, so that $B^2 - 4AC = -400$. Since $B^2 - 4AC < 0$, the equation defines an ellipse. ▬

Now work Problem 33.

6.5 EXERCISES

In Problems 1–10, identify each equation without completing the squares.

1. $x^2 + 4x + y + 3 = 0$
2. $2y^2 - 3y + 3x = 0$
3. $6x^2 + 3y^2 - 12x + 6y = 0$
4. $2x^2 + y^2 - 8x + 4y + 2 = 0$
5. $3x^2 - 2y^2 + 6x + 4 = 0$
6. $4x^2 - 3y^2 - 8x + 6y + 1 = 0$
7. $2y^2 - x^2 - y + x = 0$
8. $y^2 - 8x^2 - 2x - y = 0$
9. $x^2 + y^2 - 8x + 4y = 0$
10. $2x^2 + 2y^2 - 8x + 8y = 0$

In Problems 11–20, determine the appropriate rotation formulas to use so that the new equation contains no xy-term.

11. $x^2 + 4xy + y^2 - 3 = 0$
12. $x^2 - 4xy + y^2 - 3 = 0$
13. $5x^2 + 6xy + 5y^2 - 8 = 0$
14. $3x^2 - 10xy + 3y^2 - 32 = 0$
15. $13x^2 - 6\sqrt{3}xy + 7y^2 - 16 = 0$
16. $11x^2 + 10\sqrt{3}xy + y^2 - 4 = 0$
17. $4x^2 - 4xy + y^2 - 8\sqrt{5}x - 16\sqrt{5}y = 0$
18. $x^2 + 4xy + 4y^2 + 5\sqrt{5}y + 5 = 0$
19. $25x^2 - 36xy + 40y^2 - 12\sqrt{13}x - 8\sqrt{13}y = 0$
20. $34x^2 - 24xy + 41y^2 - 25 = 0$

In Problems 21–32, rotate the axes so that the new equation contains no xy-term. Discuss and graph the new equation. (Refer to Problems 11–20 for Problems 21–30.)

21. $x^2 + 4xy + y^2 - 3 = 0$
22. $x^2 - 4xy + y^2 - 3 = 0$
23. $5x^2 + 6xy + 5y^2 - 8 = 0$
24. $3x^2 - 10xy + 3y^2 - 32 = 0$
25. $13x^2 - 6\sqrt{3}xy + 7y^2 - 16 = 0$
26. $11x^2 + 10\sqrt{3}xy + y^2 - 4 = 0$
27. $4x^2 - 4xy + y^2 - 8\sqrt{5}x - 16\sqrt{5}y = 0$
28. $x^2 + 4xy + 4y^2 + 5\sqrt{5}y + 5 = 0$
29. $25x^2 - 36xy + 40y^2 - 12\sqrt{13}x - 8\sqrt{13}y = 0$
30. $34x^2 - 24xy + 41y^2 - 25 = 0$
31. $16x^2 + 24xy + 9y^2 - 130x + 90y = 0$
32. $16x^2 + 24xy + 9y^2 - 60x + 80y = 0$

In Problems 33–42, identify each equation without applying a rotation of axes.

33. $x^2 + 3xy - 2y^2 + 3x + 2y + 5 = 0$
34. $2x^2 - 3xy + 4y^2 + 2x + 3y - 5 = 0$
35. $x^2 - 7xy + 3y^2 - y - 10 = 0$
36. $2x^2 - 3xy + 2y^2 - 4x - 2 = 0$
37. $9x^2 + 12xy + 4y^2 - x - y = 0$
38. $10x^2 + 12xy + 4y^2 - x - y + 10 = 0$
39. $10x^2 - 12xy + 4y^2 - x - y - 10 = 0$
40. $4x^2 + 12xy + 9y^2 - x - y = 0$
41. $3x^2 - 2xy + y^2 + 4x + 2y - 1 = 0$
42. $3x^2 + 2xy + y^2 + 4x - 2y + 10 = 0$

In Problems 43–46, apply rotation formulas (5) to

$$Ax^2 + Bxy + Cy^2 + Dx + Ey + F = 0$$

to obtain the equation

$$A'x'^2 + B'x'y' + C'y'^2 + D'x' + E'y' + F' = 0$$

43. Express A', B', C', D', E', and F' in terms of A, B, C, D, E, F and the angle θ of rotation.

44. Show that $A + C = A' + C'$, and thus show that $A + C$ is **invariant**; that is, its value does not change under a rotation of axes.

45. Refer to Problem 44. Show that $B^2 - 4AC$ is invariant.

46. Prove that, except for degenerate cases, the equation

$$Ax^2 + Bxy + Cy^2 + Dx + Ey + F = 0$$

   (a) Defines a parabola if $B^2 - 4AC = 0$.
   (b) Defines an ellipse (or a circle) if $B^2 - 4AC < 0$.
   (c) Defines a hyperbola if $B^2 - 4AC > 0$.

47. Use rotation formulas (5) to show that distance is invariant under a rotation of axes. That is, show that the distance from $P_1 = (x_1, y_1)$ to

$P_2 = (x_2, y_2)$ in the xy-plane equals the distance from $P_1 = (x_1', y_1')$ to $P_2 = (x_2', y_2')$ in the $x'y'$-plane.

48. Show that the graph of the equation $x^{1/2} + y^{1/2} = a^{1/2}$ is part of the graph of a parabola.

49. Formulate a strategy for discussing and graphing an equation of the form

$$Ax^2 + Cy^2 + Dx + Ey + F = 0$$

How does your strategy change if the equation is of the form

$$Ax^2 + Bxy + Cy^2 + Dx + Ey + F = 0$$

6.6 | POLAR EQUATIONS OF CONICS

1 Discuss and Graph Polar Equations of Conics
2 Convert a Polar Equation of a Conic to a Rectangular Equation

In Sections 6.2, 6.3, and 6.4, we gave separate definitions for the parabola, ellipse, and hyperbola based on geometric properties and the distance formula. In this section, we present an alternative definition that simultaneously defines all these conics. As we shall see, this approach is well suited to polar coordinate representation. (Refer to Section 5.1.)

> Let D denote a fixed line called the **directrix**; let F denote a fixed point called the **focus**, which is not on D; and let e be a fixed positive number called the **eccentricity**. A **conic** is the set of points P in the plane such that the ratio of the distance from F to P to the distance from D to P equals e. Thus, a conic is the collection of points P for which
>
> $$\frac{d(F, P)}{d(D, P)} = e \qquad (1)$$

If $e = 1$, the conic is a **parabola.**
If $e < 1$, the conic is an **ellipse.**
If $e > 1$, the conic is a **hyperbola.**

Observe that if $e = 1$ the definition of a parabola in equation (1) is exactly the same as the definition used earlier in Section 6.2.

In the case of an ellipse, the **major axis** is a line through the focus perpendicular to the directrix. In the case of a hyperbola, the **transverse axis** is a line through the focus perpendicular to the directrix. For both an ellipse and a hyperbola, the eccentricity e satisfies

$$e = \frac{c}{a} \qquad (2)$$

where c is the distance from the center to the focus and a is the distance from the center to a vertex.

Just as we did earlier using rectangular coordinates, we derive equations for the conics in polar coordinates by choosing a convenient position for the focus F and the directrix D. The focus F is positioned at the pole, and the directrix D is either parallel to the polar axis or perpendicular to it.

Suppose that we start with the directrix D perpendicular to the polar axis at a distance p units to the left of the pole (the focus F). See Figure 52.

FIGURE 52

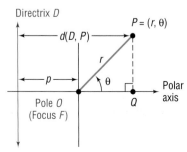

If $P = (r, \theta)$ is any point on the conic, then, by equation (1),

$$\frac{d(F, P)}{d(D, P)} = e \quad \text{or} \quad d(F, P) = e \cdot d(D, P) \tag{3}$$

Now we use the point Q obtained by dropping the perpendicular from P to the polar axis to calculate $d(D, P)$.

$$d(D, P) = p + d(O, Q) = p + r \cos \theta$$

Using this expression and the fact that $d(F, P) = d(O, P) = r$ in equation (3), we get

$$
\begin{aligned}
d(F, P) &= e \cdot d(D, P) \\
r &= e(p + r \cos \theta) \\
r &= ep + er \cos \theta \\
r - er \cos \theta &= ep \\
r(1 - e \cos \theta) &= ep \\
r &= \frac{ep}{1 - e \cos \theta}
\end{aligned}
$$

Theorem   Polar Equation of a Conic; Focus at Pole; Directrix Perpendicular to Polar Axis a Distance p to the Left of the Pole

The polar equation of a conic with focus at the pole and directrix perpendicular to the polar axis at a distance p to the left of the pole is

$$r = \frac{ep}{1 - e \cos \theta} \tag{4}$$

where e is the eccentricity of the conic.

E X A M P L E   1   Discussing and Graphing the Polar Equation of a Conic

Discuss and graph the equation: $r = \dfrac{4}{2 - \cos \theta}$

Solution   The given equation is not quite in the form of equation (4), since the first term in the denominator is 2 instead of 1. Thus, we divide the numerator and denominator by 2 to obtain

$$r = \frac{2}{1 - \frac{1}{2} \cos \theta}$$

This equation is in the form of equation (4), with

$$e = \tfrac{1}{2} \quad \text{and} \quad ep = \tfrac{1}{2}p = 2$$

Thus, $e = \tfrac{1}{2}$ and $p = 4$. We conclude that the conic is an ellipse, since $e = \tfrac{1}{2} < 1$. One focus is at the pole, and the directrix is perpendicular to the polar axis,

FIGURE 53

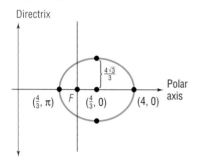

a distance of 4 units to the left of the pole. It follows that the major axis is along the polar axis. To find the vertices, we let $\theta = 0$ and $\theta = \pi$. Thus, the vertices of the ellipse are $(4, 0)$ and $(\frac{4}{3}, \pi)$. The midpoint of the vertices, $(\frac{4}{3}, 0)$ in polar coordinates, is the center of the ellipse. [Do you see why? The vertices $(4, 0)$ and $(\frac{4}{3}, \pi)$ in polar coordinates are $(4, 0)$ and $(-\frac{4}{3}, 0)$ in rectangular coordinates. The midpoint in rectangular coordinates is $(\frac{4}{3}, 0)$, which is also $(\frac{4}{3}, 0)$ in polar coordinates.] Thus, a = distance from the center to a vertex = $\frac{8}{3}$. Using $a = \frac{8}{3}$ and $e = \frac{1}{2}$ in equation (2), $e = c/a$, we find $c = \frac{4}{3}$. Finally, using $a = \frac{8}{3}$ and $c = \frac{4}{3}$ in $b^2 = a^2 - c^2$, we have

$$b^2 = a^2 - c^2 = \frac{64}{9} - \frac{16}{9} = \frac{48}{9}$$

$$b = \frac{4\sqrt{3}}{3}$$

Figure 53 shows the graph.

Check: In POLar mode with θmin $= 0$, θmax $= 2\pi$, and θstep $= \pi/24$, graph $r_1 = 4/(2 - \cos \theta)$ and compare the result with Figure 53.

Exploration   Graph $r_1 = 4/(2 + \cos \theta)$ and compare the result with Figure 53. What do you conclude? Clear the screen and graph $r_1 = 4/(2 - \sin \theta)$ and then $r_1 = 4/(2 + \sin \theta)$. Compare each of these graphs with Figure 53. What do you conclude?

Now work Problem 5.

Equation (4) was obtained under the assumption that the directrix was perpendicular to the polar axis at a distance p units to the left of the pole. A similar derivation (see Problem 37), in which the directrix is perpendicular to the polar axis at a distance p units to the right of the pole, results in the equation

$$r = \frac{ep}{1 + e \cos \theta}$$

In Problems 38 and 39 you are asked to derive the polar equations of conics with focus at the pole and directrix parallel to the polar axis. Table 5 summarizes the polar equations of conics.

TABLE 5 Polar Equations of Conics (Focus at the Pole, Eccentricity *e*)	
Equation	**Description**
(a) $r = \dfrac{ep}{1 - e \cos \theta}$	Directrix is perpendicular to the polar axis at a distance p units to the left of the pole.
(b) $r = \dfrac{ep}{1 + e \cos \theta}$	Directrix is perpendicular to the polar axis at a distance p units to the right of the pole.
(c) $r = \dfrac{ep}{1 + e \sin \theta}$	Directrix is parallel to the polar axis at a distance p units above the pole.
(d) $r = \dfrac{ep}{1 - e \sin \theta}$	Directrix is parallel to the polar axis at a distance p units below the pole.
Eccentricity	
If $e = 1$, the conic is a parabola; the axis of symmetry is perpendicular to the directrix.	
If $e < 1$, the conic is an ellipse; the major axis is perpendicular to the directrix.	
If $e > 1$, the conic is a hyperbola; the transverse axis is perpendicular to the directrix.	

E X A M P L E   2   Discussing and Graphing the Polar Equation of a Conic

Discuss and graph the equation: $r = \dfrac{6}{3 + 3 \sin \theta}$

Solution   To place the equation in proper form, we divide the numerator and denominator by 3 to get

$$r = \frac{2}{1 + \sin \theta}$$

FIGURE 54

Referring to Table 5, we conclude that this equation is in the form of equation (c) with

$$e = 1 \quad \text{and} \quad ep = 2$$

Thus, $e = 1$ and $p = 2$. The conic is a parabola with focus at the pole. The directrix is parallel to the polar axis at a distance 2 units above the pole; the axis of symmetry is perpendicular to the polar axis. The vertex of the parabola is at $(1, \pi/2)$. (Do you see why?) See Figure 54 for the graph. Notice that we plotted two additional points, $(2, 0)$ and $(2, \pi)$, to assist in graphing.

Check: Graph $r_1 = 6/(3 + 3 \sin \theta)$ and compare the result with Figure 54.

Now work Problem 7.

E X A M P L E   3   Discussing and Graphing the Polar Equation of a Conic

Discuss and graph the equation: $r = \dfrac{3}{1 + 3 \cos \theta}$

Solution   This equation is in the form of equation (b) in Table 5. We conclude that

$$e = 3 \quad \text{and} \quad ep = 3p = 3$$

FIGURE 55

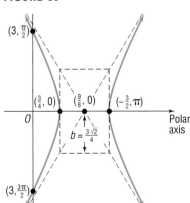

Thus, $e = 3$ and $p = 1$. This is the equation of a hyperbola with a focus at the pole. The directrix is perpendicular to the polar axis, 1 unit to the right of the pole. The transverse axis is along the polar axis. To find the vertices, we let $\theta = 0$ and $\theta = \pi$. Thus, the vertices are $(\frac{3}{4}, 0)$ and $(-\frac{3}{2}, \pi)$. The center, which is at the midpoint of $(\frac{3}{4}, 0)$ and $(-\frac{3}{2}, \pi)$, is $(\frac{9}{8}, 0)$. Thus, $c = $ distance from the center to a focus $= \frac{9}{8}$. Since $e = 3$, it follows from equation (2), $e = c/a$, that $a = \frac{3}{8}$. Finally, using $a = \frac{3}{8}$ and $c = \frac{9}{8}$ in $b^2 = c^2 - a^2$, we find

$$b^2 = c^2 - a^2 = \frac{81}{64} - \frac{9}{64} = \frac{72}{64} = \frac{9}{8}$$

$$b = \frac{3}{2\sqrt{2}} = \frac{3\sqrt{2}}{4}$$

Figure 55 shows the graph. Notice that we plotted two additional points $(3, \pi/2)$ and $(3, 3\pi/2)$, on the left branch and used symmetry to obtain the right branch. The asymptotes of this hyperbola were found in the usual way by constructing the rectangle shown.

Check: Graph $r_1 = 3/(1 + 3 \cos \theta)$ and compare the result with Figure 55.

 Now work Problem 11.

E X A M P L E 4   Converting a Polar Equation to a Rectangular Equation

Convert the polar equation

$$r = \frac{1}{3 - 3 \cos \theta}$$

to a rectangular equation.

Solution   The strategy here is first to rearrange the equation and square each side, before using the transformation equations.

$$r = \frac{1}{3 - 3 \cos \theta}$$

$$3r - 3r \cos \theta = 1$$

$$3r = 1 + 3r \cos \theta \qquad \text{Rearrange the equation.}$$

$$9r^2 = (1 + 3r \cos \theta)^2 \qquad \text{Square each side.}$$

$$9(x^2 + y^2) = (1 + 3x)^2 \qquad \text{Use the transformation equations.}$$

$$9x^2 + 9y^2 = 9x^2 + 6x + 1$$

$$9y^2 = 6x + 1$$

This is the equation of a parabola in rectangular coordinates.

 Now work Problem 19.

6.6 | EXERCISES

In Problems 1–6, identify the conic that each polar equation represents. Also, give the position of the directrix.

1. $r = \dfrac{1}{1 + \cos \theta}$ **2.** $r = \dfrac{3}{1 - \sin \theta}$ **3.** $r = \dfrac{4}{2 - 3 \sin \theta}$

4. $r = \dfrac{2}{1 + 2 \cos \theta}$ **5.** $r = \dfrac{3}{4 - 2 \cos \theta}$ **6.** $r = \dfrac{6}{8 + 2 \sin \theta}$

In Problems 7–18, identify and graph each equation.

7. $r = \dfrac{1}{1 + \cos \theta}$ **8.** $r = \dfrac{3}{1 - \sin \theta}$ **9.** $r = \dfrac{8}{4 + 3 \sin \theta}$

10. $r = \dfrac{10}{5 + 4 \cos \theta}$ **11.** $r = \dfrac{9}{3 - 6 \cos \theta}$ **12.** $r = \dfrac{12}{4 + 8 \sin \theta}$

13. $r = \dfrac{8}{2 - \sin \theta}$ **14.** $r = \dfrac{8}{2 + 4 \cos \theta}$ **15.** $r(3 - 2 \sin \theta) = 6$

16. $r(2 - \cos \theta) = 2$ **17.** $r = \dfrac{6 \sec \theta}{2 \sec \theta - 1}$ **18.** $r = \dfrac{3 \csc \theta}{\csc \theta - 1}$

In Problems 19–30, convert each polar equation to a rectangular equation.

19. $r = \dfrac{1}{1 + \cos \theta}$

20. $r = \dfrac{3}{1 - \sin \theta}$

21. $r = \dfrac{8}{4 + 3 \sin \theta}$

22. $r = \dfrac{10}{5 + 4 \cos \theta}$

23. $r = \dfrac{9}{3 - 6 \cos \theta}$

24. $r = \dfrac{12}{4 + 8 \sin \theta}$

25. $r = \dfrac{8}{2 - \sin \theta}$

26. $r = \dfrac{8}{2 + 4 \cos \theta}$

27. $r(3 - 2 \sin \theta) = 6$

28. $r(2 - \cos \theta) = 2$

29. $r = \dfrac{6 \sec \theta}{2 \sec \theta - 1}$

30. $r = \dfrac{3 \csc \theta}{\csc \theta - 1}$

In Problems 31–36, find a polar equation for each conic. For each, a focus is at the pole.

31. $e = 1$; directrix is parallel to the polar axis 1 unit above the pole

32. $e = 1$; directrix is parallel to the polar axis 2 units below the pole

33. $e = \frac{4}{5}$; directrix is perpendicular to the polar axis 3 units to the left of the pole

34. $e = \frac{2}{3}$; directrix is parallel to the polar axis 3 units above the pole

35. $e = 6$; directrix is parallel to the polar axis 2 units below the pole

36. $e = 5$; directrix is perpendicular to the polar axis 5 units to the right of the pole

37. Derive equation (b) in Table 5:

$$r = \frac{ep}{1 + e \cos \theta}$$

38. Derive equation (c) in Table 5:

$$r = \frac{ep}{1 + e \sin \theta}$$

39. Derive equation (d) in Table 5:

$$r = \frac{ep}{1 - e \sin \theta}$$

40. **Orbit of Mercury** The planet Mercury travels around the Sun in an elliptical orbit given approximately by

$$r = \frac{(3.442)10^7}{1 - 0.206 \cos \theta}$$

where r is measured in miles and the Sun is at the pole. Find the distance from Mercury to the Sun at *aphelion* (greatest distance from the Sun) and at *perihelion* (shortest distance from the Sun). See the figure.

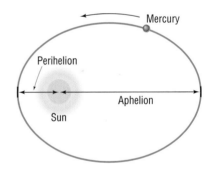

6.7 | PLANE CURVES AND PARAMETRIC EQUATIONS

1 Graph Parametric Equations

2 Find a Rectangular Equation for a Curve Defined Parametrically

3 Use Time as a Parameter in Parametric Equations

4 Find Parametric Equations for Curves Defined by Rectangular Equations

Equations of the form $y = f(x)$, where f is a function, have graphs that are intersected no more than once by any vertical line. The graphs of many of the conics and certain other, more complicated graphs do not have this characteristic. Yet each graph, like the graph of a function, is a collection of points (x, y) in the xy-plane; that is, each is a *plane curve*. In this section, we discuss another way of representing such graphs.

Let $x = f(t)$ and $y = g(t)$, where f and g are two functions whose common domain is some interval I. The collection of points defined by

$$(x, y) = (f(t), g(t))$$

is called a **plane curve**. The equations

$$x = f(t) \qquad y = g(t)$$

where t is in I, are called **parametric equations** of the curve. The variable t is called a **parameter.**

Parametric equations are particularly useful in describing movement along a curve. Suppose that a curve is defined by the parametric equations

$$x = f(t) \qquad y = g(t), \qquad a \le t \le b$$

where f and g are each defined over the interval $a \le t \le b$. For a given value of t, we can find the value of $x = f(t)$ and $y = g(t)$, thus obtaining a point (x, y) on the curve. In fact, as t varies over the interval from $t = a$ to $t = b$, successive values of t give rise to a directed movement along the curve; that is, the curve is traced out in a certain direction by the corresponding succession of points (x, y). See Figure 56. The arrows show the direction, or **orientation,** along the curve as t varies from a to b.

FIGURE 56

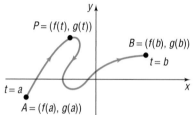

E X A M P L E   1

Discussing a Curve Defined by Parametric Equations

Discuss the curve defined by the parametric equations

$$x = 3t^2 \qquad y = 2t, \qquad -2 \le t \le 2 \tag{1}$$

Solution   For each number t, $-2 \le t \le 2$, there corresponds a number x and a number y. For example, when $t = -2$, then $x = 12$ and $y = -4$. When $t = 0$, then $x = 0$ and $y = 0$. Indeed, we can set up a table listing various choices of the parameter t and the corresponding values for x and y, as shown in Table 6. Plotting these points and connecting them with a smooth curve leads to Figure 57.

TABLE 6

t	$x = 3t^2$	$y = 2t$	(x, y)
-2	12	-4	$(12, -4)$
-1	3	-2	$(3, -2)$
0	0	0	$(0, 0)$
1	3	2	$(3, 2)$
2	12	4	$(12, 4)$

FIGURE 57

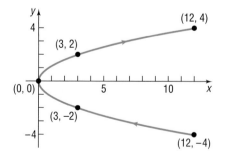

Comment   Most graphing utilities have the capability of graphing parametric equations. See Section 7 in the Appendix, page 563.

Exploration Graph the following parametric equations using a graphing utility with Xmin $= 0$, Xmax $= 15$, Ymin $= -5$, Ymax $= 5$, and Tstep $= 0.1$.

1. $x = \dfrac{3t^2}{4}$, $y = t$, $-4 \le t \le 4$

2. $x = 3t^2 + 12t + 12$, $y = 2t + 4$, $-4 \le t \le 0$

3. $x = 3t^{2/3}$, $y = 2\sqrt[3]{t}$, $-8 \le t \le 8$

Compare these graphs to the graph in Figure 57. Conclude that parametric equations defining a curve are not unique; that is, different parametric equations can represent the same graph. ▬

Now work Problem 1.

 The curve given in Example 1 should be familiar. To identify it accurately, we find the corresponding rectangular equation by eliminating the parameter t from the parametric equations (1) given in Example 1,

$$x = 3t^2 \qquad y = 2t, \qquad -2 \le t \le 2$$

Noting that we can readily solve for t in $y = 2t$, obtaining $t = y/2$, we substitute this expression in the other equation.

$$x = 3t^2 = 3\left(\frac{y}{2}\right)^2 = \frac{3y^2}{4}$$
$$\uparrow$$
$$t = \frac{y}{2}$$

This equation, $x = 3y^2/4$, is the equation of a parabola with vertex at $(0, 0)$ and axis of symmetry along the x-axis.

 Exploration In FUNCtion mode graph $x = \dfrac{3y^2}{4}\left(Y_1 = \sqrt{\dfrac{4x}{3}} \text{ and } Y_2 = -\sqrt{\dfrac{4x}{3}}\right)$ with Xmin $= 0$, Xmax $= 15$, Ymin $= -5$, and Ymax $= 5$. Compare this graph with Figure 57. Why do the graphs differ? ▬

Note that the parameterized curve defined by equation (1) and shown in Figure 57 is only a part of the parabola $x = 3y^2/4$. Thus, the graph of the rectangular equation obtained by eliminating the parameter will, in general, contain more points than the original parameterized curve. Care must therefore be taken when a parameterized curve is sketched after eliminating the parameter. Even so, the process of eliminating the parameter t of a parameterized curve in order to identify it accurately is sometimes a better approach than merely plotting points. However, the elimination process sometimes requires a little ingenuity.

E X A M P L E   2   Finding the Rectangular Equation of a Curve Defined Parametrically

Find the rectangular equation of the curve whose parametric equations are

$$x = a \cos t \qquad y = a \sin t$$

where $a > 0$ is a constant. Graph this curve, indicating its orientation.

Solution   The presence of sines and cosines in the parametric equations suggests that we use a Pythagorean identity. In fact, since

$$\cos t = \frac{x}{a} \qquad \sin t = \frac{y}{a}$$

we find that

$$\cos^2 t + \sin^2 t = 1$$

$$\left(\frac{x}{a}\right)^2 + \left(\frac{y}{a}\right)^2 = 1$$

$$x^2 + y^2 = a^2$$

FIGURE 58

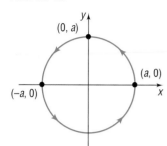

Thus, the curve is a circle with center at $(0, 0)$ and radius a. As the parameter t increases, say from $t = 0$ [the point $(a, 0)$] to $t = \pi/2$ [the point $(0, a)$] to $t = \pi$ [the point $(-a, 0)$], we see that the corresponding points are traced in a counterclockwise direction around the circle. Hence, the orientation is as indicated in Figure 58. ∎

Now work Problem 13.

Let's discuss the curve in Example 2 further. The domain of each parametric equation is $-\infty < t < \infty$. Thus, the graph in Figure 58 is actually being repeated each time that t increases by 2π.

If we wanted the curve to consist of exactly one revolution in the counterclockwise direction, we could write

$$x = a \cos t \qquad y = a \sin t, \qquad 0 \le t \le 2\pi$$

This curve starts when $t = 0$ [the point $(a, 0)$] and, proceeding counterclockwise around the circle, ends when $t = 2\pi$ [also the point $(a, 0)$].

If we wanted the curve to consist of exactly three revolutions in the counterclockwise direction, we could write

$$x = a \cos t \qquad y = a \sin t, \qquad -2\pi \le t \le 4\pi$$

or

$$x = a \cos t \qquad y = a \sin t, \qquad 0 \le t \le 6\pi$$

or

$$x = a \cos t \qquad y = a \sin t, \qquad 2\pi \le t \le 8\pi$$

E X A M P L E   3

Describing Parametric Equations

Find rectangular equations for and graph the following curves defined by parametric equations.

(a) $x = a \cos t, \quad y = a \sin t, \quad 0 \le t \le \pi, \quad a > 0$
(b) $x = -a \sin t, \quad y = -a \cos t, \quad 0 \le t \le \pi, \quad a > 0$

Solution   (a) We eliminate the parameter t using a Pythagorean identity.

$$\left(\frac{x}{a}\right)^2 + \left(\frac{y}{a}\right)^2 = \cos^2 t + \sin^2 t = 1$$

FIGURE 59

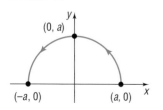

The curve defined by these parametric equations is a circle, with radius a and center at $(0, 0)$. The circle begins at the point $(a, 0)$, $(t = 0)$, passes through the point $(0, a)$, $(t = \pi/2)$, and ends at the point $(-a, 0)$, $(t = \pi)$. Thus, the parametric equations define an upper semicircle of radius a with a counterclockwise orientation. See Figure 59. The rectangular equation is

$$y = a\sqrt{1 - (x/a)^2}, \qquad -a \le x \le a$$

FIGURE 60

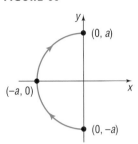

(b) We eliminate the parameter t using a Pythagorean identity.

$$\left(\frac{x}{-a}\right)^2 + \left(\frac{y}{-a}\right)^2 = \sin^2 t + \cos^2 t = 1$$

The curve defined by these parametric equations is a circle, with radius a and center at $(0,0)$. The circle begins at the point $(0,-a)$, $(t=0)$, passes through the point $(-a,0)$ $(t=\pi/2)$, and ends at the point $(0,a)$ $(t=\pi)$. Thus, the parametric equations define a left semicircle of radius a with a clockwise orientation. See Figure 60. The rectangular equation is

$$x = -a\sqrt{1 - (y/a)^2}, \qquad -a \le y \le a$$　■

Example 3 illustrates the versatility of parametric equations for replacing complicated rectangular equations, while providing additional information about orientation. These characteristics make parametric equations very useful in applications, such as projectile motion.

 Seeing the Concept　Graph $x = \cos t$, $y = \sin t$ for $0 \le t \le 2\pi$. Compare to Figure 58. Graph $x = \cos t$, $y = \sin t$ for $0 \le t \le \pi$. Compare to Figure 59. Graph $x = -\sin t$, $y = -\cos t$ for $0 \le t \le \pi$. Compare to Figure 60.　■

Time as a Parameter: Projectile Motion; Simulated Motion

 If we think of the parameter t as time, then the parametric equations $x = f(t)$ and $y = g(t)$ of a curve C specify how the x- and y-coordinates of a moving point vary with time.

For example, we can use parametric equations to describe the motion of an object, sometimes referred to as **curvilinear motion.** Using parametric equations, we can specify not only where the object travels, that is, its location (x, y), but also when it gets there, that is, the time t.

When an object is propelled upward at an inclination θ to the horizontal with initial speed v_0, the resulting motion is called **projectile motion.** See Figure 61(a).

In calculus it is shown that the parametric equations of the path of a projectile fired at an inclination θ to the horizontal, with an initial speed v_0, from a height h above the horizontal are

$$x = (v_0 \cos \theta)t \qquad y = -\frac{1}{2}gt^2 + (v_0 \sin \theta)t + h \qquad (2)$$

where t is the time and g is the constant acceleration due to gravity (approximately 32 feet per second or 9.8 meters per second). See Figure 61(b).

FIGURE 61

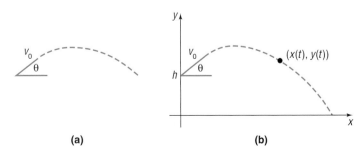

(a)　　　(b)

E X A M P L E   4   Projectile Motion

FIGURE 62

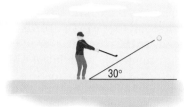

Suppose that Jim hit a golf ball with an initial velocity of 150 feet per second at an angle of 30° to the horizontal. See Figure 62.

(a) Find parametric equations that describe the position of the ball as a function of time.

(b) How long is the golf ball in the air?

(c) When is the ball at its maximum height? Determine the maximum height of the ball.

(d) Determine the distance the ball traveled.

(e) Using a graphing utility, simulate the motion of the golfball by simultaneously graphing the equations found in (a).

Solution   (a) We have $v_0 = 150$, $\theta = 30°$, $h = 0$ (the ball is on the ground), and $g = 32$ (since units are in feet and seconds). Substituting these values into equations (2), we find

$$x = (150 \cos 30°)t = 75\sqrt{3}\,t$$

$$y = \frac{-1}{2}(32)t^2 + (150 \sin 30°)t + 0 = -16t^2 + (150 \sin 30°)t$$

$$= -16t^2 + 75t$$

(b) To determine the length of time that the ball is in the air, we solve the equation $y = 0$.

$$-16t^2 + 75t = 0$$

$$t(-16t + 75) = 0$$

$$t = 0 \text{ seconds} \quad \text{or} \quad t = \frac{75}{16} = 4.6875 \text{ seconds}$$

The ball will strike the ground after 4.6875 seconds.

(c) Notice that the height y of the ball is a quadratic function of t. Thus, the maximum height of the ball can be found by determining the vertex of $y = -16t^2 + 75t$. The value of t at the vertex is

$$t = \frac{-b}{2a} = \frac{-75}{-32} = 2.34375 \text{ seconds}$$

The ball is at its maximum height after 2.34375 seconds. The maximum height of the ball is found by evaluating the function y at $t = 2.34375$ seconds.

$$\text{Maximum height} = -16(2.34375)^2 + (75)2.34375 \approx 87.89 \text{ feet}$$

FIGURE 63

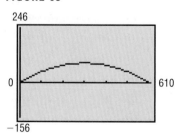

(d) Since the ball is in the air for 4.6875 seconds, the horizontal distance that the ball travels is

$$x = (75\sqrt{3})4.6875 \approx 608.92 \text{ feet}$$

(e) We enter the equation from (a) into a graphing utility with $T\text{min} = 0$, $T\text{max} = 4.7$, and $T\text{step} = 0.1$. See Figure 63.   ■

Exploration   Simulate the motion of a ball thrown straight up with an initial speed of 100 feet per second from a height of 5 feet above the ground. Use PARametric mode with $T\text{min} = 0$, $T\text{max} = 6.5$, $T\text{step} = 0.1$, $X\text{min} = 0$, $X\text{max} = 5$, $Y\text{min} = 0$, and $Y\text{max} = 180$. What happens to the speed at which the graph is

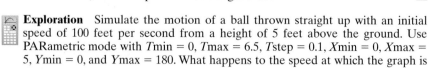

drawn as the ball goes up and then comes back down? How do you interpret this physically? Repeat the experiment using other values for *T*step. How does this affect the experiment?

Hint In the projectile motion equations, let $\theta = 90°$, $v_0 = 100$, $h = 5$, and $g = 32$. We use $x = 3$ instead of $x = 0$ to see the vertical motion better.

 Result See Figure 64. In Figure 64(a), the ball is going up. In Figure 64(b), the ball is near its highest point. Finally, in Figure 64(c), the ball is coming back down.

FIGURE 64

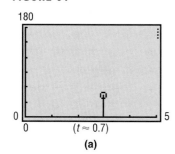

(a)

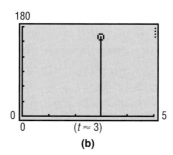

(b)

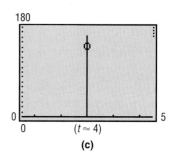

(c)

Notice that, as the ball goes up, its speed decreases, until at the highest point it is zero. Then the speed increases as the ball comes back down. ∎

A graphing utility can be used to simulate other kinds of motion as well.

EXAMPLE 5   Simulating Motion

Tanya, who is a long distance runner, runs at an average velocity of 8 miles per hour. Two hours after Tanya leaves your house, you leave in your Honda and follow the same route. If your average velocity is 40 miles per hour, how long will it be before you catch up to Tanya? See Figure 65. Use a simulation of the two motions to verify the answer.

FIGURE 65

Solution   We begin by setting up two sets of parametric equations: one to describe Tanya's motion, the other to describe the motion of the Honda. We choose time $t = 0$ to be when Tanya leaves the house. If we choose $y_1 = 2$ as Tanya's path, then we can use $y_2 = 4$ as the parallel path of the Honda. The horizontal distances traversed in time t are

$$\text{Tanya:}\quad x_1 = 8t \qquad \text{Honda:}\quad x_2 = 40(t - 2)$$

The Honda catches up to Tanya when $x_1 = x_2$.

$$8t = 40(t - 2)$$
$$8t = 40t - 80$$
$$-32t = -80$$
$$t = \frac{-80}{-32} = 2.5$$

The Honda catches up to Tanya 2.5 hours after Tanya leaves the house. In PARametric mode with Tstep $= 0.01$, we simultaneously graph

Tanya: $\quad x_1 = 8t \qquad$ Honda: $\quad x_2 = 40(t - 2)$
$\qquad\qquad y_1 = 2 \qquad\qquad\qquad\qquad y_2 = 4$

for $0 \le t \le 3$.

Figure 66 shows the relative position of Tanya and the Honda for $t = 0$, $t = 2$, $t = 2.25$, $t = 2.5$, and $t = 2.75$.

FIGURE 66

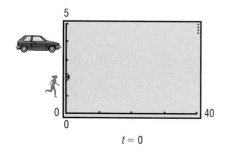

$t = 0$

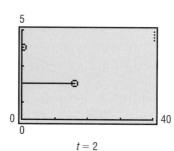

$t = 2$

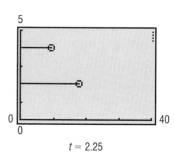

$t = 2.25$

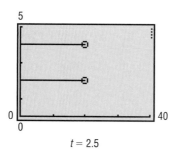

$t = 2.5$

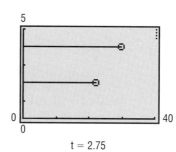

$t = 2.75$

 Now work Problem 27.

Finding Parametric Equations

We now take up the question of how to find parametric equations of a given curve.

 If a curve is defined by the equation $y = f(x)$, where f is a function, one way of finding parametric equations is simply to let $x = t$. Then $y = f(t)$. Thus,

$$x = t \qquad y = f(t) \qquad t \text{ in the domain of } f$$

are parametric equations of the curve.

E X A M P L E   6   Finding Parametric Equations for a Curve Defined by a Rectangular Equation

Find parametric equations for the equation $y = x^2 - 4$.

Solution   Let $x = t$. Then the parametric equations are

$$x = t \qquad y = t^2 - 4, \qquad -\infty < t < \infty$$

Another less obvious approach to Example 6 is to let $x = t^3$. Then the parametric equations become

$$x = t^3 \qquad y = t^6 - 4, \qquad -\infty < t < \infty$$

Care must be taken when using this approach, since the substitution for x must be a function that allows x to take on all the values stipulated by the domain of f. Thus, for example, letting $x = t^2$ so that $y = t^4 - 4$ does not result in equivalent parametric equations for $y = x^2 - 4$, since only points for which $x \geq 0$ are obtained.

E X A M P L E   7   Finding Parametric Equations for an Object in Motion

Find parametric equations for the ellipse

$$x^2 + \frac{y^2}{9} = 1$$

where the parameter t is time (in seconds) and

(a) The motion around the ellipse is clockwise, begins at the point $(0, 3)$, and requires 1 second for a complete revolution.

(b) The motion around the ellipse is counterclockwise, begins at the point $(1, 0)$, and requires 2 seconds for a complete revolution.

FIGURE 67

$x^2 + \dfrac{y^2}{9} = 1$

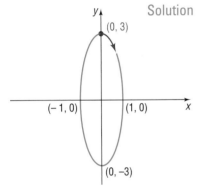

Solution   (a) See Figure 67. Since the motion begins at the point $(0, 3)$, we want $x = 0$ and $y = 3$ when $t = 0$. Since the given equation is an ellipse, we begin by letting

$$x = \sin \omega t \qquad \frac{y}{3} = \cos \omega t$$

for some constant ω. These parametric equations satisfy the equation of the ellipse. Furthermore, with this choice, when $t = 0$, we have $x = 0$ and $y = 3$.

For the motion to be clockwise, the motion will have to begin with the value of x increasing and y decreasing as t increases. This requires that $\omega > 0$. See the red part of the graph in Figure 67.

Finally, since one revolution requires 1 second, the period $2\pi/\omega = 1$, so $\omega = 2\pi$. Thus, parametric equations that satisfy the conditions stipulated are

$$x = \sin 2\pi t \qquad y = 3 \cos 2\pi t, \qquad 0 \leq t \leq 1 \qquad (3)$$

FIGURE 68

$x^2 + \dfrac{y^2}{9} = 1$

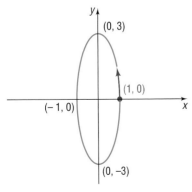

(b) See Figure 68. Since the motion begins at the point $(1, 0)$, we want $x = 1$ and $y = 0$ when $t = 0$. Since the given equation is an ellipse, we begin by letting

$$x = \cos \omega t \qquad \frac{y}{3} = \sin \omega t$$

for some constant ω. These parametric equations satisfy the equation of the ellipse. Furthermore, with this choice, when $t = 0$, we have $x = 1$ and $y = 0$.

For the motion to be counterclockwise, the motion will have to begin with the value of x decreasing and y increasing as t increases. Thus,

$\omega > 0$. Finally, since one revolution requires 2 seconds, the period is $2\pi/\omega = 2$, so $\omega = \pi$. Thus, the parametric equations that satisfy the conditions stipulated are

$$x = \cos \pi t \qquad y = 3 \sin \pi t, \qquad 0 \le t \le 2 \tag{4}$$

Now work Problem 35.

The Cycloid

Suppose that a circle of radius a rolls along a horizontal line without slipping. As the circle rolls along the line, a point P on the circle will trace out a curve called a **cycloid** (see Figure 69). We now seek parametric equations* for a cycloid.

FIGURE 69

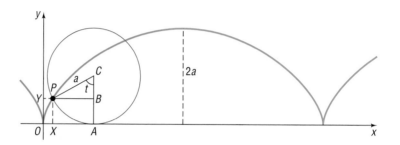

We begin with a circle of radius a and take the fixed line on which the circle rolls as the x-axis. Let the origin be one of the points at which the point P comes in contact with the x-axis. Figure 69 illustrates the position of this point P after the circle has rolled somewhat. The angle t (in radians) measures the angle through which the circle has rolled.

Since we require no slippage, it follows that

Arc $AP = d(O, A)$

Therefore,

$at = d(O, A)$

The x-coordinate of the point P is

$$d(O, X) = d(O, A) - d(X, A) = at - a \sin t = a(t - \sin t)$$

The y-coordinate of the point P is equal to

$$d(O, Y) = d(A, C) - d(B, C) = a - a \cos t = a(1 - \cos t)$$

Thus, the parametric equations of the cycloid are

$$x = a(t - \sin t) \qquad y = a(1 - \cos t) \tag{5}$$

*Any attempt to derive the rectangular equation of a cycloid would soon demonstrate how complicated the task is.

Exploration Graph $x = t - \sin t, y = 1 - \cos t, 0 \le t \le 3\pi$ using your graphing utility with Tstep $= \pi/36$ and a square screen. Compare your results with Figure 69.

Applications to Mechanics

If a is negative in equation (5), we obtain an inverted cycloid, as shown in Figure 70(a). The inverted cycloid occurs as a result of some remarkable applications in the field of mechanics. We shall mention two of them: the *brachistochrone* and the *tautochrone.**

FIGURE 70

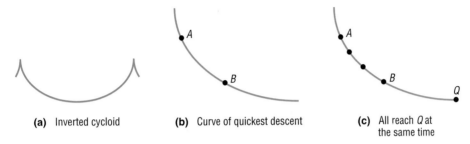

(a) Inverted cycloid   **(b)** Curve of quickest descent   **(c)** All reach Q at the same time

The **brachistochrone** is the curve of quickest descent. If a particle is constrained to follow some path from one point A to a lower point B (not on the same vertical line) and is acted on only by gravity, the time needed to make the descent is least if the path is an inverted cycloid. See Figure 70(b). This remarkable discovery, which is attributed to many famous mathematicians (including Johann Bernoulli and Blaise Pascal), was a significant step in creating the branch of mathematics known as the *calculus of variations.*

FIGURE 71
A flexible pendulum constrained by cycloids swings in a cycloid.

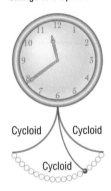

Cycloid   Cycloid

Cycloid

To define the **tautochrone,** let Q be the lowest point on an inverted cycloid. If several particles placed at various positions on an inverted cycloid simultaneously begin to slide down the cycloid, they will reach the point Q at the same time, as indicated in Figure 70(c). The tautochrone property of the cycloid was used by Christiaan Huygens (1629–1695), the Dutch mathematician, physicist, and astronomer, to construct a pendulum clock with a bob that swings along a cycloid (see Figure 71). In Huygen's clock, the bob was made to swing along a cycloid by suspending the bob on a thin wire constrained by two plates shaped like cycloids. In a clock of this design, the period of the pendulum is independent of its amplitude.

*In Greek, *brachistochrone* means "the shortest time" and *tautochrone* means "equal time."

6.7 | EXERCISES

In Problems 1–20, graph the curve whose parametric equations are given and show its orientation. Find the rectangular equation of each curve.

1. $x = 3t + 2, \quad y = t + 1; \quad 0 \le t \le 4$

2. $x = t - 3, \quad y = 2t + 4; \quad 0 \le t \le 2$

3. $x = t + 2, \quad y = \sqrt{t}; \quad t \ge 0$

4. $x = \sqrt{2t}, \quad y = 4t; \quad t \ge 0$

5. $x = t^2 + 4, \quad y = t^2 - 4; \quad -\infty < t < \infty$

6. $x = \sqrt{t} + 4, \quad y = \sqrt{t} - 4; \quad t \ge 0$

7. $x = 3t^2, \quad y = t + 1; \quad -\infty < t < \infty$

8. $x = 2t - 4, \quad y = 4t^2; \quad -\infty < t < \infty$

9. $x = 2e^t, \quad y = 1 + e^t; \quad t \ge 0$

10. $x = e^t, \quad y = e^{-t}; \quad t \ge 0$

11. $x = \sqrt{t}, \quad y = t^{3/2}; \quad t \geq 0$

12. $x = t^{3/2} + 1, \quad y = \sqrt{t}; \quad t \geq 0$

13. $x = 2 \cos t, \quad y = 3 \sin t; \quad 0 \leq t \leq 2\pi$

14. $x = 2 \cos t, \quad y = 3 \sin t; \quad 0 \leq t \leq \pi$

15. $x = 2 \cos t, \quad y = 3 \sin t; \quad -\pi \leq t \leq 0$

16. $x = 2 \cos t, \quad y = \sin t; \quad 0 \leq t \leq \pi/2$

17. $x = \sec t, \quad y = \tan t; \quad 0 \leq t \leq \pi/4$

18. $x = \csc t, \quad y = \cot t; \quad \pi/4 \leq t \leq \pi/2$

19. $x = \sin^2 t, \quad y = \cos^2 t; \quad 0 \leq t \leq \pi/2$

20. $x = t^2, \quad y = \ln t; \quad t > 0$

21. **Projectile Motion** Bob throws a ball straight up with an initial speed of 50 feet per second from a height of 6 feet.
    (a) Find parametric equations that describe the motion of the ball as a function of time.
    (b) How long is the ball in the air?
    (c) When is the ball at its maximum height? Determine the maximum height of the ball.
    (d) Simulate the motion of the ball by graphing the equations found in (a).

22. **Projectile Motion** Alice throws a ball straight up with an initial speed of 40 feet per second from a height of 5 feet.
    (a) Find parametric equations that describe the motion of the ball as a function of time.
    (b) How long is the ball in the air?
    (c) When is the ball at its maximum height? Determine the maximum height of the ball.
    (d) Simulate the motion of the ball by graphing the equations found in (a).

23. **Catching a Train** Bill's train leaves at 8:06 AM and accelerates at the rate of 2 meters per second per second. Bill, who can run 5 meters per second, arrives at the train station 5 seconds after the train had left.
    (a) Find parametric equations that describe the motion of the train and Bill as a function of time.
    (b) Determine algebraically whether Bill will catch the train. If so, when?
    (c) Simulate the motion of the train and Bill by simultaneously graphing the equations found in (a).

24. **Catching a Bus** Jodi's bus leaves at 5:03 PM and accelerates at the rate of 3 meters per second per second. Jodi, who can run 5 meters per second, arrives at the bus station 2 seconds after the bus has left.
    (a) Find parametric equations that describe the motion of the bus and Jodi as a function of time.
    (b) Determine algebraically whether Jodi will catch the bus. If so, when?
    (c) Simulate the motion of the bus and Jodi by simultaneously graphing the equations found in (a).

25. **Projectile Motion** Nolan Ryan throws a baseball with an initial speed of 145 feet per second at an angle of 20° to the horizontal. The ball leaves Nolan Ryan's hand at a height of 5 feet.
    (a) Find parametric equations that describe the position of the ball as a function of time.
    (b) How long is the ball in the air?
    (c) When is the ball at its maximum height? Determine the maximum height of the ball.
    (d) Determine the distance that the ball travels.
    (e) Using a graphing utility, simultaneously graph the equations found in (a).

26. **Projectile Motion** Mark McGuire hit a baseball with an initial speed of 180 feet per second at an angle of 40° to the horizontal. The ball was hit at a height of 3 feet off the ground.
    (a) Find parametric equations that describe the position of the ball as a function of time.
    (b) How long is the ball in the air?
    (c) When is the ball at its maximum height? Determine the maximum height of the ball.
    (d) Determine the distance that the ball travels.
    (e) Using a graphing utility, simultaneously graph the equations found in (a).

27. **Projectile Motion** Suppose that Adam throws a tennis ball off a cliff 300 meters high with an initial speed of 40 meters per second at an angle of 45° to the horizontal.
    (a) Find parametric equations that describe the position of the ball as a function of time.
    (b) How long is the ball in the air?
    (c) When is the ball at its maximum height? Determine the maximum height of the ball.
    (d) Determine the distance that the ball travels.
    (e) Using a graphing utility, simultaneously graph the equations found in (a).

28. **Projectile Motion** Suppose that Adam throws a tennis ball off a cliff 300 meters high with an initial speed of 40 meters per second at an angle of 45° to the horizontal on the Moon (gravity on the Moon is one-sixth of that on earth).
    (a) Find parametric equations that describe the position of the ball as a function of time.
    (b) How long is the ball in the air?
    (c) When is the ball at its maximum height? Determine the maximum height of the ball.
    (d) Determine the distance that the ball travels.
    (e) Using a graphing utility, simultaneously graph the equations found in (a).

29. Uniform Motion A Toyota Paseo (traveling east at 40 mph) and Pontiac Bonneville (traveling north at 30 mph) are heading toward the same intersection. The Paseo is 5 miles from the intersection when the Bonneville is 4 miles from the intersection. See the figure.

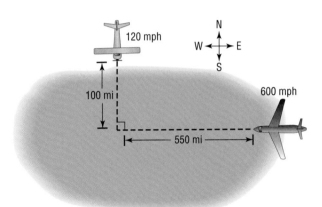

(a) Find a formula for the distance between the cars as a function of time.
(b) Graph the function in (a) using a graphing utility.
(c) What is the minimum distance between the cars? When are the cars closest?
(d) Find parametric equations that describe the motion of the Paseo and Bonneville.
(e) Simulate the motion of the cars by simultaneously graphing the equations found in (d).

30. Uniform Motion A Cessna (heading south at 120 mph) and a Boeing 747 (heading west at 600 mph) are flying toward each other at the same altitude. The Cessna is 100 miles from the point where the flight patterns intersect and the 747 is 550 miles from this intersection point. See the figure.

(a) Find a formula for the distance between the planes as a function of time.
(b) Graph the function in (a) using a graphing utility.
(c) What is the minimum distance between the planes? When are the planes closest?
(d) Find parametric equations that describe the motion of the Cessna and the 747.
(e) Simulate the motion of the planes by simultaneously graphing the equations found in (d).

In Problems 31–34, find two different parametric equations for each rectangular equation.

31. $y = x^3$ **32.** $y = x^4 + 1$ **33.** $x = y^{3/2}$ **34.** $x = \sqrt{y}$

In Problems 35–38, find parametric equations for an object that moves along the ellipse $x^2/4 + y^2/9 = 1$ with the motion described.

35. The motion begins at $(2, 0)$, is clockwise, and requires 2 seconds for a complete revolution.

36. The motion begins at $(0, 3)$, is clockwise, and requires 1 second for a complete revolution.

37. The motion begins at $(0, 3)$, is counterclockwise, and requires 1 second for a complete revolution.

38. The motion begins at $(2, 0)$, is counterclockwise, and requires 3 seconds for a complete revolution.

In Problems 39 and 40, the parametric equations of four curves are given. Graph each of them, indicating the orientation.

39. C_1: $x = t$, $y = t^2$; $-4 \le t \le 4$
C_2: $x = \cos t$, $y = 1 - \sin^2 t$; $0 \le t \le \pi$
C_3: $x = e^t$, $y = e^{2t}$; $0 \le t \le \ln 4$
C_4: $x = \sqrt{t}$, $y = t$; $0 \le t \le 16$

40. C_1: $x = t$, $y = \sqrt{1 - t^2}$; $-1 \le t \le 1$
C_2: $x = \sin t$, $y = \cos t$; $0 \le t \le 2\pi$
C_3: $x = \cos t$, $y = \sin t$; $0 \le t \le 2\pi$
C_4: $x = \sqrt{1 - t^2}$, $y = t$; $-1 \le t \le 1$

41. Show that the parametric equations for a line passing through the points (x_1, y_1) and (x_2, y_2) are

$$x = (x_2 - x_1)t + x_1 \quad y = (y_2 - y_1)t + y_1, \quad -\infty < t < \infty$$

What is the orientation of this line?

42. Projectile Motion The position of a projectile fired with an initial velocity v_0 feet per second and at an angle θ to the horizontal at the end of t seconds is given by the parametric equations

$$x = (v_o \cos \theta)t \qquad y = (v_0 \sin \theta)t - 16t^2$$

See the following illustration.

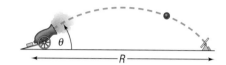

(a) Obtain the rectangular equation of the trajectory and identify the curve.

(b) Show that the projectile hits the ground ($y = 0$) when $t = \frac{1}{16}v_0 \sin \theta$.

(c) How far has the projectile traveled (horizontally) when it strikes the ground? In other words, find the range R.

(d) Find the time t when $x = y$. Then find the horizontal distance x and the vertical distance y traveled by the projectile in this time. Then compute $\sqrt{x^2 + y^2}$. This is the distance R, the range, that the projectile travels up a plane inclined at 45° to the horizontal ($x = y$). See the illustration below. (See also Problem 63 in Exercise 3.3.)

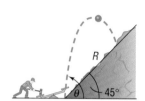

 In Problems 43–46, use a graphing utility to graph the curve defined by the given parametric equations.

43. $x = t \sin t, \quad y = t \cos t$

44. $x = \sin t + \cos t, \quad y = \sin t - \cos t$

45. $x = 4 \sin t - 2 \sin 2t$
$y = 4 \cos t - 2 \cos 2t$

46. $x = 4 \sin t + 2 \sin 2t$
$y = 4 \cos t + 2 \cos 2t$

47. The Hypocycloid The hypocycloid is a curve defined by the parametric equations

$$x(t) = \cos^3 t \qquad y(t) = \sin^3 t, \qquad 0 \le t \le 2\pi$$

(a) Graph the hypocycloid using a graphing utility.

(b) Find rectangular equations of the hypocycloid.

48. In Problem 47, we graphed the hypocycloid. Now graph the rectangular equations of the hypocycloid. Did you obtain a complete graph? If not, experiment until you do.

 49. Look up the curves called *hypocycloid* and *epicycloid*. Write a report on what you find. Be sure to draw comparisons with the cycloid.

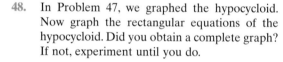

CHAPTER REVIEW

THINGS TO KNOW

Equations

Parabola	See Tables 1 and 2. (pages 386 and 388)	
Ellipse	See Table 3. (page 399)	
Hyperbola	See Table 4. (page 412)	
General equation of a conic	$Ax^2 + Bxy + Cy^2 + Dx + Ey + F = 0$	Parabola if $B^2 - 4AC = 0$ Ellipse (or circle) if $B^2 - 4AC < 0$ Hyperbola if $B^2 - 4AC > 0$
Conic in polar coordinates	$\dfrac{d(F, P)}{d(P, D)} = e$	Parabola if $e = 1$ Ellipse if $e < 1$ Hyperbola if $e > 1$
Polar equations of a conic	See Table 5. (page 428)	
Parametric equations of a curve	$x = f(t), y = g(t)$, t a parameter	

Definitions

Parabola	Set of points P in the plane for which $d(F, P) = d(P, D)$, where F is the focus and D is the directrix
Ellipse	Set of points P in the plane, the sum of whose distances from two fixed points (the foci) is a constant
Hyperbola	Set of points P in the plane, the difference of whose distances from two fixed points (the foci) is a constant

Formulas

Rotation formulas	$x = x' \cos \theta - y' \sin \theta$ $y = x' \sin \theta + y' \cos \theta$
Angle θ of rotation that eliminates the $x'y'$-term	$\cot 2\theta = \dfrac{A - C}{B} \qquad 0 < \theta < \pi/2$

HOW TO

Find the vertex, focus, and directrix of a parabola given its equation

Graph a parabola given its equation

Find an equation of a parabola given certain information about the parabola

Find the center, foci, and vertices of an ellipse given its equation

Graph an ellipse given its equation

Find an equation of an ellipse given certain information about the ellipse

Find the center, foci, vertices, and asymptotes of a hyperbola given its equation

Graph a hyperbola given its equation

Find an equation of a hyperbola given certain information about the hyperbola

Identify conics without completing the square

Identify conics without a rotation of axes

Use rotation formulas to transform second-degree equations so that no xy-term is present

Identify and graph conics given by a polar equation

Graph parametric equations

Find the rectangular equation given the parametric equations

Simulate motion problems

FILL-IN-THE-BLANK ITEMS

1. A(n) _____ is the collection of all points in the plane such that the distance from each point to a fixed point equals its distance to a fixed line.

2. A(n) _____ is the collection of all points in the plane the sum of whose distances from two fixed points is a constant.

3. A(n) _____ is the collection of all points in the plane the difference of whose distances from two fixed points is a constant.

4. For an ellipse, the foci lie on the _____ axis; for a hyperbola, the foci lie on the _____ axis.

5. For the ellipse $(x^2/9) + (y^2/16) = 1$, the major axis is along the _____.

6. The equations of the asymptotes of the hyperbola $(y^2/9) - (x^2/4) = 1$ are _____ and _____.

7. To transform the equation
$$Ax^2 + Bxy + Cy^2 + Dx + Ey + F = 0, \qquad B \neq 0$$
into one in x' and y' without an $x'y'$-term, rotate the axes through an acute angle θ that satisfies the equation
_____.

8. The polar equation
$$r = \frac{8}{4 - 2 \sin \theta}$$
is a conic whose eccentricity is _____. It is a(n) _____ whose directrix is _____ to the polar axis at a distance _____ units _____ the pole.

9. The parametric equations $x = 2 \sin t$ and $y = 3 \cos t$ represent a(n) _____.

TRUE/FALSE ITEMS

T   F   **1.**   On a parabola, the distance from any point to the focus equals the distance from that point to the directrix.

T   F   **2.**   The foci of an ellipse lie on its minor axis.

T   F   **3.**   The foci of a hyperbola lie on its transverse axis.

T   F   **4.**   Hyperbolas always have asymptotes, and ellipses never have asymptotes.

T   F   **5.**   A hyperbola never intersects its conjugate axis.

T   F   **6.**   A hyperbola always intersects its transverse axis.

T   F   **7.**   The equation $ax^2 + 6y^2 - 12y = 0$ defines an ellipse if $a > 0$.

T   F   **8.**   The equation $3x^2 + bxy + 12y^2 = 10$ defines a parabola if $b = -12$.

T   F   **9.**   If (r, θ) are polar coordinates, the equation $r = 2/(2 + 3 \sin \theta)$ defines a hyperbola.

T   F   **10.**   Parametric equations defining a curve are unique.

REVIEW EXERCISES

Blue problem numbers indicate the authors' suggestions for use in a Practice Test.

In Problems 1–20, identify each equation. If it is a parabola, give its vertex, focus, and directrix; if it is an ellipse, give its center, vertices, and foci; if it is a hyperbola, give its center, vertices, foci, and asymptotes.

1. $y^2 = -16x$

2. $16x^2 = y$

3. $\dfrac{x^2}{25} - y^2 = 1$

4. $\dfrac{y^2}{25} - x^2 = 1$

5. $\dfrac{y^2}{25} + \dfrac{x^2}{16} = 1$

6. $\dfrac{x^2}{9} + \dfrac{y^2}{16} = 1$

7. $x^2 + 4y = 4$

8. $3y^2 - x^2 = 9$

9. $4x^2 - y^2 = 8$

10. $9x^2 + 4y^2 = 36$

11. $x^2 - 4x = 2y$

12. $2y^2 - 4y = x - 2$

13. $y^2 - 4y - 4x^2 + 8x = 4$

14. $4x^2 + y^2 + 8x - 4y + 4 = 0$

15. $4x^2 + 9y^2 - 16x - 18y = 11$

16. $4x^2 + 9y^2 - 16x + 18y = 11$

17. $4x^2 - 16x + 16y + 32 = 0$

18. $4y^2 + 3x - 16y + 19 = 0$

19. $9x^2 + 4y^2 - 18x + 8y = 23$

20. $x^2 - y^2 - 2x - 2y = 1$

In Problems 21–36, obtain an equation of the conic described. Graph the equation.

21.   Parabola; focus at $(-2, 0)$; directrix the line $x = 2$

22.   Ellipse; center at $(0, 0)$; focus at $(0, 3)$; vertex at $(0, 5)$

23.   Hyperbola; center at $(0, 0)$; focus at $(0, 4)$; vertex at $(0, -2)$

24.   Parabola; vertex at $(0, 0)$; directrix the line $y = -3$

25.   Ellipse; foci at $(-3, 0)$ and $(3, 0)$; vertex at $(4, 0)$

26.   Hyperbola; vertices at $(-2, 0)$ and $(2, 0)$; focus at $(4, 0)$

27.   Parabola; vertex at $(2, -3)$; focus at $(2, -4)$

28.   Ellipse; center at $(-1, 2)$; focus at $(0, 2)$; vertex at $(2, 2)$

29.   Hyperbola; center at $(-2, -3)$; focus at $(-4, -3)$; vertex at $(-3, -3)$

30.   Parabola; focus at $(3, 6)$; directrix the line $y = 8$

31.   Ellipse; foci at $(-4, 2)$ and $(-4, 8)$; vertex at $(-4, 10)$

32.   Hyperbola; vertices at $(-3, 3)$ and $(5, 3)$; focus at $(7, 3)$

33.   Center at $(-1, 2)$; $a = 3$; $c = 4$; transverse axis parallel to the x-axis

34.   Center at $(4, -2)$; $a = 1$; $c = 4$; transverse axis parallel to the y-axis

35.   Vertices at $(0, 1)$ and $(6, 1)$; asymptote the line $3y + 2x - 9 = 0$

36.   Vertices at $(4, 0)$ and $(4, 4)$; asymptote the line $y + 2x - 10 = 0$

In Problems 37–46, identify each conic without completing the squares and without applying a rotation of axes.

37. $y^2 + 4x + 3y - 8 = 0$

38. $2x^2 - y + 8x = 0$

39. $x^2 + 2y^2 + 4x - 8y + 2 = 0$

40. $x^2 - 8y^2 - x - 2y = 0$

41. $9x^2 - 12xy + 4y^2 + 8x + 12y = 0$

42. $4x^2 + 4xy + y^2 - 8\sqrt{5}x + 16\sqrt{5}y = 0$

43. $4x^2 + 10xy + 4y^2 - 9 = 0$

44. $4x^2 - 10xy + 4y^2 - 9 = 0$

45. $x^2 - 2xy + 3y^2 + 2x + 4y - 1 = 0$

46. $4x^2 + 12xy - 10y^2 + x + y - 10 = 0$

In Problems 47–52, rotate the axes so that the new equation contains no xy-term. Discuss and graph the new equation.

47. $2x^2 + 5xy + 2y^2 - \frac{9}{2} = 0$

48. $2x^2 - 5xy + 2y^2 - \frac{9}{2} = 0$

49. $6x^2 + 4xy + 9y^2 - 20 = 0$

50. $x^2 + 4xy + 4y^2 + 16\sqrt{5}x - 8\sqrt{5}y = 0$

51. $4x^2 - 12xy + 9y^2 + 12x + 8y = 0$

52. $9x^2 - 24xy + 16y^2 + 80x + 60y = 0$

In Problems 53–58, identify the conic that each polar equation represents and graph it.

53. $r = \dfrac{4}{1 - \cos\theta}$

54. $r = \dfrac{6}{1 + \sin\theta}$

55. $r = \dfrac{6}{2 - \sin\theta}$

56. $r = \dfrac{2}{3 + 2\cos\theta}$

57. $r = \dfrac{8}{4 + 8\cos\theta}$

58. $r = \dfrac{10}{5 + 20\sin\theta}$

In Problems 59–62, convert each polar equation to a rectangular equation.

59. $r = \dfrac{4}{1 - \cos\theta}$

60. $r = \dfrac{6}{2 - \sin\theta}$

61. $r = \dfrac{8}{4 + 8\cos\theta}$

62. $r = \dfrac{2}{3 + 2\cos\theta}$

In Problems 63–68, graph the curve whose parametric equations are given and show its orientation. Find the rectangular equation of each curve.

63. $x = 4t - 2, \quad y = 1 - t; \quad -\infty < t < \infty$

64. $x = 2t^2 + 6, \quad y = 5 - t; \quad -\infty < t < \infty$

65. $x = 3\sin t, \quad y = 4\cos t + 2; \quad 0 \le t \le 2\pi$

66. $x = \ln t, \quad y = t^3; \quad t > 0$

67. $x = \sec^2 t, \quad y = \tan^2 t; \quad 0 \le t \le \pi/4$

68. $x = t^{3/2}, \quad y = 2t + 4; \quad t \ge 0$

69. Find an equation of the hyperbola whose foci are the vertices of the ellipse $4x^2 + 9y^2 = 36$ and whose vertices are the foci of this ellipse.

70. Find an equation of the ellipse whose foci are the vertices of the hyperbola $x^2 - 4y^2 = 16$ and whose vertices are the foci of this hyperbola.

71. Describe the collection of points in a plane so that the distance from each point to the point $(3, 0)$ is three-fourths of its distance from the line $x = \frac{16}{3}$.

72. Describe the collection of points in a plane so that the distance from each point to the point $(5, 0)$ is five-fourths of its distance from the line $x = \frac{16}{5}$.

73. Mirrors A mirror is shaped like a paraboloid of revolution. If a light source is located 1 foot from the base along the axis of symmetry and the opening is 2 feet across, how deep should the mirror be?

74. Parabolic Arch Bridge A bridge is built in the shape of a parabolic arch. The bridge has a span of 60 feet and a maximum height of 20 feet. Find

the height of the arch at distance of 5, 10, and 20 feet from the center.

75. Semielliptical Arch Bridge A bridge is built in the shape of a semielliptical arch. The bridge has a span of 60 feet and a maximum height of 20 feet. Find the height of the arch at distances of 5, 10, and 20 feet from the center.

76. Whispering Galleries The figure shows the specifications for an elliptical ceiling in a hall designed to be a whispering gallery. Where in the hall are the foci located?

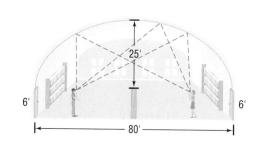

77. **LORAN** Two LORAN stations are positioned 150 miles apart along a straight shore.
    (a) A ship records a time difference of 0.00032 second between the LORAN signals. Set up an appropriate rectangular coordinate system to determine where the ship would reach shore if it were to follow the hyperbola corresponding to this time difference.
    (b) If the ship wants to enter a harbor located between the two stations 15 miles from the master station, what time difference should it be looking for?
    (c) If the ship is 20 miles offshore when the desired time difference is obtained, what is the approximate location of the ship?

    [**Note:** The speed of each radio signal is 186,000 miles per second.]

78. **Uniform Motion** Mary's train leaves at 7:15 AM and accelerates at the rate of 3 meters per second per second. Mary, who can run 6 meters per second, arrives at the train station 2 seconds after the train has left.
    (a) Find parametric equations that describe the motion of the train and Mary as a function of time.

    (b) Determine algebraically whether Mary will catch the train. If so, when?
     (c) Simulate the motion of the train and Mary by simultaneously graphing the equations found in (a).

79. **Projectile Motion** Drew Bledsoe throws a football with an initial speed of 100 feet per second at an angle of 35° to the horizontal. The ball leaves Drew Bledsoe's hand at a height of 6 feet.
    (a) Find parametric equations that describe the position of the ball as a function of time.
    (b) How long is the ball in the air?
    (c) When is the ball at its maximum height? Determine the maximum height of the ball.
    (d) Determine the distance that the ball travels.
    (e) Using a graphing utility, simultaneously graph the equations found in (a).

 80. Formulate a strategy for discussing and graphing an equation of the form

$$Ax^2 + Bxy + Cy^2 + Dx + Ey + F = 0$$

CHAPTER 7

Exponential and Logarithmic Functions

Most people view bees as pests, but farmers and scientists view bees rather differently. On the following page you will find your project as a research entomologist and will need to use the Sullivan website at:

www.prenhall.com/sullivan

to help link you to the Internet resources you will need to prepare your report.

PREPARING FOR THIS CHAPTER

Before getting started on this chapter, review the following concepts:

Exponents *(pp. 7–10)*

Functions *(Sections 1.3–1.6)*

OUTLINE

BEE PARASITIC MITE SYNDROME [BPMS]

The Bee Crisis. If you've noticed a lack of bees buzzing around your home, you're not alone. In recent years, there has been a *serious loss* of bee colonies in the United States. The cause is from two parasitic mites: Acarapis woodi and Varroa jacobsoni. Varroa first showed up in 1987. Both mites now infest bees in every state except Hawaii. There is no official wild bee census, but estimates go as high as 90 percent of the wild bee colonies have been wiped out! Scientists expect that all the bee colonies in the United States will be exposed to the Varroa mite.

Varroa mites are a serious threat to the United States honeybee industry. In 1986, more than 211,000 beekeepers produced about 200 million pounds of honey valued at $103.1 million, according to the National Honey Board. To agriculture, in general, the mite poses an even bigger threat. Crops valued at $20 billion, ranging from blueberries in Maine to almonds in California, depend on bees for pollination each year. Serious economic losses could result if Varroa mites were to devastate the United States bee population.

As a research entomologist, you want to find alternative methods of control other than the use of pesticides. Your research into Varroa resistant hybrids may offer some hope. You need to create some mathematical models that reflect possible outcomes of your research. These models will help you to determine a focus for your genetic work.

1. What are some of the *difficulties* in fighting the Varroa mite? What are likely results unless some control measures are used?
2. Pesticides are recommended to curb the Varroa epidemic. This is the main defense currently in use. How effective are the pesticides? Does the short-term gain result in any long-term costs?
3. If your research in bee hybridization allows you to breed a mite-resistant honey bee such that only one-third of your bees die from Varroa per year, would this be considered a success?
4. If you study the graphs of bee-mite interaction, you will notice some fairly well defined *logistic curves*. You search the Internet for references to logistic growth models. Using a graphing utility, determine a logistic growth function of best fit that models the situation.
5. Collect and assemble the data on *honey production* for the 1990's. Can the Varroa infestation be seen in this data?
6. Given the results of questions 4 and 5, could you justify removing the use of pesticides for the treatment of a colony?

Functions that can be expressed in terms of sums, differences, products, quotients, powers, or roots of polynomials are called **algebraic functions.** Functions that are not algebraic are termed **transcendental** (they transcend, or go beyond, algebraic functions). The trigonometric functions are examples of transcendental functions.

In this chapter, we study two transcendental functions: the *exponential* and *logarithmic functions.* These functions occur frequently in a wide variety of applications, such as biology, chemistry, economics, and psychology.

7.1 | EXPONENTIAL FUNCTIONS

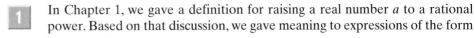

1. Evaluate Exponential Functions
2. Graph Exponential Functions
3. Define the Number e

In Chapter 1, we gave a definition for raising a real number a to a rational power. Based on that discussion, we gave meaning to expressions of the form

$$a^r$$

where the base a is a positive real number and the exponent r is a rational number.

But what is the meaning of a^x, where the base a is a positive real number and the exponent x is an irrational number? Although a rigorous definition requires methods discussed in calculus, the basis for the definition is easy to follow: Select a rational number r that is formed by truncating (removing) all but a finite number of digits from the irrational number x. Then it is reasonable to expect that

$$a^x \approx a^r$$

For example, take the irrational number $\pi = 3.14159\ldots$. Then, an approximation to a^π is

$$a^\pi \approx a^{3.14}$$

where the digits after the hundredths position have been removed from the value for π. A better approximation would be

$$a^\pi \approx a^{3.14159}$$

where the digits after the hundred-thousandths position have been removed. Continuing this way, we can obtain approximations to a^π to any desired degree of accuracy.

Most calculators have an $\boxed{x^y}$ key or a caret key $\boxed{\wedge}$ for working with exponents. To evaluate expressions of the form a^x, enter the base a, then press the $\boxed{x^y}$ key (or the $\boxed{\wedge}$ key), enter the exponent x, and press $\boxed{=}$ (or $\boxed{\text{enter}}$).

E X A M P L E   1   Using a Calculator to Evaluate Powers of 2

Using a calculator, evaluate:

(a) $2^{1.4}$ (b) $2^{1.41}$ (c) $2^{1.414}$ (d) $2^{1.4142}$ (e) $2^{\sqrt{2}}$

Solution   (a) $2^{1.4} \approx 2.639015822$ (b) $2^{1.41} \approx 2.657371628$
(c) $2^{1.414} \approx 2.66474965$ (d) $2^{1.4142} \approx 2.665119089$
(e) $2^{\sqrt{2}} \approx 2.665144143$

Now work Problem 1.

It can be shown that the familiar laws of exponents hold for real exponents.

Theorem Laws of Exponents

If s, t, a, and b are real numbers with $a > 0$ and $b > 0$, then

$$a^s \cdot a^t = a^{s+t} \qquad (a^s)^t = a^{st} \qquad (ab)^s = a^s \cdot b^s$$

$$1^s = 1 \qquad a^{-s} = \frac{1}{a^s} = \left(\frac{1}{a}\right)^s \qquad a^0 = 1 \tag{1}$$

We are now ready for the following definition.

An **exponential function** is a function of the form

$$f(x) = a^x$$

where a is a positive real number ($a > 0$) and $a \neq 1$. The domain of f is the set of all real numbers.

We exclude the base $a = 1$, because this function is simply the constant function $f(x) = 1^x = 1$. We also need to exclude the bases that are negative, because, otherwise, we would have to exclude many values of x from the domain, such as $x = \frac{1}{2}$, $x = \frac{3}{4}$, and so on. [Recall that $(-2)^{1/2}$, $(-3)^{3/4}$, and so on, are not defined in the system of real numbers.]

Graphs of Exponential Functions

2 First, we graph the exponential function $f(x) = 2^x$.

E X A M P L E   2   Graphing an Exponential Function

Graph the exponential function: $f(x) = 2^x$.

Solution   The domain of $f(x) = 2^x$ consists of all real numbers. We begin by locating some points on the graph of $f(x) = 2^x$, as listed in Table 1.

Since $2^x > 0$ for all x, the range of f is $(0, \infty)$. From this, we conclude that the graph has no x-intercepts, and, in fact, the graph will lie above the x-axis. As Table 1 indicates, the y-intercept is 1. Table 1 also indicates that as $x \to -\infty$ the value of $f(x) = 2^x$ gets closer and closer to 0. Thus, the x-axis is a horizontal asymptote to the graph as $x \to -\infty$. This gives us the end behavior of the graph for x large and negative.

To determine the end behavior for x large and positive, look again at Table 1. As $x \to \infty$, $f(x) = 2^x$ grows very quickly, causing the graph of $f(x) = 2^x$ to rise very rapidly. Thus, it is apparent that f is an increasing function and hence is one-to-one.

Using all this information, we plot some of the points from Table 1 and connect them with a smooth, continuous curve, as shown in Figure 1.

TABLE 1	
x	$f(x) = 2^x$
-10	$2^{-10} \approx 0.00098$
-3	$2^{-3} = \frac{1}{8}$
-2	$2^{-2} = \frac{1}{4}$
-1	$2^{-1} = \frac{1}{2}$
0	$2^0 = 1$
1	$2^1 = 2$
2	$2^2 = 4$
3	$2^3 = 8$
10	$2^{10} = 1024$

FIGURE 1
$y = 2^x$

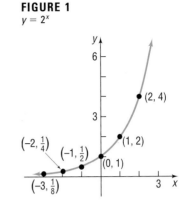

As we shall see, graphs that look like the one in Figure 1 occur very frequently in a variety of situations. For example, look at the graph in Figure 2, which illustrates the population in Ethiopia. Researchers might conclude from this graph that the population in Ethiopia is "behaving exponentially"; that is, the graph exhibits "rapid, or exponential, growth." We shall have more to say about situations that lead to exponential growth later in this chapter. For now, we continue to seek properties of the exponential functions.

FIGURE 2
Ethiopia's population is growing at one of the fastest paces in the world, exacerbating continuing shortfalls in food needed to feed its millions of people. The nation, which is only about three-quarters the size of Alaska, grows with more than 2 million births each year.

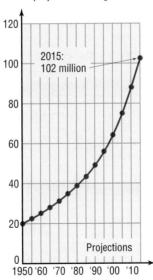

Ethiopia, in millions of people, with projections through 2015

2015:
102 million

Projections

Sources: News reports, World Bank, U.S. Agency for International Development, UN Food and Agriculture Organization, Population Reference Bureau, UNICEF, U.S. Department of Agriculture, Human Nutrition Information Service.

FIGURE 3

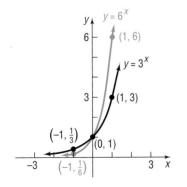

The graph of $f(x) = 2^x$ in Figure 1 is typical of all exponential functions that have a base larger than 1. Such functions are increasing functions and hence are one-to-one. Their graphs lie above the x-axis, pass through the point $(0, 1)$, and thereafter rise rapidly as $x \to \infty$. As $x \to -\infty$, the x-axis is a horizontal asymptote. There are no vertical asymptotes. Finally, the graphs are smooth and continuous, with no corners or gaps.

Figure 3 illustrates the graphs of two more exponential functions whose bases are larger than 1. Notice that for the larger base the graph is steeper when $x > 0$ and is closer to the x-axis when $x < 0$.

🖥 **Seeing the Concept** Graph $y = 2^x$ and compare what you see to Figure 1. Clear the screen and graph $y = 3^x$ and $y = 6^x$ and compare what you see to Figure 3. Clear the screen and graph $y = 10^x$ and $y = 100^x$. What viewing rectangle seems to work best? ▬

The display summarizes the information we have about $f(x) = a^x, a > 1$:

$$f(x) = a^x \qquad a > 1$$
Domain: $(-\infty, \infty)$ Range: $(0, \infty)$
x-Intercepts: None y-Intercept: 1
Horizontal asymptote: x-Axis, as $x \to -\infty$
f is an increasing function
f is one-to-one
The graph of f passes through the points $(0, 1)$ and $(1, a)$

Now we consider $f(x) = a^x$ when $0 < a < 1$.

EXAMPLE 3   Graphing an Exponential Function

Graph the exponential function: $f(x) = \left(\frac{1}{2}\right)^x$.

Solution   The domain of $f(x) = \left(\frac{1}{2}\right)^x$ consists of all real numbers. As before, we locate some points on the graph, as listed in Table 2. Since $\left(\frac{1}{2}\right)^x > 0$ for all x, the range of f is $(0, \infty)$. Thus, the graph lies above the x-axis and so has no x-intercepts. The y-intercept is 1. As $x \to -\infty$, $f(x) = \left(\frac{1}{2}\right)^x$ grows very quickly. As $x \to \infty$, the values of $f(x)$ approach 0. Thus, the x-axis ($y = 0$) is a horizontal asymptote as $x \to \infty$. It is apparent that f is a decreasing function and hence is one-to-one. Figure 4 illustrates the graph.

TABLE 2	
x	$f(x) = \left(\frac{1}{2}\right)^x$
-10	$\left(\frac{1}{2}\right)^{-10} = 1024$
-3	$\left(\frac{1}{2}\right)^{-3} = 8$
-2	$\left(\frac{1}{2}\right)^{-2} = 4$
-1	$\left(\frac{1}{2}\right)^{-1} = 2$
0	$\left(\frac{1}{2}\right)^{0} = 1$
1	$\left(\frac{1}{2}\right)^{1} = \frac{1}{2}$
2	$\left(\frac{1}{2}\right)^{2} = \frac{1}{4}$
3	$\left(\frac{1}{2}\right)^{3} = \frac{1}{8}$
10	$\left(\frac{1}{2}\right)^{10} \approx 0.00098$

FIGURE 4

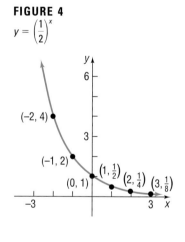

$y = \left(\frac{1}{2}\right)^x$

Note that we could have obtained the graph of $y = \left(\frac{1}{2}\right)^x$ from the graph of $y = 2^x$. If $f(x) = 2^x$, then $f(-x) = 2^{-x} = \frac{1}{2^x} = \left(\frac{1}{2}\right)^x$. Thus, the graph of $y = \left(\frac{1}{2}\right)^x = 2^{-x}$ is a reflection about the y-axis of the graph of $y = 2^x$. Compare Figures 1 and 4.

The graph of $f(x) = \left(\frac{1}{2}\right)^x$ in Figure 4 is typical of all exponential functions that have a base between 0 and 1. Such functions are decreasing, one-to-one functions. Their graphs lie above the x-axis and pass through the point

FIGURE 5

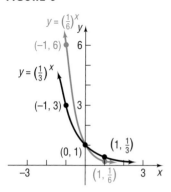

(0, 1). The graphs rise rapidly as $x \to -\infty$. As $x \to \infty$, the x-axis is a horizontal asymptote. There are no vertical asymptotes. Finally, the graphs are smooth and continuous, with no corners or gaps.

Figure 5 illustrates the graphs of two more exponential functions whose bases are between 0 and 1. Notice that the choice of a base closer to 0 results in a graph that is steeper when $x < 0$ and closer to the x-axis when $x > 0$.

Seeing the Concept Graph $y = \left(\frac{1}{2}\right)^x$ and compare what you see to Figure 4. Clear the screen and graph $y = \left(\frac{1}{3}\right)^x$ and $y = \left(\frac{1}{6}\right)^x$ and compare what you see to Figure 5. Clear the screen and graph $y = \left(\frac{1}{10}\right)^x$ and $y = \left(\frac{1}{100}\right)^x$. What viewing rectangle seems to work best? ■

The display summarizes the information we have about $f(x) = a^x$, $0 < a < 1$:

$$
\begin{array}{cc}
f(x) = a^x & 0 < a < 1 \\
\text{Domain:} \quad (-\infty, \infty) & \text{Range:} \quad (0, \infty) \\
x\text{-Intercepts:} \quad \text{None} & y\text{-Intercept:} \quad 1 \\
\text{Horizontal asymptote:} & x\text{-Axis, as } x \to \infty \\
\end{array}
$$
f is a decreasing function
f is one-to-one
The graph of f passes through the points $(0, 1)$ and $(1, a)$

Transformations (shifting, compression, stretching, and reflection) may be used to graph many functions that are basically exponential functions.

E X A M P L E   4

Graphing Exponential Functions Using Transformations

Graph $f(x) = 2^{-x} - 3$ and determine the domain, range, and horizontal asymptote of f.

Solution   We begin with the graph of $y = 2^x$. Figure 6 shows the various steps.

FIGURE 6

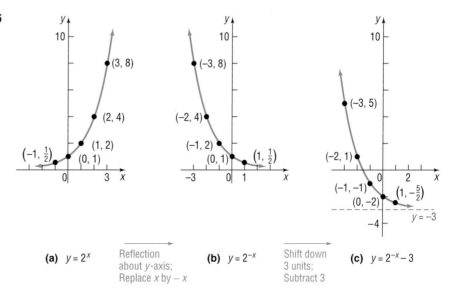

(a) $y = 2^x$ Reflection about y-axis; Replace x by $-x$ **(b)** $y = 2^{-x}$ Shift down 3 units; Subtract 3   **(c)** $y = 2^{-x} - 3$

As Figure 6(c) illustrates, the domain of $f(x) = 2^{-x} - 3$ is $(-\infty, \infty)$ and the range is $(-3, \infty)$. The horizontal asymptote of f is the line $y = -3$. ■

E X A M P L E 5   Graphing Exponential Functions Using Transformations

Graph $f(x) = -(2^{x-3})$ and determine the domain, range, and horizontal asymptote of f.

Solution   We begin with the graph of $y = 2^x$. Figure 7 shows the various steps.

FIGURE 7

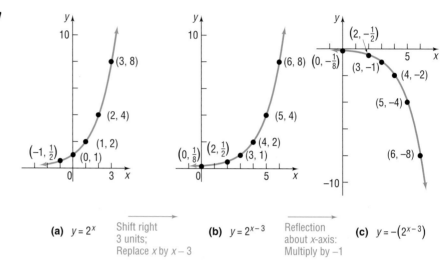

(a) $y = 2^x$ Shift right 3 units; Replace x by $x-3$ (b) $y = 2^{x-3}$ Reflection about x-axis: Multiply by -1 (c) $y = -(2^{x-3})$

The domain of $f(x) = -(2^{x-3})$ is $(-\infty, \infty)$ and the range is $(-\infty, 0)$. The horizontal asymptote of f is $y = 0$.

 Now work Problems 11 and 13.

The Base e

 As we shall see shortly, many problems that occur in nature require the use of an exponential function whose base is a certain irrational number, symbolized by the letter e.

Let's look now at one way of arriving at this important number e.

The **number e** is defined as the number that the expression

$$\left(1 + \frac{1}{n}\right)^n \tag{2}$$

approaches as $n \to \infty$. In calculus, this is expressed using limit notation as

$$e = \lim_{n \to \infty} \left(1 + \frac{1}{n}\right)^n$$

Table 3 illustrates what happens to the defining expression (2) as n takes on increasingly large values. The last number in the last column in the table is correct to nine decimal places and is the same as the entry given for e on your calculator (if expressed correct to nine decimal places).

TABLE 3

n	$\dfrac{1}{n}$	$1 + \dfrac{1}{n}$	$\left(1 + \dfrac{1}{n}\right)^{n}$
1	1	2	2
2	0.5	1.5	2.25
5	0.2	1.2	2.48832
10	0.1	1.1	2.59374246
100	0.01	1.01	2.704813829
1,000	0.001	1.001	2.716923932
10,000	0.0001	1.0001	2.718145927
100,000	0.00001	1.00001	2.718268237
1,000,000	0.000001	1.000001	2.718280469
1,000,000,000	10^{-9}	$1 + 10^{-9}$	2.718281827

The exponential function $f(x) = e^x$, whose base is the number e, occurs with such frequency in applications that it is usually referred to as *the* exponential function. Indeed, most calculators have the key* $\boxed{e^x}$ or $\boxed{exp(x)}$, which may be used to evaluate the exponential function for a given value of x.

Now use your calculator to find e^x for $x = -2$, $x = -1$, $x = 0$, $x = 1$, and $x = 2$, as we have done to create Table 4. The graph of the exponential function $f(x) = e^x$ is given in Figure 8. Since $2 < e < 3$, the graph of $y = e^x$ lies between the graphs of $y = 2^x$ and $y = 3^x$. (Refer to Figures 1 and 3.)

TABLE 4

x	e^x
-2	0.14
-1	0.37
0	1
1	2.72
2	7.39

FIGURE 8
$y = e^x$

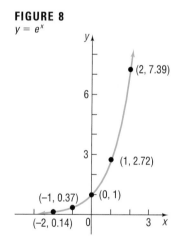

Seeing the Concept Graph $Y_1 = e^x$ and compare what you see to Figure 8. Use eVALUEate to verify the points on the graph shown in Figure 8. Now graph $Y_2 = 2^x$ and $Y_3 = 3^x$ on the same screen as $Y_1 = e^x$. Notice that the graph of $Y_1 = e^x$ lies between these two graphs. ▬

*If your calculator does not have this key but does have a $\boxed{\text{SHIFT}}$ key (or $\boxed{\text{2}^{\text{nd}}}$ key) and an $\boxed{\text{ln}}$ key, you can display the number e as follows:

Keystrokes: $\boxed{1}$ $\boxed{\text{SHIFT}}$ $\boxed{\text{ln}}$

Display: $\boxed{1}$ $\boxed{2.7182818}$

The reason this works will become clear in Section 7.2.

Now work Problem 19.

There are many applications involving the exponential function. Let's look at one.

E X A M P L E   6   Response to Advertising

Suppose that the percent R of people who respond to a newspaper advertisement for a new product and purchase the item advertised after t days is found using the formula

$$R = 50 - 50e^{-0.1t}$$

(a) What percent has responded and purchased after 5 days?
(b) What percent has responded and purchased after 10 days?
(c) What is the highest percent of people expected to respond and purchase?
 (d) Graph $R = 50 - 50e^{-0.1t}, t > 0$. Use eVALUEate to compare the values of R for $t = 5$ and $t = 10$ to the ones obtained in parts (a) and (b). Use TRACE to determine how many days are required for R to exceed 40%?

Solution   (a) After 5 days, we have $t = 5$. The corresponding percent R of people responding and purchasing is

$$R = 50 - 50e^{(-0.1)(5)} = 50 - 50e^{-0.5}$$

We use a calculator to evaluate this expression:

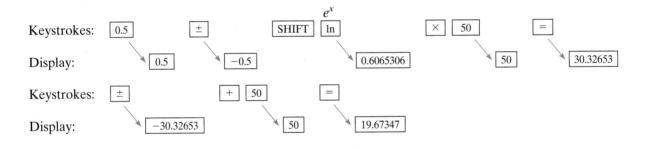

Thus, about 20% will have responded after 5 days.

(b) After 10 days, we have $t = 10$. The corresponding percent R of people responding and purchasing is

$$R = 50 - 50e^{(-0.1)(10)} = 50 - 50e^{-1} \approx 31.606$$

About 32% will have responded and purchased after 10 days.

(c) As time passes, more people are expected to respond and purchase. The highest percent expected is therefore found for the value of R as $t \to \infty$. Since $e^{-0.1t} = 1/e^{0.1t}$, it follows that $e^{-0.1t} \to 0$ as $t \to \infty$. Thus, the highest percent expected is 50%.

FIGURE 9

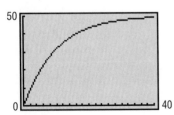

(d) See Figure 9. It will require just over 16 days to exceed 40%.   ∎

Now work Problem 43.

SUMMARY

Properties of the Exponential Function

$f(x) = a^x$, $a > 1$ Domain: $(-\infty, \infty)$;   range: $(0, \infty)$; x-intercepts: none; y-intercept: 1;   horizontal asymptote: x-axis as $x \to -\infty$;   increasing;   one-to-one
See Figure 1 for a typical graph.

$f(x) = a^x$, $0 < a < 1$ Domain: $(-\infty, \infty)$;   range: $(0, \infty)$; x-intercepts: none; y-intercept: 1;   horizontal asymptote: x-axis as $x \to \infty$;   decreasing;   one-to-one
See Figure 4 for a typical graph.

7.1 | EXERCISES

In Problems 1–10, approximate each number using a calculator. Express your answer rounded to three decimal places.

1. (a) $3^{2.2}$ (b) $3^{2.23}$ (c) $3^{2.236}$ (d) $3^{\sqrt{5}}$

2. (a) $5^{1.7}$ (b) $5^{1.73}$ (c) $5^{1.732}$ (d) $5^{\sqrt{3}}$

3. (a) $2^{3.14}$ (b) $2^{3.141}$ (c) $2^{3.1415}$ (d) 2^{π}

4. (a) $2^{2.7}$ (b) $2^{2.71}$ (c) $2^{2.718}$ (d) 2^{e}

5. (a) $3.1^{2.7}$ (b) $3.14^{2.71}$ (c) $3.141^{2.718}$ (d) π^{e}

6. (a) $2.7^{3.1}$ (b) $2.71^{3.14}$ (c) $2.718^{3.141}$ (d) e^{π}

7. $e^{1.2}$ 8. $e^{-1.3}$ 9. $e^{-0.85}$ 10. $e^{2.1}$

In Problems 11–18, the graph of an exponential function is given. Match each graph to one of the following functions:

A. $y = 3^x$ B. $y = 3^{-x}$ C. $y = -3^x$ D. $y = -3^{-x}$

E. $y = 3^x - 1$ F. $y = 3^{x-1}$ G. $y = 3^{1-x}$ H. $y = 1 - 3^x$

11.
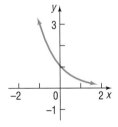

12.
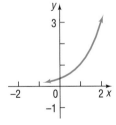

13.
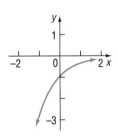

14.
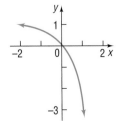

15.
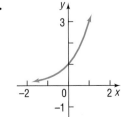

16.
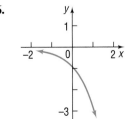

17.

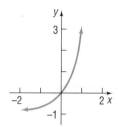

18.

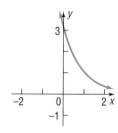

In Problems 19–26, begin with the graph of $y = e^x$ (Figure 8) and use transformations to graph each function. Determine the domain, range, and horizontal asymptote of each function.

19. $f(x) = e^{-x}$ **20.** $f(x) = -e^x$ **21.** $f(x) = e^{x+2}$ **22.** $f(x) = e^x - 1$

23. $f(x) = 5 - e^{-x}$ **24.** $f(x) = 9 - 3e^{-x}$ **25.** $f(x) = 2 - e^{-x/2}$ **26.** $f(x) = 7 - 3e^{-2x}$

In Problems 27–30, graph each function f. Based on the graph, state the domain, range, and intercepts, if any, of f.

27. $f(x) = \begin{cases} e^{-x} & \text{if } x < 0 \\ e^x & \text{if } x \geq 0 \end{cases}$ **28.** $f(x) = \begin{cases} e^x & \text{if } x < 0 \\ e^{-x} & \text{if } x \geq 0 \end{cases}$

29. $f(x) = \begin{cases} -e^x & \text{if } x < 0 \\ -e^{-x} & \text{if } x \geq 0 \end{cases}$ **30.** $f(x) = \begin{cases} -e^{-x} & \text{if } x < 0 \\ -e^x & \text{if } x \geq 0 \end{cases}$

31. If $4^x = 7$, what does 4^{-2x} equal?   **32.** If $2^x = 3$, what does 4^{-x} equal?

33. If $3^{-x} = 2$, what does 3^{2x} equal?   **34.** If $5^{-x} = 3$, what does 5^{3x} equal?

35. Optics If a single pane of glass obliterates 3% of the light passing through it, then the percent p of light that passes through n successive panes is given approximately by the equation

$$p = 100e^{-0.03n}$$

(a) What percent of light will pass through 10 panes?

(b) What percent of light will pass through 25 panes?

36. Atmospheric Pressure The atmospheric pressure p on a balloon or plane decreases with increasing height. This pressure, measured in millimeters of mercury, is related to the number of kilometers h above sea level by the formula

$$p = 760e^{-0.145h}$$

(a) Find the atmospheric pressure at a height of 2 kilometers (over a mile).

(b) What is it at a height of 10 kilometers (over 30,000 feet)?

37. Space Satellites The number of watts w provided by a space satellite's power supply after d days is given by the formula

$$w = 50e^{-0.004d}$$

(a) How much power will be available after 30 days?

(b) How much power will be available after 1 year (365 days)?

38. Healing of Wounds The normal healing of wounds can be modeled by an exponential function. If A_0 represents the original area of the wound and if A equals the area of the wound after n days, then the formula

$$A = A_0 e^{-0.35n}$$

describes the area of a wound on the nth day following an injury when no infection is present to retard the healing. Suppose that a wound initially had an area of 100 square centimeters.

(a) If healing is taking place, how large should the area of the wound be after 3 days?

(b) How large should it be after 10 days?

39. Drug Medication The formula

$$D = 5e^{-0.4h}$$

can be used to find the number of milligrams D of a certain drug that is in a patient's bloodstream h hours after the drug has been administered. How many milligrams will be present after 1 hour? After 6 hours?

40. Spreading of Rumors A model for the number of people N in a college community who have heard a certain rumor is

$$N = P(1 - e^{-0.15d})$$

where P is the total population of the community and d is the number of days that have elapsed since the rumor began. In a community of 1000 students, how many students will have heard the rumor after 3 days?

41. Exponential Probability Between 12:00 PM and 1:00 PM, cars arrive at Citibank's drive-through at the rate of 6 cars per hour (0.1 car per minute). The following formula from statistics can be used to determine the probability that a car will arrive within t minutes of 12:00 PM.

$$F(t) = 1 - e^{-0.1t}$$

(a) Determine the probability that a car will arrive within 10 minutes of 12:00 PM (that is, before 12:10 PM).

(b) Determine the probability that a car will arrive within 40 minutes of 12:00 PM (before 12:40 PM).

(c) Graph F using your graphing utility.

(d) Using TRACE, determine how many minutes are needed for the probability to reach 50%?

(e) What value does F approach as t becomes unbounded in the positive direction?

42. Exponential Probability Between 5:00 PM and 6:00 PM, cars arrive at Jiffy Lube at the rate of 9 cars per hour (0.15 car per minute). The following formula from statistics can be used to determine the probability that a car will arrive within t minutes of 5:00 PM:

$$F(t) = 1 - e^{-0.15t}$$

(a) Determine the probability that a car will arrive within 15 minutes of 5:00 PM (that is, before 5:15 PM).

(b) Determine the probability that a car will arrive within 30 minutes of 5:00 PM (before 5:30 PM).

(c) Graph F using your graphing utility.

(d) Using TRACE, determine how many minutes are needed for the probability to reach 60%?

(e) What value does F approach as t becomes unbounded in the positive direction?

43. Response to TV Advertising The percent R of viewers who respond to a television commercial for a new product after t days is found by using the formula

$$R = 70 - 70e^{-0.2t}$$

(a) What percent is expected to respond after 10 days?

(b) What percent has responded after 20 days?

(c) What is the highest percent of people expected to respond?

(d) Graph $R = 70 - 70e^{-0.2t}$, $t > 0$. Use eVALUEate to compare the values of R for $t = 10$ and $t = 20$ to the ones obtained in parts (a) and (b). Use TRACE to determine how many days are required for R to exceed 40%.

44. Profit The annual profit P of a company due to the sales of a particular item after it has been on the market x years is determined to be

$$P = \$100{,}000 - \$60{,}000\left(\tfrac{1}{2}\right)^x$$

(a) What is the annual profit after 5 years?

(b) What is the annual profit after 10 years?

(c) What is the most annual profit the company can expect from this product?

(d) Graph the profit function. Use eVALUEate to compare the values of P for $x = 5$ and $x = 10$ to the ones obtained in parts (a) and (b). Use TRACE to determine how many years it takes before an annual profit of $65,000 is obtained.

45. Alternating Current in an RL Circuit The equation governing the amount of current I (in amperes) after time t (in seconds) in a single RL circuit consisting of a resistance R (in ohms), an inductance L (in henrys), and an electromotive force E (in volts) is

$$I = \frac{E}{R}[1 - e^{-(R/L)t}]$$

(a) If $E = 120$ volts, $R = 10$ ohms, and $L = 5$ henrys, how much current I_1 is flowing after 0.3 second? After 0.5 second? After 1 second?

(b) What is the maximum current?

(c) Graph this function $I = I_1(t)$, measuring I along the y-axis and t along the x-axis.

(d) If $E = 120$ volts, $R = 5$ ohms, and $L = 10$ henrys, how much current I_2 is flowing after 0.3 second? After 0.5 second? After 1 second?

(e) What is the maximum current?

(f) Graph this function $I = I_2(t)$ on the same coordinate axes as $I_1(t)$.

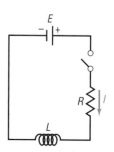

46. Alternating Current in an *RC* Circuit The equation governing the amount of current *I* (in amperes) after time *t* (in microseconds) in a single *RC* circuit consisting of a resistance *R* (in ohms), a capacitance *C* (in microfarads), and an electromotive force *E* (in volts) is

$$I = \frac{E}{R}e^{-t/(RC)}$$

(a) If $E = 120$ volts, $R = 2000$ ohms, and $C = 1.0$ microfarad, how much current I_1 is flowing initially $(t = 0)$? After 1000 microseconds? After 3000 microseconds?

(b) What is the maximum current?

(c) Graph this function $I = I_1(t)$, measuring I along the *y*-axis and *t* along the *x*-axis.

(d) If $E = 120$ volts, $R = 1000$ ohms, and $C = 2.0$ microfarads, how much current I_2 is flowing initially? After 1000 microseconds? After 3000 microseconds?

(e) What is the maximum current?

(f) Graph this function $I = I_2(t)$ on the same coordinate axes as $I_1(t)$.

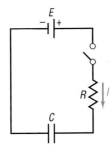

47. The Challenger Disaster* After the *Challenger* disaster in 1986, a study of the 23 launches that preceded the fatal flight was made. A mathematical model was developed involving the relationship between the Fahrenheit temperature *x* around the O-rings and the number *y* of eroded or leaky primary O-rings. The model stated that

$$y = \frac{6}{1 + e^{-(5.085 - 0.1156x)}}$$

where the number 6 indicates the 6 primary O-rings on the spacecraft.

(a) What is the predicted number of eroded or leaky primary O-rings at a temperature of 100°F?

(b) What is the predicted number of eroded or leaky primary O-rings at a temperature of 60°F?

(c) What is the predicted number of eroded or leaky primary O-rings at a temperature of 30°F?

 (d) Graph the equation and TRACE. At what temperature is the predicted number of eroded or leaky O-rings 1? 3? 5?

48. Postage Stamps* The cumulative number *y* of different postage stamps (regular and commemorative only) issued by the U.S. Post Office can be approximated (modeled) by the exponential function

$$y = 78e^{0.025x}$$

where *x* is the number of years since 1848.

(a) What is the predicted cumulative number of stamps that will have been issued by the year 1998? Check with the Postal Service and comment on the accuracy of using the function.

(b) What is the predicted cumulative number of stamps that will have been issued by the year 2000?

 (c) The cumulative number of stamps actually issued by the United States was 2 in 1848, 88 in 1868, 218 in 1888, and 341 in 1908. What conclusion can you draw about using the given function as a model over the first few decades in which stamps were issued?

*Linda Tappin, "Analyzing Data Relating to the *Challenger* Disaster," *Mathematics Teacher*, Vol. 87, No. 6, September 1994, pp. 423–426.

*David Kullman, "Patterns of Postage-stamp Production," *Mathematics Teacher*, Vol. 85, No. 3, March 1992, pp. 188–189.

49. Another Formula for _e_ Use a calculator to compute the value of

$$2 + \frac{1}{2!} + \frac{1}{3!} + \cdots + \frac{1}{n!}$$

for $n = 4, 6, 8,$ and 10. Compare each result with e.

[**Hint:** $1! = 1$, $2! = 2 \cdot 1$, $3! = 3 \cdot 2 \cdot 1$, $n! = n(n - 1) \cdot \ldots \cdot (3)(2)(1)$]

50. Another Formula for _e_ Use a calculator to compute the various values of the expression

$$2 + \cfrac{1}{1 + \cfrac{1}{2 + \cfrac{2}{3 + \cfrac{3}{4 + \cfrac{4}{\text{etc.}}}}}}$$

Compare the values to e.

51. If $f(x) = a^x$, show that

$$\frac{f(x + h) - f(x)}{h} = a^x \left(\frac{a^h - 1}{h} \right).$$

52. If $f(x) = a^x$, show that $f(A + B) = f(A) \cdot f(B)$.

53. If $f(x) = a^x$, show that $f(-x) = \dfrac{1}{f(x)}$.

54. If $f(x) = a^x$, show that $f(\alpha x) = [f(x)]^{\alpha}$.

Problems 55 and 56 provide definitions for two other transcendental functions.

55. The **hyperbolic sine function,** designated by $\sinh x$, is defined as

$$\sinh x = \frac{1}{2}(e^x - e^{-x})$$

(a) Show that $f(x) = \sinh x$ is an odd function.

(b) Graph $y = e^x$ and $y = e^{-x}$ on the same set of coordinate axes, and use the method of subtracting y-coordinates to obtain a graph of $f(x) = \sinh x$.

56. The **hyperbolic cosine function,** designated by $\cosh x$, is defined as

$$\cosh x = \frac{1}{2}(e^x + e^{-x})$$

(a) Show that $f(x) = \cosh x$ is an even function.

(b) Graph $y = e^x$ and $y = e^{-x}$ on the same set of coordinate axes, and use the method of adding y-coordinates to obtain a graph of $f(x) = \cosh x$.

(c) Refer to Problem 55. Show that, for every x,

$$(\cosh x)^2 - (\sinh x)^2 = 1$$

57. Historical Problem Pierre de Fermat (1601–1665) conjectured that the function

$$f(x) = 2^{(2^x)} + 1$$

for $x = 1, 2, 3, \ldots$, would always have a value equal to a prime number. But Leonhard Euler (1707–1783) showed that this formula fails for $x = 5$. Use a calculator to determine the prime numbers produced by f for $x = 1, 2, 3, 4$. Then show that $f(5) = 641 \times 6{,}700{,}417$, which is not prime.

58. The bacteria in a 4-liter container double every minute. After 60 minutes the container is full. How long did it take to fill half the container?

59. Explain in your own words what the number e is. Provide at least two applications that require the use of this number.

60. Do you think there is a power function that increases more rapidly than an exponential function whose base is greater than 1? Explain.

7.2 | LOGARITHMIC FUNCTIONS

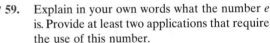

1 Change Exponential Expressions to Logarithmic Expressions
2 Change Logarithmic Expressions to Exponential Expressions
3 Evaluate Logarithmic Functions
4 Determine the Domain of a Logarithmic Function
5 Graph Logarithmic Functions

Recall (Section 1.6) that a one-to-one function $y = f(x)$ has an inverse that is defined (implicitly) by the equation $x = f(y)$. In particular, the exponential function $y = f(x) = a^x$, $a > 0$, $a \neq 1$, is one-to-one and hence has an inverse function that is defined implicitly by the equation

$$x = a^y \qquad a > 0, a \neq 1$$

This inverse function is so important that it is given a name, the *logarithmic function*.

The **logarithmic function to the base a,** where $a > 0$ and $a \neq 1$, is denoted by $y = \log_a x$ (read as "y is the logarithm to the base a of x") and is defined by

$$y = \log_a x \quad \text{if and only if} \quad x = a^y$$

The domain of the function $y = \log_a x$ is $x > 0$.

E X A M P L E   1   Relating Logarithms to Exponents

(a) If $y = \log_3 x$, then $x = 3^y$. Thus, if $x = 9$, then $y = 2$, so $9 = 3^2$ is equivalent to $2 = \log_3 9$.

(b) If $y = \log_5 x$, then $x = 5^y$. Thus, if $x = \frac{1}{5} = 5^{-1}$, then $y = -1$, so $\frac{1}{5} = 5^{-1}$ is equivalent to $-1 = \log_5 \left(\frac{1}{5}\right)$.

E X A M P L E   2   Changing Exponential Expressions to Logarithmic Expressions

 Change each exponential expression to an equivalent expression involving a logarithm.

(a) $1.2^3 = m$ (b) $e^b = 9$ (c) $a^4 = 24$

Solution   We use the fact that $y = \log_a x$ and $x = a^y$, $a > 0$, $a \neq 1$, are equivalent.

(a) If $1.2^3 = m$, then $3 = \log_{1.2} m$.
(b) If $e^b = 9$, then $b = \log_e 9$.
(c) If $a^4 = 24$, then $4 = \log_a 24$.

Now work Problem 1.

E X A M P L E   3   Changing Logarithmic Expressions to Exponential Expressions

 Change each logarithmic expression to an equivalent expression involving an exponent.

(a) $\log_a 4 = 5$ (b) $\log_e b = -3$ (c) $\log_3 5 = c$

Solution   (a) If $\log_a 4 = 5$, then $a^5 = 4$.
(b) If $\log_e b = -3$, then $e^{-3} = b$.
(c) If $\log_3 5 = c$, then $3^c = 5$.

Now work Problem 13.

To find the exact value of a logarithm, we write the logarithm in exponential notation and use the following fact:

$$\text{If } a^u = a^v, \quad \text{then} \quad u = v. \tag{1}$$

The result (1) is a consequence of the fact that exponential functions are one-to-one.

EXAMPLE 4

Finding the Exact Value of a Logarithmic Function

Find the exact value of

(a) $\log_2 16$ (b) $\log_3 \frac{1}{3}$ (c) $\log_5 25$

Solution

(a) For $y = \log_2 16$, we have the equivalent exponential equation $2^y = 16 = 2^4$, so, by (1), $y = 4$. Thus, $\log_2 16 = 4$.

(b) For $y = \log_3 \frac{1}{3}$, we have $3^y = \frac{1}{3} = 3^{-1}$, so $y = -1$. Thus, $\log_3 \frac{1}{3} = -1$.

(c) For $y = \log_5 25$, we have $5^y = 25 = 5^2$, so $y = 2$. Thus, $\log_5 25 = 2$.   ■

Now work Problem 25.

Domain of a Logarithmic Function

The logarithmic function $y = \log_a x$ has been defined as the inverse of the exponential function $y = a^x$. That is, if $f(x) = a^x$, then $f^{-1}(x) = \log_a x$. Based on the discussion given in Section 1.6 on inverse functions, we know that for a function f and its inverse f^{-1}

Domain f^{-1} = Range f and   Range f^{-1} = Domain f

Consequently, it follows that

> Domain of logarithmic function = Range of exponential function = $(0, \infty)$
> Range of logarithmic function = Domain of exponential function = $(-\infty, \infty)$

In the next box, we summarize some properties of the logarithmic function:

> $y = \log_a x$ (defining equation: $x = a^y$)
> Domain: $0 < x < \infty$ Range: $-\infty < y < \infty$

Notice that the domain of a logarithmic function consists of the *positive* real numbers.

EXAMPLE 5

Finding the Domain of a Logarithmic Function

Find the domain of each logarithmic function:

(a) $F(x) = \log_2 (1 - x)$
(b) $h(x) = \log_{1/2}|x|$

Solution (a) The domain of F consists of all x for which $(1 - x) > 0$; that is, all x for which $x < 1$, or $(-\infty, 1)$.

(b) Since $|x| > 0$ provided $x \neq 0$, the domain of h consists of all nonzero real numbers.

Now work Problem 39.

Graphs of Logarithmic Functions

Since exponential functions and logarithmic functions are inverses of each other, the graph of a logarithmic function $y = \log_a x$ is the reflection about the line $y = x$ of the graph of the exponential function $y = a^x$, as shown in Figure 10.

FIGURE 10

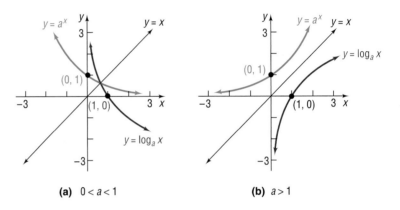

(a) $0 < a < 1$ **(b)** $a > 1$

Facts about the Graph of a Logarithmic Function $f(x) = \log_a x$

1. The x-intercept of the graph is 1. There is no y-intercept.
2. The y-axis is a vertical asymptote of the graph.
3. A logarithmic function is decreasing if $0 < a < 1$ and increasing if $a > 1$.
4. The graph is smooth and continuous, with no corners or gaps.

If the base of a logarithmic function is the number e, then we have the **natural logarithm function.** This function occurs so frequently in applications that it is given a special symbol, **ln** (from the Latin, *logarithmic naturalis*). Thus,

---

$$y = \ln x \quad \text{if and only if} \quad x = e^y$$

Since $y = \ln x$ and the exponential function $y = e^x$ are inverse functions, we can obtain the graph of $y = \ln x$ by reflecting the graph of $y = e^x$ about the line $y = x$. See Figure 11.

FIGURE 11

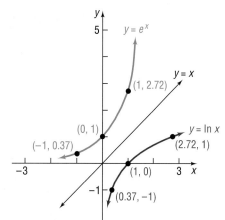

TABLE 5

x	$\ln x$
$\frac{1}{2}$	-0.69
2	0.69
3	1.10

Using a calculator with an $\boxed{ln}$ key, we can obtain other points on the graph of $f(x) = \ln x$. See Table 5.

 Check: Graph $Y_1 = e^x$ and $Y_2 = \ln x$ on the same square screen. Use eVALUEate to verify the points on the graph given in Figure 11. Do you see the symmetry of the two graphs with respect to the line $y = x$? ▬

E X A M P L E   6   Graphing Logarithmic Functions Using Transformations

Graph $f(x) = -\ln x$ by starting with the graph of $y = \ln x$. Determine the domain, range, and vertical asymptote of f.

Solution   The graph of $f(x) = -\ln x$ is obtained by a reflection about the x-axis of the graph of $y = \ln x$. See Figure 12.

FIGURE 12

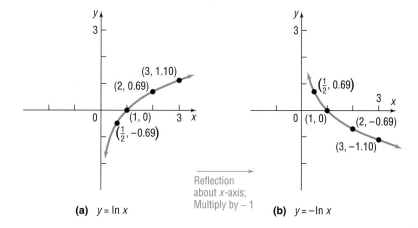

(a) $y = \ln x$

Reflection about x-axis; Multiply by -1

(b) $y = -\ln x$

The domain of $f(x) = -\ln x$ is $(0, \infty)$, the range is $(-\infty, \infty)$, and the vertical asymptote is $x = 0$. ▬

EXAMPLE 7

Graphing Logarithmic Functions Using Transformations

Determine the domain, range, and vertical asymptote of $f(x) = \ln(x + 2)$. Graph $f(x) = \ln(x + 2)$.

Solution   The domain consists of all x for which

$$x + 2 > 0 \quad \text{or} \quad x > -2$$

The graph is obtained by applying a horizontal shift to the left 2 units, as shown in Figure 13.

FIGURE 13

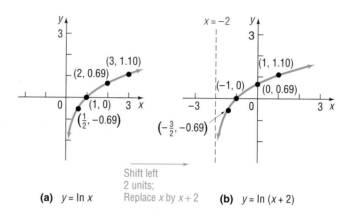

(a) $y = \ln x$ Shift left 2 units; Replace x by $x + 2$ **(b)** $y = \ln(x + 2)$

The range is $(-\infty, \infty)$, and $x = -2$ is the vertical asymptote.

EXAMPLE 8

Graphing Logarithmic Functions Using Transformations

Graph $f(x) = \ln(1 - x)$. Determine the domain, range, and vertical asymptote of f.

Solution   The domain consists of all x for which

$$1 - x > 0 \quad \text{or} \quad x < 1$$

To obtain the graph of $y = \ln(1 - x)$, we use the steps illustrated in Figure 14.

FIGURE 14

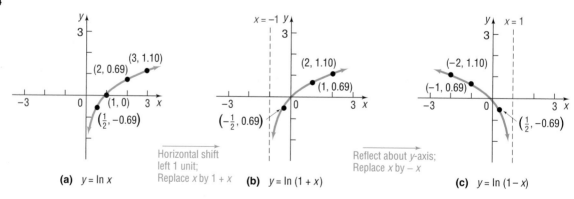

(a) $y = \ln x$ Horizontal shift left 1 unit; Replace x by $1 + x$ **(b)** $y = \ln(1 + x)$ Reflect about y-axis; Replace x by $-x$ **(c)** $y = \ln(1 - x)$

The range of $f(x) = \ln(1 - x)$ is $(-\infty, \infty)$ and the vertical asymptote is $x = 1$.

Now work Problem 67.

EXAMPLE 9

Alcohol and Driving

The concentration of alcohol in a person's blood is measurable. Recent medical research suggests that the risk R (given as a percent) of having an accident while driving a car can be modeled by the equation

$$R = 6e^{kx}$$

where x is the variable concentration of alcohol in the blood and k is a constant.

(a) Suppose that a concentration of alcohol in the blood of 0.04 results in a 10% risk ($R = 10$) of an accident. Find the constant k in the equation.
(b) Using this value of k, what is the risk if the concentration is 0.17?
(c) Using the same value of k, what concentration of alcohol corresponds to a risk of 100%?
(d) If the law asserts that anyone with a risk of having an accident of 20% or more should not have driving privileges, at what concentration of alcohol in the blood should a driver be arrested and charged with a DUI (Driving Under the Influence)?

Solution

(a) For a concentration of alcohol in the blood of 0.04 and a risk of 10%, we let $x = 0.04$ and $R = 10$ in the equation and solve for k.

$$R = 6e^{kx}$$
$$10 = 6e^{k(0.04)}$$
$$\frac{10}{6} = e^{0.04k} \qquad \text{Change to a logarithmic expression.}$$
$$0.04k = \ln\frac{10}{6} = 0.5108256$$
$$k = 12.77$$

(b) Using $k = 12.77$ and $x = 0.17$ in the equation, we find the risk R to be

$$R = 6e^{kx} = 6e^{(12.77)(0.17)} = 52.6$$

For a concentration of alcohol in the blood of 0.17, the risk of an accident is about 52.6%.

(c) Using $k = 12.77$ and $R = 100$ in the equation, we find the concentration x of alcohol in the blood to be

$$R = 6e^{kx}$$
$$100 = 6e^{12.77x}$$
$$\frac{100}{6} = e^{12.77x} \qquad \text{Change to a logarithmic expression.}$$
$$12.77x = \ln\frac{100}{6} = 2.8134$$
$$x = 0.22$$

For a concentration of alcohol in the blood of 0.22, the risk of an accident is 100%.

(d) Using $k = 12.77$ and $R = 20$ in the equation, we find the concentration x of alcohol in the blood to be

$$R = 6e^{kx}$$

$$20 = 6e^{12.77x}$$

$$\frac{20}{6} = e^{12.77x}$$

$$12.77x = \ln \frac{20}{6} = 1.204$$

$$x = 0.094$$

A driver with a concentration of alcohol in the blood of 0.094 or more should be arrested and charged with DUI. ▬

Note: Many states use 0.10 as the blood alcohol content at which a DUI citation is given. More and more states are lowering the blood alcohol content to 0.08.

SUMMARY

Properties of the Logarithmic Function

$f(x) = \log_a x, \quad a > 1$ Domain: $(0, \infty)$;   Range: $(-\infty, \infty)$; x-intercept: 1; y-intercept: none;
$(y = \log_a x$ means $x = a^y)$ vertical asymptote: y-axis;   increasing;   one-to-one
                                     See Figure 10(b) for a typical graph.

$f(x) = \log_a x, \quad 0 < a < 1$ Domain: $(0, \infty)$;   Range: $(-\infty, \infty)$; x-intercept: 1; y-intercept: none;
$(y = \log_a x$ means $x = a^y)$ vertical asymptote: y-axis;   decreasing;   one-to-one
                                       See Figure 10(a) for a typical graph.

7.2 EXERCISES

In Problems 1–12, change each exponential expression to an equivalent expression involving a logarithm.

1. $9 = 3^2$ **2.** $16 = 4^2$ **3.** $a^2 = 1.6$ **4.** $a^3 = 2.1$ **5.** $1.1^2 = M$ **6.** $2.2^3 = N$

7. $2^x = 7.2$ **8.** $3^x = 4.6$ **9.** $x^{\sqrt{2}} = \pi$ **10.** $x^\pi = e$ **11.** $e^x = 8$ **12.** $e^{2.2} = M$

In Problems 13–24, change each logarithmic expression to an equivalent expression involving an exponent.

13. $\log_2 8 = 3$ **14.** $\log_3 \left(\frac{1}{9}\right) = -2$ **15.** $\log_a 3 = 6$ **16.** $\log_b 4 = 2$

17. $\log_3 2 = x$ **18.** $\log_2 6 = x$ **19.** $\log_2 M = 1.3$ **20.** $\log_3 N = 2.1$

21. $\log_{\sqrt{2}} \pi = x$ **22.** $\log_\pi x = \frac{1}{2}$ **23.** $\ln 4 = x$ **24.** $\ln x = 4$

In Problems 25–36, find the exact value of each logarithm without using a calculator.

25. $\log_2 1$ **26.** $\log_8 8$ **27.** $\log_5 25$ **28.** $\log_3 \left(\frac{1}{9}\right)$ **29.** $\log_{1/2} 16$ **30.** $\log_{1/3} 9$

31. $\log_{10} \sqrt{10}$ **32.** $\log_5 \sqrt[3]{25}$ **33.** $\log_{\sqrt{2}} 4$ **34.** $\log_{\sqrt{3}} 9$ **35.** $\ln \sqrt{e}$ **36.** $\ln e^3$

In Problems 37–44, find the domain and the intercepts, if any, of each function.

37. $f(x) = \ln(3 - x)$ **38.** $g(x) = \ln(x - 1)$ **39.** $F(x) = \log_2 x^2$ **40.** $H(x) = \log_5 x^3$

41. $h(x) = \log_{1/2}(x^2 - 2x + 1)$ **42.** $G(x) = \log_{1/2}(x^2 - 1)$ **43.** $g(x) = \log_5 \left(\frac{x + 1}{x}\right)$ **44.** $h(x) = \log_3 \left(\frac{x}{x - 1}\right)$

In Problems 45–48, use a calculator to evaluate each expression. Round your answer to three decimal places.

45. $\ln \dfrac{5}{3}$

46. $\dfrac{\ln 5}{3}$

47. $\dfrac{\ln (10/3)}{0.04}$

48. $\dfrac{\ln (2/3)}{-0.1}$

49. Find a such that the graph of $f(x) = \log_a x$ contains the point $(2, 2)$.

50. Find a such that the graph of $f(x) = \log_a x$ contains the point $(\frac{1}{2}, -4)$.

In Problems 51–58, the graph of a logarithmic function is given. Match each graph to one of the following functions:

A. $y = \log_3 x$

B. $y = \log_3 (-x)$

C. $y = -\log_3 x$

D. $y = -\log_3 (-x)$

E. $y = \log_3 x - 1$

F. $y = \log_3 (x - 1)$

G. $y = \log_3 (1 - x)$

H. $y = 1 - \log_3 x$

51.
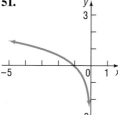

52.
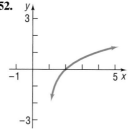

53.
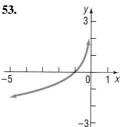

54.
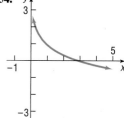

55.
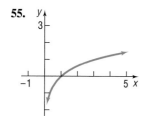

56.
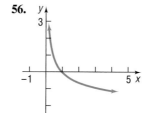

57.
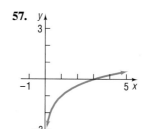

58.
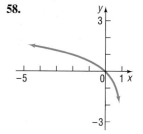

In Problems 59–70, begin with the graph of $y = \ln x$ and use transformations to graph each function. State the domain, range, and vertical asymptote of f.

59. $f(x) = \ln(x + 4)$

60. $f(x) = \ln(x - 3)$

61. $f(x) = \ln(-x)$

62. $f(x) = -\ln(-x)$

63. $g(x) = \ln 2x$

64. $h(x) = \ln \frac{1}{2}x$

65. $f(x) = 3 \ln x$

66. $f(x) = -2 \ln x$

67. $g(x) = \ln(3 - x)$

68. $h(x) = \ln(4 - x)$

69. $f(x) = -\ln(x - 1)$

70. $f(x) = 2 - \ln x$

In Problems 71–74, graph each function f. Based on the graph, state the domain, range, and intercepts, if any, of f.

71. $f(x) = \begin{cases} \ln(-x) & \text{if } x < 0 \\ \ln x & \text{if } x > 0 \end{cases}$

72. $f(x) = \begin{cases} \ln(-x) & \text{if } x \leq -1 \\ -\ln(-x) & \text{if } -1 < x < 0 \end{cases}$

73. $f(x) = \begin{cases} -\ln x & \text{if } 0 < x < 1 \\ \ln x & \text{if } x \geq 1 \end{cases}$

74. $f(x) = \begin{cases} \ln x & \text{if } 0 < x < 1 \\ -\ln x & \text{if } x \geq 1 \end{cases}$

75. **Optics** If a single pane of glass obliterates 10% of the light passing through it, then the percent P of light that passes through n successive panes is given approximately by the equation

$$P = 100e^{-0.1n}$$

  (a) How many panes are necessary to block at least 50% of the light?

  (b) How many panes are necessary to block at least 75% of the light?

76. **Chemistry** The pH of a chemical solution is given by the formula

$$pH = -\log_{10} [H^+]$$

where $[H^+]$ is the concentration of hydrogen ions in moles per liter. Values of pH range from 0 (acidic) to 14 (alkaline).

  (a) Find the pH of a 1-liter container of water with 0.0000001 mole of hydrogen ion.

  (b) Find the hydrogen ion concentration of a mildly acidic solution with a pH of 4.2.

77. Space Satellites   The number of watts w provided by a space satellite's power supply after d days is given by the formula

$$w = 50e^{-0.004d}$$

(a) How long will it take for the available power to drop to 30 watts?
(b) How long will it take for the available power to drop to only 5 watts?

78. Healing of Wounds   The normal healing of wounds can be modeled by an exponential function. If A_0 represents the original area of the wound and if A equals the area of the wound after n days, then the formula

$$A = A_0e^{-0.35n}$$

describes the area of a wound on the nth day following an injury when no infection is present to retard the healing. Suppose that a wound initially had an area of 100 square centimeters.

(a) If healing is taking place, how many days should pass before the wound is one-half its original size?
(b) How long before the wound is 10% of its original size?

79. Drug Medication   The formula

$$D = 5e^{-0.4h}$$

can be used to find the number of milligrams D of a certain drug that is in a patient's bloodstream h hours after the drug has been administered. When the number of milligrams reaches 2, the drug is to be administered again. What is the time between injections?

80. Spreading of Rumors   A model for the number of people N in a college community who have heard a certain rumor is

$$N = P(1 - e^{-0.15d})$$

where P is the total population of the community and d is the number of days that have elapsed since the rumor began. In a community of 1,000 students, how many days will elapse before 450 students have heard the rumor?

81. Current in an *RL* Circuit   The equation governing the amount of current I (in amperes) after time t (in seconds) in a simple RL circuit

consisting of a resistance R (in ohms), an inductance L (in henrys), and an electromotive force E (in volts) is

$$I = \frac{E}{R}[1 - e^{-(R/L)t}]$$

If $E = 12$ volts, $R = 10$ ohms, and $L = 5$ henrys, how long does it take to obtain a current of 0.5 ampere? Of 1.0 ampere? Graph the equation.

82. Learning Curve   Psychologists sometimes use the function

$$L(t) = A(1 - e^{-kt})$$

to measure the amount L learned at time t. The number A represents the amount to be learned, and the number k measures the rate of learning. Suppose that a student has an amount A of 200 vocabulary words to learn. A psychologist determines that the student learned 20 vocabulary words after 5 minutes.

(a) Determine the rate of learning k.
(b) Approximately how many words will the student have learned after 10 minutes?
(c) After 15 minutes?
(d) How long does it take for the student to learn 180 words?

83. Alcohol and Driving   The concentration of alcohol in a person's blood is measurable. Suppose that the risk R (given as a percent) of having an accident while driving a car can be modeled by the equation

$$R = 3e^{kx}$$

where x is the variable concentration of alcohol in the blood and k is a constant.

(a) Suppose that a concentration of alcohol in the blood of 0.06 results in a 10% risk ($R = 10$) of an accident. Find the constant k in the equation.
(b) Using this value of k, what is the risk if the concentration is 0.17?
(c) Using the same value of k, what concentration of alcohol corresponds to a risk of 100%?
(d) If the law asserts that anyone with a risk of having an accident of 15% or more should not have driving privileges, at what concentration of alcohol in the blood should a driver be arrested and charged with a DUI?
 (e) Compare this situation with that of Example 9. If you were a lawmaker, which situation would you support? Give your reasons.

84. Is there any function of the form $y = x^{\alpha}, 0 < \alpha < 1$, that increases more slowly than a logarithmic function whose base is greater than 1? Explain.

85. Constructing a Function   Look back at Figure 2. Assuming that the points (1950, 20) and (1990, 50) are on the graph, find an exponential equation $y = Ae^{bt}$ that fits the data.

[**Hint:** Let $t = 0$ correspond to the year 1950. Then show that $A = 20$. Now find b.]

Is the projection of 102 million in 2015 confirmed by your model? Try to obtain similar data about the United States birthrate and construct a function to fit those data.

86. Critical Thinking   In buying a new car, one consideration might be how well the price of the car holds up over time. Different makes of cars have different depreciation rates. One way to compute a depreciation rate for a car is given

here. Suppose that the current prices of a certain Mercedes automobile are as follows:

New	1 Year Old	2 Years Old	3 Years Old	4 Years Old	5 Years Old
$38,000	$36,600	$32,400	$28,750	$25,400	$21,200

Use the formula New = Old(e^{Rt}) to find R, the annual depreciation rate, for a specific time t. When might be the best time to trade in the car? Consult the NADA ("blue") book and compare two like models that you are interested in. Which has the better depreciation rate?

7.3 PROPERTIES OF LOGARITHMS; CURVE FITTING

1 Write a Logarithmic Expression as a Sum or Difference of Logarithms
2 Write a Logarithmic Expression as a Single Logarithm
3 Evaluate Logarithms Whose Base Is Neither 10 nor e
4 Draw Scatter Diagrams of Exponential and Logarithmic Data

Logarithms have some very useful properties that can be derived directly from the definition and the laws of exponents.

EXAMPLE 1   Establishing Properties of Logarithms

(a) Show that $\log_a 1 = 0$.
(b) Show that $\log_a a = 1$.

Solution   (a) This fact was established when we graphed $y = \log_a x$ (see Figure 20). Algebraically, for $y = \log_a 1$, we have $a^y = 1 = a^0$, so $y = 0$.
(b) For $y = \log_a a$, we have $a^y = a = a^1$, so $y = 1$. ∎

To summarize

$$\log_a 1 = 0 \qquad \log_a a = 1$$

Theorem   Properties of Logarithms

In the properties given next, M and a are positive real numbers, with $a \neq 1$, and r is any real number.

The number $\log_a M$ is the exponent to which a must be raised to obtain M. That is,

$$a^{\log_a M} = M \tag{1}$$

The logarithm to the base a of a raised to a power equals that power. That is,

$$\log_a a^r = r \qquad (2)$$

Proof of Property (1) Let $x = \log_a M$. Change this logarithmic expression to the equivalent exponential expression:

$$a^x = M$$

But $x = \log_a M$, so

$$a^{\log_a M} = M$$

Proof of Property (2) Let $x = a^r$. Change this exponential expression to the equivalent logarithmic expression

$$\log_a x = r$$

But $x = a^r$, so

$$\log_a a^r = r$$

E X A M P L E   2   Using Properties (1) and (2)

(a) $2^{\log_2 \pi} = \pi$ (b) $\log_{0.2} 0.2^{-\sqrt{2}} = -\sqrt{2}$ (c) $\ln e^{kt} = kt$

Other useful properties of logarithms are given now.

Theorem   Properties of Logarithms

In the following properties, M, N, and a are positive real numbers, with $a \neq 1$, and r is any real number.

The Log of a Product Equals the Sum of the Logs

$$\log_a (MN) = \log_a M + \log_a N \qquad (3)$$

The Log of a Quotient Equals the Difference of the Logs

$$\log_a\left(\frac{M}{N}\right) = \log_a M - \log_a N \qquad (4)$$

$$\log_a\left(\frac{1}{N}\right) = -\log_a N \qquad (5)$$

$$\log_a M^r = r \log_a M \qquad (6)$$

We shall derive properties (3) and (6) and leave the derivations of properties (4) and (5) as exercises (see Problems 76 and 77).

Proof of Property (3)   Let $A = \log_a M$ and let $B = \log_a N$. These expressions are equivalent to the exponential expressions

$$a^A = M \quad \text{and} \quad a^B = N$$

Now

$$
\begin{aligned}
\log_a (MN) = \log_a (a^A a^B) &= \log_a a^{A+B} &&\text{Law of exponents} \\
&= A + B &&\text{Property (2) of logarithms} \\
&= \log_a M + \log_a N
\end{aligned}
$$

Proof of Property (6)   Let $A = \log_a M$. This expression is equivalent to

$$a^A = M$$

Now

$$
\begin{aligned}
\log_a M^r = \log_a (a^A)^r &= \log_a a^{rA} &&\text{Law of exponents} \\
&= rA &&\text{Property (2) of logarithms} \\
&= r \log_a M
\end{aligned}
$$

1 Logarithms can be used to transform products into sums, quotients into differences, and powers into factors. Such transformations prove useful in certain types of calculus problems.

EXAMPLE 3

Writing a Logarithmic Expression as a Sum of Logarithms

Write $\log_a(x\sqrt{x^2 + 1})$ as a sum of logarithms. Express all powers as factors.

Solution

$$
\begin{aligned}
\log_a(x\sqrt{x^2 + 1}) &= \log_a x + \log_a \sqrt{x^2 + 1} &&\text{Property (3)} \\
&= \log_a x + \log_a (x^2 + 1)^{1/2} \\
&= \log_a x + \tfrac{1}{2} \log_a (x^2 + 1) &&\text{Property (6)}
\end{aligned}
$$

EXAMPLE 4

Writing a Logarithmic Expression as a Difference of Logarithms

Write

$$\log_a \frac{x^2}{(x - 1)^3}$$

as a difference of logarithms. Express all powers as factors.

Solution

$$\log_a \frac{x^2}{(x - 1)^3} = \underset{\substack{\uparrow \\ \text{Property (4)}}}{\log_a x^2 - \log_a (x - 1)^3} = \underset{\substack{\uparrow \\ \text{Property (6)}}}{2 \log_a x - 3 \log_a (x - 1)}$$

 Now work Problem 13.

EXAMPLE 5   Writing a Logarithmic Expression as a Sum and Difference of Logarithms

Write

$$\log_a \frac{x^3\sqrt{x^2 + 1}}{(x + 1)^4}$$

as a sum and difference of logarithms. Express all powers as factors.

Solution

$$\log_a \frac{x^3\sqrt{x^2 + 1}}{(x + 1)^4} = \log_a (x^3\sqrt{x^2 + 1}) - \log_a (x + 1)^4$$

$$= \log_a x^3 + \log_a \sqrt{x^2 + 1} - \log_a (x + 1)^4$$

$$= \log_a x^3 + \log_a (x^2 + 1)^{1/2} - \log_a (x + 1)^4$$

$$= 3 \log_a x + \frac{1}{2} \log_a (x^2 + 1) - 4 \log_a (x + 1)$$

■

2 Another use of properties (3) through (6) is to write sums and/or differences of logarithms with the same base as a single logarithm.

EXAMPLE 6   Writing Expressions as a Single Logarithm

Write each of the following as a single logarithm:

(a) $\log_a 7 + 4 \log_a 3$
(b) $\frac{2}{3} \log_a 8 - \log_a (3^4 - 8)$
(c) $\log_a x + \log_a 9 + \log_a (x^2 + 1) - \log_a 5$

Solution

(a) $\log_a 7 + 4 \log_a 3 = \log_a 7 + \log_a 3^4$ Property (6)

$$= \log_a 7 + \log_a 81$$

$$= \log_a (7 \cdot 81)$$ Property (3)

$$= \log_a 567$$

(b) $\frac{2}{3} \log_a 8 - \log_a (3^4 - 8) = \log_a 8^{2/3} - \log_a (81 - 8)$ Property (6)

$$= \log_a 4 - \log_a 73$$

$$= \log_a \left(\frac{4}{73}\right)$$ Property (4)

(c) $\log_a x + \log_a 9 + \log_a (x^2 + 1) - \log_a 5 = \log_a 9x + \log_a (x^2 + 1) - \log_a 5$

$$= \log_a [9x(x^2 + 1)] - \log_a 5$$

$$= \log_a \left[\frac{9x(x^2 + 1)}{5}\right]$$

■

Warning   A common error made by some students is to express the logarithm of a sum as the sum of logarithms:

$$\log_a(M + N) \quad \text{is not equal to} \quad \log_a M + \log_a N$$

Correct statement $\log_a (MN) = \log_a M + \log_a N$ Property (3)

Another common error is to express the difference of logarithms as the quotient of logarithms:

$$\log_a M - \log_a N \quad \text{is not equal to} \quad \frac{\log_a M}{\log_a N}$$

Correct statement $\quad \log_a M - \log_a N = \log_a\left(\dfrac{M}{N}\right) \quad$ Property (4)

 Now work Problem 23.

There remain two other properties of logarithms that we need to know. They are a consequence of the fact that the logarithmic function $y = \log_a x$ is one-to-one.

Theorem

In the following properties, M, N, and a are positive real numbers, with $a \neq 1$:

$$\text{If } M = N, \quad \text{then} \quad \log_a M = \log_a N. \tag{7}$$
$$\text{If } \log_a M = \log_a N, \quad \text{then} \quad M = N. \tag{8}$$

Properties (7) and (8) are useful for solving *logarithmic equations*, a topic discussed in the next section.

Using a Calculator to Evaluate Logarithms with Bases Other Than e or 10

3 Logarithms to the base 10, called **common logarithms,** were used to facilitate arithmetic computations before the widespread use of calculators. (See the Historical Feature at the end of this section.) Natural logarithms, that is, logarithms whose base is the number e, remain very important because they arise frequently in the study of natural phenomena.

Common logarithms are usually abbreviated by writing **log,** with the base understood to be 10, just as natural logarithms are abbreviated by **ln,** with the base understood to be e.

Most calculators have both $\boxed{\log}$ and $\boxed{\ln}$ keys to calculate the common logarithm and natural logarithm of a number. Let's look at an example to see how to calculate logarithms having a base other than 10 or e.

E X A M P L E   7   Evaluating Logarithms Whose Base Is Neither 10 Nor e

Evaluate $\log_2 7$.

Solution   Let $y = \log_2 7$. Then $2^y = 7$, so

$$2^y = 7$$
$$\ln 2^y = \ln 7 \qquad \text{Property (7)}$$
$$y \ln 2 = \ln 7 \qquad \text{Property (6)}$$
$$y = \frac{\ln 7}{\ln 2} \qquad \text{Solve for } y$$
$$\approx 2.8074 \qquad \text{Use calculator } (\boxed{\text{ln}} \text{ key}). \qquad \blacksquare$$

Example 7 shows how to change the base from 2 to e. In general, to change from the base b to the base a, we use the **Change-of-Base Formula.**

Theorem   Change-of-Base Formula

If $a \neq 1$, $b \neq 1$, and M are positive real numbers, then

$$\log_a M = \frac{\log_b M}{\log_b a} \qquad (9)$$

Proof   We derive this formula as follows: Let $y = \log_a M$. Then $a^y = M$, so

$$\log_b a^y = \log_b M \qquad \text{Property (7)}$$
$$y \log_b a = \log_b M \qquad \text{Property (6)}$$
$$y = \frac{\log_b M}{\log_b a} \qquad \text{Solve for } y$$
$$\log_a M = \frac{\log_b M}{\log_b a} \qquad y = \log_a M \qquad \blacksquare$$

Since calculators have only keys for $\boxed{\text{log}}$ and $\boxed{\text{ln}}$, in practice, the Change-of-Base Formula uses either $b = 10$ or $b = e$. Thus,

---

$$\log_a M = \frac{\log M}{\log a} \quad \text{and} \quad \log_a M = \frac{\ln M}{\ln a} \qquad (10)$$

Comment   To graph logarithmic functions when the base is different from e or 10 requires the Change-of-Base Formula. For example, to graph $y = \log_2 x$, we would instead graph $y = (\ln x)/(\ln 2)$. Try it. $\blacksquare$

E X A M P L E   8   Using the Change-of-Base Formula

Calculate:

(a) $\log_5 89$ (b) $\log_{\sqrt{2}} \sqrt{5}$

Solution   (a) $\log_5 89 = \dfrac{\log 89}{\log 5} \approx \dfrac{1.94939}{0.69897} = 2.7889$

or

$\log_5 89 = \dfrac{\ln 89}{\ln 5} \approx \dfrac{4.4886}{1.6094} = 2.7889$

(b) $\log_{\sqrt{2}} \sqrt{5} = \dfrac{\log \sqrt{5}}{\log \sqrt{2}} = \dfrac{\frac{1}{2}\log 5}{\frac{1}{2}\log 2} \approx \dfrac{0.69897}{0.30103} = 2.3219$

or

$\log_{\sqrt{2}} \sqrt{5} = \dfrac{\ln \sqrt{5}}{\ln \sqrt{2}} = \dfrac{\frac{1}{2}\ln 5}{\frac{1}{2}\ln 2} \approx \dfrac{1.6094}{0.6931} = 2.3219$

Now work Problem 37.

Exponential and Logarithmic Curve Fitting

 When placed in a scatter diagram, data sometimes indicate a relation between the variables that is exponential or logarithmic. Figure 15 shows typical scatter diagrams for data that are exponential or logarithmic.

FIGURE 15

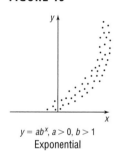

$y = ab^x, a > 0, b > 1$
Exponential

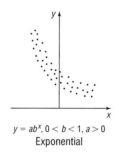

$y = ab^x, 0 < b < 1, a > 0$
Exponential

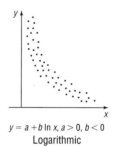

$y = a + b \ln x, a > 0, b < 0$
Logarithmic

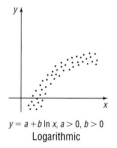

$y = a + b \ln x, a > 0, b > 0$
Logarithmic

E X A M P L E   9   Fitting a Curve to an Exponential Model

Day	Weight (in Grams)
0	100
1	91.6
2	84.1
3	74.2
4	68.3
5	63.7
6	58.5
7	53.8
8	50.2
9	47.3

Beth is interested in finding a function that explains the relation between the amount of an unknown radioactive material and time. She obtains 100 grams of the radioactive material and every day for 9 days measures the amount of radioactive material in the sample; she obtains the following data:

(a) Draw a scatter diagram using day as the independent variable.
(b) The exponential function of best fit to the data is found to be

$$y = 98(0.92)^x$$

Express the function of best fit in the form $A = A_0 e^{kt}$.
(c) Use the solution to (b) to estimate the time it takes until 50 grams of material is left. (Since 50 grams is one-half of the original amount, this time is referred to as the **half-life** of the radioactive material).
(d) Use the solution to (b) to predict how much of the material remains after 25 days.

(e) Use a graphing utility to verify the exponential function of best fit.

(f) Use a graphing utility to draw a scatter diagram of the data and then graph the exponential function of best fit on it.

Solution   (a) See Figure 16.

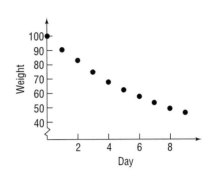

FIGURE 16

FIGURE 17

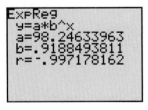

FIGURE 18

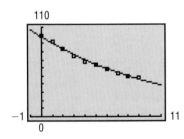

(b) To express $y = ab^x$ in the form $A = A_0 e^{kt}$, we proceed as follows:

$$ab^x = A_0 e^{kt}$$
$$a = A_0 \qquad b^x = e^{kt} = (e^k)^t$$
$$b = e^k$$

Since $y = ab^x = 98(0.92)^x$, we find $a = 98$ and $b = 0.92$. Thus,

$$a = A_0 = 98 \quad \text{and} \quad b = 0.92 = e^k$$
$$k = \ln(0.92) = -0.083$$

As a result, $A = A_0 e^{kt} = 98e^{-0.083t}$.

(c) We set $A = 50$ and solve the equation $50 = 98e^{-0.083t}$ for t. The result is that the half-life is $t = 8.1$ days.

(d) After $t = 25$ days, the amount A of the radioactive material left is

$$A = 98e^{-0.083(25)} = 12.3 \text{ grams}$$

(e) A graphing utility fits the data in Figure 16 to an exponential model of the form $y = ab^x$ by using the EXPonential REGression option.* See Figure 17. Notice that $r = -0.997178162$, so $|r|$ is close to 1, indicating a good fit.

(f) See Figure 18.   ■

Now work Problem 71.

*If your utility does not have such an option but does have a LINear REGression option, you can transform the exponential model to a linear model by the following technique:

$y = ab^x$	Exponential form
$\ln y = \ln(ab^x)$	Take natural logs of each side.
$\ln y = \ln a + \ln b^x$	Property of logarithms
$\ln y = \ln a + (\ln b)x$	Property of logarithms
$Y = \ln a + (\ln b)x$	

Now apply the LINear REGression techniques discussed in Appendix A.2. Be careful though. The dependent variable is now $Y = \ln y$, while the independent variable remains x. Once the line of best fit is obtained, you can find a and b and obtain the exponential curve $y = ab^x$.

Many relations between variables do not follow an exponential model but instead, the independent variable is related to the dependent variable using a logarithmic model.

E X A M P L E   10

Fitting a Curve to a Logarithmic Model

Jodi, a meteorologist, is interested in finding a function that explains the relation between the height of a weather balloon (in kilometers) and the atmospheric pressure (measured in millimeters of mercury) on the balloon. She collects the data in the table to the left:

Atmospheric Pressure, p	Height, h
760	0
740	0.184
725	0.328
700	0.565
650	1.079
630	1.291
600	1.634
580	1.862
550	2.235

(a) Draw a scatter diagram of the data with atmospheric pressure as the independent variable.

(b) The logarithmic function of best fit to the data is found to be

$$h = 45.8 - 6.9 \ln p$$

where h is the height of the weather balloon and p is the atmospheric pressure. Use the logarithmic function of best fit to predict the height of the balloon if the atmospheric pressure is 560 millimeters of mercury.

(c) Use a graphing utility to verify the logarithmic function of best fit.

(d) Use a graphing utility to draw a scatter diagram of the data, and then graph the logarithmic function of best fit on it.

Solution

(a) See Figure 19.

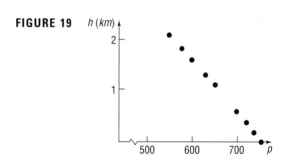

FIGURE 19

(b) Using the given function, Jodi predicts the height of the weather balloon when the atmospheric pressure is 560 to be

$$h = 45.8 - 6.9 \ln 560 \approx 2.14 \text{ kilometers}$$

(c) A graphing utility fits the data in Figure 19 to a logarithmic model of the form $y = a + b \ln x$ by using the Logarithm REGression option. See Figure 20. Notice that $|r|$ is close to 1, indicating a good fit.

(d) Figure 21 shows the graph of $h = 45.8 - 6.9 \ln p$ on the scatter diagram.

FIGURE 20

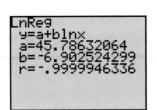

FIGURE 21

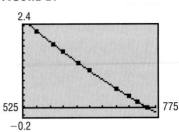

SUMMARY OF PROPERTIES OF LOGARITHMS

In the list that follows, $a > 0$, $a \neq 1$, and $b > 0$, $b \neq 1$; also, $M > 0$ and $N > 0$.

Definition $\qquad$ $y = \log_a x$ means $x = a^y$

Properties of logarithms $\qquad$ $\log_a 1 = 0$; $\quad \log_a a = 1$

$$a^{\log_a M} = M; \quad \log_a a^r = r$$

$$\log_a (MN) = \log_a M + \log_a N$$

$$\log_a \left(\frac{M}{N} \right) = \log_a M - \log_a N$$

$$\log_a \left(\frac{1}{N} \right) = -\log_a N$$

$$\log_a M^r = r \log_a M$$

Change-of-Base Formula $\qquad$ $\log_a M = \dfrac{\log_b M}{\log_b a}$

HISTORICAL FEATURE Logarithms were invented about 1590 by John Napier (1550–1617) and Jobst Bürgi (1552–1632), working independently. Napier, whose work had the greater influence, was a Scottish lord, a secretive man whose neighbors were inclined to believe him to be in league with the devil. His approach to logarithms was quite different from ours; it was based on the relationship between arithmetic and geometric sequences and not on the inverse function relationship of logarithms to exponential functions (described in Section 7.2). Napier's tables, published in 1614, listed what would now be called *natural logarithms* of sines and were rather difficult to use. A London professor, Henry Briggs, became interested in the tables and visited Napier. In their conversations, they developed the idea of common logarithms, which were published in 1617. Their importance for calculation was immediately recognized, and by 1650 they were being printed as far away as China. They remained an important calculation tool until the advent of the inexpensive hand-held calculator about 1972, which has decreased their calculational, but not their theoretical, importance.

A side effect of the invention of logarithms was the popularization of the decimal system of notation for real numbers.

7.3 | EXERCISES

In Problems 1–12, suppose that ln 2 = a and ln 3 = b. Use properties of logarithms to write each logarithm in terms of a and b.

1. $\ln 6$

2. $\ln \frac{2}{3}$

3. $\ln 1.5$

4. $\ln 0.5$

5. $\ln 2e$

6. $\ln\left(\dfrac{3}{e}\right)$

7. $\ln 12$

8. $\ln 24$

9. $\ln \sqrt[5]{18}$

10. $\ln \sqrt[4]{48}$

11. $\log_2 3$

12. $\log_3 2$

In Problems 13–22, write each expression as a sum and/or difference of logarithms. Express powers as factors.

13. $\ln(x^2\sqrt{1-x})$

14. $\ln(x^4\sqrt{x^2+1})$

15. $\log_2\left(\dfrac{x^3}{x-3}\right)$

16. $\log_5\left(\dfrac{\sqrt[3]{x^2+1}}{x^2-1}\right)$

17. $\log\left[\dfrac{x(x+2)}{(x+3)^2}\right]$

18. $\log\dfrac{x^3\sqrt{x+1}}{(x-2)^2}$

19. $\ln\left[\dfrac{x^2-x-2}{(x+4)^2}\right]^{1/3}$

20. $\ln\left[\dfrac{(x-4)^2}{x^2-1}\right]^{2/3}$

21. $\ln\dfrac{5x\sqrt{1-3x}}{(x-4)^3}$

22. $\ln\left[\dfrac{5x^2\sqrt[3]{1-x}}{4(x+1)^2}\right]$

In Problems 23–32, write each expression as a single logarithm.

23. $3\log_5 u + 4\log_5 v$

24. $\log_3 u^2 - \log_3 v$

25. $\log_{1/2}\sqrt{x} - \log_{1/2} x^3$

26. $\log_2\left(\dfrac{1}{x}\right) + \log_2\left(\dfrac{1}{x^2}\right)$

27. $\ln\left(\dfrac{x}{x-1}\right) + \ln\left(\dfrac{x+1}{x}\right) - \ln(x^2-1)$

28. $\log\left(\dfrac{x^2+2x-3}{x^2-4}\right) - \log\left(\dfrac{x^2+7x+6}{x+2}\right)$

29. $8\log_2\sqrt{3x-2} - \log_2\left(\dfrac{4}{x}\right) + \log_2 4$

30. $21\log_3\sqrt[3]{x} + \log_3 9x^2 - \log_5 25$

31. $2\log_a 5x^3 - \frac{1}{2}\log_a(2x+3)$

32. $\frac{1}{3}\log(x^3+1) + \frac{1}{2}\log(x^2+1)$

In Problems 33–36, find the exact value of each expression.

33. $3^{\log_3 5 - \log_3 4}$

34. $5^{\log_5 6 + \log_5 7}$

35. $e^{\log_{e^2} 16}$

36. $e^{\log_{e^2} 9}$

In Problems 37–44, use the Change-of-Base Formula and a calculator to evaluate each logarithm. Round your answer to three decimal places.

37. $\log_3 21$

38. $\log_5 18$

39. $\log_{1/3} 71$

40. $\log_{1/2} 15$

41. $\log_{\sqrt{2}} 7$

42. $\log_{\sqrt{5}} 8$

43. $\log_\pi e$

44. $\log_\pi \sqrt{2}$

45. Show that $\log_a(x+\sqrt{x^2-1}) + \log_a(x-\sqrt{x^2-1}) = 0$.

46. Show that $\log_a(\sqrt{x}+\sqrt{x-1}) + \log_a(\sqrt{x}-\sqrt{x-1}) = 0$.

47. Show that $\ln(1+e^{2x}) = 2x + \ln(1+e^{-2x})$.

48. If $f(x) = \log_a x$, show that $\dfrac{f(x+h)-f(x)}{h} = \log_a\left(1+\dfrac{h}{x}\right)^{1/h}, h \neq 0$.

49. If $f(x) = \log_a x$, show that $-f(x) = \log_{1/a} x$.

50. If $f(x) = \log_a x$, show that $f(1/x) = -f(x)$.

51. If $f(x) = \log_a x$, show that $f(AB) = f(A) + f(B)$.

52. If $f(x) = \log_a x$, show that $f(x^\alpha) = \alpha f(x)$.

For Problems 53–56, graph each function using a graphing utility and the Change-of-Base Formula.

53. $y = \log_4 x$

54. $y = \log_5 x$

55. $y = \log_2 (x + 2)$

56. $y = \log_4 (x - 3)$

In Problems 57–66, express y as a function of x. The constant C is a positive number.

57. $\ln y = \ln x + \ln C$

58. $\ln y = \ln(x + C)$

59. $\ln y = \ln x + \ln(x + 1) + \ln C$

60. $\ln y = 2 \ln x - \ln(x + 1) + \ln C$

61. $\ln y = 3x + \ln C$

62. $\ln y = -2x + \ln C$

63. $\ln(y - 3) = -4x + \ln C$

64. $\ln(y + 4) = 5x + \ln C$

65. $3 \ln y = \frac{1}{2} \ln(2x + 1) - \frac{1}{3} \ln(x + 4) + \ln C$

66. $2 \ln y = -\frac{1}{2} \ln x + \frac{1}{3} \ln(x^2 + 1) + \ln C$

67. Find the value of $\log_2 3 \cdot \log_3 4 \cdot \log_4 5 \cdot \log_5 6 \cdot \log_6 7 \cdot \log_7 8$.

68. Find the value of $\log_2 4 \cdot \log_4 6 \cdot \log_6 8$.

69. Find the value of $\log_2 3 \cdot \log_3 4 \cdot \ldots \cdot \log_n (n + 1) \cdot \log_{n+1} 2$.

70. Find the value of $\log_2 2 \cdot \log_2 4 \cdot \ldots \cdot \log_2 2^n$.

71. **Chemistry** A chemist has a 100-gram sample of a radioactive material. He records the amount of radioactive material every week for 6 weeks and obtains the following data:

Week	Weight (in Grams)
0	100.0
1	88.3
2	75.9
3	69.4
4	59.1
5	51.8
6	45.5

(a) Draw a scatter diagram with week as the independent variable.

(b) The exponential function of best fit to the data is found to be

$$y = 100(0.88)^x$$

Express the function of best fit in the form $A = A_0 e^{kt}$.

(c) Use the solution to (b) to estimate the time it takes until 50 grams of material is left. (Since 50 grams is one-half of the original

amount, this time is referred to as the **half-life** of the radioactive material).

(d) Use the solution to (b) to predict how much radioactive material will be left after 50 weeks.

 (e) Use a graphing utility to verify the exponential function of best fit.

(f) Use a graphing utility to draw a scatter diagram of the data and then graph the exponential function of best fit on it.

72. **Chemistry** A chemist has a 1,000-gram sample of a radioactive material. She records the amount of radioactive material remaining in the sample every day for a week and obtains the following data:

Day	Weight (in Grams)
0	1000.0
1	897.1
2	802.5
3	719.8
4	651.1
5	583.4
6	521.7
7	468.3

(a) Draw a scatter diagram with day as the independent variable.

(b) The exponential function of best fit to the data is found to be

$$y = 999(0.898)^x$$

Express the function of best fit in the form $A = A_0 e^{kt}$.

(c) Use the solution to (b) to estimate the time it takes until 500 grams of material is left. (Since 500 grams is one-half of the original amount, this time is referred to as the half-life of the radioactive material.)

(d) Use the solution to (b) to predict how much radioactive material is left after 20 days.

 (e) Use a graphing utility to verify the exponential function of best fit.

(f) Use a graphing utility to draw a scatter diagram of the data and then graph the exponential function of best fit on it.

73. **Economics and Marketing** A store manager collected the following data regarding price and quantity demanded of shoes:

Price ($/Unit)	Quantity Demanded
79	10
67	20
54	30
46	40
38	50
31	60

(a) Draw a scatter diagram with quantity demanded on the x-axis and price on the y-axis.

(b) The exponential function of best fit to the data is found to be

$$y = 96(0.98)^x$$

Express the function of best fit in the form $y = A_0 e^{kx}$.

(c) Use the solution to (b) to predict the quantity demanded if the price is $60.

(d) Use a graphing utility to verify the exponential function of best fit.

(e) Use a graphing utility to draw a scatter diagram of the data and then graph the exponential function of best fit on it.

74. **Economics and Marketing** A store manager collected the following data regarding price and quantity supplied of dresses:

Price ($/Unit)	Quantity Supplied
25	10
32	20
40	30
46	40
60	50
74	60

(a) Draw a scatter diagram with quantity supplied on the x-axis and price on the y-axis.

(b) The exponential function of best fit to the data is found to be

$$y = 20.5(1.02)^x$$

Express the function of best fit in the form $A = A_0 e^{kx}$.

(c) Use the solution to (b) to predict the quantity supplied if the price is $45.

(d) Use a graphing utility to verify the exponential function of best fit.

(e) Use a graphing utility to draw a scatter diagram of the data and then graph the exponential function of best fit on it.

75. **Economics and Marketing** The data at the top of page 486 represent the price and quantity demanded in 1997 for IBM personal computers at Best Buy.

(a) Draw a scatter diagram of the data with price as the dependent variable.

(b) The logarithmic function of best fit to the data is found to be

$$y = 32,741 - 6071 \ln x$$

Use it to predict the number of IBM personal computers that would be demanded if the price were $1650.

(c) Use a graphing utility to verify the logarithmic function of best fit.

(d) Use a graphing utility to draw a scatter diagram of the data and then graph the logarithmic function of best fit on it.

Price ($/Computer)	Quantity Demanded
2300	152
2000	159
1700	164
1500	171
1300	176
1200	180
1000	189

76. Show that $\log_a (M/N) = \log_a M - \log_a N$, where a, M, and N are positive real numbers, with $a \neq 1$.

77. Show that $\log_a (1/N) = -\log_a N$, where a and N are positive real numbers, with $a \neq 1$.

 78. Find the domain of $f(x) = \log_a x^2$ and the domain of $g(x) = 2 \log_a x$. Since $\log_a x^2 = 2 \log_a x$, how do you reconcile the fact that the domains are not equal? Write a brief explanation.

7.4 LOGARITHMIC AND EXPONENTIAL EQUATIONS

> 1 Solve Logarithmic Equations
> 2 Solve Exponential Equations
> 3 Solve Logarithmic and Exponential Equations Using a Graphing Utility

Logarithmic Equations

1 Equations that contain terms of the form $\log_a x$, where a is a positive real number, with $a \neq 1$, are often called **logarithmic equations.**

Our practice will be to solve equations, whenever possible, by finding exact solutions using algebraic methods. When algebraic methods cannot be used, approximate solutions will be obtained using a graphing utility. The reader is encouraged to pay particular attention to the form of the equations for which exact solutions are possible.

EXAMPLE 1   Solving a Logarithmic Equation

Solve: $\log_3 (4x - 7) = 2$

Solution   We can obtain an exact solution by changing the logarithm to exponential form.

$$\log_3 (4x - 7) = 2$$
$$4x - 7 = 3^2$$
$$4x - 7 = 9$$
$$4x = 16$$
$$x = 4$$

EXAMPLE 2   Solving a Logarithmic Equation

Solve: $2 \log_5 x = \log_5 9$

Solution   Because each logarithm is to the same base, 5, we can obtain an exact solution, as follows:

$$2 \log_5 x = \log_5 9$$
$$\log_5 x^2 = \log_5 9 \qquad \text{Property (6), Section 7.3}$$
$$x^2 = 9 \qquad \text{Property (8), Section 7.3}$$
$$x = 3 \quad \text{or} \quad \cancel{x = -3} \qquad \text{Recall that logarithms of negative}$$

numbers are not defined, so, in the expression $2 \log_5 x$, x must be positive. Therefore, -3 is extraneous and we discard it.

The equation has only one solution, 3.   ▬

Now work Problem 1.

E X A M P L E   3   Solving a Logarithmic Equation

Solve: $\log_4 (x + 3) + \log_4 (2 - x) = 1$

Solution   To obtain an exact solution, we need to express the left side as a single logarithm. Then we will change the expression to exponential form.

$$\log_4 (x + 3) + \log_4 (2 - x) = 1$$
$$\log_4 [(x + 3)(2 - x)] = 1 \qquad \text{Property (3), Section 7.3}$$
$$(x + 3)(2 - x) = 4^1 = 4$$
$$-x^2 - x + 6 = 4$$
$$x^2 + x - 2 = 0$$
$$(x + 2)(x - 1) = 0$$
$$x = -2 \quad \text{or} \quad x = 1 \qquad\qquad ▬$$

You should verify that both of these are solutions.

Now work Problem 11.

     Care must be taken when solving logarithmic equations. Be sure to check each apparent solution in the original equation and discard any that are extraneous. In the expression $\log_a M$, remember that a and M are positive and $a \neq 1$.

Exponential Equations

2   Equations that involve terms of the form a^x, $a > 0$, $a \neq 1$, are often referred to as **exponential equations.** Such equations sometimes can be solved by appropriately applying the laws of exponents and equations (1), that is,

$$\text{If } a^u = a^v, \quad \text{then } u = v. \qquad\qquad (1)$$

     To use equation (1), each side of the equation must be written with the same base.

E X A M P L E   4   Solving an Exponential Equation

Solve: $3^{x+1} = 81$

Solution   Since $81 = 3^4$, we can write the equation as

$$3^{x+1} = 81 = 3^4$$

Now we have the same base, 3, on each side, so we can apply (1) to obtain

$$x + 1 = 4$$
$$x = 3$$

 Now work Problem 19.

E X A M P L E   5   Solving an Exponential Equation

Solve: $e^{-x^2} = (e^x)^2 \cdot \dfrac{1}{e^3}$

Solution   We use some Laws of Exponents first to express the right-hand side with the base e.

$$(e^x)^2 \cdot \frac{1}{e^3} = e^{2x} \cdot e^{-3} = e^{2x-3}$$

Now we have $e^{-x^2} = e^{2x-3}$, so we can apply (1) to get

$$-x^2 = 2x - 3$$
$$x^2 + 2x - 3 = 0$$
$$(x + 3)(x - 1) = 0$$
$$x = -3 \quad \text{or} \quad x = 1$$

E X A M P L E   6   Solving an Exponential Equation

Solve: $4^x - 2^x - 12 = 0$

Solution   We note that $4^x = (2^2)^x = 2^{2x} = (2^x)^2$, so the equation is actually quadratic in form, and we can write it as

$$(2^x)^2 - 2^x - 12 = 0 \quad \text{If } u = 2^x, \text{ then } u^2 - u - 12 = 0.$$

Now we can factor as usual:

$$(2^x - 4)(2^x + 3) = 0 \qquad (u - 4)(u + 3) = 0$$
$$2^x - 4 = 0 \quad \text{or} \quad 2^x + 3 = 0 \qquad u - 4 = 0 \quad \text{or} \quad u + 3 = 0$$
$$2^x = 4 \qquad\qquad 2^x = -3 \qquad u = 2^x = 4 \qquad u = 2^x = -3$$

The equation on the left has the solution $x = 2$, since $4 = 2^2$; the equation on the right has no solution, since $2^x > 0$ for all x.

In each of the preceding three examples, we were able to write each exponential expression using the same base, obtaining exact solutions to the

MISSION POSSIBLE

McNewton's Coffee

Your team has been called in to solve a problem encountered by a fast food restaurant. They believe that their coffee should be brewed at 170° Fahrenheit. However, at that temperature it is too hot to drink, and a customer who accidentally spills the coffee on himself might receive third-degree burns.

What they need is a special container that will heat the water to 170°, brew the coffee at that temperature, then cool it quickly to a drinkable temperature, say 140°F, and hold it there, or at least keep it at or above 120°F for a reasonable period of time without further cooking. To cool down the coffee, three companies have submitted proposals with these specifications:

(a) The CentiKeeper Company has a container that will reduce the temperature of a liquid from 200°F to 100°F in 90 minutes by maintaining a constant temperature of 70°F.

(b) The TempControl Company has a container that will reduce the temperature of a liquid from 200°F to 110°F in 60 minutes by maintaining a constant temperature of 60°F.

(c) The Hot'n'Cold, Inc., has a container that will reduce the temperature of a liquid from 210°F to 90°F in 30 minutes by maintaining a constant temperature of 50°F.

Your job is to make a recommendation as to which container to purchase. For this you will need Newton's Law of Cooling, which follows:

$$u(t) = T + (u_0 - T)e^{kt} \quad k < 0$$

In this formula, T represents the temperature of the surrounding medium, u_0 is the initial temperature of the heated object, t is the length of time in minutes, k is a negative constant, and u represents the temperature at time t.

1. Use Newton's Law of Cooling to find the constant k of the formula for each container.
2. Use a graphing utility to graph each relation.
3. How long does it take each container to lower the coffee temperature from 170°F to 140°F?
4. How long will the coffee temperature remain between 120°F and 140°F?
5. On the basis of this information, which company should get the contract with McNewton's? What are your reasons?
6. Define "capital cost" and "operating cost". How might they affect your choice?

equation. When this is not possible, logarithms can sometimes be used to obtain the solution.

E X A M P L E 7   Solving an Exponential Equation

Solve: $2^x = 5$

Solution   Since we cannot express 5 as a power of 2, we write the exponential equation as the equivalent logarithmic equation:

$$2^x = 5$$

$$x = \log_2 5 = \frac{\ln 5}{\ln 2}$$

↑

Change-of-Base Formula (10)

Alternatively, we can solve the equation $2^x = 5$ by taking the natural logarithm (or common logarithm) of each side. Taking the natural logarithm,

$$2^x = 5$$
$$\ln 2^x = \ln 5$$
$$x \ln 2 = \ln 5$$
$$x = \frac{\ln 5}{\ln 2}$$

Using a calculator, the solution, rounded to three decimal places, is

$$x = \frac{\ln 5}{\ln 2} \approx 2.322$$

Now work Problem 33.

E X A M P L E 8   Solving an Exponential Equation

Solve: $8 \cdot 3^x = 5$

Solution
$$8 \cdot 3^x = 5 \qquad \text{Isolate } 3^x \text{ on the left side.}$$
$$3^x = \tfrac{5}{8} \qquad \text{Proceed as in Example 7.}$$
$$x = \log_3 \left(\tfrac{5}{8}\right) = \frac{\ln \tfrac{5}{8}}{\ln 3}$$

Using a calculator, the solution, rounded to three decimal places, is

$$x = \frac{\ln \left(\tfrac{5}{8}\right)}{\ln 3} \approx -0.428$$

E X A M P L E 9

Solving an Exponential Equation

Solve: $5^{x-2} = 3^{3x+2}$

Solution

Because the bases on each side are different, we take the natural logarithm of each side and apply appropriate properties of logarithms. The result is an equation in x that we can solve.

$$5^{x-2} = 3^{3x+2}$$
$$\ln 5^{x-2} = \ln 3^{3x+2} \qquad \text{Property (7)}$$
$$(x-2)\ln 5 = (3x+2)\ln 3 \qquad \text{Property (6)}$$
$$(\ln 5)x - 2\ln 5 = (3\ln 3)x + 2\ln 3 \qquad \text{Remove parentheses}$$
$$(\ln 5 - 3\ln 3)x = 2\ln 3 + 2\ln 5 \qquad \text{Isolate } x \text{ terms on the left}$$
$$x = \frac{2(\ln 3 + \ln 5)}{\ln 5 - 3\ln 3} \approx -3.212$$

 Now work Problem 41.

 ## Graphing Utility Solutions

The techniques introduced in this section only apply to certain types of logarithmic and exponential equations. Solutions for other types are usually studied in calculus, using numerical methods. However, we can use a graphing utility to approximate the solution of other types.

E X A M P L E 10

Solving Equations Using a Graphing Utility

Solve: $\log_3 x + \log_4 x = 4$
Express the solution(s) correct to two decimal places.

FIGURE 22

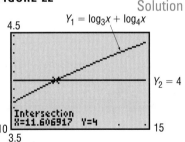

Solution

The solution is found by graphing

$$Y_1 = \log_3 x + \log_4 x = \frac{\log x}{\log 3} + \frac{\log x}{\log 4} \quad \text{and} \quad Y_2 = 4$$

(Remember that you must use the Change-of-Base Formula to graph Y_1.) Y_1 is an increasing function (do you know why?), and so there is only one point of intersection for Y_1 and Y_2. Figure 22 shows the graphs of Y_1 and Y_2. Using the INTERSECT* command, the solution 11.60, correct to two decimal places, is obtained.

Can you discover an algebraic solution to Example 10?

[**Hint:** Factor $\log x$ from Y_1.]

E X A M P L E 11

Solving Equations Using a Graphing Utility

Solve: $x + e^x = 2$
Express the solution(s) correct to two decimal places.

*If your graphing utility does not have an INTERSECT command, you will have to use BOX or ZOOM-IN with TRACE to find the point of intersection.

Solution   The solution is found by graphing $Y_1 = x + e^x$ and $Y_2 = 2$. Y_1 is an increasing function (do you know why?), and so there is only one point of intersection for Y_1 and Y_2. Figure 23 shows the graphs of Y_1 and Y_2. Using the INTERSECT command, the solution 0.44, correct to two decimal places, is obtained.

FIGURE 23

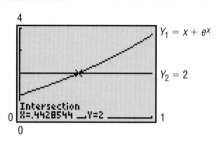

7.4 | EXERCISES

In Problems 1–58, solve each equation.

1. $\log_2 (2x + 1) = 3$
2. $\log_3 (3x - 2) = 2$
3. $\log_3 (x^2 + 1) = 2$
4. $\log_5 (x^2 + x + 4) = 2$
5. $\frac{1}{2} \log_3 x = 2 \log_3 2$
6. $-2 \log_4 x = \log_4 9$
7. $2 \log_5 x = 3 \log_5 4$
8. $3 \log_2 x = -\log_2 27$
9. $3 \log_2 (x - 1) + \log_2 4 = 5$
10. $2 \log_3 (x + 4) - \log_3 9 = 2$
11. $\log x + \log (x + 15) = 2$
12. $\log_4 x + \log_4 (x - 3) = 1$
13. $\log_x 4 = 2$
14. $\log_x \left(\frac{1}{8}\right) = 3$
15. $\log_3 (x - 1)^2 = 2$
16. $\log_2 (x + 4)^3 = 6$
17. $\log_{1/2} (3x + 1)^{1/3} = -2$
18. $\log_{1/3} (1 - 2x)^{1/2} = -1$
19. $2^{2x+1} = 4$
20. $5^{1-2x} = \frac{1}{5}$
21. $3^{x^3} = 9^x$
22. $4^{x^2} = 2^x$
23. $8^{x^2-2x} = \frac{1}{2}$
24. $9^{-x} = \frac{1}{3}$
25. $2^x \cdot 8^{-x} = 4^x$
26. $\left(\frac{1}{2}\right)^{1-x} = 4$
27. $2^{2x} + 2^x - 12 = 0$
28. $3^{2x} + 3^x - 2 = 0$
29. $3^{2x} + 3^{x+1} - 4 = 0$
30. $4^x - 2^x = 0$
31. $4^x = 8$
32. $9^{2x} = 27$
33. $2^x = 10$
34. $3^x = 14$
35. $8^{-x} = 1.2$
36. $2^{-x} = 1.5$
37. $3^{1-2x} = 4^x$
38. $2^{x+1} = 5^{1-2x}$
39. $\left(\frac{3}{5}\right)^x = 7^{1-x}$
40. $\left(\frac{4}{3}\right)^{1-x} = 5^x$
41. $1.2^x = (0.5)^{-x}$
42. $(0.3)^{1+x} = 1.7^{2x-1}$
43. $\pi^{1-x} = e^x$
44. $e^{x+3} = \pi^x$
45. $5(2^{3x}) = 8$
46. $0.3(4^{0.2x}) = 0.2$
47. $400e^{0.2x} = 600$
48. $500e^{0.3x} = 600$
49. $\log_a (x - 1) - \log_a (x + 6) = \log_a (x - 2) - \log_a (x + 3)$
50. $\log_a x + \log_a (x - 2) = \log_a (x + 4)$
51. $\log_{1/3} (x^2 + x) - \log_{1/3} (x^2 - x) = -1$
52. $\log_4 (x^2 - 9) - \log_4 (x + 3) = 3$
53. $\log_2 8^x = -3$
54. $\log_3 3^x = -1$
55. $\log_2 (x^2 + 1) - \log_4 x^2 = 1$
56. $\log_2 (3x + 2) - \log_4 x = 3$
    [**Hint:** Change $\log_4 x^2$ to base 2.]
57. $\log_{16} x + \log_4 x + \log_2 x = 7$
58. $\log_9 x + 3 \log_3 x = 14$
59. $(\sqrt[3]{2})^{2-x} = 2^{x^2}$
60. $\log_2 x^{\log_2 x} = 4$

In Problems 61–76, use a graphing utility to solve each equation. Express your answer correct to two decimal places.

61. $\log_5 x + \log_3 x = 1$
62. $\log_2 x + \log_6 x = 3$
63. $\log_5 (x + 1) - \log_4 (x - 2) = 1$
64. $\log_2 (x - 1) - \log_6 (x + 2) = 2$
65. $e^x = -x$
66. $e^{2x} = x + 2$
67. $e^x = x^2$
68. $e^x = x^3$
69. $\ln x = -x$
70. $\ln 2x = -x + 2$
71. $\ln x = x^3 - 1$
72. $\ln x = -x^2$
73. $e^x + \ln x = 4$
74. $e^x - \ln x = 4$
75. $e^{-x} = \ln x$
76. $e^{-x} = -\ln x$

7.5 COMPOUND INTEREST

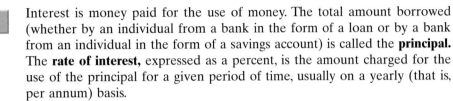

1 Determine the Future Value of a Lump Sum of Money
2 Calculate Effective Rates of Return
3 Determine the Present Value of a Lump Sum of Money
4 Determine the Time Required to Double or Triple a Lump Sum of Money

1 Interest is money paid for the use of money. The total amount borrowed (whether by an individual from a bank in the form of a loan or by a bank from an individual in the form of a savings account) is called the **principal.** The **rate of interest,** expressed as a percent, is the amount charged for the use of the principal for a given period of time, usually on a yearly (that is, per annum) basis.

If a principal of P dollars is borrowed for a period of t years at a per annum interest rate r, expressed as a decimal, the interest I charged is

> Simple Interest Formula
>
> $$I = Prt \qquad\qquad (1)$$

Interest charged according to formula (1) is called **simple interest.**

In working with problems involving interest, we use the term **payment period** as follows:

Annually	Once per year	Monthly	12 times per year
Semiannually	Twice per year	Daily	365 times per year*
Quarterly	4 times per year		

When the interest due at the end of a payment period is added to the principal so that the interest computed at the end of the next payment period is based on this new principal amount (old principal + interest), the interest is said to have been **compounded.** Thus, **compound interest** is interest paid on previously earned interest.

E X A M P L E   1   Computing Compound Interest

A credit union pays interest of 8% per annum compounded quarterly on a certain savings plan. If $1000 is deposited in such a plan and the interest is left to accumulate, how much is in the account after 1 year?

Solution   We use the simple interest formula, $I = Prt$. The principal P is $1000 and the rate of interest is 8% $= 0.08$. After the first quarter of a year, the time t is $\frac{1}{4}$ year, so the interest earned is

$$I = Prt = (\$1000)(0.08)(\tfrac{1}{4}) = \$20$$

---

*Most banks use a 360 day "year." Why do you think they do?

The new principal is $P + I = \$1000 + \$20 = \$1020$. At the end of the second quarter, the interest on this principal is

$$I = (\$1020)(0.08)\left(\tfrac{1}{4}\right) = \$20.40$$

At the end of the third quarter, the interest on the new principal of $\$1020 + \$20.40 = \$1040.40$ is

$$I = (\$1040.40)(0.08)\left(\tfrac{1}{4}\right) = \$20.81$$

Finally, after the fourth quarter, the interest is

$$I = (\$1061.21)(0.08)\left(\tfrac{1}{4}\right) = \$21.22$$

Thus, after 1 year the account contains $\$1082.43$.

The pattern of the calculations performed in Example 1 leads to a general formula for compound interest. To fix our ideas, let P represent the principal to be invested at a per annum interest rate r, which is compounded n times per year. (For computing purposes, r is expressed as a decimal.) The interest earned after each compounding period is the principal times r/n. Thus, the amount A after one compounding period is

$$A = P + P\left(\frac{r}{n}\right) = P\left(1 + \frac{r}{n}\right)$$

After two compounding periods, the amount A, based on the new principal $P(1 + r/n)$, is

$$A = \underbrace{P\left(1 + \frac{r}{n}\right)}_{\substack{\text{New} \\ \text{principal}}} + \underbrace{P\left(1 + \frac{r}{n}\right)\left(\frac{r}{n}\right)}_{\substack{\text{Interest on} \\ \text{new principal}}} = P\left(1 + \frac{r}{n}\right)\left(1 + \frac{r}{n}\right) = P\left(1 + \frac{r}{n}\right)^2$$

After three compounding periods,

$$A = P\left(1 + \frac{r}{n}\right)^2 + P\left(1 + \frac{r}{n}\right)^2\left(\frac{r}{n}\right) = P\left(1 + \frac{r}{n}\right)^2\left(1 + \frac{r}{n}\right) = P\left(1 + \frac{r}{n}\right)^3$$

Continuing this way, after n compounding periods (1 year),

$$A = P\left(1 + \frac{r}{n}\right)^n$$

Because t years will contain $n \cdot t$ compounding periods, after t years we have

$$A = P\left(1 + \frac{r}{n}\right)^{nt}$$

Theorem **Compound Interest Formula**

The amount A after t years due to a principal P invested at an annual interest rate r compounded n times per year is

$$A = P\left(1 + \frac{r}{n}\right)^{nt} \tag{2}$$

E X A M P L E   2

Comparing Investments Using Different Compounding Periods

Investing $1000 at an annual rate of 10% compounded annually, quarterly, monthly, and daily will yield the following amounts after 1 year:

Annual compounding:
$$A = P(1 + r)$$
$$= (\$1000)(1 + 0.10) = \$1100.00$$

Quarterly compounding:
$$A = P\left(1 + \frac{r}{4}\right)^4$$
$$= (\$1000)(1 + 0.025)^4 = \$1103.81$$

Monthly compounding:
$$A = P\left(1 + \frac{r}{12}\right)^{12}$$
$$= (\$1000)(1 + 0.00833)^{12} = \$1104.71$$

Daily compounding:
$$A = P\left(1 + \frac{r}{365}\right)^{365}$$
$$= (\$1000)(1 + 0.000274)^{365} = \$1105.16$$

 Now work Problem 1.

From Example 2, we can see that the effect of compounding more frequently is that the account after 1 year is higher: $1000 compounded 4 times a year at 10% results in $1103.81; $1000 compounded 12 times a year at 10% results in $1104.71; and $1000 compounded 365 times a year at 10% results in $1105.16. This leads to the following question: What would happen to the amount after 1 year if the number of times that the interest is compounded were increased without bound?

Let's find the answer. Suppose that P is the principal, r is the per annum interest rate, and n is the number of times that the interest is compounded each year. The amount after 1 year is

$$A = P\left(1 + \frac{r}{n}\right)^n$$

Now suppose that the number n of times that the interest is compounded per year gets larger and larger; that is, suppose that $n \to \infty$. Then,

$$A = P\left(1 + \frac{r}{n}\right)^n = P\left[1 + \frac{1}{n/r}\right]^n = P\left(\left[1 + \frac{1}{n/r}\right]^{n/r}\right)^r = P\left[\left(1 + \frac{1}{h}\right)^h\right]^r \quad (3)$$

$$\uparrow$$
$$h = \frac{n}{r}$$

In (3), as $n \to \infty$, then $h = n/r \to \infty$, and the expression in brackets equals e. [Refer to equation (2) on p. 456.] Thus, $A \to Pe^r$. Table 6 compares $(1 + r/n)^n$, for large values of n, to e^r for $r = 0.05, r = 0.10, r = 0.15,$ and $r = 1$. The larger

n gets, the closer $(1 + r/n)^n$ gets to e^r. Thus, no matter how frequent the compounding, the amount after 1 year has the definite ceiling Pe^r.

TABLE 6

	$\left(1 + \dfrac{r}{n}\right)^n$			
	$n = 100$	$n = 1000$	$n = 10{,}000$	e^r
$r = 0.05$	1.0512579	1.05127	1.051271	1.0512711
$r = 0.10$	1.1051157	1.1051654	1.1051703	1.1051709
$r = 0.15$	1.1617037	1.1618212	1.1618329	1.1618342
$r = 1$	2.7048138	2.7169239	2.7181459	2.7182818

When interest is compounded so that the amount after 1 year is Pe^r, we say the interest is **compounded continuously.**

Theorem   Continuous Compounding

The amount A after t years due to a principal P invested at an annual interest rate r compounded continuously is

$$A = Pe^{rt} \qquad (4)$$

E X A M P L E   3   Using Continuous Compounding

The amount A that results from investing a principal P of \$1000 at an annual rate r of 10% compounded continuously for a time t of 1 year is

$$A = \$1000e^{0.10} = (\$1000)(1.10517) = \$1105.17$$

Now work Problem 9.

The **effective rate of interest** is the equivalent annual simple rate of interest that would yield the same amount as compounding after 1 year. For example, based on Example 3, a principal of \$1000 will result in \$1105.17 at a rate of 10% compounded continuously. To get this same amount using a simple rate of interest would require that interest of \$1105.17 − \$1000.00 = \$105.17 be earned on the principal. Since \$105.17 is 10.517% of \$1000, a simple rate of interest of 10.517% is needed to equal 10% compounded continuously. Thus, the effective rate of interest of 10% compounded continuously is 10.517%.

Based on the results of Examples 2 and 3, we find the following comparisons:

	Annual Rate	Effective Rate
Annual compounding	10%	10%
Quarterly compounding	10%	10.381%
Monthly compounding	10%	10.471%
Daily compounding	10%	10.516%
Continuous compounding	10%	10.517%

 Now work Problem 21.

E X A M P L E 4   Computing the Value of an IRA

On January 2, 1998, $2000 is placed in an Individual Retirement Account (IRA) that will pay interest of 10% per annum compounded continuously. What will the IRA be worth on January 1, 2018?

Solution   The amount A after 20 years is

$$A = Pe^{rt} = \$2000e^{(0.10)(20)} = \$14{,}778.11$$

 Exploration   How long will it be until $A = \$40{,}000$?

[**Hint:** Graph $Y_1 = 2000e^{0.1x}$ and $Y_2 = 40{,}000$. Use INTERSECT to find x.]

FIGURE 24

Time is money

 When people engaged in finance speak of the "time value of money," they are usually referring to the **present value** of money. The present value of A dollars to be received at a future date is the principal you would need to invest now so that it would grow to A dollars in the specified time period. Thus, the present value of money to be received at a future date is always less than the amount to be received, since the amount to be received will equal the present value (money invested now) *plus* the interest accrued over the time period.

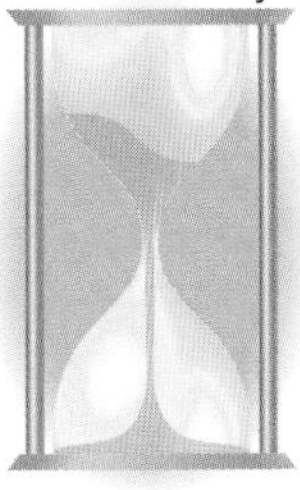

We use the compound interest formula (2) to get a formula for present value. If P is the present value of A dollars to be received after t years at a per annum interest rate r compounded n times per year, then, by formula (2),

$$A = P\left(1 + \frac{r}{n}\right)^{nt}$$

To solve for P, we divide both sides by $(1 + r/n)^{nt}$, and the result is

$$\frac{A}{(1 + \frac{r}{n})^{nt}} = P \quad \text{or} \quad P = A\left(1 + \frac{r}{n}\right)^{-nt}$$

Theorem   Present Value Formulas

The present value P of A dollars to be received after t years, assuming a per annum interest rate r compounded n times per year, is

$$P = A\left(1 + \frac{r}{n}\right)^{-nt} \tag{5}$$

If the interest is compounded continuously, then

$$P = Ae^{-rt} \qquad (6)$$

To prove (6), solve formula (4) for P.

E X A M P L E 5   Computing the Value of a Zero-Coupon Bond

A zero-coupon (noninterest-bearing) bond can be redeemed in 10 years for $1000. How much should you be willing to pay for it now if you want a return of

(a) 8% compounded monthly?

(b) 7% compounded continuously?

Solution   (a) We are seeking the present value of $1000. Thus, we use formula (5) with $A = \$1000$, $n = 12$, $r = 0.08$, and $t = 10$.

$$P = A\left(1 + \frac{r}{n}\right)^{-nt}$$

$$= \$1000\left(1 + \frac{0.08}{12}\right)^{-12(10)}$$

$$= \$450.52$$

For a return of 8% compounded monthly, you should pay $450.52 for the bond.

(b) Here, we use formula (6) with $A = \$1000$, $r = 0.07$, and $t = 10$:

$$P = Ae^{-rt}$$

$$= \$1000e^{-(0.07)(10)}$$

$$= \$496.59$$

For a return of 7% compounded continuously, you should pay $496.59 for the bond.

Now work Problem 11.

E X A M P L E 6   Rate of Interest Required to Double on Investment

What annual rate of interest compounded annually should you seek if you want to double your investment in 5 years?

Solution   If P is the principal and we want P to double, the amount A will be $2P$. We use the compound interest formula with $n = 1$ and $t = 5$ to find r.

$$2P = P(1 + r)^5$$
$$2 = (1 + r)^5$$
$$1 + r = \sqrt[5]{2}$$
$$r = \sqrt[5]{2} - 1 = 1.148698 - 1 = 0.148698$$

The annual rate of interest needed to double the principal in 5 years is 14.87%.

Now work Problem 23.

E X A M P L E   7   Doubling and Tripling Time for an Investment

(a) How long will it take for an investment to double in value if it earns 5% compounded continuously?

(b) How long will it take to triple at this rate?

Solution   (a) If P is the initial investment and we want P to double, the amount A will be $2P$. We use formula (4) for continuously compounded interest with $r = 0.05$. Then

$$A = Pe^{rt}$$
$$2P = Pe^{0.05t}$$
$$2 = e^{0.05t}$$
$$0.05t = \ln 2$$
$$t = \frac{\ln 2}{0.05} = 13.86$$

It will take about 14 years to double the investment.

(b) To triple the investment, we set $A = 3P$ in formula (4).

$$A = Pe^{rt}$$
$$3P = Pe^{0.05t}$$
$$3 = e^{0.05t}$$
$$0.05t = \ln 3$$
$$t = \frac{\ln 3}{0.05} = 21.97$$

It will take about 22 years to triple the investment.

 Now work Problem 29.

7.5 | EXERCISES

In Problems 1–10, find the amount that results from each investment.

1. $100 invested at 4% compounded quarterly after a period of 2 years

2. $50 invested at 6% compounded monthly after a period of 3 years

3. $500 invested at 8% compounded quarterly after a period of $2\frac{1}{2}$ years

4. $300 invested at 12% compounded monthly after a period of $1\frac{1}{2}$ years

5. $600 invested at 5% compounded daily after a period of 3 years

6. $700 invested at 6% compounded daily after a period of 2 years

7. $10 invested at 11% compounded continuously after a period of 2 years

8. $40 invested at 7% compounded continuously after a period of 3 years

9. $100 invested at 10% compounded continuously after a period of $2\frac{1}{4}$ years

10. $100 invested at 12% compounded continuously after a period of $3\frac{3}{4}$ years

In Problems 11–20, find the principal needed now to get each amount; that is, find the present value.

11. To get $100 after 2 years at 6% compounded monthly

12. To get $75 after 3 years at 8% compounded quarterly

13. To get $1000 after $2\frac{1}{2}$ years at 6% compounded daily

14. To get $800 after $3\frac{1}{2}$ years at 7% compounded monthly

15. To get $600 after 2 years at 4% compounded quarterly

16. To get $300 after 4 years at 3% compounded daily

17. To get $80 after $3\frac{1}{4}$ years at 9% compounded continuously

18. To get $800 after $2\frac{1}{2}$ years at 8% compounded continuously

19. To get $400 after 1 year at 10% compounded continuously

20. To get $1000 after 1 year at 12% compounded continuously

21. Find the effective rate of interest for $5\frac{1}{4}$% compounded quarterly

22. What interest rate compounded quarterly will give an effective interest rate of 7%?

23. What annual rate of interest is required to double an investment in 3 years?

24. What annual rate of interest is required to double an investment in 10 years?

In Problems 25–28, which of the two rates would yield the larger amount in 1 year?
[**Hint:** Start with a principal of $10,000 in each instance.]

25. 6% compounded quarterly or $6\frac{1}{4}$% compounded annually?

26. 9% compounded quarterly or $9\frac{1}{4}$% compounded annually?

27. 9% compounded monthly or 8.8% compounded daily?

28. 8% compounded semiannually or 7.9% compounded daily?

29. How long does it take for an investment to double in value if it is invested at 8% per annum compounded monthly? Compounded continuously?

30. How long does it take for an investment to double in value if it is invested at 10% per annum compounded monthly? Compounded continuously?

31. If Tanisha has $100 to invest at 8% per annum compounded monthly, how long will it be before she has $150? If the compounding is continuous, how long will it be?

32. If Angela has $100 to invest at 10% per annum compounded monthly, how long will it be before she has $175? If the compounding is continuous, how long will it be?

33. How many years will it take for an initial investment of $10,000 to grow to $25,000? Assume a rate of interest of 6% compounded continuously.

34. How many years will it take for an initial investment of $25,000 to grow to $80,000? Assume a rate of interest of 7% compounded continuously.

35. What will a $90,000 house cost 5 years from now if the inflation rate over that period averages 3% compounded annually?

36. Sears charges 1.25% per month on the unpaid balance for customers with charge accounts (in-terest is compounded monthly). A customer charges $200 and does not pay her bill for 6 months. What is the bill at that time?

37. Jerome will be buying a new car for $15,000 in 3 years. How much money should he ask his parents for now so that, if he invests it at 5% compounded continuously, he will have enough to buy the car?

38. John will require $3000 in 6 months to pay off a loan that has no prepayment privileges. If he has the $3000 now, how much of it should he save in an account paying 3% compounded monthly so that in 6 months he will have exactly $3000?

39. George is contemplating the purchase of 100 shares of a stock selling for $15 per share. The stock pays no dividends. The history of the stock indicates that it should grow at an annual rate of 15% per year. How much will the 100 shares of stock be worth in 5 years?

40. Tracy is contemplating the purchase of 100 shares of stock selling for $15 per share. The stock pays no dividends. Her broker says the stock will be worth $20 per share in 2 years. What is the annual rate of return on this investment?

41. A business purchased for $650,000 in 1995 is sold in 1998 for $850,000. What is the annual rate of return for this investment?

42. Tanya has just inherited a diamond ring appraised at $5000. If diamonds have appreciated in value at an annual rate of 8%, what was the value of the ring 10 years ago when the ring was purchased?

43. Jim places $1000 in a bank account that pays 5.6% compounded continuously. After 1 year, will he have enough money to buy a computer system that costs $1060? If another bank will pay Jim 5.9% compounded monthly, which is the better deal?

44. On January 1, Kim places $1000 in a Certificate of Deposit (CD) that pays 6.8% compounded continuously and matures in 3 months. Then Kim places the $1000 and the interest in a passbook account that pays 5.25% compounded monthly. How much does Kim have in the passbook account on May 1?

45. Will invests $2000 in a bond trust that pays 9% interest compounded semiannually. His friend Henry invests $2000 in a Certificate of Deposit that pays $8\frac{1}{2}$% compounded continuously. Who has more money after 20 years, Will or Henry?

46. Suppose that April has access to an investment that will pay 10% interest compounded continuously. Which is better: To be given $1000 now so that she can take advantage of this investment opportunity or to be given $1325 after 3 years?

47. Colleen and Bill have just purchased a house for $150,000, with the seller holding a second mortgage of $50,000. They promise to pay the seller $50,000 plus all accrued interest 5 years from now. The seller offers them three interest options on the second mortgage:

(a) Simple interest at 12% per annum
(b) $11\frac{1}{2}$% interest compounded monthly
(c) $11\frac{1}{4}$% interest compounded continuously

Which option is best; that is, which results in the least interest on the loan?

48. The First National Bank advertises that it pays interest on savings accounts at the rate of 4.25% compounded daily. Find the effective rate if the bank uses (a) 360 days or (b) 365 days in determining the daily rate.

Problems 49–52 involve zero-coupon bonds. A zero-coupon bond is a bond that is sold now at a discount and will pay its face value at some time in the future when it matures; no interest payments are made.

49. A zero-coupon bond can be redeemed in 20 years for $10,000. How much should you be willing to pay for it now if you want a return of:
(a) 10% compounded monthly?
(b) 10% compounded continuously?

50. A child's grandparents are considering buying a $40,000 face value zero-coupon bond at birth so that she will have enough money for her college education 17 years later. If thay want a rate of return of 8% compounded annually, what should they pay for the bond?

51. How much should a $10,000 face value zero-coupon bond, maturing in 10 years, be sold for now if its rate of return is to be 8% compounded annually?

52. If Pat pays $12,485.52 for a $25,000 face value zero-coupon bond that matures in 8 years, what is his annual rate of return?

53. Explain in your own words what the term *compound interest* means. What does *continuous compounding* mean?

54. Explain in your own words the meaning of present value.

55. Write a program that will calculate the amount after *n* years if a principal *P* is invested at *r*% per annum compounded quarterly. Use it to verify your answers to Problems 1 and 3.

56. Write a program that will calculate the principal needed now to get the amount *A* in *n* years at *r*% per annum compounded daily. Use it to verify your answer to Problem 13.

57. Write a program that will calculate the annual rate of interest required to double an investment in *n* years. Use it to verify your answers to Problems 23 and 24.

58. Write a program that will calculate the number of years required for an initial investment of *x* dollars to grow to *y* dollars at *r*% per annum compounded continuously. Use it to verify your answer to Problems 33 and 34.

59. **Time to Double or Triple an Investment** The formula

$$y = \frac{\ln m}{n \ln\left(1 + \dfrac{r}{n}\right)}$$

can be used to find the number of years *y* required to multiply an investment *m* times when *r* is the per annum interest rate compounded *n* times a year.

(a) How many years will it take to double the value of an IRA that compounds annually at the rate of 12%?
(b) How many years will it take to triple the value of a savings account that compounds quarterly at an annual rate of 6%?
(c) Give a derivation of this formula.

60.   Time to Reach an Investment Goal   The formula

$$y = \frac{\ln A - \ln P}{r}$$

can be used to find the number of years y required for an investment P to grow to a value A when compounded continuously at an annual rate r.

(a) How long will it take to increase an initial investment of $1000 to $8000 at an annual rate of 10%?

(b) What annual rate is required to increase the value of a $2000 IRA to $30,000 in 35 years?

(c) Give a derivation of this formula.

61.   Critical Thinking   You have just contracted to buy a house and will seek financing in the amount of $100,000. You go to several banks. Bank 1 will lend you $100,000 at the rate of 8.75% amortized over 30 years with a loan origination fee of 1.75%. Bank 2 will lend you $100,000 at the rate of 8.375% amortized over 15 years with a loan origination fee of 1.5%. Bank 3 will lend you $100,000 at the rate of 9.125% amortized over 30 years with no loan origination fee. Bank 4 will lend you $100,000 at the rate of 8.625% amortized over 15 years with no loan origination fee. Which loan would you take? Why? Be sure to have sound reasons for your choice. If the amount of the monthly payment does not matter to you, which loan would you take? Again, have sound reasons for your choice. Use the information in the table to assist you. Compare your final decision with others in the class. Discuss.

	Monthly Payment	Loan Origination Fee
Bank 1	$786.70	$1,750.00
Bank 2	$977.42	$1,500.00
Bank 3	$813.63	$0.00
Bank 4	$990.68	$0.00

7.6 | GROWTH AND DECAY

1 Find Equations of Populations That Obey the Law of Uninhibited Growth
2 Find Equations of Populations That Obey the Law of Decay
3 Use Newton's Law of Cooling
4 Use Logistic Growth Models

1 Many natural phenomena have been found to follow the law that an amount A varies with time t according to

$$A(t) = A_0 e^{kt} \tag{1}$$

where $A_0 = A(0)$ is the original amount ($t = 0$) and $k \neq 0$ is a constant.

If $k > 0$, then equation (1) states that the amount A is increasing over time; if $k < 0$, the amount A is decreasing over time. In either case, when an amount A varies over time according to equation (1), it is said to follow the **exponential law** or the **law of uninhibited growth** ($k > 0$) **or decay** ($k < 0$). See Figure 25.

FIGURE 25

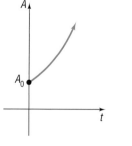

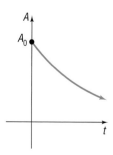

(a) $A(t) = A_0 e^{kt}, \; k > 0$

(b) $A(t) = A_0 e^{kt}, \; k < 0$

For example, we saw in Section 7.5 that continuously compounded interest follows the law of uninhibited growth. In this section we shall look at three additional phenomena that follow the exponential law.

Biology

Cell division is a universal process in the growth of living organisms such as amoebas, plants, human skin cells, and many others. Based on an ideal situation in which no cells die and no by-products are produced, the number of cells present at a given time follows the law of uninhibited growth. Actually, however, after enough time has passed, growth at an exponential rate will cease due to the influence of factors such as lack of living space, dwindling food supply, and so on. The law of uninhibited growth accurately reflects only the early stages of the cell division process.

The cell division process begins with a culture containing N_0 cells. Each cell in the culture grows for a certain period of time and then divides into two identical cells. We assume that the time needed for each cell to divide in two is constant and does not change as the number of cells increases. These cells then grow, and eventually each divides in two, and so on.

A model that gives the number N of cells in the culture after a time t has passed (in the early stages of growth) is

Uninhibited Growth of Cells

$$N(t) = N_0 e^{kt} \qquad k > 0 \qquad\qquad (2)$$

where k is a positive constant and N_0 is the initial number of cells.

In using equation (2) to model the growth of cells, we are using a function that yields positive real numbers, even though we are counting the number of cells, which must be an integer. This is a common practice in many applications.

E X A M P L E   1   Bacterial Growth

A colony of bacteria increases according to the law of uninhibited growth.

(a) If the number of bacteria doubles in 3 hours, find k in equation (2) and express N as a function of t.

(b) How long will it take for the size of the colony to triple?

(c) How long does it take for the population to double a second time (that is, increase four times).

Solution (a) Using formula (2), the number N of cells at a time t is

$$N(t) = N_0 e^{kt}$$

where N_0 is the initial number of bacteria present and k is a positive number. We seek the number k. The number of cells doubles in 3 hours; thus, we have

$$N(3) = 2N_0$$

But $N(3) = N_0 e^{k(3)}$, so

$$N_0 e^{k(3)} = 2N_0$$
$$e^{3k} = 2$$
$$3k = \ln 2 \qquad \text{Write the exponential equation as a logarithm.}$$
$$k = \tfrac{1}{3}\ln 2 \approx \tfrac{1}{3}(0.6931) = 0.2310$$

Formula (2) for this growth process is therefore

$$N(t) = N_0 e^{0.2310t}$$

(b) The time t needed for the size of the colony to triple requires that $N = 3N_0$. Thus, we substitute $3N_0$ for N to get

$$3N_0 = N_0 e^{0.2310t}$$
$$3 = e^{0.2310t}$$
$$0.2310t = \ln 3$$
$$t = \frac{\ln 3}{0.2310} \approx \frac{1.0986}{0.2310} = 4.756 \text{ hours}$$

It will take about 4.756 hours for the size of the colony to triple.

(c) If the number of bacteria doubles in 3 hours, it will double again in 6 hours. ∎

Now work Problem 1.

Radioactive Decay

Radioactive materials follow the law of uninhibited decay. Thus, the amount A of a radioactive material present at time t is given by the model

> **Uninhibited Radioactive Decay**
>
> $$A(t) = A_0 e^{kt} \quad k < 0 \tag{3}$$

where $A_0 = A(0)$ is the original amount of radioactive material and k is a negative number.

All radioactive substances have a specific **half-life,** which is the time required for half of the radioactive substance to decay. In **carbon-dating,** we use the fact that all living organisms contain two kinds of carbon, carbon-12 (a stable carbon) and carbon-14 (a radioactive carbon, with a half-life of 5600 years). While an organism is living, the ratio of carbon-12 to carbon-14 is constant. But when an organism dies, the original amount of carbon-12 present remains unchanged, whereas the amount of carbon-14 begins to decrease. This change in the amount of carbon-14 present relative to the amount of carbon-12 present makes it possible to calculate when an organism died.

E X A M P L E   2   Estimating the Age of Ancient Tools

Traces of burned wood found along with ancient stone tools in an archaeo-
logical dig in Chile. The wood was found to contain approximately 1.67% of
the original amount of carbon-14. If the half-life of carbon-14 is 5600 years,
approximately when was the tree cut and burned?

Solution   Using equation (3), the amount A of carbon-14 present at time t is

$$A(t) = A_0 e^{kt}$$

where A_0 is the original amount of carbon-14 present and k is a negative
number. We first seek the number k. To find it, we use the fact that after 5600
years half of the original amount of carbon-14 remains. Thus,

$$\frac{1}{2} A_0 = A_0 e^{k(5600)}$$

$$\frac{1}{2} = e^{5600k}$$

$$5600k = \ln \frac{1}{2}$$

$$k = \frac{1}{5600} \ln \frac{1}{2} \approx -0.000124$$

Formula (3) therefore becomes

$$A(t) = A_0 e^{-0.000124t}$$

If the amount A of carbon-14 now present is 1.67% of the original amount,
it follows that

$$0.0167 A_0 = A_0 e^{-0.000124t}$$
$$0.0167 = e^{-0.000124t}$$
$$-0.000124t = \ln 0.0167$$

$$t = \frac{1}{-0.000124} \ln 0.0167 \approx 33,000 \text{ years}$$

The tree was cut and burned about 33,000 years ago. Some archaeologists
use this conclusion to argue that humans lived in the Americas 33,000 years
ago, much earlier than is generally accepted.   ▬

    Now work Problem 3.

Newton's Law of Cooling

Newton's Law of Cooling* states that the temperature of a heated object decreases exponentially over time toward the temperature of the surrounding
medium. That is, the temperature u of a heated object at a given time t can
be modeled by the function

*Named after Sir Isaac Newton (1642–1727), one of the cofounders of calculus.

Newton's Law of Cooling

$$u(t) = T + (u_0 - T)e^{kt} \qquad k < 0 \tag{4}$$

where T is the constant temperature of the surrounding medium, u_0 is the initial temperature of the heated object, and k is a negative constant.

EXAMPLE 3

Using Newton's Law of Cooling

An object is heated to 100°C and is then allowed to cool in a room whose air temperature is 30°C.

(a) If the temperature of the object is 80°C after 5 minutes, when will its temperature be 50°C?

(b) Determine the elapsed time before the object is 35°C.

(c) What do you notice about $u(t)$, the temperature, as t, time, increases?

Solution (a) Using equation (4) with $T = 30$ and $u_0 = 100$, the temperature (in degrees Celsius) of the object at time t (in minutes) is

$$u(t) = 30 + (100 - 30)e^{kt} = 30 + 70e^{kt} \tag{5}$$

where k is a negative constant. To find k, we use the fact that $u = 80$ when $t = 5$. Then

$$80 = 30 + 70e^{k(5)}$$
$$50 = 70e^{5k}$$
$$e^{5k} = \frac{50}{70}$$
$$5k = \ln\frac{5}{7}$$
$$k = \frac{1}{5}\ln\frac{5}{7} \approx -0.0673$$

Formula (5) therefore becomes

$$u(t) = 30 + 70e^{-0.0673t} \tag{6}$$

Now, we want to find t when $u = 50$°C, so

$$50 = 30 + 70e^{-0.0673t}$$
$$20 = 70e^{-0.0673t}$$
$$e^{-0.0673t} = \frac{20}{70}$$
$$-0.0673t = \ln\frac{2}{7}$$
$$t = \frac{1}{-0.0673}\ln\frac{2}{7} \approx 18.6 \text{ minutes}$$

Thus, the temperature of the object will be 50°C after about 18.6 minutes.

(b) If $u = 35°C$, then based on equation (6), we have

$$35 = 30 + 70e^{-0.0673t}$$
$$5 = 70e^{-0.0673t}$$
$$e^{-0.0673t} = \frac{5}{70}$$
$$-0.0673t = \ln \frac{5}{70}$$
$$t = \frac{\ln \frac{5}{70}}{-0.0673} = 39.2 \text{ minutes}$$

The object will reach a temperature of 35°C after about 39.2 minutes.

(c) Refer to equation (6). As t increases, the value of $e^{-0.0673t}$ approaches zero, so the value of $u(t)$ approaches 30°C. ∎

Logistic Models

4 The exponential growth model $A(t) = A_0 e^{kt}$, $k > 0$ assumes uninhibited growth, meaning that the value of the function grows without limit. Recall that we stated that cell division could be modeled using this function, assuming that no cells die and no by-products are produced. However, cell division would eventually be limited by factors such as living space, food supply, and so forth. The **logistic growth model** is an exponential function that can model situations where the growth of the dependent variable is limited. The logistic growth model is given by the function

> ### Logistic Growth Model
>
> $$P(t) = \frac{c}{1 + ae^{-bt}}$$
>
> where a, b, and c are constants with $c > 0$ and $b > 0$.

The number c is called the **carrying capacity** because the value $P(t)$ approaches c as t approaches infinity; that is, $\lim_{t \to \infty} P(t) = c$. Thus, c represents the maximum value that the function can attain.

EXAMPLE 4   Fruit Fly Population

Fruit flies are placed in a half-pint milk bottle with a banana (for food) and yeast plants (for food and to provide a stimulus to lay eggs). Suppose that the fruit fly population after t days is given by

$$P(t) = \frac{230}{1 + 56.5e^{-0.37t}}$$

(a) What is the carrying capacity of the half-pint bottle? That is, what is $P(t)$ as $t \to \infty$?

(b) How many fruit flies were initially placed in the half-pint bottle?

(c) When will the population of fruit flies be 180?

(d) Using a graphing utility, graph $P(t)$.

Solution (a) As $t \to \infty$, $e^{-0.37t} \to 0$, and $P(t) \to 230/1$. The carrying capacity of the half-pint bottle is 230 fruit flies.

(b) To find the initial number of fruit flies in the half-pint bottle, we evaluate $P(0)$.

$$P(0) = \frac{230}{1 + 56.5e^{-0.37(0)}}$$

$$= \frac{230}{1 + 56.5}$$

$$= 4$$

Initially, there were four fruit flies in the half-pint bottle.

(c) To determine when the population of fruit flies will be 180, we solve the equation

$$\frac{230}{1 + 56.5e^{-0.37t}} = 180$$

$$230 = 180(1 + 56.5e^{-0.37t})$$

$$1.2778 = 1 + 56.5e^{-0.37t} \qquad \text{Divide each side by 180}$$

$$0.2778 = 56.5e^{-0.37t}$$

$$0.0049 = e^{-0.37t} \qquad \text{Divide each side by 56.5}$$

$$\ln(0.0049) = -0.37t$$

$$t \approx 14.4 \text{ days}$$

It will take approximately 14.4 days for the population to reach 180 fruit flies.

(d) See Figure 26.

FIGURE 26

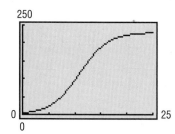

250

0

0   25

Exploration On the same viewing rectangle, graph $Y_1 = \dfrac{500}{1 + 24e^{-0.03t}}$ and $Y_2 = \dfrac{500}{1 + 24e^{-0.08t}}$. What effect does b have on the logistic growth function?

7.6 EXERCISES

1. **Growth of an Insect Population** The size P of a certain insect population at time t (in days) obeys the equation $P = 500e^{0.02t}$.
    (a) After how many days will the population reach 1000?
    (b) 2000?

2. **Growth of Bacteria** The number N of bacteria present in a culture at time t (in hours) obeys the equation $N = 1000e^{0.01t}$.
    (a) After how many hours will the population equal 1500?
    (b) 2000?

3. **Radioactive Decay** Strontium-90 is a radioactive material that decays according to the equation $A = A_0 e^{-0.0244t}$, where A_0 is the initial

amount present and A is the amount present at time t (in years).
    (a) What is the half-life of strontium-90?
    (b) Determine how long it takes for 100 grams of strontium-90 to decay to 10 grams.

4. **Radioactive Decay** Iodine-131 is a radioactive material that decays according to the equation $A = A_0 e^{-0.087t}$, where A_0 is the initial amount present and A is the amount present at time t (in days).
    (a) What is the half-life of iodine-131?
    (b) How long does it take for 100 grams of iodine-131 to decay to 10 grams?

5. **Growth of a Colony of Mosquitoes** The population of a colony of mosquitoes obeys the law

of uninhibited growth. If there are 1000 mosquitoes initially and there are 1800 after 1 day, what is the size of the colony after 3 days? How long is it until there are 10,000 mosquitoes?

6. **Bacterial Growth**   A culture of bacteria obeys the law of uninhibited growth. If 500 bacteria are present initially and there are 800 after 1 hour, how many will be present in the culture after 5 hours? How long is it until there are 20,000 bacteria?

7. **Population Growth**   The population of a southern city follows an exponential model. If the population doubled in size over an 18-month period and the current population is 10,000, what will the population be 2 years from now?

8. **Population Growth**   The population of a midwestern city follows an exponential model. If the population decreased from 900,000 to 800,000 from 1995 to 1997, what will the population be in 1999?

9. **Radioactive Decay**   The half-life of radium is 1690 years. If 10 grams are present now, how much will be present in 50 years?

10. **Radioactive Decay**   The half-life of radioactive potassium is 1.3 billion years. If 10 grams is present now, how much will be present in 100 years? In 1000 years?

11. **Estimating the Age of a Tree**   A piece of charcoal is found to contain 30% of the carbon-14 that it originally had. When did the tree from which the charcoal came die? Use 5600 years as the half-life of carbon-14.

12. **Estimating the Age of a Fossil**   A fossilized leaf contains 70% of its normal amount of carbon-14. How old is the fossil?

13. **Cooling Time of a Pizza**   A pizza baked at 450°F is removed from the oven at 5:00 PM into a room that is a constant 70°F. After 5 minutes, the pizza is at 300°F.
    (a) At what time can you begin eating the pizza if you want its temperature to be 135°F?
    (b) Determine the time that needs to elapse before the pizza is 160°F?
    (c) What do you notice about the temperature as time passes?

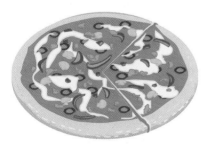

14. **Newton's Law of Cooling**   A thermometer reading 72°F is placed in a refrigerator where the temperature is a constant 38°F. If the thermometer reads 60°F after 2 minutes, what will it read after 7 minutes?
    (a) How long will it take before the thermometer reads 39°F?
    (b) Determine the time needed to elapse before the thermometer reads 45°F.
    (c) What do you notice about the temperature as time passes?

15. **Newton's Law of Cooling**   A thermometer reading 8°C is brought into a room with a constant temperature of 35°C. If the thermometer reads 15°C after 3 minutes, what will it read after being in the room for 5 minutes? For 10 minutes?

    [**Hint:** You need to construct a formula similar to equation (4).]

16. **Thawing Time of a Steak**   A frozen steak has a temperature of 28°F. It is placed in a room with a constant temperature of 70°F. After 10 minutes, the temperature of the steak has risen to 35°F. What will the temperature of the steak be after 30 minutes? How long will it take the steak to thaw to a temperature of 45°F? [See the hint given for Problem 15.]

17. **Decomposition of Salt in Water**   Salt (NaCl) decomposes in water into sodium (Na^+) and chloride (Cl^-) ions according to the law of uninhibited decay. If the initial amount of salt is 25 kilograms and, after 10 hours, 15 kilograms of salt is left, how much salt is left after 1 day? How long does it take until $\frac{1}{2}$ kilogram of salt is left?

18. **Voltage of a Conductor**   The voltage of a certain conductor decreases over time according to the law of uninhibited decay. If the initial voltage is 40 volts and 2 seconds later it is 10 volts, what is the voltage after 5 seconds?

19. **Radioactivity from Chernobyl**   After the release of radioactive material into the atmosphere from a nuclear power plant at Chernobyl (Ukraine) in 1986, the hay in Austria was contaminated by iodine-131 (half-life of 8 days). If it is all right to feed the hay to cows when 10% of the iodine-131 remains, how long do the farmers need to wait to use this hay?

20. **Pig Roasts**   The hotel Bora-Bora is having a pig roast. At noon, the chef put the pig in a large earthen oven. The pig's original temperature was 75°F. At 2:00 PM, the chef checked the pig's temperature and was upset because it had reached only 100°F. If the oven's temperature remains a constant 325°F, at what time may the

hotel serve its guests, assuming that pork is done when it reaches 175°F?

21. Proportion of the Population That Owns a VCR The logistic growth model

$$P(t) = \frac{0.9}{1 + 6e^{-0.32t}}$$

relates the proportion of U.S. households that own a VCR to the year. Let $t = 0$ represent 1984, $t = 1$ represent 1985, and so on.

(a) What proportion of U.S. households owned a VCR in 1984?

(b) Determine the maximum proportion of households that will own a VCR.

(c) When will 0.8 (80%) of U.S. households own a VCR?

22. Market Penetration of Intel's Coprocessor The logistic growth model

$$P(t) = \frac{0.90}{1 + 3.5e^{-0.339t}}$$

relates the proportion of new personal computers sold at Best Buy that have Intel's latest coprocessor t months after it has been introduced.

(a) What proportion of new personal computers sold at Best Buy will have Intel's latest coprocessor when it is first introduced (that is, at $t = 0$)?

(b) Determine the maximum proportion of new personal computers sold at Best Buy that will have Intel's latest coprocessor.

(c) When will 0.75 (75%) of new personal computers sold at Best Buy have Intel's latest coprocessor?

23. Population of a Bacteria Culture The logistic growth model

$$P(t) = \frac{1000}{1 + 32.33e^{-0.439t}}$$

represents the population of a bacteria after t hours.

(a) What is the carrying capacity of the environment?

(b) What was the initial amount of bacteria in the population?

(c) When will the amount of bacteria be 800?

24. Population of an Endangered Species Often environmentalists will capture an endangered species and transport the species to a controlled environment where the species can produce offspring and regenerate its population. Suppose that six American Bald Eagles are captured and transported to Montana and set free. Based on experience, the environmentalists expect the population to grow according to the model

$$P(t) = \frac{500}{1 + 83.33e^{-0.162t}}$$

(a) What is the carrying capacity of the environment?

(b) What is the predicted population of the American Bald Eagle in 20 years?

(c) When will the population be 300?

7.7 | LOGARITHMIC SCALES

1 Decibels (Loudness of Sound)
2 Richter Scale (Earthquake Magnitude)

Common logarithms often appear in the measurement of quantities, because they provide a way to scale down positive numbers that vary from very small to very large. For example, if a certain quantity can take on values from $0.0000000001 = 10^{-10}$ to $10,000,000,000 = 10^{10}$, the common logarithms of such numbers would be between -10 and 10, respectively.

Loudness of Sound

1 Our first application utilizes a logarithmic scale to measure the loudness of a sound. Physicists define the **intensity of a sound wave** as the amount of energy that the wave transmits through a given area. For example, the least intense sound that a human ear can detect at a frequency of 100 hertz is about 10^{-12} watt per square meter. The **loudness** $L(x)$, measured in **decibels** (named in honor of Alexander Graham Bell), of a sound of intensity x (measured in watts per square meter) is defined as

$$L(x) = 10 \log \frac{x}{I_0} \qquad (1)$$

where $I_0 = 10^{-12}$ watt per square meter is the least intense sound that a human ear can detect. If we let $x = I_0$ in equation (1), we get

$$L(I_0) = 10 \log \frac{I_0}{I_0} = 10 \log 1 = 0$$

Thus, at the threshold of human hearing, the loudness is 0 decibels. Figure 27 gives the loudness of some common sounds.

FIGURE 27
Loudness of common sounds (in decibels)

Decibels

140	Shotgun blast, jet 100 feet away at takeoff	Pain
130	Motor test chamber	Human ear pain threshold
120	Firecrackers, severe thunder, pneumatic jackhammer, hockey crowd	Uncomfortably loud
110	Amplified rock music	
100	Textile loom, subway train, elevated train, farm tractor, power lawn mower, newspaper press	Loud
90	Heavy city traffic, noisy factory	
80	Diesel truck going 40 mi/hr 50 feet away, crowded restaurant, garbage disposal, average factory, vacuum cleaner	Moderately loud
70	Passenger car going 50 mi/hr 50 feet away	
60	Quiet typewriter, singing birds, window air conditioner, quiet automobile	Quiet
50	Normal conversation, average office	
40	Household refrigerator, quiet office	Very quiet
30	Average home, dripping faucet, whisper 5 feet away	
20	Light rainfall, rustle of leaves	Average person's threshold of hearing
10	Whisper across room	Just audible
0		Threshold for acute hearing

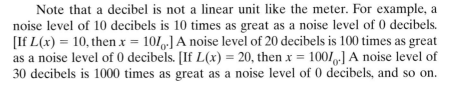

Note that a decibel is not a linear unit like the meter. For example, a noise level of 10 decibels is 10 times as great as a noise level of 0 decibels. [If $L(x) = 10$, then $x = 10I_0$.] A noise level of 20 decibels is 100 times as great as a noise level of 0 decibels. [If $L(x) = 20$, then $x = 100I_0$.] A noise level of 30 decibels is 1000 times as great as a noise level of 0 decibels, and so on.

EXAMPLE 1   Finding the Intensity of a Sound

Use Figure 27 to find the intensity of the sound of a dripping faucet.

Solution   From Figure 27, we see that the loudness of the sound of a dripping faucet is 30 decibels. Thus, by equation (1), its intensity x may be found as follows:

$$30 = 10 \log \left(\frac{x}{I_0}\right)$$

$$3 = \log \left(\frac{x}{I_0}\right) \qquad \text{Divide by 10.}$$

$$\frac{x}{I_0} = 10^3 \qquad \text{Write in exponential form.}$$

$$x = 1000 I_0$$

where $I_0 = 10^{-12}$ watt per square meter. Thus, the intensity of the sound of a dripping faucet is 1000 times as great as a noise level of 0 decibels; that is, such a sound has an intensity of $1000 \cdot 10^{-12} = 10^{-9}$ watt per square meter. ■

Now work Problem 5.

E X A M P L E   2   Finding the Loudness of a Sound

Use Figure 27 to determine the loudness of a subway train if it is known that this sound is 10 times as intense as the sound due to heavy city traffic.

Solution   The sound due to heavy city traffic has a loudness of 90 decibels. Its intensity, therefore, is the value of x in the equation

$$90 = 10 \log \left(\frac{x}{I_0}\right)$$

A sound 10 times as intense as x has loudness $L(10x)$. Thus, the loudness of the subway train is

$$L(10x) = 10 \log \left(\frac{10x}{I_0}\right) \qquad \text{Substitute 10x for x.}$$

$$= 10 \log \left(10 \cdot \frac{x}{I_0}\right)$$

$$= 10 \left[\log 10 + \log \left(\frac{x}{I_0}\right)\right] \qquad \text{Log of product = sum of logs}$$

$$= 10 \log 10 + 10 \log \left(\frac{x}{I_0}\right) \qquad \text{log 10 = 1; 90 = 10 log (x/I}_0\text{)}$$

$$= 10 + 90 = 100 \text{ decibels} \qquad ■$$

Magnitude of an Earthquake

Our second application uses a logarithmic scale to measure the magnitude of an earthquake.

The **Richter scale*** is one way of converting seismographic readings into numbers that provide an easy reference for measuring the magnitude M of an earthquake. All earthquakes are compared to a **zero-level earthquake** whose seismographic reading measures 0.001 millimeter at a distance of 100 kilometers from the epicenter. An earthquake whose seismographic reading measures x millimeters has **magnitude** $M(x)$, given by

*Named after the American scientist, C. F. Richter, who devised it in 1935.

$$M(x) = \log\left(\frac{x}{x_0}\right) \qquad (2)$$

where $x_0 = 10^{-3}$ is the reading of a zero-level earthquake the same distance from its epicenter.

E X A M P L E   3   Finding the Magnitude of an Earthquake

What is the magnitude of an earthquake whose seismographic reading is 0.1 millimeter at a distance of 100 kilometers from its epicenter?

Solution   If $x = 0.1$, the magnitude $M(x)$ of this earthquake is

$$M(0.1) = \log\left(\frac{x}{x_0}\right) = \log\left(\frac{0.1}{0.001}\right) = \log\left(\frac{10^{-1}}{10^{-3}}\right) = \log 10^2 = 2$$

This earthquake thus measures 2.0 on the Richter scale.   ▬

 Now work Problem 7.

Based on formula (2), we define the **intensity of an earthquake** as the ratio of x to x_0. For example, the intensity of the earthquake described in Example 3 is $\frac{0.1}{0.001} = 10^2 = 100$. That is, it is 100 times as intense as a zero-level earthquake.

E X A M P L E   4   Comparing the Intensity of Two Earthquakes

The devastating San Francisco earthquake of 1906 measured 6.9 on the Richter scale. How did the intensity of that earthquake compare to the Papua, New Guinea, earthquake of 1988, which measured 6.7 on the Richter scale?

FIGURE 28

Solution   Let x_1 and x_2 denote the seismographic readings, respectively, of the 1906 San Francisco earthquake and the 1988 Papua, New Guinea, earthquake. Then, based on formula (2),

$$6.9 = \log\left(\frac{x_1}{x_0}\right) \qquad 6.7 = \log\left(\frac{x_2}{x_0}\right)$$

Consequently,

$$\frac{x_1}{x_0} = 10^{6.9} \qquad \frac{x_2}{x_0} = 10^{6.7}$$

The 1906 San Francisco earthquake was $10^{6.9}$ times as intense as a zero-level earthquake. The Papua, New Guinea, earthquake was $10^{6.7}$ times as intense as a zero-level earthquake. Thus,

$$\frac{x_1}{x_2} = \frac{10^{6.9}x_0}{10^{6.7}x_0} = 10^{0.2} \approx 1.58$$
$$x_1 \approx 1.58x_2$$

Hence, the San Francisco earthquake was 1.58 times as intense as the Papua, New Guinea, earthquake. ▬

Example 4 demonstrates that the relative intensity of two earthquakes can be found by raising 10 to a power equal to the difference of their readings on the Richter scale.

7.7 | EXERCISES

1. **Loudness of a Dishwasher** Find the loudness of a dishwasher that operates at an intensity of 10^{-5} watt per square meter. Express your answer in decibels.

2. **Loudness of a Diesel Engine** Find the loudness of a diesel engine that operates at an intensity of 10^{-3} watt per square meter. Express your answer in decibels.

3. **Loudness of a Jet Engine** With engines at full throttle, a Boeing 727 jetliner produces noise at an intensity of 0.15 watt per square meter. Find the loudness of the engines in decibels.

4. **Loudness of a Whisper** A whisper produces noise at an intensity of $10^{-9.8}$ watt per square meter. What is the loudness of a whisper in decibels?

5. **Intensity of a Sound at the Threshold of Pain** For humans, the threshold of pain due to sound averages 130 decibels. What is the intensity of such a sound in watts per square meter?

6. **Comparing Sounds** If one sound is 50 times as intense as another, what is the difference in the loudness of the two sounds? Express your answer in decibels.

7. **Magnitude of an Earthquake** Find the magnitude of an earthquake whose seismographic reading is 10.0 millimeters at a distance of 100 kilometers from its epicenter.

8. **Magnitude of an Earthquake** Find the magnitude of an earthquake whose seismographic reading is 1210 millimeters at a distance of 100 kilometers from its epicenter.

9. **Comparing Earthquakes** The Mexico City earthquake of 1985 registered 8.1 on the Richter scale. What would a seismograph 100 kilometers from the epicenter have measured for this earthquake? How does this earthquake compare in intensity to the 1906 San Francisco earthquake, which registered 6.9 on the Richter scale?

10. **Comparing Earthquakes** Two earthquakes differ by 1.0 when measured on the Richter scale. How would the seismographic readings differ at a distance of 100 kilometers from the epicenter? How do their intensities compare?

11. **NBA Finals 1997** In game 5 of the NBA Finals between the Chicago Bulls and the Utah Jazz at the Delta Center, the crowd noise measured 110 decibels. NBA guidelines say sound levels are not to exceed 95 decibels. Compute the ratio of the intensities of these two sounds to determine by how much the crowd noise exceeded guidelines.

CHAPTER REVIEW

THINGS TO KNOW

Properties of the exponential function

$f(x) = a^x, \quad a > 1$ Domain: $(-\infty, \infty)$; Range: $(0, \infty)$; x-intercepts: none; y-intercept: 1; horizontal asymptote: x-axis as $x \to -\infty$; increasing; one-to-one

See Figure 1 for a typical graph.

$f(x) = a^x, \quad 0 < a < 1$ Domain: $(-\infty, \infty)$; Range: $(0, \infty)$; x-intercepts: none; y-intercept: 1; horizontal asymptote: x-axis as $x \to \infty$; decreasing; one-to-one

See Figure 4 for a typical graph.

Properties of the logarithmic function

$f(x) = \log_a x, \quad a > 1$
$(y = \log_a x \text{ means } x = a^y)$ Domain: $(0, \infty)$; Range: $(-\infty, \infty)$; x-intercept: 1; y-intercept: none; vertical asymptote: y-axis; increasing; one-to-one

See Figure 10(b) for a typical graph.

$f(x) = \log_a x, \quad 0 < a < 1$
$(y = \log_a x \text{ means } x = a^y)$ Domain: $(0, \infty)$; Range: $(-\infty, \infty)$; x-intercept: 1; y-intercept: none; vertical asymptote: y-axis; decreasing; one-to-one

See Figure 10(a) for a typical graph.

Number e

Value approached by the expression $\left(1 + \dfrac{1}{n}\right)^n$ as $n \to \infty$; that is, $\lim\limits_{n \to \infty} \left(1 + \dfrac{1}{n}\right)^n = e$

Natural logarithm

$y = \ln x$ means $x = e^y$

Properties of logarithms

$\log_a 1 = 0 \qquad \log_a a = 1 \qquad a^{\log_a M} = M \qquad \log_a a^r = r$

$\log_a (MN) = \log_a M + \log_a N \qquad \log_a \left(\dfrac{M}{N}\right) = \log_a M - \log_a N \qquad \log_a \left(\dfrac{1}{N}\right) = -\log_a N \qquad \log_a M^r = r \log_a M$

FORMULAS

Change-of-Base Formula $\log_a M = \dfrac{\log_b M}{\log_b a}$

Compound interest $A = P\left(1 + \dfrac{r}{n}\right)^{nt}$

Continuous compounding $A = Pe^{rt}$

Present value $P = A\left(1 + \dfrac{r}{n}\right)^{-nt}$ or $P = Ae^{-rt}$

Growth and decay $A(t) = A_0 e^{kt}$

HOW TO

Graph exponential and logarithmic functions.

Solve certain exponential equations.

Solve certain logarithmic equations.

Solve problems involving compound interest.

Solve problems involving growth and decay.

Solve problems involving intensity of sound and intensity of earthquakes.

FILL-IN-THE-BLANK ITEMS

1. The graph of every exponential function $f(x) = a^x$, $a > 0$, $a \neq 1$, passes through the two points _____.

2. If the graph of an exponential function $f(x) = a^x$, $a > 0$, $a \neq 1$, is decreasing, then its base must be less than _____.

3. If $3^x = 3^4$, then $x =$ _____.

4. The logarithm of a product equals the _____ of the logarithms.

5. For every base, the logarithm of _____ equals 0.

6. If $\log_8 M = \log_5 7/\log_5 8$, then $M =$ _____.

7. The domain of the logarithmic function $f(x) = \log_a x$ consists of _____.

8. The graph of every logarithmic function $f(x) = \log_a x$, $a > 0$, $a \neq 1$, passes through the two points _____.

9. If the graph of a logarithmic function $f(x) = \log_a x$, $a > 0$, $a \neq 1$, is increasing, then its base must be larger than _____.

10. If $\log_3 x = \log_3 7$, then $x =$ _____.

TRUE/FALSE ITEMS

T F   1. The graph of every exponential function $f(x) = a^x$, $a > 0$, $a \neq 1$, will contain the points $(0, 1)$ and $(1, a)$.

T F   2. The graphs of $y = 3^{-x}$ and $y = \left(\frac{1}{3}\right)^x$ are identical.

T F   3. The present value of $1000 to be received after 2 years at 10% per annum compounded continuously is approximately $1205.

T F   4. If $y = \log_a x$, then $y = a^x$.

T F   5. The graph of every logarithmic function $f(x) = \log_a x$, $a > 0$, $a \neq 1$, will contain the points $(1, 0)$ and $(a, 1)$.

T F   6. $a^{\log_M a} = M$, where $a > 0$, $a \neq 1$, $M > 0$.

T F   7. $\log_a (M + N) = \log_a M + \log_a N$, where $a > 0$, $a \neq 1$, $M > 0$, $N > 0$.

T F   8. $\log_a M - \log_a N = \log_a(M/N)$, where $a > 0$, $a \neq 1$, $M > 0$, $N > 0$.

REVIEW EXERCISES

Blue problem numbers indicate the author's suggestions for use in a Practice Test.

In Problems 1–6, evaluate each expression. Do not use a calculator.

1. $\log_2 \left(\frac{1}{8}\right)$ 2. $\log_3 81$ 3. $\ln e^{\sqrt{2}}$ 4. $e^{\ln 0.1}$ 5. $2^{\log_2 0.4}$ 6. $\log_2 2^{\sqrt{3}}$

In Problems 7–12, write each expression as a single logarithm.

7. $3 \log_4 x^2 + \frac{1}{2} \log_4 \sqrt{x}$

8. $-2 \log_3 \left(\frac{1}{x}\right) + \frac{1}{3} \log_3 \sqrt{x}$

9. $\ln \left(\frac{x - 1}{x}\right) + \ln \left(\frac{x}{x + 1}\right) - \ln(x^2 - 1)$

10. $\log (x^2 - 9) - \log (x^2 + 7x + 12)$

11. $2 \log 2 + 3 \log x - \frac{1}{2}[\log(x + 3) + \log(x - 2)]$

12. $\frac{1}{2} \ln(x^2 + 1) - 4 \ln \frac{1}{2} - \frac{1}{2}[\ln(x - 4) + \ln x]$

In Problems 13–20, find y as a function of x. The constant C is a positive number.

13. $\ln y = 2x^2 + \ln C$

14. $\ln(y - 3) = \ln 2x^2 + \ln C$

15. $\frac{1}{2}\ln y = 3x^2 + \ln C$

16. $\ln 2y = \ln(x + 1) + \ln(x + 2) + \ln C$

17. $\ln(y - 3) + \ln(y + 3) = x + C$

18. $\ln(y - 1) + \ln(y + 1) = -x + C$

19. $e^{y + C} = x^2 + 4$

20. $e^{3y - C} = (x + 4)^2$

In Problems 21–30, use transformations to graph each function. Determine the domain, range, and any asymptotes.

21. $f(x) = e^{-x}$

22. $f(x) = \ln(-x)$

23. $f(x) = 1 - e^x$

24. $f(x) = 3 + \ln x$

25. $f(x) = 3e^x$

26. $f(x) = \frac{1}{2}\ln x$

27. $f(x) = e^{|x|}$

28. $f(x) = \ln|x|$

29. $f(x) = 3 - e^{-x}$

30. $f(x) = 4 - \ln(-x)$

In Problems 31–50, solve each equation.

31. $4^{1-2x} = 2$

32. $8^{6+3x} = 4$

33. $3^{x^2+x} = \sqrt{3}$

34. $4^{x-x^2} = \frac{1}{2}$

35. $\log_x 64 = -3$

36. $\log_{\sqrt{2}} x = -6$

37. $5^x = 3^{x+2}$

38. $5^{x+2} = 7^{x-2}$

39. $9^{2x} = 27^{3x-4}$

40. $25^{2x} = 5^{x^2-12}$

41. $\log_3 \sqrt{x - 2} = 2$

42. $2^{x+1} \cdot 8^{-x} = 4$

43. $8 = 4^{x^2} \cdot 2^{5x}$

44. $2^x \cdot 5 = 10^x$

45. $\log_6 (x + 3) + \log_6 (x + 4) = 1$

46. $\log_{10} (7x - 12) = 2 \log_{10} x$

47. $e^{1-x} = 5$

48. $e^{1-2x} = 4$

49. $2^{3x} = 3^{2x+1}$

50. $2^{x^3} = 3^{x^2}$

In Problems 51–54, use the following result: If x is the atmospheric pressure (measured in millimeters of mercury), then the formula for the altitude h(x) (measured in meters above sea level) is

$$h(x) = (30T + 8000)\log\left(\frac{P_0}{x}\right)$$

where T is the temperature (in degrees Celsius) and P_0 is the atmospheric pressure at sea level, which is approximately 760 millimeters of mercury.

51. **Finding the Altitude of an Airplane** At what height is a Piper Cub whose instruments record an outside temperature of 0°C and a barometric pressure of 300 millimeters of mercury?

52. **Finding the Height of a Mountain** How high is a mountain if instruments placed on its peak record a temperature of 5°C and a barometric pressure of 500 millimeters of mercury?

53. **Atmospheric Pressure Outside an Airplane** What is the atmospheric pressure outside a Boeing 737 flying at an altitude of 10,000 meters if the outside air temperature is −100°C?

54. **Atmospheric Pressure at High Altitudes** What is the atmospheric pressure (in millimeters of mercury) on Mt. Everest, which has an altitude of approximately 8900 meters, if the air temperature is 5°C?

55. **Amplifying Sound** An amplifier's power output P (in watts) is related to its decibel voltage gain d by the formula $P = 25e^{0.1d}$.

(a) Find the power output for a decibel voltage gain of 4 decibels.

(b) For a power output of 50 watts, what is the decibel voltage gain?

56. **Limiting Magnitude of a Telescope** A telescope is limited in its usefulness by the brightness of the star it is aimed at and by the diameter of its lens. One measure of a star's brightness is its *magnitude:* the dimmer the star, the larger its magnitude. A formula for the limiting magnitude L of a telescope, that is, the magnitude of the dimmest star that it can be used to view, is given by

$$L = 9 + 5.1 \log d$$

where d is the diameter (in inches) of the lens.

(a) What is the limiting magnitude of a 3.5-inch telescope?

(b) What diameter is required to view a star of magnitude 14?

57. **Product Demand** The demand for a new product increases rapidly at first and then levels off. The percent P of actual purchases of this product after it has been on the market t months is

$$P = 90 - 80\left(\frac{3}{4}\right)^t$$

(a) What is the percent of purchases of the product after 5 months?
(b) What is the percent of purchases of the product after 10 months?
(c) What is the maximum percent of purchases of the product?
(d) How many months does it take before 40% of purchases occur?
(e) How many months before 70% of purchases occur?

58. Disseminating Information A survey of a certain community of 10,000 residents shows that the number of residents N who have heard a piece of information after m months is given by the formula

$$m = 55.3 - 6 \ln(10{,}000 - N)$$

How many months will it take for half of the citizens to learn about a community program of free blood pressure readings?

59. Salvage Value The number of years n for a piece of machinery to depreciate to a known salvage value can be found using the formula

$$n = \frac{\log s - \log i}{\log(1 - d)}$$

where s is the salvage value of the machinery, i is its initial value, and d is the annual rate of depreciation.
(a) How many years will it take for a piece of machinery to decline in value from $90,000 to $10,000 if the annual rate of depreciation is 0.20 (20%)?
(b) How many years will it take for a piece of machinery to lose half of its value if the annual rate of depreciation is 15%?

60. Funding a College Education A child's grandparents purchase a $10,000 bond fund that matures in 18 years to be used for her college education. The bond fund pays 4% interest compounded semiannually. How much will the bond fund be worth at maturity?

61. Funding a College Education A child's grandparents wish to purchase a bond fund that matures in 18 years to be used for her college education. The bond fund pays 4% interest compounded semiannually. How much should they purchase so that the bond fund will be worth $85,000 at maturity?

62. Funding an IRA First Colonial Bankshares Corporation advertised the following IRA investment plans.

Target IRA Plans

For each $5000 Maturity Value Desired	
Deposit:	At a Term of:
$620.17	20 Years
$1045.02	15 Years
$1760.92	10 Years
$2967.26	5 Years

(a) Assuming continuous compounding, what was the annual rate of interest they offered?
(b) First Colonial Bankshares claims that $4000 invested today will have a value of over $32,000 in 20 years. Use the answer found in part (a) to find the actual value of $4000 in 20 years. Assume continuous compounding.

63. Loudness of a Garbage Disposal Find the loudness of a garbage disposal unit that operates at an intensity of 10^{-4} watt per square meter. Express your answer in decibels.

64. Comparing Earthquakes On September 9, 1985, the western suburbs of Chicago experienced a mild earthquake that registered 3.0 on the Richter scale. How did this earthquake compare in intensity to the great San Francisco earthquake of 1906, which registered 6.9 on the Richter scale?

65. Estimating the Date on Which a Prehistoric Man Died The bones of a prehistoric man found in the desert of New Mexico contain approximately 5% of the original amount of carbon-14. If the half-life of carbon-14 is 5600 years, approximately how long ago did the man die?

66. Temperature of a Skillet A skillet is removed from an oven whose temperature is 450°F and placed in a room whose temperature is 70°F. After 5 minutes, the temperature of the skillet is 400°F. How long will it be until its temperature is 150°F?

67. Biology A certain bacteria initially increases according to the law of uninhibited growth. A biologist collects the following data for this bacteria:

Time (Hours)	Population
0	1000
1	1415
2	2000
3	2828
4	4000
5	5656
6	8000

(a) Draw a scatter diagram.
(b) The exponential function of best fit to the data is found to be
$$y = 1000(\sqrt{2})^x$$
Express the function of best fit in the form $N = N_0 e^{kt}$.
(c) Use the solution to (b) to predict the population at $t = 7$ hours.

(d) Use a graphing utility to verify the exponential function of best fit.
(e) Use a graphing utility to draw a scatter diagram of the data and then graph the exponential function of best fit on it.

68. Finance The following data represent the amount of money an investor has in an investment account each year for 10 years. She wishes to determine the effective rate of return on her investment.

Year	Value of Account
1985	$10,000
1986	$10,573
1987	$11,260
1988	$11,733
1989	$12,424
1990	$13,269
1991	$13,968
1992	$14,823
1993	$15,297
1994	$16,539

(a) Draw a scatter diagram with time as the independent variable and the value of the account as the dependent variable.
(b) The exponential function of best fit to the data is found to be
$$y = 10,014(1.057)^x$$
where x is the number of years since 1985. Express the function of best fit in the form $A = A_0 e^{kt}$.
(c) Use the solution to (b) to estimate the value of the account in the year 2020.
(d) Use a graphing utility to verify the exponential function of best fit.
(e) Use a graphing utility to draw a scatter diagram of the data and then graph the exponential function of best fit on it.

Algebra Topics

A.1 | LINES

1   Calculate and Interpret the Slope of a Line
2   Graph Lines
3   Find the Equation of Vertical Lines
4   Use the Point–Slope Form of a Line; Identify Horizontal Lines
5   Write the Equation of a Line from Two Points
6   Write the Equation of a Line in Slope–Intercept Form
7   Identify the Slope and y-Intercept of a Line from Its Equation
8   Write the Equation of a Line in General Form
9   Find Equations of Parallel Lines
10 Find Equations of Perpendicular Lines

In this section we study a certain type of equation that contains two variables, called a *linear equation,* and its graph, a *line.*

Slope of a Line

FIGURE 1

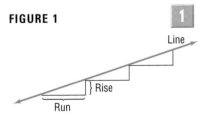

Line

} Rise

Run

1   Consider the staircase illustrated in Figure 1. Each step contains exactly the same horizontal **run** and the same vertical **rise.** The ratio of the rise to the run, called the *slope,* is a numerical measure of the steepness of the staircase. For example, if the run is increased and the rise remains the same, the staircase becomes less steep. If the run is kept the same, but the rise is increased, the staircase becomes more steep. This important characteristic of a line is best defined using rectangular coordinates.

Let $P = (x_1, y_1)$ and $Q = (x_2, y_2)$ be two distinct points with $x_1 \neq x_2$. The **slope m** of the nonvertical line L containing P and Q is defined by the formula

$$m = \frac{y_2 - y_1}{x_2 - x_1} \qquad x_1 \neq x_2 \qquad (1)$$

If $x_1 = x_2$, L is a **vertical line** and the slope m of L is **undefined** (since this results in division by 0).

Figure 2(a) provides an illustration of the slope of a nonvertical line; Figure 2(b) illustrates a vertical line.

FIGURE 2

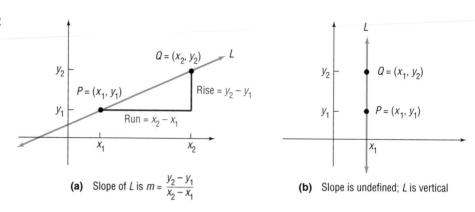

(a) Slope of L is $m = \dfrac{y_2 - y_1}{x_2 - x_1}$ **(b)** Slope is undefined; L is vertical

As Figure 2(a) illustrates, the slope m of a nonvertical line may be viewed as

$$m = \frac{y_2 - y_1}{x_2 - x_1} = \frac{\text{rise}}{\text{run}}$$

We can also express the slope m of a nonvertical line as

$$m = \frac{y_2 - y_1}{x_2 - x_1} = \frac{\text{change in } y}{\text{change in } x} = \frac{\Delta y}{\Delta x}$$

That is, the slope m of a nonvertical line L measures the amount y changes as x changes from x_1 to x_2. This is called the **average rate of change** of y with respect to x.

Two comments about computing the slope of a nonvertical line may prove helpful.

1. Any two distinct points on the line can be used to compute the slope of the line. (See Figure 3 for justification.)

FIGURE 3
Triangles *ABC* and *PQR* are similar (equal angles). Hence, ratios of corresponding sides are equal. Thus:

Slope using P and $Q = \dfrac{y_2 - y_1}{x_2 - x_1}$

$\qquad = $ slope using A and $B = \dfrac{d(B, C)}{d(A, C)}$

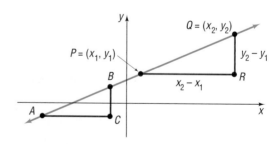

2. The slope of a line may be computed from $P = (x_1, y_1)$ to $Q = (x_2, y_2)$ or from Q to P because

$$\frac{y_2 - y_1}{x_2 - x_1} = \frac{y_1 - y_2}{x_1 - x_2}$$

EXAMPLE 1

Finding and Interpreting the Slope of a Line Joining Two Points

The slope m of the line joining the points $(1,2)$ and $(5,-3)$ may be computed as

$$m = \frac{-3-2}{5-1} = \frac{-5}{4} \quad \text{or as} \quad m = \frac{2-(-3)}{1-5} = \frac{5}{-4} = \frac{-5}{4}$$

For every 4-unit change in x, y will change by -5 units. That is, if x increases by 4 units, then y decreases by 5 units. The average rate of change of y with respect to x is $\frac{-5}{4}$.

Now work Problem 7.

To get a better idea of the meaning of the slope m of a line L, consider the following example.

EXAMPLE 2

Finding the Slopes of Various Lines Containing the Same Point (2, 3)

Compute the slopes of the lines L_1, L_2, L_3, and L_4 containing the following pairs of points. Graph all four lines on the same set of coordinate axes.

$$
\begin{array}{lll}
L_1\colon & P = (2,3) & Q_1 = (-1,-2) \\
L_2\colon & P = (2,3) & Q_2 = (3,-1) \\
L_3\colon & P = (2,3) & Q_3 = (5,3) \\
L_4\colon & P = (2,3) & Q_4 = (2,5)
\end{array}
$$

FIGURE 4

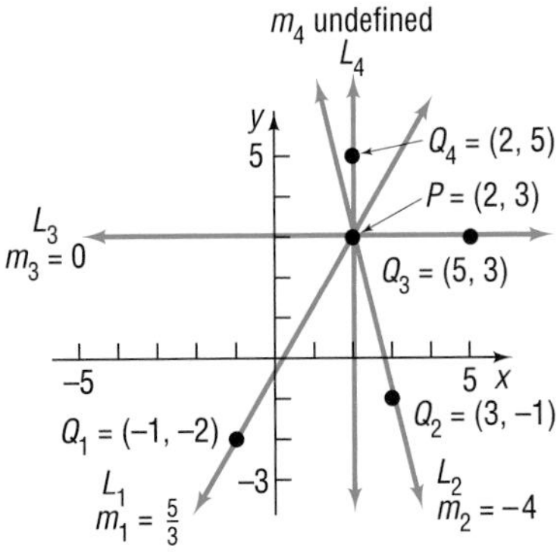

Solution

Let m_1, m_2, m_3, and m_4 denote the slopes of the lines L_1, L_2, L_3, and L_4, respectively. Then

$$m_1 = \frac{-2-3}{-1-2} = \frac{-5}{-3} = \frac{5}{3} \quad \text{A rise of 5 divided by a run of 3.}$$

$$m_2 = \frac{-1-3}{3-2} = \frac{-4}{1} = -4$$

$$m_3 = \frac{3-3}{5-2} = \frac{0}{3} = 0$$

m_4 is undefined

The graphs of these lines are given in Figure 4.

Figure 4 illustrates the following facts:

1. When the slope of a line is positive, the line slants upward from left to right (L_1).
2. When the slope of a line is negative, the line slants downward from left to right (L_2).
3. When the slope is 0, the line is horizontal (L_3).
4. When the slope is undefined, the line is vertical (L_4).

FIGURE 5

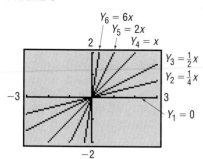

Seeing the Concept On the same square screen, graph the following equations:

$Y_1 = 0$	Slope of line is 0.
$Y_2 = \frac{1}{4}x$	Slope of line is $\frac{1}{4}$.
$Y_3 = \frac{1}{2}x$	Slope of line is $\frac{1}{2}$.
$Y_4 = x$	Slope of line is 1.
$Y_5 = 2x$	Slope of line is 2.
$Y_6 = 6x$	Slope of line is 6.

See Figure 5.

FIGURE 6

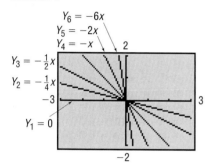

Seeing the Concept On the same square screen, graph the following equations:

$Y_1 = 0$	Slope of line is 0.
$Y_2 = -\frac{1}{4}x$	Slope of line is $-\frac{1}{4}$.
$Y_3 = -\frac{1}{2}x$	Slope of line is $-\frac{1}{2}$.
$Y_4 = -x$	Slope of line is -1.
$Y_5 = -2x$	Slope of line is -2.
$Y_6 = -6x$	Slope of line is -6.

See Figure 6.

Figures 5 and 6 illustrate that the closer the line is to the vertical position the greater the magnitude of the slope.

The next example illustrates how the slope of a line can be used to graph the line.

E X A M P L E   3

Graphing a Line Given a Point and a Slope

Draw a graph of the line that passes through the point $(3, 2)$ and has a slope of:

(a) $\frac{3}{4}$ (b) $-\frac{4}{5}$

Solution

(a) Slope = rise/run. The fact that the slope is $\frac{3}{4}$ means that for every horizontal movement (run) of 4 units to the right there will be a vertical movement (rise) of 3 units. If we start at the given point $(3, 2)$ and move 4 units to the right and 3 units up, we reach the point $(7, 5)$. By drawing the line through this point and the point $(3, 2)$, we have the graph. See Figure 7.

(b) The fact that the slope is

$$-\frac{4}{5} = \frac{-4}{5} = \frac{\text{rise}}{\text{run}}$$

means that for every horizontal movement of 5 units to the right there will be a corresponding vertical movement of -4 units (a downward movement). If we start at the given point $(3, 2)$ and move 5 units to the right and then 4 units down, we arrive at the point $(8, -2)$. By drawing the line through these points, we have the graph. See Figure 8.

FIGURE 7
Slope $= \frac{3}{4}$

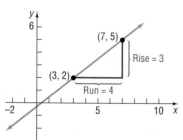

FIGURE 8
Slope $= -\frac{4}{5}$

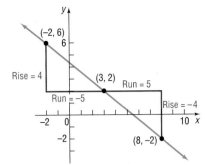

Alternatively, we could start at the point $(3, 2)$ and move 5 units to the left (Run $= -5$), then 4 units up (Rise $= 4$), to arrive at the point $(-2, 6)$. Figure 66 also shows this approach.

Now work Problem 15.

Equations of Lines

Now that we have discussed the slope of a line, we are ready to derive equations of lines. As we shall see, there are several forms of the equation of a line. Let's start with an example.

E X A M P L E   4

Graphing a Line

FIGURE 9

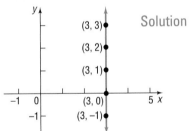

Graph the equation: $x = 3$

Solution   We are looking for all points (x, y) in the plane for which $x = 3$. Thus, no matter what y-coordinate is used, the corresponding x-coordinate always equals 3. Consequently, the graph of the equation $x = 3$ is a vertical line with x-intercept 3 and undefined slope. See Figure 9.

As suggested by Example 4, we have the following result:

Theorem   Equation of a Vertical Line

A vertical line is given by the equation of the form

$$x = a$$

where a is the x-intercept.

Comment   To graph an equation using a graphing utility, we need to express the equation in the form $y =$ expression in x. But $x = 3$ cannot be put in this form. To overcome this, most graphing utilities have special ways for drawing vertical lines. LINE, PLOT, and VERT are among the more common ones. Consult your manual to determine the correct methodology for your graphing utility.

FIGURE 10

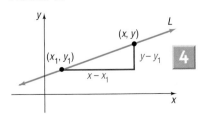

Now let L be a nonvertical line with slope m containing the point (x_1, y_1). See Figure 10. For any other point (x, y) on L, we have

$$m = \frac{y - y_1}{x - x_1} \quad \text{or} \quad y - y_1 = m(x - x_1)$$

Theorem   Point–Slope Form of an Equation of a Line

An equation of a nonvertical line of slope m that passes through the point (x_1, y_1) is

$$y - y_1 = m(x - x_1) \qquad (2)$$

E X A M P L E   5

Using the Point–Slope Form of a Line

An equation of the line with slope 4 and passing through the point $(1, 2)$ can be found by using the point–slope form with $m = 4$, $x_1 = 1$, and $y_1 = 2$.

$$y - y_1 = m(x - x_1) \qquad m = 4,\ x_1 = 1,\ y_1 = 2$$
$$y - 2 = 4(x - 1)$$
$$y = 4x - 2$$

See Figure 11.

FIGURE 11
$y = 4x - 2$

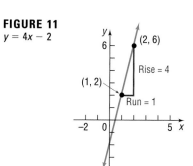

E X A M P L E   6

Finding the Equation of a Horizontal Line

Find an equation of the horizontal line passing through the point $(3, 2)$.

FIGURE 12
$y = 2$

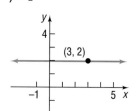

Solution   The slope of a horizontal line is 0. To get an equation, we use the point–slope form with $m = 0$, $x_1 = 3$, and $y_1 = 2$.

$$y - y_1 = m(x - x_1) \qquad m = 0,\ x_1 = 3,\ y_1 = 2$$
$$y - 2 = 0 \cdot (x - 3)$$
$$y - 2 = 0$$
$$y = 2$$

See Figure 12 for the graph.

As suggested by Example 6, we have the following result:

Theorem   Equation of a Horizontal Line

A horizontal line is given by an equation of the form

$$y = b$$

where b is the y-intercept.

E X A M P L E   7

5

Finding an Equation of a Line Given Two Points

Find an equation of the line L passing through the points $(2, 3)$ and $(-4, 5)$. Graph the line L.

Solution

Since two points are given, we first compute the slope of the line.

$$m = \frac{5-3}{-4-2} = \frac{2}{-6} = -\frac{1}{3}$$

FIGURE 13

$y - 3 = -\dfrac{1}{3}(x - 2)$

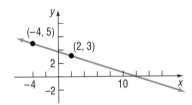

We use the point $(2, 3)$ and the fact that the slope $m = -\dfrac{1}{3}$ to get the point–slope form of the equation of the line.

$$y - 3 = -\frac{1}{3}(x - 2)$$

See Figure 13 for the graph.

In the solution to Example 7, we could have used the other point, $(-4, 5)$, instead of the point $(2, 3)$. The equation that results, although it looks different, is equivalent to the equation we obtained in the example. (Try it for yourself.)

6

Another useful equation of a line is obtained when the slope m and y-intercept b are known. In this event, we know both the slope m of the line and a point $(0, b)$ on the line; thus, we may use the point–slope form, equation (2), to obtain the following equation:

$$y - b = m(x - 0) \quad \text{or} \quad y = mx + b$$

Theorem   Slope–Intercept Form of an Equation of a Line

An equation of a line L with slope m and y-intercept b is

$$y = mx + b \tag{3}$$

FIGURE 14

$y = mx + 2$

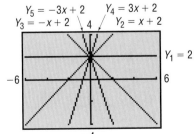

Seeing the Concept   To see the role that the slope m plays, graph the following lines on the same square screen:

$$Y_1 = 2$$
$$Y_2 = x + 2$$
$$Y_3 = -x + 2$$
$$Y_4 = 3x + 2$$
$$Y_5 = -3x + 2$$

See Figure 14. What do you conclude about the lines $y = mx + 2$?

FIGURE 15
$y = 2x + b$

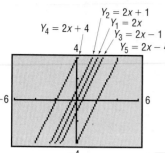

$Y_2 = 2x + 1$
$Y_1 = 2x$
$Y_4 = 2x + 4$
$Y_3 = 2x - 1$
$Y_5 = 2x - 4$

Seeing the Concept To see the role of the y-intercept b, graph the following lines on the same square screen:

$$Y_1 = 2x$$
$$Y_2 = 2x + 1$$
$$Y_3 = 2x - 1$$
$$Y_4 = 2x + 4$$
$$Y_5 = 2x - 4$$

See Figure 15. What do you conclude about the lines $y = 2x + b$?

7 When the equation of a line is written in slope–intercept form, it is easy to find the slope m and y-intercept b of the line. For example, suppose that the equation of a line is $y = -2x + 3$. Compare this equation to $y = mx + b$.

$$y = -2x + 3$$
$$\qquad\uparrow\quad\uparrow$$
$$y = \quad mx + b$$

The slope of this line is -2 and its y-intercept is 3.

E X A M P L E 8

Finding the Slope and y-Intercept

Find the slope m and y-intercept b of the equation $2x + 4y - 8 = 0$. Graph the equation.

Solution   To obtain the slope and y-intercept, we transform the equation into its slope–intercept form. Thus, we need to solve for y.

$$2x + 4y - 8 = 0$$
$$4y = -2x + 8$$
$$y = -\tfrac{1}{2}x + 2 \qquad y = mx + b$$

FIGURE 16
$2x + 4y - 8 = 0$

The coefficient of x, $-\tfrac{1}{2}$, is the slope, and the y-intercept is 2. We can graph the line in two ways:

1. Use the fact that the y-intercept is 2 and the slope is $-\tfrac{1}{2}$. Then, starting at the point $(0, 2)$, go to the right 2 units and then down 1 unit to the point $(2, 1)$. See Figure 16.

Or:

FIGURE 17
$2x + 4y - 8 = 0$

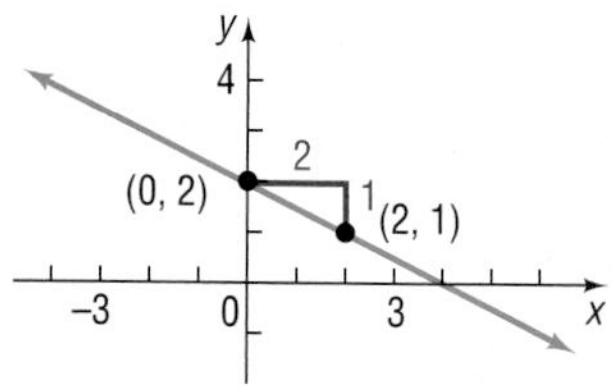

2. Locate the intercepts. Because the y-intercept is 2, we know that one intercept is $(0, 2)$. To obtain the x-intercept, let $y = 0$ and solve for x. When $y = 0$, we have

$$2x + 4 \cdot 0 - 8 = 0$$
$$2x - 8 = 0$$
$$x = 4$$

Thus, the intercepts are $(4, 0)$ and $(0, 2)$. See Figure 17.

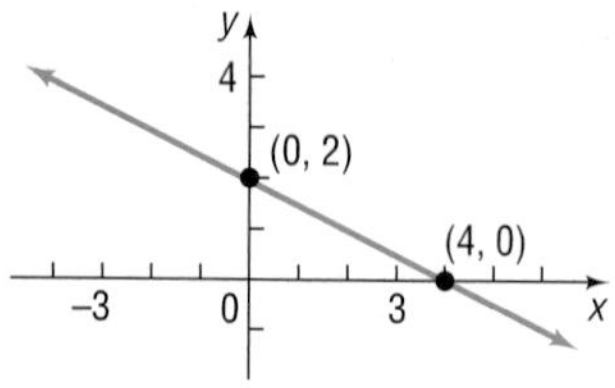

Now work Problem 61.

8 The form of the equation of the line in Example 8, $2x + 4y - 8 = 0$, is called the *general form*.

The equation of a line L is in **general form** when it is written as

$$Ax + By + C = 0 \qquad (4)$$

where A, B, and C are three real numbers and A and B are not both 0.

Every line has an equation that is equivalent to an equation written in general form. For example, a vertical line whose equation is

$$x = a$$

can be written in the general form

$$1 \cdot x + 0 \cdot y - a = 0 \quad A = 1, B = 0, C = -a.$$

A horizontal line whose equation is

$$y = b$$

can be written in the general form

$$0 \cdot x + 1 \cdot y - b = 0 \quad A = 0, B = 1, C = -b.$$

Lines that are neither vertical nor horizontal have general equations of the form

$$Ax + By + C = 0 \quad A \neq 0 \text{ and } B \neq 0.$$

Because the equation of every line can be written in general form, any equation equivalent to (4) is called a **linear equation.**

 Now work Problem 39.

Parallel and Perpendicular Lines

FIGURE 18
The lines are parallel if and only if their slopes are equal.

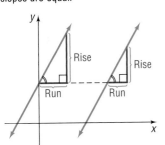

When two lines (in the plane) have no points in common, they are said to be **parallel.** Look at Figure 18. There we have drawn two lines and have constructed two right triangles by drawing sides parallel to the coordinate axes. These lines are parallel if and only if the right triangles are similar. (Do you see why? Two angles are equal.) But the triangles are similar if and only if the ratios of corresponding sides are equal.

This suggests the following result:

> **Theorem   Criterion for Parallel Lines**
>
> Two distinct nonvertical lines are parallel if and only if their slopes are equal.

The use of the words "if and only if" in the preceding theorem means that actually two statements are being made, one the converse of the other.

If two distinct nonvertical lines are parallel, then their slopes are equal.

If two distinct nonvertical lines have equal slopes, then they are parallel.

EXAMPLE 9

Showing That Two Lines Are Parallel

Show that the lines given by the following equations are parallel:

$$L: \quad 2x + 3y - 6 = 0 \qquad M: \quad 4x + 6y = 0$$

FIGURE 19
Parallel lines.

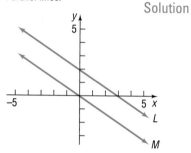

Solution   To determine whether these lines have equal slopes, we write each equation in slope–intercept form:

$$L: \quad 2x + 3y - 6 = 0 \qquad\qquad M: \quad 4x + 6y = 0$$
$$3y = -2x + 6 \qquad\qquad\qquad 6y = -4x$$
$$y = -\tfrac{2}{3}x + 2 \qquad\qquad\qquad y = -\tfrac{2}{3}x$$
$$\text{Slope} = -\tfrac{2}{3} \qquad\qquad\qquad \text{Slope} = -\tfrac{2}{3}$$

Because these lines have the same slope, $-\tfrac{2}{3}$, but different y-intercepts, the lines are parallel. See Figure 19.

EXAMPLE 10

Finding a Line That Is Parallel to a Given Line

Find an equation for the line that contains the point $(2, -3)$ and is parallel to the line $2x + y - 6 = 0$.

FIGURE 20

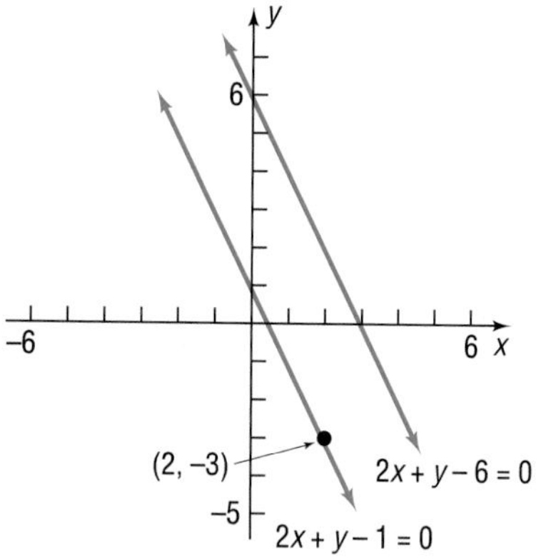

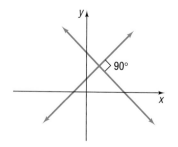

Solution   The slope of the line we seek equals the slope of the line $2x + y - 6 = 0$, since the two lines are to be parallel. Thus, we begin by writing the equation of the line $2x + y - 6 = 0$ in slope–intercept form:

$$2x + y - 6 = 0$$
$$y = -2x + 6$$

The slope is -2. Since the line we seek contains the point $(2, -3)$, we use the point–slope form to obtain

$$y + 3 = -2(x - 2) \qquad \text{Point–slope form.}$$
$$2x + y - 1 = 0 \qquad\qquad \text{General form.}$$
$$y = -2x + 1 \qquad\qquad \text{Slope–intercept form.}$$

This line is parallel to the line $2x + y - 6 = 0$ and contains the point $(2, -3)$. See Figure 20.

 Now work Problem 43.

When two lines intersect at a right angle (90°), they are said to be **perpendicular.** See Figure 21.

The following result gives a condition, in terms of their slopes, for two lines to be perpendicular:

FIGURE 21
Perpendicular lines.

Theorem   Criterion for Perpendicular Lines

Two nonvertical lines are perpendicular if and only if the product of their slopes is -1.

Here, we shall prove the "only if" part of the statement:

> If two nonvertical lines are perpendicular, then the product of their slopes is -1.

In Problem 102, you are asked to prove the "if" part of the theorem; that is,

> If two nonvertical lines have slopes whose product is -1, then the lines are perpendicular.

FIGURE 22

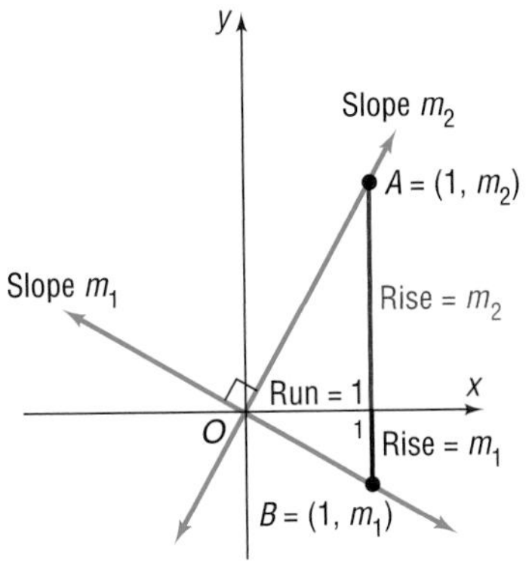

Slope m_2
$A = (1, m_2)$
Slope m_1
Rise $= m_2$
Run $= 1$
Rise $= m_1$
$B = (1, m_1)$

Proof   Let m_1 and m_2 denote the slopes of the two lines. There is no loss in generality (that is, neither the angle nor the slopes are affected) if we situate the lines so that they meet at the origin. See Figure 22. The point $A = (1, m_2)$ is on the line having slope m_2, and the point $B = (1, m_1)$ is on the line having slope m_1. (Do you see why this must be true?)

Suppose that the lines are perpendicular. Then triangle OAB is a right triangle. As a result of the Pythagorean Theorem, it follows that

$$[d(O, A)]^2 + [d(O, B)]^2 = [d(A, B)]^2 \tag{5}$$

By the distance formula, we can write each of these distances as

$$[d(O, A)]^2 = (1 - 0)^2 + (m_2 - 0)^2 = 1 + m_2^2$$
$$[d(O, B)]^2 = (1 - 0)^2 + (m_1 - 0)^2 = 1 + m_1^2$$
$$[d(A, B)]^2 = (1 - 1)^2 + (m_2 - m_1)^2 = m_2^2 - 2m_1m_2 + m_1^2$$

Using these facts in equation (5), we get

$$(1 + m_2^2) + (1 + m_1^2) = m_2^2 - 2m_1m_2 + m_1^2$$

which, upon simplification, can be written as

$$m_1m_2 = -1$$

Thus, if the lines are perpendicular, the product of their slopes is -1.   ∎

You may find it easier to remember the condition for two nonvertical lines to be perpendicular by observing that the equality $m_1m_2 = -1$ means that m_1 and m_2 are negative reciprocals of each other; that is, either $m_1 = -1/m_2$ or $m_2 = -1/m_1$.

E X A M P L E   11   Finding the Slope of a Line Perpendicular to a Given Line

If a line has slope $\frac{3}{2}$, any line having slope $-\frac{2}{3}$ is perpendicular to it.   ∎

E X A M P L E   12   Finding the Equation of a Line Perpendicular to a Given Line

10 Find an equation of the line passing through the point $(1, -2)$ and perpendicular to the line $x + 3y - 6 = 0$. Graph the two lines.

Solution   We first write the equation of the given line in slope–intercept form to find its slope.

$$x + 3y - 6 = 0$$
$$3y = -x + 6$$
$$y = -\tfrac{1}{3}x + 2$$

FIGURE 23

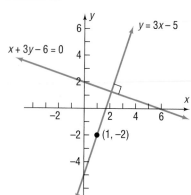

The given line has slope $-\frac{1}{3}$. Any line perpendicular to this line will have slope 3. Because we require the point $(1, -2)$ to be on this line with slope 3, we use the point–slope form of the equation of a line.

$$y - (-2) = 3(x - 1) \quad \text{Point–slope form.}$$
$$y + 2 = 3(x - 1)$$

This equation is equivalent to the forms

$$3x - y - 5 = 0 \qquad \text{General form.}$$
$$y = 3x - 5 \qquad \text{Slope–intercept form.}$$

Figure 23 shows the graphs.

Now work Problem 49.

A.1 | EXERCISES

In Problems 1–4 (a) find the slope of the line and (b) interpret the slope.

1. 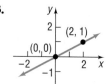   **2.** 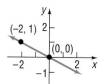   **3.** 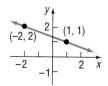   **4.**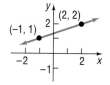

In Problems 5–14, plot each pair of points and determine the slope of the line containing them. Graph the line.

5. $(2, 3); (4, 0)$ **6.** $(4, 2); (3, 4)$ **7.** $(-2, 3); (2, 1)$ **8.** $(-1, 1); (2, 3)$

9. $(-3, -1); (2, -1)$ **10.** $(4, 2); (-5, 2)$ **11.** $(-1, 2); (-1, -2)$ **12.** $(2, 0); (2, 2)$

13. $(\sqrt{2}, 3); (1, \sqrt{3})$ **14.** $(-2\sqrt{2}, 0); (4, \sqrt{5})$

In Problems 15–22, graph the line passing through the point P and having slope m.

15. $P = (1, 2); \quad m = 3$ **16.** $P = (2, 1); \quad m = 4$ **17.** $P = (2, 4); \quad m = \frac{-3}{4}$

18. $P = (1, 3); \quad m = \frac{-2}{5}$ **19.** $P = (-1, 3); \quad m = 0$ **20.** $P = (2, -4); \quad m = 0$

21. $P = (0, 3); \quad$ slope undefined   **22.** $P = (-2, 0); \quad$ slope undefined

In Problems 23–30, find an equation of each line. Express your answer using either the general form or the slope–intercept form of the equation of a line, whichever you prefer.

23. 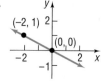   **24.** 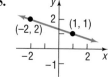   **25.** 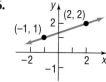   **26.**

27.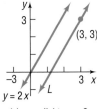

$y = 2x$

L is parallel to $y = 2x$

28.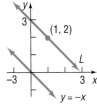

$y = -x$

L is parallel to $y = -x$

29.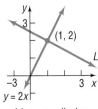

$y = 2x$

L is perpendicular to $y = 2x$

30.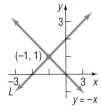

$y = -x$

L is perpendicular to $y = -x$

In Problems 31–54, find an equation for the line with the given properties. Express your answer using either the general form or the slope–intercept form of the equation of a line, whichever you prefer.

31. Slope $= 3$;   passing through $(-2, 3)$

32. Slope $= 2$;   passing through $(4, -3)$

33. Slope $= -\frac{2}{3}$;   passing through $(1, -1)$

34. Slope $= \frac{1}{2}$;   passing through $(3, 1)$

35. Passing through $(1, 3)$ and $(-1, 2)$

36. Passing through $(-3, 4)$ and $(2, 5)$

37. Slope $= -3$; y-intercept $= 3$

38. Slope $= -2$; y-intercept $= -2$

39. x-intercept $= 2$; y-intercept $= -1$

40. x-intercept $= -4$; y-intercept $= 4$

41. Slope undefined;   passing through $(2, 4)$

42. Slope undefined;   passing through $(3, 8)$

43. Parallel to the line $y = 2x$;
passing through $(-1, 2)$

44. Parallel to the line $y = -3x$;
passing through $(-1, 2)$

45. Parallel to the line $2x - y + 2 = 0$;
passing through $(0, 0)$

46. Parallel to the line $x - 2y + 5 = 0$;
passing through $(0, 0)$

47. Parallel to the line $x = 5$;
passing through $(4, 2)$

48. Parallel to the line $y = 5$;
passing through $(4, 2)$

49. Perpendicular to the line $y = \frac{1}{2}x + 4$;
passing through $(1, -2)$

50. Perpendicular to the line $y = 2x - 3$;
passing through $(1, -2)$

51. Perpendicular to the line $2x + y - 2 = 0$;
passing through $(-3, 0)$

52. Perpendicular to the line $x - 2y + 5 = 0$;
passing through $(0, 4)$

53. Perpendicular to the line $x = 8$;
passing through $(3, 4)$

54. Perpendicular to the line $y = 8$;
passing through $(3, 4)$

In Problems 55–74, find the slope and y-intercept of each line. Graph the line.

55. $y = 2x + 3$

56. $y = -3x + 4$

57. $\frac{1}{2}y = x - 1$

58. $\frac{1}{3}x + y = 2$

59. $y = \frac{1}{2}x + 2$

60. $y = 2x + \frac{1}{2}$

61. $x + 2y = 4$

62. $-x + 3y = 6$

63. $2x - 3y = 6$

64. $3x + 2y = 6$

65. $x + y = 1$

66. $x - y = 2$

67. $x = -4$

68. $y = -1$

69. $y = 5$

70. $x = 2$

71. $y - x = 0$

72. $x + y = 0$

73. $2y - 3x = 0$

74. $3x + 2y = 0$

75. Find an equation of the x-axis.

76. Find an equation of the y-axis.

In Problems 77–80, match each graph with the correct equation:
(a) $y = x$ (b) $y = 2x$ (c) $y = x/2$ (d) $y = 4x$

77.

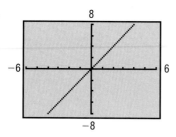

78.

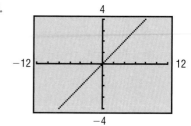

79.

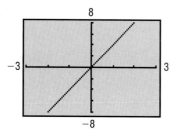

80.

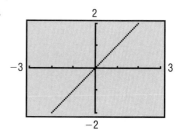

In Problems 81–86, write an equation of each line. Express your answer using either the general form or the slope–intercept form of the equation of a line, whichever you prefer.

81.

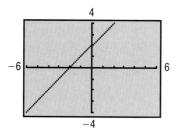

82.

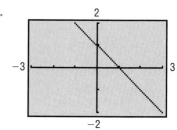

83.

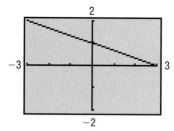

84.

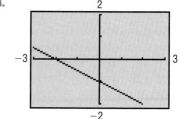

85.

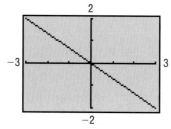

86.

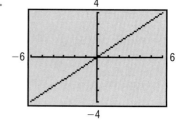

87. Measuring Temperature   The relationship between Celsius (°C) and Fahrenheit (°F) degrees for measuring temperature is linear. Find an equation relating °C and °F if 0°C corresponds to 32°F and 100°C corresponds to 212°F. Use the equation to find the Celsius measure of 70°F.

88. Measuring Temperature   The Kelvin (K) scale for measuring temperature is obtained by adding 273 to the Celsius temperature.
   (a) Write an equation relating K and °C.
   (b) Write an equation relating K and °F (see Problem 87).

89. Business: Computing Profit   Each Sunday a newspaper agency sells x copies of a newspaper for $1.00 per copy. The cost to the agency of each newspaper is $0.50. The agency pays a fixed cost for storage, delivery, and so on, of $100 per Sunday.
   (a) Write an equation that relates the profit P, in dollars, to the number x of copies sold. Graph this equation.
   (b) What is the profit to the agency if 1,000 copies are sold?
   (c) What is the profit to the agency if 5,000 copies are sold?

90. Business: Computing Profit   Repeat Problem 89 if the cost to the agency is $0.45 per copy and the fixed cost is $125 per Sunday.

91. Cost of Electricity   In 1997, Florida Power and Light Company supplied electricity to residential customers for a monthly customer charge of $5.65 plus 6.543¢ per kilowatt-hour supplied in the month for the first 750 kWhr used.* Write an equation that relates the monthly charge C, in dollars, to the number x of kilowatt-hours used in the month. Graph this equation. What is the monthly charge for using 300 kilowatt-hours? For using 750 kilowatt-hours?

92. Show that an equation for a line with nonzero x- and y-intercepts can be written as

$$\frac{x}{a} + \frac{y}{b} = 1$$

where a is the x-intercept and b is the y-intercept. This is called the **intercept form** of the equation of a line.

93. The **tangent line** to a circle may be defined as the line that intersects the circle in a single point, called the **point of tangency** (see the figure). If the equation of the circle is $x^2 + y^2 = r^2$ and the equation of the tangent line is $y = mx + b$, show that
   (a) $r^2(1 + m^2) = b^2$.

Source: Florida Power and Light Co., Miami, Florida, 1997.

[**Hint:** The quadratic equation $x^2 + (mx + b)^2 = r^2$ has exactly one solution.]
   (b) The point of tangency is $(-r^2m/b, r^2/b)$.
   (c) The tangent line is perpendicular to the line containing the center of the circle and the point of tangency.

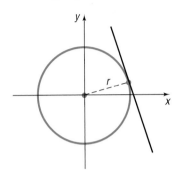

94. The **Greek method** for finding the equation of the tangent line to a circle uses the fact that at any point on a circle the line containing the radius and the tangent line are perpendicular (see Problem 93). Use this method to find an equation of the tangent line to the circle $x^2 + y^2 = 9$ at the point $(1, 2\sqrt{2})$.

95. Use the Greek method described in Problem 94 to find an equation of the tangent line to the circle $x^2 + y^2 - 4x + 6y + 4 = 0$ at the point $(3, 2\sqrt{2} - 3)$.

96. Refer to Problem 93. The line $x - 2y + 4 = 0$ is tangent to a circle at $(0, 2)$. The line $y = 2x - 7$ is tangent to the same circle at $(3, -1)$. Find the center of the circle.

97. Find an equation of the line containing the centers of the two circles

$$x^2 + y^2 - 4x + 6y + 4 = 0$$

and

$$x^2 + y^2 + 6x + 4y + 9 = 0$$

98. Show that the line containing the points (a, b) and (b, a) is perpendicular to the line $y = x$. Also show that the midpoint of (a, b) and (b, a) lies on the line $y = x$.

99. The equation $2x - y + C = 0$ defines a **family of lines,** one line for each value of C. On one set of coordinate axes, graph the members of the family when $C = -4$, $C = 0$, and $C = 2$. Can you draw a conclusion from the graph about each member of the family?

100. Rework Problem 99 for the family of lines $Cx + y + 4 = 0$.

101. If a circle of radius 2 is made to roll along the x-axis, what is an equation for the path of the center of the circle?

102. Prove that if two nonvertical lines have slopes whose product is −1 then the lines are perpendicular.

[**Hint:** Refer to Figure 22, and use the converse of the Pythagorean Theorem.]

103. Which form of the equation of a line do you prefer to use? Justify your position with an example that shows that your choice is better than another. Have reasons.

104. Can every line be written in slope–intercept form? Explain.

105. Does every line have two distinct intercepts? Explain. Are there lines that have no intercepts? Explain.

106. What can you say about two lines that have equal slopes and equal y-intercepts?

107. What can you say about two lines with the same x-intercept and the same y-intercept? Assume that the x-intercept is not 0.

108. If two lines have the same slope, but different x-intercepts, can they have the same y-intercept?

109. If two lines have the same y-intercept, but different slopes, can they have the same x-intercept? What is the only way that this can happen?

110. The accepted symbol used to denote the slope of a line is the letter *m*. Investigate the origin of this symbolism. Begin by consulting a French dictionary and looking up the French word *monter*. Write a brief essay on your findings.

111. Grade of a Road The term *grade* is used to describe the inclination of a road. How does this term relate to the notion of slope of a line? Is a 4% grade very steep? Investigate the grades of some mountainous roads and determine their slopes. Write a brief essay on your findings.

112. Carpentry Carpenters use the term *pitch* to describe the steepness of staircases and roofs. How does pitch relate to slope? Investigate typical pitches used for stairs and for roofs. Write a brief essay on your findings.

A.2 LINEAR CURVE FITTING

> **1** Distinguish between Linear and Nonlinear Relations
> **2** Use a Graphing Utility to Find the Line of Best Fit

Curve fitting is an area of statistics in which a relation between two or more variables is explained through an equation. For example, the equation $S = \$100{,}000 + 12A$, where S is sales (in dollars) and A is the advertising expenditure (in dollars), implies that if advertising expenditures were $0, sales would be $\$100{,}000 + 12(0) = \$100{,}000$, and if advertising expenditures were $10,000, sales would be $\$100{,}000 + 12(\$10{,}000) = \$220{,}000$. In this model, the variable A is called the predictor (independent) variable and S is called the response (dependent) variable because, if the level of advertising is known, it can be used to predict sales. Curve fitting is used to find an equation that relates two or more variables using observed or experimental data.

Steps for Curve Fitting

> STEP 1: Obtain data that relate two variables. Then plot ordered pairs of the variables as points to obtain a **scatter diagram.**
> STEP 2: Find an equation that describes this relation.

Scatter diagrams are used to help us to see the type of relation that exists between two variables. In this text, we will discuss a variety of different relations that may exist between two variables. For now, we concentrate on distinguishing between linear and nonlinear relations. See Figure 24.

FIGURE 24
Linear relations

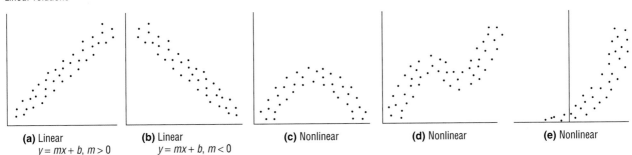

(a) Linear
$y = mx + b, m > 0$

(b) Linear
$y = mx + b, m < 0$

(c) Nonlinear

(d) Nonlinear

(e) Nonlinear

E X A M P L E   1

Distinguishing between Linear and Nonlinear Relations

Determine whether the relation between two variables shown in Figure 25 is linear or nonlinear.

FIGURE 25

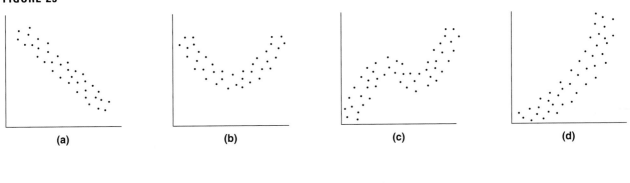

(a)

(b)

(c)

(d)

Solution   (a) Linear   (b) Nonlinear   (c) Nonlinear   (d) Nonlinear   ■

We will study linearly related data in this section. Fitting an equation to nonlinear data will be discussed in other chapters.

Suppose that the scatter diagram of a set of data appears to be linearly related. One way to obtain an equation for such data is to draw a line through two points on the scatter diagram and estimate the equation of the line.

EXAMPLE 2

Finding an Equation for Linearly Related Data

A farmer collected the following data, which show crop yields for various amounts of fertilizer used.

Plot		1	2	3	4	5	6	7	8	9	10	11	12
Fertilizer, X (pounds/100 ft²)		0	0	5	5	10	10	15	15	20	20	25	25
Yield, Y (bushels)		4	6	10	7	12	10	15	17	18	21	23	22

(a) Draw a scatter diagram of the data.
(b) Select two points from the data and find an equation of the line containing the points.
(c) Graph the line on the scatter diagram.

Solution

(a) The data collected indicate that a relation exists between the amount of fertilizer used and crop yield. To draw a scatter diagram, we plot points, using fertilizer as the x-coordinate and yield as the y-coordinate. See Figure 26. From the scatter diagram, it appears that a linear relation exists between the amount of fertilizer used and yield.

FIGURE 26

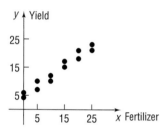

(b) Select two points, say $(0, 4)$ and $(25, 23)$. (You should select your own two points and complete the solution.) The slope of the line joining the points $(0, 4)$ and $(25, 23)$ is

$$m = \frac{23 - 4}{25 - 0} = \frac{19}{25} = 0.76$$

The equation of the line with slope 0.76 and passing through $(0, 4)$ is found using the point–slope form with $m = 0.76$, $x_1 = 0$, and $y_1 = 4$.

FIGURE 27

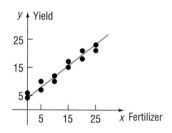

$$y - y_1 = m(x - x_1)$$
$$y - 4 = 0.76(x - 0)$$
$$y = 0.76x + 4$$

(c) Figure 27 shows the scatter diagram with the graph of the line found in part (b). ∎

Now work Problem 7(a), (b), and (c).

The line obtained in Example 2 depends on the selection of points, which will vary from person to person. So the line we found might be different from the line you found. Although the line we found in Example 2 appears to "fit" the data well, there may be a line that "fits it better." Do you think your line "fits" the data better? Is there a line of "best fit"? As it turns out, there is a method for finding the line that best fits linearly related data.* This line is called the **line of best fit.**

*We shall not discuss in this book the underlying mathematics of lines of best fit. Most books in statistics and many in linear algebra discuss this topic.

Lines of Best Fit

E X A M P L E 3

Finding the Line of Best Fit

 With the data from Example 2, find the line of best fit using a graphing utility.

Solution   Graphing utilities contain built-in programs that find the linear equation of "best fit" for a collection of points in a scatter diagram. (Look in your owner's manual under Linear Regression or Line of Best Fit for details on how to execute the program.) Upon executing the LINear REGression program, we obtain the results shown in Figure 28. The output that the utility provides shows us the equation, $y = ax + b$, where a is the slope of the line and b is the y-intercept. The line of best fit that relates fertilizer and yield is

$$\text{Yield} = 0.717(\text{fertilizer}) + 4.786$$

Figure 29 shows the graph of the line of best fit, along with the scatter diagram.

FIGURE 28   **FIGURE 29**

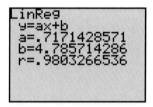

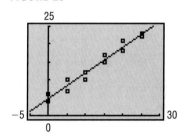

Now work Problem 7(d) and (e).

Does the line of best fit appear to be a good "fit"? In other words, does the line appear to accurately describe the relation between yield and fertilizer?

And just how "good" is this line of "best fit"? The answers are given by what is called the *correlation coefficient.* Look again at Figure 28. The last line of output is $r = 0.98$. This number, called the **correlation coefficient, r,** $0 \le |r| \le 1$, is a measure of the strength of the *linear relation* that exists between two variables. The closer $|r|$ is to 1 the more perfect the linear relationship is. If r is close to 0, there is little or no *linear* relationship between the variables. A negative value of r, $r < 0$, indicates that as x increases y decreases; a positive value of r, $r > 0$, indicates that as x increases y does also. Thus, the data given in Example 2, having a correlation coefficient of 0.98, are strongly indicative of a linear relationship.

A.2 | EXERCISES

In Problems 1–6, examine the scatter diagram and determine whether the type of relation, if any, that may exist is linear or nonlinear.

1.
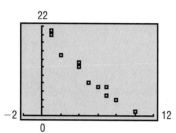

2.
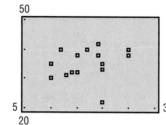

3.
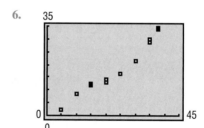

4.

5.

6.

In Problems 7–14:
(a) Draw a scatter diagram.
(b) Select two points from the scatter diagram and find the equation of the line passing through the points selected. *
(c) Graph the line found in (b) on the scatter diagram.
(d) Use a graphing utility to find the line of best fit.
(e) Use a graphing utility to graph the line of best fit on the scatter diagram.

7.

x	3	4	5	6	7	8	9
y	4	6	7	10	12	14	16

8.

x	3	5	7	9	11	13
y	0	2	3	6	9	11

9.

x	-2	-1	0	1	2
y	-4	0	1	4	5

10.

x	-2	-1	0	1	2
y	7	6	3	2	0

11.

x	20	30	40	50	60
y	100	95	91	83	70

12.

x	5	10	15	20	25
y	2	4	7	11	18

13.

x	-20	-17	-15	-14	-10
y	100	120	118	130	140

14.

x	-30	-27	-25	-20	-14
y	10	12	13	13	18

*Answers will vary. We'll use the first and last data points.

15. Consumption and Disposable Income An economist wishes to estimate a line that relates personal consumption expenditures (C) and disposable income (I). Both C and I are in thousands of dollars. She interviews 8 heads of households for families of size 4 and obtains the following data:

C (000)	I (000)
16	20
18	20
13	18
21	27
27	36
26	37
36	45
39	50

Let I represent the independent variable and C the dependent variable.
(a) Draw a scatter diagram.
(b) Find a line that fits the data.*
(c) Interpret the slope. The slope of this line is called the **marginal propensity to consume.**
(d) Predict the consumption of a family whose disposable income is $42,000.
(e) Use a graphing utility to find the line of best fit to the data.

16. Marginal Propensity to Save The same economist as the one in Problem 15 wants to estimate a line that relates savings (S) and disposable income (I). Let $S = I - C$ be the dependent variable and I the independent variable. The slope of this line is called the **marginal propensity to save.**
(a) Draw a scatter diagram.
(b) Find a line that fits the data.
(c) Interpret the slope.
(d) Predict the savings of a family whose income is $42,000.
(e) Use a graphing utility to find the line of best fit to the data.

17. Average Speed of a Car An individual wanted to determine the relation that might exist between speed and miles per gallon of an automobile. Let X be the average speed of a car on the highway measured in miles per hour, and let Y

represent the miles per gallon of the automobile. The following data are collected:

X	50	55	55	60	60	62	65	65
Y	28	26	25	22	20	20	17	15

(a) Draw a scatter diagram.
(b) Find a line that fits the data.*
(c) Interpret the slope.
(d) Predict the miles per gallon of a car traveling 61 miles per hour.
(e) Use a graphing utility to find the line of best fit to the data.

18. Height versus Weight A doctor wished to determine whether a relation exists between the height of a female and her weight. She obtained the heights and weights of 10 females aged 18 to 24. Let height be the independent variable, X, measured in inches, and weight be the dependent variable, Y, measured in pounds.

X	60	61	62	62	64	65	65	67	68	68
Y	105	110	115	120	120	125	130	135	135	145

(a) Draw a scatter diagram.
(b) Find a line that fits the data.
(c) Interpret the slope.
(d) Predict the weight of a female aged 18 to 24 whose height is 66 inches.
(e) Use a graphing utility to find the line of best fit to the data.

19. Sales Data versus Income The following data represent sales and net income before taxes (both are in billions of dollars) for all manufacturing firms within the United States for 1980–1989. Treat sales as the independent variable and net income before taxes as the dependent variable.

Year	Sales	Net Income Before Taxes
1980	1912.8	92.6
1981	2144.7	101.3
1982	2039.4	70.9
1983	2114.3	85.8
1984	2335.0	107.6
1985	2331.4	87.6
1986	2220.9	83.1
1987	2378.2	115.6
1988	2596.2	154.6
1989	2745.1	136.3

*Answers will vary. We'll use the first and last data points.

Source: Economic Report of the President, February 1995.

(a) Draw a scatter diagram.

(b) Find a line that fits the data.

(c) Interpret the slope.

(d) Predict the net income before taxes of manufacturing firms in 1990 if sales are $2456.4 billion.

 (e) Use a graphing utility to find the line of best fit to the data.

20. **Employment and the Labor Force**   The following data represent the civilian labor force (people aged 16 years and older, excluding those serving in the military) and the number of employed people in the United States for the years 1981–1991. Treat the size of the labor force as the independent variable and the number employed as the dependent variable. Both the size of the labor force and the number employed are measured in thousands of people.

(a) Draw a scatter diagram.

(b) Find a line that fits the data.

(c) Interpret the slope.

(d) Predict the number of employed people if the civilian labor force is 122,340,000 people.

(e) Use a graphing utility to find the line of best fit to the data.

Year	Civilian Labor Force	Number Employed
1981	108,670	100,397
1982	110,204	99,526
1983	111,550	100,834
1984	113,544	105,005
1985	115,461	107,150
1986	117,834	109,597
1987	119,865	112,440
1988	121,669	114,968
1989	123,869	117,342
1990	124,787	117,914
1991	125,303	116,877

Source: Business Statistics, 1963–1991, U.S. Department of Commerce, Economics and Statistics Administration, Bureau of Economic Analysis, June 1992.

A.3 | COMPLETING THE SQUARE

1 Complete the Square of an Equation

1 The idea behind the method of **completing the square** is to "adjust" the left side of a second-degree polynomial, $ax^2 + bx + c$, so that it becomes a perfect square (the square of a first-degree polynomial). For example, $x^2 + 6x + 9$ and $x^2 - 4x + 4$ are perfect squares because

$$x^2 + 6x + 9 = (x + 3)^2 \quad \text{and} \quad x^2 - 4x + 4 = (x - 2)^2$$

How do we adjust the second-degree polynomial? We do it by adding the appropriate number to create a perfect square. For example, to make $x^2 + 6x$ a perfect square, we add 9.

Let's look at several examples of completing the square when the coefficient of x^2 is 1.

Start	Add	Result
$x^2 + 4x$	4	$x^2 + 4x + 4 = (x + 2)^2$
$x^2 + 12x$	36	$x^2 + 12x + 36 = (x + 6)^2$
$x^2 - 6x$	9	$x^2 - 6x + 9 = (x - 3)^2$
$x^2 + x$	$\frac{1}{4}$	$x^2 + x + \frac{1}{4} = (x + \frac{1}{2})^2$

Do you see the pattern? Provided the coefficient of x^2 is 1, we complete the square by adding the square of one-half the coefficient of x.

Start	Add	Result
$x^2 + mx$	$\left(\dfrac{m}{2}\right)^2$	$x^2 + mx + \left(\dfrac{m}{2}\right)^2 = \left(x + \dfrac{m}{2}\right)^2$

 Now work Problem 1.

E X A M P L E   1   Completing the Square of an Equation Containing Two Variables

Complete the squares of x and y in the equation

$$x^2 + y^2 - 2x + 4y - 4 = 0$$

Solution   We rearrange the equation, grouping the terms involving the variable x and the variable y.

$$(x^2 - 2x) + (y^2 + 4y) = 4$$

Next, we complete the square of each parenthetical expression. Of course, since we want an equivalent equation, whatever we add to the left side, we also add to the right side.

$$(x^2 - 2x + 1) + (y^2 + 4y + 4) = 4 + 1 + 4$$
$$(x - 1)^2 + (y + 2)^2 = 9$$

The terms involving the variables x and y now appear as perfect squares. ▬

 Now work Problem 7.

A.3 | EXERCISES

In Problems 1–6, tell what number should be added to complete the square of each expression.

1. $x^2 - 4x$ 2. $x^2 - 2x$ 3. $x^2 + \frac{1}{2}x$

4. $x^2 - \frac{1}{3}x$ 5. $x^2 - \frac{2}{3}x$ 6. $x^2 - \frac{2}{5}x$

In Problems 7–12, complete the squares of x and y in each equation.

7. $x^2 + y^2 - 4x + 4y - 1 = 0$ 8. $x^2 + y^2 + 4x + 4y - 8 = 0$ 9. $x^2 + y^2 + 6x - 2y + 1 = 0$

10. $x^2 + y^2 - 8x + 2y + 1 = 0$ 11. $x^2 + y^2 + x - y - \frac{1}{2} = 0$ 12. $x^2 + y^2 - x + y - \frac{3}{2} = 0$

A.4 | THE QUADRATIC FORMULA

> **1** Use the Quadratic Formula

We begin with a preliminary result. Suppose that we wish to solve the quadratic equation

$$x^2 = p \tag{1}$$

where $p \geq 0$ is a nonnegative number. We proceed as follows:

$x^2 - p = 0$	Put in standard form.
$(x - \sqrt{p})(x + \sqrt{p}) = 0$	Factor (over the real numbers).
$x = \sqrt{p}$ or $x = -\sqrt{p}$	Solve

Thus, we have the following result:

$$\text{If } x^2 = p \text{ and } p \geq 0, \text{ then } x = \sqrt{p} \text{ or } x = -\sqrt{p}. \tag{2}$$

Note that if $p > 0$ the equation $x^2 = p$ has two solutions: $x = \sqrt{p}$ and $x = -\sqrt{p}$. We usually abbreviate these solutions as $x = \pm\sqrt{p}$, read as "x equals plus or minus the square root of p." For example, the two solutions of the equation

$$x^2 = 4$$

are

$$x = \pm\sqrt{4}$$

and, since $\sqrt{4} = 2$, we have

$$x = \pm 2$$

The solution set is $\{-2, 2\}$.

Do not confuse the two solutions of the equation $x^2 = 4$ with the value of the principal square root of 4, which is $\sqrt{4} = 2$. The principal square root of a positive number is unique, while the equation $x^2 = p, p > 0$, has two solutions.

The next example illustrates how the procedure of completing the square can be used to solve a quadratic equation.

E X A M P L E 1

Solving a Quadratic Equation by Completing the Square

Solve by completing the square: $x^2 + 5x + 4 = 0$

Solution

We always begin this procedure by rearranging the equation so that the constant is on the right side.

$$x^2 + 5x + 4 = 0$$
$$x^2 + 5x = -4$$

Since the coefficient of x^2 is 1, we can complete the square on the left side by adding $(\frac{1}{2} \cdot 5)^2 = \frac{25}{4}$. Of course, in an equation, whatever we add to the left side must also be added to the right side. Thus, we add $\frac{25}{4}$ to *both* sides.

$$x^2 + 5x + \tfrac{25}{4} = -4 + \tfrac{25}{4}$$
$$(x + \tfrac{5}{2})^2 = \tfrac{9}{4}$$
$$x + \tfrac{5}{2} = \pm\sqrt{\tfrac{9}{4}} \qquad \text{Apply (2)}$$
$$x + \tfrac{5}{2} = \pm\tfrac{3}{2}$$
$$x = -\tfrac{5}{2} \pm \tfrac{3}{2}$$
$$x = -\tfrac{5}{2} + \tfrac{3}{2} = -1 \quad \text{or} \quad x = -\tfrac{5}{2} - \tfrac{3}{2} = -4$$

The solution set is $\{-4, -1\}$.

Now work Problem 1.

We can use the method of completing the square to obtain a general formula for solving the quadratic equation

$$ax^2 + bx + c = 0, \qquad a \neq 0$$

As in Example 2, we rearrange the term as

$$ax^2 + bx = -c$$

Since $a \neq 0$, we can divide both sides by a to get

$$x^2 + \frac{b}{a}x = -\frac{c}{a}$$

Now the coefficient of x^2 is 1. To complete the square on the left side, add the square of one-half the coefficient of x; that is, add

$$\left(\frac{1}{2} \cdot \frac{b}{a}\right)^2 = \frac{b^2}{4a^2}$$

to each side. Then

$$x^2 + \frac{b}{a}x + \frac{b^2}{4a^2} = \frac{b^2}{4a^2} - \frac{c}{a}$$
$$\left(x + \frac{b}{2a}\right)^2 = \frac{b^2 - 4ac}{4a^2} \qquad (3)$$

Provided $b^2 - 4ac \geq 0$, we now can apply the result in equation (2) to get

$$x + \frac{b}{2a} = \pm\sqrt{\frac{b^2 - 4ac}{4a^2}}$$
$$x = -\frac{b}{2a} \pm \frac{\sqrt{b^2 - 4ac}}{2a} = \frac{-b \pm \sqrt{b^2 - 4ac}}{2a}$$

What if $b^2 - 4ac$ is negative? Then equation (3) states that the left expression (a real number squared) equals the right expression (a negative number). Since this occurrence is impossible for real numbers, we conclude that if $b^2 - 4ac < 0$ the quadratic equation has no real solution.

We now state the *quadratic formula*.

1

Theorem   Quadratic Formula

Consider the quadratic equation

$$ax^2 + bx + c = 0 \qquad a \neq 0 \tag{4}$$

If $b^2 - 4ac < 0$, this equation has no real solution.
If $b^2 - 4ac \geq 0$, the real solution(s) of this equation is (are) given by the **quadratic formula:**

$$x = \frac{-b \pm \sqrt{b^2 - 4ac}}{2a} \tag{5}$$

The quantity $b^2 - 4ac$ is called the **discriminant** of the quadratic equation, because its value tells us whether the equation has real solutions. In fact, it also tells us how many solutions to expect.

Discriminant of a
Quadratic Equation

For a quadratic equation $ax^2 + bx + c = 0$:

1. If $b^2 - 4ac > 0$, there are two unequal real solutions.
2. If $b^2 - 4ac = 0$, there is a repeated real solution.
3. If $b^2 - 4ac < 0$, there is no real solution.

Thus, when asked to find the real solutions, if any, of a quadratic equation, always evaluate the discriminant first to see how many real solutions there are.

E X A M P L E   2

Solving a Quadratic Equation Using the Quadratic Formula

Use the quadratic formula to find the real solutions, if any, of the equation

$$3x^2 - 5x + 1 = 0$$

Solution   The equation is in standard form, so we compare it to $ax^2 + bx + c = 0$ to find a, b, and c.

$$3x^2 - 5x + 1 = 0$$
$$ax^2 + bx + c = 0$$

With $a = 3, b = -5$, and $c = 1$, we evaluate the discriminant $b^2 - 4ac$.

$$b^2 - 4ac = (-5)^2 - 4(3)(1) = 25 - 12 = 13$$

Since $b^2 - 4ac > 0$, there are two real solutions, which can be found using the quadratic formula.

$$x = \frac{-b \pm \sqrt{b^2 - 4ac}}{2a} = \frac{5 \pm \sqrt{13}}{6}$$

The solution set is $\left\{ \dfrac{5 - \sqrt{13}}{6}, \dfrac{5 + \sqrt{13}}{6} \right\}$.

 Now work Problem 7.

E X A M P L E   3   Solving Quadratic Equations Using the Quadratic Formula

Use the quadratic formula to find the real solutions, if any, of the equation

$$3x^2 + 2 = 4x$$

Solution   The equation, as given, is not in standard form.

$$3x^2 + 2 = 4x$$
$$3x^2 - 4x + 2 = 0 \qquad \text{Put in standard form.}$$
$$ax^2 + bx + c = 0 \qquad \text{Compare to standard form.}$$

With $a = 3, b = -4$, and $c = 2$, we find

$$b^2 - 4ac = 16 - 24 = -8$$

Since $b^2 - 4ac < 0$, the equation has no real solution.

A.4 | EXERCISES

In Problems 1–6, solve each equation by completing the square.

1. $x^2 + 4x - 21 = 0$

2. $x^2 - 6x = 13$

3. $x^2 - \frac{1}{2}x = \frac{3}{16}$

4. $x^2 + \frac{2}{3}x = \frac{1}{3}$

5. $3x^2 + x - \frac{1}{2} = 0$

6. $2x^2 - 3x = 1$

In Problems 7–18, find the real solutions, if any, of each equation. Use the quadratic formula.

7. $x^2 - 4x + 2 = 0$

8. $x^2 + 4x + 2 = 0$

9. $x^2 - 4x - 1 = 0$

10. $x^2 + 6x + 1 = 0$

11. $2x^2 - 5x + 3 = 0$

12. $2x^2 + 5x + 3 = 0$

13. $4y^2 - y + 2 = 0$

14. $4t^2 + t + 1 = 0$

15. $4x^2 = 1 - 2x$

16. $2x^2 = 1 - 2x$

17. $9t^2 - 6t + 1 = 0$

18. $4u^2 - 6u + 9 = 0$

Graphing Utilities

B.1   THE VIEWING RECTANGLE

FIGURE 1
$y = 2x$

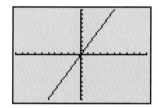

FIGURE 2

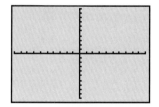

FIGURE 3

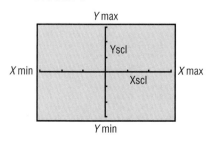

All graphing utilities, that is, all graphing calculators and all computer software graphing packages, graph equations by plotting points on a screen. The screen itself actually consists of small rectangles, called **pixels.** The more pixels the screen has the better the resolution. Most graphing calculators have 48 pixels per square inch; most computer screens have 32 to 108 pixels per square inch. When a point to be plotted lies inside a pixel, the pixel is turned on (lights up). Thus, the graph of an equation is a collection of pixels. Figure 1 shows one way the graph of $y = 2x$ might look on a TI-83 graphing calculator.

The screen of a graphing utility will display the coordinate axes of a rectangular coordinate system. However, you must set the scale on each axis. You must also include the smallest and largest values of x and y that you want included in the graph. This is called **setting the RANGE** and it gives the **viewing rectangle** or **WINDOW.**

Figure 2 illustrates a typical viewing rectangle.

To select the viewing rectangle, we must give values to the following expressions:

Xmin:	the smallest value of x
Xmax:	the largest value of x
Xscl:	the number of units per tick mark on the x-axis
Ymin:	the smallest value of y
Ymax:	the largest value of y
Yscl:	the number of units per tick mark on the y-axis

Figure 3 illustrates these settings for a typical screen.

If the scale used on each axis is known, we can determine the minimum and maximum values of x and y shown on the screen by counting the tick marks. Look again at Figure 2. For a scale of 1 on each axis, the minimum and maximum values of x are -10 and 10, respectively; the minimum and maximum values of y are also -10 and 10. If the scale is 2 on each axis, then the minimum and maximum values of x are -20 and 20, respectively; the minimum and maximum values of y are -20 and 20, respectively.

Conversely, if we know the minimum and maximum values of x and y, we can determine the scales being used by counting the tick marks displayed. We shall follow the practice of showing the minimum and maximum values of x and y in our illustrations so that you will know how the WINDOW was set. See Figure 4.

FIGURE 4

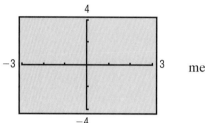

means

$$Xmin = -3 \qquad Ymin = -4$$
$$Xmax = 3 \qquad Ymax = 4$$
$$Xscl = 1 \qquad Yscl = 2$$

E X A M P L E   1

Finding the Coordinates of a Point Shown on a Graphing Utility Screen

Find the coordinates of the point shown in Figure 5.

FIGURE 5

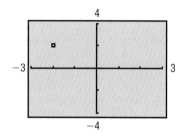

Solution   First, we note that the WINDOW used in Figure 5 is

$$Xmin = -3 \qquad Xmax = 3 \qquad Xscl = 1$$
$$Ymin = -4 \qquad Ymax = 4 \qquad Yscl = 2$$

The point shown is 2 tick units to the left on the horizontal axis (scale = 1) and 1 tick up on the vertical scale (scale = 2). Thus, the coordinates of the point shown are $(-2, 2)$. ∎

Now work Problems 1 and 11.

B.1 | **EXERCISES**

In Problems 1–4, determine the coordinates of the points shown. Tell in which quadrant each point lies.

1.

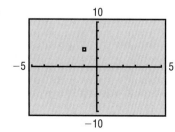

2.

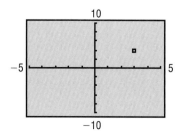

3.

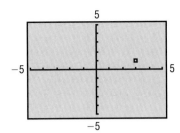

4.

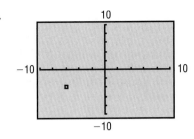

In Problems 5–10, select a WINDOW so that each of the given points will lie within the viewing rectangle.

5. $(-10, 5), (3, -2), (4, -1)$

6. $(5, 0), (6, 8), (-2, -3)$

7. $(40, 20), (-20, -80), (10, 40)$

8. $(-80, 60), (20, -30), (-20, -40)$

9. $(0, 0), (100, 5), (5, 150)$

10. $(0, -1), (100, 50), (-10, 30)$

In Problems 11–20, determine the WINDOW used for each viewing rectangle.

11.

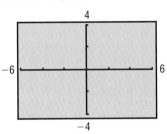

12.

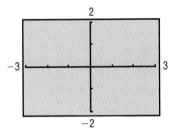

13.

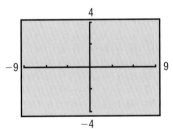

14.

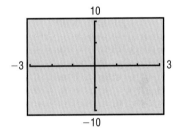

15.

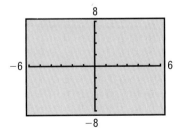

16.

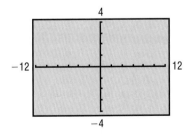

17.

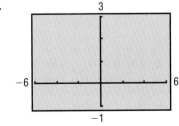

18.

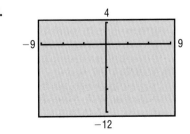

19.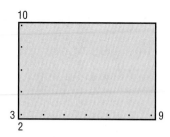

20.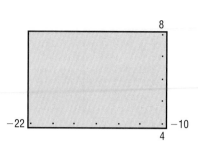

B.2 | USING A GRAPHING UTILITY TO GRAPH EQUATIONS

Examples 5 and 6 of Section 1.2 demonstrate that a graph can be obtained by plotting points in a rectangular coordinate system and connecting them. Graphing utilities perform these same steps when graphing an equation. For example, the TI-83 determines 95 evenly spaced input values,* uses the equation to determine the output values, plots these points on the screen, and, finally (if in the connected mode), draws a line segment between consecutive points.

Most graphing utilities require the following steps in order to obtain the graph of an equation:

Steps for Graphing an Equation Using a Graphing Utility

STEP 1: Solve the equation for y in terms of x.

STEP 2: Get into the graphing mode of your graphing utility. The screen will usually display $y =$ _____, prompting you to enter the expression involving x that you found in Step 1. (Consult your manual for the correct way to enter the expression; for example, $y = x^2$ might be entered as $x\wedge 2$ or as $x * x$ or as $x \, x^Y \, 2$).

STEP 3: Select the viewing rectangle. Without prior knowledge about the behavior of the graph of the equation, it is common to select the **standard viewing rectangle** initially. The viewing rectangle is then adjusted based on the graph that appears. In this text, the standard viewing rectangle will be

$$X\text{min} = -10 \qquad X\text{max} = 10 \qquad X\text{scl} = 1$$
$$Y\text{min} = -10 \qquad Y\text{max} = 10 \qquad Y\text{scl} = 1$$

STEP 4: Execute (or graph).

STEP 5: Adjust the viewing rectangle until a complete graph is obtained.

E X A M P L E   1   Graphing an Equation Using a Graphing Utility

Graph the equation: $6x^2 + 3y = 36$.

Solution   STEP 1: We solve for y in terms of x.

$$6x^2 + 3y = 36$$
$$3y = -6x^2 + 36 \qquad \text{Subtract } 6x^2 \text{ from both sides of the equation.}$$
$$y = -2x^2 + 12 \qquad \text{Divide both sides of the equation by 3.}$$

*These input values depend on the values of Xmin and Xmax. For example, if Xmin = −10 and Xmax = 10, then the first input value will be −10 and the next input value will be −10 + (10 − (−10))/94 ≈ −9.7872, and so on.

STEP 2: From the graphing mode, enter the expression $-2x^2 + 12$ after the prompt $y =$.

STEP 3: Set the viewing rectangle to the standard viewing rectangle.

STEP 4: Execute (or graph). The screen should look like Figure 6.

FIGURE 6

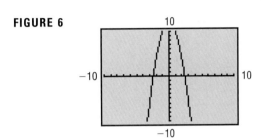

STEP 5: The graph of $y = -2x^2 + 12$ is not complete. The value of Ymax must be increased so that the top portion of the graph is visible. After increasing the value of Ymax to 12, we obtain the graph in Figure 7. The graph is now complete.

FIGURE 7

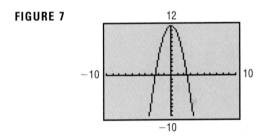

Look again at Figure 7. Although a complete graph is shown, the graph might be improved by adjusting the values of Xmin and Xmax. Figure 8 shows the graph of $y = -2x^2 + 12$ using Xmin $= -4$ and Xmax $= 4$. Do you think this is a better choice for the viewing rectangle?

FIGURE 8

Now work Problems 11(a)–11(d).

EXAMPLE 2 Creating a Table and Graphing an Equation

Create a table and graph the equation: $y = x^3$

Solution Most graphing utilities have the capability of creating a table of values for an equation. (Check your manual to see if your graphing utility has this capability.) Table 1 illustrates a table of values for $y = x^3$ on a TI-83. See Figure 9 for the graph.

TABLE 1

X	Y1	
-3	-27	
-2	-8	
-1	-1	
0	0	
1	1	
2	8	
3	27	

Y1=X^3

FIGURE 9

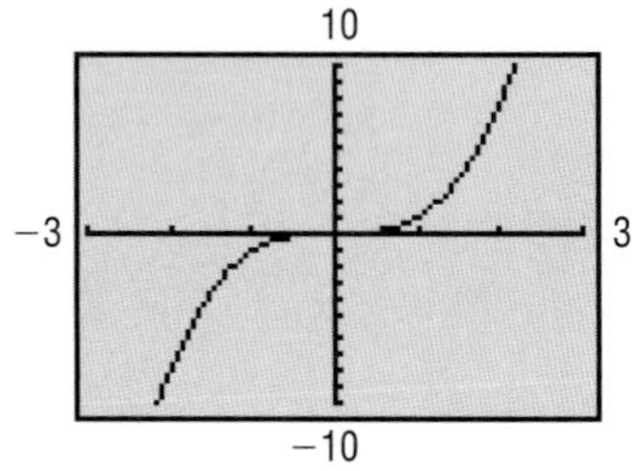

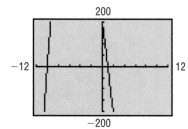

EXAMPLE 3 Using ZOOM-OUT to Obtain a Complete Graph

Graph the equation $y = x^3 - 11x^2 - 190x + 200$ using the following settings for the first viewing rectangle.

Xmin: −12	Xmax: 12	Xscl: 2
Ymin: −200	Ymax: 200	Yscl: 50

Solution Figure 10 shows the graph.

FIGURE 10

Notice how ragged the graph looks. The y-scale is clearly not adequate. The graph appears to have two x-intercepts. The graph is not complete. We can use the ZOOM-OUT function to help obtain a complete graph.

After the first ZOOM-OUT, we obtain Figure 11.

FIGURE 11

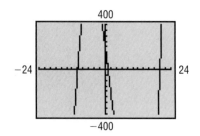

Notice that the min/max settings have been doubled, while the scales remained the same.* Also notice that now there are three x-intercepts, the most a polynomial of degree 3 can have. However, we still cannot see the top or bottom portion of the graph. ZOOM-OUT again to obtain the graph in Figure 12.

FIGURE 12

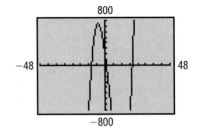

Now we can see the entire graph with the exception of the bottom portion. Therefore, ZOOMing-OUT further may not provide us with a complete graph. To get a complete graph, we instead choose to adjust the WINDOW. After some experimentation, we obtain Figure 13, a complete graph.

FIGURE 13
$y = x^3 - 11x^2 - 190x + 200$

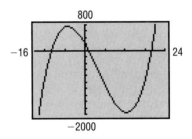

*On some graphing utilities, the default factor for the ZOOM-OUT function is 4, meaning that the original min/max setting will be multiplied by 4. Also, you can set the factor yourself, if you want; furthermore, the factor need not be the same for x and y.

B.2 | EXERCISES

In Problems 1–20, graph each equation using the following WINDOW's:

(a) Xmin $= -5$	(b) Xmin $= -10$	(c) Xmin $= -10$	(d) Xmin $= -5$
Xmax $= 5$	Xmax $= 10$	Xmax $= 10$	Xmax $= 5$
Xscl $= 1$	Xscl $= 1$	Xscl $= 2$	Xscl $= 1$
Ymin $= -4$	Ymin $= -8$	Ymin $= -8$	Ymin $= -20$
Ymax $= 4$	Ymax $= 8$	Ymax $= 8$	Ymax $= 20$
Yscl $= 1$	Yscl $= 1$	Yscl $= 2$	Yscl $= 5$

1. $y = x + 2$ **2.** $y = x - 2$ **3.** $y = -x + 2$ **4.** $y = -x - 2$

5. $y = 2x + 2$ **6.** $y = 2x - 2$ **7.** $y = -2x + 2$ **8.** $y = -2x - 2$

9. $y = x^2 + 2$ **10.** $y = x^2 - 2$ **11.** $y = -x^2 + 2$ **12.** $y = -x^2 - 2$

13. $y = 2x^2 + 2$ **14.** $y = 2x^2 - 2$ **15.** $y = -2x^2 + 2$ **16.** $y = -2x^2 - 2$

17. $3x + 2y = 6$ **18.** $3x - 2y = 6$ **19.** $-3x + 2y = 6$ **20.** $-3x - 2y = 6$

In Problems 21–40, use a graphing utility to graph each equation.

21. $3x + 5y = 75$ **22.** $3x - 5y = 75$ **23.** $3x + 5y = -75$ **24.** $3x - 5y = -75$

25. $y = (x - 10)^2$ **26.** $y = (x + 10)^2$ **27.** $y = x^2 - 100$ **28.** $y = x^2 + 100$

29. $x^2 + y^2 = 100$ **30.** $x^2 + y^2 = 64$ **31.** $3x^2 + y^2 = 900$ **32.** $4x^2 + y^2 = 1600$

33. $x^2 + 3y^2 = 900$ **34.** $x^2 + 4y^2 = 1600$ **35.** $y = x^2 - 10x$ **36.** $y = x^2 + 10x$

37. $y = x^2 - 18x$ **38.** $y = x^2 + 18x$ **39.** $y = x^2 - 36x$ **40.** $y = x^2 + 36x$

B.3 | USING A GRAPHING UTILITY TO LOCATE INTERCEPTS

Four tools that can be used to locate intercepts using a graphing utility are TRACE, BOX, VALUE, and ROOT (or ZERO).

TRACE

Most graphing utilities allow you to move from point to point along the graph, displaying on the screen the coordinates of each point. This feature is called TRACE.

E X A M P L E 1   Using TRACE to Locate Intercepts

Graph the equation: $y = x^3 - 8$. Use TRACE to locate the intercepts.

Solution   Figure 14 shows the graph of $y = x^3 - 8$.

FIGURE 14

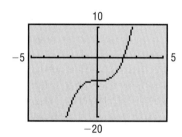

Activate the TRACE feature. As you move the cursor along the graph, you will see the coordinates of each point displayed. Just before you get to the x-axis, the display will look like the one in Figure 15(a). (Due to differences in graphing utilities, your display may be slightly different from the one shown here.)

FIGURE 15

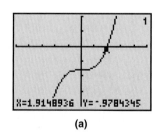

(a)

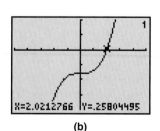

(b)

In Figure 15(a), the negative value of the y-coordinate indicates that we are still below the x-axis. The next position of the cursor is shown in Figure 15(b).

The positive value of the y-coordinate indicates that we are now above the x-axis. This means that between these two points the x-axis was crossed. The x-intercept lies between 1.9148936 and 2.0212766.

Using TRACE, we find that the y-intercept is −8. See Figure 16.

FIGURE 16

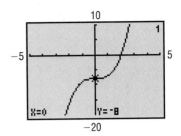

BOX

Most graphing utilities have a BOX feature that allows you to box in a specific part of the graph of an equation.

E X A M P L E   2

Using BOX

Graph the equation $y = x^3 - 8$ and use the BOX feature to improve on the approximation found in Example 1 for the x-intercept.

Solution   Using the viewing rectangle of Figure 14, graph the equation. Activate the BOX feature. (With some graphing utilities, this requires positioning the cursor at one corner of the box and then tracing out the sides of the box to the diagonal corner.) See Figure 17. Once executed, the box becomes the viewing rectangle. The result is shown in Figure 18.

FIGURE 17   **FIGURE 18**

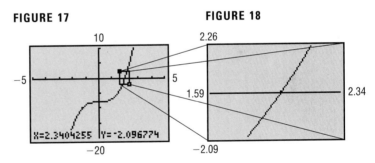

Now you can TRACE to get the approximation shown in Figure 19.

FIGURE 19

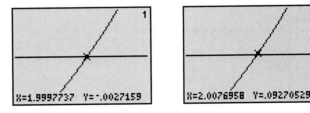

Now we know that the x-intercept lies between 1.9997737 and 2.0076958.

VALUE and ROOT (ZERO)

Most graphing utilities have an eVALUEate feature that, given a value of x, determines the value of y for an equation. We can use this feature to evaluate an equation at $x = 0$ to determine the y-intercept. Using the VALUE feature, we find the y-intercept of $y = x^3 - 8$ is −8.

Most graphing utilities also have a ROOT (or ZERO) feature that can be used to determine the x-intercept of an equation. Using ROOT on a TI-82, we find that the x-intercept of $y = x^3 - 8$ is 2. Consult your owner's manual to determine the appropriate keystrokes for these functions.

If your utility has no ROOT or ZERO feature, you can use BOX to approximate the intercepts of an equation.

Approximating x-Intercepts Using BOX

To approximate the x-intercepts of an equation, we use an algorithm in which each successive application results in one more decimal place of accuracy.

We begin with an example that illustrates the meaning of writing a number "correct to n decimal places."

E X A M P L E 3   Writing a Number Correct to n Decimal Places

Writing the Number	1/7	$\sqrt{2}$	π
Correct to 1 decimal place	0.1	1.4	3.1
Correct to 2 decimal places	0.14	1.41	3.14
Correct to 3 decimal places	0.142	1.414	3.141
Correct to 4 decimal places	0.1428	1.4142	3.1415
Correct to 5 decimal places	0.14285	1.41421	3.14159
Correct to 6 decimal places	0.142857	1.414213	3.141592

As the example illustrates, **correct to n decimal places** means the decimal that results from truncation after the nth decimal.

The following example illustrates the steps to use to approximate the x-intercepts of an equation using a graphing utility.

E X A M P L E 4   Using BOX to Approximate the x-Intercepts of an Equation

Find the smaller of the two x-intercepts of the equation $y = x^2 - 6x + 7$. Express the answer correct to two decimal places.

Solution   We begin by graphing the equation using a scale of 1 on each axis. Figure 20 shows the graph.

The smaller of the two x-intercepts lies between 1 and 2. (Remember that Xscl = 1.) Thus, correct to 0 decimal places, the smaller x-intercept is $x = 1$.

Next, we BOX the graph from approximately $x = 1$ to $x = 2$ and from $y = -1$ to $y = 1$. See Figure 21. Adjust the WINDOW so that Xscl = 0.1 and Yscl = 0.1. Graph the equation again. See Figure 22.

FIGURE 20

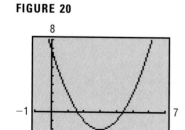

FIGURE 21   **FIGURE 22**

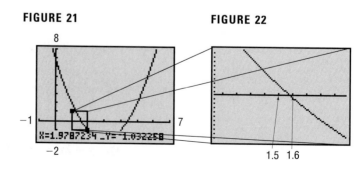

The x-intercept lies between 1.5 and 1.6. (Remember that the BOX was from $x = 1$ to $x = 2$ and Xscl $= 0.1$.) Thus, correct to one decimal place, the smaller x-intercept is $x = 1.5$.

Next, BOX again from $x = 1.5$ to $x = 1.6$ and from $y = -0.1$ to $y = 0.1$. See Figure 23. Adjust the WINDOW so that Xscl $= 0.01$ and Yscl $= 0.01$ and graph. See Figure 24.

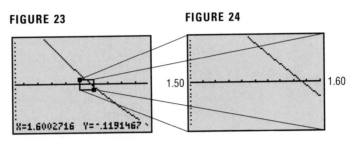

FIGURE 23   **FIGURE 24**

The x-intercept lies between 1.58 and 1.59. Remember that the BOX was from $x = 1.5$ to $x = 1.6$ with Xscl $= 0.01$. Thus, correct to two decimal places, the smaller of the two x-intercepts is $x = 1.58$. ∎

The steps to follow for approximating x-intercepts of equations correct to any desired number of decimals are given next.

Steps for Approximating the x-Intercepts of an Equation

> STEP 1: Write the equation in the form {expression in x} $= 0$.
> STEP 2: Graph the equation $y = $ {expression in x} using Xscl $= 1$ and Yscl $= 1$.
> STEP 3: BOX the graph so that the x-intercept is within the BOX. Adjust Xscl and Yscl to one-tenth of their current value and graph the equation again.
> STEP 4: If additional accuracy is desired, repeat step 3.

In these steps, each repetition of step 3 gives one more decimal place of accuracy. Thus, when Xscl $= 0.01$, the x-intercept will be known correct to two decimal places.

If the x-intercept that you are trying to approximate appears to fall on a tick mark, then you must evaluate the equation at that tick mark. If the value is zero, then there is an exact solution. If the value is not zero, then you know whether the graph is above or below the x-axis at the tick mark by noting the sign of the value at the tick mark.

B.3 EXERCISES

In Problems 1–6, write each expression as a decimal correct to three decimal places.

1. $\frac{3}{7}$ **2.** $\frac{2}{11}$ **3.** $\sqrt{5}$ **4.** $\sqrt{6}$ **5.** $\sqrt[3]{2}$ **6.** $\sqrt[3]{3}$

In Problems 7–12, use a graphing utility to approximate the smaller of the two x-intercepts of each equation. Express the answer correct to two decimal places.

7. $y = x^2 + 4x + 2$ **8.** $y = x^2 + 4x - 3$ **9.** $y = 2x^2 + 4x + 1$

10. $y = 3x^2 + 5x + 1$ **11.** $y = 2x^2 - 3x - 1$ **12.** $y = 2x^2 - 4x - 1$

In Problems 13–20, use a graphing utility to approximate the **positive** x-intercepts of each equation. Express the answer correct to two decimal places.

13. $y = x^3 + 3.2x^2 - 16.83x - 5.31$

14. $y = x^3 + 3.2x^2 - 7.25x - 6.3$

15. $y = x^4 - 1.4x^3 - 33.71x^2 + 23.94x + 292.41$

16. $y = x^4 + 1.2x^3 - 7.46x^2 - 4.692x + 15.2881$

17. $y = \pi x^3 - (8.88\pi + 1)x^2 - (42.066\pi - 8.88)x + 42.066$

18. $y = \pi x^3 - (5.63\pi + 2)x^2 - (108.392\pi - 11.26)x + 216.784$

19. $y = x^3 + 19.5x^2 - 1021x + 1000.5$

20. $y = x^3 + 14.2x^2 - 4.8x - 12.4$

B.4 SQUARE SCREENS

FIGURE 25

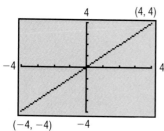

Most graphing utilities have a rectangular screen. Because of this, using the same settings for both x and y will result in a distorted view. For example, Figure 25 shows the graph of the line $y = x$ connecting the points $(-4, -4)$ and $(4, 4)$.

We expect the line to bisect the first and third quadrants, but it doesn't. We need to adjust the selections for Xmin, Xmax, Ymin, and Ymax so that a **square screen** results. On most graphing utilities, this is accomplished by setting the ratio of x to y at 3:2.* In other words,

$$2(X\text{max} - X\text{min}) = 3(Y\text{max} - Y\text{min})$$

E X A M P L E 1

FIGURE 26

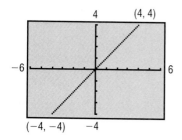

Examples of Viewing Rectangles That Result in Square Screens

(a) Xmin = -3
   Xmax = 3
   Xscl = 1
   Ymin = -2
   Ymax = 2
   Yscl = 1

(b) Xmin = -6
   Xmax = 6
   Xscl = 1
   Ymin = -4
   Ymax = 4
   Yscl = 1

(c) Xmin = -6
   Xmax = 6
   Xscl = 2
   Ymin = -4
   Ymax = 4
   Yscl = 1

Figure 26 shows the graph of the line $y = x$ on a square screen using the viewing rectangle given in Example 1(b). Notice that the line now bisects the first and third quadrants. Compare this illustration to Figure 25.

*Some graphing utilities have a built-in function that automatically squares the screen. For example, the TI-85 has a ZSQR function that does this. Some graphing utilities require a ratio other than 3:2 to square the screen. For example, the HP 48G requires the ratio of x to y to be 2:1 for a square screen. Consult your manual.

B.4 EXERCISES

In Problems 1–8, determine which of the given WINDOW's result in a square screen.

1. Xmin = -3
   Xmax = 3
   Xscl = 1
   Ymin = -2
   Ymax = 2
   Yscl = 1

2. Xmin = -5
   Xmax = 5
   Xscl = 1
   Ymin = -4
   Ymax = 4
   Yscl = 1

3. Xmin = 0
   Xmax = 9
   Xscl = 3
   Ymin = -2
   Ymax = 4
   Yscl = 2

4. Xmin = -6
   Xmax = 6
   Xscl = 2
   Ymin = -4
   Ymax = 4
   Yscl = 1

5. $X\text{min} = -6$
$X\text{max} = 6$
$X\text{scl} = 1$
$Y\text{min} = -2$
$Y\text{max} = 2$
$Y\text{scl} = .5$

6. $X\text{min} = -6$
$X\text{max} = 6$
$X\text{scl} = 1$
$Y\text{min} = -4$
$Y\text{max} = 4$
$Y\text{scl} = 1$

7. $X\text{min} = 0$
$X\text{max} = 9$
$X\text{scl} = 1$
$Y\text{min} = -2$
$Y\text{max} = 4$
$Y\text{scl} = 1$

8. $X\text{min} = -6$
$X\text{max} = 6$
$X\text{scl} = 2$
$Y\text{min} = -4$
$Y\text{max} = 4$
$Y\text{scl} = 2$

9. If $X\text{min} = -4$, $X\text{max} = 8$, and $X\text{scl} = 1$, how should $Y\text{min}$, $Y\text{max}$, and $Y\text{scl}$ be selected so that the viewing rectangle contains the point $(4, 8)$ and the screen is square?

10. If $X\text{min} = -6$, $X\text{max} = 12$, and $X\text{scl} = 2$, how should $Y\text{min}$, $Y\text{max}$, and $Y\text{scl}$ be selected so that the viewing rectangle contains the point $(4, 8)$ and the screen is square?

B.5 | USING A GRAPHING UTILITY TO GRAPH INEQUALITIES

It is easiest to begin with an example.

E X A M P L E   1   Graphing an Inequality Using a Graphing Utility

Use a graphing utility to graph: $3x + y - 6 \le 0$.

Solution   We begin by graphing the equation $3x + y - 6 = 0$ ($Y_1 = -3x + 6$). See Figure 27.

FIGURE 27

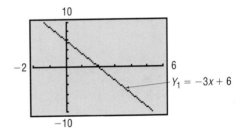

$Y_1 = -3x + 6$

As with graphing by hand, we need to test points selected from each region and determine whether they satisfy the inequality. To test the point $(-1, 2)$, for example, enter $3*-1+2-6 \le 0$. See Figure 28(a). The 1 that appears indicates that the statement entered (the inequality) is true. When the point $(5, 5)$ is tested, a 0 appears, indicating that the statement entered is false. Thus, $(-1, 2)$ is a part of the graph of the inequality and $(5, 5)$ is not. Figure 28(b) shows the graph of the inequality on a TI-83.*

FIGURE 28

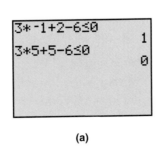

(a)

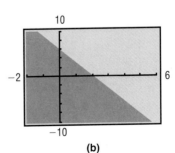

(b)

*Consult your owner's manual for shading techniques.

The steps to follow to graph an inequality using a graphing utility are given next.

Steps for Graphing an Inequality Using a Graphing Utility

STEP 1: Replace the inequality symbol by an equal sign and graph the resulting equation, $y = \quad$.

STEP 2: In each of the regions, select a test point P.
  (a) Use a graphing utility to determine if the test point P satisfies the inequality. If the test point satisfies the inequality, then so do all the points in the region. Indicate this by using the graphing utility to shade the region.
  (b) If the coordinates of P do not satisfy the inequality, then none of the points in that region do.

B.6   USING A GRAPHING UTILITY TO GRAPH A POLAR EQUATION

Most graphing utilities require the following steps in order to obtain the graph of a polar equation. Be sure to be in POLar mode.

Graphing a Polar Equation Using a Graphing Utility

STEP 1: Set the mode to POLar. Solve the equation for r in terms of θ.

STEP 2: Select the viewing rectangle in polar mode. Besides setting Xmin, Xmax, Xscl, and so forth, the viewing rectangle in polar mode requires setting the minimum and maximum values for θ and an increment setting for θ (θstep). In addition, a square screen and radian measure should be used.

STEP 3: Enter the expression involving θ that you found in Step 1. (Consult your manual for the correct way to enter the expression.)

STEP 4: Execute.

EXAMPLE 1

Graphing a Polar Equation Using a Graphing Utility

Use a graphing utility to graph the polar equation $r \sin \theta = 2$.

Solution   STEP 1: We solve the equation for r in terms of θ.

$$r \sin \theta = 2$$

$$r = \frac{2}{\sin \theta}$$

STEP 2: From the POLar mode, select the viewing rectangle. We will use the one given next.

θmin $= 0$	Xmin $= -9$	Ymin $= -6$
θmax $= 2\pi$	Xmax $= 9$	Ymax $= 6$
θstep $= \pi/24$	Xscl $= 1$	Yscl $= 1$

θstep determines the number of points that the graphing utility will plot. For example, if θstep is $\pi/24$, then the graphing utility will evaluate r at $\theta = 0$(θmin), $\pi/24$, $2\pi/24$, $3\pi/24$, and so forth, up to 2π(θmax). The smaller θstep is, the more points that the graphing

FIGURE 29

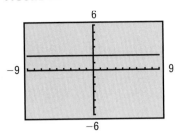

utility will plot. The student is encouraged to experiment with different values for θmin, θmax, and θstep to see how the graph is affected.

STEP 3: Enter the expression $\dfrac{2}{\sin \theta}$ after the prompt $r =$.

STEP 4: Execute.

The graph is shown in Figure 29.   ▬

B.7 USING A GRAPHING UTILITY TO GRAPH PARAMETRIC EQUATIONS

Most graphing utilities have the capability of graphing parametric equations. The following steps are usually required in order to obtain the graph of parametric equations. Check your owner's manual to see how your utility works.

Graphing Parametric Equations Using a Graphing Utility

> STEP 1: Set the mode to PARametric. Enter $x(t)$ and $y(t)$.
> STEP 2: Select the viewing rectangle. In addition to setting Xmin, Xmax, Xscl, and so on, the viewing rectangle in parametric mode requires setting minimum and maximum values for the parameter t and an increment setting for t (Tstep).
> STEP 3: Execute.

E X A M P L E   1

Graphing a Curve Defined by Parametric Equations Using a Graphing Utility

Graph the curve defined by the parametric equations

$$x = 3t^2 \qquad y = 2t \qquad -2 \le t \le 2$$

Solution   STEP 1: With the graphing utility in PARametric mode, enter the equations $x(t) = 3t^2$ and $y(t) = 2t$.

STEP 2: Select the viewing rectangle. The interval I is $-2 \le t \le 2$, so we select the following square viewing rectangle:

Tmin $= -2$	Xmin $= 0$	Ymin $= -5$
Tmax $= 2$	Xmax $= 15$	Ymax $= 5$
Tstep $= 0.1$	Xscl $= 1$	Yscl $= 1$

We choose Tmin $= -2$ and Tmax $= 2$ because $-2 \le t \le 2$. Finally, the choice for Tstep will determine the number of points that the graphing utility will plot. For example, with Tstep at 0.1, the graphing utility will evaluate x and y at $t = -2, -1.9, -1.8$, and so on. The smaller the Tstep is, the more points that the graphing utility will plot. The reader is encouraged to experiment with different values of Tstep to see how the graph is affected.

STEP 3: Execute. Notice the direction that the graph is drawn. This direction shows the orientation of the curve.

The graph shown in Figure 30 is complete.   ▬

FIGURE 30

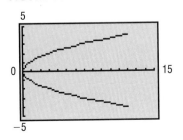

ANSWERS

C H A P T E R 1   1.1 Exercises

1. $>$ **3.** $>$ **5.** $>$ **7.** $=$ **9.** $<$ **11.**

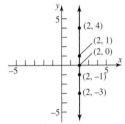

13. $x > 0$ **15.** $x < 2$ **17.** $x \leq 1$ **19.** $2 < x < 5$

21. $[0, 4]$

23. $[4, 6)$

25. $2 \leq x \leq 5$

27. $x \geq 4$ or $4 \leq x < \infty$

29. 1 **31.** 5 **33.** 1 **35.** $\dfrac{1}{16}$ **37.** $\dfrac{4}{9}$ **39.** $\dfrac{1}{9}$ **41.** $\dfrac{9}{4}$ **43.** 324 **45.** 27 **47.** 16 **49.** $4\sqrt{2}$ **51.** $-\dfrac{2}{3}$ **53.** y^2 **55.** $\dfrac{y}{x^2}$ **57.** $\dfrac{1}{x^3 y}$ **59.** $\dfrac{25y^2}{16x^2}$

61. $\dfrac{1}{x^3 y^3 z^2}$ **63.** $\dfrac{-8x^3}{9yz^2}$ **65.** $\dfrac{y^2}{x^2 + y^2}$ **67.** $\dfrac{16x^2}{9y^2}$ **69.** 13 **71.** 26 **73.** 25 **75.** 4 **77.** 24 **79.** Yes; 5 **81.** Not a right triangle

83. Yes; 25 **85.** 72.42 **87.** 30.66 **89.** 111.17 **91.** 11.03 **93.** -0.49 **95.** 0.59 **97.** 9.4 in. **99.** 58.3 ft **101.** 46.7 mi **103.** 3 mi

105. $a \leq b, c > 0; a - b \leq 0$

$$(a - b)c \leq 0(c)$$
$$ac - bc \leq 0$$
$$ac \leq bc$$

107. $\dfrac{a + b}{2} - a = \dfrac{a + b - 2a}{2} = \dfrac{b - a}{2} > 0;$ therefore, $a < \dfrac{a + b}{2}$

$b - \dfrac{a + b}{2} = \dfrac{2b - a - b}{2} = \dfrac{b - a}{2} > 0;$ therefore, $b > \dfrac{a + b}{2}$

109. No; No **111.** 1 **113.** 3.15 or 3.16 **115.** The light can be seen on the horizon 23.3 miles distant. Planes flying at 10,000 feet can see it 146 miles away. The ship would need to be 185.9 ft tall for the data to be correct.

1.2 Exercises

1. (a) Quadrant II **(b)** Positive x-axis **(c)** Quadrant III **(d)** Quadrant I **(e)** Negative y-axis **(f)** Quadrant IV

3. The points form a vertical line that is 2 units to the right of the y-axis.

5. $\sqrt{5}; \left(1, \dfrac{1}{2}\right)$ **7.** $\sqrt{10}; \left(-\dfrac{1}{2}, \dfrac{3}{2}\right)$ **9.** $5; \left(3, -\dfrac{3}{2}\right)$ **11.** $\sqrt{85}; \left(\dfrac{3}{2}, 1\right)$ **13.** $2\sqrt{5}; (5, -1)$ **15.** $2.625; (1.05, 0.7)$ **17.** $\sqrt{a^2 + b^2}; \left(\dfrac{a}{2}, \dfrac{b}{2}\right)$

19. $d(A, B) = \sqrt{13}$

$d(B, C) = \sqrt{13}$

$d(A, C) = \sqrt{26}$

$(\sqrt{13})^2 + (\sqrt{13})^2 = (\sqrt{26})^2$

Area $= \dfrac{13}{2}$ square units

21. $d(A, B) = \sqrt{130}$

$d(B, C) = \sqrt{26}$

$d(A, C) = 2\sqrt{26}$

$(\sqrt{26})^2 + (2\sqrt{26})^2 = (\sqrt{130})^2$

Area $= 26$ square units

23. $(2, -4); (2, 2)$ **25.** $(-2, 0); (6, 0)$

27.
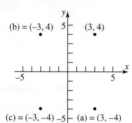

29.
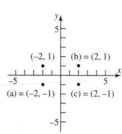

31.
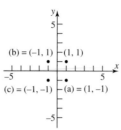

33.
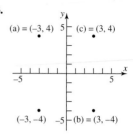

35. (a) $(-1, 0), (1, 0)$ **(b)** x-axis, y-axis, origin **37. (a)** $\left(-\frac{\pi}{2}, 0\right), (0, 1), \left(\frac{\pi}{2}, 0\right)$ **(b)** y-axis

39. (a) $(0, 0)$ **(b)** x-axis **41. (a)** $(-1.5, 0), (0, -2), (1.5, 0)$ **(b)** y-axis **43. (a)** none **(b)** origin

45. $x^2 + (y - 2)^2 = 4; x^2 + y^2 - 4y = 0$ **47.** $(x - 4)^2 + (y + 3)^2 = 25; x^2 + y^2 - 8x + 6y = 0$

49. $x^2 + y^2 = 4; x^2 + y^2 - 4 = 0$ **51.** Center $(2, 1)$; radius 2; $(x - 2)^2 + (y - 1)^2 = 4$

53. Center $\left(\frac{5}{2}, 2\right)$; radius $\frac{3}{2}$; $\left(x - \frac{5}{2}\right)^2 + (y - 2)^2 = \frac{9}{4}$ **55. (b)** **57. (c)**

59. $r = 2; (h, k) = (0, 0)$ **61.** $r = 2; (h, k) = (3, 0)$ **63.** $r = 3; (h, k) = (-2, 2)$ **65.** $r = \frac{1}{2}; (h, k) = \left(\frac{1}{2}, -1\right)$

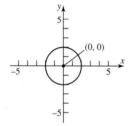

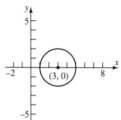

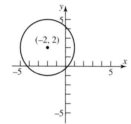

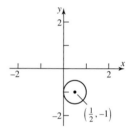

67. $r = 5; (h, k) = (3, -2)$

69.

71.

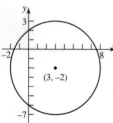

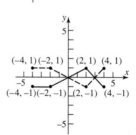

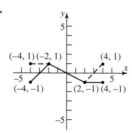

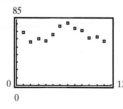

73. $(0, 0)$; symmetric with respect to the y-axis **75.** $(0, 0)$; symmetric with respect to the origin
77. $(0, 9), (3, 0), (-3, 0)$; symmetric with respect to the y-axis
79. $(-3, 0), (3, 0), (0, -2), (0, 2)$; symmetric with respect to the x-axis, y-axis, and origin
81. $(0, -27), (3, 0)$; no symmetry **83.** $(0, -4), (4, 0), (-1, 0)$; no symmetry **85.** $(0, 0)$; symmetric with respect to the origin
87. (a) $(90, 0), (90, 90), (0, 90)$ **(b)** 232.4 ft **(c)** 366.2 ft **89.** $d = 50t$
91. (a)

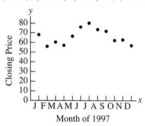

Month of 1997

(b)

(c) The price of the stock is decreasing, increasing, and then decreasing over time.

93. (a)

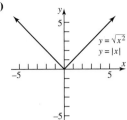

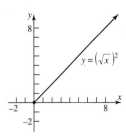

(b) Since $\sqrt{x^2} = |x|$, for all x, the graphs of $y = \sqrt{x^2}$ and $y = |x|$ are the same.

(c) For $y = (\sqrt{x})^2$, the domain is $x \geq 0$; for $y = x$, the domain is all real numbers. Thus, $(\sqrt{x})^2 = x$ only for $x \geq 0$.

(d) For $y = \sqrt{x^2}$, the range is $y \geq 0$; for $y = x$, the range is all real numbers. Also, $\sqrt{x^2} = |x|$, which equals x only if $x \geq 0$.

1.3 Exercises

1. Function **3.** Not a function **5.** Function **7.** Function **9.** Not a function **11.** Function
13. (a) -4 **(b)** -5 **(c)** -9 **(d)** $-3x^2 - 2x - 4$ **(e)** $3x^2 - 2x + 4$ **(f)** $-3x^2 - 4x - 5$
15. (a) 0 **(b)** $\dfrac{1}{2}$ **(c)** $-\dfrac{1}{2}$ **(d)** $\dfrac{-x}{x^2 + 1}$ **(e)** $\dfrac{-x}{x^2 + 1}$ **(f)** $\dfrac{x + 1}{x^2 + 2x + 2}$ **17. (a)** 4 **(b)** 5 **(c)** 5 **(d)** $|x| + 4$ **(e)** $-|x| - 4$
(f) $|x + 1| + 4$ **19. (a)** $-\dfrac{1}{5}$ **(b)** $-\dfrac{3}{2}$ **(c)** $\dfrac{1}{8}$ **(d)** $\dfrac{-2x + 1}{-3x - 5}$ **(e)** $\dfrac{-2x - 1}{3x - 5}$ **(f)** $\dfrac{2x + 3}{3x - 2}$
21. $f(0) = 3; f(-6) = -3$ **23.** Positive **25.** $-3, 6$ and 10 **27.** $\{x|\ -6 \leq x \leq 11\}$ **29.** $(-3,0), (6,0), (10,0)$ **31.** 3 times

33. (a) No **(b)** $-3; (4, -3)$ **(c)** $14; (14, 2)$ **(d)** $\{x|\ x \neq 6\}$

35. (a) Yes **(b)** $\dfrac{8}{17}; \left(2, \dfrac{8}{17}\right)$ **(c)** $-1, 1; (-1, 1), (1, 1)$ **(d)** All real numbers **37.** Not a function

39. Function **(a)** Domain: $\{x|\ -\pi \leq x \leq \pi\}$; Range: $\{y|\ -1 \leq y \leq 1\}$ **(b)** Intercepts: $\left(-\dfrac{\pi}{2}, 0\right), \left(\dfrac{\pi}{2}, 0\right), (0, 1)$ **(c)** y-axis

41. Not a function
43. Function **(a)** Domain: $\{x|x > 0\}$; Range: All real numbers **(b)** Intercept: $(1, 0)$ **(c)** None
45. Function **(a)** Domain: all real numbers; Range: $\{y|y \leq 2\}$ **(b)** Intercepts: $(-3, 0), (3, 0), (0, 2)$ **(c)** y-axis
47. Function **(a)** Domain: $\{x|x \neq 2\}$; Range: $\{y|y \neq 1\}$ **(b)** Intercept: $(0, 0)$ **(c)** None **49.** All real numbers **51.** All real numbers
53. $\{x|x \neq -1, x \neq 1\}$ **55.** $\{x|x \neq 0\}$ **57.** $\{x|x \geq 4\}$ **59.** $\{x|x > 9\}$ **61.** $\{x|x < 1 \text{ or } x \geq 2\}$

63. (a) III **(b)** IV **(c)** I **(d)** V **(e)** II **65.** $A = -\dfrac{7}{2}$ **67.** $A = -4$ **69.** $A = 8$; undefined at $x = 3$

71. (a) No **(b)**

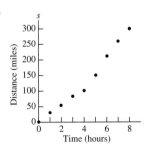

Price (dollars/pair)

(c) For the points $(20, 60)$ and $(30, 44)$, $D = -1.6p + 92$
(d) If the price increases \$1, the quantity demanded decreases by 1.6.
(e) $D(p) = -1.6p + 92$
(f) $\{p|\ 0 < p \leq 57.5\}$
(g) $D(28) = 47.2$; about 47 pairs
(h) $D = -1.34p + 86.20$

73. (a) Yes **(b)**

Time (hours)

(c) For the points $(0, 0)$ and $(8, 300)$, $s = \dfrac{75}{2}t$
(d) If t increases by 1 hour, then distance increases by 37.5 miles
(e) $s(t) = \dfrac{75}{2}t$
(f) $\{t|\ t \geq 0\}$
(g) 412.5 miles
(h) $s(t) = 37.78t - 19.13$

75. $A(x) = \dfrac{1}{2}x^2$ **77.** $G(x) = 10x$

79. (a) $A(x) = (8.5 - 2x)(11 - 2x)$
(b) Domain: $\{x|\ 0 \le x \le 4.25\}$; Range: $\{A|\ 0 \le A \le 93.5\}$
(c) $A(1) = 58.5\ \text{in.}^2$, $A(1.2) = 52.46\ \text{in.}^2$, $A(1.5) = 44\ \text{in.}^2$
(d)

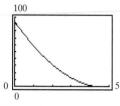

81. (a)

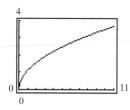

(b) The period of T varies from 1.107 seconds to 3.501 seconds.
(c) 81.56 feet

(e) $A(x) = 70$ when $x \approx 0.65$ in.; $A(x) = 50$ when $x \approx 1.28$ in.
83. Function   **85.** Function   **87.** Not a function   **89.** Function   **91.** Only $h(x) = 2x$ **93.** No
95. Any number of x-intercepts; at most 1 y-intercept   **97.** Yes, $f(x) = 0$

1.4 Exercises

1. C   **3.** E   **5.** B   **7.** F   **9. (a)** Domain: $\{x|\ -3 \le x \le 4\}$; Range: $\{y|\ 0 \le y \le 3\}$
(b) Increasing on $(-3, 0)$ and on $(2, 4)$; Decreasing on $(0, 2)$ **(c)** Neither   **(d)** $(-3, 0), (0, 3), (2, 0)$
11. (a) Domain: all real numbers; Range: $\{y|y > 0\}$ **(b)** Increasing on $(-\infty, \infty)$ **(c)** Neither   **(d)** $(0, 1)$
13. (a) Domain: $\{x|-\pi \le x \le \pi\}$; Range: $\{y|\ -1 \le y \le 1\}$ **(b)** Increasing on $\left(-\dfrac{\pi}{2}, \dfrac{\pi}{2}\right)$; Decreasing on $\left(-\pi, -\dfrac{\pi}{2}\right)$ and on $\left(\dfrac{\pi}{2}, \pi\right)$
(c) Odd   **(d)** $(-\pi, 0), (0, 0), (\pi, 0)$ **15. (a)** Domain: $\{x|x \ne 2\}$; Range: $\{y|y \ne 1\}$ **(b)** Decreasing on $(-\infty, 2)$ and on $(2, \infty)$
(c) Neither   **(d)** $(0, 0)$ **17. (a)** Domain: $\{x|x \ne 0\}$; Range: all real numbers   **(b)** Increasing on $(-\infty, 0)$ and on $(0, \infty)$ **(c)** Odd
(d) $(-1, 0), (1, 0)$ **19. (a)** Domain: $\{x|x \ne -2, x \ne 2\}$; Range: $\{y|y \le 0$ or $y > 1\}$
(b) Increasing on $(-\infty, -2)$ and on $(-2, 0)$; Decreasing on $(0, 2)$ and on $(2, \infty)$ **(c)** Even   **(d)** $(0, 0)$
21. (a) Domain: $\{x|-4 \le x \le 4\}$; Range: $\{y|0 \le y \le 2\}$ **(b)** Increasing on $(-2, 0)$ and on $(2, 4)$; Decreasing on $(-4, -2)$ and on $(0, 2)$
(c) Even   **(d)** $(-2, 0), (0, 2), (2, 0)$ **23. (a)** 4   **(b)** 2   **(c)** 5   **25. (a)** 2   **(b)** 3   **(c)** -4 **27.** 3   **29.** -3 **31.** $3x + 1$ **33.** $x^2 + x$
35. $-\dfrac{1}{x + 1}$ **37.** $\dfrac{1}{\sqrt{x} + 1} = \dfrac{\sqrt{x} - 1}{x - 1}$ **39.** Odd   **41.** Even   **43.** Odd   **45.** Neither   **47.** Even   **49.** Odd   **51.** At most one

53. (a) All real numbers
(b) $(0, 0)$
(c)

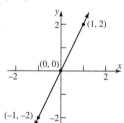

55. (a) All real numbers
(b) $(-1, 0), (0, 0)$
(c)

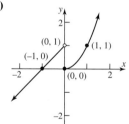

57. (a) $\{x|x \ge -2\}$
(b) $(0, 1)$
(c)

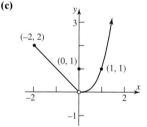

(d) All real numbers

(d) All real numbers

(d) $\{y|y > 0\}$

59. (a) $-2x + 5$ **(b)** $-2x - 5$ **(c)** $4x + 5$ **(d)** $2x - 1$ **(e)** $\dfrac{2}{x} + 5$ **(f)** $\dfrac{1}{2x + 5}$

61. (a) $2x^2 - 4$ **(b)** $-2x^2 + 4$ **(c)** $8x^2 - 4$ **(d)** $2x^2 - 12x + 14$ **(e)** $\dfrac{2}{x^2} - 4$ **(f)** $\dfrac{1}{2x^2 - 4}$

63. (a) $-x^3 + 3x$ **(b)** $-x^3 + 3x$ **(c)** $8x^3 - 6x$ **(d)** $x^3 - 9x^2 + 24x - 18$ **(e)** $\dfrac{1}{x^3} - \dfrac{3}{x}$ **(f)** $\dfrac{1}{x^3 - 3x}$

65. (a) $|x|$ **(b)** $-|x|$ **(c)** $2|x|$ **(d)** $|x - 3|$ **(e)** $\dfrac{1}{|x|}$ **(f)** $\dfrac{1}{|x|}$ **67.** 2   **69.** $2x + h + 2$

71. $f(x) = \begin{cases} -x & \text{if } -1 \le x \le 0 \\ \frac{1}{2}x & \text{if } 0 < x \le 2 \end{cases}$ (Other answers are possible.)

73. $f(x) = \begin{cases} -x & \text{if } x \le 0 \\ -x + 2 & \text{if } 0 < x \le 2 \end{cases}$ (Other answers are possible.)

75. Not even; $f(2) \ne f(-2)$

77. (a), (b), (e)

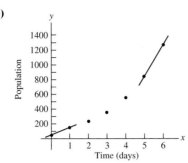

(c) 103 bacteria/day
(d) The population is increasing at an average rate of 103 bacteria per day between day 0 and day 1.
(f) 441 bacteria/day
(g) The population is increasing at an average rate of 441 bacteria per day between day 5 and day 6.
(h) The average rate of change is increasing.

79. (a), (b), (e)

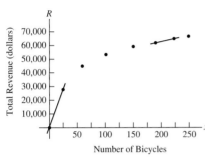

(c) 1120 dollars/bicycle
(d) For each additional bicycle sold between 0 and 25 bicycles, total revenue increases by $1120.
(f) 75 dollars/bicycle
(g) For each additional bicycle sold between 190 and 223 bicycles, total revenue increases by $75.
(h) As the number of bicycles sold increases, marginal revenue decreases.
(i) 150 bicycles

81. (a) $43.47   **(b)** $241.49

(c) $C = \begin{cases} 9 + 0.68935x & \text{if } 0 \le x \le 50 \\ 21.465 + 0.44005x & \text{if } x > 50 \end{cases}$

(d)

83. Each graph is that of $y = x^2$, but shifted vertically. If $y = x^2 + k, k > 0$, the shift is up k units; if $y = x^2 + k$, $k < 0$, the shift is down $|k|$ units.
85. Each graph is that of $y = |x|$, but either compressed or stretched. If $y = k|x|$ and $k > 1$, the graph is stretched vertically; if $y = k|x|, 0 < k < 1$, the graph is compressed vertically.
87. The graph of $y = f(-x)$ is the reflection about the y-axis of the graph of $y = f(x)$.
89. They are all ∪-shaped and open upward. All three go through the points $(-1, 1)$, $(0,0)$ and $(1,1)$. As the exponent increases, the steepness of the curve increases (except near $x = 0$).

1.5 Exercises

1. B   **3.** H   **5.** I   **7.** L   **9.** F   **11.** G   **13.** $y = (x - 4)^3$ **15.** $y = x^3 + 4$ **17.** $y = -x^3$ **19.** $y = 4x^3$ **21.** $y = -\sqrt{-x} - 2$
23. $y = -\sqrt{x + 3} + 2$
25.

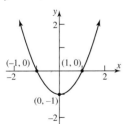

27.

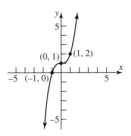

29.

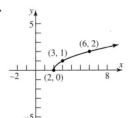

31.

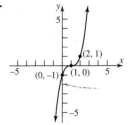

33.

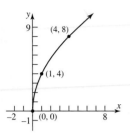

35.

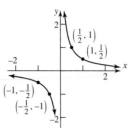

37.

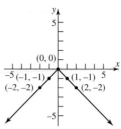

39.

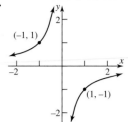

41.

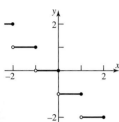

43.

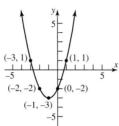

45.

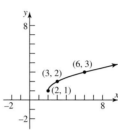

47.

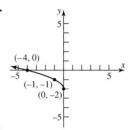

49.

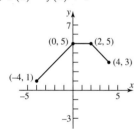

51.

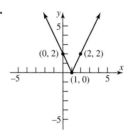

53.

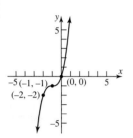

55. (a) $F(x) = f(x) + 3$

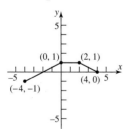

(b) $G(x) = f(x + 2)$

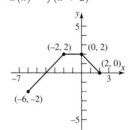

(c) $P(x) = -f(x)$

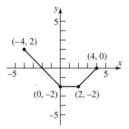

(d) $Q(x) = \dfrac{1}{2}f(x)$

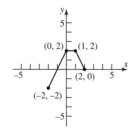

(e) $g(x) = f(-x)$

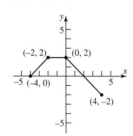

(f) $h(x) = f(2x)$

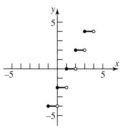

57. (a) $F(x) = f(x) + 3$

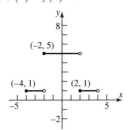

(b) $G(x) = f(x + 2)$

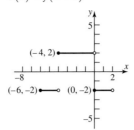

(c) $P(x) = -f(x)$

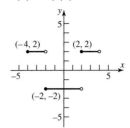

(d) $Q(x) = \frac{1}{2}f(x)$

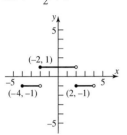

(e) $g(x) = f(-x)$

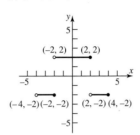

(f) $h(x) = f(2x)$

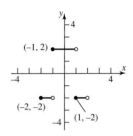

59. (a) $F(x) = f(x) + 3$

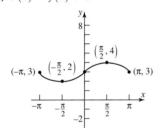

(b) $G(x) = f(x + 2)$

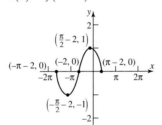

(c) $P(x) = -f(x)$

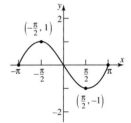

(d) $Q(x) = \frac{1}{2}f(x)$

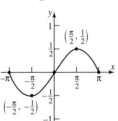

(e) $g(x) = f(-x)$

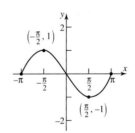

(f) $h(x) = f(2x)$

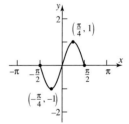

61. (a)

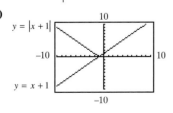

(b)

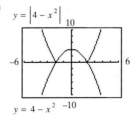

(c)

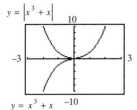

(d) Any part of the graph of $y = f(x)$ that lies below the x-axis is reflected about the x-axis to obtain the graph of $y = |f(x)|$.

63. (a)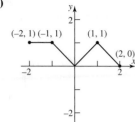

(b)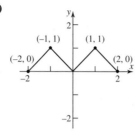

65. $f(x) = (x + 1)^2 - 1$

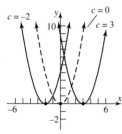

67. $f(x) = (x - 4)^2 - 15$

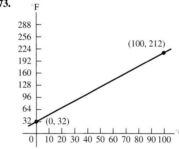

69. $f(x) = \left(x + \dfrac{1}{2}\right)^2 + \dfrac{3}{4}$

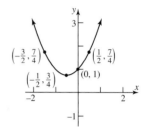

71.

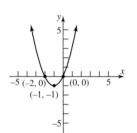

73.

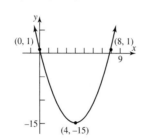
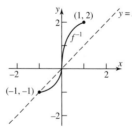

1.6 Exercises

1. (a)

Domain	Range
$200 →	20 hours
$300 →	25 hours
$350 →	30 hours
$425 →	40 hours

(b) Inverse is a function.

3. (a)

Domain	Range
$200 →	20 hours
	25 hours
$350 →	30 hours
$425 →	40 hours

(b) Inverse is not a function.

5. (a) $\{(6, 2), (6, -3), (9, 4), (10, 1)\}$ **(b)** Inverse is not a function.
7. (a) $\{(0, 0), (1, 1), (16, 2), (81, 3)\}$ **(c)** Inverse is a function.
9. One-to-one **11.** Not one-to-one **13.** One-to-one

15.

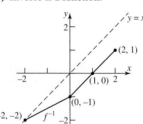

17.

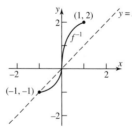

19.

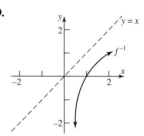

21. $f(g(x)) = f\left(\dfrac{1}{3}(x - 4)\right) = 3\left[\dfrac{1}{3}(x - 4)\right] + 4 = x; g(f(x)) = g(3x + 4) = \dfrac{1}{3}[(3x + 4) - 4] = x$

23. $f(g(x)) = 4\left[\dfrac{x}{4} + 2\right] - 8 = x; g(f(x)) = \dfrac{4x - 8}{4} + 2 = x$ **25.** $f(g(x)) = (\sqrt[3]{x + 8})^3 - 8 = x; g(f(x)) = \sqrt[3]{(x^3 - 8) + 8} = x$

27. $f(g(x)) = \dfrac{1}{\left(\dfrac{1}{x}\right)} = x; g(f(x)) = \dfrac{1}{\left(\dfrac{1}{x}\right)} = x$ **29.** $f(g(x)) = \dfrac{2\left(\dfrac{4x-3}{2-x}\right)+3}{\dfrac{4x-3}{2-x}+4} = x; g(f(x)) = \dfrac{4\left(\dfrac{2x+3}{x+4}\right)-3}{2-\dfrac{2x+3}{x+4}} = x$

31. $f^{-1}(x) = \dfrac{1}{3}x$

$f(f^{-1}(x)) = 3\left(\dfrac{1}{3}x\right) = x$

$f^{-1}(f(x)) = \dfrac{1}{3}(3x) = x$

Domain f = Range $f^{-1} = (-\infty, \infty)$
Range f = Domain $f^{-1} = (-\infty, \infty)$

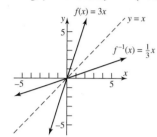

33. $f^{-1}(x) = \dfrac{x}{4} - \dfrac{1}{2}$

$f(f^{-1}(x)) = 4\left(\dfrac{x}{4} - \dfrac{1}{2}\right) + 2 = x$

$f^{-1}(f(x)) = \dfrac{4x+2}{4} - \dfrac{1}{2} = x$

Domain f = Range $f^{-1} = (-\infty, \infty)$
Range f = Domain $f^{-1} = (-\infty, \infty)$

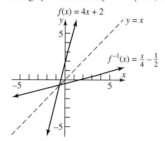

35. $f^{-1}(x) = \sqrt[3]{x+1}$

$f(f^{-1}(x)) = (\sqrt[3]{x+1})^3 - 1 = x$

$f^{-1}(f(x)) = \sqrt[3]{(x^3-1)+1} = x$

Domain f = Range $f^{-1} = (-\infty, \infty)$
Range f = Domain $f^{-1} = (-\infty, \infty)$

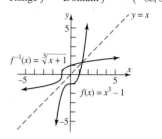

37. $f^{-1}(x) = \sqrt{x-4}$

$f(f^{-1}(x)) = (\sqrt{x-4})^2 + 4 = x$

$f^{-1}(f(x)) = \sqrt{(x^2+4)-4} = \sqrt{x^2} = |x| = x, x \geq 0$

Domain f = Range $f^{-1} = [0, \infty)$
Range f = Domain $f^{-1} = [4, \infty)$

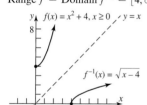

39. $f^{-1}(x) = \dfrac{4}{x}$

$f(f^{-1}(x)) = \dfrac{4}{4/x}$

$f^{-1}(f(x)) = \dfrac{4}{4/x}$

Domain f = Range f^{-1} = All real numbers except 0
Range f = Domain f^{-1} = All real numbers except 0

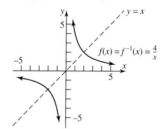

41. $f^{-1}(x) = \dfrac{2x+1}{x}$

$f(f^{-1}(x)) = \dfrac{1}{\dfrac{2x+1}{x}-2} = x$

$f^{-1}(f(x)) = \dfrac{2\left(\dfrac{1}{x-2}\right)+1}{\dfrac{1}{x-2}} = x$

Domain f = Range f^{-1} = All real numbers except 2
Range f = Domain f^{-1} = All real numbers except 0

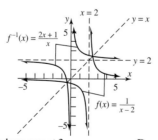

43. $f^{-1}(x) = \dfrac{2-3x}{x}$

$f(f^{-1}(x)) = \dfrac{2}{3+\dfrac{2-3x}{x}} = x$

$f^{-1}(f(x)) = \dfrac{2-3\left(\dfrac{2}{3+x}\right)}{\dfrac{2}{3+x}} = x$

Domain f = Range f^{-1} = All real numbers except -3
Range f = Domain f^{-1} = All real numbers except 0

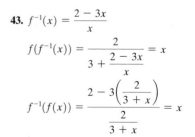

45. $f^{-1}(x) = \sqrt{x} - 2$
$f(f^{-1}(x)) = (\sqrt{x} - 2 + 2)^2 = x$
$f^{-1}(f(x)) = \sqrt{(x+2)^2} - 2 = |x+2| - 2 = x, x \geq -2$
Domain f = Range $f^{-1} = [-2, \infty)$; Range f = Domain $f^{-1} = [0, \infty)$

47. $f^{-1}(x) = \dfrac{x}{x-2}$

$f(f^{-1}(x)) = \dfrac{2\left(\dfrac{x}{x-2}\right)}{\dfrac{x}{x-2}-1} = x$

$f^{-1}(f(x)) = \dfrac{\dfrac{2x}{x-1}}{\dfrac{2x}{x-1}-2} = x$

Domain f = Range f^{-1} = All real numbers except 1

Range f = Domain f^{-1} = All real numbers except 2

49. $f^{-1}(x) = \dfrac{3x+4}{2x-3}$

$f(f^{-1}(x)) = \dfrac{3\left(\dfrac{3x+4}{2x-3}\right)+4}{2\left(\dfrac{3x+4}{2x-3}\right)-3} = x$

$f^{-1}(f(x)) = \dfrac{3\left(\dfrac{3x+4}{2x-3}\right)+4}{2\left(\dfrac{3x+4}{2x-3}\right)-3} = x$

Domain f = Range f^{-1} = All real numbers except $\dfrac{3}{2}$

Range f = Domain f^{-1} = All real numbers except $\dfrac{3}{2}$

51. $f^{-1}(x) = \dfrac{-2x+3}{x-2}$

$f(f^{-1}(x)) = \dfrac{2\left(\dfrac{-2x+3}{x-2}\right)+3}{\dfrac{-2x+3}{x-2}+2} = x$

$f^{-1}(f(x)) = \dfrac{-2\left(\dfrac{2x+3}{x+2}\right)+3}{\dfrac{2x+3}{x+2}-2} = x$

Domain f = Range f^{-1} = All real numbers except -2
Range f = Domain f^{-1} = All real numbers except 2

53. $f^{-1}(x) = \dfrac{x^3}{8}$

$f(f^{-1}(x)) = 2\sqrt[3]{\dfrac{x^3}{8}} = x$

$f^{-1}(f(x)) = \dfrac{(2\sqrt[3]{x})^3}{8} = x$

Domain f = Range f^{-1} = $(-\infty, \infty)$
Range f = Domain f^{-1} = $(-\infty, \infty)$

55. $f^{-1}(x) = \dfrac{1}{m}(x-b), m \neq 0$ **59.** Quadrant I **61.** $f(x) = |x|, x \geq 0$, is one-to-one; $f^{-1}(x) = x, x \geq 0$

63. $f(g(x)) = \dfrac{9}{5}\left[\dfrac{5}{9}(x-32)\right] + 32 = x; g(f(x)) = \dfrac{5}{9}\left[\left(\dfrac{9}{5}x + 32\right) - 32\right] = x$ **65.** $l(T) = \dfrac{gT^2}{4\pi^2}, T > 0$

67. $f^{-1}(x) = \dfrac{-dx+b}{cx-a}; f = f^{-1}$ if $a = -d$

Fill-in-the-Blank Items

1. y-axis **3.** independent; dependent **5.** even; odd **7.** one-to-one

True/False Items

1. T **3.** F **5.** T **7.** T **9.** T

Review Exercises

1. Center $(1, -2)$, radius $= 3$ **3.** Center $(1, -2)$, radius $= \sqrt{5}$ **5.** Intercept: $(0, 0)$; symmetric with respect to the x-axis

7. Intercepts: $(0, -1), (0, 1), \left(\dfrac{1}{2}, 0\right), \left(-\dfrac{1}{2}, 0\right)$; symmetric with respect to the x-axis, y-axis and origin

9. Intercept: $(0, 1)$; symmetric with respect to the y-axis **11.** Intercepts: $(0, 0), (0, -2), (-1, 0)$; no symmetry
13. $f(x) = -2x + 3$ **15.** $A = 11$ **17.** b, c, d
19. (a) $f(-x) = \dfrac{-3x}{x^2-4}$ (b) $-f(x) = \dfrac{-3x}{x^2-4}$ (c) $f(x+2) = \dfrac{3x+6}{x^2+4x}$ (d) $f(x-2) = \dfrac{3x-6}{x^2-4x}$
21. (a) $f(-x) = \sqrt{x^2-4}$ (b) $-f(x) = -\sqrt{x^2-4}$ (c) $f(x+2) = \sqrt{x^2+4x}$ (d) $f(x-2) = \sqrt{x^2-4x}$
23. (a) $f(-x) = \dfrac{x^2-4}{x^2}$ (b) $-f(x) = -\dfrac{x^2-4}{x^2}$ (c) $f(x+2) = \dfrac{x^2+4x}{x^2+4x+4}$ (d) $f(x-2) = \dfrac{x^2-4x}{x^2-4x+4}$
25. Odd **27.** Even **29.** Neither **31.** $\{x|x \neq -3, x \neq 3\}$ **33.** $\{x|x \leq 2\}$ **35.** $\{x|x > 0\}$ **37.** $\{x|x \neq -3, x \neq 1\}$
39. $\{x|x \geq -1\}$ **41.** $\{x|x \geq 0\}$ **43.** -5 **45.** $3 - 4x$

47.

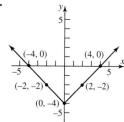

Intercepts: $(-4, 0)$, $(0, -4)$, $(4, 0)$
Domain: all real numbers
Range: $\{y | y \geq -4\}$

49.

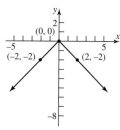

Intercept: $(0, 0)$
Domain: all real numbers
Range: $\{y | y \leq 0\}$

51.

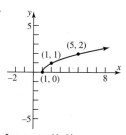

Intercept: $(1, 0)$
Domain: $\{x | x \geq 1\}$
Range: $\{y | y \geq 0\}$

53.

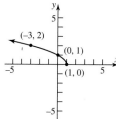

Intercepts: $(0, 1)$, $(1, 0)$
Domain: $\{x | x \leq 1\}$
Range: $\{y | y \geq 0\}$

55.

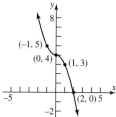

Intercepts: $(0, 4)$, $(2, 0)$
Domain: all real numbers
Range: all real numbers

57.

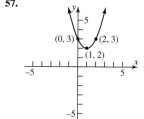

Intercept: $(0, 3)$
Domain: all real numbers
Range: $\{y | y \geq 2\}$

59.

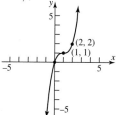

Intercept: $(0, 0)$
Domain: all real numbers
Range: all real numbers

61.

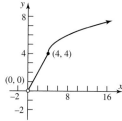

Intercept: none
Domain: $\{x | x > 0\}$
Range: $\{y | y > 0\}$

63.

Intercept: $(0, 0)$
Domain: $\{x | x \neq 1\}$
Range: $\{y | y \neq 1\}$

65.

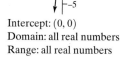

Intercepts: $(x, 0)$, where $-1 < x \leq 0$
Domain: all real numbers
Range: the set of integers

67. $f^{-1}(x) = \dfrac{2x + 3}{x - 2}; f(f^{-1}(x)) = \dfrac{2\left(\dfrac{2x + 3}{x - 2}\right) + 3}{\dfrac{2x + 3}{x - 2} - 2} = x; f^{-1}(f(x)) = \dfrac{2\left(\dfrac{2x + 3}{x - 2}\right) + 3}{\dfrac{2x + 3}{x - 2} - 2} = x;$

Domain f = Range f^{-1} = All real numbers except 2; Range f = Domain f^{-1} = All real numbers except 2

69. $f^{-1}(x) = \dfrac{x + 1}{x}; f(f^{-1}(x)) = \dfrac{1}{\dfrac{x + 1}{x} - 1} = x; f^{-1}(f(x)) = \dfrac{\dfrac{1}{x - 1} + 1}{\dfrac{1}{x - 1}} = x;$

Domain f = Range f^{-1} = All real numbers except 1; Range f = Domain f^{-1} = All real numbers except 0

71. $f^{-1}(x) = \dfrac{27}{x^3}; f(f^{-1}(x)) = \dfrac{3}{\left(\dfrac{27}{x^3}\right)^{1/3}} = x; f^{-1}(f(x)) = \dfrac{27}{\left(\dfrac{3}{x^{1/3}}\right)^3} = x;$

Domain f = Range f^{-1} = All real numbers except 0; Range f = Domain f^{-1} = All real numbers except 0

73. (a)

(b)

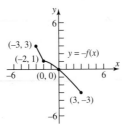

(c)

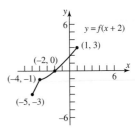

(d)

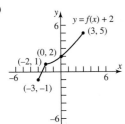

(e)

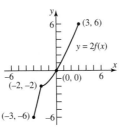

(f)

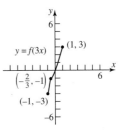

75. $T(h) = -0.0025h + 30, 0 \le x \le 10{,}000$ **77.** $S(x) = kx(36 - x^2)^{3/2}$; Domain: $\{x \mid 0 < x < 6\}$

C H A P T E R   2   2.1 Exercises

1.

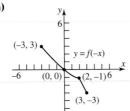

3.

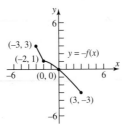

135°

5.

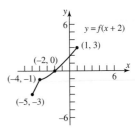

450°

7.

$\frac{3\pi}{4}$

9.

$-\frac{\pi}{6}$

11.

$\frac{16\pi}{3}$

13. $\frac{\pi}{6}$ **15.** $\frac{4\pi}{3}$ **17.** $-\frac{\pi}{3}$ **19.** π **21.** $-\frac{3\pi}{4}$ **23.** $-\frac{\pi}{2}$ **25.** 60° **27.** −225° **29.** 90° **31.** 15° **33.** −90° **35.** −30° **37.** 5 m

39. 6 ft **41.** 0.6 radian **43.** $\frac{\pi}{3} \approx 1.047$ in. **45.** 0.30 **47.** −0.70 **49.** 2.18 **51.** 5.93 **53.** 179.91° **55.** 587.28° **57.** 114.59°

59. 362.11° **61.** 40.17° **63.** 1.03° **65.** 9.15° **67.** 40°19′12″ **69.** 18°15′18″ **71.** 19°59′24″ **73.** $3\pi \approx 9.4248$ in.; $5\pi \approx 15.7080$ in.

75. $\omega = \frac{1}{60}$ radian/sec; $v = \frac{1}{12}$ cm/sec **77.** approximately 452.5 rpm **79.** approximately 37.7 in. **81.** approximately 2292 mph

83. $\frac{3}{4}$ rpm **85.** approximately 2.86 mph **87.** approximately 31.47 revolutions/minute **89.** approximately 1037 mi/hr

91. $v_1 = r_1 w_1$, $v_2 = r_2 w_2$, and $v_1 = v_2$ so $r_1 w_1 = r_2 w_2 \Rightarrow \dfrac{r_1}{r_2} = \dfrac{w_2}{w_1}$.

2.2 Exercises

1. $\sin \theta = \dfrac{4}{5}$; $\cos \theta = -\dfrac{3}{5}$; $\tan \theta = -\dfrac{4}{3}$; $\csc \theta = \dfrac{5}{4}$; $\sec \theta = -\dfrac{5}{3}$; $\cot \theta = -\dfrac{3}{4}$

3. $\sin \theta = -\dfrac{3\sqrt{13}}{13}$; $\cos \theta = \dfrac{2\sqrt{13}}{13}$; $\tan \theta = -\dfrac{3}{2}$; $\csc \theta = -\dfrac{\sqrt{13}}{3}$; $\sec \theta = \dfrac{\sqrt{13}}{2}$; $\cot \theta = -\dfrac{2}{3}$

5. $\sin \theta = -\dfrac{\sqrt{2}}{2}$; $\cos \theta = -\dfrac{\sqrt{2}}{2}$; $\tan \theta = 1$; $\csc \theta = -\sqrt{2}$; $\sec \theta = -\sqrt{2}$; $\cot \theta = 1$

7. $\sin \theta = -\dfrac{2\sqrt{13}}{13}$; $\cos \theta = -\dfrac{3\sqrt{13}}{13}$; $\tan \theta = \dfrac{2}{3}$; $\csc \theta = -\dfrac{\sqrt{13}}{2}$; $\sec \theta = -\dfrac{\sqrt{13}}{3}$; $\cot \theta = \dfrac{3}{2}$

9. $\sin\theta = -\dfrac{3}{5}$; $\cos\theta = \dfrac{4}{5}$; $\tan\theta = -\dfrac{3}{4}$; $\csc\theta = -\dfrac{5}{3}$; $\sec\theta = \dfrac{5}{4}$; $\cot\theta = -\dfrac{4}{3}$ **11.** $\dfrac{1}{2}(\sqrt{2}+1)$ **13.** 2 **15.** $\dfrac{1}{2}$ **17.** $\sqrt{6}$ **19.** 4

21. 0 **23.** 0 **25.** $2\sqrt{2}+\dfrac{4\sqrt{3}}{3}$ **27.** -1 **29.** 1

31. $\sin\left(\dfrac{2\pi}{3}\right) = \dfrac{\sqrt{3}}{2}$; $\cos\left(\dfrac{2\pi}{3}\right) = -\dfrac{1}{2}$; $\tan\left(\dfrac{2\pi}{3}\right) = -\sqrt{3}$; $\csc\left(\dfrac{2\pi}{3}\right) = \dfrac{2\sqrt{3}}{3}$; $\sec\left(\dfrac{2\pi}{3}\right) = -2$; $\cot\left(\dfrac{2\pi}{3}\right) = -\dfrac{\sqrt{3}}{3}$

33. $\sin 150° = \dfrac{1}{2}$; $\cos 150° = -\dfrac{\sqrt{3}}{2}$; $\tan 150° = -\dfrac{\sqrt{3}}{3}$; $\csc 150° = 2$; $\sec 150° = -\dfrac{2\sqrt{3}}{3}$; $\cot 150° = -\sqrt{3}$

35. $\sin\left(-\dfrac{\pi}{6}\right) = -\dfrac{1}{2}$; $\cos\left(-\dfrac{\pi}{6}\right) = \dfrac{\sqrt{3}}{2}$; $\tan\left(-\dfrac{\pi}{6}\right) = -\dfrac{\sqrt{3}}{3}$; $\csc\left(-\dfrac{\pi}{6}\right) = -2$; $\sec\left(-\dfrac{\pi}{6}\right) = \dfrac{2\sqrt{3}}{3}$; $\cot\left(-\dfrac{\pi}{6}\right) = -\sqrt{3}$

37. $\sin 225° = -\dfrac{\sqrt{2}}{2}$, $\cos 225° = -\dfrac{\sqrt{2}}{2}$; $\tan 225° = 1$; $\csc 225° = -\sqrt{2}$; $\sec 225° = -\sqrt{2}$; $\cot 225° = 1$

39. $\sin\left(\dfrac{5\pi}{2}\right) = 1$; $\cos\left(\dfrac{5\pi}{2}\right) = 0$; $\tan\left(\dfrac{5\pi}{2}\right)$ is not defined; $\csc\left(\dfrac{5\pi}{2}\right) = 1$; $\sec\left(\dfrac{5\pi}{2}\right)$ is not defined; $\cot\left(\dfrac{5\pi}{2}\right) = 0$

41. $\sin(-180°) = 0$; $\cos(-180°) = -1$; $\tan(-180°) = 0$; $\csc(-180°)$ is not defined; $\sec(-180°) = -1$; $\cot(-180°)$ is not defined

43. $\sin\left(\dfrac{3\pi}{2}\right) = -1$; $\cos\left(\dfrac{3\pi}{2}\right) = 0$; $\tan\left(\dfrac{3\pi}{2}\right)$ is not defined; $\csc\left(\dfrac{3\pi}{2}\right) = -1$; $\sec\left(\dfrac{3\pi}{2}\right)$ is not defined; $\cot\left(\dfrac{3\pi}{2}\right) = 0$

45. $\sin 450° = 1$; $\cos 450° = 0$; $\tan 450°$ is not defined; $\csc 450° = 1$; $\sec 450°$ is not defined; $\cot 450° = 0$ **47.** 0.47 **49.** 0.38 **51.** 1.33

53. 0.36 **55.** 0.31 **57.** 3.73 **59.** 1.04 **61.** 5.67 **63.** 0.84 **65.** 0.02 **67.** 0.93 **69.** 0.31 **71.** $\dfrac{\sqrt{3}}{2}$ **73.** $\dfrac{1}{2}$ **75.** $\dfrac{3}{4}$ **77.** $\dfrac{\sqrt{3}}{2}$

79. $\sqrt{3}$ **81.** $\dfrac{\sqrt{3}}{4}$ **83.** 0 **85.** -0.1 **87.** 3 **89.** 5 **91.** $R \approx 310.56$ ft; $H \approx 77.64$ ft **93.** $R \approx 19{,}542$ m; $H \approx 2278$ m

95. (a) 1.2 sec **(b)** 1.12 sec **(c)** 1.2 sec **97. (a)** 1.9 hr; 0.57 hr **(b)** 1.69 hr; 0.75 hr **(c)** 1.63 hr; 0.86 hr **(d)** 1.67 hr

99. (a) 16.6 ft **(b)**

(c) 67.5°

2.3 Exercises

1. $\dfrac{\sqrt{2}}{2}$ **3.** 1 **5.** 1 **7.** $\sqrt{3}$ **9.** $\dfrac{\sqrt{2}}{2}$ **11.** 0 **13.** $\sqrt{2}$ **15.** $\dfrac{\sqrt{3}}{3}$ **17.** II **19.** IV **21.** IV **23.** II

25. $\tan\theta = 2$; $\cot\theta = \dfrac{1}{2}$; $\sec\theta = \sqrt{5}$; $\csc\theta = \dfrac{\sqrt{5}}{2}$ **27.** $\tan\theta = \dfrac{\sqrt{3}}{3}$; $\cot\theta = \sqrt{3}$; $\sec\theta = \dfrac{2\sqrt{3}}{3}$; $\csc\theta = 2$

29. $\tan\theta = -\dfrac{\sqrt{2}}{4}$; $\cot\theta = -2\sqrt{2}$; $\sec\theta = \dfrac{3\sqrt{2}}{4}$; $\csc\theta = -3$ **31.** $\tan\theta \approx 0.2679$; $\cot\theta \approx 3.7322$; $\sec\theta \approx 1.0353$; $\csc\theta \approx 3.8640$

33. $\cos\theta = -\dfrac{5}{13}$; $\tan\theta = -\dfrac{12}{5}$; $\csc\theta = \dfrac{13}{12}$; $\sec\theta = -\dfrac{13}{5}$; $\cot\theta = -\dfrac{5}{12}$

35. $\sin\theta = -\dfrac{3}{5}$; $\tan\theta = \dfrac{3}{4}$; $\csc\theta = -\dfrac{5}{3}$; $\sec\theta = -\dfrac{5}{4}$; $\cot\theta = \dfrac{4}{3}$

37. $\cos\theta = -\dfrac{12}{13}$; $\tan\theta = -\dfrac{5}{12}$; $\cot\theta = -\dfrac{12}{5}$; $\sec\theta = -\dfrac{13}{12}$; $\csc\theta = \dfrac{13}{5}$

39. $\sin\theta = \dfrac{2\sqrt{2}}{3}$; $\tan\theta = -2\sqrt{2}$; $\cot\theta = -\dfrac{\sqrt{2}}{4}$; $\sec\theta = -3$; $\csc\theta = \dfrac{3\sqrt{2}}{4}$

41. $\cos\theta = -\dfrac{\sqrt{5}}{3}$; $\tan\theta = -\dfrac{2\sqrt{5}}{5}$; $\cot\theta = -\dfrac{\sqrt{5}}{2}$; $\sec\theta = -\dfrac{3\sqrt{5}}{5}$; $\csc\theta = \dfrac{3}{2}$

43. $\sin\theta = -\dfrac{\sqrt{3}}{2}$; $\cos\theta = \dfrac{1}{2}$; $\tan\theta = -\sqrt{3}$; $\cot\theta = -\dfrac{\sqrt{3}}{3}$; $\csc\theta = -\dfrac{2\sqrt{3}}{3}$

45. $\sin\theta = -\dfrac{3}{5}$; $\cos\theta = -\dfrac{4}{5}$; $\cot\theta = \dfrac{4}{3}$; $\sec\theta = -\dfrac{5}{4}$; $\csc\theta = -\dfrac{5}{3}$

47. $\sin\theta = \dfrac{\sqrt{10}}{10}$; $\cos\theta = -\dfrac{3\sqrt{10}}{10}$; $\cot\theta = -3$; $\sec\theta = -\dfrac{\sqrt{10}}{3}$; $\csc\theta = \sqrt{10}$ **49.** $-\dfrac{\sqrt{3}}{2}$ **51.** $-\dfrac{\sqrt{3}}{3}$ **53.** 2   **55.** -1 **57.** -1

59. $\dfrac{\sqrt{2}}{2}$ **61.** 0   **63.** $-\sqrt{2}$ **65.** $\dfrac{2\sqrt{3}}{3}$ **67.** -1 **69.** -2 **71.** $\dfrac{2-\sqrt{2}}{2}$ **73.** 1   **75.** 1   **77.** 0   **79.** 0.9   **81.** 9

83. All real numbers   **85.** Odd multiples of $\dfrac{\pi}{2}$ **87.** Odd multiples of $\dfrac{\pi}{2}$ **89.** $-1 \le y \le 1$ **91.** All real numbers   **93.** $|y| \ge 1$

95. Odd; Yes; Origin   **97.** Odd; Yes; Origin   **99.** Even; Yes; y-axis   **101. (a)** $-\dfrac{1}{3}$ **(b)** 1   **103. (a)** -2 **(b)** 6   **105. (a)** -4 **(b)** -12

107. 15.8 min

109. Let $P = (x, y)$ be the point on the unit circle that corresponds to θ. Consider the equation $\tan\theta = \dfrac{y}{x} = a$. Then $y = ax$. But

$x^2 + y^2 = 1$ so that $x^2 + a^2 x^2 = 1$. Thus, $x = \pm\dfrac{1}{\sqrt{1+a^2}}$ and $y = \pm\dfrac{a}{\sqrt{1+a^2}}$; that is, for any real number a, there is a point

$P = (x, y)$ in the unit circle for which $\tan\theta = a$. In other words, $-\infty < \tan\theta < \infty$, and the range of the tangent function is the set of all
real numbers.

111. Suppose there is a number $p, 0 < p < 2\pi$, for which $\sin(\theta + p) = \sin\theta$ for all θ. If $\theta = 0$, then $\sin(0 + p) = \sin p = \sin 0 = 0$; so

that $p = \pi$. If $\theta = \dfrac{\pi}{2}$, then $\sin\left(\dfrac{\pi}{2} + p\right) = \sin\left(\dfrac{\pi}{2}\right)$. But $p = \pi$. Thus, $\sin\left(\dfrac{3\pi}{2}\right) = -1 = \sin\left(\dfrac{\pi}{2}\right) = 1$. This is impossible.

The smallest positive number p for which $\sin(\theta + p) = \sin\theta$ for all θ is therefore $p = 2\pi$.

113. $\sec\theta = \dfrac{1}{(\cos\theta)}$; since $\cos\theta$ has period 2π, so does $\sec\theta$

115. If $P = (a, b)$ is the point on the unit circle corresponding to θ, then $Q = (-a, -b)$ is the point on the unit circle corresponding to

$\theta + \pi$. Thus $\tan(\theta + \pi) = \dfrac{(-b)}{(-a)} = \dfrac{b}{a} = \tan\theta$; that is, the period of the tangent function is π.

117. Let $P = (a, b)$ be the point on the unit circle corresponding to θ. Then $\csc\theta = \dfrac{1}{b} = \dfrac{1}{(\sin\theta)}$; $\sec\theta = \dfrac{1}{a} = \dfrac{1}{(\cos\theta)}$;

$\cot\theta = \dfrac{a}{b} = \dfrac{1}{b/a} = \dfrac{1}{(\tan\theta)}$.

119. $(\sin\theta\cos\phi)^2 + (\sin\theta\sin\phi)^2 + \cos^2\theta = \sin^2\theta\cos^2\phi + \sin^2\theta\sin^2\phi + \cos^2\theta$
$= \sin^2\theta(\cos^2\phi + \sin^2\phi) + \cos^2\theta = \sin^2\theta + \cos^2\theta = 1$

2.4 Exercises

1. $\sin\theta = \dfrac{5}{13}$; $\cos\theta = \dfrac{12}{13}$; $\tan\theta = \dfrac{5}{12}$; $\csc\theta = \dfrac{13}{5}$; $\sec\theta = \dfrac{13}{12}$; $\cot\theta = \dfrac{12}{5}$

3. $\sin\theta = \dfrac{2\sqrt{13}}{13}$; $\cos\theta = \dfrac{3\sqrt{13}}{13}$; $\tan\theta = \dfrac{2}{3}$; $\csc\theta = \dfrac{\sqrt{13}}{2}$; $\sec\theta = \dfrac{\sqrt{13}}{3}$; $\cot\theta = \dfrac{3}{2}$

5. $\sin\theta = \dfrac{\sqrt{3}}{2}$; $\cos\theta = \dfrac{1}{2}$; $\tan\theta = \sqrt{3}$; $\csc\theta = \dfrac{2\sqrt{3}}{3}$; $\sec\theta = 2$; $\cot\theta = \dfrac{\sqrt{3}}{3}$

7. $\sin\theta = \dfrac{\sqrt{6}}{3}$; $\cos\theta = \dfrac{\sqrt{3}}{3}$; $\tan\theta = \sqrt{2}$; $\csc\theta = \dfrac{\sqrt{6}}{2}$; $\sec\theta = \sqrt{3}$; $\cot\theta = \dfrac{\sqrt{2}}{2}$

9. $\sin\theta = \dfrac{\sqrt{5}}{5}$; $\cos\theta = \dfrac{2\sqrt{5}}{5}$; $\tan\theta = \dfrac{1}{2}$; $\csc\theta = \sqrt{5}$; $\sec\theta = \dfrac{\sqrt{5}}{2}$; $\cot\theta = 2$ **11.** $30°$ **13.** $60°$ **15.** $30°$ **17.** $\dfrac{\pi}{4}$ **19.** $\dfrac{\pi}{3}$ **21.** $45°$

23. $\dfrac{\pi}{3}$ **25.** $60°$ **27.** $\dfrac{1}{2}$ **29.** $\dfrac{\sqrt{2}}{2}$ **31.** -2 **33.** $-\sqrt{3}$ **35.** $\dfrac{\sqrt{2}}{2}$ **37.** $\sqrt{3}$ **39.** $\dfrac{1}{2}$ **41.** $-\dfrac{\sqrt{3}}{2}$ **43.** $-\sqrt{3}$ **45.** $\sqrt{2}$ **47.** 0 **49.** 1

51. 0 **53.** 0 **55.** 1 **57. (a)** $\dfrac{1}{3}$ **(b)** $\dfrac{8}{9}$ **(c)** 3 **(d)** 3 **59. (a)** 17 **(b)** $\dfrac{1}{4}$ **(c)** 4 **(d)** $\dfrac{17}{16}$ **61. (a)** $\dfrac{1}{4}$ **(b)** 15 **(c)** 4 **(d)** $\dfrac{16}{15}$

63. 0.6 **65.** 0 **67.** $20°$

69. (a) $T(\theta) = 1 + \dfrac{2}{3\sin\theta} - \dfrac{1}{4\tan\theta}$ **(b)** 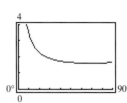 $\theta = 67.97°$ for least time; the least time is $T = 1.61$ hrs. Sally is on the road 0.9 hr.

71. (a) 10 min   **(b)** 20 min   **(c)** $T(\theta) = 5\left(1 - \dfrac{1}{3\tan\theta} + \dfrac{1}{\sin\theta}\right)$ **(d)** 10.4 min

(e) T is least for $\theta = 70.5°$; the least time is 9.7 min; $x = 177$ ft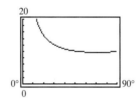

75.

θ	0.5	0.4	0.2	0.1	0.01	0.001	0.0001	0.00001
$\sin\theta$	0.4794	0.3894	0.1987	0.0998	0.0100	0.0010	0.0001	0.00001
$\dfrac{\sin\theta}{\theta}$	0.9589	0.9735	0.9933	0.9983	1.0000	1.0000	1.0000	1.0000

$\dfrac{(\sin\theta)}{\theta}$ approaches 1 as θ approaches 0.

77. (a) $|OA| = |OC| = 1$; angle OAC + angle OCA + $180° - \theta = 180°$; angle OAC = angle OCA, so angle $OAC = \dfrac{\theta}{2}$

(b) $\sin\theta = \dfrac{|CD|}{|OC|} = |CD|$; $\cos\theta = \dfrac{|OD|}{|OC|} = |OD|$ **(c)** $\tan\dfrac{\theta}{2} = \dfrac{|CD|}{|AD|} = \dfrac{\sin\theta}{1 + |OD|} = \dfrac{\sin\theta}{1 + \cos\theta}$

79. $h = x\tan\theta$ and $h = (1 - x)\tan n\theta$; thus, $x\tan\theta = (1 - x)\tan n\theta$
$$x = \dfrac{\tan n\theta}{\tan\theta + \tan n\theta}$$

81. (a) Area $\triangle OAC = \dfrac{1}{2}|OC||AC| = \dfrac{1}{2}\cdot\dfrac{|OC|}{1}\cdot\dfrac{|AC|}{1} = \dfrac{1}{2}\sin\alpha\cos\alpha$

(b) Area $\triangle OCB = \dfrac{1}{2}|BC||OC| = \dfrac{1}{2}|OB|^2\dfrac{|BC|}{|OB|}\cdot\dfrac{|OC|}{|OB|} = \dfrac{1}{2}|OB|^2\sin\beta\cos\beta$

(c) Area $\triangle OAB = \dfrac{1}{2}|BD||OA| = \dfrac{1}{2}|OB|\dfrac{|BD|}{|OB|} = \dfrac{1}{2}|OB|\sin(\alpha + \beta)$

(d) $\dfrac{\cos\alpha}{\cos\beta} = \dfrac{\dfrac{|OC|}{1}}{\dfrac{|OC|}{|OB|}} = |OB|$ **(e)** Use the hint and above results.

83. $\sin \alpha = \tan \alpha \cos \alpha = \cos \beta \cos \alpha = \cos \beta \tan \beta = \sin \beta$;

$$\sin^2\alpha + \cos^2 \alpha = 1$$
$$\sin^2 \alpha + \tan^2 \beta = 1$$
$$\sin^2 \alpha + \frac{\sin^2 \beta}{\cos^2 \beta} = 1$$
$$\sin^2 \alpha + \frac{\sin^2 \alpha}{1 - \sin^2\alpha} = 1$$
$$\sin^2 \alpha - \sin^4 \alpha + \sin^2 \alpha = 1 - \sin^2 \alpha$$
$$\sin^4 \alpha - 3 \sin^2 \alpha + 1 = 0$$
$$\sin^2 \alpha = \frac{3 \pm \sqrt{5}}{2}$$
$$\sin^2 \alpha = \frac{3 - \sqrt{5}}{2}, \text{ since } \frac{3 + \sqrt{5}}{2} > 1$$
$$\sin \alpha = \sqrt{\frac{3 - \sqrt{5}}{2}}$$

2.5 Exercises

1. 0   **3.** $-\dfrac{\pi}{2} \le x \le \dfrac{\pi}{2}$ **5.** 1   **7.** $0, \pi, 2\pi$ **9.** $\sin x = 1$ for $x = -\dfrac{3\pi}{2}, \dfrac{\pi}{2}$; $\sin x = -1$ for $x = -\dfrac{\pi}{2}, \dfrac{3\pi}{2}$ **11.** 0   **13.** 1

15. $\sec x = 1$ for $x = -2\pi, 0, 2\pi$; $\sec x = -1$ for $x = -\pi, \pi$ **17.** $-\dfrac{3\pi}{2}, -\dfrac{\pi}{2}, \dfrac{\pi}{2}, \dfrac{3\pi}{2}$ **19.** $-\dfrac{3\pi}{2}, -\dfrac{\pi}{2}, \dfrac{\pi}{2}, \dfrac{3\pi}{2}$

21. B, C, F   **23.** C   **25.** D   **27.** D   **29.** B

31.

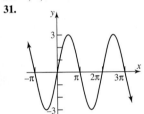

33.

35.

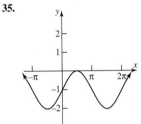

37.

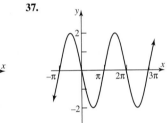

39.

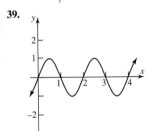

41.

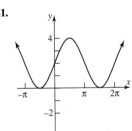

43.

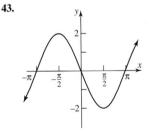

45.

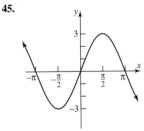

47.

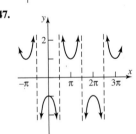

49.

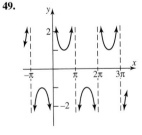

51.

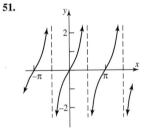

53.

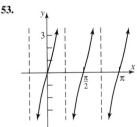

55.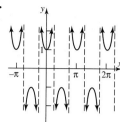

57.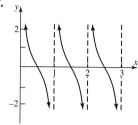

59.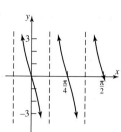

61.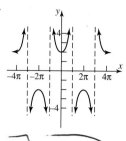

63.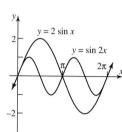

65.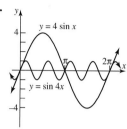

67.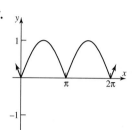

69. (a) $L = \dfrac{3}{\cos \theta} + \dfrac{4}{\sin \theta} = 3 \sec \theta + 4 \csc \theta$ **(b)** 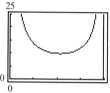   **(c)** L is the least when $\theta = 0.83$ **(d)** $L(0.83) = 9.87$ ft.

71. $y = \sin \omega x$ has period $\dfrac{2\pi}{\omega}$.   **73.** The graphs are the same; yes.

2.6 Exercises

1. Amplitude $= 2$; Period $= 2\pi$ **3.** Amplitude $= 4$; Period $= \pi$ **5.** Amplitude $= 6$; Period $= 2$ **7.** Amplitude $= \dfrac{1}{2}$; Period $= \dfrac{4\pi}{3}$

9. Amplitude $= \dfrac{5}{3}$; Period $= 3$ **11.** F   **13.** A   **15.** H   **17.** C   **19.** J   **21.** A   **23.** D   **25.** B

27.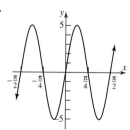

29.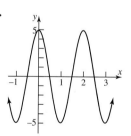

31.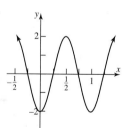

33.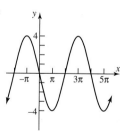

35.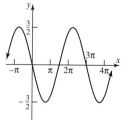

37. $y = 5 \cos \dfrac{\pi}{4}x$ **39.** $y = -3 \cos \dfrac{1}{2}x$ **41.** $y = \dfrac{3}{4} \sin 2\pi x$ **43.** $y = -\sin \dfrac{3}{2}x$ **45.** $y = -2 \cos \dfrac{3\pi}{2}x$

47. $y = 3 \sin \dfrac{\pi}{2}x$ **49.** $y = -4 \cos 3x$

51. Amplitude $= 4$

Period $= \pi$

Phase shift $= \dfrac{\pi}{2}$

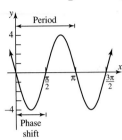

53. Amplitude $= 2$

Period $= \dfrac{2\pi}{3}$

Phase shift $= -\dfrac{\pi}{6}$

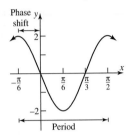

55. Amplitude $= 3$

Period $= \pi$

Phase shift $= -\dfrac{\pi}{4}$

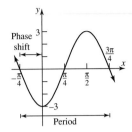

57. Amplitude $= 4$

Period $= 2$

Phase shift $= -\dfrac{2}{\pi}$

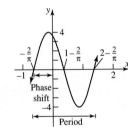

59. Amplitude $= 3$

Period $= 2$

Phase shift $= \dfrac{2}{\pi}$

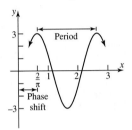

61. Amplitude $= 3$

Period $= \pi$

Phase shift $= \dfrac{\pi}{4}$

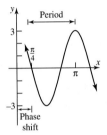

63. $y = 3 \sin 2x$ **65.** $y = 3 \sin \pi x$ **67.** $y = 2 \sin 2\left(x - \dfrac{1}{2}\right)$ **69.** $y = 3 \sin \dfrac{2}{3}\left(x + \dfrac{1}{3}\right)$

71. Period $= \dfrac{1}{30}$, Amplitude $= 220$

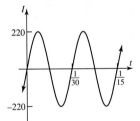

73. Period $= \dfrac{1}{15}$, Amplitude $= 120$, Phase shift $= \dfrac{1}{90}$

75. (a) Amplitude $= 220$, period $= \dfrac{1}{60}$

(b), (e)

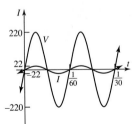

(c) $I = 22 \sin 120\pi t$

(d) Amplitude $= 22$, Period $= \dfrac{1}{60}$

77. (a) $P = \dfrac{(V_0 \sin 2\pi f t)^2}{R} = \dfrac{V_0^2}{R} \sin^2 2\pi f t$ **(b)** $P = \dfrac{V_0^2}{R} \dfrac{1}{2}(1 - \cos 4\pi f t)$

79. (a)

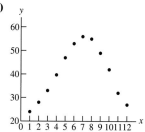

(b) $y = 15.9 \sin\left(\dfrac{\pi}{6}x - \dfrac{2\pi}{3}\right) + 40.1$ **(c)**

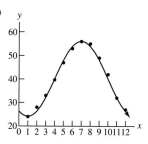

(d) $y = 15.62 \sin(0.517x - 2.096) + 40.377$ **(e)**

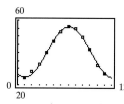

81. (a)

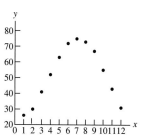

(b) $y = 24.95 \sin\left(\dfrac{\pi}{6}x - \dfrac{2\pi}{3}\right) + 50.45$ **(c)**

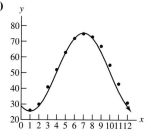

(d) $y = 25.693 \sin(0.476x - 1.814) + 49.854$ **(e)**

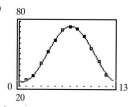

83. (a) 4:08 PM   **(b)** $y = 4.4 \sin\left(\dfrac{\pi}{6.25}x - 6.6643\right) + 3.8$ **(c)**   **(d)** 8.2 feet

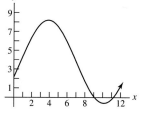

85. (a) $y = 1.0835 \sin\left(\dfrac{2\pi}{365}x - 2.45\pi\right) + 11.6665$ **(b)**   **(c)** 11.83 hours

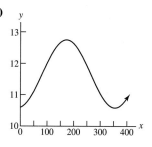

87. (a) $y = 5.3915 \sin\left(\dfrac{2\pi}{365} x - 2.45\pi\right) + 10.8415$ **(b)** **(c)** 11.66 hours

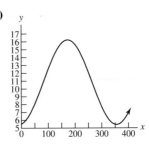

Fill-in-the-Blank Items

1. angle; initial side; terminal side   **3.** π **5.** cosine   **7.** $45°$ **9.** $y = \cos x, y = \sec x$ **11.** $y = 3 \sin \pi x$

True/False Items

1. F   **3.** F   **5.** F   **7.** T   **9.** F

Review Exercises

1. $\dfrac{3\pi}{4}$ **3.** $\dfrac{\pi}{10}$ **5.** $135°$ **7.** $-450°$ **9.** $\dfrac{1}{2}$ **11.** $\dfrac{3\sqrt{2}}{2} - \dfrac{4\sqrt{3}}{3}$ **13.** $-3\sqrt{2} - 2\sqrt{3}$ **15.** 3   **17.** 0   **19.** 0   **21.** 1   **23.** 1   **25.** 1   **27.** -1

29. 1   **31.** $\cos\theta = \dfrac{3}{5}$; $\tan\theta = -\dfrac{4}{3}$; $\csc\theta = -\dfrac{5}{4}$; $\sec\theta = \dfrac{5}{3}$; $\cot\theta = -\dfrac{3}{4}$

33. $\sin\theta = -\dfrac{12}{13}$; $\cos\theta = -\dfrac{5}{13}$; $\csc\theta = -\dfrac{13}{12}$; $\sec\theta = -\dfrac{13}{5}$; $\cot\theta = \dfrac{5}{12}$ **35.** $\sin\theta = \dfrac{3}{5}$; $\cos\theta = -\dfrac{4}{5}$; $\tan\theta = -\dfrac{3}{4}$; $\csc\theta = \dfrac{5}{3}$; $\cot\theta = -\dfrac{4}{3}$

37. $\cos\theta = -\dfrac{5}{13}$; $\tan\theta = -\dfrac{12}{5}$; $\csc\theta = \dfrac{13}{12}$; $\sec\theta = -\dfrac{13}{5}$; $\cot\theta = -\dfrac{5}{12}$

39. $\cos\theta = \dfrac{12}{13}$; $\tan\theta = -\dfrac{5}{12}$; $\csc\theta = -\dfrac{13}{5}$; $\sec\theta = \dfrac{13}{12}$; $\cot\theta = -\dfrac{12}{5}$

41. $\sin\theta = -\dfrac{\sqrt{10}}{10}$; $\cos\theta = -\dfrac{3\sqrt{10}}{10}$; $\csc\theta = -\sqrt{10}$; $\sec\theta = -\dfrac{\sqrt{10}}{3}$; $\cot\theta = 3$

43. $\sin\theta = -\dfrac{2\sqrt{2}}{3}$; $\cos\theta = \dfrac{1}{3}$; $\tan\theta = -2\sqrt{2}$; $\csc = -\dfrac{3\sqrt{2}}{4}$; $\cot\theta = -\dfrac{\sqrt{2}}{4}$

45. $\sin\theta = \dfrac{\sqrt{5}}{5}$; $\cos\theta = -\dfrac{2\sqrt{5}}{5}$; $\tan\theta = -\dfrac{1}{2}$; $\csc\theta = \sqrt{5}$; $\sec\theta = -\dfrac{\sqrt{5}}{2}$

47.   **49.**   **51.**   **53.**

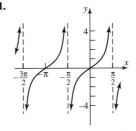

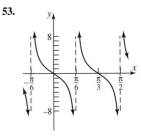

55. Amplitude $= 4$; Period $= 2\pi$ **57.** Amplitude $= 8$; Period $= 4$

59. Amplitude $= 4$

Period $= \dfrac{2\pi}{3}$

Phase shift $= 0$

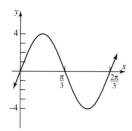

61. Amplitude $= 2$

Period $= 4$

Phase shift $= -\dfrac{1}{\pi}$

63. Amplitude $= \dfrac{1}{2}$

Period $= \dfrac{4\pi}{3}$

Phase shift $= \dfrac{2\pi}{3}$

65. Amplitude $= \dfrac{2}{3}$

Period $= 2$

Phase shift $= \dfrac{6}{\pi}$

67. $y = 5 \cos \dfrac{x}{4}$ **69.** $y = -6 \cos \dfrac{\pi}{4}x$ **71.** $\dfrac{\pi}{3}$ ft   **73.** approximately 114.59 revolutions per hour

75. 0.1 revolutions per second $= \dfrac{\pi}{5}$ radians/second   **77. (a)** 120   **(b)** $\dfrac{1}{60}$ **(c)**

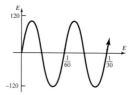

79. (a)

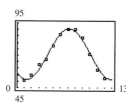

(b) $y = 19.5 \sin\left(\dfrac{\pi}{6}x - \dfrac{2\pi}{3}\right) + 70.5$ **(c)**

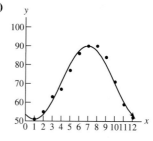

(d) $y = 19.52 \sin(0.54x - 2.28) + 71.01$ **(e)**

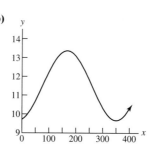

81. (a) $y = 1.85 \sin\left(\dfrac{2\pi}{365}x - 2.45\pi\right) + 11.517$ **(b)**   **(c)** 11.80 hours

C H A P T E R   3   3.1 Exercises

1. $\csc\theta \cdot \cos\theta = \dfrac{1}{\sin\theta} \cdot \cos\theta = \dfrac{\cos\theta}{\sin\theta} = \cot\theta$ **3.** $1 + \tan^2(-\theta) = 1 + (-\tan\theta)^2 = 1 + \tan^2\theta = \sec^2\theta$

5. $\cos\theta(\tan\theta + \cot\theta) = \cos\theta\left(\dfrac{\sin\theta}{\cos\theta} + \dfrac{\cos\theta}{\sin\theta}\right) = \cos\theta\left(\dfrac{\sin^2\theta + \cos^2\theta}{\cos\theta\sin\theta}\right) = \dfrac{1}{\sin\theta} = \csc\theta$

7. $\tan\theta\cot\theta - \cos^2\theta = \dfrac{\sin\theta}{\cos\theta} \cdot \dfrac{\cos\theta}{\sin\theta} - \cos^2\theta = 1 - \cos^2\theta = \sin^2\theta$ **9.** $(\sec\theta - 1)(\sec\theta + 1) = \sec^2\theta - 1 = \tan^2\theta$

11. $(\sec\theta + \tan\theta)(\sec\theta - \tan\theta) = \sec^2\theta - \tan^2\theta = 1$ **13.** $\cos^2\theta(1 + \tan^2\theta) = \cos^2\theta + \cos^2\theta \cdot \dfrac{\sin^2\theta}{\cos^2\theta} = \cos^2\theta + \sin^2\theta = 1$

15. $(\sin\theta + \cos\theta)^2 + (\sin\theta - \cos\theta)^2 = \sin^2\theta + 2\sin\theta\cos\theta + \cos^2\theta + \sin^2\theta - 2\sin\theta\cos\theta + \cos^2\theta$
$$= \sin^2\theta + \cos^2\theta + \sin^2\theta + \cos^2\theta = 1 + 1 = 2$$

17. $\sec^4\theta - \sec^2\theta = \sec^2\theta(\sec^2\theta - 1) = (1 + \tan^2\theta)\tan^2\theta = \tan^4\theta + \tan^2\theta$

19. $\sec\theta - \tan\theta = \dfrac{1}{\cos\theta} - \dfrac{\sin\theta}{\cos\theta} = \dfrac{1 - \sin\theta}{\cos\theta} \cdot \dfrac{1 + \sin\theta}{1 + \sin\theta} = \dfrac{1 - \sin^2\theta}{\cos\theta(1 + \sin\theta)} = \dfrac{\cos^2\theta}{\cos\theta(1 + \sin\theta)} = \dfrac{\cos\theta}{1 + \sin\theta}$

21. $3\sin^2\theta + 4\cos^2\theta = 3\sin^2\theta + 3\cos^2\theta + \cos^2\theta = 3(\sin^2\theta + \cos^2\theta) + \cos^2\theta = 3 + \cos^2\theta$

23. $1 - \dfrac{\cos^2\theta}{1 + \sin\theta} = 1 - \dfrac{1 - \sin^2\theta}{1 + \sin\theta} = 1 - (1 - \sin\theta) = \sin\theta$ **25.** $\dfrac{1 + \tan\theta}{1 - \tan\theta} = \dfrac{1 + \dfrac{1}{\cot\theta}}{1 - \dfrac{1}{\cot\theta}} = \dfrac{\dfrac{\cot\theta + 1}{\cot\theta}}{\dfrac{\cot\theta - 1}{\cot\theta}} = \dfrac{\cot\theta + 1}{\cot\theta - 1}$

27. $\dfrac{\sec\theta}{\csc\theta} + \dfrac{\sin\theta}{\cos\theta} = \dfrac{\dfrac{1}{\cos\theta}}{\dfrac{1}{\sin\theta}} + \tan\theta = \dfrac{\sin\theta}{\cos\theta} + \tan\theta = \tan\theta + \tan\theta = 2\tan\theta$ **29.** $\dfrac{1 + \sin\theta}{1 - \sin\theta} = \dfrac{1 + \dfrac{1}{\csc\theta}}{1 - \dfrac{1}{\csc\theta}} = \dfrac{\dfrac{\csc\theta + 1}{\csc\theta}}{\dfrac{\csc\theta - 1}{\csc\theta}} = \dfrac{\csc\theta + 1}{\csc\theta - 1}$

31. $\dfrac{1 - \sin\theta}{\cos\theta} + \dfrac{\cos\theta}{1 - \sin\theta} = \dfrac{(1 - \sin\theta)^2 + \cos^2\theta}{\cos\theta(1 - \sin\theta)} = \dfrac{1 - 2\sin\theta + \sin^2\theta + \cos^2\theta}{\cos\theta(1 - \sin\theta)} = \dfrac{2 - 2\sin\theta}{\cos\theta(1 - \sin\theta)} = \dfrac{2(1 - \sin\theta)}{\cos\theta(1 - \sin\theta)} = \dfrac{2}{\cos\theta}$
$$= 2\sec\theta$$

33. $\dfrac{\sin\theta}{\sin\theta - \cos\theta} = \dfrac{1}{\dfrac{\sin\theta - \cos\theta}{\sin\theta}} = \dfrac{1}{1 - \dfrac{\cos\theta}{\sin\theta}} = \dfrac{1}{1 - \cot\theta}$

35. $(\sec\theta - \tan\theta)^2 = \sec^2\theta - 2\sec\theta\tan\theta + \tan^2\theta = \dfrac{1}{\cos^2\theta} - \dfrac{2\sin\theta}{\cos^2\theta} + \dfrac{\sin^2\theta}{\cos^2\theta} = \dfrac{1 - 2\sin\theta + \sin^2\theta}{\cos^2\theta} = \dfrac{(1 - \sin\theta)^2}{1 - \sin^2\theta}$
$$= \dfrac{(1 - \sin\theta)^2}{(1 - \sin\theta)(1 + \sin\theta)} = \dfrac{1 - \sin\theta}{1 + \sin\theta}$$

37. $\dfrac{\cos\theta}{1 - \tan\theta} + \dfrac{\sin\theta}{1 - \cot\theta} = \dfrac{\cos\theta}{1 - \dfrac{\sin\theta}{\cos\theta}} + \dfrac{\sin\theta}{1 - \dfrac{\cos\theta}{\sin\theta}} = \dfrac{\cos\theta}{\dfrac{\cos\theta - \sin\theta}{\cos\theta}} + \dfrac{\sin\theta}{\dfrac{\sin\theta - \cos\theta}{\sin\theta}} = \dfrac{\cos^2\theta}{\cos\theta - \sin\theta} + \dfrac{\sin^2\theta}{\sin\theta - \cos\theta}$
$$= \dfrac{\cos^2\theta - \sin^2\theta}{\cos\theta - \sin\theta} = \dfrac{(\cos\theta - \sin\theta)(\cos\theta + \sin\theta)}{\cos\theta - \sin\theta} = \sin\theta + \cos\theta$$

39. $\tan\theta + \dfrac{\cos\theta}{1 + \sin\theta} = \dfrac{\sin\theta}{\cos\theta} + \dfrac{\cos\theta}{1 + \sin\theta} = \dfrac{\sin\theta(1 + \sin\theta) + \cos^2\theta}{\cos\theta(1 + \sin\theta)} = \dfrac{\sin\theta + \sin^2\theta + \cos^2\theta}{\cos\theta(1 + \sin\theta)} = \dfrac{\sin\theta + 1}{\cos\theta(1 + \sin\theta)} = \dfrac{1}{\cos\theta} = \sec\theta$

41. $\dfrac{\tan\theta + \sec\theta - 1}{\tan\theta - \sec\theta + 1} = \dfrac{\tan\theta + (\sec\theta - 1)}{\tan\theta - (\sec\theta - 1)} \cdot \dfrac{\tan\theta + (\sec\theta - 1)}{\tan\theta + (\sec\theta - 1)} = \dfrac{\tan^2\theta + 2\tan\theta(\sec\theta - 1) + \sec^2\theta - 2\sec\theta + 1}{\tan^2\theta - (\sec^2\theta - 2\sec\theta + 1)}$
$$= \dfrac{\sec^2\theta - 1 + 2\tan\theta(\sec\theta - 1) + \sec^2\theta - 2\sec\theta + 1}{\sec^2\theta - 1 - \sec^2\theta + 2\sec\theta - 1} = \dfrac{2\sec^2\theta - 2\sec\theta + 2\tan\theta(\sec\theta - 1)}{-2 + 2\sec\theta}$$
$$= \dfrac{2\sec\theta(\sec\theta - 1) + 2\tan\theta(\sec\theta - 1)}{2(\sec\theta - 1)} = \dfrac{2(\sec\theta - 1)(\sec\theta + \tan\theta)}{2(\sec\theta - 1)} = \tan\theta + \sec\theta$$

43. $\dfrac{\tan\theta - \cot\theta}{\tan\theta + \cot\theta} = \dfrac{\dfrac{\sin\theta}{\cos\theta} - \dfrac{\cos\theta}{\sin\theta}}{\dfrac{\sin\theta}{\cos\theta} + \dfrac{\cos\theta}{\sin\theta}} = \dfrac{\dfrac{\sin^2\theta - \cos^2\theta}{\cos\theta\sin\theta}}{\dfrac{\sin^2\theta + \cos^2\theta}{\cos\theta\sin\theta}} = \dfrac{\sin^2\theta - \cos^2\theta}{1} = \sin^2\theta - \cos^2\theta$

45. $\dfrac{\tan\theta - \cot\theta}{\tan\theta + \cot\theta} + 1 = \dfrac{\dfrac{\sin\theta}{\cos\theta} - \dfrac{\cos\theta}{\sin\theta}}{\dfrac{\sin\theta}{\cos\theta} + \dfrac{\cos\theta}{\sin\theta}} + 1 = \dfrac{\dfrac{\sin^2\theta - \cos^2\theta}{\cos\theta\sin\theta}}{\dfrac{\sin^2\theta + \cos^2\theta}{\cos\theta\sin\theta}} + 1 = \sin^2\theta - \cos^2\theta + 1 = \sin^2\theta + (1 - \cos^2\theta) = 2\sin^2\theta$

47. $\dfrac{\sec\theta + \tan\theta}{\cot\theta + \cos\theta} = \dfrac{\dfrac{1}{\cos\theta} + \dfrac{\sin\theta}{\cos\theta}}{\dfrac{\cos\theta}{\sin\theta} + \dfrac{\cos\theta\sin\theta}{\sin\theta}} = \dfrac{\dfrac{1+\sin\theta}{\cos\theta}}{\dfrac{\cos\theta + \cos\theta\sin\theta}{\sin\theta}} = \dfrac{1+\sin\theta}{\cos\theta}\cdot\dfrac{\sin\theta}{\cos\theta(1+\sin\theta)} = \dfrac{\sin\theta}{\cos\theta}\cdot\dfrac{1}{\cos\theta} = \tan\theta\sec\theta$

49. $\dfrac{1-\tan^2\theta}{1+\tan^2\theta} + 1 = \dfrac{1-\tan^2\theta}{\sec^2\theta} + 1 = \dfrac{1}{\sec^2\theta} - \dfrac{\tan^2\theta}{\sec^2\theta} + 1 = \cos^2\theta - \dfrac{\dfrac{\sin^2\theta}{\cos^2\theta}}{\dfrac{1}{\cos^2\theta}} + 1 = \cos^2\theta - \sin^2\theta + 1$

$= \cos^2\theta + (1 - \sin^2\theta) = 2\cos^2\theta$

51. $\dfrac{\sec\theta - \csc\theta}{\sec\theta\csc\theta} = \dfrac{\dfrac{1}{\cos\theta} - \dfrac{1}{\sin\theta}}{\dfrac{1}{\cos\theta}\cdot\dfrac{1}{\sin\theta}} = \dfrac{\dfrac{\sin\theta - \cos\theta}{\cos\theta\sin\theta}}{\dfrac{1}{\cos\theta\sin\theta}} = \sin\theta - \cos\theta$

53. $\sec\theta - \cos\theta - \sin\theta\tan\theta = \left(\dfrac{1}{\cos\theta} - \cos\theta\right) - \sin\theta\cdot\dfrac{\sin\theta}{\cos\theta} = \dfrac{1-\cos^2\theta}{\cos\theta} - \dfrac{\sin^2\theta}{\cos\theta} = \dfrac{\sin^2\theta}{\cos\theta} - \dfrac{\sin^2\theta}{\cos\theta} = 0$

55. $\dfrac{1}{1-\sin\theta} + \dfrac{1}{1+\sin\theta} = \dfrac{1+\sin\theta + 1 - \sin\theta}{(1+\sin\theta)(1-\sin\theta)} = \dfrac{2}{1-\sin^2\theta} = \dfrac{2}{\cos^2\theta} = 2\sec^2\theta$

57. $\dfrac{\sec\theta}{1-\sin\theta} = \dfrac{\sec\theta}{1-\sin\theta}\cdot\dfrac{1+\sin\theta}{1+\sin\theta} = \dfrac{\sec\theta(1+\sin\theta)}{1-\sin^2\theta} = \dfrac{\sec\theta(1+\sin\theta)}{\cos^2\theta} = \dfrac{1+\sin\theta}{\cos^3\theta}$

59. $\dfrac{(\sec\theta - \tan\theta)^2 + 1}{\csc\theta(\sec\theta - \tan\theta)} = \dfrac{\sec^2\theta - 2\sec\theta\tan\theta + \tan^2\theta + 1}{\dfrac{1}{\sin\theta}\left(\dfrac{1}{\cos\theta} - \dfrac{\sin\theta}{\cos\theta}\right)} = \dfrac{2\sec^2\theta - 2\sec\theta\tan\theta}{\dfrac{1}{\sin\theta}\left(\dfrac{1-\sin\theta}{\cos\theta}\right)} = \dfrac{\dfrac{2}{\cos^2\theta} - \dfrac{2\sin\theta}{\cos^2\theta}}{\dfrac{1-\sin\theta}{\sin\theta\cos\theta}} = \dfrac{2-2\sin\theta}{\cos^2\theta}\cdot\dfrac{\sin\theta\cos\theta}{1-\sin\theta}$

$= \dfrac{2(1-\sin\theta)}{\cos\theta}\cdot\dfrac{\sin\theta}{1-\sin\theta} = \dfrac{2\sin\theta}{\cos\theta} = 2\tan\theta$

61. $\dfrac{\sin\theta + \cos\theta}{\cos\theta} - \dfrac{\sin\theta - \cos\theta}{\sin\theta} = \dfrac{\sin\theta(\sin\theta + \cos\theta) - \cos\theta(\sin\theta - \cos\theta)}{\cos\theta\sin\theta} = \dfrac{\sin^2\theta + \sin\theta\cos\theta - \sin\theta\cos\theta + \cos^2\theta}{\cos\theta\sin\theta}$

$= \dfrac{1}{\cos\theta\sin\theta} = \sec\theta\csc\theta$

63. $\dfrac{\sin^3\theta + \cos^3\theta}{\sin\theta + \cos\theta} = \dfrac{(\sin\theta + \cos\theta)(\sin^2\theta - \sin\theta\cos\theta + \cos^2\theta)}{\sin\theta + \cos\theta} = \sin^2\theta + \cos^2\theta - \sin\theta\cos\theta = 1 - \sin\theta\cos\theta$

65. $\dfrac{\cos^2\theta - \sin^2\theta}{1-\tan^2\theta} = \dfrac{\cos^2\theta - \sin^2\theta}{1 - \dfrac{\sin^2\theta}{\cos^2\theta}} = \dfrac{\cos^2\theta - \sin^2\theta}{\dfrac{\cos^2\theta - \sin^2\theta}{\cos^2\theta}} = \cos^2\theta$

67. $\dfrac{(2\cos^2\theta - 1)^2}{\cos^4\theta - \sin^4\theta} = \dfrac{[2\cos^2\theta - (\sin^2\theta + \cos^2\theta)]^2}{(\cos^2\theta - \sin^2\theta)(\cos^2\theta + \sin^2\theta)} = \cos^2\theta - \sin^2\theta = (1 - \sin^2\theta) - \sin^2\theta = 1 - 2\sin^2\theta$

69. $\dfrac{1+\sin\theta + \cos\theta}{1+\sin\theta - \cos\theta} = \dfrac{(1+\sin\theta) + \cos\theta}{(1+\sin\theta) - \cos\theta}\cdot\dfrac{(1+\sin\theta) + \cos\theta}{(1+\sin\theta) + \cos\theta} = \dfrac{1 + 2\sin\theta + \sin^2\theta + 2(1+\sin\theta)(\cos\theta) + \cos^2\theta}{1 + 2\sin\theta + \sin^2\theta - \cos^2\theta}$

$= \dfrac{1 + 2\sin\theta + \sin^2\theta + 2(1+\sin\theta)(\cos\theta) + (1-\sin^2\theta)}{1 + 2\sin\theta + \sin^2\theta - (1-\sin^2\theta)} = \dfrac{2 + 2\sin\theta + 2(1+\sin\theta)(\cos\theta)}{2\sin\theta + 2\sin^2\theta}$

$= \dfrac{2(1+\sin\theta) + 2(1+\sin\theta)(\cos\theta)}{2\sin\theta(1+\sin\theta)} = \dfrac{2(1+\sin\theta)(1+\cos\theta)}{2\sin\theta(1+\sin\theta)} = \dfrac{1+\cos\theta}{\sin\theta}$

71. $(a\sin\theta + b\cos\theta)^2 + (a\cos\theta - b\sin\theta)^2 = a^2\sin^2\theta + 2ab\sin\theta\cos\theta + b^2\cos^2\theta + a^2\cos^2\theta - 2ab\sin\theta\cos\theta + b^2\sin^2\theta$

$= a^2(\sin^2\theta + \cos^2\theta) + b^2(\cos^2\theta + \sin^2\theta) = a^2 + b^2$

73. $\dfrac{\tan\alpha + \tan\beta}{\cot\alpha + \cot\beta} = \dfrac{\tan\alpha + \tan\beta}{\dfrac{1}{\tan\alpha} + \dfrac{1}{\tan\beta}} = \dfrac{\tan\alpha + \tan\beta}{\dfrac{\tan\beta + \tan\alpha}{\tan\alpha\tan\beta}} = (\tan\alpha + \tan\beta)\cdot\dfrac{\tan\alpha\tan\beta}{\tan\alpha + \tan\beta} = \tan\alpha\tan\beta$

75. $(\sin\alpha + \cos\beta)^2 + (\cos\beta + \sin\alpha)(\cos\beta - \sin\alpha) = (\sin^2\alpha + 2\sin\alpha\cos\beta + \cos^2\beta) + (\cos^2\beta - \sin^2\alpha)$

$= 2\cos^2\beta + 2\sin\alpha\cos\beta = 2\cos\beta(\cos\beta + \sin\alpha)$

77. $\ln|\sec\theta| = \ln|\cos\theta|^{-1} = -\ln|\cos\theta|$

79. $\ln|1+\cos\theta| + \ln|1-\cos\theta| = \ln(|1+\cos\theta|\,|1-\cos\theta|) = \ln|1-\cos^2\theta| = \ln|\sin^2\theta| = 2\ln|\sin\theta|$

3.2 Exercises

1. $\frac{1}{4}(\sqrt{6} + \sqrt{2})$ **3.** $\frac{1}{4}(\sqrt{2} - \sqrt{6})$ **5.** $-\frac{1}{4}(\sqrt{2} + \sqrt{6})$ **7.** $\frac{\sqrt{3} - 1}{1 + \sqrt{3}} = 2 - \sqrt{3}$ **9.** $-\frac{1}{4}(\sqrt{6} + \sqrt{2})$ **11.** $\frac{4}{\sqrt{6} + \sqrt{2}} = \sqrt{6} - \sqrt{2}$

13. $\frac{1}{2}$ **15.** 0 **17.** 1 **19.** -1 **21.** $\frac{1}{2}$ **23. (a)** $\frac{2\sqrt{5}}{25}$ **(b)** $\frac{11\sqrt{5}}{25}$ **(c)** $\frac{2\sqrt{5}}{5}$ **(d)** 2 **25. (a)** $\frac{4 - 3\sqrt{3}}{10}$ **(b)** $\frac{-3 - 4\sqrt{3}}{10}$ **(c)** $\frac{4 + 3\sqrt{3}}{10}$

(d) $\frac{4 + 3\sqrt{3}}{4\sqrt{3} - 3} = \frac{25\sqrt{3} + 48}{39}$ **27. (a)** $-\frac{1}{26}(5 + 12\sqrt{3})$ **(b)** $\frac{1}{26}(12 - 5\sqrt{3})$ **(c)** $-\frac{1}{26}(5 - 12\sqrt{3})$

(d) $\frac{-5 + 12\sqrt{3}}{12 + 5\sqrt{3}} = \frac{-240 + 169\sqrt{3}}{69}$ **29. (a)** $-\frac{2\sqrt{2}}{3}$ **(b)** $\frac{-2\sqrt{2} + \sqrt{3}}{6}$ **(c)** $\frac{-2\sqrt{2} + \sqrt{3}}{6}$ **(d)** $\frac{2\sqrt{2} - 1}{2\sqrt{2} + 1} = \frac{9 - 4\sqrt{2}}{7}$

31. $\sin\left(\frac{\pi}{2} + \theta\right) = \sin\frac{\pi}{2}\cos\theta + \cos\frac{\pi}{2}\sin\theta = 1 \cdot \cos\theta + 0 \cdot \sin\theta = \cos\theta$

33. $\sin(\pi - \theta) = \sin\pi\cos\theta - \cos\pi\sin\theta = 0 \cdot \cos\theta - (-1)\sin\theta = \sin\theta$
35. $\sin(\pi + \theta) = \sin\pi\cos\theta + \cos\pi\sin\theta = 0 \cdot \cos\theta + (-1)\sin\theta = -\sin\theta$

37. $\tan(\pi - \theta) = \frac{\tan\pi - \tan\theta}{1 + \tan\pi\tan\theta} = \frac{0 - \tan\theta}{1 + 0} = -\tan\theta$

39. $\sin\left(\frac{3\pi}{2} + \theta\right) = \sin\frac{3\pi}{2}\cos\theta + \cos\frac{3\pi}{2}\sin\theta = (-1)\cos\theta + 0 \cdot \sin\theta = -\cos\theta$

41. $\sin(\alpha + \beta) + \sin(\alpha - \beta) = \sin\alpha\cos\beta + \cos\alpha\sin\beta + \sin\alpha\cos\beta - \cos\alpha\sin\beta = 2\sin\alpha\cos\beta$

43. $\frac{\sin(\alpha + \beta)}{\sin\alpha\cos\beta} = \frac{\sin\alpha\cos\beta + \cos\alpha\sin\beta}{\sin\alpha\cos\beta} = \frac{\sin\alpha\cos\beta}{\sin\alpha\cos\beta} + \frac{\cos\alpha\sin\beta}{\sin\alpha\cos\beta} = 1 + \cot\alpha\tan\beta$

45. $\frac{\cos(\alpha + \beta)}{\cos\alpha\cos\beta} = \frac{\cos\alpha\cos\beta - \sin\alpha\sin\beta}{\cos\alpha\cos\beta} = \frac{\cos\alpha\cos\beta}{\cos\alpha\cos\beta} - \frac{\sin\alpha\sin\beta}{\cos\alpha\cos\beta} = 1 - \tan\alpha\tan\beta$

47. $\frac{\sin(\alpha + \beta)}{\sin(\alpha - \beta)} = \frac{\sin\alpha\cos\beta + \cos\alpha\sin\beta}{\sin\alpha\cos\beta - \cos\alpha\sin\beta} = \frac{\dfrac{\sin\alpha\cos\beta + \cos\alpha\sin\beta}{\cos\alpha\cos\beta}}{\dfrac{\sin\alpha\cos\beta - \cos\alpha\sin\beta}{\cos\alpha\cos\beta}} = \frac{\dfrac{\sin\alpha\cos\beta}{\cos\alpha\cos\beta} + \dfrac{\cos\alpha\sin\beta}{\cos\alpha\cos\beta}}{\dfrac{\sin\alpha\cos\beta}{\cos\alpha\cos\beta} - \dfrac{\cos\alpha\sin\beta}{\cos\alpha\cos\beta}} = \frac{\tan\alpha + \tan\beta}{\tan\alpha - \tan\beta}$

49. $\cot(\alpha + \beta) = \frac{\cos(\alpha + \beta)}{\sin(\alpha + \beta)} = \frac{\cos\alpha\cos\beta - \sin\alpha\sin\beta}{\sin\alpha\cos\beta + \cos\alpha\sin\beta} = \frac{\dfrac{\cos\alpha\cos\beta - \sin\alpha\sin\beta}{\sin\alpha\cos\beta + \cos\alpha\sin\beta}}{\dfrac{\sin\alpha\sin\beta}{\sin\alpha\sin\beta}} = \frac{\dfrac{\cos\alpha\cos\beta}{\sin\alpha\sin\beta} - \dfrac{\sin\alpha\sin\beta}{\sin\alpha\sin\beta}}{\dfrac{\sin\alpha\cos\beta}{\sin\alpha\sin\beta} + \dfrac{\cos\alpha\sin\beta}{\sin\alpha\sin\beta}} = \frac{\cot\alpha\cot\beta - 1}{\cot\beta + \cot\alpha}$

51. $\sec(\alpha + \beta) = \frac{1}{\cos(\alpha + \beta)} = \frac{1}{\cos\alpha\cos\beta - \sin\alpha\sin\beta} = \frac{\dfrac{\sin\alpha\sin\beta}{\sin\alpha\sin\beta}}{\dfrac{\cos\alpha\cos\beta - \sin\alpha\sin\beta}{\sin\alpha\sin\beta}} = \frac{\dfrac{1}{\sin\alpha} \cdot \dfrac{1}{\sin\beta}}{\dfrac{\cos\alpha\cos\beta}{\sin\alpha\sin\beta} - \dfrac{\sin\alpha\sin\beta}{\sin\alpha\sin\beta}} = \frac{\csc\alpha\csc\beta}{\cot\alpha\cot\beta - 1}$

53. $\sin(\alpha - \beta)\sin(\alpha + \beta) = (\sin\alpha\cos\beta - \cos\alpha\sin\beta)(\sin\alpha\cos\beta + \cos\alpha\sin\beta) = \sin^2\alpha\cos^2\beta - \cos^2\alpha\sin^2\beta$
$= (\sin^2\alpha)(1 - \sin^2\beta) - (1 - \sin^2\alpha)(\sin^2\beta) = \sin^2\alpha - \sin^2\beta$
55. $\sin(\theta + k\pi) = \sin\theta\cos k\pi + \cos\theta\sin k\pi = (\sin\theta)(-1)^k + (\cos\theta)(0) = (-1)^k\sin\theta, k$ any integer
57. $\frac{\sin(x + h) - \sin x}{h} = \frac{\sin x\cos h + \cos x\sin h - \sin x}{h} = \frac{\cos x\sin h - \sin x(1 - \cos h)}{h} = \cos x \cdot \frac{\sin h}{h} - \sin x \cdot \frac{1 - \cos h}{h}$

59. $\tan\frac{\pi}{2}$ is not defined; $\tan\left(\frac{\pi}{2} - \theta\right) = \frac{\sin\left(\dfrac{\pi}{2} - \theta\right)}{\cos\left(\dfrac{\pi}{2} - \theta\right)} = \frac{\cos\theta}{\sin\theta} = \cot\theta$ **61.** $\tan\theta = \tan(\theta_2 - \theta_1) = \frac{\tan\theta_2 - \tan\theta_1}{1 + \tan\theta_1\tan\theta_2} = \frac{m_2 - m_1}{1 + m_1 m_2}$

63. No; $\tan\left(\frac{\pi}{2}\right)$ is undefined.

3.3 Exercises

1. (a) $\frac{24}{25}$ **(b)** $\frac{7}{25}$ **(c)** $\frac{\sqrt{10}}{10}$ **(d)** $\frac{3\sqrt{10}}{10}$ **3. (a)** $\frac{24}{25}$ **(b)** $-\frac{7}{25}$ **(c)** $\frac{2\sqrt{5}}{5}$ **(d)** $-\frac{\sqrt{5}}{5}$

5. (a) $-\frac{2\sqrt{2}}{3}$ **(b)** $\frac{1}{3}$ **(c)** $\sqrt{\frac{3 + \sqrt{6}}{6}}$ **(d)** $\sqrt{\frac{3 - \sqrt{6}}{6}}$ **7. (a)** $\frac{4\sqrt{2}}{9}$ **(b)** $-\frac{7}{9}$ **(c)** $\frac{\sqrt{3}}{3}$ **(d)** $\frac{\sqrt{6}}{3}$

9. (a) $-\frac{4}{5}$ **(b)** $\frac{3}{5}$ **(c)** $\sqrt{\frac{5 + 2\sqrt{5}}{10}}$ **(d)** $\sqrt{\frac{5 - 2\sqrt{5}}{10}}$ **11. (a)** $-\frac{3}{5}$ **(b)** $-\frac{4}{5}$ **(c)** $\frac{1}{2}\sqrt{\frac{10 - \sqrt{10}}{5}}$ **(d)** $-\frac{1}{2}\sqrt{\frac{10 + \sqrt{10}}{5}}$

13. $\frac{\sqrt{2 - \sqrt{2}}}{2}$ **15.** $1 - \sqrt{2}$ **17.** $-\frac{\sqrt{2 + \sqrt{3}}}{2}$ **19.** $\frac{2}{\sqrt{2} + \sqrt{2}} = (2 - \sqrt{2})\sqrt{2 + \sqrt{2}}$ **21.** $-\frac{\sqrt{2 - \sqrt{2}}}{2}$

23. $\sin^4\theta = (\sin^2\theta)^2 = \left(\dfrac{1-\cos 2\theta}{2}\right)^2 = \dfrac{1}{4}(1 - 2\cos 2\theta + \cos^2 2\theta) = \dfrac{1}{4} - \dfrac{1}{2}\cos 2\theta + \dfrac{1}{4}\cos^2 2\theta$

$= \dfrac{1}{4} - \dfrac{1}{2}\cos 2\theta + \dfrac{1}{4}\left(\dfrac{1+\cos 4\theta}{2}\right) = \dfrac{1}{4} - \dfrac{1}{2}\cos 2\theta + \dfrac{1}{8} + \dfrac{1}{8}\cos 4\theta = \dfrac{3}{8} - \dfrac{1}{2}\cos 2\theta + \dfrac{1}{8}\cos 4\theta$

25. $\sin 4\theta = \sin 2(2\theta) = 2\sin 2\theta \cos 2\theta = (4\sin\theta\cos\theta)(1 - 2\sin^2\theta) = 4\sin\theta\cos\theta - 8\sin^3\theta\cos\theta = (\cos\theta)(4\sin\theta - 8\sin^3\theta)$

27. $16\sin^5\theta - 20\sin^3\theta + 5\sin\theta$

29. $\cos^4\theta - \sin^4\theta = (\cos^2\theta + \sin^2\theta)(\cos^2\theta - \sin^2\theta) = \cos 2\theta$

31. $\cot 2\theta = \dfrac{1}{\tan 2\theta} = \dfrac{1}{\dfrac{2\tan\theta}{1-\tan^2\theta}} = \dfrac{1-\tan^2\theta}{2\tan\theta} = \dfrac{1 - \dfrac{1}{\cot^2\theta}}{2\left(\dfrac{1}{\cot\theta}\right)} = \dfrac{\dfrac{\cot^2\theta - 1}{\cot^2\theta}}{\dfrac{2}{\cot\theta}} = \dfrac{\cot^2\theta - 1}{\cot^2\theta} \cdot \dfrac{\cot\theta}{2} = \dfrac{\cot^2\theta - 1}{2\cot\theta}$

33. $\sec 2\theta = \dfrac{1}{\cos 2\theta} = \dfrac{1}{2\cos^2\theta - 1} = \dfrac{1}{\dfrac{2}{\sec^2\theta} - 1} = \dfrac{1}{\dfrac{2 - \sec^2\theta}{\sec^2\theta}} = \dfrac{\sec^2\theta}{2 - \sec^2\theta}$

35. $\cos^2 2\theta - \sin^2 2\theta = \cos[2(2\theta)] = \cos 4\theta$

37. $\dfrac{\cos 2\theta}{1+\sin 2\theta} = \dfrac{\cos^2\theta - \sin^2\theta}{1 + 2\sin\theta\cos\theta} = \dfrac{(\cos\theta - \sin\theta)(\cos\theta + \sin\theta)}{\sin^2\theta + \cos^2\theta + 2\sin\theta\cos\theta} = \dfrac{(\cos\theta - \sin\theta)(\cos\theta + \sin\theta)}{(\sin\theta + \cos\theta)(\sin\theta + \cos\theta)} = \dfrac{\cos\theta - \sin\theta}{\cos\theta + \sin\theta}$

$= \dfrac{\dfrac{\cos\theta - \sin\theta}{\sin\theta}}{\dfrac{\cos\theta + \sin\theta}{\sin\theta}} = \dfrac{\dfrac{\cos\theta}{\sin\theta} - \dfrac{\sin\theta}{\sin\theta}}{\dfrac{\cos\theta}{\sin\theta} + \dfrac{\sin\theta}{\sin\theta}} = \dfrac{\cot\theta - 1}{\cot\theta + 1}$

39. $\sec^2\dfrac{\theta}{2} = \dfrac{1}{\cos^2\left(\dfrac{\theta}{2}\right)} = \dfrac{1}{\dfrac{1+\cos\theta}{2}} = \dfrac{2}{1+\cos\theta}$

41. $\cot^2\dfrac{\theta}{2} = \dfrac{1}{\tan^2\left(\dfrac{\theta}{2}\right)} = \dfrac{1}{\dfrac{1-\cos\theta}{1+\cos\theta}} = \dfrac{1+\cos\theta}{1-\cos\theta} = \dfrac{1 + \dfrac{1}{\sec\theta}}{1 - \dfrac{1}{\sec\theta}} = \dfrac{\dfrac{\sec\theta + 1}{\sec\theta}}{\dfrac{\sec\theta - 1}{\sec\theta}} = \dfrac{\sec\theta + 1}{\sec\theta} \cdot \dfrac{\sec\theta}{\sec\theta - 1} = \dfrac{\sec\theta + 1}{\sec\theta - 1}$

43. $\dfrac{1 - \tan^2\left(\dfrac{\theta}{2}\right)}{1 + \tan^2\left(\dfrac{\theta}{2}\right)} = \dfrac{1 - \dfrac{1-\cos\theta}{1+\cos\theta}}{1 + \dfrac{1-\cos\theta}{1+\cos\theta}} = \dfrac{\dfrac{1 + \cos\theta - (1 - \cos\theta)}{1+\cos\theta}}{\dfrac{1 + \cos\theta + 1 - \cos\theta}{1+\cos\theta}} = \dfrac{2\cos\theta}{1+\cos\theta} \cdot \dfrac{1+\cos\theta}{2} = \cos\theta$

45. $\dfrac{\sin 3\theta}{\sin\theta} - \dfrac{\cos 3\theta}{\cos\theta} = \dfrac{\sin 3\theta\cos\theta - \cos 3\theta\sin\theta}{\sin\theta\cos\theta} = \dfrac{\sin(3\theta - \theta)}{\dfrac{1}{2}(2\sin\theta\cos\theta)} = \dfrac{2\sin 2\theta}{\sin 2\theta} = 2$

47. $\tan 3\theta = \tan(\theta + 2\theta) = \dfrac{\tan\theta + \tan 2\theta}{1 - \tan\theta\tan 2\theta} = \dfrac{\tan\theta + \dfrac{2\tan\theta}{1-\tan^2\theta}}{1 - \dfrac{\tan\theta(2\tan\theta)}{1-\tan^2\theta}} = \dfrac{\tan\theta - \tan^3\theta + 2\tan\theta}{1 - \tan^2\theta - 2\tan^2\theta} = \dfrac{3\tan\theta - \tan^3\theta}{1 - 3\tan^2\theta}$

49. $\sin 2\theta = \dfrac{4x}{4 + x^2}$ **51.** $-\dfrac{1}{4}$

53. $A = \dfrac{1}{2}h(\text{base}) = h\left(\dfrac{1}{2}\text{base}\right) = s\cos\dfrac{\theta}{2} \cdot s\sin\dfrac{\theta}{2} = \dfrac{1}{2}s^2\sin\theta$

55.

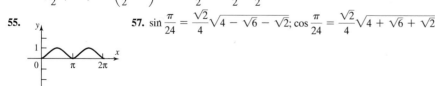

57. $\sin\dfrac{\pi}{24} = \dfrac{\sqrt{2}}{4}\sqrt{4 - \sqrt{6} - \sqrt{2}}$; $\cos\dfrac{\pi}{24} = \dfrac{\sqrt{2}}{4}\sqrt{4 + \sqrt{6} + \sqrt{2}}$

59. $\sin^3\theta + \sin^3(\theta + 120°) + \sin^3(\theta + 240°) = \sin^3\theta + (\sin\theta\cos 120° + \cos\theta\sin 120°)^3 + (\sin\theta\cos 240° + \cos\theta\sin 240°)^3$

$= \sin^3\theta + \left(-\dfrac{1}{2}\sin\theta + \dfrac{\sqrt{3}}{2}\cos\theta\right)^3 + \left(-\dfrac{1}{2}\sin\theta - \dfrac{\sqrt{3}}{2}\cos\theta\right)^3$

$= \sin^3\theta + \dfrac{1}{8}(3\sqrt{3}\cos^3\theta - 9\cos^2\theta\sin\theta + 3\sqrt{3}\cos\theta\sin^2\theta - \sin^3\theta) - \dfrac{1}{8}(\sin^3\theta + 3\sqrt{3}\sin^2\theta\cos\theta + 9\sin\theta\cos^2\theta + 3\sqrt{3}\cos^3\theta)$

$= \dfrac{3}{4}\sin^3\theta - \dfrac{9}{4}\cos^2\theta\sin\theta = \dfrac{3}{4}[\sin^3\theta - 3\sin\theta(1 - \sin^2\theta)] = \dfrac{3}{4}(4\sin^3\theta - 3\sin\theta) = -\dfrac{3}{4}\sin 3\theta \text{ (from Example 2)}$

61. $\frac{1}{2}(\ln |1 - \cos 2\theta| - \ln 2) = \ln\left(\frac{|1 - \cos 2\theta|}{2}\right)^{1/2} = \ln |\sin^2 \theta|^{1/2} = \ln |\sin \theta|$

63. (a) $R = \dfrac{v_0^2 \sqrt{2}}{16}(\sin \theta \cos \theta - \cos^2 \theta) = \dfrac{v_0^2 \sqrt{2}}{16}\left(\dfrac{1}{2} \sin 2\theta - \dfrac{1 + \cos 2\theta}{2}\right)$ **(b)**

$= \dfrac{v_0^2 \sqrt{2}}{32}(\sin 2\theta - \cos 2\theta - 1)$

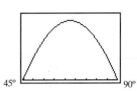

45°   90°

(c) $\theta = 67.5°$ makes R largest

3.4 Exercises

1. $\frac{1}{2}(\cos 2\theta - \cos 6\theta)$ **3.** $\frac{1}{2}(\sin 6\theta + \sin 2\theta)$ **5.** $\frac{1}{2}(\cos 8\theta + \cos 2\theta)$ **7.** $\frac{1}{2}(\cos \theta - \cos 3\theta)$ **9.** $\frac{1}{2}(\sin 2\theta + \sin \theta)$ **11.** $2 \sin \theta \cos 3\theta$

13. $2 \cos 3\theta \cos \theta$ **15.** $2 \sin 2\theta \cos \theta$ **17.** $2 \sin \theta \sin\left(\dfrac{\theta}{2}\right)$ **19.** $\dfrac{\sin \theta + \sin 3\theta}{2 \sin 2\theta} = \dfrac{2 \sin 2\theta \cos(-\theta)}{2 \sin 2\theta} = \cos(-\theta) = \cos \theta$

21. $\dfrac{\sin 4\theta + \sin 2\theta}{\cos 4\theta + \cos 2\theta} = \dfrac{2 \sin 3\theta \cos \theta}{2 \cos 3\theta \cos \theta} = \dfrac{\sin 3\theta}{\cos 3\theta} = \tan 3\theta$ **23.** $\dfrac{\cos \theta - \cos 3\theta}{\sin \theta + \sin 3\theta} = \dfrac{-2 \sin 2\theta \sin(-\theta)}{2 \sin 2\theta \cos(-\theta)} = \dfrac{\sin \theta}{\cos \theta} = \tan \theta$

25. $\sin \theta(\sin \theta + \sin 3\theta) = \sin \theta[2 \sin 2\theta \cos(-\theta)] = 2 \sin 2\theta \sin \theta \cos \theta = \cos \theta(2 \sin 2\theta \sin \theta) = \cos \theta\left[2 \cdot \dfrac{1}{2}(\cos \theta - \cos 3\theta)\right]$

$= \cos \theta(\cos \theta - \cos 3\theta)$

27. $\dfrac{\sin 4\theta + \sin 8\theta}{\cos 4\theta + \cos 8\theta} = \dfrac{2 \sin 6\theta \cos(-2\theta)}{2 \cos 6\theta \cos(-2\theta)} = \dfrac{\sin 6\theta}{\cos 6\theta} = \tan 6\theta$

29. $\dfrac{\sin 4\theta + \sin 8\theta}{\sin 4\theta - \sin 8\theta} = \dfrac{2 \sin 6\theta \cos(-2\theta)}{2 \sin(-2\theta) \cos 6\theta} = \dfrac{\sin 6\theta}{\cos 6\theta} \cdot \dfrac{\cos 2\theta}{-\sin 2\theta} = \tan 6\theta(-\cot 2\theta) = -\dfrac{\tan 6\theta}{\tan 2\theta}$

31. $\dfrac{\sin \alpha + \sin \beta}{\sin \alpha - \sin \beta} = \dfrac{2 \sin \dfrac{\alpha + \beta}{2} \cos \dfrac{\alpha - \beta}{2}}{2 \sin \dfrac{\alpha - \beta}{2} \cos \dfrac{\alpha + \beta}{2}} = \dfrac{\sin \dfrac{\alpha + \beta}{2}}{\cos \dfrac{\alpha + \beta}{2}} \cdot \dfrac{\cos \dfrac{\alpha - \beta}{2}}{\sin \dfrac{\alpha - \beta}{2}} = \tan \dfrac{\alpha + \beta}{2} \cot \dfrac{\alpha - \beta}{2}$

33. $\dfrac{\sin \alpha + \sin \beta}{\cos \alpha + \cos \beta} = \dfrac{2 \sin \dfrac{\alpha + \beta}{2} \cos \dfrac{\alpha - \beta}{2}}{2 \cos \dfrac{\alpha + \beta}{2} \cos \dfrac{\alpha - \beta}{2}} = \dfrac{\sin \dfrac{\alpha + \beta}{2}}{\cos \dfrac{\alpha + \beta}{2}} = \tan \dfrac{\alpha + \beta}{2}$

35. $1 + \cos 2\theta + \cos 4\theta + \cos 6\theta = (1 + \cos 6\theta) + (\cos 2\theta + \cos 4\theta) = 2 \cos^2 3\theta + 2 \cos 3\theta \cos(-\theta) = 2 \cos 3\theta(\cos 3\theta + \cos \theta)$
$= 2 \cos 3\theta(2 \cos 2\theta \cos \theta) = 4 \cos \theta \cos 2\theta \cos 3\theta$

37. (a) $y = 2 \sin 2061\pi t \cos 357\pi t$ **(b)** $y_{max} = 2$ **(c)**

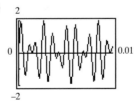

39. $\sin 2\alpha + \sin 2\beta + \sin 2\gamma = 2 \sin(\alpha + \beta) \cos(\alpha - \beta) + \sin 2\gamma = 2 \sin(\alpha + \beta) \cos(\alpha - \beta) + 2 \sin \gamma \cos \gamma$
$= 2 \sin(\pi - \gamma) \cos(\alpha - \beta) + 2 \sin \gamma \cos \gamma = 2 \sin \gamma \cos(\alpha - \beta) + 2 \sin \gamma \cos \gamma = 2 \sin \gamma[\cos(\alpha - \beta) + \cos \gamma]$
$= 2 \sin \gamma\left(2 \cos \dfrac{\alpha - \beta + \gamma}{2} \cos \dfrac{\alpha - \beta - \gamma}{2}\right) = 4 \sin \gamma \cos \dfrac{\pi - 2\beta}{2} \cos \dfrac{2\alpha - \pi}{2} = 4 \sin \gamma \cos\left(\dfrac{\pi}{2} - \beta\right) \cos\left(\alpha - \dfrac{\pi}{2}\right)$
$= 4 \sin \gamma \sin \beta \sin \alpha$

41. $\quad\quad\quad \sin(\alpha - \beta) = \sin \alpha \cos \beta - \cos \alpha \sin \beta$
$\quad\quad\quad\quad \sin(\alpha + \beta) = \sin \alpha \cos \beta + \cos \alpha \sin \beta$
$\sin(\alpha - \beta) + \sin(\alpha + \beta) = 2 \sin \alpha \cos \beta$

$\quad\quad\quad\quad \sin \alpha \cos \beta = \dfrac{1}{2}[\sin(\alpha + \beta) + \sin(\alpha - \beta)]$

43. $2 \cos \dfrac{\alpha + \beta}{2} \cos \dfrac{\alpha - \beta}{2} = 2 \cdot \dfrac{1}{2}\left[\cos\left(\dfrac{\alpha + \beta}{2} + \dfrac{\alpha - \beta}{2}\right) + \cos\left(\dfrac{\alpha + \beta}{2} - \dfrac{\alpha - \beta}{2}\right)\right] = \cos \dfrac{2\alpha}{2} + \cos \dfrac{2\beta}{2} = \cos \alpha + \cos \beta$

3.5 Exercises

1. 0 **3.** $-\dfrac{\pi}{2}$ **5.** 0 **7.** $\dfrac{\pi}{4}$ **9.** $\dfrac{\pi}{3}$ **11.** $\dfrac{5\pi}{6}$ **13.** 0.10 **15.** 1.37 **17.** 0.51 **19.** -0.38 **21.** -0.12 **23.** 1.08 **25.** $\dfrac{\sqrt{2}}{2}$ **27.** $-\dfrac{\sqrt{3}}{3}$

29. 2 **31.** $\sqrt{2}$ **33.** $-\dfrac{\sqrt{2}}{2}$ **35.** $\dfrac{2\sqrt{3}}{3}$ **37.** $\dfrac{3\pi}{4}$ **39.** $\dfrac{\pi}{6}$ **41.** $\dfrac{\sqrt{2}}{4}$ **43.** $\dfrac{\sqrt{5}}{2}$ **45.** $-\dfrac{\sqrt{14}}{2}$ **47.** $-\dfrac{3\sqrt{10}}{10}$ **49.** $\sqrt{5}$ **51.** $-\dfrac{\pi}{4}$ **53.** $\dfrac{\sqrt{3}}{2}$

55. $-\dfrac{24}{25}$ **57.** $-\dfrac{33}{65}$ **59.** $\dfrac{65}{63}$ **61.** $\dfrac{1}{39}(48 - 25\sqrt{3})$ **63.** $\dfrac{\sqrt{3}}{2}$ **65.** $\dfrac{7}{25}$ **67.** $\dfrac{24}{7}$ **69.** $\dfrac{24}{25}$ **71.** $\dfrac{1}{5}$ **73.** $\dfrac{25}{7}$ **75.** 4

77. $u\sqrt{1 - v^2} - v\sqrt{1 - u^2}$ **79.** $\dfrac{u\sqrt{1 - v^2} - v}{\sqrt{1 + u^2}}$ **81.** $\dfrac{uv - \sqrt{1 - u^2}\sqrt{1 - v^2}}{v\sqrt{1 - u^2} + u\sqrt{1 - v^2}}$

83. Let $\theta = \tan^{-1} v$. Then $\tan \theta = v$, $-\dfrac{\pi}{2} < \theta < \dfrac{\pi}{2}$. Now, $\sec \theta > 0$ and $\tan^2 \theta + 1 = \sec^2 \theta$. Thus $\sec(\tan^{-1} v) = \sec \theta = \sqrt{1 + v^2}$.

85. Let $\theta = \cos^{-1} v$. Then $\cos \theta = v$, $0 \le \theta \le \pi$, and $\tan(\cos^{-1} v) = \tan \theta = \dfrac{\sin \theta}{\cos \theta} = \dfrac{\sqrt{1 - \cos^2 \theta}}{\cos \theta} = \dfrac{\sqrt{1 - v^2}}{v}$.

87. Let $\theta = \sin^{-1} v$. Then $\sin \theta = v$, $-\dfrac{\pi}{2} \le \theta \le \dfrac{\pi}{2}$, and $\cos(\sin^{-1} v) = \cos \theta = \sqrt{1 - \sin^2 \theta} = \sqrt{1 - v^2}$.

89. Let $\alpha = \sin^{-1} v$ and $\beta = \cos^{-1} v$. Then $\sin \alpha = \cos \beta = v$, and since $\sin \alpha = \cos\left(\dfrac{\pi}{2} - \alpha\right)$, $\cos\left(\dfrac{\pi}{2} - \alpha\right) = \cos \beta$. If $v \ge 0$,

then $0 \le \alpha \le \dfrac{\pi}{2}$, so that $\left(\dfrac{\pi}{2} - \alpha\right)$ and β both lie on $\left[0, \dfrac{\pi}{2}\right]$. If $v < 0$, then $-\dfrac{\pi}{2} \le \alpha < 0$, so that $\left(\dfrac{\pi}{2} - \alpha\right)$ and β

both lie on $\left(\dfrac{\pi}{2}, \pi\right)$. Either way, $\cos\left(\dfrac{\pi}{2} - \alpha\right) = \cos \beta$ implies $\dfrac{\pi}{2} - \alpha = \beta$, or $\alpha + \beta = \dfrac{\pi}{2}$.

91. Let $\alpha = \tan^{-1} \dfrac{1}{v}$ and $\beta = \tan^{-1} v$. Because $\dfrac{1}{v}$ must be defined, $v \ne 0$ and so $\alpha, \beta \ne 0$. Then $\tan \alpha = \dfrac{1}{v} = \dfrac{1}{\tan \beta} = \cot \beta$, and since

$\tan \alpha = \cot\left(\dfrac{\pi}{2} - \alpha\right)$, $\cot\left(\dfrac{\pi}{2} - \alpha\right) = \cot \beta$. Because $v > 0$, $0 < \alpha < \dfrac{\pi}{2}$ and so $\dfrac{\pi}{2} - \alpha$ and β both lie on $\left(0, \dfrac{\pi}{2}\right)$.

Then $\cot\left(\dfrac{\pi}{2} - \alpha\right) = \cot \beta$ implies $\dfrac{\pi}{2} - \alpha = \beta$, or $\alpha + \beta = \dfrac{\pi}{2}$.

93. $\sin(\sin^{-1} v + \cos^{-1} v) = \sin(\sin^{-1} v)\cos(\cos^{-1} v) - \cos(\sin^{-1} v)\sin(\cos^{-1} v) = (v)(v) + \sqrt{1 - v^2}\sqrt{1 - v^2} = v^2 + 1 - v^2 = 1$

95. 1.32 **97.** 0.46 **99.** -0.34 **101.** 2.72 **103.** -0.73 **105.** 2.55 **107.** 2.77 minutes **109.** $-1 \le x \le 1$ **111.** $0 \le x \le \pi$

113.

3.6 Exercises

1. $\dfrac{\pi}{6}, \dfrac{5\pi}{6}$ **3.** $\dfrac{5\pi}{6}, \dfrac{11\pi}{6}$ **5.** $\dfrac{\pi}{2}, \dfrac{3\pi}{2}$ **7.** $\dfrac{\pi}{2}, \dfrac{7\pi}{6}, \dfrac{11\pi}{6}$ **9.** $\dfrac{\pi}{3}, \dfrac{2\pi}{3}, \dfrac{4\pi}{3}, \dfrac{5\pi}{3}$ **11.** $\dfrac{4\pi}{9}, \dfrac{8\pi}{9}, \dfrac{16\pi}{9}$ **13.** $\dfrac{3\pi}{4}, \dfrac{7\pi}{4}$ **15.** $\dfrac{11\pi}{6}$ **17.** 0.41, 2.73

19. 1.37, 4.51 **21.** 2.69, 3.59 **23.** 1.82, 4.46 **25.** $\dfrac{\pi}{2}, \dfrac{2\pi}{3}, \dfrac{4\pi}{3}, \dfrac{3\pi}{2}$ **27.** $\dfrac{\pi}{2}, \dfrac{7\pi}{6}, \dfrac{11\pi}{6}$ **29.** $0, \dfrac{\pi}{4}, \dfrac{5\pi}{4}$ **31.** $\dfrac{\pi}{4}, \dfrac{5\pi}{4}$ **33.** $0, \dfrac{\pi}{3}, \pi, \dfrac{5\pi}{3}$ **35.** $\dfrac{\pi}{2}, \dfrac{3\pi}{2}$

37. $0, \dfrac{2\pi}{3}, \dfrac{4\pi}{3}$ **39.** $0, \dfrac{\pi}{3}, \dfrac{\pi}{2}, \dfrac{2\pi}{3}, \pi, \dfrac{4\pi}{3}, \dfrac{3\pi}{2}, \dfrac{5\pi}{3}$ **41.** $0, \dfrac{\pi}{5}, \dfrac{2\pi}{5}, \dfrac{3\pi}{5}, \dfrac{4\pi}{5}, \pi, \dfrac{6\pi}{5}, \dfrac{7\pi}{5}, \dfrac{8\pi}{5}, \dfrac{9\pi}{5}$ **43.** $\dfrac{\pi}{6}, \dfrac{5\pi}{6}, \dfrac{3\pi}{2}$ **45.** $\dfrac{\pi}{3}, \dfrac{5\pi}{3}$

47. No real solutions **49.** No real solutions **51.** $\dfrac{\pi}{2}, \dfrac{7\pi}{6}$ **53.** $0, \dfrac{\pi}{3}, \pi, \dfrac{5\pi}{3}$ **55.** $\dfrac{\pi}{4}$

57. $-1.29, 0$ **59.** $-2.24, 0, 2.24$ **61.** $-0.82, 0.82$

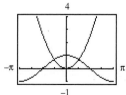

63. $-1.30, 1.97, 3.83$ **65.** 0.52 **67.** 1.25 **69.** $-1.02, 1.02$ **71.** 0, 2.14 **73.** 0.76, 1.34

75. (a) 60° **(b)** 60° **(c)** $A(60°) = 12\sqrt{3}$ sq in. **(d)**

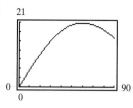

$\theta_{\max} = 60°$
Maximum Area = 20.78 sq in.

77. 2.02, 4.91
79. (a) 29.99° or 60.01° **(b)** 123.6 meters **(c)**

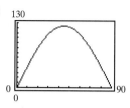

81. 28.9° **83.** Yes; it varies from 1.28 to 1.34 **85.** 1.47

87. If θ is the original angle of incidence and ϕ is the angle of refraction, then $\dfrac{\sin \theta}{\sin \phi} = n_2$. The angle of incidence of the emerging

beam is also ϕ, and the index of refraction is $\dfrac{1}{n_2}$. Thus, θ is the angle of refraction of the emerging beam.

Fill-in-the-Blank Items

1. identity; conditional **3.** + **5.** $1 - \cos \alpha$ **7.** 0

True/False Items

1. T **3.** T **5.** F **7.** F

Review Exercises

1. $\tan \theta \cot \theta - \sin^2 \theta = 1 - \sin^2 \theta = \cos^2 \theta$ **3.** $\cos^2 \theta(1 + \tan^2 \theta) = \cos^2 \theta \sec^2 \theta = 1$

5. $4\cos^2 \theta + 3\sin^2 \theta = \cos^2 \theta + 3(\cos^2 \theta + \sin^2 \theta) = 3 + \cos^2 \theta$

7. $\dfrac{1 - \cos \theta}{\sin \theta} + \dfrac{\sin \theta}{1 - \cos \theta} = \dfrac{(1 - \cos \theta)^2 + \sin^2 \theta}{\sin \theta(1 - \cos \theta)} = \dfrac{1 - 2\cos \theta + \cos^2 \theta + \sin^2 \theta}{\sin \theta(1 - \cos \theta)} = \dfrac{2(1 - \cos \theta)}{\sin \theta(1 - \cos \theta)} = 2 \csc \theta$

9. $\dfrac{\cos \theta}{\cos \theta - \sin \theta} = \dfrac{\dfrac{\cos \theta}{\cos \theta}}{\dfrac{\cos \theta - \sin \theta}{\cos \theta}} = \dfrac{1}{1 - \dfrac{\sin \theta}{\cos \theta}} = \dfrac{1}{1 - \tan \theta}$

11. $\dfrac{\csc \theta}{1 + \csc \theta} = \dfrac{\dfrac{1}{\sin \theta}}{1 + \dfrac{1}{\sin \theta}} = \dfrac{1}{1 + \sin \theta} = \dfrac{1}{1 + \sin \theta} \cdot \dfrac{1 - \sin \theta}{1 - \sin \theta} = \dfrac{1 - \sin \theta}{1 - \sin^2 \theta} = \dfrac{1 - \sin \theta}{\cos^2 \theta}$

13. $\csc \theta - \sin \theta = \dfrac{1}{\sin \theta} - \sin \theta = \dfrac{1 - \sin^2 \theta}{\sin \theta} = \dfrac{\cos^2 \theta}{\sin \theta} = \cos \theta \cdot \dfrac{\cos \theta}{\sin \theta} = \cos \theta \cot \theta$

15. $\dfrac{1 - \sin \theta}{\sec \theta} = \cos \theta(1 - \sin \theta) \cdot \dfrac{1 + \sin \theta}{1 + \sin \theta} = \dfrac{\cos \theta(1 - \sin^2 \theta)}{1 + \sin \theta} = \dfrac{\cos^3 \theta}{1 + \sin \theta}$

17. $\cot \theta - \tan \theta = \dfrac{\cos \theta}{\sin \theta} - \dfrac{\sin \theta}{\cos \theta} = \dfrac{\cos^2 \theta - \sin^2 \theta}{\sin \theta \cos \theta} = \dfrac{1 - 2\sin^2 \theta}{\sin \theta \cos \theta}$

19. $\dfrac{\cos(\alpha + \beta)}{\cos \alpha \sin \beta} = \dfrac{\cos \alpha \cos \beta - \sin \alpha \sin \beta}{\cos \alpha \sin \beta} = \dfrac{\cos \alpha \cos \beta}{\cos \alpha \sin \beta} - \dfrac{\sin \alpha \sin \beta}{\cos \alpha \sin \beta} = \cot \beta - \tan \alpha$

21. $\dfrac{\cos(\alpha - \beta)}{\cos \alpha \cos \beta} = \dfrac{\cos \alpha \cos \beta + \sin \alpha \sin \beta}{\cos \alpha \cos \beta} = \dfrac{\cos \alpha \cos \beta}{\cos \alpha \cos \beta} + \dfrac{\sin \alpha \sin \beta}{\cos \alpha \cos \beta} = 1 + \tan \alpha \tan \beta$

23. $(1 + \cos \theta)\left(\tan \dfrac{\theta}{2}\right) = \left(2 \cos^2 \dfrac{\theta}{2}\right) \dfrac{\sin\left(\dfrac{\theta}{2}\right)}{\cos\left(\dfrac{\theta}{2}\right)} = 2 \sin \dfrac{\theta}{2} \cos \dfrac{\theta}{2} = \sin \theta$

25. $2 \cot \theta \cot 2\theta = 2\left(\dfrac{\cos \theta}{\sin \theta}\right)\left(\dfrac{\cos 2\theta}{\sin 2\theta}\right) = \dfrac{2 \cos \theta(\cos^2 \theta - \sin^2 \theta)}{2 \sin^2 \theta \cos \theta} = \dfrac{\cos^2 \theta - \sin^2 \theta}{\sin^2 \theta} = \cot^2 \theta - 1$

27. $1 - 8 \sin^2 \theta \cos^2 \theta = 1 - 2(2 \sin \theta \cos \theta)^2 = 1 - 2 \sin^2 2\theta = \cos 4\theta$

29. $\dfrac{\sin 2\theta + \sin 4\theta}{\cos 2\theta + \cos 4\theta} = \dfrac{2 \sin 3\theta \cos(-\theta)}{2 \cos 3\theta \cos(-\theta)} = \tan 3\theta$

31. $\dfrac{\cos 2\theta - \cos 4\theta}{\cos 2\theta + \cos 4\theta} - \tan \theta \tan 3\theta = \dfrac{-2 \sin 3\theta \sin(-\theta)}{2 \cos 3\theta \cos(-\theta)} - \tan \theta \tan 3\theta = \tan 3\theta \tan \theta - \tan \theta \tan 3\theta = 0$

33. $\dfrac{1}{4}(\sqrt{6} - \sqrt{2})$ **35.** $\dfrac{1}{4}(\sqrt{6} - \sqrt{2})$ **37.** $\dfrac{1}{2}$ **39.** $\sqrt{\dfrac{2 - \sqrt{2}}{2 + \sqrt{2}}} = \sqrt{2} - 1$

41. (a) $-\dfrac{33}{65}$ **(b)** $-\dfrac{56}{65}$ **(c)** $-\dfrac{63}{65}$ **(d)** $\dfrac{33}{56}$ **(e)** $\dfrac{24}{25}$ **(f)** $\dfrac{119}{169}$ **(g)** $\dfrac{5\sqrt{26}}{26}$ **(h)** $\dfrac{2\sqrt{5}}{5}$

43. (a) $-\dfrac{16}{65}$ **(b)** $-\dfrac{63}{65}$ **(c)** $-\dfrac{56}{65}$ **(d)** $\dfrac{16}{63}$ **(e)** $\dfrac{24}{25}$ **(f)** $\dfrac{119}{169}$ **(g)** $\dfrac{\sqrt{26}}{26}$ **(h)** $-\dfrac{\sqrt{10}}{10}$

45. (a) $-\dfrac{63}{65}$ **(b)** $\dfrac{16}{65}$ **(c)** $\dfrac{33}{65}$ **(d)** $-\dfrac{63}{16}$ **(e)** $\dfrac{24}{25}$ **(f)** $-\dfrac{119}{169}$ **(g)** $\dfrac{2\sqrt{13}}{13}$ **(h)** $-\dfrac{\sqrt{10}}{10}$

47. (a) $\dfrac{(-\sqrt{3} - 2\sqrt{2})}{6}$ **(b)** $\dfrac{(1 - 2\sqrt{6})}{6}$ **(c)** $\dfrac{(-\sqrt{3} + 2\sqrt{2})}{6}$ **(d)** $\dfrac{(-\sqrt{3} - 2\sqrt{2})}{(1 - 2\sqrt{6})} = \dfrac{(8\sqrt{2} + 9\sqrt{3})}{23}$ **(e)** $-\dfrac{\sqrt{3}}{2}$ **(f)** $-\dfrac{7}{9}$ **(g)** $\dfrac{\sqrt{3}}{3}$

(h) $\dfrac{\sqrt{3}}{2}$ **49. (a)** 1 **(b)** 0 **(c)** $-\dfrac{1}{9}$ **(d)** Not defined **(e)** $\dfrac{4\sqrt{5}}{9}$ **(f)** $-\dfrac{1}{9}$ **(g)** $\dfrac{\sqrt{30}}{6}$ **(h)** $-\dfrac{\sqrt{6}\sqrt{3} - \sqrt{5}}{6}$

51. $\dfrac{\pi}{2}$ **53.** $\dfrac{\pi}{4}$ **55.** $\dfrac{5\pi}{6}$ **57.** $\dfrac{\sqrt{2}}{2}$ **59.** $-\sqrt{3}$ **61.** $\dfrac{2\sqrt{3}}{3}$ **63.** $\dfrac{3}{5}$ **65.** $-\dfrac{4}{3}$ **67.** $-\dfrac{\pi}{6}$ **69.** $-\dfrac{\pi}{4}$ **71.** $\dfrac{4 + 3\sqrt{3}}{10}$

73. $\dfrac{3\sqrt{3} + 4}{3 - 4\sqrt{3}} = \dfrac{48 + 25\sqrt{3}}{-39}$ **75.** $-\dfrac{24}{25}$ **77.** $\dfrac{\pi}{3}, \dfrac{5\pi}{3}$ **79.** $\dfrac{3\pi}{4}, \dfrac{5\pi}{4}$ **81.** $\dfrac{3\pi}{4}, \dfrac{7\pi}{4}$ **83.** $0, \dfrac{\pi}{2}, \pi, \dfrac{3\pi}{2}$ **85.** $1.12, \pi - 1.12$

87. $0, \pi$ **89.** $0, \dfrac{2\pi}{3}, \pi, \dfrac{4\pi}{3}$ **91.** $0, \dfrac{\pi}{6}, \dfrac{5\pi}{6}$ **93.** $\dfrac{\pi}{6}, \dfrac{\pi}{2}, \dfrac{5\pi}{6}$ **95.** $\dfrac{\pi}{2}, \pi$ **97.** 1.11 **99.** 0.86 **101.** 2.21

C H A P T E R   4   4.1 Exercises

1. $a \approx 13.74, c \approx 14.62, \alpha = 70°$ **3.** $b \approx 5.03, c \approx 7.83, \alpha = 50°$ **5.** $a \approx 0.71, c \approx 4.06, \beta = 80°$ **7.** $b \approx 10.72, c \approx 11.83, \beta = 65°$
9. $b \approx 3.08, a \approx 8.46, \alpha = 70°$ **11.** $c \approx 5.83, \alpha \approx 59.0°, \beta = 31.0°$ **13.** $b \approx 4.58, \alpha \approx 23.6°, \beta = 66.4°$ **15.** 4.59 in., 6.55 in.
17. 5.52 in. or 11.83 in.   **19.** $23.6°$ and $66.4°$ **21.** 70.02 ft   **23.** 985.91 ft   **25.** 137.37 m   **27.** 20.67 ft   **29.** 449.36 ft   **31.** $80.5°$ **33.** 30 ft
35. 530 ft   **37.** 555 ft   **39. (a)** 112 ft/sec or 76.3 mph   **(b)** 82.41 ft/sec or 56.2 mph   **(c)** under $18.8°$ **41.** S76.6°E   **43.** $14.9°$
45. (a) 3.1 mi **(b)** 3.2 mi **(c)** 3.8 mi
47. (a) $\cos \dfrac{\theta}{2} = \dfrac{3960}{3960 + h}$ **(b)** $d = 3960\,\theta$ **(c)** $\cos \dfrac{d}{7920} = \dfrac{3960}{3960 + h}$ **(d)** 206 mi **(e)** 2990 miles
49. No. Move the tripod 1 foot back.

4.2 Exercises

1. $a = 3.23, b = 3.55, \alpha = 40°$ **3.** $a = 3.25, c = 4.23, \beta = 45°$ **5.** $\gamma = 95°, c = 9.86, a = 6.36$ **7.** $\alpha = 40°, a = 2, c = 3.06$
9. $\gamma = 120°, b = 1.06, c = 2.69$ **11.** $\alpha = 100°, a = 5.24, c = 0.92$ **13.** $\beta = 40°, a = 5.64, b = 3.86$ **15.** $\gamma = 100°, a = 1.31, b = 1.31$
17. One triangle; $\beta = 30.7°, \gamma = 99.3°, c = 3.86$ **19.** One triangle; $\gamma = 36.2°, \alpha = 43.8°, a = 3.51$ **21.** No triangle
23. Two triangles; $\gamma_1 = 30.9°, \alpha_1 = 129.1°, a_1 = 9.08$ or $\gamma_2 = 149.1°, \alpha_2 = 10.9°, a_2 = 2.21$ **25.** No triangle
27. Two triangles; $\alpha_1 = 57.7°, \beta_1 = 97.3°, b_1 = 2.35$ or $\alpha_2 = 122.3°, \beta_2 = 32.7°, b_2 = 1.28$
29. (a) Station Able is 143.3 mi from the ship; Station Baker is 135.6 mi from the ship.   **(b)** Approx. 41 min   **31.** 1490.5 ft   **33.** 381.7 ft
35. (a) 169 mi **(b)** 161.3°   **37.** 84.7°; 183.7 ft   **39.** 2.64 mi   **41.** 1.88 mi   **43. (a)** 1818 m   **45.** 39.4 ft

47. $\dfrac{a + b}{c} = \dfrac{a}{c} + \dfrac{b}{c} = \dfrac{\sin \alpha}{\sin \gamma} + \dfrac{\sin \beta}{\sin \gamma} = \dfrac{\sin \alpha + \sin \beta}{\sin \gamma} = \dfrac{2 \sin \dfrac{\alpha + \beta}{2} \cos \dfrac{\alpha - \beta}{2}}{2 \sin \dfrac{\gamma}{2} \cos \dfrac{\gamma}{2}} = \dfrac{\sin\left(\dfrac{\pi}{2} - \dfrac{\gamma}{2}\right) \cos \dfrac{\alpha - \beta}{2}}{\sin \dfrac{\gamma}{2} \cos \dfrac{\gamma}{2}} = \dfrac{\cos \dfrac{1}{2}(\alpha - \beta)}{\sin \dfrac{1}{2}\gamma}$

49. $a = \dfrac{b \sin \alpha}{\sin \beta} = \dfrac{b \sin[180° - (\beta + \gamma)]}{\sin \beta} = \dfrac{b}{\sin \beta}(\sin \beta \cos \gamma + \cos \beta \sin \gamma) = b \cos \gamma + \dfrac{b \sin \gamma}{\sin \beta}\cos \beta = b \cos \gamma + c \cos \beta$

51. $\sin \beta = \sin(\text{Angle } ABC) = \sin(\text{Angle } AB'C) = \dfrac{b}{2r}; \dfrac{\sin \beta}{b} = \dfrac{1}{2r}$; the result follows using the Law of Sines.

4.3 Exercises

1. $b = 2.95, \alpha = 28.7°, \gamma = 106.3°$ **3.** $c = 3.75, \alpha = 32.1°, \beta = 52.9°$ **5.** $\alpha = 48.5°, \beta = 38.6°, \gamma = 92.9°$
7. $\alpha = 127.2°, \beta = 32.1°, \gamma = 20.7°$ **9.** $c = 2.57, \alpha = 48.6°, \beta = 91.4°$ **11.** $a = 2.99, \beta = 19.2°, \gamma = 80.8°$
13. $b = 4.14, \alpha = 43.0°, \gamma = 27.0°$ **15.** $c = 1.69, \alpha = 65.0°, \beta = 65.0°$ **17.** $\alpha = 67.4°, \beta = 90°, \gamma = 22.6°$
19. $\alpha = 60°, \beta = 60°, \gamma = 60°$ **21.** $\alpha = 33.6°, \beta = 62.2°, \gamma = 84.3°$ **23.** $\alpha = 97.9°, \beta = 52.4°, \gamma = 29.7°$
25. 70.75 ft   **27. (a)** 12°   **(b)** 220.8 mph   **29. (a)** 63.7 ft   **(b)** 66.8 ft   **(c)** 92.5°   **31. (a)** 492.6 ft   **(b)** 269.3 ft   **33.** 342.3 ft
35. Using the Law of Cosines:
$L^2 = x^2 + r^2 - 2rx \cos \theta$
$x^2 - 2rx \cos \theta + r^2 - L^2 = 0$
Then, using the quadratic formula:
$x = r \cos \theta + \sqrt{r^2 \cos^2 \theta + L^2 - r^2}$

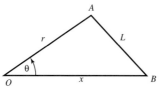

37. $\cos \dfrac{\gamma}{2} = \sqrt{\dfrac{1 + \cos \gamma}{2}} = \sqrt{\dfrac{1 + \dfrac{a^2 + b^2 - c^2}{2ab}}{2}} = \sqrt{\dfrac{2ab + a^2 + b^2 - c^2}{4ab}} = \sqrt{\dfrac{(a + b)^2 - c^2}{4ab}} = \sqrt{\dfrac{(a + b + c)(a + b - c)}{4ab}}$
$= \sqrt{\dfrac{2s(2s - 2c)}{4ab}} = \sqrt{\dfrac{s(s - c)}{ab}}$

39. $\dfrac{\cos \alpha}{a} + \dfrac{\cos \beta}{b} + \dfrac{\cos \gamma}{c} = \dfrac{b^2 + c^2 - a^2}{2abc} + \dfrac{a^2 + c^2 - b^2}{2abc} + \dfrac{a^2 + b^2 - c^2}{2abc} = \dfrac{b^2 + c^2 - a^2 + a^2 + c^2 - b^2 + a^2 + b^2 - c^2}{2abc}$
$= \dfrac{a^2 + b^2 + c^2}{2abc}$

4.4 Exercises

1. 2.83   **3.** 2.99   **5.** 14.98   **7.** 9.56   **9.** 3.86   **11.** 1.48   **13.** 2.82   **15.** 1.53   **17.** 30   **19.** 1.73   **21.** 19.90   **23.** 19.81   **25.** 9.03 sq. ft.
27. $5446.38   **29.** 9.26 sq. cm.   **31.** $A = \dfrac{1}{2}ab \sin \gamma = \dfrac{1}{2}a \sin \gamma \left(\dfrac{a \sin \beta}{\sin \alpha}\right) = \dfrac{a^2 \sin \beta \sin \gamma}{2 \sin \alpha}$ **33.** 0.92   **35.** 2.27   **37.** 5.44

39. $A = \dfrac{1}{2}r^2(\theta + \sin \theta)$ **41.** 31,144 sq ft

43. $h_1 = 2\dfrac{K}{a}, h_2 = 2\dfrac{K}{b}, h_3 = 2\dfrac{K}{c}$. Then $\dfrac{1}{h_1} + \dfrac{1}{h_2} + \dfrac{1}{h_3} = \dfrac{a}{2K} + \dfrac{b}{2K} + \dfrac{c}{2K} = \dfrac{a + b + c}{2K} = \dfrac{2s}{2K} = \dfrac{s}{K}$.

47. $\cot \dfrac{\alpha}{2} + \cot \dfrac{\beta}{2} + \cot \dfrac{\gamma}{2} = \dfrac{s - a}{r} + \dfrac{s - b}{r} + \dfrac{s - c}{r} = \dfrac{3s - (a + b + c)}{r} = \dfrac{3s - 2s}{r} = \dfrac{s}{r}$

4.5 Exercises

1. $d = -5 \cos \pi t$ **3.** $d = -6 \cos 2t$ **5.** $d = -5 \sin \pi t$ **7.** $d = -6 \sin 2t$ **9. (a)** Simple harmonic   **(b)** 5 m   **(c)** $\dfrac{2\pi}{3}$ sec
(d) $\dfrac{3}{2\pi}$ oscillation/sec   **11. (a)** Simple harmonic   **(b)** 6 m   **(c)** 2 sec   **(d)** $\dfrac{1}{2}$ oscillation/sec   **13. (a)** Simple harmonic   **(b)** 3 m
(c) 4π sec   **(d)** $\dfrac{1}{4\pi}$ oscillation/sec   **15. (a)** Simple harmonic   **(b)** 2 m   **(c)** 1 sec   **(d)** 1 oscillation/sec

17.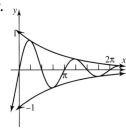

19.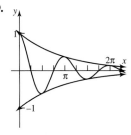

21. (a) 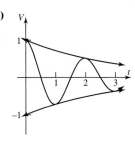 **(b)** At $t = 0, 1, 2,$ and 3.
(c) After 2.75 seconds.

Fill-in-the Blank Items

1. Sines **3.** Heron's

True/False Items

1. F **3.** T

Review Exercises

1. $\alpha = 70°, b \approx 3.42, a \approx 9.4$ **3.** $a \approx 4.58, \alpha \approx 66.4°, \beta \approx 23.6°$ **5.** $\gamma = 100°, b = 0.65, c = 1.29$ **7.** $\beta = 56.8°, \gamma = 23.2°, b = 4.25$
9. No triangle **11.** $b = 3.32, \alpha = 62.8°, \gamma = 17.2°$ **13.** No triangle **15.** $c = 2.32, \alpha = 16.1°, \beta = 123.9°$
17. $\beta = 36.2°, \gamma = 63.8°, c = 4.56$ **19.** $\alpha = 39.6°, \beta = 18.6°, \gamma = 121.8°$
21. Two triangles: $\beta_1 = 13.4°, \gamma_1 = 156.6°, c_1 = 6.86; \beta_2 = 166.6°, \gamma_2 = 3.4°, c_2 = 1.02$
23. $a = 5.23, \beta = 46°, \gamma = 64°$ **25.** 1.93 **27.** 18.79 **29.** 6 **31.** 3.80 **33.** 0.32 **35.** 839 ft **37.** 23.32 ft **39.** 2.15 mi **41.** 204.1 mi
43. (a) 2.59 mi **(b)** 2.92 mi **(c)** 2.53 mi **45. (a)** 131.8 mi **(b)** 23.1° **(c)** 0.2 hr **47.** 8799 sq ft. **49.** 1.92 sq. in. **51.** 76.9 in.

53. (a) Simple harmonic **(b)** 6 ft **(c)** π sec **(d)** $\dfrac{1}{\pi}$ oscillation/sec **55. (a)** Simple harmonic **(b)** 2 ft **(c)** 2 sec **(d)** $\dfrac{1}{2}$ oscillation/sec

57.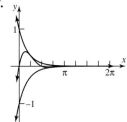

C H A P T E R   5   5.1 Exercises

1. A **3.** C **5.** B **7.** A
9.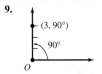

11.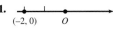

13.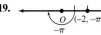

15.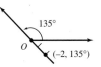

17.

19.

21.

23.
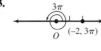

25.

27.

(a) $\left(5, -\dfrac{4\pi}{3}\right)$

(a) $(2, -2\pi)$

(a) $\left(1, -\dfrac{3\pi}{2}\right)$

(a) $\left(3, -\dfrac{5\pi}{4}\right)$

(b) $\left(-5, \dfrac{5\pi}{3}\right)$

(b) $(-2, \pi)$

(b) $\left(-1, \dfrac{3\pi}{2}\right)$

(b) $\left(-3, \dfrac{7\pi}{4}\right)$

(c) $\left(5, \dfrac{8\pi}{3}\right)$

(c) $(2, 2\pi)$

(c) $\left(1, \dfrac{5\pi}{2}\right)$

(c) $\left(3, \dfrac{11\pi}{4}\right)$

29. $(0, 3)$ **31.** $(-2, 0)$ **33.** $(-3\sqrt{3}, 3)$ **35.** $(\sqrt{2}, -\sqrt{2})$ **37.** $\left(-\dfrac{1}{2}, \dfrac{\sqrt{3}}{2}\right)$ **39.** $(2, 0)$ **41.** $(-2.57, 7.05)$ **43.** $(-4.98, -3.85)$

45. $(3, 0)$ **47.** $(1, \pi)$ **49.** $\left(\sqrt{2}, -\dfrac{\pi}{4}\right)$ **51.** $\left(2, \dfrac{\pi}{6}\right)$ **53.** $(2.47, -1.02)$ **55.** $(9.30, 0.47)$ **57.** $r^2 = \dfrac{3}{2}$ **59.** $r^2 \cos^2\theta - 4r\sin\theta = 0$

61. $r^2 \sin 2\theta = 1$ **63.** $r\cos\theta = 4$ **65.** $x^2 + y^2 - x = 0$ or $\left(x - \dfrac{1}{2}\right)^2 + y^2 = \dfrac{1}{4}$ **67.** $(x^2 + y^2)^{3/2} - x = 0$ **69.** $x^2 + y^2 = 4$

71. $y^2 = 8(x + 2)$ **73.** $d = \sqrt{(r_2 \cos\theta_2 - r_1 \cos\theta_1)^2 + (r_2 \sin\theta_2 - r_1 \sin\theta_1)^2} = \sqrt{r_1^2 + r_2^2 - 2r_1 r_2 \cos(\theta_2 - \theta_1)}$

5.2 Exercises

1. Circle, radius 4, center at pole

3. Line through pole, making an angle of $\dfrac{\pi}{3}$ with polar axis

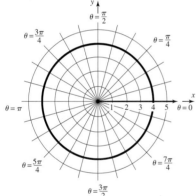

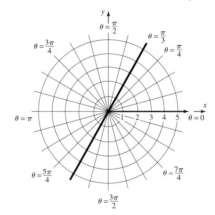

5. Horizontal line 4 units above the pole

7. Vertical line 2 units to the left of the pole

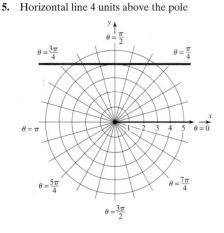

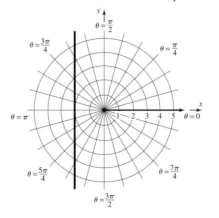

9. Circle, radius 1, center (1, 0) in rectangular coordinates

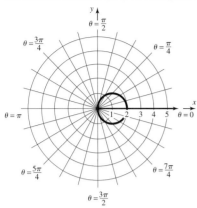

11. Circle, radius 2, center at (0, −2) in rectangular coordinates

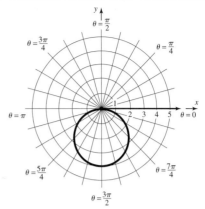

13. Circle, radius 2, center at (2, 0) in rectangular coordinates

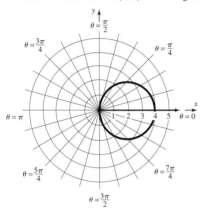

15. Circle, radius 1, center (0, −1) in rectangular coordinates

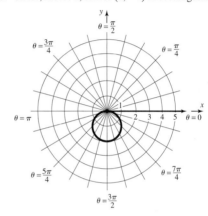

17. E   **19.** F   **21.** H   **23.** D

25. Cardioid

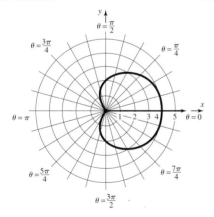

27. Cardioid

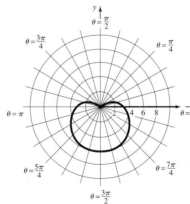

29. Limaçon without inner loop

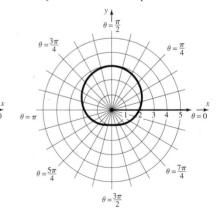

31. Limaçon without inner loop

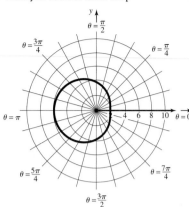

33. Limaçon with inner loop

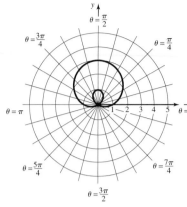

35. Limaçon with inner loop

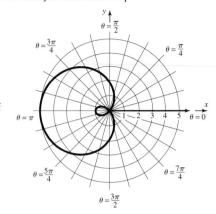

37. Rose

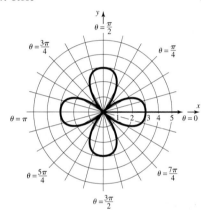

39. Rose

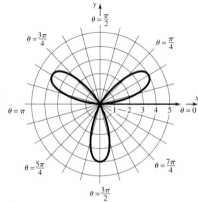

41. Lemniscate

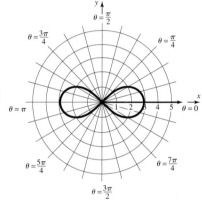

43. Spiral

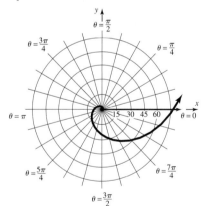

45. Cardioid

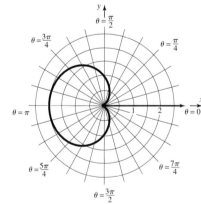

47. Limaçon with inner loop

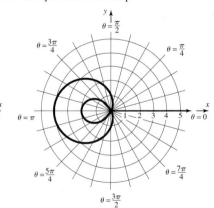

49.

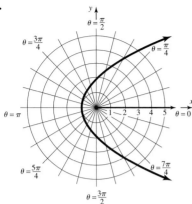

51.

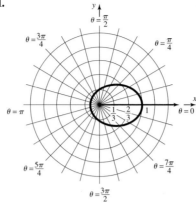

53.

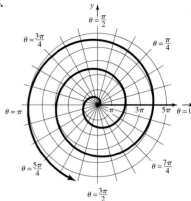

55.

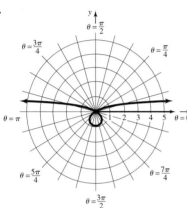

57.

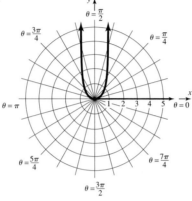

59. $r \sin \theta = a$
$y = a$

61. $r = 2a \sin \theta$
$r^2 = 2ar \sin \theta$
$x^2 + y^2 = 2ay$
$x^2 + y^2 - 2ay = 0$
$x^2 + (y - a)^2 = a^2$
Circle, radius a, center at $(0, a)$
in rectangular coordinates

63. $r = 2a \cos \theta$
$r^2 = 2ar \cos \theta$
$x^2 + y^2 = 2ax$
$x^2 - 2ax + y^2 = 0$
$(x - a)^2 + y^2 = a^2$
Circle, radius a, center at $(a, 0)$
in rectangular coordinates

65. (a) $r^2 = \cos \theta$: $r^2 = \cos(\pi - \theta)$
                $r^2 = -\cos \theta$
             Not equivalent; test fails
       (b) $r^2 = \sin \theta$: $r^2 = \sin(\pi - \theta)$
                $r^2 = \sin \theta$
             Test works

               $(-r)^2 = \cos(-\theta)$
                 $r^2 = \cos \theta$
             New test works
               $(-r)^2 = \sin(-\theta)$
                 $r^2 = -\sin \theta$
             Not equivalent; new test fails

5.3 Exercises

1. $8 + 5i$ **3.** $-7 + 6i$ **5.** $-6 - 11i$ **7.** $6 - 18i$ **9.** $6 + 4i$ **11.** $10 - 5i$ **13.** 37 **15.** $\dfrac{6}{5} + \dfrac{8}{5}i$ **17.** $1 - 2i$ **19.** $\dfrac{5}{2} - \dfrac{7}{2}i$

21. $-\dfrac{1}{2} + \dfrac{\sqrt{3}}{2}i$ **23.** $2i$ **25.** $-i$ **27.** i **29.** -6 **31.** $-10i$ **33.** $-2 + 2i$ **35.** 0 **37.** 0 **39.** $2i$ **41.** $5i$ **43.** $5i$ **45.** $\{-2i, 2i\}$

47. $\{-4, 4\}$ **49.** 6 **51.** 25 **53.** $z + \bar{z} = (a + bi) + (a - bi) = 2a; z - \bar{z} = (a + bi) - (a - bi) = 2bi$

55. $\overline{z + w} = \overline{(a + bi) + (c + di)} = \overline{(a + c) + (b + d)i} = (a + c) - (b + d)i = (a - bi) + (c - di) = \bar{z} + \bar{w}$

5.4 Exercises

1.
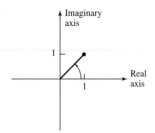
$\sqrt{2}(\cos 45° + i \sin 45°)$

3.
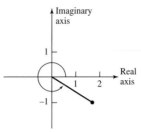
$2(\cos 330° + i \sin 330°)$

5.
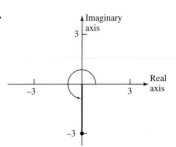
$3(\cos 270° + i \sin 270°)$

7.
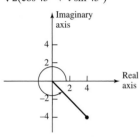
$4\sqrt{2}(\cos 315° + i \sin 315°)$

9.
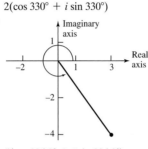
$5(\cos 306.9° + i \sin 306.9°)$

11.
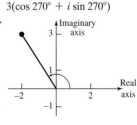
$\sqrt{13}(\cos 123.7° + i \sin 123.7°)$

13. $-1 + \sqrt{3}i$ **15.** $2\sqrt{2} - 2\sqrt{2}i$ **17.** $-3i$ **19.** $-0.035 + 0.197i$ **21.** $1.970 + 0.347i$

23. $zw = 8(\cos 60° + i \sin 60°); \dfrac{z}{w} = \dfrac{1}{2}(\cos 20° + i \sin 20°)$ **25.** $zw = 12(\cos 40° + i \sin 40°); \dfrac{z}{w} = \dfrac{3}{4}(\cos 220° + i \sin 220°)$

27. $zw = 4\left(\cos \dfrac{9\pi}{40} + i \sin \dfrac{9\pi}{40}\right); \dfrac{z}{w} = \cos \dfrac{\pi}{40} + i \sin \dfrac{\pi}{40}$ **29.** $zw = 4\sqrt{2}(\cos 15° + i \sin 15°); \dfrac{z}{w} = \sqrt{2}(\cos 75° + i \sin 75°)$

31. $-32 + 32\sqrt{3}i$ **33.** $32i$ **35.** $\dfrac{27}{2} + \dfrac{27\sqrt{3}}{2}i$ **37.** $-\dfrac{25\sqrt{2}}{2} + \dfrac{25\sqrt{2}}{2}i$ **39.** $-4 + 4i$ **41.** $-23 + 14.14i$ or $-23 + 10\sqrt{2}i$

43. $\sqrt[6]{2}(\cos 15° + i \sin 15°)$, $\sqrt[6]{2}(\cos 135° + i \sin 135°)$, $\sqrt[6]{2}(\cos 255° + i \sin 255°)$

45. $\sqrt[4]{8}(\cos 75° + i \sin 75°)$, $\sqrt[4]{8}(\cos 165° + i \sin 165°)$, $\sqrt[4]{8}(\cos 255° + i \sin 255°)$, $\sqrt[4]{8}(\cos 345° + i \sin 345°)$

47. $2(\cos 67.5° + i \sin 67.5°)$, $2(\cos 157.5° + i \sin 157.5°)$, $2(\cos 247.5° + i \sin 247.5°)$, $2(\cos 337.5° + i \sin 337.5°)$

49. $\cos 18° + i \sin 18°$, $\cos 90° + i \sin 90°$, $\cos 162° + i \sin 162°$, $\cos 234° + i \sin 234°$, $\cos 306° + i \sin 306°$

51. $1, i, -1, -i$ **53.** Look at formula (8); $|z_k| = \sqrt[n]{r}$ for all k.

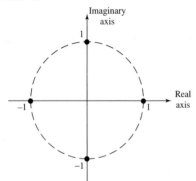

55. Look at formula (8). The z_k are spaced apart by an angle of $\dfrac{2\pi}{n}$.

5.5 Exercises

1.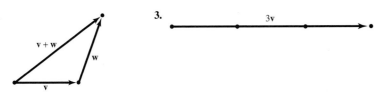

3.

5.

7.

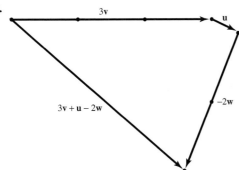

9. T **11.** F **13.** F **15.** T **17.** 12 **19.** $\mathbf{v} = 3\mathbf{i} + 4\mathbf{j}$ **21.** $\mathbf{v} = 2\mathbf{i} + 4\mathbf{j}$ **23.** $\mathbf{v} = 8\mathbf{i} - \mathbf{j}$ **25.** $\mathbf{v} = -\mathbf{i} + \mathbf{j}$ **27.** 5 **29.** $\sqrt{2}$ **31.** $\sqrt{13}$

33. $-\mathbf{j}$ **35.** $\sqrt{89}$ **37.** $\sqrt{34} - \sqrt{13}$ **39.** $\mathbf{i}$ **41.** $\frac{3}{5}\mathbf{i} - \frac{4}{5}\mathbf{j}$ **43.** $\frac{\sqrt{2}}{2}\mathbf{i} - \frac{\sqrt{2}}{2}\mathbf{j}$ **45.** $\mathbf{v} = \frac{8\sqrt{5}}{5}\mathbf{i} + \frac{4\sqrt{5}}{5}\mathbf{j}$ or $\mathbf{v} = -\frac{8\sqrt{5}}{5}\mathbf{i} - \frac{4\sqrt{5}}{5}\mathbf{j}$

47. $\{-2 + \sqrt{21}, -2 - \sqrt{21}\}$ **49.** 460 kph **51.** 288 mph **53.**

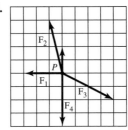

5.6 Exercises

1. $0; 90°$ **3.** $4; 37°$ **5.** $\sqrt{3} - 1; 75°$ **7.** $24; 16°$ **9.** $0; 90°$ **11.** $\frac{3}{2}$ **13.** $\mathbf{v}_1 = \text{proj}_\mathbf{w}\, \mathbf{v} = \frac{5}{2}\mathbf{i} - \frac{5}{2}\mathbf{j}, \mathbf{v}_2 = -\frac{1}{2}\mathbf{i} - \frac{1}{2}\mathbf{j}$

15. $\mathbf{v}_1 = \text{proj}_\mathbf{w}\, \mathbf{v} = -\frac{1}{5}\mathbf{i} - \frac{2}{5}\mathbf{j}, \mathbf{v}_2 = \frac{6}{5}\mathbf{i} - \frac{3}{5}\mathbf{j}$ **17.** $\mathbf{v}_1 = \text{proj}_\mathbf{w}\, \mathbf{v} = \frac{14}{5}\mathbf{i} + \frac{7}{5}\mathbf{j}, \mathbf{v}_2 = \frac{1}{5}\mathbf{i} - \frac{2}{5}\mathbf{j}$ **19.** 496.7 mph; 38.5° west of south

21. 8.6° off direct heading across the current, upstream; 1.52 min **23.** 60°; 17.32 min

25. $\frac{6\sqrt{5}}{5}$ ft-lb ≈ 2.68 ft-lb **27.** $1000\sqrt{3}$ ft-lb ≈ 1732 ft-lb

29. Let $\mathbf{u} = a_1\mathbf{i} + b_1\mathbf{j}, \mathbf{v} = a_2\mathbf{i} + b_2\mathbf{j}, \mathbf{w} = a_3\mathbf{i} + b_3\mathbf{j}$. Compute $\mathbf{u} \cdot (\mathbf{v} + \mathbf{w})$ and $\mathbf{u} \cdot \mathbf{v} + \mathbf{u} \cdot \mathbf{w}$.

31. $\cos \alpha = \frac{\mathbf{v} \cdot \mathbf{i}}{\|\mathbf{v}\| \|\mathbf{i}\|} = \mathbf{v} \cdot \mathbf{i}$; if $\mathbf{v} = x\mathbf{i} + y\mathbf{j}$, then $\mathbf{v} \cdot \mathbf{i} = x = \cos \alpha$ and $\mathbf{v} \cdot \mathbf{j} = y = \cos\left(\frac{\pi}{2} - \alpha\right) = \sin \alpha$.

33. $\mathbf{v} = a\mathbf{i} + b\mathbf{j}$; $\text{proj}_\mathbf{i}\, \mathbf{v} = \frac{\mathbf{v} \cdot \mathbf{i}}{\|\mathbf{i}\|^2}\mathbf{i} = (\mathbf{v} \cdot \mathbf{i})\mathbf{i}$; $\mathbf{v} \cdot \mathbf{i} = a, \mathbf{v} \cdot \mathbf{j} = b$, so $\mathbf{v} = (\mathbf{v} \cdot \mathbf{i})\mathbf{i} + (\mathbf{v} \cdot \mathbf{j})\mathbf{j}$.

35. $(\mathbf{v} - \alpha\mathbf{w}) \cdot \mathbf{w} = \mathbf{v} \cdot \mathbf{w} - \alpha\mathbf{w} \cdot \mathbf{w} = \mathbf{v} \cdot \mathbf{w} - \alpha\|\mathbf{w}\|^2 = \mathbf{v} \cdot \mathbf{w} - \frac{\mathbf{v} \cdot \mathbf{w}}{\|\mathbf{w}\|^2}\|\mathbf{w}\|^2 = 0$ **37.** 0

5.7 Exercises

1. A plane consisting of all points of the form $(x, 0, z)$. **3.** A plane consisting of all points of the form $(x, y, 2)$.
5. A plane consisting of all points of the form $(-4, y, z)$. **7.** A line consisting of all points of the form $(1, 2, z)$. **9.** $\sqrt{21}$ **11.** $\sqrt{33}$
13. $\sqrt{26}$ **15.** $(2, 0, 0); (2, 1, 0); (0, 1, 0); (2, 0, 3); (0, 1, 3); (0, 0, 3)$ **17.** $(1, 4, 3); (3, 2, 3); (3, 4, 3); (3, 2, 5); (1, 4, 5); (1, 2, 5)$
19. $(-1, 2, 2); (4, 0, 2); (4, 2, 2); (-1, 2, 5); (4, 0, 5); (-1, 0, 5)$ **21.** $\mathbf{v} = 3\mathbf{i} + 4\mathbf{j} - \mathbf{k}$ **23.** $\mathbf{v} = 2\mathbf{i} + 4\mathbf{j} + \mathbf{k}$ **25.** $\mathbf{v} = 8\mathbf{i} - \mathbf{j}$ **27.** 7

29. $\sqrt{3}$ **31.** $\sqrt{22}$ **33.** $-\mathbf{j} - 2\mathbf{k}$ **35.** $\sqrt{105}$ **37.** $\sqrt{38} - \sqrt{17}$ **39.** $\mathbf{i}$ **41.** $\frac{3}{7}\mathbf{i} - \frac{6}{7}\mathbf{j} - \frac{2}{7}\mathbf{k}$ **43.** $\frac{\sqrt{3}}{3}\mathbf{i} + \frac{\sqrt{3}}{3}\mathbf{j} + \frac{\sqrt{3}}{3}\mathbf{k}$ **45.** $0; 90°$

47. $-2; 100°$ **49.** $0; 90°$ **51.** $52; 0°$ **53.** $\alpha \approx 65°; \beta \approx 149°; \gamma \approx 107°; \mathbf{v} = 7(\cos 65°\mathbf{i} + \cos 149°\mathbf{j} + \cos 107°\mathbf{k})$

55. $\alpha \approx 55°; \beta \approx 55°; \gamma \approx 55°; \mathbf{v} = \sqrt{3}(\cos 55°\mathbf{i} + \cos 55°\mathbf{j} + \cos 55°\mathbf{k})$

57. $\alpha = 45°; \beta = 45°; \gamma = 90°; \mathbf{v} = \sqrt{2}(\cos 45°\mathbf{i} + \cos 45°\mathbf{j} + \cos 90°\mathbf{k})$

59. $\alpha \approx 61°; \beta \approx 144°; \gamma \approx 71°; \mathbf{v} = \sqrt{38}(\cos 61°\mathbf{i} + \cos 144°\mathbf{j} + \cos 71°\mathbf{k})$

61. $d(P_0, P) = \sqrt{(x - x_0)^2 + (y - y_0)^2 + (z - z_0)^2}$ **63.** $(x - 1)^2 + (y - 2)^2 + (z - 2)^2 = 4$ **65.** Center: $(-1, 1, 0)$; Radius: 2
$R = \sqrt{(x - x_0)^2 + (y - y_0)^2 + (z - z_0)^2}$
$R^2 = (x - x_0)^2 + (y - y_0)^2 + (z - z_0)^2$

67. Center: $(2, -2, -1)$; Radius: 3 **69.** Center $(2, 0, -1)$; Radius: $\frac{3\sqrt{2}}{2}$ **71.** 2 newton-meters **73.** 9

Fill-in-the-Blank Items

1. pole; polar axis **3.** $r = 2\cos\theta$ **5.** real; imaginary; imaginary unit **7.** unit

True/False Items

1. F **3.** F **5.** T **7.** T **9.** T

Review Exercises

1. $\left(\dfrac{3\sqrt{3}}{2}, \dfrac{3}{2}\right)$

3. $(1, \sqrt{3})$

5. $(0, 3)$

7. $\left(3\sqrt{2}, \dfrac{3\pi}{4}\right), \left(-3\sqrt{2}, -\dfrac{\pi}{4}\right)$ **9.** $\left(2, -\dfrac{\pi}{2}\right), \left(-2, \dfrac{\pi}{2}\right)$ **11.** $(5, 0.93), (-5, 4.07)$ **13.** $3r^2 - 6r\sin\theta = 0$

15. $r^2(2 - 3\sin^2\theta) - \tan\theta = 0$ **17.** $r^3\cos\theta = 4$ **19.** $x^2 + y^2 - 2y = 0$ **21.** $x^2 + y^2 = 25$ **23.** $x + 3y = 6$

25. Circle: radius 2, center at $(2, 0)$ in rectangular coordinates

27. Cardioid

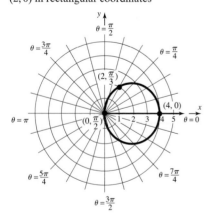

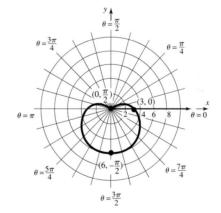

29. Limaçon without inner loop

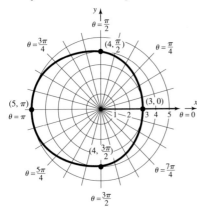

31. $4 + 7i$ **33.** $11 - 10i$ **35.** $\dfrac{9}{10} - \dfrac{1}{10}i$ **37.** -1 **39.** $\sqrt{2}(\cos 225° + i \sin 225°)$ **41.** $5(\cos 323.1° + i \sin 323.1°)$

43. $-\sqrt{3} + i$ **45.** $-\dfrac{3}{2} + \left(\dfrac{3\sqrt{3}}{2}\right)i$ **47.** $0.098 - 0.017i$ **49.** $zw = \cos 130° + i \sin 130°; \dfrac{z}{w} = \cos 30° + i \sin 30°$

51. $zw = 6(\cos 0 + i \sin 0); \dfrac{z}{w} = \dfrac{3}{2}\left(\cos \dfrac{8\pi}{5} + i \sin \dfrac{8\pi}{5}\right)$ **53.** $zw = 5(\cos 5° + i \sin 5°); \dfrac{z}{w} = 5(\cos 15° + i \sin 15°)$

55. $\dfrac{27}{2} + \dfrac{27\sqrt{3}}{2}i$ **57.** $4i$ **59.** 64 **61.** $-527 - 336i$ **63.** $3, 3(\cos 120° + i \sin 120°), 3(\cos 240° + i \sin 240°)$

65. $\mathbf{v} = 2\mathbf{i} - 4\mathbf{j}; \|\mathbf{v}\| = 2\sqrt{5}$ **67.** $\mathbf{v} = -\mathbf{i} + 3\mathbf{j}; \|\mathbf{v}\| = \sqrt{10}$ **69.** $\mathbf{v} = -3\mathbf{i} - 2\mathbf{j} + \mathbf{k}; \|\mathbf{v}\| = \sqrt{14}$ **71.** $\mathbf{v} = 3\mathbf{i} - \mathbf{k}; \|\mathbf{v}\| = \sqrt{10}$

73. $-20\mathbf{i} + 13\mathbf{j}$ **75.** $\sqrt{5}$ **77.** $\sqrt{5} + 5 \approx 7.24$ **79.** $\dfrac{-2\sqrt{5}}{5}\mathbf{i} + \dfrac{\sqrt{5}}{5}\mathbf{j}$ **81.** $21\mathbf{i} - 2\mathbf{j} - 5\mathbf{k}$ **83.** $\sqrt{38}$ **85.** 0

87. $\dfrac{3\sqrt{14}}{14}\mathbf{i} + \dfrac{\sqrt{14}}{14}\mathbf{j} - \dfrac{\sqrt{14}}{7}\mathbf{k}; -\dfrac{3\sqrt{14}}{14}\mathbf{i} - \dfrac{\sqrt{14}}{14}\mathbf{j} + \dfrac{\sqrt{14}}{7}\mathbf{k}$ **89.** $\mathbf{v} \cdot \mathbf{w} = -11; \theta \approx 169.7°$ **91.** $\mathbf{v} \cdot \mathbf{w} = -4; \theta \approx 153.4°$

93. $\mathbf{v} \cdot \mathbf{w} = 1; \theta \approx 70.5°$ **95.** $\mathbf{v} \cdot \mathbf{w} = 0; \theta = 90°$ **97.** $\text{proj}_\mathbf{w}\mathbf{v} = \dfrac{9}{10}(3\mathbf{i} + \mathbf{j})$ **99.** $\alpha \approx 56°; \beta \approx 138°; \gamma \approx 68°$

101. $\sqrt{29} \approx 5.39$ mph; 0.4 mi **103.** At an angle of 70.5° to the shore

C H A P T E R   6   6.2 Exercises

1. B **3.** E **5.** H **7.** C
9. $y^2 = 16x$

11. $x^2 = -12y$

13. $y^2 = -8x$

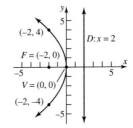

15. $x^2 = 2y$

17. $(x - 2)^2 = -8(y + 3)$

19. $x^2 = \dfrac{4}{3}y$

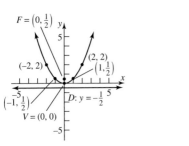

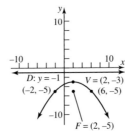

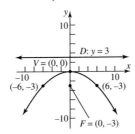

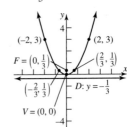

21. $(x + 3)^2 = 4(y - 3)$

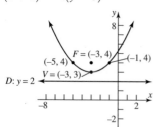

23. $(y + 2)^2 = -8(x + 1)$

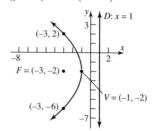

25. Vertex: $(0, 0)$; Focus: $(0, 1)$;
Directrix: $y = -1$

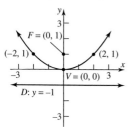

27. Vertex: $(0, 0)$; Focus: $(-4, 0)$;
Directrix: $x = 4$

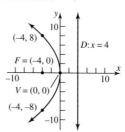

29. Vertex: $(-1, 2)$; Focus: $(1, 2)$;
Directrix: $x = -3$

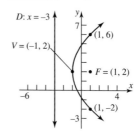

31. Vertex: $(3, -1)$; Focus: $\left(3, -\dfrac{5}{4} \right)$;

Directrix: $y = -\dfrac{3}{4}$

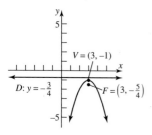

33. Vertex: $(2, -3)$; Focus: $(4, -3)$;

Directrix: $x = 0$

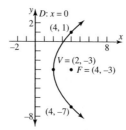

35. Vertex: $(0, 2)$; Focus: $(-1, 2)$;

Directrix: $x = 1$

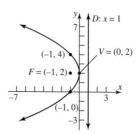

37. Vertex: $(-4, -2)$; Focus: $(-4, -1)$;

Directrix: $y = -3$

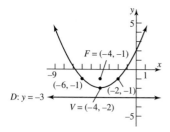

39. Vertex: $(-1, -1)$; Focus: $\left(-\dfrac{3}{4}, -1 \right)$;

Directrix: $x = -\dfrac{5}{4}$

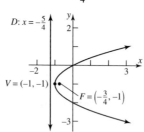

41. Vertex: $(2, -8)$; Focus: $\left(2, -\dfrac{31}{4} \right)$;

Directrix: $y = -\dfrac{33}{4}$

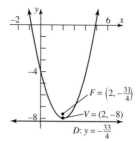

43. $(y - 1)^2 = x$ **45.** $(y - 1)^2 = -(x - 2)$ **47.** $x^2 = 4(y - 1)$ **49.** $y^2 = \dfrac{1}{2}(x + 2)$

51. 1.5625 ft from the base of the dish, along the axis of symmetry   **53.** 1 in. from the vertex   **55.** 20 ft   **57.** 0.78125 ft
59. 4.17 ft from the base along the axis of symmetry   **61.** 24.31 ft, 18.75 ft, 7.64 ft

63. $Ax^2 + Ey = 0$ This is the equation of a parabola with vertex at $(0,0)$ and axis of symmetry the y-axis. The focus is

$x^2 = -\dfrac{E}{A}y$ $\left(0, -\dfrac{E}{4A}\right)$; the directrix is the line $y = \dfrac{E}{4A}$. The parabola opens up if $-\dfrac{E}{A} > 0$ and down if $-\dfrac{E}{A} < 0$.

65. $Ax^2 + Dx + Ey + F = 0, A \neq 0$

$Ax^2 + Dx = -Ey - F$

$x^2 + \dfrac{D}{A}x = -\dfrac{E}{A}y - \dfrac{F}{A}$

$\left(x + \dfrac{D}{2A}\right)^2 = -\dfrac{E}{A}y - \dfrac{F}{A} + \dfrac{D^2}{4A^2}$

$\left(x + \dfrac{D}{2A}\right)^2 = -\dfrac{E}{A}y + \dfrac{D^2 - 4AF}{4A^2}$

(a) If $E \neq 0$, then the equation may be written as

$\left(x + \dfrac{D}{2A}\right)^2 = -\dfrac{E}{A}\left(y - \dfrac{D^2 - 4AF}{4AE}\right)$

This is the equation of a parabola with vertex at

$\left(-\dfrac{D}{2A}, \dfrac{D^2 - 4AF}{4AE}\right)$ and axis of symmetry parallel to the y-axis.

(b)–(d) If $E = 0$, the graph of the equation contains no points if $D^2 - 4AF < 0$, is a single vertical line if $D^2 - 4AF = 0$, and is two vertical lines if $D^2 - 4AF > 0$.

6.3 Exercises

1. C   **3.** B

5. Vertices: $(-5, 0), (5, 0)$

Foci: $(-\sqrt{21}, 0), (\sqrt{21}, 0)$

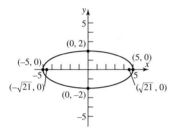

7. Vertices: $(0, -5), (0, 5)$

Foci: $(0, -4), (0, 4)$

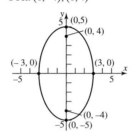

9. $\dfrac{x^2}{4} + \dfrac{y^2}{16} = 1$

Vertices: $(0, -4), (0, 4)$

Foci: $(0, -2\sqrt{3}), (0, 2\sqrt{3})$

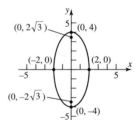

11. $\dfrac{x^2}{8} + \dfrac{y^2}{2} = 1$

Vertices: $(-2\sqrt{2}, 0), (2\sqrt{2}, 0)$

Foci: $(-\sqrt{6}, 0), (\sqrt{6}, 0)$

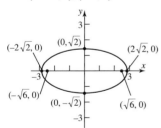

13. Vertices: $(-4, 0), (4, 0), (0, -4), (0, 4)$

Focus: $(0, 0)$

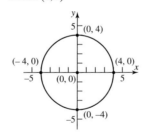

15. $\dfrac{x^2}{25} + \dfrac{y^2}{16} = 1$

17. $\dfrac{x^2}{9} + \dfrac{y^2}{25} = 1$

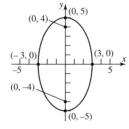

19. $\dfrac{x^2}{9} + \dfrac{y^2}{5} = 1$

21. $\dfrac{x^2}{4} + \dfrac{y^2}{13} = 1$

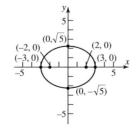

23. $x^2 + \dfrac{y^2}{16} = 1$

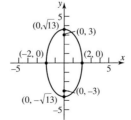

25. $\dfrac{(x+1)^2}{4} + (y-1)^2 = 1$ **27.** $(x-1)^2 + \dfrac{y^2}{4} = 1$ **29.** Center: $(3,-1)$; Vertices: $(3,-4)$, $(3,2)$

Foci: $(3, -1 - \sqrt{5})$, $(3, -1 + \sqrt{5})$

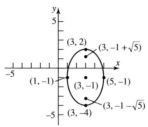

31. $\dfrac{(x+5)^2}{16} + \dfrac{(y-4)^2}{4} = 1$

Center: $(-5,4)$; Vertices: $(-9,4)$, $(-1,4)$
Foci: $(-5 - 2\sqrt{3}, 4)$, $(-5 + 2\sqrt{3}, 4)$

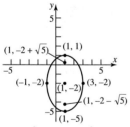

33. $\dfrac{(x+2)^2}{4} + (y-1)^2 = 1$

Center: $(-2,1)$; Vertices: $(-4,1)$, $(0,1)$
Foci: $(-2 - \sqrt{3}, 1)$, $(-2 + \sqrt{3}, 1)$

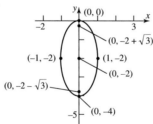

35. $\dfrac{(x-2)^2}{3} + \dfrac{(y+1)^2}{2} = 1$,

Center: $(2,-1)$; Vertices: $(2 - \sqrt{3}, -1)$,
$(2 + \sqrt{3}, -1)$; Foci: $(1,-1)$, $(3,-1)$

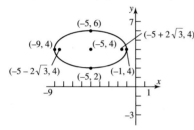

37. $\dfrac{(x-1)^2}{4} + \dfrac{(y+2)^2}{9} = 1$

Center: $(1,-2)$; Vertices: $(1,-5)$, $(1,1)$
Foci: $(1, -2 - \sqrt{5})$, $(1, -2 + \sqrt{5})$

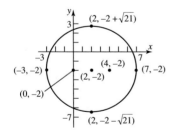

39. $x^2 + \dfrac{(y+2)^2}{4} = 1$

Center: $(0,-2)$; Vertices: $(0,-4)$, $(0,0)$
Foci: $(0, -2 - \sqrt{3})$, $(0, -2 + \sqrt{3})$

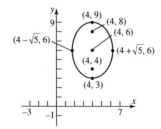

41. $\dfrac{(x-2)^2}{25} + \dfrac{(y+2)^2}{21} = 1$

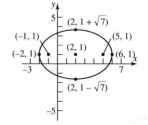

43. $\dfrac{(x-4)^2}{5} + \dfrac{(y-6)^2}{9} = 1$

45. $\dfrac{(x-2)^2}{16} + \dfrac{(y-1)^2}{7} = 1$

47. $\dfrac{(x-1)^2}{10} + (y-2)^2 = 1$

49. $\dfrac{(x-1)^2}{9} + (y-2)^2 = 1$

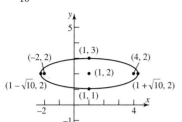

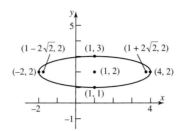

51.

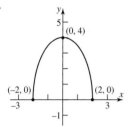

53.

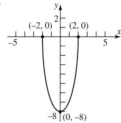

55. $\dfrac{x^2}{100} + \dfrac{y^2}{36} = 1$ **57.** 43.3 ft

59. 24.65 ft, 21.65 ft, 13.82 ft

61. 0 ft, 12.99 ft, 15 ft, 12.99 ft, 0 ft   **63.** 91.5 million miles; $\dfrac{x^2}{(93)^2} + \dfrac{y^2}{8646.75} = 1$

65. perihelion: 460.6 million miles; mean distance: 483.8 million miles; $\dfrac{x^2}{(483.8)^2} + \dfrac{y^2}{233{,}524.2} = 1$ **67.** 30 ft

69. (a) $Ax^2 + Cy^2 + F = 0$ If A and C are of the same and F is of opposite sign, then the equation takes the form

$Ax^2 + Cy^2 = -F$ $\dfrac{x^2}{\left(-\dfrac{F}{A}\right)} + \dfrac{y^2}{\left(-\dfrac{F}{C}\right)} = 1$, where $-\dfrac{F}{A}$ and $-\dfrac{F}{C}$ are positive. This is the equation of an ellipse

with center at $(0,0)$.

(b) If $A = C$, the equation may be written as $x^2 + y^2 = -\dfrac{F}{A}$. This is the equation of a circle with center at $(0,0)$ and radius equal

to $\sqrt{-\dfrac{F}{A}}$.

6.4 Exercises

1. B   **3.** A

5. $x^2 - \dfrac{y^2}{8} = 1$

7. $\dfrac{y^2}{16} - \dfrac{x^2}{20} = 1$

9. $\dfrac{x^2}{9} - \dfrac{y^2}{16} = 1$

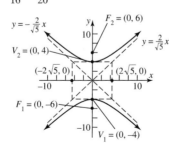

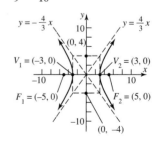

11. $\dfrac{y^2}{36} - \dfrac{x^2}{9} = 1$

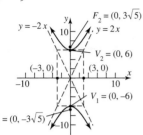

13. $\dfrac{x^2}{8} - \dfrac{y^2}{8} = 1$

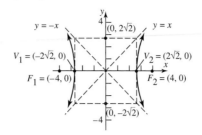

15. $\dfrac{x^2}{25} - \dfrac{y^2}{9} = 1$

Center: $(0, 0)$
Transverse axis: x-axis
Vertices: $(-5, 0), (5, 0)$
Foci: $(-\sqrt{34}, 0), (\sqrt{34}, 0)$

Asymptotes: $y = \pm\dfrac{3}{5}x$

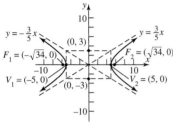

17. $\dfrac{x^2}{4} - \dfrac{y^2}{16} = 1$

Center: $(0, 0)$
Transverse axis: x-axis
Vertices: $(-2, 0), (2, 0)$
Foci: $(-2\sqrt{5}, 0), (2\sqrt{5}, 0)$

Asymptotes: $y = \pm 2x$

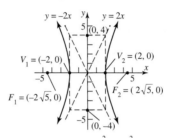

19. $\dfrac{y^2}{9} - x^2 = 1$

Center: $(0, 0)$
Transverse axis: y-axis
Vertices: $(0, -3), (0, 3)$
Foci: $(0, -\sqrt{10}), (0, \sqrt{10})$

Asymptotes: $y = \pm 3x$

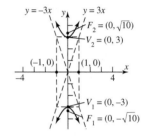

21. $\dfrac{y^2}{25} - \dfrac{x^2}{25} = 1$

Center: $(0, 0)$
Transverse axis: y-axis
Vertices: $(0, -5), (0, 5)$
Foci: $(0, -5\sqrt{2}), (0, 5\sqrt{2})$
Asymptotes: $y = \pm x$

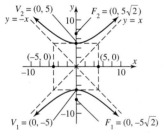

23. $x^2 - y^2 = 1$ **25.** $\dfrac{y^2}{36} - \dfrac{x^2}{9} = 1$

27. $\dfrac{(x-4)^2}{4} - \dfrac{(y+1)^2}{5} = 1$

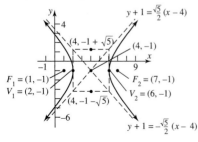

29. $\dfrac{(y+4)^2}{4} - \dfrac{(x+3)^2}{12} = 1$

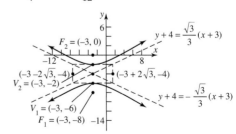

31. $(x - 5)^2 - \dfrac{(y - 7)^2}{3} = 1$

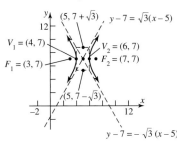

33. $\dfrac{(x - 1)^2}{4} - \dfrac{(y + 1)^2}{9} = 1$

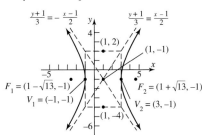

35. $\dfrac{(x - 2)^2}{4} - \dfrac{(y + 3)^2}{9} = 1$

Center: $(2, -3)$
Transverse axis: Parallel to x-axis
Vertices: $(0, -3), (4, -3)$
Foci: $(2 - \sqrt{13}, -3), (2 + \sqrt{13}, -3)$
Asymptotes: $y + 3 = \pm\dfrac{3}{2}(x - 2)$

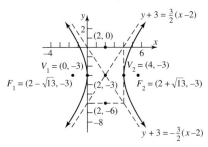

37. $\dfrac{(y - 2)^2}{4} - (x + 2)^2 = 1$

Center: $(-2, 2)$
Transverse axis: Parallel to y-axisis
Vertices: $(-2, 0), (-2, 4)$
Foci: $(-2, 2 - \sqrt{5}), (-2, 2 + \sqrt{5})$
Asymptotes: $y - 2 = \pm2(x + 2)$

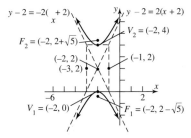

39. $\dfrac{(x + 1)^2}{4} - \dfrac{(y + 2)^2}{4} = 1$

Center: $(-1, -2)$
Transverse axis: Parallel to x-axis
Vertices: $(-3, -2), (1, -2))$
Foci: $(-1 - 2\sqrt{2}, -2), (-1 + 2\sqrt{2}, -2)$
Asymptotes: $y + 2 = \pm(x + 1)$

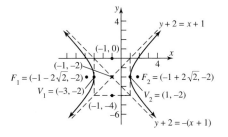

41. $(x - 1)^2 - (y + 1)^2 = 1$

Center: $(1, -1)$
Transverse axis: Parallel to x-axis
Vertices: $(0, -1), (2, -1)$
Foci: $(1 - \sqrt{2}, -1), (1 + \sqrt{2}, -1)$
Asymptotes: $y + 1 = \pm(x - 1)$

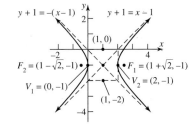

43. $\dfrac{(y-2)^2}{4} - (x+1)^2 = 1$

Center: $(-1, 2)$
Transverse axis: Parallel to y-axis
Vertices: $(-1, 0), (-1, 4)$
Foci: $(-1, 2 - \sqrt{5}), (-1, 2 + \sqrt{5})$
Asymptotes: $y - 2 = \pm 2(x + 1)$

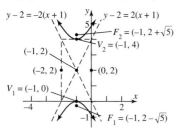

45. $\dfrac{(x-3)^2}{4} - \dfrac{(y+2)^2}{16} = 1$

Center: $(3, -2)$
Transverse axis: Parallel to x-axis
Vertices: $(1, -2), (5, -2)$
Foci: $(3 - 2\sqrt{5}, -2), (3 + 2\sqrt{5}, -2)$
Asymptotes: $y + 2 = \pm 2(x - 3)$

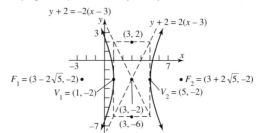

47. $\dfrac{(y-1)^2}{4} - (x+2)^2 = 1$

Center: $(-2, 1)$
Transverse axis: Parallel to y-axis
Vertices: $(-2, -1), (-2, 3)$
Foci: $(-2, 1 - \sqrt{5}), (-2, 1 + \sqrt{5})$

Asymptotes: $y - 1 = \pm 2(x + 2)$

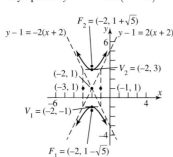

49.

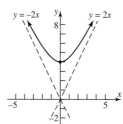

51.

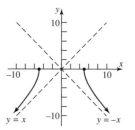

53. (a) The ship will reach shore at a point 64.66 miles from the master station.   **(b)** 0.00086 sec   **(c)** $(104, 50)$
55. (a) 450 ft   **57.** If e is close to 1, narrow hyperbola; if e is very large, wide hyperbola
59. $\dfrac{x^2}{4} - y^2 = 1$; asymptotes $y = \pm\dfrac{1}{2}x$, $y^2 - \dfrac{x^2}{4} = 1$; asymptotes $y = \pm\dfrac{1}{2}x$

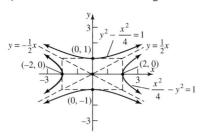

61. $Ax^2 + Cy^2 + F = 0$ If A and C are of opposite sign and $F \neq 0$, this equation may be written as $\dfrac{x^2}{\left(-\dfrac{F}{A}\right)} + \dfrac{y^2}{\left(-\dfrac{F}{C}\right)} = 1$,

$Ax^2 + Cy^2 = -F$ where $-\dfrac{F}{A}$ and $-\dfrac{F}{C}$ are opposite in sign. This is the equation of a hyperbola with center $(0, 0)$.

The transverse axis is the x-axis if $-\dfrac{F}{A} > 0$; the transverse axis is the y-axis if $-\dfrac{F}{A} < 0$.

6.5 Exercises

1. Parabola   **3.** Ellipse   **5.** Hyperbola   **7.** Hyperbola   **9.** Circle   **11.** $x = \dfrac{\sqrt{2}}{2}(x' - y'),\ y = \dfrac{\sqrt{2}}{2}(x' + y')$

13. $x = \dfrac{\sqrt{2}}{2}(x' - y'),\ y = \dfrac{\sqrt{2}}{2}(x' + y')$ **15.** $x = \dfrac{1}{2}(x' - \sqrt{3}y'),\ y = \dfrac{1}{2}(\sqrt{3}x' + y')$ **17.** $x = \dfrac{\sqrt{5}}{5}(x' - 2y'),\ y = \dfrac{\sqrt{5}}{5}(2x' + y')$

19. $x = \dfrac{\sqrt{13}}{13}(3x' - 2y'),\ y = \dfrac{\sqrt{13}}{13}(2x' + 3y')$

21. $\theta = 45°$ (see Problem 11)

$$x'^2 - \frac{y'^2}{3} = 1$$

Hyperbola
Center at origin
Transverse axis is the x'-axis
Vertices at $(\pm 1, 0)$

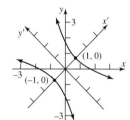

23. $\theta = 45°$ (see Problem 13)

$$x'^2 + \frac{y'^2}{4} = 1$$

Ellipse
Center at $(0, 0)$
Major axis is the y'-axis.
Vertices at $(0, \pm 2)$

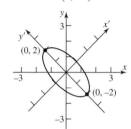

25. $\theta = 60°$ (see Problem 15)

$$\frac{x'^2}{4} + y'^2 = 1$$

Ellipse
Center at $(0, 0)$
Major axis is the x'-axis.
Vertices at $(\pm 2, 0)$

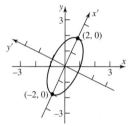

27. $\theta \approx 63°$ (see Problem 17)

$$y'^2 = 8x'$$

Parabola

Vertex at $(0, 0)$

Focus at $(2, 0)$

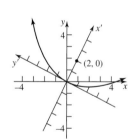

29. $\theta \approx 34°$ (see Problem 19)

$$\frac{(x' - 2)^2}{4} + y'^2 = 1$$

Ellipse

Center at $(2, 0)$

Major axis is the x'-axis.

Vertices at $(4, 0)$ and $(0, 0)$

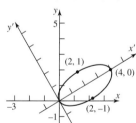
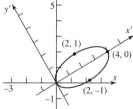

31. $\cot 2\theta = \dfrac{7}{24}; \theta \approx 37°$

$$(x' - 1)^2 = -6\left(y' - \frac{1}{6}\right)$$

Parabola

Vertex at $\left(1, \dfrac{1}{6}\right)$

Focus at $\left(1, -\dfrac{4}{3}\right)$

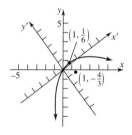

33. Hyperbola   **35.** Hyperbola   **37.** Parabola   **39.** Ellipse   **41.** Ellipse
43. Refer to equation (6):
$$A' = A\cos^2\theta + B\sin\theta\cos\theta + C\sin^2\theta$$
$$B' = B(\cos^2\theta - \sin^2\theta) + 2(C - A)(\sin\theta\cos\theta)$$
$$C' = A\sin^2\theta - B\sin\theta\cos\theta + C\cos^2\theta$$
$$D' = D\cos\theta + E\sin\theta$$
$$E' = -D\sin\theta + E\cos\theta$$
$$F' = F$$
45. Use Problem 43 to find $B'^2 - 4A'C'$. After much cancellation, $B'^2 - 4A'C' = B^2 - 4AC$.
47. Use formulas (5) and find $d^2 = (x_2 - x_1)^2 + (y_2 - y_1)^2$. After simplifying, $(x_2 - x_1)^2 + (y_2 - y_1)^2 = (x'_2 - x'_1)^2 + (y'_2 - y'_1)^2$.

6.6 Exercises

1. Parabola; directrix is perpendicular to the polar axis 1 unit to the right of the pole.

3. Hyperbola; directrix is parallel to the polar axis $\dfrac{4}{3}$ units below the pole.

5. Ellipse; directrix is perpendicular to the polar axis $\frac{3}{2}$ units to the left of the pole.

7. Parabola; directrix is perpendicular to the polar axis 1 unit to the right of the pole.

9. Ellipse; directrix is parallel to the polar axis $\frac{8}{3}$ units above the pole.

11. Hyperbola; directrix is perpendicular to the polar axis $\frac{3}{2}$ units to the left of the pole.

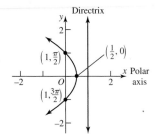

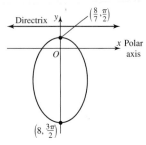

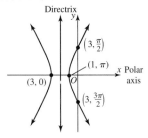

13. Ellipse; directrix is parallel to the polar axis 8 units below the pole; vertices are at $\left(8, \frac{\pi}{2}\right)$ and $\left(\frac{8}{3}, \frac{3\pi}{2}\right)$.

15. Ellipse; directrix is parallel to the polar axis 3 units below the pole; vertices are at $\left(6, \frac{\pi}{2}\right)$ and $\left(\frac{6}{5}, \frac{3\pi}{2}\right)$.

17. Ellipse; directrix is perpendicular too the polar axis 6 units to the left of the pole; vertices are at $(6, 0)$ and $(2, \pi)$.

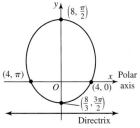

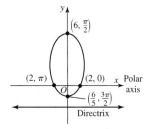

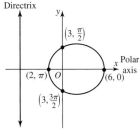

19. $y^2 + 2x - 1 = 0$ **21.** $16x^2 + 7y^2 + 48y - 64 = 0$ **23.** $3x^2 - y^2 + 12x + 9 = 0$ **25.** $4x^2 + 3y^2 - 16y - 64 = 0$

27. $9x^2 + 5y^2 - 24y - 36 = 0$ **29.** $3x^2 + 4y^2 - 12x - 36 = 0$ **31.** $r = \dfrac{1}{1 + \sin\theta}$ **33.** $r = \dfrac{12}{5 - 4\cos\theta}$ **35.** $r = \dfrac{12}{1 - 6\sin\theta}$

37. Use $d(D, P) = p - r\cos\theta$ in the derivation of equation (a) in Table 5.
39. Use $d(D, P) = p + r\sin\theta$ in the derivation of equation (a) in Table 5.

6.7 Exercises

1. $x - 3y + 1 = 0$

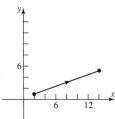

3. $y = \sqrt{x - 2}$

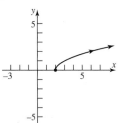

5. $x = y + 8$

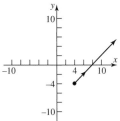

7. $x = 3(y - 1)^2$

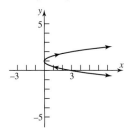

9. $2y = 2 + x$

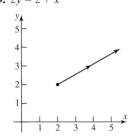

11. $y = x^3$

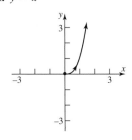

13. $\dfrac{x^2}{4} + \dfrac{y^2}{9} = 1$

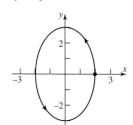

15. $\dfrac{x^2}{4} + \dfrac{y^2}{9} = 1$

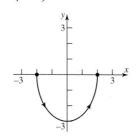

17. $x^2 - y^2 = 1$

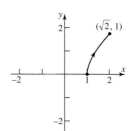
$(\sqrt{2}, 1)$

19. $x + y = 1$

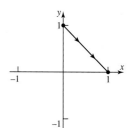

21. **(a)** $x = 3$
$y = -16t^2 + 50t + 6$

(b) 3.24 seconds

(c) 1.5625 seconds; 45.0625 feet

(d)
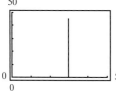

23. **(a)** Train: $x_1 = t^2$ Bill: $x_2 = 5(t - 5)$
$y_1 = 1$ $y_2 = 3$

(b) Bill won't catch the train.

(c)
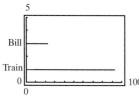

25. **(a)** $x = (145 \cos 20°)t$
$y = -16t^2 + (145 \sin 20°)t + 5$

(b) 3.197 seconds

(c) 1.55 seconds; 43.43 feet

(d) 435.61 feet

(e)
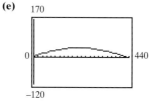

27. **(a)** $x = (40 \cos 45°)t$
$y = -4.9t^2 + (40 \sin 45°)t + 300$

(b) 11.226 seconds

(c) 2.886 seconds; 340.8 meters

(d) 317.5 meters

(e)
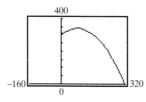

29. (a) $d = \sqrt{(40t - 5)^2 + (30t - 4)^2}$ **(b)**

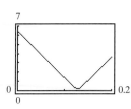

(c) 0.2 miles; 0.128 seconds

(d) Paseo: $x = 40t - 5$ Bonneville: $x = 0$
$\quad\quad\quad\quad\; y = 0$ $\quad\quad\quad\quad\quad\; y = 30t - 4$

(e) Turn axes off to see the graph:

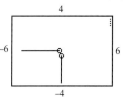

31. $x = t, y = t^3; x = \sqrt[3]{t}, y = t$ **33.** $x = t, y = t^{2/3}; x = t^{3/2}, y = t$ **35.** $x = 2 \cos \pi t, y = -3 \sin \pi t, 0 \le t \le 2$
37. $x = -2 \sin 2\pi t, y = 3 \cos 2\pi t, 0 \le t \le 1$
39.

41. The orientation is from (x_1, y_1) to (x_2, y_2).
43.   **45.**   **47. (a)**   **(b)** $x^{2/3} + y^{2/3} = 1$

Fill-in-the-Blank Items

1. parabola   **3.** hyperbola   **5.** y-axis   **7.** $\cot 2\theta = \dfrac{A - C}{B}$ **9.** ellipse

True/False Items

1. T   **3.** T   **5.** T   **7.** T   **9.** T

Review Exercises

1. Parabola; vertex $(0, 0)$, focus $(-4, 0)$, directrix $x = 4$

3. Hyperbola; center $(0, 0)$, vertices $(5, 0)$ and $(-5, 0)$, foci $(\sqrt{26}, 0)$ and $(-\sqrt{26}, 0)$, asymptotes $y = \dfrac{1}{5}x$ and $y = -\dfrac{1}{5}x$

5. Ellipse; center $(0, 0)$, vertices $(0, 5)$ and $(0, -5)$, foci $(0, 3)$ and $(0, -3)$

7. $x^2 = -4(y - 1)$: Parabola; vertex $(0, 1)$, focus $(0, 0)$, directrix $y = 2$

9. $\dfrac{x^2}{2} - \dfrac{y^2}{8} = 1$: Hyperbola; center $(0, 0)$, vertices $(\sqrt{2}, 0)$ and $(-\sqrt{2}, 0)$, foci $(\sqrt{10}, 0)$ and $(-\sqrt{10}, 0)$, asymptotes $y = 2x$ and $y = -2x$

11. $(x - 2)^2 = 2(y + 2)$: Parabola; vertex $(2, -2)$, focus $\left(2, -\dfrac{3}{2}\right)$, directrix $y = -\dfrac{5}{2}$

13. $\dfrac{(y-2)^2}{4} - (x-1)^2 = 1$: Hyperbola; center $(1, 2)$, vertices $(1, 4)$ and $(1, 0)$, foci $(1, 2 + \sqrt{5})$ and $(1, 2 - \sqrt{5})$, asymptotes $y - 2 = \pm 2(x - 1)$

15. $\dfrac{(x-2)^2}{9} + \dfrac{(y-1)^2}{4} = 1$: Ellipse; center $(2, 1)$, vertices $(5, 1)$ and $(-1, 1)$, foci $(2 + \sqrt{5}, 1)$ and $(2 - \sqrt{5}, 1)$

17. $(x - 2)^2 = -4(y + 1)$: Parabola; vertex $(2, -1)$, focus $(2, -2)$, directrix $y = 0$

19. $\dfrac{(x-1)^2}{4} + \dfrac{(y+1)^2}{9} = 1$: Ellipse; center $(1, -1)$, vertices $(1, 2)$ and $(1, -4)$, foci $(1, -1 + \sqrt{5})$ and $(1, -1 - \sqrt{5})$

21. $y^2 = -8x$

23. $\dfrac{y^2}{4} - \dfrac{x^2}{12} = 1$

25. $\dfrac{x^2}{16} + \dfrac{y^2}{7} = 1$

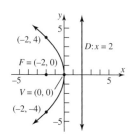

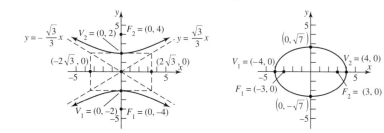

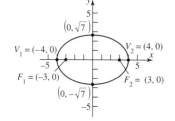

27. $(x - 2)^2 = -4(y + 3)$

29. $(x + 2)^2 - \dfrac{(y + 3)^2}{3} = 1$

31. $\dfrac{(x + 4)^2}{16} + \dfrac{(y - 5)^2}{25} = 1$

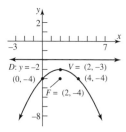

33. $\dfrac{(x + 1)^2}{9} - \dfrac{(y - 2)^2}{7} = 1$

35. $\dfrac{(x - 3)^2}{9} - \dfrac{(y - 1)^2}{4} = 1$

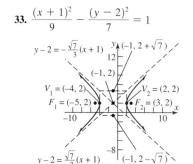

37. Parabola **39.** Ellipse **41.** Parabola **43.** Hyperbola **45.** Ellipse

47. $x'^2 - \dfrac{y'^2}{9} = 1$

Hyperbola
Center at the origin

Transverse axis the x'-axis

Vertices at $(\pm 1, 0)$

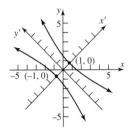

49. $\dfrac{x'^2}{2} + \dfrac{y'^2}{4} = 1$

Ellipse
Center at origin

Major axis the y'-axis

Vertices at $(0, \pm 2)$

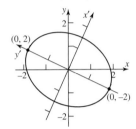

51. $y'^2 = -\dfrac{4\sqrt{13}}{13}x'$

Parabola
Vertex at the origin

Focus on the x'-axis at $\left(-\dfrac{\sqrt{13}}{13}, 0\right)$

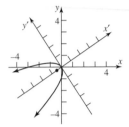

53. Parabola; directrix is perpendicular to the polar axis 4 units to the left of the pole.

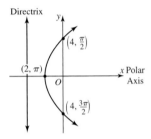

55. Ellipse; directrix is parallel to the polar axis 6 units below the pole; vertices are $\left(6, \dfrac{\pi}{2}\right)$ and $\left(2, \dfrac{3\pi}{2}\right)$.

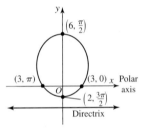

57. Hyperbola; directrix is perpendicular to the polar axis 1 unit to the right of the pole; vertices are at $\left(\dfrac{2}{3}, 0\right)$ and $(-2, \pi)$.

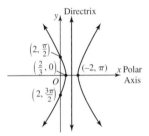

59. $y^2 - 8x - 16 = 0$ **61.** $3x^2 - y^2 - 8x + 4 = 0$ 0

63. $x + 4y = 2$

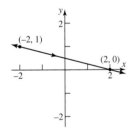

65. $\dfrac{x^2}{9} + \dfrac{(y-2)^2}{16} = 1$

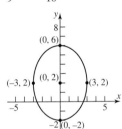

67. $1 + y = x$

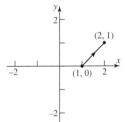

69. $\dfrac{x^2}{5} - \dfrac{y^2}{4} = 1$ **71.** The ellipse $\dfrac{x^2}{16} + \dfrac{y^2}{7} = 1$ **73.** $\dfrac{1}{4}$ ft or 3 in.   **75.** 19.72 ft, 18.86 ft, 14.91 ft

77. (a) 45.24 miles from the Master Station   **(b)** 0.000645 second   **(c)** (66, 20)

79. (a) $x = (100 \cos 35°)t$
$\quad\;\; y = -16t^2 + (100 \sin 35°)t + 6$

(d) 302 feet

(b) 3.6866 seconds

(e)

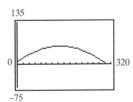

(c) 1.7924 seconds; 57.4 feet

C H A P T E R 7 7.1 Exercises

1. (a) 11.212 **(b)** 11.587 **(c)** 11.664 **(d)** 11.665 **3. (a)** 8.815 **(b)** 8.821 **(c)** 8.824 **(d)** 8.825
5. (a) 21.217 **(b)** 22.217 **(c)** 22.440 **(d)** 22.459 **7.** 3.320 **9.** 0.427 **11.** B **13.** D **15.** A **17.** E
19.

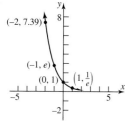

Domain: $(-\infty, \infty)$
Range: $(0, \infty)$
Horizontal asymptote: $y = 0$

21.

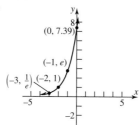

Domain: $(-\infty, \infty)$
Range: $(0, \infty)$
Horizontal asymptote: $y = 0$

23.

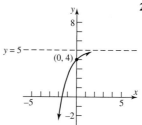

Domain: $(-\infty, \infty)$
Range: $(-\infty, 5)$
Horizontal asymptote: $y = 5$

25.

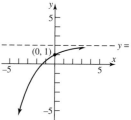

Domain: $(-\infty, \infty)$
Range: $(-\infty, 2)$
Horizontal asymptote: $y = 2$

27.

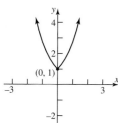

Domain: $(-\infty, \infty)$
Range: $[1, \infty)$
Intercept: $(0, 1)$

29.

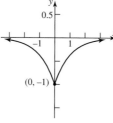

Domain: $(-\infty, \infty)$
Range: $[-1, 0)$
Intercept: $(0, -1)$

31. $\dfrac{1}{49}$ **33.** $\dfrac{1}{4}$ **35. (a)** 74% **(b)** 47% **37. (a)** 44 watts **(b)** 11.6 watts **39.** 3.35 milligrams; 0.45 milligrams

41. (a) 0.63 **(b)** 0.98

(c)

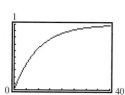

43. (a) 60.5% **(b)** 68.7% **(c)** 70% **(d)** About $4\dfrac{1}{4}$ days

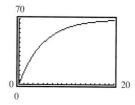

(d) About 7 minutes **(e)** 1
45. (a) 5.414 amperes, 7.585 amperes, 10.376 amperes **(b)** 12 amperes
(d) 3.343 amperes, 5.309 amperes, 9.443 amperes **(e)** 24 amperes
(c), (f)

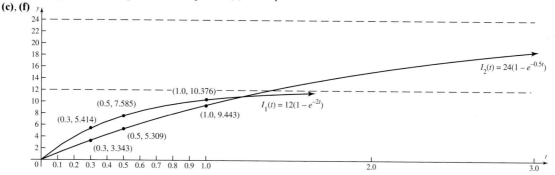

47. (a) 9.23×10^{-3}, or about 0
(b) 0.81, or about 1 **(c)** 5.01, or about 5
(d) 57.91°, 43.99°, 30.07°

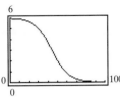

49. $n = 4$: 2.7083; $n = 6$: 2.7181; $n = 8$: 2.7182788; $n = 10$: 2.7182818

51. $\dfrac{f(x + h) - f(x)}{h} = \dfrac{a^{x+h} - a^x}{h} = \dfrac{a^x a^h - a^x}{h} = \dfrac{a^x(a^h - 1)}{h}$ **53.** $f(-x) = a^{-x} = \dfrac{1}{a^x} = \dfrac{1}{f(x)}$

55. a. $f(-x) = \dfrac{1}{2}(e^{-x} - e^{-(-x)}) = \dfrac{1}{2}(e^{-x} - e^x) = -\dfrac{1}{2}(e^x - e^{-x}) = -f(x)$

b.

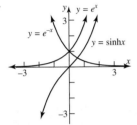

57. $f(1) = 5, f(2) = 17, f(3) = 257, f(4) = 65,537, f(5) = 4,294,967,297 = 641 \times 6,700,417$

7.2 Exercises

1. $2 = \log_3 9$ **3.** $2 = \log_a 1.6$ **5.** $2 = \log_{1.1} M$ **7.** $x = \log_2 7.2$ **9.** $\sqrt{2} = \log_x \pi$ **11.** $x = \ln 8$ **13.** $2^3 = 8$ **15.** $a^6 = 3$ **17.** $3^x = 2$
19. $2^{1.3} = M$ **21.** $(\sqrt{2})^x = \pi$ **23.** $e^x = 4$ **25.** 0 **27.** 2 **29.** -4 **31.** $\dfrac{1}{2}$ **33.** 4 **35.** $\dfrac{1}{2}$ **37.** $\{x | x < 3\}$; $(0, \ln 3), (2, 0)$
39. All real numbers except 0; $(-1, 0), (1, 0)$ **41.** All real numbers except 1; $(0, 0), (2, 0)$ **43.** $\{x | x < -1 \text{ or } x > 0\}$; No intercepts
45. 0.511 **47.** 30.099 **49.** $\sqrt{2}$ **51.** B **53.** D **55.** A **57.** E

59.

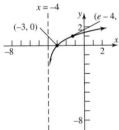

Domain: $(-4, \infty)$
Range: $(-\infty, \infty)$
Vertical asymptote: $x = -4$

61.

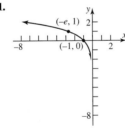

Domain: $(-\infty, 0)$
Range: $(-\infty, \infty)$
Vertical asymptote: $x = 0$

63.

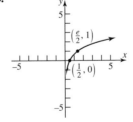

Domain: $(0, \infty)$
Range: $(-\infty, \infty)$
Vertical asymptote: $x = 0$

65.

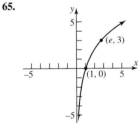

Domain: $(0, \infty)$
Range: $(-\infty, \infty)$
Vertical asymptote: $x = 0$

67.

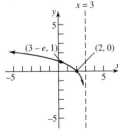

Domain: $(-\infty, 3)$
Range: $(-\infty, \infty)$
Vertical asymptote: $x = 3$

69.

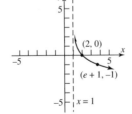

Domain: $(1, \infty)$
Range: $(-\infty, \infty)$
Vertical asymptote: $x = 1$

71.

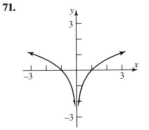

Domain: $\{x | x \neq 0\}$
Range: $(-\infty, \infty)$
Intercepts: $(-1, 0), (1, 0)$

73.

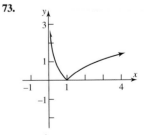

Domain: $\{x | x > 0\}$
Range: $\{y | y \geq 0\}$
Intercept: $(1, 0)$

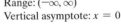

75. (a) $n \approx 6.93$ so 7 panes are necessary   **(b)** $n \approx 13.86$ so 14 panes are necessary
77. (a) $d \approx 127.7$ so it takes about 128 days   **(b)** $d \approx 575.6$ so it takes about 576 days
79. $h \approx 2.29$ so the time between injections is about 2 hours, 17 minutes
81. 0.2695 sec; 0.8959 sec

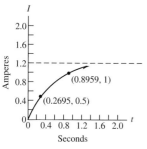

83. (a) $k = 20.07$
(b) 91%
(c) 0.175
(d) 0.08

85. $y = 20e^{0.023t}$; $y = 89.2$ is predicted

7.3 Exercises

1. $a + b$ **3.** $b - a$ **5.** $a + 1$ **7.** $2a + b$ **9.** $\frac{1}{5}(a + 2b)$ **11.** $\frac{b}{a}$ **13.** $2 \ln x + \frac{1}{2} \ln(1 - x)$ **15.** $3 \log_2 x - \log_2(x - 3)$

17. $\log x + \log(x + 2) - 2 \log(x + 3)$ **19.** $\frac{1}{3} \ln(x - 2) + \frac{1}{3} \ln(x + 1) - \frac{2}{3} \ln(x + 4)$

21. $\ln 5 + \ln x + \frac{1}{2} \ln(1 - 3x) - 3 \ln(x - 4)$ **23.** $\log_5 u^3 v^4$ **25.** $-\frac{5}{2} \log_{1/2} x$ **27.** $-2 \ln(x - 1)$ **29.** $\log_2[x(3x - 2)^4]$

31. $\log_a\left(\frac{25x^6}{\sqrt{2x + 3}}\right)$ **33.** $\frac{5}{4}$ **35.** 4   **37.** 2.771   **39.** -3.880 **41.** 5.615   **43.** 0.874

45. $\log_a(x + \sqrt{x^2 - 1}) + \log_a(x - \sqrt{x^2 - 1}) = \log_a[(x + \sqrt{x^2 - 1})(x - \sqrt{x^2 - 1})] = \log_a[x^2 - (x^2 - 1)] = \log_a 1 = 0$

47. $\ln(1 + e^{2x}) = \ln(e^{2x} \cdot e^{-2x} + e^{2x}) = \ln[e^{2x}(e^{-2x} + 1)] = \ln e^{2x} + \ln(e^{-2x} + 1) = 2x + \ln(1 + e^{-2x})$

49. If $f(x) = \log_a x$, then $x = a^{f(x)} = \frac{1}{a^{-f(x)}} = \left(\frac{1}{a}\right)^{-f(x)}$ so $-f(x) = \log_{1/a} x$

51. $f(AB) = \log_a(AB) = \log_a A + \log_a B = f(A) + f(B)$

53.

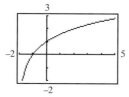

55.

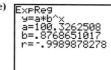

57. $y = Cx$ **59.** $y = Cx(x + 1)$ **61.** $y = Ce^{3x}$ **63.** $y = Ce^{-4x} + 3$ **65.** $y = \frac{\sqrt[3]{C}(2x + 1)^{1/6}}{(x + 4)^{1/9}}$ **67.** 3   **69.** 1

71. (a)

(graph: Weight (grams) vs Week)

(b) $A = 100e^{-0.128t}$
(c) 5.4 weeks
(d) 0.17 grams
(e)
```
ExpReg
y=a*b^x
a=100.3262508
b=.8768651017
r=-.9989878278
```
(f)

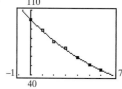

73. (a)

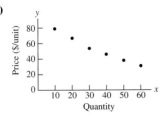

(b) $y = 96e^{-0.02x}$

(c) 23.3 units

(d)

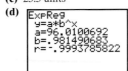

(e)

75. (a)

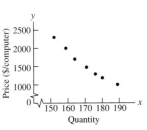

(b) 168 computers

(c)

```
LnReg
y=a+blnx
a=32741.02369
b=-6070.956754
r=-.9917146001
```

(d)

77. $\log_a \left(\dfrac{1}{N} \right) = \log_a 1 - \log_a N = -\log_a N$

7.4 Exercises

1. $\dfrac{7}{2}$ **3.** $\{-2\sqrt{2}, 2\sqrt{2}\}$ **5.** 16   **7.** 8   **9.** 3   **11.** 5   **13.** 2   **15.** $\{-2, 4\}$ **17.** 21   **19.** $\dfrac{1}{2}$ **21.** $\{-\sqrt{2}, 0, \sqrt{2}\}$ **23.** $\left\{1 - \dfrac{\sqrt{6}}{3}, 1 + \dfrac{\sqrt{6}}{3}\right\}$

25. 0   **27.** $\dfrac{\ln 3}{\ln 2} \approx 1.585$ **29.** 0   **31.** $\dfrac{3}{2}$ **33.** $\dfrac{\ln 10}{\ln 2} \approx 3.322$ **35.** $-\dfrac{\ln 1.2}{\ln 8} \approx -0.088$ **37.** $\dfrac{\ln 3}{2 \ln 3 + \ln 4} \approx 0.307$

39. $\dfrac{\ln 7}{\ln 0.6 + \ln 7} \approx 1.356$ **41.** 0   **43.** $\dfrac{\ln \pi}{1 + \ln \pi} \approx 0.534$ **45.** $\dfrac{\ln 1.6}{3 \ln 2} \approx 0.226$ **47.** $5 \ln 1.5 \approx 2.027$ **49.** $\dfrac{9}{2}$ **51.** 2   **53.** −1   **55.** 1

57. 16   **59.** $\left\{-1, \dfrac{2}{3}\right\}$ **61.** 1.92   **63.** 2.78   **65.** −0.56   **67.** −0.70   **69.** 0.56   **71.** $\{0.39, 1.00\}$ **73.** 1.31   **75.** 1.30

7.5 Exercises

1. $108.29   **3.** $609.50   **5.** $697.09   **7.** $12.46   **9.** $125.23   **11.** $88.72   **13.** $860.72   **15.** $554.09   **17.** $59.71   **19.** $361.93   **21.** 5.35%

23. 26%   **25.** $6\dfrac{1}{4}$% compounded annually   **27.** 9% compounded monthly   **29.** 104.32 mo; 103.97 mo   **31.** 61.02 mo; 60.82 mo

33. 15.27 yrs or 15 yrs, 4 mo   **35.** $104,335   **37.** $12,910.62   **39.** About $3017 or $30.17 per share   **41.** 9.35%
43. Not quite. Jim will have $1057.60. The second bank gives a better deal, since Jim will have $1060.62 after 1 year.
45. Will has $11,632.73; Henry has $10,947.89. Will has more money.
47. (a) Interest is $30,000   **(b)** Interest is $38,613.59   **(c)** Interest is $37,752.73. Simple interest at 12% is best.
49. (a) $1364.62   **(b)** $1353.35   **51.** $4631.93

59. (a) 6.1 yrs   **(b)** 18.45 yrs

(c) $mP = P\left(1 + \dfrac{r}{n}\right)^{nt}$

$m = \left(1 + \dfrac{r}{n}\right)^{nt}$

$\ln m = \ln\left(1 + \dfrac{r}{n}\right)^{nt} = nt \ln\left(1 + \dfrac{r}{n}\right)$

$t = \dfrac{\ln m}{n \ln\left(1 + \dfrac{r}{n}\right)}$

7.6 Exercises

1. (a) 34.7 days   **(b)** 69.3 days   **3. (a)** 28.4 yrs   **(b)** 94.4 yrs   **5.** 5832; 3.9 days   **7.** 25,198   **9.** 9.797 g   **11.** 9727 yrs ago
13. (a) 5:18 PM   **(b)** 14.3 min   **(c)** As time passes, the temperature of the pizza gets closer to 70°F.
15. 18.63°C; 25.1°C   **17.** 7.34 kg; 76.6 hr   **19.** 26.5 days   **21. (a)** 0.1286   **(b)** 0.9   **(c)** 1996   **23. (a)** 1000   **(b)** 30   **(c)** After 11.076 hrs

7.7 Exercises

1. 70 decibels   **3.** 111.76 decibels   **5.** 10 watts/m^2 **7.** 4.0 on the Richter scale
9. 125,892.54 mm; the Mexico City earthquake was 15.85 times as intense as the one in San Francisco.
11. Crowd noise was 31.6 times as intense as guidelines allow.

Fill-in-the-Blank Items

1. $(0, 1)$ and $(1, a)$ **3.** 4   **5.** 1   **7.** $\{x \mid x > 0\}$ **9.** 1

True/False Items

1. T   **3.** F   **5.** T   **7.** F

Review Exercises

1. -3 **3.** $\sqrt{2}$ **5.** 0.4   **7.** $\dfrac{25}{4}\log_4 x$ **9.** $-2\ln(x + 1)$ **11.** $\log\left(\dfrac{4x^3}{[(x + 3)(x - 2)]^{1/2}}\right)$ **13.** $y = Ce^{2x^2}$ **15.** $y = (Ce^{3x^2})^2$
17. $y = \sqrt{e^{x+C} + 9}$ **19.** $y = \ln(x^2 + 4) - C$

21.
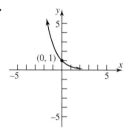
Domain: $(-\infty, \infty)$
Range: $(0, \infty)$
Asymptote: $y = 0$

23.
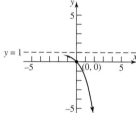
Domain: $(-\infty, \infty)$
Range: $(-\infty, 1)$
Asymptote: $y = 1$

25.
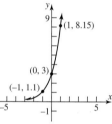
Domain: $(-\infty, \infty)$
Range: $(0, \infty)$
Asymptote: $y = 0$

27.
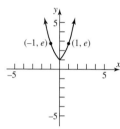
Domain: $(-\infty, \infty)$
Range: $[1, \infty)$
Asymptote: None

29.
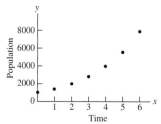
Domain: $(-\infty, \infty)$
Range: $(-\infty, 3)$
Asymptote: $y = 3$

31. $\dfrac{1}{4}$ **33.** $\left\{\dfrac{-1 - \sqrt{3}}{2}, \dfrac{-1 + \sqrt{3}}{2}\right\}$ **35.** $\dfrac{1}{4}$ **37.** $\dfrac{2\ln 3}{\ln 5 - \ln 3} \approx 4.301$
39. $\dfrac{12}{5}$ **41.** 83   **43.** $\left\{-3, \dfrac{1}{2}\right\}$ **45.** -1 **47.** $1 - \ln 5 \approx -0.609$
49. $\dfrac{\ln 3}{3\ln 2 - 2\ln 3} \approx -9.327$ **51.** 3229.5 m   **53.** 7.6 mm of mercury
55. (a) 37.3 watts   (b) 6.9 decibels
57. (a) 71%   (b) 85.5%   (c) 90%   (d) About 1.6 mo   (e) About 4.8 mo
59. (a) 9.85 yrs   (b) 4.27 yrs   **61.** \$41,669   **63.** 80 decibels   **65.** 24,203 yrs ago

67. (a)

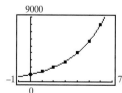

(b) $N = 1000e^{0.346574t}$
(c) 11,314
(d)
```
ExpReg
y=a*b^x
a=1000.187781
b=1.41414215
r=.9999999592
```
(e)

A P P E N D I X   A   A.1 Exercises

1. (a) $\dfrac{1}{2}$ **(b)** For every 2 unit change in x, y will change 1 unit.

3. (a) $-\dfrac{1}{3}$ **(b)** For every 3 unit change in x, y will decrease by 1 unit.

5. Slope $= -\dfrac{3}{2}$

7. Slope $= -\dfrac{1}{2}$

9. Slope $= 0$

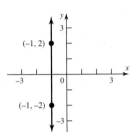

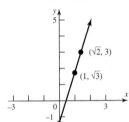

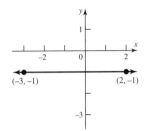

11. Slope undefined

13. Slope $= \dfrac{\sqrt{3} - 3}{1 - \sqrt{2}} \approx 3.06$

15.

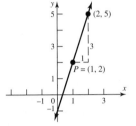

17.

19.

21.

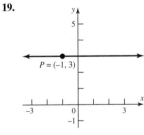

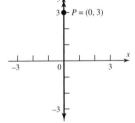

23. $x - 2y = 0$ or $y = \dfrac{1}{2}x$ **25.** $x + 3y - 4 = 0$ or $y = -\dfrac{1}{3}x + \dfrac{4}{3}$ **27.** $2x - y - 3 = 0$ or $y = 2x - 3$

29. $x + 2y - 5 = 0$ or $y = -\dfrac{1}{2}x + \dfrac{5}{2}$

31. $3x - y + 9 = 0$ or $y = 3x + 9$ **33.** $2x + 3y + 1 = 0$ or $y = -\dfrac{2}{3}x - \dfrac{1}{3}$ **35.** $x - 2y + 5 = 0$ or $y = \dfrac{1}{2}x + \dfrac{5}{2}$

37. $3x + y - 3 = 0$ or $y = -3x + 3$ **39.** $x - 2y - 2 = 0$ or $y = \dfrac{1}{2}x - 1$ **41.** $x - 2 = 0$; no slope-intercept form

43. $2x - y + 4 = 0$ or $y = 2x + 4$ **45.** $2x - y = 0$ or $y = 2x$ **47.** $x - 4 = 0$; no y-intercept form   **49.** $2x + y = 0$ or $y = -2x$

51. $x - 2y + 3 = 0$ or $y = \dfrac{1}{2}x + \dfrac{3}{2}$ **53.** $y - 4 = 0$ or $y = 4$

55. Slope $= 2$; y-intercept $= 3$

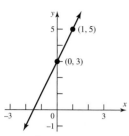

57. $y = 2x - 2$; Slope $= 2$; y-intercept $= -2$

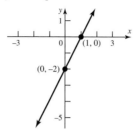

59. Slope $= \dfrac{1}{2}$; y-intercept $= 2$

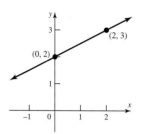

61. $y = -\dfrac{1}{2}x + 2$; Slope $= -\dfrac{1}{2}$ y-intercept $= 2$

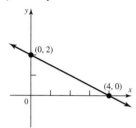

63. $y = \dfrac{2}{3}x - 2$; Slope $= \dfrac{2}{3}$ y-intercept $= -2$

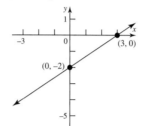

65. $y = -x + 1$; Slope $= -1$ y-intercept $= 1$

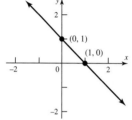

67. Slope undefined; no y-intercept

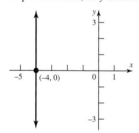

69. Slope $= 0$; y-intercept $= 5$

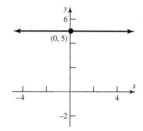

71. $y = x$; Slope $= 1$; y-intercept $= 0$

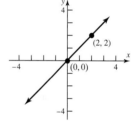

73. $y = \dfrac{3}{2}x$; Slope $= \dfrac{3}{2}$ y-intercept $= 0$

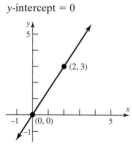

75. $y = 0$ **77.** (b)   **79.** (d)

81. $x - y + 2 = 0$ or $y = x + 2$ **83.** $x + 3y - 3 = 0$ or $y = -\dfrac{1}{3}x + 1$ **85.** $2x + 3y = 0$ or $y = -\dfrac{2}{3}x$

87. $°C = \dfrac{5}{9}(°F - 32)$; approx. $21°C$

89. (a) $P = 0.5x - 100$ **(b)** \$400
(c) \$2400

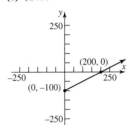

91. $C = 0.06543x + 5.65; C = $25.28;$
$C = 54.72

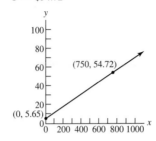

93. (a) $x^2 + (mx + b)^2 = r^2$
$(1 + m^2)x^2 + 2mbx + b^2 - r^2 = 0$
One solution if and only if discriminant $= 0$
$(2mb)^2 - 4(1 + m^2)(b^2 - r^2) = 0$
$-4b^2 + 4r^2 + 4m^2r^2 = 0$
$r^2(1 + m^2) = b^2$

(b) $x = \dfrac{-2mb}{2(1 + m^2)} = \dfrac{-2mb}{2b^2/r^2} = -\dfrac{r^2m}{b}$

$y = m\left(-\dfrac{r^2m}{b}\right) + b = -\dfrac{r^2m^2}{b} + b = \dfrac{-r^2m^2 + b^2}{b} = \dfrac{-r^2m^2 + r^2(1 + m^2)}{b} = \dfrac{r^2}{b}$

(c) Slope of tangent line $= m$

Slope of line joining center to point of tangency $= \dfrac{r^2/b}{-r^2m/b} = -\dfrac{1}{m}$

95. $\sqrt{2}x + 4y - 11\sqrt{2} + 12 = 0$ **97.** $x + 5y + 13 = 0$
99. All have the same slope, 2;   **101.** $y = 2$
the lines are parallel.

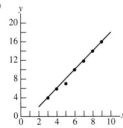

A.2 Exercises

1. Linear   **3.** Linear   **5.** Nonlinear
7. (a), (c)

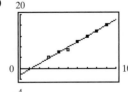

(b) Using $(3, 4)$ and $(8, 14)$, $y = 2x - 2$
(d) $y = 2.0357x - 2.3571$ **(e)**

9. (a), (c)

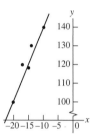

(b) Using $(0, 1)$ and $(2, 5)$, $y = 2x + 1$.

(d) $y = 2.2x + 1.2$ **(e)**

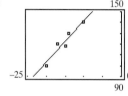

11. (a), (c)

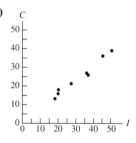

(b) Using $(30, 95)$ and $(50, 83)$, $y = -\dfrac{3}{5}x + 113$.   **(d)** $y = -0.72x + 116.6$

(e)

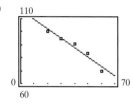

13. (a), (c)

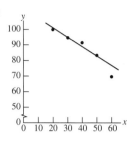

(b) Using $(-20, 100)$ and $(-10, 140)$, $y = 4x + 180$.   **(d)** $y = 3.8613x + 180.292$

(e)

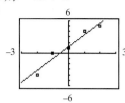

15. (a)

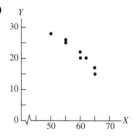

(b) Using $(20, 16)$ and $(50, 39)$, $C - 16 = \dfrac{23}{30}(I - 20)$

(c) As disposable income increases by \$1, consumption increases by about \$0.77.

(d) \$32,867

(e) $C = 0.755I + 0.6266$

17. (a)

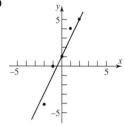

(b) Using $(50, 28)$ and $(65, 15)$, $Y - 28 = -\dfrac{13}{15}(X - 50)$

(c) As speed increases by 1 mph, mpg decreases by 0.87.

(d) 18.5 mpg

(e) $Y = -0.8265X + 70.3903$

19. (a)

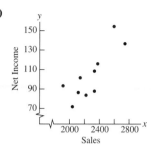

(b) Using $(1912.8, 92.6)$ and $(2745.1, 136.3)$, $y - 92.6 = 0.05(x - 1912.8)$
(c) As sales increase by \$1, the net income before taxes increases by \$0.05.
(d) \$119.8 billion
(e) $y = 0.0842x - 88.5776$

A.3 Exercises

1. 4 **3.** $\dfrac{1}{16}$ **5.** $\dfrac{1}{9}$ **7.** $(x - 2)^2 + (y + 2)^2 = 9$ **9.** $(x + 3)^2 + (y - 1)^2 = 9$ **11.** $\left(x + \dfrac{1}{2}\right)^2 + \left(y - \dfrac{1}{2}\right)^2 = 1$

A.4 Exercises

1. $\{-7, 3\}$ **3.** $\left\{-\dfrac{1}{4}, \dfrac{3}{4}\right\}$ **5.** $\left\{\dfrac{-1 - \sqrt{7}}{6}, \dfrac{-1 + \sqrt{7}}{6}\right\}$ **7.** $\{2 - \sqrt{2}, 2 + \sqrt{2}\}$ **9.** $\{2 - \sqrt{5}, 2 + \sqrt{5}\}$ **11.** $\left\{1, \dfrac{3}{2}\right\}$
13. No real solutions **15.** $\left\{\dfrac{-1 - \sqrt{5}}{4}, \dfrac{-1 + \sqrt{5}}{4}\right\}$ **17.** $\left\{\dfrac{1}{3}\right\}$

A P P E N D I X B B.1 Exercises

1. $(-1, 4)$ **3.** $(3, 1)$ **5.** $X\min = -11$, $X\max = 5$, $X\text{scl} = 1$, $Y\min = -3$, $Y\max = 6$, $Y\text{scl} = 1$ **7.** $X\min = -30$, $X\max = 50$, $X\text{scl} = 10$, $Y\min = -90$, $Y\max = 50$, $Y\text{scl} = 10$ **9.** $X\min = -10$, $X\max = 110$, $X\text{scl} = 10$, $Y\min = -10$, $Y\max = 160$, $Y\text{scl} = 10$
11. $X\min = -6$, $X\max = 6$, $X\text{scl} = 2$, $Y\min = -4$, $Y\max = 4$, $Y\text{scl} = 2$ **13.** $X\min = -9$, $X\max = 9$, $X\text{scl} = 3$, $Y\min = -4$, $Y\max = 4$, $Y\text{scl} = 2$ **15.** $X\min = -6$, $X\max = 6$, $X\text{scl} = 1$, $Y\min = -8$, $Y\max = 8$, $Y\text{scl} = 2$ **17.** $X\min = -6$, $X\max = 6$, $X\text{scl} = 2$, $Y\min = -1$, $Y\max = 3$, $Y\text{scl} = 1$ **19.** $X\min = 3$, $X\max = 9$, $X\text{scl} = 1$, $Y\min = 2$, $Y\max = 10$, $Y\text{scl} = 2$

B.2 Exercises

1. (a) 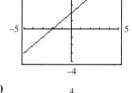 **(b)** 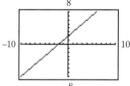 **(c)** 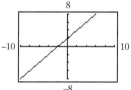 **(d)**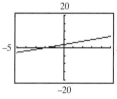

3. (a)  **(b)** 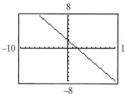 **(c)** 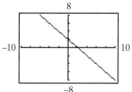 **(d)**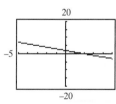

5. (a) 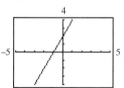 **(b)** 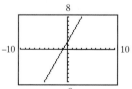 **(c)** 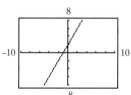 **(d)**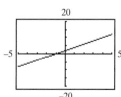

7. (a) 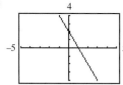   **(b)** 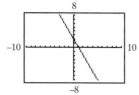   **(c)** 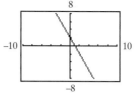   **(d)**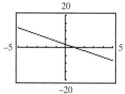

9. (a) 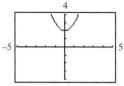   **(b)** 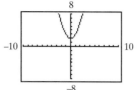   **(c)** 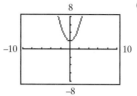   **(d)**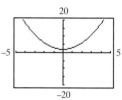

11. (a) 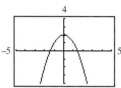   **(b)** 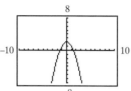   **(c)** 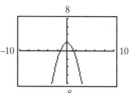   **(d)**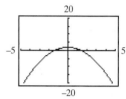

13. (a) 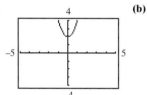   **(b)** 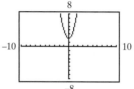   **(c)** 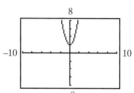   **(d)**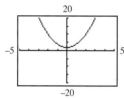

15. (a) 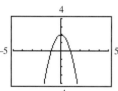   **(b)** 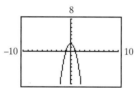   **(c)** 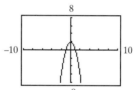   **(d)**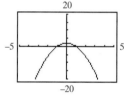

17. (a) 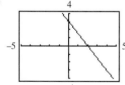   **(b)** 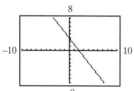   **(c)** 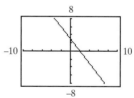   **(d)**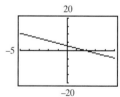

19. (a) 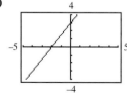   **(b)** 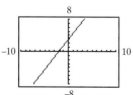   **(c)** 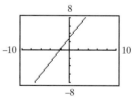   **(d)**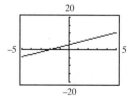

21.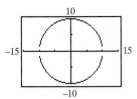

23.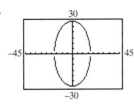

25.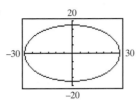

27.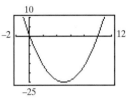

29.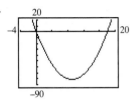

31.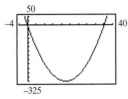

33.

35.

37.

39.

B.3 Exercises

1. 0.428 **3.** 2.236 **5.** 1.259 **7.** −3.41 **9.** −1.70 **11.** −0.28 **13.** 3.00 **15.** 4.50 **17.** 0.31, 12.30 **19.** 1.00, 23.00

B.4 Exercises

1. Yes **3.** Yes **5.** No **7.** Yes **9.** ymin = 4   Other answers are possible
ymax = 12
yscl = 1

INDEX